Annotated Instructor's Edition

INTRODUCTORY ALGEBRA

K. Elayn Martin-Gay

University of New Orleans

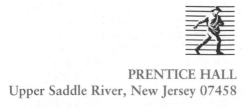

PRENTICE HALL
Upper Saddle River, New Jersey 07458

Library of Congress Cataloging-in-Publication Data

Martin-Gay, K. Elayn
   Introductory algebra / K. Elayn Martin-Gay.
     p. cm.
   Includes index.
   ISBN 0-13-228834-6 (student ed.).—ISBN 0-13-862467-4 (annotated
instructor's ed.)
   1. Algebra. I. Title.
QA152.2.M36883 1999   98-4121
512.9—dc21   CIP

To my mother, Barbara M. Miller,
and her husband, Leo Miller,
and to the memory
of my father, Robert J. Martin

Acquisitions Editor: Karin E. Wagner
Editor-in-Chief: Jerome Grant
Editorial Director: Tim Bozik
Associate Editor-in-Chief, Development: Carol Trueheart
Senior Managing Editor: Linda Mihatov Behrens
Executive Managing Editor: Kathleen Schiaparelli
Assistant Vice President of Production and Manufacturing: David W. Riccardi
Marketing Manager: Jolene Howard
Manufacturing Buyer: Alan Fischer
Manufacturing Manager: Trudy Pisciotti
Editorial Assistant/Supplements Editor: Kate Marks
Supplements Editor: Audra J. Walsh
Art Director/Cover Designer: Maureen Eide
Associate Creative Director: Amy Rosen
Director of Creative Services: Paula Maylahn
Assistant to Art Director: John Christiana
Art Manager: Gus Vibal
Art Editors: Karen Branson, Grace Hazeldine
Cover image: Fritz Prenzel/Tony Stone Images
Test Design and Project Management: Elm Street Publishing Services, Inc.
Photo Researcher: Diana Gongora
Photo Research Administrator: Melinda Reo
Art Studio: Academy Artworks

© 1999 by Prentice-Hall, Inc.
Simon & Schuster/A Viacom Company
Upper Saddle River, New Jersey 07458

Photo Credits appear on page P-1, which constitutes a continuation of the copyright page.

Printed in the United States of America
10 9 8 7 6 5 4 3 2 1

ISBN: 0-13-228834-6   (Student Edition 0-13-228834-6)

Prentice-Hall International (UK) Limited, *London*
Prentice-Hall of Australia Pty. Limited, *Sydney*
Prentice-Hall Canada Inc., *Toronto*
Prentice-Hall Hispanoamericana, S.A., *Mexico*
Prentice-Hall of India Private Limited, *New Delhi*
Prentice-Hall of Japan, Inc., *Tokyo*
Simon & Schuster Asia Pte. Ltd., *Singapore*
Editora Prentice-Hall do Brasil, Ltda., *Rio de Janeiro*

Contents

Preface

ABOUT THIS BOOK

This worktext was written to provide a solid foundation in algebra for students who might have had no previous experience in algebra. Specific care was taken to ensure that students have the most up-to-date relevant text preparation for their next mathematics course or for nonmathematical courses that require an understanding of algebraic fundamentals. I have tried to achieve this by writing a user-friendly text that is keyed to objectives and contains many worked-out examples. As suggested by the AMATYC Crossroads Document and the NCTM Standards (plus Addenda), real-life and real-data applications, data interpretation, conceptual understanding, problem solving, writing, cooperative learning, appropriate use of technology, mental mathematics, number sense, critical thinking, and geometric concepts are emphasized and integrated throughout the book.

KEY PEDAGOGICAL FEATURES

Readability and Connections   I have tried to make the writing style as clear as possible while still retaining the mathematical integrity of the content. When a new topic is presented, an effort has been made to relate the new ideas to those that students may already know. Constant reinforcement and connections within problem-solving strategies, data interpretation, geometry, patterns, graphs, and situations from everyday life can help students gradually master both new and old information. In addition, each section begins with a list of objectives covered in the section. Clear organization of section material based on objectives further enhances readability.

Problem-Solving Process   This is formally introduced in Chapter 2 with a four-step process that is integrated throughout the text. The four steps are Understand, Translate, Solve, and Interpret. The repeated use of these steps in a variety of examples shows their wide applicability. Reinforcing the steps can increase students' comfort level and confidence in tackling problems.

Applications and Connections   Every effort was made to include as many interesting and relevant real-life applications as possible throughout the text in both worked-out examples and exercise sets. The applications help to motivate students and strengthen their understanding of mathematics in the real world. They show connections to a wide range of fields including agriculture, allied health, anthropology, art, astronomy, biology, business, chemistry, construction, consumer affairs, earth science, education, entertainment, environmental issues, finance, geography, government, history, medicine, music, nutrition, physics, sports, travel, and weather. Many of the applications are based on recent real data. Sources for data include newspapers, magazines, publicly held companies, government agencies, special-interest groups, research organizations, and reference books. Opportunities for obtaining your own real data are also included.

Exercise Sets   Each text section ends with an Exercise Set. Each skill-based exercise in the set is keyed to one of the objectives of the section. Wherever possible, a specific example is also given. In addition to the approximately 4000 exercises in end-of-section exercise sets, exercises may also be found in the Pretests, Integrated Reviews, Chapter Reviews, Chapter Tests, and Cumulative Reviews. Each Exercise Set contains one or more of the following features:

Mental Math   Found at the beginning of an exercise set, these mental warmups reinforce concepts found in the accompanying section and increase students' confidence before they tackle an exercise set. By relying on their own mental skills, students increase not only their confidence in themselves but also their number sense and estimation ability.

Review and Preview   These exercises occur in each exercise set (except for those in Chapters R and 1) after the exercises keyed to the objectives of the section. Review and Preview problems are keyed to earlier sections and review concepts learned earlier in the text that are needed in the next section or in the next chapter. These exercises show the links between earlier topics and later material.

Combining Concepts   These exercises are found at the end of each exercise set after the Review and Preview exercises. Combining Concepts exercises require students to combine several concepts from that section or to take the concepts of the section a step further by combining them with concepts learned in previous sections. For instance, sometimes students are required to combine the concepts of the section with the problem-solving process they learned in Chapter 2 to try their hand at solving an application problem.

Internet Excursions   These exercises occur once per chapter. Internet Excursions require students to use the Internet as a data-collection tool to complete the exercises, allowing students first-hand experience with manipulating and working with real data.

Conceptual and Writing Exercises   These exercises occur in almost every exercise set and are marked with the icon ✏. They require students to show an understanding of a concept learned in the corresponding section. This is accomplished by asking students questions that require them to use two or more concepts together. Some require students to stop, think, and explain in their own words the concept(s) used in the exercises they have just completed. Guidelines recommended by the American Mathematical Association of Two Year Colleges (AMATYC) and other professional groups recommend incorporating writing in mathematics courses to reinforce concepts.

Data and Graphical Interpretation   There is an emphasis on data interpretation in exercises via tables and graphs. The ability to interpret data and read and create a variety of types of graphs is developed gradually so students become comfortable with it. In addition, an appendix on mean, median, and mode together with exercises is included.

Practice Problems   Throughout the text, each worked-out example has a parallel Practice Problem placed next to the example in the margin. These invite students to be actively involved in the learning process before beginning the end-of-section exercise set. Practice Problems immediately reinforce a skill after it is developed.

Concept Checks   These margin exercises are appropriately placed in many sections of the text. They allow students to gauge their grasp of an idea as it is being explained in the text. Concept Checks stress conceptual understanding at point of use and help suppress misconceived notions before they start.

Integrated Reviews These "mid-chapter reviews" are appropriately placed once per chapter. Integrated Reviews allow students to review and assimilate the many different skills learned separately over several sections before moving on to related material in the chapter.

Helpful Hints Helpful Hints contain practical advice on applying mathematical concepts. These are found throughout the text and strategically placed where students are most likely to need immediate reinforcement. Helpful Hints are highlighted for quick reference.

Focus On Appropriately placed throughout each chapter, these are divided into Focus on Study Skills, Focus on Mathematical Connections, Focus on Business and Career, Focus on the Real World, and Focus on History. They are written to help students develop effective habits for studying mathematics, engage in investigations of other branches of mathematics, understand the importance of mathematics in various careers and in the world of business, and see the relevance of mathematics in both the present and past through critical thinking exercises and group activities.

Calculator and Graphing Calculator Explorations These optional explorations offer point-of-use instruction, through examples and exercises, on the proper use of scientific and graphing calculators as tools in the mathematical problem-solving process. Placed appropriately throughout the text, Calculator and Graphing Calculator Explorations also reinforce concepts learned in the corresponding section and motivate discovery-based learning.

Additional exercises building on the skill developed in the Explorations may be found in exercise sets throughout the text. Exercises requiring a calculator are marked with the ▦ icon. Exercises requiring a graphing calculator are marked with the ▦ icon.

Chapter Activity These features occur once per chapter at the end of the chapter, often serving as a chapter wrap-up. For individual or group completion, the Chapter Activity, usually hands-on or data-based, complements and extends to concepts of the chapter, allowing students to make decisions and interpretations and to think and write about algebra.

Visual Reinforcement of Concepts The text contains a wealth of graphics, models, photographs, and illustrations to visually clarify and reinforce concepts. These include bar graphs, line graphs, calculator screens, application illustrations, and geometric figures.

Pretests Each chapter begins with a pretest that is designed to help students identify areas where they need to pay special attention in the upcoming chapter.

Chapter Highlights Found at the end of each chapter, these contain key definitions, concepts, and examples to help students understand and retain what they have learned.

Chapter Review and Test The end of each chapter contains a review of topics introduced in the chapter. The Chapter Review offers exercises that are keyed to sections of the chapter. The Chapter Test is a practice test and is not keyed to sections of the chapter.

Cumulative Review These features are found at the end of each chapter (except Chapters R and 1). Each problem contained in the cumulative review is actually an earlier worked example in the text that is referenced in the back of the book along with the answer. Students who need to see a complete worked-out solution, with explanation, can do so by turning to the appropriate example in the text.

Student Resource Icons   At the beginning of each section, videotape, software, and solutions manual icons are displayed. These icons help reinforce that these learning aids are available should students wish to use them to help them review concepts and skills at their own pace. These items have direct correlation to the text and emphasize the text's methods of solution. In addition, a videotape index can be found at the end of the book.

Functional Use of Color and Design   Elements of the text are highlighted with color or design to make it easier for students to read and study. Color is also used to clarify the problem-solving process in worked examples.

SUPPLEMENTS FOR THE INSTRUCTOR

Printed Supplements

Annotated Instructor's Edition (0-13-862467-4)

▲ Answers to all exercises printed on the same text page.
▲ Teaching Tips throughout the text placed at key points in the margin.

Instructor's Solution Manual (0-13-862483-6)

▲ Solutions to even-numbered section exercises.
▲ Solutions to every (even and odd) Mental Math exercise.
▲ Solutions to every (even and odd) Practice Problem (margin exercise).
▲ Solutions to every (even and odd) exercise found in the Chapter Pretests, Integrated Reviews (mid-chapter reviews), Chapter Reviews, Chapter Tests, Cumulative Reviews.

Instructor's Resource Manual with Tests (0-13-862475-5)

▲ Notes to the Instructor that includes an introduction to Interactive Learning, Interpreting Graphs and Data, Alternative Assessment, Using Technology and Helping Students Succeed.
▲ Two free-response Pretests per chapter.
▲ Eight Chapter Tests per chapter (3 multiple-choice, 5 free-response).
▲ Two Cumulative Review Tests (one multiple-choice, one free-response) every two chapters (after chapters 2, 4, 6, 9).
▲ Eight Final Exams (3 multiple-choice, 5 free-response).
▲ Twenty additional exercises per section for added test exercises if needed.

Media Supplements

TestPro4 Computerized Testing

▲ Algorithmically driven, text-specific testing program.
▲ Networkable for administering tests and capturing grades on-line.
▲ Edit and add your own questions—create nearly unlimited number of tests and drill worksheets.

Companion Web site

▲ www.prenhall.com/martin-gay
▲ Links related to the Internet Excursions in each chapter allow you to collect data to solve specific internet exercises.
▲ Additional links to helpful, generic sites include Fun Math and For Additional Help.

SUPPLEMENTS FOR THE STUDENT

Printed Supplements

Student's Solution Manual (0-13-862525-5)

▲ Solutions to odd-numbered section exercises.
▲ Solutions to every (even and odd) Mental Math exercise.
▲ Solutions to every (even and odd) Practice Problem (margin exercise).
▲ Solutions to every (even and odd) exercise found in the Chapter Pretests, Integrated Reviews (mid-chapter reviews), Chapter Reviews, Chapter Tests, Cumulative Reviews.

New York Times *Themes of the Times*

▲ Have your instructor contact the local Prentice Hall sales representative.

How to Study Mathematics

▲ Have your instructor contact the local Prentice Hall sales representative.

Internet Guide

▲ Have your instructor contact the local Prentice Hall sales representative.

Media Supplements

MathPro4 computerized tutorial

▲ Keyed to each section of the text for text-specific tutorial exercises and instruction.
▲ Includes Warm-up exercises and graded Practice Problems.
▲ Algorithmically driven and fully networkable.
▲ Have your instructor contact the local Prentice Hall sales representative—also available for purchase for home use.

Videotape Series (0-13-862517-4)

▲ Written and presented by Elayn Martin-Gay.
▲ Keyed to each section of the text.
▲ Step-by-step solutions to exercises from each section of the text. Exercises that are worked in the videos are marked with a video icon (▭).

Companion Web site

▲ www.prenhall.com/martin-gay
▲ Links related to the Internet Excursions in each chapter allow you to collect data to solve specific internet exercises.
▲ Additional links to generic sites include Fun Math, For Additional Help . . .

ACKNOWLEDGMENTS

First, as usual, I would like to thank my husband, Clayton, for his constant encouragement. I would also like to thank my children, Eric and Bryan, for their sense of humor and especially for their suggestion of letting Dad cook the bacon that I always used to burn.

I would also like to thank my extended family for their invaluable help and also their sense of humor. Their contributions are too numerous to list. They are Rod and Karen Pasch; Michael, Christopher, Matthew, and Jessica Callac; Stuart, Earline, Melissa, Mandy and Bailey Martin; Mark, Sabrina, and Madison Martin; Leo and Barbara Miller; and Jewett Gay.

I would like to thank the following reviewers for their input and suggestions:

Peter Arvanites, *Rockland Community College*
David Boni, *Monroe Community College*
Paulette Botley, *Edmonds Community College*

James W. Brewer, *Florida Atlantic University*
Joan Burke, *Montclair State University*
Marjorie Darrah, *Alderson-Broaddus College*
Mary Ellen Gallegos, *Santa Fe Community College*
James W. Harris, *John A. Logan College*
Brian Hayes, *Triton College*
Kayana Hoagland, *South Puget Sound Community College*
Deanna Li, *North Seattle Community College*
Marcel Maupin, *Oklahoma State University, Oklahoma City*
Christopher McNally, *Tallahassee Community College*
Julie Miller, *Daytona Beach Community College*
Cameron Neal, Jr., *Temple College*
Ellen O'Connell, *Triton College*
Ted Panitz, *Cape Cod Community College*
Marilyn Garrett Platt, *Gaston College*
Lynne Sage, *Bellevue Community College*
Sue Sharkey, *Waukesha County Technical College*

There were many people who helped me develop this text and I will attempt to thank some of them here. Penny Korn did an excellent job of providing answers. Cheryl Roberts Cantwell was invaluable for contributing to the overall accuracy of this text. Emily Keaton was also invaluable for her many suggestions and contributions during the development and writing of this first edition. Ingrid Mount at Elm Street Publishing Services provided guidance throughout the production process. I thank Terri Bittner, Cindy Trimble, Jeff Rector, and Teri Lovelace at Laurel Technical Services for all their work on the supplements, text and thorough accuracy check. Lastly, a special thank you to my editor, Karin Wagner, for her support and assistance throughout the development and production of this text and to all the staff at Prentice Hall: Linda Behrens, Alan Fischer, Maureen Eide, Karen Branson, Grace Hazeldine, Gus Vibal, Kate Marks, Jolene Howard, John Tweeddale, Jerome Grant, and Tim Bozik.

K. Elayn Martin-Gay

ABOUT THE AUTHOR

K. Elayn Martin-Gay has taught mathematics at the University of New Orleans for more than 19 years. Her numerous teaching awards include the local University Alumni Association's Award for Excellence in Teaching, and Outstanding Developmental Educator at University of New Orleans, presented by the Louisiana Association of Developmental Educators.

Elayn is the author of an entire product line of highly successful textbooks. The Martin-Gay library includes the exciting new worktext series: *Basic College Mathematics*, *Introductory Algebra* and *Intermediate Algebra*, the successful hardbound series; *Beginning Algebra* 2e, *Intermediate Algebra* 2e, *Intermediate Algebra A Graphing Approach* (co-authored with Margaret Greene), *Prealgebra* 2e, and *Introductory and Intermediate Algebra*, a combined algebra text.

Prior to writing textbooks, Elayn developed an acclaimed series of lecture videos to support developmental mathematics students in their quest for success. These highly successful videos originally served as the foundation material for her texts. Today the tapes specifically support each book in the Martin-Gay series.

Index of Applications

How to Use This Worktext: A Guide for Students

Introductory Algebra has been written and designed to help you succeed in this course. The goals that the author will help you achieve include:

▲ Exposure to real-world applications you will encounter in life and other courses

▲ An organized, integrated system of learning that combines the text with a comprehensive set of print and media supplements

▲ Better preparation relevant to the *next* course you will take in mathematics

Take a few moments now to acquaint yourself with some of the features that have been built into *Introductory Algebra* to help you excel.

Equations, Inequalities, and Problem Solving

CHAPTER 2

In this chapter, we solve equations and inequalities. Once we know how to solve equations and inequalities, we may solve problems. Of course, problem solving is an integral topic in algebra and its discussion is continued throughout this text.

2.1 Simplifying Expressions

2.2 The Addition Property of Equality

2.3 The Multiplication Property of Equality

2.4 Further Solving Linear Equations

　　Integrated Review—Solving Linear Equations

2.5 An Introduction to Problem Solving

2.6 Formulas and Problem Solving

2.7 Percent, Ratio, and Proportion

2.8 Solving Linear Inequalities

A glacier is a giant mass of rocks and ice that flows downh[ill] like a river. Alaska alone has an estimated 100,000 glacie[rs]. They form high in the mountains where snow does not me[lt]. After years of accumulated snowfall, the weight of the co[m]pacted ice causes it to slide downhill, and a glacier is formed. [In] Example 1 on page 101, we will find the time it takes for ice [at] the beginning of a glacier to reach its face, or ending wall.

　　Glaciers account for approximately 77 percent of the worl[d's] fresh water. They also contain records of past environments be-tween their layers of snow. Scientists study glaciers to learn about past global climates and then use this information to pre-dict future environmental changes.

(*Source: Blue Ice in Motion—The Story of Alaska's Glaciers*, by Sally D. Wiley, published in 1990 by The Alaska Natural History Association and Judith Ann Rose)

The photo applications at the opening of every chapter, and applications through-out, introduce you to situations that you might encounter in the real-world and are applicable to the mathematics you will learn in the upcoming chapter. These applications are often discussed later in the chapter as well.

Page 89

Connect the Concepts

Learning and succeeding in a math course require practice and a broader understanding of how everything works together. As you study, make connections using this text's organization to help you. All of these features have been included to enhance your understanding of algebraic concepts.

Concept Checks are special margin exercises found in most sections. Work these to gauge your grasp of the idea being explained in the text.

✓ CONCEPT CHECK

Suppose you have simplified several equations and obtain the following results. What can you conclude about the solutions to the original equation?
a. $7 = 7$ b. $x = 0$ c. $7 = -4$

Page 121

▚ COMBINING CONCEPTS

Solve.

59. $0.07x - 5.06 = -4.92$

60. $0.06y + 2.63 = 2.5562$

61. The equation $3x + 6 = 2x + 10 + x - 4$ is true for all real numbers. Substitute a few real numbers for x to see that this is so and then try solving the equation.

62. The equation $6x + 2 - 2x = 4x + 1$ has no solution. Try solving this equation for x and see what happens.

Look for **Combining Concepts** exercises at the end of each exercise set. Solving these exercises will expose you to the way mathematical ideas build upon each other.

Page 116

Practice Problem 3

Use the circle graph and part (c) of Example 3 to answer each question.

a. What percent of homeowners spend $250–$4999 on yearly home maintenance?

b. What percent of homeowners spend $250 or more on ye tenance?

c. How many of the the town of Fairvi pect to spend fro year on home mai

Example 3

The circle graph below shows how much money homeowners in the United States spend annually on maintaining their homes. Use this graph to answer the questions below.

Yearly Home Maintenance in the U.S.

45% Under $250

38% $250–$999

16% $1000–$499

1% $5000 or more

Practice Problems occur in the margins next to every Example. Work these problems after you read an Example to reinforce your understanding.

Page 152

Test Yourself and Check Your Understanding

Good exercise sets are an essential ingredient for a solid introductory algebra textbook. The exercises you will find in this worktext are intended to help you build skills and understand concepts as well as motivate and challenge you. Note that the features like Chapter Highlights, Tests, and Cumulative Reviews are found at the end of each chapter to help you study and organize your notes.

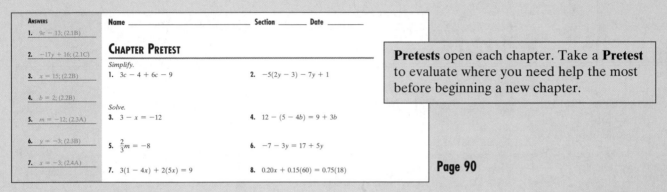

ANSWERS	
1. $9c - 13$; (2.1B)	
2. $-17y + 16$; (2.1C)	
3. $x = 15$; (2.2B)	
4. $b = 2$; (2.2B)	
5. $m = -12$; (2.3A)	
6. $y = -3$; (2.3B)	
7. $x = -3$; (2.4A)	

Name _____ Section _____ Date _____

CHAPTER PRETEST

Simplify.

1. $3c - 4 + 6c - 9$ **2.** $-5(2y - 3) - 7y + 1$

Solve.

3. $3 - x = -12$ **4.** $12 - (5 - 4b) = 9 + 3b$

5. $\frac{2}{3}m = -8$ **6.** $-7 - 3y = 17 + 5y$

7. $3(1 - 4x) + 2(5x) = 9$ **8.** $0.20x + 0.15(60) = 0.75(18)$

Pretests open each chapter. Take a **Pretest** to evaluate where you need help the most before beginning a new chapter.

Page 90

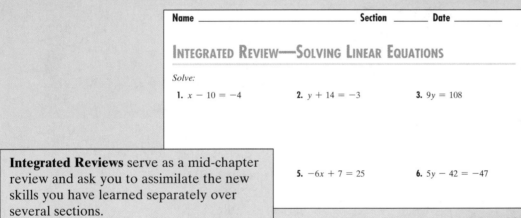

Name _____ Section _____ Date _____

INTEGRATED REVIEW—SOLVING LINEAR EQUATIONS

Solve:

1. $x - 10 = -4$ **2.** $y + 14 = -3$ **3.** $9y = 108$

5. $-6x + 7 = 25$ **6.** $5y - 42 = -47$

Integrated Reviews serve as a mid-chapter review and ask you to assimilate the new skills you have learned separately over several sections.

Page 127

MENTAL MATH

Solve each equation mentally. See Examples 1 and 2.

1. $x + 4 = 6$ **2.** $x + 7 = 10$ **3.** $n + 18 = 30$

4. $z + 22 = 40$ **5.** $b - 11 = 6$ **6.** $d - 16 = 5$

Confidence-building **Mental Math** problems are in most sections.

Page 105

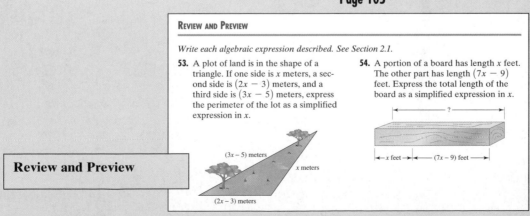

REVIEW AND PREVIEW

Write each algebraic expression described. See Section 2.1.

53. A plot of land is in the shape of a triangle. If one side is x meters, a second side is $(2x - 3)$ meters, and a third side is $(3x - 5)$ meters, express the perimeter of the lot as a simplified expression in x.

54. A portion of a board has length x feet. The other part has length $(7x - 9)$ feet. Express the total length of the board as a simplified expression in x.

$(3x - 5)$ meters

x meters

$(2x - 3)$ meters

x feet $(7x - 9)$ feet

Review and Preview

Page 125

Get Involved!

Discover how algebra relates to and appears in the world around you. Evaluate and interpret real data in graphs, tables, and charts—make an educated guess on the answers and outcomes! Knowing how to use data and graphs to estimate and predict is a valuable skill in the workplace as well as in other courses.

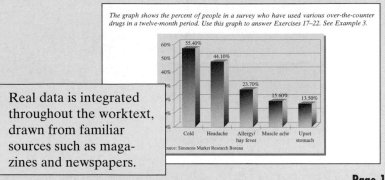

The graph shows the percent of people in a survey who have used various over-the-counter drugs in a twelve-month period. Use this graph to answer Exercises 17–22. See Example 3.

Source: Simmons Market Research Bureau

Real data is integrated throughout the worktext, drawn from familiar sources such as magazines and newspapers.

Page 158

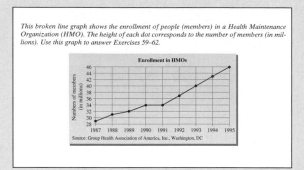

This broken line graph shows the enrollment of people (members) in a Health Maintenance Organization (HMO). The height of each dot corresponds to the number of members (in millions). Use this graph to answer Exercises 59–62.

Source: Group Health Association of America, Inc., Washington, DC

Page 172

Name _____

Fill in the percent column in each table. See Examples 1 through 3.

25.

THE WALT DISNEY COMPANY 1996 OPERATING INCOME		
	Millions of Dollars	Percent of Total (Round to nearest whole percent)
Creative Content	1596	$\frac{1596}{3333} \approx 48\%$
Theme Parks	990	
Broadcasting	747	
TOTALS	3333	

Source: Hoover's Company Profiles

26.

NIKE, INC. 1996 SALES		
	Millions of Dollars	Percent of Total (Round to nearest whole percent)
U.S. Footwear	2773	$\frac{2773}{6471} \approx 43\%$
International Footwear	1682	
U.S. Apparel	843	
International Apparel	651	
Other Brands	522	
TOTALS	6471	

Source: NIKE, Inc.

Graphics, models and illustrations provide visual reinforcement.

Page 159

CALCULATOR EXPLORATIONS
CHECKING EQUATIONS

We can use a calculator to check possible solutions of equations. To do this, replace the variable by the possible solution and evaluate both sides of the equation separately.

Equation: $3x - 4 = 2(x + 6)$ Solution: $x = 16$

$3x - 4 = 2(x + 6)$ Original equation

$3(16) - 4 \stackrel{?}{=} 2(16 + 6)$ Replace x with 16.

Now evaluate each side with your calculator.

Evaluate left side: $\boxed{3} \boxed{\times} \boxed{16} \boxed{-} \boxed{4} \boxed{=}$ Display: $\boxed{44}$

or

$\boxed{\text{ENTER}}$

Evaluate right side: $\boxed{2} \boxed{(} \boxed{16} \boxed{+} \boxed{6} \boxed{)} \boxed{=}$ Display: $\boxed{44}$

or

$\boxed{\text{ENTER}}$

Since the left side equals the right side, the equation checks.

Use a calculator to check the possible solutions to each equation.

1. $2x = 48 + 6x$; $x = -12$ solution

2. $-3x - 7 = 3x - 1$; $x = -1$ solution

3. $5x - 2.6 = 2(x + 0.8)$; $x = 4.4$ not a solution

4. $-1.6x - 3.9 = -6.9x - 25.6$; $x = 5$ not a solution

5. $\frac{564x}{4} = 200x - 11(649)$; $x = 121$ solution

6. $20(x - 39) = 5x - 432$; $x = 23.2$ solution

Scientific and Graphing Calculator Explorations and exercises are woven into appropriate sections.

Page 122

Focus On boxes found throughout each chapter help you see the relevance of math through critical thinking exercises and group activities. Try these on your own or with another student.

Focus on Study Skills

Focus On Study Skills

STUDY TIPS

Have you wondered what you can do to be successful in your algebra course? If so, that may well be your first step to success in algebra! Here are some tips on how to use this text and how to study mathematics in general.

Using this Text

1. Each example in the section has a parallel Practice Problem. As you read a section, try each Practice Problem after you've finished the corresponding example. This "learn-by-doing" approach will help you grasp ideas before you move on to other concepts.
2. The main section of exercises in an exercise set are referenced by an objective, such as **A** or **B** and also an example(s). Use this referencing in case you have trouble completing an assignment from the exercise set.
3. If you need extra help in a particular section, check at the beginning of the section to see what videotapes and software are available.
4. Integrated Reviews in each chapter offer you a chance to practice—in one place—the many concepts that you have learned separately over several sections.
5. There are many opportunities at the end of each chapter to help you understand the concepts of the chapter.

Page 40

Focus on History

Focus On History

THE GOLDEN RECTANGLE IN ART

The golden rectangle is a rectangle whose length is approximately 1.6 times its width. The early Greeks thought that a rectangle with these dimensions was the most pleasing to the eye. Examples of the golden rectangle are found in many ancient, as well as modern, works of art. For example, the Parthenon in Athens, Greece, shows the golden rectangle in many aspects of its design. Modern-era artists, including Piet Mondrian (1872–1944) and Georges Seurat (1859–1891), also frequently used the proportions of a golden rectangle in their paintings.

Page 104

Focus on Mathematical Connections

Focus On Mathematical Connections

SEQUENCES

The ratio of the lengths of the sides of the golden rectangle is approximately 1.6. This value is also known as the golden ratio. The actual value of the golden ratio is $\frac{1 + \sqrt{5}}{2} \approx 1.618033989\ldots$, which is an irrational number. Interestingly enough, there is a very simple sequence, or ordered list of numbers, that can be used to approximate the golden ratio. The sequence is called the Fibonacci sequence, and it is very easy to construct. Start with the first two terms of the sequence, the numbers 1 and 1. The next term in the sequence is the sum of the two previous terms, and so on. For example,

		$1+1=$	$1+2=$	$2+3=$	$3+5=$	$5+8=$	
1,	1,	2,	3,	5,	8,	13,	

		$8+13=$	$13+21=$	$21+34=$	
		21,	34,	55, ...	

Now, let's look at the ratios of a Fibonacci sequence term to the previous one:

$\frac{1}{1} = 1,\ \frac{2}{1} = 2,\ \frac{3}{2} = 1.5,\ \frac{5}{3} \approx 1.66667,\ \frac{8}{5} = 1.6,\ \frac{13}{8} = 1.625,\ \frac{21}{13} \approx 1.615,\ \frac{34}{21} \approx 1.619,$

Page 168

Focus on the Real World

Focus On The Real World

SURVEYS

Recall that the golden rectangle is a rectangle whose length is approximately 1.6 times its width. It is thought that for about 75% of adults, a rectangle in the shape of the golden rectangle is most pleasing to the eye.

(a)   (b)   (c)

GROUP ACTIVITIES

Page 113

Focus on the Business and Career

Focus On Business and Career

FAST-GROWING CAREERS

According to U.S. Bureau of Labor Statistics projections, the careers listed below will have the largest job growth into the next century.

		Employment [Numbers in thousands]		
	Occupation	1994	2005	Change
1.	Cashiers	3,005	3,567	+562
2.	Janitors and cleaners, including maids and housekeeping cleaners	3,043	3,602	+559
3.	Salespersons, retail	3,842	4,374	+532
4.	Waiters and waitresses	1,847	2,326	+479
5.	Registered nurses	1,906	2,379	+473
6.	General managers and top executives	3,046	3,512	+466
7.	Systems analysts	483	928	+445
8.	Home health aides	420	848	+428
9.	Guards	867	1,282	+415

Page 378

Enrich Your Learning

Seek out these additional Student Resources to match your personal learning style.

Text-specific videos hosted by the award-winning teacher and author of *Introductory Algebra* cover each objective in every chapter section as a supplementary review.

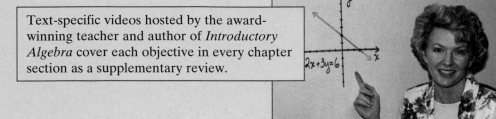

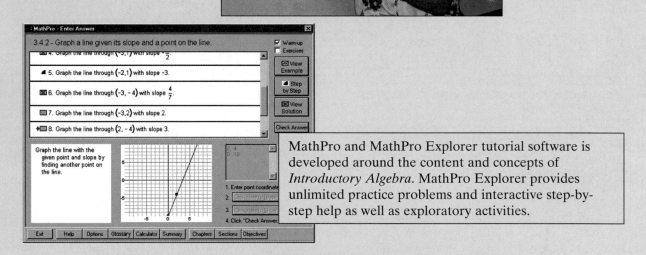

MathPro and MathPro Explorer tutorial software is developed around the content and concepts of *Introductory Algebra*. MathPro Explorer provides unlimited practice problems and interactive step-by-step help as well as exploratory activities.

Also available:

STUDENT SOLUTIONS MANUAL
Laurel Technical Services

Basic College Mathematics

K. Elayn Martin-Gay

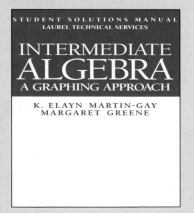

STUDENT SOLUTIONS MANUAL
LAUREL TECHNICAL SERVICES

INTERMEDIATE ALGEBRA
A GRAPHING APPROACH

K. ELAYN MARTIN-GAY
MARGARET GREENE

Ask your instructor or bookstore about these additional study aids.

Prealgebra Review

Mathematics is an important tool for everyday life. Knowing basic mathematical skills can simplify many tasks. For example, we use fractions to represent parts of a whole, such as "half an hour" or "third of a cup." Understanding decimals helps us work efficiently in our money system. Percent is a concept used virtually every day in ordinary and business life.

This optional review chapter covers basic topics and skills from prealgebra. Knowledge of these topics is needed for success in algebra.

R.1   Factors and the Least Common Multiple

R.2   Fractions

R.3   Decimals and Percents

Donald and Doris Fisher opened the first The Gap store (named after the "generation gap") in 1969 near the campus of what is now the San Francisco State University. This single store, which sold mostly Levi's blue jeans, has evolved into an international clothing giant. The Gap, Inc., now sells a variety of clothing through its store chains that include babyGap, Banana Republic, The Gap, GapKids, and Old Navy Clothing. In 1997, The Gap, Inc., had approximately $5.3 billion in revenue. In Exercise 64 on page R-17, we will look at the fraction of the total number of The Gap, Inc., stores accounted for by each store chain.

1. 1, 2, 3, 4, 6, 12 (R.1A)

2. $2 \cdot 3 \cdot 5 \cdot 5$ (R.1B)

3. 280 (R.1C)

4. $\frac{35}{40}$ (R.2A)

5. $\frac{3}{5}$ (R.2B)

6. $\frac{12}{25}$ (R.2B)

7. $\frac{1}{12}$ (R.2C)

8. $\frac{13}{12}$ (R.2D)

9. $\frac{30}{49}$ (R.2C)

10. $\frac{1}{9}$ (R.2D)

11. 78.53 (R.3B)

12. 5.33 (R.3B)

13. 2.432 (R.3B)

14. 34.9 (R.3B)

15. $\frac{716}{100}$ (R.3A)

16. 0.1875 (R.3D)

17. $0.8\overline{3}$ (R.3D)

18. 78.6 (R.3C)

19. 78.62 (R.3C)

20. 0.806 (R.3E)

21. 30% (R.3E)

Name _____ Section _____ Date _____

CHAPTER R PRETEST

1. List the factors of 12.

2. Write the prime factorization of 150.

3. Find the LCM of 8, 14, and 20.

4. Write $\frac{7}{8}$ as an equivalent expression with a denominator of 40.

Simplify each fraction.

5. $\frac{24}{40}$

6. $\frac{120}{250}$

Perform each indicated operation and simplify.

7. $\frac{2}{9} \cdot \frac{3}{8}$

8. $\frac{1}{4} + \frac{5}{6}$

9. $\frac{3}{7} \div \frac{7}{10}$

10. $\frac{2}{3} - \frac{5}{9}$

Perform each indicated operation.

11. $76 + 0.5 + 2.03$

12. $18 - 12.67$

13. $\begin{array}{r} 12.8 \\ \times\ 0.19 \\ \hline \end{array}$

14. $7.5\overline{)261.75}$

15. Write 7.16 as a fraction. Do not simplify.

16. Write $\frac{3}{16}$ as a decimal.

17. Write $\frac{5}{6}$ as a decimal.

18. Round 78.6159 to the nearest tenth.

19. Round 78.6159 to the nearest hundredth.

20. Write 80.6% as a decimal.

21. Write 0.3 as a percent.

R.1   Factors and the Least Common Multiple

A   Factoring Numbers

> To factor means to write as a product.

In arithmetic we factor numbers, and in algebra we factor expressions containing variables. Throughout this text, you will encounter the word *factor* often. Always remember that factoring means writing as a product

Since $2 \cdot 3 = 6$, we say that 2 and 3 are **factors** of 6. Also, $2 \cdot 3$ is a **factorization** of 6.

Example 1   List the factors of 6.

Solution: First we write the different factorizations of 6.

$$6 = 1 \cdot 6, \quad 6 = 2 \cdot 3$$

The factors of 6 are 1, 2, 3, and 6.

Example 2   List the factors of 20.

Solution: $20 = 1 \cdot 20, \; 20 = 2 \cdot 10, \; 20 = 4 \cdot 5, \; 20 = 2 \cdot 2 \cdot 5$

The factors of 20 are 1, 2, 4, 5, 10, and 20.

In this section, we will concentrate on **natural numbers** only. The natural numbers (also called counting numbers) are

Natural Numbers: 1, 2, 3, 4, 5, 6, 7, and so on

Every natural number except 1 is either a prime number or a composite number.

> **Prime and Composite Numbers**
>
> A **prime number** is a natural number greater than 1 whose only factors are 1 and itself. The first few prime numbers are 2, 3, 5, 7, 11, 13, 17, 19, 23, 29, . . .
> A **composite number** is a natural number greater than 1 that is not prime.

Example 3   Identify each number as prime or composite:

3, 20, 7, 4

Solution: 3 is a prime number. Its factors are 1 and 3 only.
20 is a composite number. Its factors are 1, 2, 4, 5, 10, and 20.
7 is a prime number. Its factors are 1 and 7 only.
4 is a composite number. Its factors are 1, 2, and 4.

B   Writing Prime Factorizations

When a number is written as a product of primes, this product is called the **prime factorization** of the number. For example, the prime factorization of 12 is $2 \cdot 2 \cdot 3$ since

$$12 = 2 \cdot 2 \cdot 3$$

and all the factors are prime.

Objectives

A Write the factors of a number.
B Write the prime factorization of a number.
C Find the LCM of a list of numbers.

SSM   CD-ROM   Video
R.1

Practice Problem 1

List the factors of 4.

Practice Problem 2

List the factors of 18.

Practice Problem 3

Identify each number as prime or composite:

5, 18, 11, 6

Answers

1. 1, 2, 4,   **2.** 1, 2, 3, 6, 9, 18,   **3.** 5, 11 prime; 6, 18 composite

Practice Problem 4

Write the prime factorization of 28.

Example 4

Write the prime factorization of 45.

Solution: We can begin by writing 45 as the product of two numbers, say 9 and 5.

$$45 = 9 \cdot 5$$

The number 5 is prime, but 9 is not. So we write 9 as $3 \cdot 3$.

$$45 = 9 \cdot 5$$
$$= 3 \cdot 3 \cdot 5$$

Each factor is now a prime number, so the prime factorization of 45 is $3 \cdot 3 \cdot 5$.

HELPFUL HINT

Recall that order is not important when multiplying numbers. For example,

$$3 \cdot 3 \cdot 5 = 3 \cdot 5 \cdot 3 = 5 \cdot 3 \cdot 3 = 45$$

For this reason, any of the products shown can be called *the* prime factorization of 45.

Practice Problem 5

Write the prime factorization of 60.

Example 5

Write the prime factorization of 80.

Solution: We first write 80 as a product of two numbers. We continue this process until all factors are prime.

$$80 = 8 \cdot 10$$
$$4 \cdot 2 \cdot 2 \cdot 5$$
$$= 2 \cdot 2 \cdot 2 \cdot 2 \cdot 5$$

All factors are now prime, so the prime factorization of 80 is

$$2 \cdot 2 \cdot 2 \cdot 2 \cdot 5$$

TRY THE CONCEPT CHECK IN THE MARGIN.

✓ CONCEPT CHECK

Suppose that you choose $80 = 4 \cdot 20$ as your first step in Example 5 and another student chooses $80 = 5 \cdot 16$. Will both end up with the same prime factorization as in Example 5? Explain.

HELPFUL HINT

There are a few quick **divisibility tests** to determine if a number is divisible by the primes 2, 3, or 5.
A whole number is divisible by

▲ **2** if the ones digit is 0, 2, 4, 6, or 8.

  132 is divisible by 2

▲ **3** if the sum of the digits is divisible by 3.

  144 is divisible by 3 since $1 + 4 + 4 = 9$ is divisible by 3.

▲ **5** if the ones digit is 0 or 5.

  1115 is divisible by 5

Answers

4. $28 = 2 \cdot 2 \cdot 7$, **5.** $60 = 2 \cdot 2 \cdot 3 \cdot 5$

✓ **Concept Check:** yes; answers may vary

Copyright 1000 Prentice Hall Inc

When finding the prime factorization of larger numbers, you may want to use the procedure shown in Example 6.

Example 6   Write the prime factorization of 252.

Solution: Since the ones digit of 252 is 2, we know that 252 is divisible by 2.

$$\frac{126}{2)\overline{252}}$$

126 is divisible by 2 also.

$$\begin{array}{r} 63 \\ 2)\overline{126} \\ 2)\overline{252} \end{array}$$

63 is not divisible by 2 but is divisible by 3. We divide 63 by 3 and continue in this same manner until the quotient is a prime number.

$$\begin{array}{r} 7 \\ 3)\overline{21} \\ 3)\overline{63} \\ 2)\overline{126} \\ 2)\overline{252} \end{array}$$

The prime factorization of 252 is $2 \cdot 2 \cdot 3 \cdot 3 \cdot 7$.

C FINDING THE LEAST COMMON MULTIPLE

A **multiple** of a number is the product of that number and any natural number. For example, the multiples of 3 are

$$\underline{3 \cdot 1} \quad \underline{3 \cdot 2} \quad \underline{3 \cdot 3} \quad \underline{3 \cdot 4} \quad \underline{3 \cdot 5} \quad \underline{3 \cdot 6} \quad \underline{3 \cdot 7}$$

3,   6,   9,   12,   15,   18,   21, and so on

The multiples of 2 are

$$\underline{2 \cdot 1} \quad \underline{2 \cdot 2} \quad \underline{2 \cdot 3} \quad \underline{2 \cdot 4} \quad \underline{2 \cdot 5} \quad \underline{2 \cdot 6} \quad \underline{2 \cdot 7}$$

2,   4,   6,   8,   10,   12,   14, and so on

Notice that 2 and 3 have multiples that are common to both.

Multiples of 2:   2, 4, 6 , 8, 10, 12 , 14, 16, 18 , and so on

Multiples of 3:   3, 6 , 9, 12 , 15, 18 , 21, and so on

The least or smallest common multiple of 2 and 3 is 6. The number 6 is called the **least common multiple** or **LCM** of 2 and 3. It is the smallest number that is a multiple of both 2 and 3.

Finding the LCM by the method above can sometimes be time-consuming. Let's look at another method that uses prime factorization.

To find the LCM of 4 and 10, for example, we write the prime factorization of each.

$$4 = 2 \cdot 2$$
$$10 = 2 \cdot 5$$

If the LCM is to be a multiple of 4, it must contain the factors $2 \cdot 2$. If the LCM is to be a multiple of 10, it must contain the factors $2 \cdot 5$. Since we decide whether the LCM is a multiple of 4 and 10 separately, the LCM does not need to contain three factors of 2. The LCM only needs to contain

Practice Problem 6

Write the prime factorization of 297.

TEACHING TIP

Consider suggesting that students adopt a procedure that can be used each time they look for a prime factorization. For instance, they might try looking first for factors of 2, then 3, then 5, and so on in order of consecutive primes.

Answer

6. $3 \cdot 3 \cdot 3 \cdot 11$

a factor the greatest number of times that the factor appears in any **one** prime factorization.

The LCM is a
multiple of 4.

$$\text{LCM} = \overbrace{2 \cdot \underbrace{2 \cdot 5}} = 20$$

The LCM is a
multiple of 10.

The number 2 is a factor twice since that is the greatest number of times that 2 is a factor in either of the prime factorizations.

TO FIND THE LCM OF A LIST OF NUMBERS

Step 1. Write the prime factorization of each number.

Step 2. Write the product containing each different prime factor (from Step 1) the greatest number of times that it appears in any one factorization. This product is the LCM.

Practice Problem 7

Find the LCM of 14 and 35.

Example 7   Find the LCM of 18 and 24.

Solution: First we write the prime factorization of each number.

$$18 = 2 \cdot 3 \cdot 3$$
$$24 = 2 \cdot 2 \cdot 2 \cdot 3$$

Now we write each factor the greatest number of times that it appears in any **one** prime factorization.

The greatest number of times that 2 appears is **3** times.
The greatest number of times that 3 appears is **2** times.

$$\text{LCM} = \underbrace{2 \cdot 2 \cdot 2}_{\substack{2 \text{ is a factor} \\ 3 \text{ times.}}} \cdot \underbrace{3 \cdot 3}_{\substack{3 \text{ is a factor} \\ 2 \text{ times.}}} = 72$$

Practice Problem 8

Find the LCM of 5 and 9.

TEACHING TIP

You may want to challenge your students to find all the possible solutions to the following: 60 is the least common multiple of 12 and _____

Example 8   Find the LCM of 11 and 6.

Solution: 11 is a prime number, so we simply rewrite it. Then we write the prime factorization of 6.

$$11 = 11$$
$$6 = 2 \cdot 3$$
$$\text{LCM} = 2 \cdot 3 \cdot 11 = 66.$$

Practice Problem 9

Find the LCM of 4, 15, and 10.

Example 9   Find the LCM of 5, 6, and 12.

Solution:
$$5 = 5$$
$$6 = 2 \cdot 3$$
$$12 = 2 \cdot 2 \cdot 3$$
$$\text{LCM} = 2 \cdot 2 \cdot 3 \cdot 5 = 60.$$

Answers

7. 70,   **8.** 45,   **9.** 60

Exercise Set R.1

A *List the factors of each number. See Examples 1 and 2.*

1. 9   **2.** 8   **3.** 24   **4.** 36

5. 42   **6.** 50   **7.** 80   **8.** 63

Identify each number as prime or composite. See Example 3.

9. 13   **10.** 17   **11.** 21   **12.** 39

13. 37   **14.** 41   **15.** 2065   **16.** 1798

B *Write each prime factorization. See Examples 4 through 6.*

17. 18   **18.** 12   **19.** 20   **20.** 30

21. 56   **22.** 48   **23.** 300   **24.** 500

25. 81   **26.** 64   **27.** 588   **28.** 315

C *Find the LCM of each list of numbers. See Examples 7 through 9.*

29. 6, 14   **30.** 9, 15   **31.** 3, 4   **32.** 4, 5

Answers

1. 1, 3, 9
2. 1, 2, 4, 8
3. 1, 2, 3, 4, 6, 8, 12, 24
4. 1, 2, 3, 4, 6, 9, 12, 18, 36
5. 1, 2, 3, 6, 7, 14, 21, 42
6. 1, 2, 5, 10, 25, 50
7. 1, 2, 4, 5, 8, 10, 16, 20, 40, 80
8. 1, 3, 7, 9, 21, 63
9. prime
10. prime
11. composite
12. composite
13. prime
14. prime
15. composite
16. composite
17. $2 \cdot 3 \cdot 3$
18. $2 \cdot 2 \cdot 3$
19. $2 \cdot 2 \cdot 5$
20. $2 \cdot 3 \cdot 5$
21. $2 \cdot 2 \cdot 2 \cdot 7$
22. $2 \cdot 2 \cdot 2 \cdot 2 \cdot 3$
23. $2 \cdot 2 \cdot 3 \cdot 5 \cdot 5$
24. $2 \cdot 2 \cdot 5 \cdot 5 \cdot 5$
25. $3 \cdot 3 \cdot 3 \cdot 3$
26. $2 \cdot 2 \cdot 2 \cdot 2 \cdot 2 \cdot 2$
27. $2 \cdot 2 \cdot 3 \cdot 7 \cdot 7$
28. $3 \cdot 3 \cdot 5 \cdot 7$
29. 42
30. 45
31. 12
32. 20

33. 60

34. 120

35. 35

36. 22

37. 12

38. 18

39. 60

40. 90

41. 350

42. 180

43. 72

44. 126

45. 60

46. 180

47. 30

48. 105

49. 360

50. 140

51. 24

52. 45

53. 2520

54. 9000

55. every 35 days

56. every 140 days

Name _____

33. 20, 30   **34.** 30, 40   **35.** 5, 7   **36.** 2, 11

37. 6, 12   **38.** 6, 18   **39.** 12, 20   **40.** 18, 30

41. 50, 70   **42.** 20, 90   **43.** 24, 36   **44.** 18, 21

45. 5, 10, 12   **46.** 3, 9, 20   **47.** 2, 3, 5   **48.** 3, 5, 7

49. 8, 18, 30   **50.** 4, 14, 35   **51.** 4, 8, 24   **52.** 5, 15, 45

COMBINING CONCEPTS

Find the LCM of each pair of numbers.

53. 315, 504

54. 1000, 1125

55. Craig Campanella and Edie Hall both have night jobs. Craig has every fifth night off and Edie has every seventh night off. How often will they have the same night off?

56. Elizabeth Kaster and Lori Sypher are both publishing company representatives in Louisiana. Elizabeth spends a day in New Orleans every 35 days, and Lori spends a day in New Orleans every 20 days. How often are they in New Orleans on the same day?

R.2 FRACTIONS

A quotient of two numbers such as $\frac{2}{9}$ is called a **fraction**. The parts of a fraction are:

Fraction bar $\rightarrow \dfrac{2}{9} \begin{array}{l} \leftarrow \text{Numerator} \\ \leftarrow \text{Denominator} \end{array}$

$\frac{2}{9}$ of the circle is shaded.

A fraction may be used to refer to part of a whole. For example, $\frac{2}{9}$ of the circle in the figure is shaded. The denominator 9 tells us how many equal parts the whole circle is divided into and the numerator 2 tells us how many equal parts are shaded.

In this section, we will use **whole numbers**. The whole numbers consist of 0 and the natural numbers.

Whole Numbers: 0, 1, 2, 3, 4, 5, and so on

A WRITING EQUIVALENT FRACTIONS

More than one fraction can be used to name the same part of a whole. Such fractions are called **equivalent fractions**.

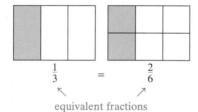

$$\frac{1}{3} = \frac{2}{6}$$

equivalent fractions

> **EQUIVALENT FRACTIONS**
>
> Fractions that represent the same portion of a whole are called **equivalent fractions**.

To write equivalent fractions, we use the **fundamental principle of fractions**. This principle guarantees that, if we multiply both the numerator and the denominator by the same nonzero number, the result is an equivalent fraction. For example, if we multiply the numerator and denominator of $\frac{1}{3}$ by the same number, 2, the result is the equivalent fraction $\frac{2}{6}$.

$$\frac{1 \cdot 2}{3 \cdot 2} = \frac{2}{6}$$

> **FUNDAMENTAL PRINCIPLE OF FRACTIONS**
>
> If a, b, and c are numbers, then
> $$\frac{a}{b} = \frac{a \cdot c}{b \cdot c} \quad \text{or} \quad \frac{a \cdot c}{b \cdot c} = \frac{a}{b}$$
> as long as b and c are not 0.

Objectives

A Write equivalent fractions.
B Write fractions in simplest form.
C Multiply and divide fractions.
D Add and subtract fractions.

SSM   CD-ROM   Video R.2

TEACHING TIP

When describing the Fundamental Principle of Fractions, consider discussing the following problem. Suppose we have two cakes of equal size. One is cut into 6 equal pieces and the other is cut into 12 equal pieces. How many pieces of the 12-piece cake would equal one piece of the 6-piece cake? We can describe this situation using the fractions: $\frac{1}{6} = \frac{1 \cdot 2}{6 \cdot 2} = \frac{2}{12}$.

Practice Problem 1

Write $\frac{1}{4}$ as an equivalent fraction with a denominator of 20.

Practice Problem 2

Simplify: $\frac{20}{35}$

✓ CONCEPT CHECK

Explain the error in the following steps.

a. $\frac{15}{55} = \frac{1\boxed{5}}{5\boxed{5}} = \frac{1}{5}$

b. $\frac{6}{7} = \frac{\boxed{5+1}}{\boxed{5+2}} = \frac{1}{2}$

Practice Problems 3–4

Simplify each fraction.

3. $\frac{7}{20}$

4. $\frac{12}{40}$

Example 1

Write $\frac{2}{5}$ as an equivalent fraction with a denominator of 15.

Solution: Since $5 \cdot 3 = 15$, we use the fundamental principle of fractions and multiply the numerator and denominator of $\frac{2}{5}$ by 3.

$$\frac{2}{5} = \frac{2 \cdot 3}{5 \cdot 3} = \frac{6}{15}$$

Then $\frac{2}{5}$ is equivalent to $\frac{6}{15}$. They both represent the same part of a whole.

B SIMPLIFYING FRACTIONS

A fraction is said to be **simplified** or in **lowest terms** when the numerator and the denominator have no factors in common other than 1. For example, the fraction $\frac{5}{11}$ is in lowest terms since 5 and 11 have no common factors other than 1.

One way to simplify fractions is to write both the numerator and the denominator as a product of primes and then apply the fundamental principle of fractions.

Example 2

Simplify: $\frac{42}{49}$

Solution: We write the numerator and the denominator as products of primes. Then we apply the fundamental principle of fractions to the common factor 7.

$$\frac{42}{49} = \frac{2 \cdot 3 \cdot \boxed{7}}{7 \cdot \boxed{7}} = \frac{2 \cdot 3}{7} = \frac{6}{7}$$

TRY THE CONCEPT CHECK IN THE MARGIN.

Examples

Simplify each fraction.

3. $\frac{11}{27} = \frac{11}{3 \cdot 3 \cdot 3}$ There are no common factors other than 1, so $\frac{11}{27}$ is already simplified.

4. $\frac{88}{20} = \frac{\boxed{2} \cdot \boxed{2} \cdot 2 \cdot 11}{\boxed{2} \cdot \boxed{2} \cdot 5} = \frac{22}{5}$

The improper fraction $\frac{22}{5}$ from Example 4 may be written as the mixed number $4\frac{2}{5}$, but in this text, we will not do so.

Some fractions may be simplified by recalling that the fraction bar means division.

$$\frac{6}{6} = 6 \div 6 = 1 \quad \text{and} \quad \frac{3}{1} = 3 \div 1 = 3$$

Answers

1. $\frac{5}{20}$, 2. $\frac{4}{7}$, 3. $\frac{7}{20}$, 4. $\frac{3}{10}$

✓ Concept Check: answers may vary

Examples   Simplify by dividing the numerator by the denominator.

5. $\dfrac{3}{3} = 3 \div 3 = 1$

6. $\dfrac{4}{2} = 4 \div 2 = 2$

7. $\dfrac{7}{7} = 7 \div 7 = 1$

8. $\dfrac{8}{1} = 8 \div 1 = 8$

▬

In general, if the numerator and the denominator are the same, the fraction is equivalent to 1. Also, if the denominator of a fraction is 1, the fraction is equivalent to the numerator.

If a is any number other than 0, then $\dfrac{a}{a} = 1$.

Also, if a is any number, $\dfrac{a}{1} = a$.

C MULTIPLYING AND DIVIDING FRACTIONS

To multiply two fractions, we multiply numerator times numerator to obtain the numerator of the product. Then we multiply denominator times denominator to obtain the denominator of the product.

MULTIPLYING FRACTIONS

$\dfrac{a}{b} \cdot \dfrac{c}{d} = \dfrac{a \cdot c}{b \cdot d}$,   if $b \neq 0$ and $d \neq 0$

Example 9   Multiply: $\dfrac{2}{15} \cdot \dfrac{5}{13}$. Simplify the product if possible.

Solution: $\dfrac{2}{15} \cdot \dfrac{5}{13} = \dfrac{2 \cdot 5}{15 \cdot 13}$ Multiply numerators.
Multiply denominators.

To simplify the product, we divide the numerator and the denominator by any common factors.

$= \dfrac{2 \cdot 5}{3 \cdot 5 \cdot 13}$

$= \dfrac{2}{39}$

▬

TEACHING TIP

Before Example 9, help students visualize multiplying fractions of the form $\dfrac{1}{a} \cdot \dfrac{1}{b}$. For instance, $\dfrac{1}{3} \cdot \dfrac{1}{4}$ can be visualized by letting a square represent 1 whole. Represent $\dfrac{1}{4}$ by dividing the square into 4 equal vertical strips. Represent $\dfrac{1}{3}$ of $\dfrac{1}{4}$ by dividing one of the vertical strips into 3 equal horizontal strips. Now ask your students how much of the whole is $\dfrac{1}{3}$ of $\dfrac{1}{4}$?

Before we divide fractions, we first define **reciprocals**. Two numbers are reciprocals of each other if their product is 1.

The reciprocal of $\dfrac{2}{3}$ is $\dfrac{3}{2}$ because $\dfrac{2}{3} \cdot \dfrac{3}{2} = \dfrac{6}{6} = 1$.

The reciprocal of 5 is $\dfrac{1}{5}$ because $5 \cdot \dfrac{1}{5} = \dfrac{5}{1} \cdot \dfrac{1}{5} = \dfrac{5}{5} = 1$.

To divide fractions, we multiply the first fraction by the reciprocal of the second fraction. For example,

$$\frac{1}{2} \div \frac{5}{7} = \frac{1}{2} \cdot \frac{7}{5} = \frac{1 \cdot 7}{2 \cdot 5} = \frac{7}{10}$$

HELPFUL HINT   To divide, multiply by the reciprocal.

DIVIDING FRACTIONS

$$\frac{a}{b} \div \frac{c}{d} = \frac{a}{b} \cdot \frac{d}{c}, \qquad \text{if } b \neq 0, d \neq 0, \text{ and } c \neq 0$$

Practice Problems 10–12

Divide and simplify.

10. $\dfrac{2}{9} \div \dfrac{3}{4}$

11. $\dfrac{8}{11} \div 24$

12. $\dfrac{5}{4} \div \dfrac{5}{8}$

Examples   Divide and simplify.

10. $\dfrac{4}{5} \div \dfrac{5}{16} = \dfrac{4}{5} \cdot \dfrac{16}{5} = \dfrac{4 \cdot 16}{5 \cdot 5} = \dfrac{64}{25}$

11. $\dfrac{7}{10} \div 14 = \dfrac{7}{10} \div \dfrac{14}{1} = \dfrac{7}{10} \cdot \dfrac{1}{14} = \dfrac{7 \cdot 1}{2 \cdot 5 \cdot 2 \cdot 7} = \dfrac{1}{20}$

12. $\dfrac{3}{8} \div \dfrac{3}{10} = \dfrac{3}{8} \cdot \dfrac{10}{3} = \dfrac{3 \cdot 2 \cdot 5}{2 \cdot 2 \cdot 2 \cdot 3} = \dfrac{5}{4}$

D   ADDING AND SUBTRACTING FRACTIONS

To add or subtract fractions with the same denominator, we combine numerators and place the sum or difference over the common denominator.

ADDING AND SUBTRACTING FRACTIONS WITH THE SAME DENOMINATOR

$$\frac{a}{b} + \frac{c}{b} = \frac{a+c}{b}, \qquad \text{if } b \neq 0$$

$$\frac{a}{b} - \frac{c}{b} = \frac{a-c}{b}, \qquad \text{if } b \neq 0$$

Answers

10. $\dfrac{8}{27}$,   11. $\dfrac{1}{33}$,   12. 2

Examples Add or subtract as indicated. Then simplify if possible.

13. $\dfrac{2}{7} + \dfrac{4}{7} = \dfrac{2+4}{7} = \dfrac{6}{7}$

14. $\dfrac{3}{10} + \dfrac{2}{10} = \dfrac{3+2}{10} = \dfrac{5}{10} = \dfrac{5}{2 \cdot 5} = \dfrac{1}{2}$

15. $\dfrac{9}{7} - \dfrac{2}{7} = \dfrac{9-2}{7} = \dfrac{7}{7} = 1$

16. $\dfrac{5}{3} - \dfrac{1}{3} = \dfrac{5-1}{3} = \dfrac{4}{3}$

To add or subtract with different denominators, we first write the fractions as **equivalent fractions** with the same denominator. We will use the smallest or least common denominator, the LCD. The LCD is the same as the least common multiple we reviewed in Section R.1.

Example 17 Add: $\dfrac{2}{5} + \dfrac{1}{4}$

Solution: We first must find the least common denominator before the fractions can be added. The least common multiple for the denominators 5 and 4 is 20. This is the LCD we will use.

We write both fractions as equivalent fractions with denominators of 20. Since

$$\dfrac{2}{5} = \dfrac{2 \cdot 4}{5 \cdot 4} = \dfrac{8}{20} \quad \text{and} \quad \dfrac{1}{4} = \dfrac{1 \cdot 5}{4 \cdot 5} = \dfrac{5}{20}$$

then

$$\dfrac{2}{5} + \dfrac{1}{4} = \dfrac{8}{20} + \dfrac{5}{20} = \dfrac{13}{20}$$

Example 18 Subtract and simplify: $\dfrac{19}{6} - \dfrac{23}{12}$

Solution: The LCD is 12. We write both fractions as equivalent fractions with denominators of 12.

$$\dfrac{19}{6} - \dfrac{23}{12} = \dfrac{19 \cdot 2}{6 \cdot 2} - \dfrac{23}{12}$$
$$= \dfrac{38}{12} - \dfrac{23}{12}$$
$$= \dfrac{15}{12} = \dfrac{3 \cdot 5}{2 \cdot 2 \cdot 3} = \dfrac{5}{4}$$

Add or subtract as indicated. Then simplify if possible.

13. $\dfrac{2}{11} + \dfrac{5}{11}$

14. $\dfrac{1}{8} + \dfrac{3}{8}$

15. $\dfrac{13}{10} - \dfrac{3}{10}$

16. $\dfrac{7}{6} - \dfrac{2}{6}$

Add: $\dfrac{3}{8} + \dfrac{1}{20}$

Subtract and simplify: $\dfrac{8}{15} - \dfrac{1}{3}$

TEACHING TIP

Consider ending this lesson with the following activities:

▲ Will the sum of two fractions which are less than 1 ever be greater than 1? If so, give an example.
▲ Will the product of two fractions which are less than 1 ever be greater than 1? If so, give an example.
▲ Will the quotient of two fractions which are less than 1 ever be greater than 1? If so, give an example.

Answers

13. $\dfrac{7}{11}$, 14. $\dfrac{1}{2}$, 15. 1, 16. $\dfrac{5}{6}$, 17. $\dfrac{17}{40}$,

18. $\dfrac{1}{5}$

Focus On Study Skills

CRITICAL THINKING

What is Critical Thinking?

Although exact definitions often vary, thinking critically usually refers to evaluating, analyzing, and interpreting information in order to make a decision, draw a conclusion, reach a goal, make a prediction, or form an opinion. Critical thinking often involves problem solving, communication, and reasoning skills. Critical thinking is more than a technique that helps students pass their courses. Critical thinking skills are life skills. Developing these skills can help you solve problems in your workplace and in everyday life. For instance, well-developed critical thinking skills would be useful in the following situation:

> Suppose you work as a medical lab technician. Your lab supervisor has decided that some lab equipment should be replaced. She asks you to collect information on several different models from equipment manufacturers. Your assignment is to study the data and then make a recommendation on which model the lab should buy.

How Can Critical Thinking be Developed?

Just as physical exercise can help to develop and strengthen certain muscles of the body, mental exercise can help to develop critical thinking skills. Mathematics is ideal for helping to develop critical thinking skills because it requires using logic and reasoning, recognizing patterns, making conjectures and educated guesses, and drawing conclusions. You will find many opportunities to build your critical thinking skills throughout *Introductory Algebra*:

▲ In real-life application problems (see Exercise 24 in Section 2.5)
▲ In conceptual and writing exercises marked with the ◻ icon (see Exercise 43 in Section 1.2)
▲ In the Combining Concepts subsection of exercise sets (see Exercise 57 in Section 3.4)
▲ In the Chapter Activities (see the Chapter 7 Activity)
▲ In the Critical Thinking and Group Activities found in Focus On features like this one throughout the book (see pages 46 and 404).

Name _____ Section _____ Date _____

ANSWERS

1. $\frac{21}{30}$

2. $\frac{6}{9}$

3. $\frac{4}{18}$

4. $\frac{64}{56}$

5. $\frac{16}{20}$

6. $\frac{20}{25}$

7. $\frac{1}{2}$

8. $\frac{1}{2}$

9. $\frac{2}{3}$

10. $\frac{3}{4}$

11. $\frac{3}{7}$

12. $\frac{5}{9}$

13. 1

14. 1

15. 5

16. 7

17. $\frac{3}{5}$

18. $\frac{14}{15}$

19. $\frac{4}{5}$

20. $\frac{1}{5}$

21. $\frac{11}{8}$

22. $\frac{8}{3}$

23. $\frac{30}{61}$

24. $\frac{18}{35}$

25. $\frac{3}{8}$

26. 1

27. $\frac{1}{2}$

28. $\frac{1}{8}$

EXERCISE SET R.2

A *Write each fraction as an equivalent fraction with the given denominator. See Example 1.*

1. $\frac{7}{10}$ with a denominator of 30   2. $\frac{2}{3}$ with a denominator of 9

3. $\frac{2}{9}$ with a denominator of 18   4. $\frac{8}{7}$ with a denominator of 56

5. $\frac{4}{5}$ with a denominator of 20   6. $\frac{4}{5}$ with a denominator of 25

B *Simplify each fraction. See Examples 2 through 8.*

7. $\frac{2}{4}$ 8. $\frac{3}{6}$ 9. $\frac{10}{15}$ 10. $\frac{15}{20}$

11. $\frac{3}{7}$ 12. $\frac{5}{9}$ 13. $\frac{20}{20}$ 14. $\frac{24}{24}$

15. $\frac{35}{7}$ 16. $\frac{42}{6}$ 17. $\frac{18}{30}$ 18. $\frac{42}{45}$

19. $\frac{16}{20}$ 20. $\frac{8}{40}$ 21. $\frac{66}{48}$ 22. $\frac{64}{24}$

23. $\frac{120}{244}$ 24. $\frac{360}{700}$

C *Multiply or divide as indicated. See Examples 9 through 12.*

25. $\frac{1}{2} \cdot \frac{3}{4}$ 26. $\frac{10}{6} \cdot \frac{3}{5}$ 27. $\frac{2}{3} \cdot \frac{3}{4}$ 28. $\frac{7}{8} \cdot \frac{3}{21}$

29. $\dfrac{6}{7}$

30. $\dfrac{7}{6}$

31. 15

32. $\dfrac{2}{3}$

33. $\dfrac{1}{6}$

34. $\dfrac{2}{147}$

35. $\dfrac{3}{80}$

36. $\dfrac{5}{72}$

37. 1

38. 1

39. $\dfrac{3}{5}$

40. $\dfrac{5}{7}$

41. $\dfrac{9}{35}$

42. $\dfrac{4}{11}$

43. $\dfrac{1}{3}$

44. $\dfrac{1}{5}$

45. $\dfrac{23}{21}$

46. $\dfrac{11}{12}$

47. $\dfrac{65}{21}$

48. $\dfrac{52}{35}$

49. $\dfrac{5}{7}$

50. $\dfrac{2}{5}$

51. $\dfrac{5}{66}$

52. $\dfrac{1}{6}$

53. $\dfrac{7}{5}$

54. $\dfrac{13}{8}$

55. $\dfrac{17}{18}$

56. $\dfrac{43}{44}$

57. $\dfrac{1}{5}$

58. $\dfrac{6}{11}$

59. $\dfrac{3}{8}$

60. $\dfrac{1}{12}$

Name _____

29. $\dfrac{1}{2} \div \dfrac{7}{12}$ 30. $\dfrac{7}{12} \div \dfrac{1}{2}$ 31. $\dfrac{3}{4} \div \dfrac{1}{20}$ 32. $\dfrac{3}{5} \div \dfrac{9}{10}$

33. $\dfrac{7}{10} \cdot \dfrac{5}{21}$ 34. $\dfrac{3}{35} \cdot \dfrac{10}{63}$ 35. $\dfrac{9}{20} \div 12$ 36. $\dfrac{25}{36} \div 10$

D *Add or subtract as indicated. See Examples 13 through 18.*

37. $\dfrac{4}{5} + \dfrac{1}{5}$ 38. $\dfrac{6}{7} + \dfrac{1}{7}$ 39. $\dfrac{4}{5} - \dfrac{1}{5}$ 40. $\dfrac{6}{7} - \dfrac{1}{7}$

41. $\dfrac{23}{105} + \dfrac{4}{105}$ 42. $\dfrac{13}{132} + \dfrac{35}{132}$ 43. $\dfrac{17}{21} - \dfrac{10}{21}$ 44. $\dfrac{18}{35} - \dfrac{11}{35}$

45. $\dfrac{2}{3} + \dfrac{3}{7}$ 46. $\dfrac{3}{4} + \dfrac{1}{6}$ 47. $\dfrac{10}{3} - \dfrac{5}{21}$ 48. $\dfrac{11}{7} - \dfrac{3}{35}$

49. $\dfrac{10}{21} + \dfrac{5}{21}$ 50. $\dfrac{11}{35} + \dfrac{3}{35}$ 51. $\dfrac{5}{22} - \dfrac{5}{33}$ 52. $\dfrac{7}{10} - \dfrac{8}{15}$

53. $\dfrac{12}{5} - 1$ 54. $2 - \dfrac{3}{8}$ 55. $\dfrac{2}{3} - \dfrac{5}{9} + \dfrac{5}{6}$ 56. $\dfrac{8}{11} - \dfrac{1}{4} + \dfrac{1}{2}$

COMBINING CONCEPTS

Each circle below represents a whole, or 1. Determine the unknown part of the circle.

57.   58.   59.   60.

Name _____

61. In 1985, United States citizens needed $8\frac{1}{2}$ or $\frac{17}{2}$ years of work credit to have fully insured status to receive Social Security benefits. In 1988, $9\frac{1}{4}$ or $\frac{37}{4}$ years of work credit were necessary to receive fully insured status. How many more years of work credit were needed in 1988 than in 1985? (*Source:* Social Security Administration)

62. Approximately $\frac{29}{50}$ of Americans either do not exercise regularly or do not exercise at all. What fraction of Americans exercise regularly? (*Source:* Random House. Information Resources, Inc.)

63. Land use in the United States is summarized in the graph shown, called a circle graph or pie chart. Use the graph to answer the questions. (*Source:* The Central Intelligence Agency, *1996 World Factbook*)

Land Use in the United States

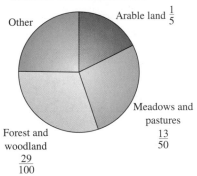

Other

Arable land $\frac{1}{5}$

Meadows and pastures $\frac{13}{50}$

Forest and woodland $\frac{29}{100}$

a. What fraction of land in the United States is arable?

b. Meadows and pastures make up what fraction of land in the United States?

c. What fraction of land falls in the "other" category?

d. What fractional part of land is classified as either meadows and pastures or forest and woodland?

64. The Gap, Inc., operated a total of 1883 stores in 1997. The table lists the various stores operated by The Gap, Inc., and the fraction of the total number of stores accounted for by each type. (*Source:* The Gap, Inc.)

Store Name	Fraction of Total
The Gap	$\frac{1}{2}$
GapKids	$\frac{13}{50}$
Banana Republic	$\frac{3}{25}$
Old Navy Clothing	$\frac{11}{100}$
babyGap	$\frac{1}{100}$
Total:	1

a. What fraction of the stores in operation in 1997 contained the word "Gap"?

b. What fraction of the stores in operation in 1997 were *not* aimed primarily at infants and children?

61. $\frac{3}{4}$ year

62. $\frac{21}{50}$

63. a. $\frac{1}{5}$

b. $\frac{13}{50}$

c. $\frac{1}{4}$

d. $\frac{11}{20}$

64. a. $\frac{77}{100}$

b. $\frac{73}{100}$

Name _____

 Internet Excursions

Go to http://www.prenhall.com/martin-gay

Publicly held corporations sell shares of their company's stock on a stock exchange such as the New York Stock Exchange. Stock prices are frequently reported as fractions or mixed numbers. Many sites on the World Wide Web allow you to look up stock prices if you know its ticker symbol. For instance, by visiting the given World Wide Web address you will have access to the CNN Financial Network Web site, or a related site, where you can find today's opening, high, and low prices and the previous day's closing price for any stock by entering its ticker symbol. You can also find the stock's highest price and lowest price in the past 52 weeks.

65. Look up current stock prices for Coca-Cola Co. (ticker symbol: KO). Use the information on the Website to complete the table below. What is the difference between today's high price and today's low price?

Date	
Time	
Today's high	
Today's low	

66. Look up the current stock prices for Wal-Mart Stores, Inc. (ticker symbol: WMT). Use the information on the Website to complete the table below. How much more was the 52-week high price than the 52-week low price?

Date	
Time	
52-week high	
52-week low	

R.3 DECIMALS AND PERCENTS

A WRITING DECIMALS AS FRACTIONS

Like fractional notation, decimal notation is used to denote a part of a whole. Below is a place value chart that shows the value of each place.

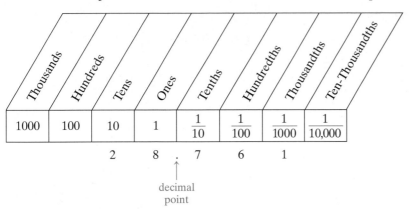

TRY THE CONCEPT CHECK IN THE MARGIN.

The next chart shows decimals written as fractions.

Decimal Form	Fractional Form
0.1	$\dfrac{1}{10}$
tenths	
0.07	$\dfrac{7}{100}$
hundredths	
2.31	$\dfrac{231}{100}$
hundredths	
0.9862	$\dfrac{9862}{10,000}$
ten-thousandths	

Examples Write each decimal as a fraction. Do not simplify.

1. $0.37 = \dfrac{37}{100}$
   2 decimal   2 zeros
   places

2. $1.3 = \dfrac{13}{10}$
   1 decimal   1 zero
   place

3. $2.649 = \dfrac{2649}{1000}$
   3 decimal   3 zeros
   places

TEACHING TIP

It may be helpful to point out to students the symmetry about the ones column on both sides of the place value chart.

✓ CONCEPT CHECK

Fill in the blank: In the number 52.634, the 3 is in the _____ place.

a. Tens

b. Ones

c. Tenths

d. Hundredths

e. Thousandths

Practice Problems 1–3

Write each decimal as a fraction. Do not simplify.

1. 0.27

2. 5.1

3. 7.685

Answers
✓Concept Check: d
1. $\dfrac{27}{100}$, 2. $\dfrac{51}{10}$, 3. $\dfrac{7685}{1000}$

B ADDING, SUBTRACTING, MULTIPLYING, AND DIVIDING DECIMALS

To **add** or **subtract** decimals, we write the numbers vertically with decimal points lined up. We then add the like place values from right to left. We place the decimal point in the answer directly below the decimal points in the problem.

Practice Problem 4

Add.

a. $7.19 + 19.782 + 1.006$

b. $12 + 0.79 + 0.03$

Example 4   Add.

a. $5.87 + 23.279 + 0.003$ b. $7 + 0.23 + 0.6$

Solution: a.
$$\begin{array}{r} 5.87 \\ 23.279 \\ +\ 0.003 \\ \hline 29.152 \end{array}$$

b.
$$\begin{array}{r} 7. \\ 0.23 \\ +\ 0.6 \\ \hline 7.83 \end{array}$$

Practice Problem 5

Subtract.

a. $84.23 - 26.982$

b. $90 - 0.19$

Example 5   Subtract.

a. $32.15 - 11.237$ b. $70 - 0.48$

Solution: a.
$$\begin{array}{r} 1\ \ \ 11\ \ 4\ \ 10 \\ 3\ \cancel{2}\ .\ \cancel{1}\ \cancel{5}\ \cancel{0} \\ -\ 1\ 1\ .\ 2\ 3\ 7 \\ \hline 2\ 0\ .\ 9\ 1\ 3 \end{array}$$

b.
$$\begin{array}{r} 6\ 9\ \ \ 9\ 10 \\ \cancel{7}\ \cancel{0}\ .\ \cancel{0}\ \cancel{0} \\ -\ \ \ 0\ .\ 4\ 8 \\ \hline 6\ 9\ .\ 5\ 2 \end{array}$$

Now let's study the following product of decimals. Notice the pattern in the decimal points.

$$0.\underset{\substack{\uparrow \\ 2\ \text{decimal} \\ \text{places}}}{03} \times 0.\underset{\substack{\uparrow \\ 1\ \text{decimal} \\ \text{place}}}{6} = \frac{3}{100} \times \frac{6}{10} = \frac{18}{1000} \quad \text{or} \quad 0.\underset{\substack{\uparrow \\ 3\ \text{decimal} \\ \text{places}}}{018}$$

In general, to **multiply** decimals we multiply the numbers as if they were whole numbers. The decimal point in the product is placed so that the number of decimal places in the product is the same as the *sum* of the number of decimal places in the factors.

Practice Problem 6

Multiply.

a. 0.31×4.6

b. 1.26×0.03

Example 6   Multiply.

a. 0.072×3.5 b. 0.17×0.02

Solution: a.
$$\begin{array}{rl} 0.072 & \text{3 decimal places} \\ \times\ \ \ 3.5 & \text{1 decimal place} \\ \hline 360 & \\ 216\ \ \ & \\ \hline 0.2520 & \text{4 decimal places} \end{array}$$

b.
$$\begin{array}{rl} 0.17 & \text{2 decimal places} \\ \times\ 0.02 & \text{2 decimal places} \\ \hline 0.0034 & \text{4 decimal places} \end{array}$$

To divide a decimal by a whole number using long division, we place the decimal point in the quotient directly above the decimal point in the dividend. For example,

$$\begin{array}{r} 2.47 \\ 3\overline{\smash{)}7.41} \\ -6 \\ \hline 1\ 4 \\ -1\ 2 \\ \hline 2\ 1 \\ -2\ 1 \\ \hline 0 \end{array}$$

To check, see that

$2.47 \times 3 = 7.41$

TEACHING TIP

Before discussing adding and subtracting decimals, consider using the following activity: Ask students to explain in writing, as if to someone who knows nothing about our money system, how to add 53 cents and 17 dollars and 12 cents. Ask them to share their explanation. Then ask them to use decimals to represent the problem.

Answers

4. a. 27.978,   **b.** 12.82,   **5. a.** 57.248,
b. 89.81,   **6. a.** 1.426,   **b.** 0.0378

In general, to **divide** decimals we move the decimal point in the divisor to the right until the divisor is a whole number. Then we move the decimal point in the dividend the same number of places that the decimal point in the divisor was moved. The decimal point in the quotient lies directly above the decimal point in the dividend.

Example 7   Divide.

a. $9.46 \div 0.04$ **b.** $31.5 \div 0.007$

Solution: **a.**
$$
\begin{array}{r}
236.5 \\
04.\overline{)946.0} \\
-8 \\
\hline
14 \\
-12 \\
\hline
26 \\
-24 \\
\hline
20 \\
-20 \\
\hline
0
\end{array}
$$

b.
$$
\begin{array}{r}
4500. \\
0007.\overline{)31500.} \\
-28 \\
\hline
35 \\
-35 \\
\hline
0
\end{array}
$$

Practice Problem 7

Divide.

a. $21.75 \div 0.5$

b. $15.6 \div 0.006$

C ROUNDING DECIMALS

We **round** the decimal part of a decimal number in nearly the same way as we round the whole numbers. The only difference is that we drop digits to the right of the rounding place, instead of replacing these digits by 0 s. For example,

24.954   rounded to the nearest hundredth is   24.95
   ↑

TO ROUND DECIMALS TO A PLACE VALUE TO THE RIGHT OF THE DECIMAL POINT

Step 1. Locate the digit to the right of the given place value.

Step 2. ▲ If this digit is 5 or greater, add 1 to the digit in the given place value and drop all digits to its right.

  ▲ If this digit is less than 5, drop all digits to the right of the given place.

Example 8   Round 7.8265 to the nearest hundredth.

Solution:
     ┌ hundredths place
7.8265
   ↑___ Step 1. Locate the digit to the right of the hundredths place.

     Step 2. This digit is 5 or greater, so we add 1 to the hundredths place digit and drop all digits to its right.

Thus, 7.8265 rounded to the nearest hundredth is 7.83.

Practice Problem 8

Round 12.9187 to the nearest hundredth.

Example 9   Round 19.329 to the nearest tenth.

Solution:
     ┌ tenths place
19.329
   ↑___ Step 1. Locate the digit to the right of the tenths place.

     Step 2. This digit is less than 5, so we drop this digit and all digits to its right.

Thus, 19.329 rounded to the nearest tenth is 19.3.

Practice Problem 9

Round 245.348 to the nearest tenth.

Answers

7. a. 43.5, **b.** 2600, **8.** 12.92, **9.** 245.3

D WRITING FRACTIONS AS DECIMALS

To write fractions as decimals, interpret the fraction bar as division and find the quotient.

> ### WRITING FRACTIONS AS DECIMALS
>
> To write fractions as decimals, divide the numerator by the denominator.

Practice Problem 10

Write $\frac{2}{5}$ as a decimal.

Example 10 Write $\frac{1}{4}$ as a decimal.

Solution:

$$
\begin{array}{r}
0.25 \\
4\overline{)1.00} \\
-8 \\
\hline
20 \\
-20 \\
\hline
0
\end{array}
$$

$$\frac{1}{4} = 0.25$$

Practice Problem 11

Write $\frac{5}{6}$ as a decimal.

Example 11 Write $\frac{2}{3}$ as a decimal.

Solution:

$$
\begin{array}{r}
0.666 \\
3\overline{)2.000} \\
-1\,8 \\
\hline
20 \\
-18 \\
\hline
20 \\
-18 \\
\hline
2
\end{array}
$$

This pattern will continue so that $\frac{2}{3} = 0.6666\ldots$.

A bar can be placed over the digit 6 to indicate that it repeats.

$$\frac{2}{3} = 0.666\ldots = 0.\overline{6}$$

We can also write a decimal approximation for $\frac{2}{3}$. For example, $\frac{2}{3}$ rounded to the nearest hundredth is 0.67. This can be written as $\frac{2}{3} \approx 0.67$. The $\approx$ sign means "is approximately equal to."

✓ CONCEPT CHECK

The notation $0.5\overline{2}$ is the same as

a. $\frac{52}{100}$

b. $\frac{52\ldots}{100}$

c. $0.52222222\ldots$

TRY THE CONCEPT CHECK IN THE MARGIN.

Practice Problem 12

Write $\frac{1}{9}$ as a decimal. Round to the nearest thousandth.

Example 12 Write $\frac{22}{7}$ as a decimal. Round to the nearest hundredth.

(The fraction $\frac{22}{7}$ is an approximation for π.)

Answers

10. 0.4, **11.** 0.8$\overline{3}$, **12.** 0.111

✓Concept Check: c

Solution:

$$\begin{array}{r} 3.142 \approx 3.14 \\ 7\overline{)22.000} \\ -21 \\ \hline 10 \\ -7 \\ \hline 30 \\ -28 \\ \hline 20 \\ -14 \\ \hline 6 \end{array}$$

If rounding to the nearest hundredth, carry the division process out to one more decimal place, the thousandths place.

The fraction $\dfrac{22}{7}$ in decimal form is approximately 3.14.

E WRITING PERCENTS AS DECIMALS AND DECIMALS AS PERCENTS

The word **percent** comes from the Latin phrase *per centum*, which means **"per 100."** Thus, 53% means 53 per 100, or

$$53\% = \frac{53}{100}$$

When solving problems containing percents, it is often necessary to write a percent as a decimal. To see how this is done, study the chart below.

Percent	Fraction	Decimal
7%	$\dfrac{7}{100}$	0.07
63%	$\dfrac{63}{100}$	0.63
109%	$\dfrac{109}{100}$	1.09

To convert directly from a percent to a decimal, notice that

$$7\% = 0.07$$

TO WRITE A PERCENT AS A DECIMAL

Drop the percent symbol and move the decimal point two places to the left.

Example 13 Write each percent as a decimal.

   a. 25%   **b.** 2.6%   **c.** 195%

Solution: We drop the % and move the decimal point two places to the left. Recall that the decimal point of a whole number is to the right of the ones place digit.

   a. $25\% = 25.\% = 0.25$

   b. $2.6\% = 02.6\% = 0.026$

   c. $195\% = 195.\% = 1.95$

Practice Problem 13

Write each percent as a decimal.

a. 20%

b. 1.2%

c. 465%

Answers

13. a. 0.20, **b.** 0.12, **c.** 4.65

To write a decimal as a percent, we simply reverse the preceding steps. That is, we move the decimal point two places to the right and attach the percent symbol, %.

TO WRITE A DECIMAL AS A PERCENT

Move the decimal point two places to the right and attach the percent symbol,%.

Practice Problem 14

Write each decimal as a percent.

a. 0.42

b. 0.003

c. 2.36

d. 0.7

Example 14

Write each decimal as a percent.

a. 0.85   **b.** 1.25   **c.** 0.012   **d.** 0.6

Solution: We move the decimal point two places to the right and attach the percent symbol, %.

a. $0.85 = 0.85 = 85\%$

b. $1.25 = 1.25 = 125\%$

c. $0.012 = 0.012 = 1.2\%$

d. $0.6 = 0.60 = 60\%$

Answers

14. a. 42%, **b.** 0.3%, **c.** 236%, **d.** 70%

Name _____ **Section** _____ **Date** _____

Exercise Set R.3

A *Write each decimal as a fraction. Do not simplify. See Examples 1 through 3.*

1. 0.6   **2.** 0.9   **3.** 1.86   **4.** 7.23

5. 0.114   **6.** 0.239   **7.** 123.1   **8.** 892.7

B *Add or subtract as indicated. See Examples 4 and 5.*

9. $5.7 + 1.13$ **10.** $2.31 + 6.4$ **11.** $24.6 + 2.39 + 0.0678$

12. $32.4 + 1.58 + 0.0934$ **13.** $8.8 - 2.3$ **14.** $7.6 - 2.1$

15. $18 - 2.78$ **16.** $28 - 3.31$ **17.**
$$
\begin{array}{r}
45.02 \\
3.006 \\
+ \ 8.405 \\
\hline
\end{array}
$$

18.
$$
\begin{array}{r}
65.0028 \\
5.0903 \\
+ \ 6.9 \\
\hline
\end{array}
$$
19.
$$
\begin{array}{r}
654.9 \\
- \ 56.67 \\
\hline
\end{array}
$$
20.
$$
\begin{array}{r}
863.2 \\
- \ 39.45 \\
\hline
\end{array}
$$

Multiply or divide as indicated. See Examples 6 and 7.

21.
$$
\begin{array}{r}
0.2 \\
\times \ 0.6 \\
\hline
\end{array}
$$
22.
$$
\begin{array}{r}
0.7 \\
\times \ 0.9 \\
\hline
\end{array}
$$
23.
$$
\begin{array}{r}
6.75 \\
\times \ 10 \\
\hline
\end{array}
$$
24.
$$
\begin{array}{r}
8.91 \\
\times \ 100 \\
\hline
\end{array}
$$

25.
$$
\begin{array}{r}
5.62 \\
\times \ 7.7 \\
\hline
\end{array}
$$
26.
$$
\begin{array}{r}
8.03 \\
\times \ 5.5 \\
\hline
\end{array}
$$
27.
$$
\begin{array}{r}
16.003 \\
\times \ 5.31 \\
\hline
\end{array}
$$
28.
$$
\begin{array}{r}
31.006 \\
\times \ 3.71 \\
\hline
\end{array}
$$

29. $5\overline{)0.47}$ **30.** $2\overline{)11.7}$ **31.** $0.6\overline{)42}$ **32.** $0.9\overline{)36}$

33. $0.82\overline{)4.756}$ **34.** $0.92\overline{)3.312}$ **35.** $0.063\overline{)52.92}$ **36.** $0.054\overline{)51.84}$

37. 0.6

38. 0.6

39. 0.23

40. 0.45

41. 0.594

42. 63.452

43. 98,207.2

44. 68,936.5

45. 12.35

46. 42.988

47. 0.75

48. 0.36

49. $0.\overline{3} \approx 0.33$

50. $0.\overline{7} \approx 0.78$

51. 0.4375

52. 0.625

53. $0.\overline{54} \approx 0.55$

54. $0.1\overline{6} \approx 0.17$

55. 0.28

56. 0.36

57. 0.031

58. 0.022

59. 1.35

60. 4.17

61. 0.9655

62. 0.8149

63. 0.61

64. 0.823; 0.816

Name _____

C *Round each decimal to the given place value. See Examples 8 and 9.*

37. 0.57, nearest tenth

38. 0.58, nearest tenth

39. 0.234, nearest hundredth

40. 0.452, nearest hundredth

41. 0.5942, nearest thousandth

42. 63.4523, nearest thousandth

43. 98,207.23, nearest tenth

44. 68,936.543, nearest tenth

45. 12.347, nearest tenth

46. 42.9878, nearest thousandth

D *Write each fraction as a decimal. If the decimal is a repeating decimal, write using the bar notation and then round to the nearest hundredth. See Examples 10 through 12.*

47. $\dfrac{3}{4}$

48. $\dfrac{9}{25}$

49. $\dfrac{1}{3}$

50. $\dfrac{7}{9}$

51. $\dfrac{7}{16}$

52. $\dfrac{5}{8}$

53. $\dfrac{6}{11}$

54. $\dfrac{1}{6}$

E *Write each percent as a decimal. See Example 13.*

55. 28%

56. 36%

57. 3.1%

58. 2.2%

59. 135%

60. 417%

61. 96.55%

62. 81.49%

63. In a recent telephone survey, approximately 61% of the respondents said that they are better off financially than their parents were at the same age. Write this percent as a decimal. (*Source: Reader's Digest*, December 1996)

64. The average one-year survival rate for a heart transplant recipient is 82.3%. The average one-year survival rate for a liver transplant patient is 81.6%. Write each percent as a decimal. (*Source*: Bureau of Health Resources Development)

Write each decimal as a percent. See Example 14.

65. 0.68   **66.** 0.32   **67.** 0.876   **68.** 0.521

69. 1   **70.** 3   **71.** 0.5   **72.** 0.1

▧ **COMBINING CONCEPTS**

73. The estimated life expectancy at birth for female Canadians was 82.65 years in 1996. The estimated life expectancy at birth for male Canadians was only 75.67 years in 1996. How much longer is a female Canadian born in 1996 expected to live than a male Canadian born in the same year? (*Source*: The Central Intelligence Agency, *1996 World Factbook*)

74. The chart shows the average number of pounds of meats consumed by each United States citizen in 1995. (*Source*: National Agricultural Statistics Service)

Meat	Pounds
Chicken	71.3
Turkey	17.9
Beef	67.3
Pork	52.5
Veal	1.0
Lamb/Mutton	1.2

a. How much more beef than pork did the average U.S. citizen consume in 1995?

b. How much poultry (chicken and turkey) did the average U.S. citizen consume in 1995?

c. What was the total amount of meat consumed by the average U.S. citizen in 1995?

75. An estimated $\frac{16}{25}$ of Americans own at least one credit card. What percent of Americans own credit cards? (*Sources*: *Bank Advertising News* and The Gallup Organization, 1996)

65. 68%

66. 32%

67. 87.6%

68. 52.1%

69. 100%

70. 300%

71. 50%

72. 10%

73. 6.98 years

74. a. 14.8 pounds

b. 89.2 pounds

c. 211.2 pounds

75. 64%

Focus On History

FACTORING MACHINE

Small numbers can be broken down into their prime factors relatively easily. However, factoring larger numbers can be difficult and time-consuming. The first known successful attempt to automate the process of factoring whole numbers is credited to a French infantry officer and mathematics enthusiast, Eugène Olivier Carissan. In 1919, he designed and built a machine that uses gears and a hand crank to factor numbers.

Carissan's factoring machine had been all but forgotten after his death in 1925. In 1989, a Canadian researcher came across a description of the machine in an article printed in an obscure French journal in 1920. This led to a five-year search for traces of the machine. Eventually, the factoring machine was found in a French astronomical observatory which had received the invention from Carissan's family after his death.

Mathematical historians agree that the factoring machine was a remarkable achievement in its precomputer era. Up to 40 numbers per second could be processed by the machine while its operator turned the crank at two revolutions per minute. Carissan was able to use his machine to prove that the number 708,158,977 was prime in under 10 minutes. He could also find the prime factorizations of up to 13-digit numbers with the machine.

CHAPTER R ACTIVITY
INTERPRETING SURVEY RESULTS

This activity may be completed by working in groups or individually.

Conduct the following survey with 12 students in one of your classes and record the results.

a. What is your age?
Under 20   20s   30s   40s   50s   60 and older
b. What is your gender?
Female   Male
c. How did you arrive on campus today?
Walked   Drove   Bicycled
Took public transportation   Other

1. For each survey question, tally the results for each category.   answers may vary

Age

Category	Tally
Under 20	
20s	
30s	
40s	
50s	
60+	
Total	

Gender

Category	Tally
Female	
Male	
Total	

Mode of Transportation

Category	Tally
Walk	
Drive	
Bicycle	
Public Transit	
Other	
Total	

2. For each survey question, find the fraction of the total number of responses that fall in each answer category. Use the tallies from Question 1 to complete the Fraction columns of the tables at the right.   answers may vary

3. For each survey question, convert the fraction of the total number of responses that fall in each answer category to a decimal number. Use the fractions from Question 2 to complete the Decimal columns of the tables below.
answers may vary

4. For each survey question, find the percent of the total number of responses that fall in each answer category. Complete the Percent columns of the tables below.
answers may vary

5. Study the tables. What may you conclude from them? What do they tell you about your survey respondents? Write a paragraph summarizing your findings.   answers may vary

Age

Category	Fraction	Decimal	Percent
Under 20			
20s			
30s			
40s			
50s			
60+			

Gender

Category	Fraction	Decimal	Percent
Female			
Male			

Mode of Transportation

Category	Fraction	Decimal	Percent
Walk			
Drive			
Bicycle			
Public Transit			
Other			

CHAPTER R HIGHLIGHTS

DEFINITIONS AND CONCEPTS	EXAMPLES

SECTION R.1 FACTORS AND THE LEAST COMMON MULTIPLE

To **factor** means to write as a product.	The factors of 12 are 1, 2, 3, 4, 6, 12
When a number is written as a product of primes, this product is called the **prime factorization** of a number.	Write the prime factorization of 60. $$60 = 6 \cdot 10$$ $$2 \cdot 3 \cdot 2 \cdot 5$$ The prime factorization of 60 is $2 \cdot 2 \cdot 3 \cdot 5$.
The least common multiple (LCM) of a list of numbers is the smallest number that is a multiple of all the numbers in the list. TO FIND THE LCM OF A LIST OF NUMBERS Step 1. Write the prime factorization of each number. Step 2. Write the product containing each different prime factor (from Step 1) the greatest number of times that it appears in any one factorization. This product is the LCM.	Find the LCM of 12 and 40. $$12 = 2 \cdot 2 \cdot 3$$ $$40 = 2 \cdot 2 \cdot 2 \cdot 5$$ $$\text{LCM} = 2 \cdot 2 \cdot 2 \cdot 3 \cdot 5 = 120$$

SECTION R.2 FRACTIONS

Fractions that represent the same quantity are called **equivalent fractions**.	$$\frac{1}{5} = \frac{1 \cdot 4}{5 \cdot 4} = \frac{4}{20}$$ $\frac{1}{5}$ and $\frac{4}{20}$ are equivalent fractions.
FUNDAMENTAL PRINCIPLE OF FRACTIONS If a, b, and c are numbers, then $$\frac{a}{b} = \frac{a \cdot c}{b \cdot c} \quad \text{or} \quad \frac{a \cdot c}{b \cdot c} = \frac{a}{b}$$ as long as b and c are not 0.	
A fraction is **simplified** when the numerator and the denominator have no factors in common other than 1.	$\frac{13}{17}$ is simplified.
To simplify a fraction, factor the numerator and the denominator; then apply the fundamental principle of fractions.	Simplify. $$\frac{6}{14} = \frac{2 \cdot 3}{2 \cdot 7} = \frac{3}{7}$$

Two fractions are **reciprocals** if their product is 1. The reciprocal of $\frac{a}{b}$ is $\frac{b}{a}$, as long as a and b are not 0.

The reciprocal of $\frac{6}{25}$ is $\frac{25}{6}$.

To multiply fractions, numerator times numerator is the numerator of the product and denominator times denominator is the denominator of the product.

$$\frac{2}{5} \cdot \frac{3}{7} = \frac{6}{35}$$

To divide fractions, multiply the first fraction by the reciprocal of the second fraction.

$$\frac{5}{9} \div \frac{2}{7} = \frac{5}{9} \cdot \frac{7}{2} = \frac{35}{18}$$

To add fractions with the same denominator, add the numerators and place the sum over the common denominator.

$$\frac{5}{11} + \frac{3}{11} = \frac{8}{11}$$

To subtract fractions with the same denominator, subtract the numerators and place the difference over the common denominator.

$$\frac{13}{15} - \frac{3}{15} = \frac{10}{15} = \frac{2}{3}$$

To add or subtract fractions with different denominators, first write each fraction as an equivalent fraction with the LCD as denominator.

$$\frac{2}{9} + \frac{3}{6} = \frac{2 \cdot 2}{9 \cdot 2} + \frac{3 \cdot 3}{6 \cdot 3} = \frac{4 + 9}{18} = \frac{13}{18}$$

SECTION R.3   DECIMALS AND PERCENTS

To write decimals as fractions, use place values.

$$0.12 = \frac{12}{100}$$

TO ADD OR SUBTRACT DECIMALS

Step 1. Write the decimals so that the decimal points line up vertically.

Step 2. Add or subtract as for whole numbers.

Step 3. Place the decimal point in the sum or difference so that it lines up vertically with the decimal points in the problem.

Subtract: $2.8 - 1.04$ Add: $25 + 0.02$

$$
\begin{array}{r}
{}^{7\ 10} \\
2.8\cancel{0} \\
-1.04 \\
\hline
1.76
\end{array}
\qquad
\begin{array}{r}
25. \\
+\ 0.02 \\
\hline
25.02
\end{array}
$$

TO MULTIPLY DECIMALS

Step 1. Multiply the decimals as though they are whole numbers.

Step 2. The decimal point in the product is placed so that the number of decimal places in the product is equal to the **sum** of the number of decimal places in the factors.

Multiply: 1.48×5.9

$$
\begin{array}{r}
1.4\,8 \quad \leftarrow\text{2 decimal places} \\
\times \quad 5.9 \quad \leftarrow\text{1 decimal place} \\
\hline
1\,3\,3\,2 \\
7\,4\,0 \\
\hline
8.7\,3\,2 \quad \leftarrow\text{3 decimal places}
\end{array}
$$

To Divide Decimals

Step 1. Move the decimal point in the divisor to the right until the divisor is a whole number.

Step 2. Move the decimal point in the dividend to the right the **same number of places** as the decimal point was moved in Step 1.

Step 3. Divide. The decimal point in the quotient is directly over the moved decimal point in the dividend.

To write fractions as decimals, divide the numerator by the denominator.

Divide: $1.118 \div 2.6$

$$
\begin{array}{r}
0.43 \\
2.6\overline{)1.118} \\
-104 \\
\hline
78 \\
-78 \\
\hline
0
\end{array}
$$

Write $\dfrac{3}{8}$ as a decimal.

$$
\begin{array}{r}
0.375 \\
8\overline{)3.000} \\
-24 \\
\hline
60 \\
-56 \\
\hline
40 \\
-40 \\
\hline
0
\end{array}
$$

To write a percent as a decimal, drop the % symbol and move the decimal point two places to the left.

To write a decimal as a percent, move the decimal point two places to the right and attach the % symbol.

$25\% = 25.\% = 0.25$

$0.7 = 0.70 = 70\%$

CHAPTER R REVIEW

(R.1) *Write the prime factorization of each number.*

1. 42 $2 \cdot 3 \cdot 7$

2. 800 $2 \cdot 2 \cdot 2 \cdot 2 \cdot 2 \cdot 5 \cdot 5$

Find the least common multiple (LCM) of each list of numbers.

3. 12, 30   60

4. 7, 42   42

5. 4, 6, 10   60

6. 2, 5, 7   70

(R.2) *Write each fraction as an equivalent fraction with the given denominator.*

7. $\frac{5}{8}$ with a denominator of 24 $\frac{15}{24}$

8. $\frac{2}{3}$ with a denominator of 60 $\frac{40}{60}$

Simplify each fraction.

9. $\frac{8}{20}$ $\frac{2}{5}$

10. $\frac{15}{100}$ $\frac{3}{20}$

11. $\frac{12}{6}$ 2

12. $\frac{8}{8}$ 1

Perform each indicated operation and simplify.

13. $\frac{1}{7} \cdot \frac{8}{11}$ $\frac{8}{77}$

14. $\frac{5}{12} + \frac{2}{15}$ $\frac{11}{20}$·

15. $\frac{3}{10} \div 6$ $\frac{1}{20}$

16. $\frac{7}{9} - \frac{1}{6}$ $\frac{11}{18}$

The area of a plane figure is a measure of the amount of surface of the figure. Find the area of each figure below. (The area of a rectangle is the product of its length and width. The area of a triangle is $\frac{1}{2}$ the product of its base and height.)

17. 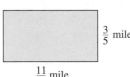 $\frac{11}{20}$ sq. mile

$\frac{3}{5}$ mile

$\frac{11}{12}$ mile

18. 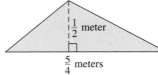 $\frac{5}{16}$ sq. meter

$\frac{1}{2}$ meter

$\frac{5}{4}$ meters

(R.3) *Write each decimal as a fraction. Do not simplify.*

19. 1.81 $\frac{181}{100}$

20. 0.035 $\frac{35}{1000}$

Perform each indicated operation.

21. $\begin{array}{r} 76.358 \\ +18.76 \\ \hline 95.118 \end{array}$

22. $35 + 0.02 + 1.765$ 36.785

23. $18 - 4.62$ 13.38

24.   804.062
        −112.489
        ‾‾‾‾‾‾‾‾
         691.573

25.   7.6
        ×   12
        ‾‾‾‾‾‾‾
         91.2

26.   14.63
        ×   3.2
        ‾‾‾‾‾‾‾‾
         46.816

27. $27\overline{)772.2}$ 28.6

28. $0.06\overline{)13.8}$ 230

Round each decimal to the given place value.

29. 0.7652, nearest hundredth   0.77

30. 25.6293, nearest tenth   25.6

Write each fraction as a decimal. If the decimal is a repeating decimal, write it using the bar notation and then round to the nearest thousandth.

31. $\frac{1}{2}$ 0.5

32. $\frac{3}{8}$ 0.375

33. $\frac{4}{11}$ $0.\overline{36} \approx 0.364$

34. $\frac{5}{6}$ $0.8\overline{3} \approx 0.833$

Write each percent as a decimal.

35. 29%   0.29

36. 1.4%   0.014

Write each decimal as a percent.

37. 0.39   39%

38. 1.2   120%

39. In 1996, 70.8% of women in the workforce had children under 18. Write this percent as a decimal. (*Source: USA Today*, 10/22/97)   0.708

40. Choose the true statement.
  a. $2.3\% = 0.23$
  b. $5 = 500\%$
  c. $40\% = 4$ b

Name _____ **Section** _____ **Date** _____

Chapter R Test

1. Write the prime factorization of 72.

2. Find the LCM of 5, 18, 20.

3. Write $\frac{5}{12}$ as an equivalent fraction with a denominator of 60.

Simplify each fraction.

4. $\frac{15}{20}$

5. $\frac{48}{100}$

6. Write 1.3 as a fraction.

Perform each indicated operation and simplify.

7. $\frac{5}{8} + \frac{7}{10}$

8. $\frac{2}{3} \cdot \frac{27}{49}$

9. $\frac{9}{10} \div 18$

10. $\frac{8}{9} - \frac{1}{12}$

Perform each indicated operation.

11. $43 + 0.21 + 1.9$

12. $123.6 - 57.72$

13. $\begin{array}{r} 7.93 \\ \times\ 1.6 \\ \hline \end{array}$

14. $0.25\overline{)80}$

15. Round 23.7272 to the nearest hundredth.

16. Write $\frac{7}{8}$ as a decimal.

17. Write $\frac{1}{6}$ as a repeating decimal. Then approximate the result to the nearest thousandth.

18. Write 58.1% as a decimal.

19. Write 0.07 as a percent.

20. Write $\frac{3}{4}$ as a percent. (*Hint:* Write $\frac{3}{4}$ as a decimal, then write the decimal as a percent.)

Answers

1. $2 \cdot 2 \cdot 2 \cdot 3 \cdot 3$

2. 180

3. $\frac{25}{60}$

4. $\frac{3}{4}$

5. $\frac{12}{25}$

6. $\frac{13}{10}$

7. $\frac{53}{40}$

8. $\frac{18}{49}$

9. $\frac{1}{20}$

10. $\frac{29}{36}$

11. 45.11

12. 65.88

13. 12.688

14. 320

15. 23.73

16. 0.875

17. $0.1\overline{6} \approx 0.167$

18. 0.581

19. 7%

20. 75%

Name _____

Most of the water on Earth is in the form of oceans. Only a small part is fresh water. The graph below is called a circle graph or pie chart. This particular circle graph shows the distribution of fresh water on Earth. Use this graph to answer Questions 21–24.

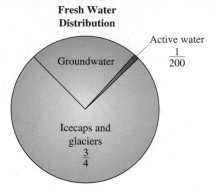

Fresh Water Distribution

Active water
$\dfrac{1}{200}$

Groundwater

Icecaps and glaciers
$\dfrac{3}{4}$

21. What fractional part of fresh water is icecaps and glaciers?

22. What fractional part of fresh water is active water?

23. What fractional part of fresh water is groundwater?

24. What fractional part of fresh water is groundwater or icecaps and glaciers?

Real Numbers and Introduction to Algebra

The power of mathematics is its flexibility. We apply numbers to almost every aspect of our lives. The power of algebra is its generality. In algebra, we use letters to represent numbers.

In this chapter, we begin with a review of the basic symbols—the language—of arithmetic. We then introduce the use of a variable in place of a number. From there, we translate phrases to algebraic expressions and sentences to equations. This is the beginning of problem solving, which we formally study in Chapter 2.

The stars have been a source of interest to different cultures for centuries. Polaris, the North Star, guided ancient sailors. The Egyptians honored Sirius, the brightest star in the sky, in temples. Around 150 B.C., a Greek astronomer, Hipparchus, devised a system of classifying the brightness of stars. He called the brightest stars "first magnitude" and the faintest stars "sixth magnitude." Hipparchus's system is the basis of the apparent magnitude scale used by modern astronomers. This modern scale has been modified to include negative numbers. In Exercises 81–86 on page 13, we shall see how this scale is used to describe the brightness of objects such as the sun, the moon, and some planets.

ANSWERS

1. $>$ (1.1A)

2. $<$ (1.1A)

3. $>$ (1.1A)

4. 5 (1.1D)

5. 1.2 (1.1D)

6. 0 (1.1D)

7. 6 (1.2B)

8. $2x - 10$ (1.2D)

9. 64 (1.2A)

10. -9 (1.5A)

11. $\dfrac{3}{5}$ (1.3C)

12. 8 (1.6A)

13. 53 (1.2A)

14. 3 (1.3A)

15. -27 (1.4A)

16. 56 (1.5A)

17. -70 (1.6B)

18. 4 (1.6B)

19. -40 (1.5B)

20. Not a solution (1.4C)

21. Solution (1.6D)

22. $36°$ (1.4D)

23. $5 + 2y$ (1.7A)

24. $12 + 8t$ (1.7B)

25. Additive inverse property (1.7C)

CHAPTER 1 PRETEST

Insert $<$, $>$, or $=$ to form a true statement.

1. 0 -3
2. -10 -8
3. 1.7   1.07

Find the absolute value.

4. $|5|$
5. $|-1.2|$
6. $|0|$

7. Evaluate $xy - x^2$ when $x = 2$ and $y = 5$.

8. Write the following phrase as an algebraic expression. Let x represent the unknown number.
Twice a number decreased by 10.

Evaluate the following.

9. 4^3
10. -3^2

11. Find the opposite of $-\dfrac{3}{5}$.

12. Find the reciprocal of $\dfrac{1}{8}$.

Perform the indicated operations and simplify.

13. $3 + 2 \cdot 5^2$
14. $-10 + 13$
15. $-6 - 21$
16. $(-7)(-8)$

17. $-2.8 \div 0.04$

18. $\dfrac{-4 - 6^2}{5(-2)}$

19. Evaluate $x^2 - 2xy$ when $x = -4$ and $y = -7$.

Decide whether the given number is a solution to the given equation.

20. $x - 12 = 5;\ x = 7$

21. $\dfrac{x}{8} + 2 = 7;\ x = 40$

22. At 6:00 a.m., the temperature was $-14°$F in Syracuse, New York. By 10:00 a.m., the temperature had risen to $22°$F. How much had the temperature risen?

23. Use the commutative property of addition to complete the statement:
$2y + 5 =$ _____.

24. Use the distributive property to write the expression without parentheses
$4(3 + 2t) =$ _____.

25. Identify the property illustrated by the expression $8 + (-8) = 0$.

1.1 SYMBOLS AND SETS OF NUMBERS

We begin with a review of the set of natural numbers and the set of whole numbers and how we use symbols to compare these numbers. A **set** is a collection of objects, each of which is called a **member** or **element** of the set. A pair of brace symbols { } encloses the list of elements and is translated as "the set of" or "the set containing."

NATURAL NUMBERS

$$\{1, 2, 3, 4, 5, 6, \ldots\}$$

WHOLE NUMBERS

$$\{0, 1, 2, 3, 4, 5, 6, \ldots\}$$

HELPFUL HINT

The three dots (an ellipsis) at the end of the list of elements of a set means that the list continues in the same manner indefinitely.

These numbers can be pictured on a **number line**. To draw a number line, first draw a line. Choose a point on the line and label it 0. To the right of 0, label any other point 1. Being careful to use the same distance as from 0 to 1, mark off equally spaced distances. Label these points 2, 3, 4, 5, and so on. Since the whole numbers continue indefinitely, it is not possible to show every whole number on the number line. The arrow at the right end of the line indicates that the pattern continues indefinitely.

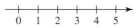

A EQUALITY AND INEQUALITY SYMBOLS

Picturing natural numbers and whole numbers on a number line helps us to see the order of the numbers. Symbols can be used to describe in writing the order of two quantities. We will use equality symbols and inequality symbols to compare quantities.

Below is a review of these symbols. The letters a and b are used to represent quantities. Letters such as a and b that are used to represent numbers or quantities are called **variables**.

		MEANING
Equality symbol:	$a = b$	a is equal to b.
Inequality symbols:	$a \neq b$	a is not equal to b.
	$a < b$	a is less than b.
	$a > b$	a is greater than b.
	$a \leq b$	a is less than or equal to b.
	$a \geq b$	a is greater than or equal to b.

These symbols may be used to form **mathematical statements** such as

$$2 = 2 \quad \text{and} \quad 2 \neq 6$$

Objectives

A Define the meaning of the symbols $=$, $\neq$, $<$, $>$, $\leq$, and $\geq$.

B Translate sentences into mathematical statements.

C Identify integers, rational numbers, irrational numbers, and real numbers.

D Find the absolute value of a real number.

SSM   CD-ROM   Video 1.1

TEACHING TIP

Encourage students to keep an organized math notebook. Remind them that they will remember the vocabulary, concepts, and hints when they write them down in their own words. Point out that the text hightlights important vocabulary in bold and gives Helpful Hints to the students in boxes.

TEACHING TIP

Students may be familiar with the term *Counting Numbers* for the term Natural Numbers.

On the number line, we see that a number **to the right of** another number is **larger**. Similarly, a number **to the left of** another number is smaller. For example, 3 is to the left of 5 on the number line, which means that 3 is less than 5, or 3 < 5. Similarly, 2 is to the right of 0 on the number line, which means 2 is greater than 0, or 2 > 0. Since 0 is to the left of 2, we can also say that 0 is less than 2, or 0 < 2.

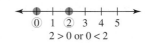

3 < 5   2 > 0 or 0 < 2

Notice that 2 > 0 has exactly the same meaning as 0 < 2. Switching the order of the numbers and reversing the "direction of the inequality symbol" does not change the meaning of the statement.

5 > 3 has the same meaning as 3 < 5.

Also notice that when the statement is true, the inequality arrow points to the smaller number.

Use the arrowheads on a number line to help students remember the meaning of > and < . Point out that the > arrowhead points to the larger number and is shorthand for *greater than*. The < arrowhead points to the smaller number and is shorthand for *less than*.

Examples

Determine whether each statement is true or false.

1. 2 < 3   True. Since 2 is to the left of 3 on the number line
2. 72 > 27   True. Since 72 is to the right of 27 on the number line
3. 8 ≥ 8   True. Since 8 = 8 is true
4. 8 ≤ 8   True. Since 8 = 8 is true
5. 23 ≤ 0   False. Since neither 23 < 0 nor 23 = 0 is true
6. 0 ≤ 23   True. Since 0 < 23 is true

Practice Problems 1–6

Determine whether each statement is true or false.

1. 8 < 6

2. 100 > 10

3. 21 ≤ 21

4. 21 ≥ 21

5. 0 ≤ 5

6. 25 ≥ 22

HELPFUL HINT

If either 3 < 3 or 3 = 3 is true, then 3 ≤ 3 is true.

B Translating Sentences into Mathematical Statements

Now, let's use the symbols discussed above to translate sentences into mathematical statements.

Practice Problem 7

Translate each sentence into a mathematical statement.

a. Fourteen is greater than or equal to fourteen.

b. Zero is less than five.

c. Nine is not equal to ten.

Example 7

Translate each sentence into a mathematical statement.

a. Nine is less than or equal to eleven.
b. Eight is greater than one.
c. Three is not equal to four.

Solution: **a.**

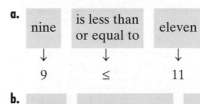

b.

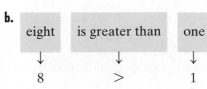

Answers

1. false, **2.** true, **3.** true, **4.** true, **5.** true,
6. true, **7. a.** 14 ≥ 14, **b.** 0 < 5, **c.** 9 ≠ 10

c.

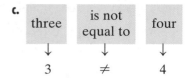

| three | is not equal to | four |

↓   ↓   ↓

3   ≠   4

C IDENTIFYING COMMON SETS OF NUMBERS

Whole numbers are not sufficient to describe many situations in the real world. For example, quantities smaller than zero must sometimes be represented, such as temperatures, less than 0 degrees.

We can place numbers less than zero on the number line as follows: Numbers less than 0 are to the left of 0 and are labeled $-1, -2, -3$, and so on. The numbers we have labeled on the number line below are called the set of **integers**.

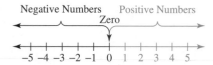

Negative Numbers   Positive Numbers
Zero

−5 −4 −3 −2 −1 0 1 2 3 4 5

Integers to the left of 0 are called **negative integers**; integers to the right of 0 are called **positive integers**. The integer 0 is neither positive nor negative.

INTEGERS

$$\{\ldots, -3, -2, -1, 0, 1, 2, 3, \ldots\}$$

Example 8   Use an integer to express the number in the following. "Pole of Inaccessibility, Antarctica, is the coldest location in the world, with an average annual temperature of 72 degrees below zero." (*Source: The Guinness Book of Records*)

Solution: The integer -72 represents 72 degrees below zero.

A problem with integers in real-life settings arises when quantities are smaller than some integer but greater than the next smallest integer. On the number line, these quantities may be visualized by points between integers. Some of these quantities between integers can be represented as a quotient of integers. For example,

The point on the number line halfway between 0 and 1 can be represented by $\frac{1}{2}$, a quotient of integers.

Practice Problem 8

Use an integer to express the number in the following. "The lowest altitude in North America is found in Death Valley, California. Its altitude is 282 feet below sea level." (*Source: The World Almanac, 1997*)

Answer

8. -282

Use a number line to help students develop a number sense of fractions. Show them that just as $\frac{3}{1}$ is three whole units from zero, $\frac{7}{4}$ is 7 one-fourth units from zero.

⌐ HELPFUL HINT

We commonly refer to rational numbers as fractions.

Practice Problem 9

Graph the numbers on the number line.

$$-2.5, \quad -\frac{2}{3}, \quad \frac{1}{5}, \quad \frac{5}{4}, \quad 2.25$$

⟵┼──┼──┼──┼──┼──┼──┼──┼──┼──┼──┼⟶
−5 −4 −3 −2 −1 0 1 2 3 4 5

Answer

9.

⟵┼──┼──┼──┼──┼──┼──┼──┼──┼──┼──┼⟶
−5 −4 −3 −2 −1 0 1 2 3 4 5

The point on the number line halfway between 0 and −1 can be represented by $-\frac{1}{2}$. Other quotients of integers and their graphs are shown in the margin.

These numbers, each of which can be represented as a quotient of integers, are examples of rational numbers. It's not possible to list the set of rational numbers using the notation that we have been using. For this reason, we will use a different notation.

RATIONAL NUMBERS

$$\left\{ \frac{a}{b} \,\middle|\, a \text{ and } b \text{ are integers and } b \neq 0 \right\}$$

We read this set as "the set of numbers $\frac{a}{b}$ such that a and b are integers and **b is not 0**."

Notice that every integer is also a rational number since each integer can be written as a quotient of integers. For example, the integer 5 is also a rational number since $5 = \frac{5}{1}$. In this rational number, $\frac{5}{1}$, recall that the top number, 5, is called the numerator and the bottom number, 1, is called the denominator.

Let's practice graphing numbers on a number line.

Example 9   Graph the numbers on the number line.

$$-\frac{4}{3}, \quad \frac{1}{4}, \quad \frac{3}{2}, \quad 2\frac{1}{8}, \quad 3.5$$

Solution: To help graph the improper fractions in the list, we first write them as mixed numbers.

$-\frac{4}{3}$ or $-1\frac{1}{3}$ $\frac{1}{4}$ $\frac{3}{2}$ or $1\frac{1}{2}$ $2\frac{1}{8}$ 3.5
⟵●┼──┼●─┼●─┼──┼●─┼──⟶
−2 −1   0   1   2   3   4

Every rational number has a point on the number line that corresponds to it. But not every point on the number line corresponds to a rational number. Those points that do not correspond to rational numbers correspond instead to **irrational numbers**.

IRRATIONAL NUMBERS

{nonrational numbers that correspond to points on the number line}

An irrational number that you have probably seen is π. Also, $\sqrt{2}$, the length of the diagonal of the square shown, is an irrational number.

Both rational and irrational numbers can be written as decimal numbers. The decimal equivalent of a rational number will either terminate or repeat in a pattern. For example, upon dividing we find that

$$\frac{3}{4} = 0.75 \qquad \text{(decimal number terminates or ends)}$$

$$\frac{2}{3} = 0.66666\ldots \qquad \text{(decimal number repeats in a pattern)}$$

The decimal representation of an irrational number will neither terminate nor repeat. (For further review of decimals, see Section R.3.)

The set of numbers, each of which corresponds to a point on the number line, is called the set of **real numbers**. One and only one point on the number line corresponds to each real number.

REAL NUMBERS

{numbers that correspond to points on the number line}

Several different sets of numbers have been discussed in this section. The following diagram shows the relationships among these sets of real numbers. Notice that, together, the rational numbers and the irrational numbers make up the real numbers.

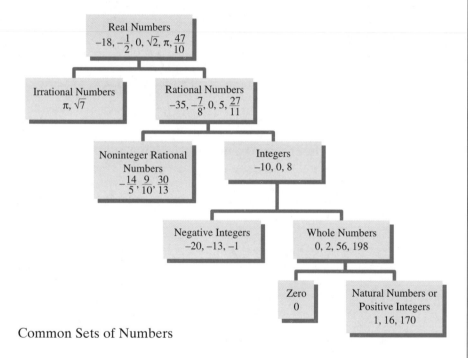

Common Sets of Numbers

Example 10   Given the set $\{-2, 0, \frac{1}{4}, 112, -3, 11, \sqrt{2}\}$, list the numbers in this set that belong to the set of:

    a. Natural numbers   **b.** Whole numbers
    c. Integers   **d.** Rational numbers
    e. Irrational numbers   **f.** Real numbers

Solution: **a.** The natural numbers are 11 and 112.
             b. The whole numbers are 0, 11, and 112.
             c. The integers are $-3, -2, 0, 11,$ and 112.
             d. Recall that integers are rational numbers also. The rational numbers are $-3, -2, 0, \frac{1}{4}, 11,$ and 112.
             e. The irrational number is $\sqrt{2}$.
             f. The real numbers are all numbers in the given set.

Practice Problem 10

Given the set $\{-100, -\frac{2}{5}, 0, \pi, 6, 913\}$, list the numbers in this set that belong to the set of:

a. Natural numbers

b. Whole numbers

c. Integers

d. Rational numbers

e. Irrational numbers

f. Real numbers

Answers

10. a. 6, 913,   **b.** 0, 6, 913,   **c.** $-100, 0, 6, 913,$
d. $-100, -\frac{2}{5}, 0, 6, 913,$ **e.** π,   **f.** all numbers in the given set

D FINDING THE ABSOLUTE VALUE OF A NUMBER

The number line not only gives us a picture of the real numbers, it also helps us visualize the distance between numbers. The distance between a real number a and 0 is given a special name called the **absolute value** of a. "The absolute value of a" is written in symbols as $|a|$.

> ### ABSOLUTE VALUE
>
> The **absolute value** of a real number a, denoted by $|a|$, is the distance between a and 0 on a number line.

For example, $|3| = 3$ and $|-3| = 3$ since both 3 and -3 are a distance of 3 units from 0 on the number line.

> ### HELPFUL HINT
>
> Since $|a|$ is a distance, $|a|$ is always either positive or 0, never negative. That is, **for any real number a, $|a| \geq 0$.**

Practice Problem 11

Find the absolute value of each number.

a. $|7|$, b. $|-8|$, c. $\left|-\frac{2}{3}\right|$

Example 11

Find the absolute value of each number.

a. $|4|$ **b.** $|-5|$ **c.** $|0|$

Solution:

a. $|4| = 4$ since 4 is 4 units from 0 on the number line.

b. $|-5| = 5$ since -5 is 5 units from 0 on the number line.

c. $|0| = 0$ since 0 is 0 units from 0 on the number line.

Practice Problem 12

Insert $<$, $>$, or $=$ in the appropriate space to make each statement true.

a. $|-4|$ 4, b. -3 $|0|$,

c. $|-2.7|$ $|-2|$, d. $|6|$ $|16|$,

e. $|-6|$ $|-16|$

Example 12

Insert $<$, $>$, or $=$ in the appropriate space to make each statement true.

a. $|0|$ 2   **b.** $|-5|$ 5   **c.** $|-3|$ $|-2|$

d. $|5|$ $|6|$ **e.** $|-7|$ $|6|$

Solution:

a. $|0| < 2$ since $|0| = 0$ and $0 < 2$.

b. $|-5| = 5$.

c. $|-3| > |-2|$ since $3 > 2$.

d. $|5| < |6|$ since $5 < 6$.

e. $|-7| > |6|$ since $7 > 6$.

Answers

11. a. 7, **b.** 8, **c.** $\frac{2}{3}$ **12. a.** $=$, **b.** $<$, **c.** $>$, **d.** $<$, **e.** $<$

Name _____ **Section** _____ **Date** _____

EXERCISE SET 1.1

A _Insert_ $<$, $>$, _or_ $=$ _in the space between the paired numbers to make each statement true. See Examples 1 through 6._

1. 4   10   **2.** 8   5   **3.** 7   3   **4.** 9   15

5. 6.26   6.26   **6.** 2.13   1.13   **7.** 0   7   **8.** 20   0

9. The freezing point of water is 32° Fahrenheit. The boiling point of water is 212° Fahrenheit. Write an inequality statement using $<$ or $>$ comparing the numbers 32 and 212.

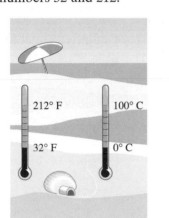

10. The freezing point of water is 0° Celsius. The boiling point of water is 100° Celsius. Write an inequality statement using $<$ or $>$ comparing the numbers 0 and 100.

Determine whether each statement is true or false. See Examples 1 through 6.

11. $11 \leq 11$ **12.** $8 \geq 9$ **13.** $10 > 11$ **14.** $17 > 16$

15. $3 + 8 \geq 3(8)$ **16.** $8 \cdot 8 \leq 8 \cdot 7$ **17.** $7 > 0$ **18.** $4 < 7$

19. An angle measuring 30° and an angle measuring 45° are shown. Use the inequality symbols $\leq$ or $\geq$ to write a statement comparing the numbers 30 and 45.

30°   45°

20. The sum of the measures of the angles of a triangle is 180°. The sum of the measures of the angles of a parallelogram is 360°. Use the inequality symbols $\leq$ or $\geq$ to write a statement comparing the numbers 360 and 180.

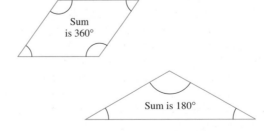

1. $<$
2. $>$
3. $>$
4. $<$
5. $=$
6. $>$
7. $<$
8. $>$
9. $32 < 212$
10. $0 < 100$
11. true
12. false
13. false
14. true
15. false
16. false
17. true
18. true
19. $30 \leq 45$
20. $360 \geq 180$

21. $20 \leq 25$

22. $13 \geq -13$

23. $6 > 0$

24. $3 < 5$

25. $-12 < -10$

26. $-2 > -4$

27. $8 < 12$

28. $15 > 5$

29. $5 \geq 4$

30. $-10 \leq 37$

31. $15 \neq -2$

32. $-7 \neq 7$

33. $535; -8$

34. $23; -12$

35. $-398,000$

36. $308,000$

37. $350; -126$

38. $30; -50$

39. see number line

40. see number line

41. see number line

42. see number line

10

Name _____

Rewrite each inequality so that the inequality symbol points in the opposite direction and the resulting statement has the same meaning as the given one.

21. $25 \geq 20$ **22.** $-13 \leq 13$ **23.** $0 < 6$

24. $5 > 3$ **25.** $-10 > -12$ **26.** $-4 < -2$

B *Write each sentence as a mathematical statement. See Example 7.*

27. Eight is less than twelve.   **28.** Fifteen is greater than five.

29. Five is greater than or equal to four.   **30.** Negative ten is less than or equal to thirty-seven.

31. Fifteen is not equal to negative two.   **32.** Negative seven is not equal to seven.

C *Use integers to represent the values in each statement. See Example 8.*

33. Driskill Mountain, in Louisiana, has an altitude of 535 feet. New Orleans, Louisiana lies 8 feet below sea level. (*Source*: U.S. Geological Survey)

34. During a Green Bay Packers football game, the team gained 23 yards and then lost 12 yards on consecutive plays.

35. From 1989 to 1990, major league baseball attendance decreased by 398,000. (*Source*: The National League of Professional Baseball Clubs)

36. From 1990 to 1991, major league baseball attendance increased by 308,000. (*Source*: The National League of Professional Baseball Clubs)

37. Aaron Miller deposited $350 in his savings account. He later withdrew $126.

38. Aris Peña was deep-sea diving. During her dive, she ascended 30 feet and later descended 50 feet.

Graph each set of numbers on the number line. See Example 9.

39. $-4, 0, 2, 5$

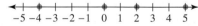

40. $-3, 0, 1, 5$

41. $-2, 4, \frac{1}{2}, -\frac{1}{4}$

42. $-5, 3, -\frac{1}{2}, \frac{1}{4}$

Name _____

43. $-2.5, \dfrac{7}{4}, 3.25, -\dfrac{3}{2}$

44. $4.5, -\dfrac{9}{4}, 1.75, \dfrac{5}{2}$

Tell which set or sets each number belongs to: natural numbers, whole numbers, integers, rational numbers, irrational numbers, and real numbers. See Example 10.

 **45.** 0

46. $\dfrac{1}{4}$

47. -2

48. $-\dfrac{1}{2}$

49. 6

50. 5

51. $\dfrac{2}{3}$

52. $\sqrt{3}$

Determine whether each statement is true or false.

53. Every rational number is also an integer.

54. Every negative number is also a rational number.

55. Every natural number is positive.

56. Every rational number is also a real number.

57. 0 is a real number.

58. Every real number is also a rational number.

59. Every whole number is an integer.

60. $\dfrac{1}{2}$ is an integer.

D *Insert* $<$, $>$, *or* $=$ *in the appropriate space to make each statement true. See Examples 11 and 12.*

61. $|-5|$ ___ -4

62. 0 ___ $|0|$

63. $|-1|$ ___ $|1|$

64. $\left|\dfrac{2}{5}\right|$ ___ $\left|-\dfrac{2}{5}\right|$

65. $|-2|$ ___ $|-3|$

66. -500 ___ $|-50|$

67. $|0|$ ___ $|-8|$

68. $|-12|$ ___ $\dfrac{24}{2}$

◆ COMBINING CONCEPTS

Tell whether each statement is true or false.

69. $\dfrac{1}{2} < \dfrac{1}{3}$

70. $\dfrac{3}{6} \geq \dfrac{1}{2}$

71. $|-5.3| \geq |5.3|$

72. $-1\dfrac{1}{2} > -\dfrac{1}{2}$

73. $-9.6 > -9.1$

74. $-7.3 < -7.1$

75. $-\dfrac{2}{3} \leq -\dfrac{1}{5}$

76. $-\dfrac{5}{6} > -\dfrac{1}{6}$

43. see number line

44. see number line

45. whole, integers, rational, real

46. rational, real

47. integers, rational, real

48. rational, real

49. natural, whole, integers, rational, real

50. natural, whole, integers, rational, real

51. rational, real

52. irrational, real

53. false

54. false

55. true

56. true

57. true

58. false

59. true

60. false

61. $>$

62. $=$

63. $=$

64. $=$

65. $<$

66. $<$

67. $<$

68. $=$

69. false

70. true

71. true

72. false

73. false

74. true

75. true

76. false

11

77. 90

78. 60

79. 70 ≤ 90

80. His quiz scores are improving

81. −0.04 > −26.7

82. 0.96 < 0.98

83. sun

84. Spica

85. sun

86. Regulus

87. answers may vary

88. answers may vary

This graph shows the first three quiz scores for Bill Seggerson in his anatomy class. Each bar represents a different quiz and the height of each bar represents Bill's score for that particular quiz. Use this graph to answer Exercises 77–80.

77. What is Bill's highest quiz score?

78. What is Bill's lowest quiz score?

79. Write an inequality statement using ≤ or ≥ comparing the scores for Quiz 2 and Quiz 3.

80. Do you notice any trends shown by this bar graph?

The apparent magnitude of a star is the measure of its brightness as seen by someone on Earth. The smaller the apparent magnitude, the brighter the star. Below, the apparent magnitudes of some stars are listed. Use this table to answer Exercises 81–86.

Star	Apparent Magnitude	Star	Apparent Magnitude
Arcturus	−0.04	Spica	0.98
Sirius	−1.46	Rigel	0.12
Vega	0.03	Regulus	1.35
Antares	0.96	Canopus	−0.72
Sun	−26.7	Hadar	0.61

(*Source: Norton's 2000.0: Star Atlas and Reference Handbook*, 18th ed., Longman Group, UK, 1989)

81. The apparent magnitude of the sun is −26.7. The apparent magnitude of the star Arcturus is −0.04. Write an inequality statement comparing the numbers −0.04 and −26.7.

82. The apparent magnitude of Antares is 0.96. The apparent magnitude of Spica is 0.98. Write an inequality statement comparing the numbers 0.96 and 0.98.

83. Which is brighter, the sun or Arcturus?

84. Which is dimmer, Antares or Spica?

85. Which star listed is the brightest?

86. Which star listed is the dimmest?

87. In your own words, explain how to find the absolute value of a number.

88. Give an example of a real-life situation that can be described with integers but not with whole numbers.

1.2 INTRODUCTION TO VARIABLE EXPRESSIONS AND EQUATIONS

A EXPONENTS AND THE ORDER OF OPERATIONS

Frequently in algebra, products occur that contain repeated multiplication of the same factor. For example, the volume of a cube whose sides each measure 2 centimeters is $(2 \cdot 2 \cdot 2)$ cubic centimeters. We may use **exponential notation** to write such products in a more compact form. For example.

$2 \cdot 2 \cdot 2$ may be written as 2^3

2 cm

Volume is $(2 \cdot 2 \cdot 2)$ cubic centimeters.

The 2 in 2^3 is called the **base;** it is the repeated factor. The 3 in 2^3 is called the **exponent** and is the number of times the base is used as a factor. The expression 2^3 is called an **exponential expression**.

exponent

$2^3 = 2 \cdot 2 \cdot 2 = 8$

base —↑   2 is a factor 3 times.

Example 1   Evaluate (find the value of) each expression.

a. 3^2 [read as "3 squared" or as "3 to the second power"]
b. 5^3 [read as "5 cubed" or as "5 to the third power"]
c. 2^4 [read as "2 to the fourth power"]
d. 7^1 **e.** $\left(\dfrac{3}{7}\right)^2$

Solution: **a.** $3^2 = 3 \cdot 3 = 9$
b. $5^3 = 5 \cdot 5 \cdot 5 = 125$
c. $2^4 = 2 \cdot 2 \cdot 2 \cdot 2 = 16$
d. $7^1 = 7$
e. $\left(\dfrac{3}{7}\right)^2 = \left(\dfrac{3}{7}\right)\left(\dfrac{3}{7}\right) = \dfrac{3 \cdot 3}{7 \cdot 7} = \dfrac{9}{49}$

⌐HELPFUL HINT

$2^3 \neq 2 \cdot 3$ since 2^3 indicates repeated **multiplication** of the same factor.
$2^3 = 2 \cdot 2 \cdot 2 = 8$, whereas $2 \cdot 3 = 6$

Using symbols for mathematical operations is a great convenience. The more operation symbols presented in an expression, the more careful we must be when performing the indicated operation. For example, in the expression $2 + 3 \cdot 7$, do we add first or multiply first? To eliminate confusion, **grouping symbols** are used. Examples of grouping symbols are parentheses ( ), brackets [ ], braces { }, and the fraction bar. If we wish $2 + 3 \cdot 7$ to be simplified by adding first, we enclose $2 + 3$ in parentheses.

$(2 + 3) \cdot 7 = 5 \cdot 7 = 35$

Objectives

A Define and use exponents and the order of operations.

B Evaluate algebraic expressions, given replacement values for variables.

C Determine whether a number is a solution of a given equation.

D Translate phrases into expressions and sentences into equations.

SSM   CD-ROM   Video 1.2

TEACHING TIP

Consider beginning this lesson with the following activity:
Have students write an expression for 4 added to itself 5 times:
$4 + 4 + 4 + 4 + 4$. Ask if anyone knows a mathematical shorthand for the expression. $(4 \cdot 5)$ Now write an expression for 4 multiplied by itself 5 times $(4 \cdot 4 \cdot 4 \cdot 4 \cdot 4)$. Ask if anyone knows a mathematical shorthand for this expression. (4^5).

Practice Problem 1

Evaluate each expression.

a. 4^2

b. 2^2

c. 3^4

d. 9^1

e. $\left(\dfrac{2}{5}\right)^2$

Answers

1. a. 16, **b.** 4, **c.** 81, **d.** 9, **e.** $\dfrac{4}{25}$

If we wish to multiply first, $3 \cdot 7$ may be enclosed in parentheses.

$$2 + (3 \cdot 7) = 2 + 21 = 23$$

To eliminate confusion when no grouping symbols are present, we use the following agreed-upon order of operations.

TEACHING TIP

Help students remember the rules for order of operations with a mnemonic device such as **P**lease **e**xcuse **m**y **d**ear **A**unt **S**ally but be sure they apply it correctly. A common mistake is to apply multiplication before division and addition before subtraction. Check if they apply it correctly by having them evaluate $12 \div 3 \cdot 2$ and $10 - 4 + 2$.

ORDER OF OPERATIONS

Simplify expressions using the following order. If grouping symbols such as parentheses are present, simplify expressions within those first, starting with the innermost set. If fraction bars are present, simplify the numerator and the denominator separately.

1. Evaluate exponential expressions.
2. Perform multiplications or divisions in order from left to right.
3. Perform additions or subtractions in order from left to right.

Using this order of operations, we now simplify $2 + 3 \cdot 7$. There are no grouping symbols and no exponents, so we multiply and then add.

$$2 + 3 \cdot 7 = 2 + 21 \qquad \text{Multiply.}$$
$$= 23 \qquad \text{Add.}$$

Practice Problems 2–4

Simplify each expression.

2. $3 + 2 \cdot 4^2$

3. $\dfrac{9}{5} \cdot \dfrac{1}{3} - \dfrac{1}{3}$

4. $8[2(6 + 3) - 9]$

Examples   Simplify each expression.

2. $6 \div 3 + 5^2 = 6 \div 3 + 25 \qquad$ Evaluate 5^2.
$\qquad\qquad\quad\ = 2 + 25 \qquad\qquad$ Divide.
$\qquad\qquad\quad\ = 27 \qquad\qquad\quad$ Add.

3. $\dfrac{3}{2} \cdot \dfrac{1}{2} - \dfrac{1}{2} = \dfrac{3}{4} - \dfrac{1}{2} \qquad$ Multiply.

$\qquad\qquad\quad = \dfrac{3}{4} - \dfrac{2}{4} \qquad$ The least common denominator is 4.

$\qquad\qquad\quad = \dfrac{1}{4} \qquad\qquad$ Subtract.

4. $3[4(5 + 2) - 10] = 3[4(7) - 10] \qquad$ Simplify the expression in parentheses. They are the innermost grouping symbols.
$\qquad\qquad\qquad\quad = 3[28 - 10] \qquad$ Multiply 4 and 7.
$\qquad\qquad\qquad\quad = 3[18] \qquad\quad$ Subtract inside the brackets.
$\qquad\qquad\qquad\quad = 54 \qquad\qquad$ Multiply.

In the next example, the fraction bar serves as a grouping symbol and separates the numerator and denominator. Simplify each separately.

Practice Problem 5

Simplify.

$$\dfrac{1 + |7 - 4| + 3^2}{8 - 5}$$

Example 5   Simplify: $\dfrac{3 + |4 - 3| + 2^2}{6 - 3}$

Solution:

$$\dfrac{3 + |4 - 3| + 2^2}{6 - 3} = \dfrac{3 + |1| + 2^2}{6 - 3} \qquad \text{Simplify the expression inside the absolute value bars.}$$

$$= \dfrac{3 + 1 + 2^2}{3} \qquad \text{Find the absolute value and simplify the denominator.}$$

$$= \dfrac{3 + 1 + 4}{3} \qquad \text{Evaluate the exponential expression.}$$

$$= \dfrac{8}{3} \qquad \text{Simplify the numerator.}$$

Answers

2. 35,   **3.** $\dfrac{4}{15}$,   **4.** 72,   **5.** $\dfrac{13}{3}$

HELPFUL HINT

Be careful when evaluating an exponential expression.

$$3 \cdot 4^2 = 3 \cdot 16 = 48 \qquad (3 \cdot 4)^2 = (12)^2 = 144$$

↑   ↑

Base is 4.   Base is $3 \cdot 4$.

B EVALUATING ALGEBRAIC EXPRESSIONS

An **algebraic expression** is a collection of numbers, variables, operation symbols, and grouping symbols. For example,

$$2x, \qquad -3, \qquad 2x - 10, \qquad 5(p^2 + 1), \qquad \text{and} \qquad \frac{3y^2 - 6y + 1}{5}$$

are algebraic expressions. The expression $2x$ means $2 \cdot x$. Also, $5(p^2 + 1)$ means $5 \cdot (p^2 + 1)$ and $3y^2$ means $3 \cdot y^2$. If we give a specific value to a variable, we can **evaluate an algebraic expression**. To evaluate an algebraic expression means to find its numerical value once we know the value of the variables.

Algebraic expressions often occur during problem solving. For example, the expression

$$16t^2$$

gives the distance in feet (neglecting air resistance) that an object will fall in t seconds. (See Exercise 63 in this section.)

Example 6   Evaluate each expression when $x = 3$ and $y = 2$.

a. $2x - y$ **b.** $\dfrac{3x}{2y}$ **c.** $\dfrac{x}{y} + \dfrac{y}{2}$ **d.** $x^2 - y^2$

Solution: **a.** Replace x with 3 and y with 2.

$$2x - y = 2(3) - 2 \quad \text{Let } x = 3 \text{ and } y = 2.$$
$$= 6 - 2 \qquad \text{Multiply.}$$
$$= 4 \qquad \text{Subtract.}$$

b. $\dfrac{3x}{2y} = \dfrac{3 \cdot 3}{2 \cdot 2} = \dfrac{9}{4} \quad$ Let $x = 3$ and $y = 2$.

c. Replace x with 3 and y with 2. Then simplify.

$$\frac{x}{y} + \frac{y}{2} = \frac{3}{2} + \frac{2}{2} = \frac{5}{2}$$

d. Replace x with 3 and y with 2.

$$x^2 - y^2 = 3^2 - 2^2 = 9 - 4 = 5$$

Practice Problem 6

Evaluate each expression when $x = 1$ and $y = 4$.

a. $2y - x$

b. $\dfrac{8x}{3y}$

c. $\dfrac{x}{y} + \dfrac{5}{y}$

d. $y^2 - x^2$

Answers

6. a. 7, **b.** $\dfrac{2}{3}$, **c.** $\dfrac{3}{2}$, **d.** 15

C SOLUTIONS OF EQUATIONS

Many times a problem-solving situation is modeled by an equation. An **equation** is a mathematical statement that two expressions have equal value. The equal symbol "=" is used to equate the two expressions. For example, $3 + 2 = 5$, $7x = 35$, $\frac{2(x-1)}{3} = 0$, and $I = PRT$ are all equations.

> **HELPFUL HINT**
>
> An equation contains the equal symbol "=". An algebraic expression does not.

TRY THE CONCEPT CHECK IN THE MARGIN.

When an equation contains a variable, deciding which values of the variable make an equation a true statement is called **solving** an equation for the variable. A **solution** of an equation is a value for the variable that makes the equation true. For example, 3 is a solution of the equation $x + 4 = 7$, because if x is replaced with 3 the statement is true.

$$x + 4 = 7$$
$$\downarrow$$
$$3 + 4 \overset{?}{=} 7 \quad \text{Replace } x \text{ with 3.}$$
$$7 = 7 \quad \text{True.}$$

Similarly, 1 is not a solution of the equation $x + 4 = 7$, because $1 + 4 = 7$ is **not** a true statement.

Example 7   Decide whether 2 is a solution of $3x + 10 = 8x$.

Solution: Replace x with 2 and see if a true statement results.

$$3x + 10 = 8x \quad \text{Original equation}$$
$$3(2) + 10 \overset{?}{=} 8(2) \quad \text{Replace } x \text{ with 2.}$$
$$6 + 10 \overset{?}{=} 16 \quad \text{Simplify each side.}$$
$$16 = 16 \quad \text{True.}$$

Since we arrived at a true statement after replacing x with 2 and simplifying both sides of the equation, 2 is a solution of the equation.

D TRANSLATING WORDS TO SYMBOLS

Now that we know how to represent an unknown number by a variable, let's practice translating phrases into algebraic expressions and sentences into equations. Oftentimes solving problems involves the ability to translate word phrases and sentences into symbols. Below is a list of key words and phrases to help us translate.

Addition (+)	Subtraction(−)	Multiplication (·)	Division (÷)	Equality (=)
Sum	Difference of	Product	Quotient	Equals
Plus	Minus	Times	Divide	Gives
Added to	Subtracted from	Multiply	Into	Is/was/should be
More than	Less than	Twice	Ratio	Yields
Increased by	Decreased by	Of	Divided by	Amounts to
Total	Less			Represents
				Is the same as

✓ **CONCEPT CHECK**

Which of the following are equations? Which are expressions?

a. $5x = 8$
b. $5x - 8$
c. $12y + 3x$
d. $12y = 3x$

Practice Problem 7

Decide whether 3 is a solution of $5x - 10 = x + 2$.

Answers

✓ **Concept Check**

equations: a, d; expressions: b, c

7. It is a solution.

Example 8   Write an algebraic expression that represents each phrase. Let the variable *x* represent the unknown number.

   a. The sum of a number and 3
   b. The product of 3 and a number
   c. Twice a number
   d. 10 decreased by a number
   e. 5 times a number increased by 7

Solution: **a.** $x + 3$ since "sum" means to add
   b. $3 \cdot x$ and $3x$ are both ways to denote the product of 3 and x
   c. $2 \cdot x$ or $2x$
   d. $10 - x$ because "decreased by" means to subtract
   e. $\underbrace{5x}_{\substack{\text{5 times}\\ \text{a number}}} + 7$

Practice Problem 8

Write an algebraic expression that represents each phrase. Let the variable *x* represent the unknown number.

a. The product of a number and 5

b. A number added to 7

c. Three times a number

d. A number subtracted from 8

e. Twice a number plus 1

HELPFUL HINT

Make sure you understand the difference when translating phrases containing "decreased by," "subtracted from" and "less than."

Phrase	Translation	
A number decreased by 10	$x - 10$	
A number subtracted from 10	$10 - x$	Notice the order.
10 less than a number	$x - 10$	

Now let's practice translating sentences into equations.

Example 9   Write each sentence as an equation. Let *x* represent the unknown number.

   a. The quotient of 15 and a number is 4.
   b. Three subtracted from 12 is a number.
   c. Four times a number added to 17 is 21.

Solution: **a.** In words: | the quotient of 15 and a number | is | 4 |
                     ↓   ↓ ↓

          Translate: $\dfrac{15}{x}$ $= 4$

   b. In words: | three subtracted **from** 12 | is | a number |
                        ↓   ↓   ↓

          Translate: $12 - 3$ $=$ x

          Care must be taken when the operation is subtraction. The expression $3 - 12$ would be incorrect. Notice that $3 - 12 \neq 12 - 3$.

   c. In words: | four times a number | added to | 17 | is | 21 |
                      ↓   ↓   ↓ ↓ ↓

          Translate: $4x$ $+$ $17 = 21$

Practice Problem 9

Write each sentence as an equation. Let *x* represent the unknown number.

a. The product of a number and 6 is 24.

b. The difference of 10 and a number is 18.

c. Twice a number decreased by 1 is 99.

Answers

8. a. $5x$, **b.** $7 + x$, **c.** $3x$, **d.** $8 - x$,
e. $2x + 1$, **9. a.** $6x = 24$, **b.** $10 - x = 18$,
c. $2x - 1 = 99$

CALCULATOR EXPLORATIONS
EXPONENTS

To evaluate exponential expressions on a calculator, find the key marked y^x or $\wedge$. To evaluate, for example, 3^5, press the following keys: $\boxed{3}$ $\boxed{y^x}$ $\boxed{5}$ $\boxed{=}$ or $\boxed{3}$ $\boxed{\wedge}$ $\boxed{5}$ $\boxed{=}$.

⇕ or
$\boxed{\text{ENTER}}$

The display should read $\boxed{243}$.

ORDER OF OPERATIONS

Some calculators follow the order of operations, and others do not. To see whether or not your calculator has the order of operations built in, use your calculator to find $2 + 3 \cdot 4$. To do this, press the following sequence of keys:

$\boxed{2}$ $\boxed{+}$ $\boxed{3}$ $\boxed{\times}$ $\boxed{4}$ $\boxed{=}$.

⇕ or
$\boxed{\text{ENTER}}$

The correct answer is 14 because the order of operations is to multiply before we add. If the calculator displays $\boxed{14}$, then it has the order of operations built in.

Even if the order of operations is built in, parentheses must sometimes be inserted. For example, to simplify $\dfrac{5}{12 - 7}$, press the keys

$\boxed{5}$ $\boxed{\div}$ $\boxed{(}$ $\boxed{1}$ $\boxed{2}$ $\boxed{-}$ $\boxed{7}$ $\boxed{)}$ $\boxed{=}$.

⇕ or
$\boxed{\text{ENTER}}$

The display should read $\boxed{1}$

Use a calculator to evaluate each expression.

1. 5^3 125

2. 7^4 2401

3. 9^5 59,049

4. 8^6 262,144

5. $2(20 - 5)$ 30

6. $3(14 - 7) + 21$ 42

7. $24(862 - 455) + 89$ 9857

8. $99 + (401 + 962)$ 1462

9. $\dfrac{4623 + 129}{36 - 34}$ 2376

10. $\dfrac{956 - 452}{89 - 86}$ 168

Name _____ Section _____ Date _____

EXERCISE SET 1.2

A *Evaluate. See Example 1.*

1. 3^5 **2.** 5^3 **3.** 3^3 **4.** 4^4

5. 1^5 **6.** 1^8 **7.** 5^1 **8.** 8^1

9. $\left(\dfrac{1}{5}\right)^3$ **10.** $\left(\dfrac{6}{11}\right)^2$ **11.** $\left(\dfrac{2}{3}\right)^4$ **12.** $\left(\dfrac{1}{2}\right)^5$

13. 7^2 **14.** 9^2 **15.** $(1.2)^2$ **16.** $(0.07)^2$

17. The area of a square whose sides each measure 5 meters is $(5 \cdot 5)$ square meters. Write this area using exponential notation.

5 meters →

18. The volume of a solid is a measure of the space it encloses. The volume of a sphere whose radius is 5 meters is $\left(\dfrac{4}{3}\pi \cdot 5 \cdot 5 \cdot 5\right)$ cubic meters. Write this volume using exponential notation.

5 m

Simplify each expression. See Examples 2 through 5.

19. $5 + 6 \cdot 2$ **20.** $8 + 5 \cdot 3$ **21.** $4 \cdot 8 - 6 \cdot 2$

22. $12 \cdot 5 - 3 \cdot 6$ **23.** $2(8 - 3)$ **24.** $5(6 - 2)$

25. $2 + (5 - 2) + 4^2$ **26.** $6 - 2 \cdot 2 + 2^5$ **27.** $5 \cdot 3^2$

28. $2 \cdot 5^2$ **29.** $\dfrac{1}{4} \cdot \dfrac{2}{3} - \dfrac{1}{6}$ **30.** $\dfrac{3}{4} \cdot \dfrac{1}{2} + \dfrac{2}{3}$

20

Name _____

31. $\dfrac{6-4}{9-2}$

32. $\dfrac{8-5}{24-20}$

33. $2[5+2(8-3)]$

34. $3[4+3(6-4)]$

35. $\dfrac{19-3\cdot 5}{6-4}$

36. $\dfrac{4\cdot 3+2}{4+3\cdot 2}$

37. $\dfrac{|6-2|+3}{8+2\cdot 5}$

38. $\dfrac{15-|3-1|}{12-3\cdot 2}$

39. $\dfrac{3+3(5+3)}{3^2+1}$

40. $\dfrac{3+6(8-5)}{4^2+2}$

41. $\dfrac{6+|8-2|+3^2}{18-3}$

42. $\dfrac{16+|13-5|+4^2}{17-5}$

43. Are parentheses necessary in the expression $2+(3\cdot 5)$? Explain your answer.

44. Are parentheses necessary in the expression $(2+3)\cdot 5$? Explain your answer.

B *Evaluate each expression when x = 1, y = 3, and z = 5. See Example 6.*

45. $3y$

46. $4x$

47. $\dfrac{z}{5x}$

48. $\dfrac{y}{2z}$

49. $3x-2$

50. $6y-8$

51. $|2x+3y|$

52. $|5z-2y|$

53. $xy+z$

54. $yz-x$

55. $5y^2$

56. $2z^2$

Evaluate each expression when x = 2, y = 6, and z = 3. See Example 6.

57. $5z$

58. $7x$

59. $\dfrac{y}{x}$

60. $\dfrac{y}{x\cdot z}$

61. $\dfrac{y}{x}+\dfrac{y}{x}$

62. $\dfrac{9}{z}+\dfrac{4z}{y}$

Name _____

Neglecting air resistance, the expression $16t^2$ gives the distance in feet an object will fall in t seconds.

63. Complete the chart below. To evaluate $16t^2$, remember to first find t^2, then multiply by 16.

Time t (in seconds)	Distance $16t^2$ (in feet)
1	
2	
3	
4	

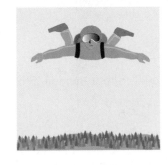

64. Does an object fall the same distance *during* each second? Why or why not? (See Exercise 63.)

C *Decide whether the given number is a solution of the given equation. See Example 7.*

65. $3x - 6 = 9$; 5

66. $2x + 7 = 3x$; 6

67. $2x + 6 = 5x - 1$; 0

68. $4x + 2 = x + 8$; 2

69. $2x - 5 = 5$; 8

70. $3x - 10 = 8$; 6

71. $x + 6 = x + 6$; 2

72. $x + 6 = x + 6$; 10

73. $x = 5x + 15$; 0

74. $4 = 1 - x$; 1

D *Write each phrase as an algebraic expression. Let x represent the unknown number. See Example 8.*

75. Fifteen more than a number

76. One-half times a number

77. Five subtracted from a number

78. The quotient of a number and 9

79. Three times a number increased by 22

80. The product of 8 and a number

63. 16, 64, 144, 256

64. No

65. solution

66. not a solution

67. not a solution

68. solution

69. not a solution

70. solution

71. solution

72. solution

73. not a solution

74. not a solution

75. $x + 15$

76. $\frac{1}{2}x$

77. $x - 5$

78. $\frac{x}{9}$

79. $3x + 22$

80. $8x$

82. $8 - 4 = 2^2$

83. $3 \neq 4 \div 2$

84. $16 - 4 > 10$

85. $5 + x = 20$

86. $2x = 17$

87. $13 - 3x = 13$

88. $x - 7 = 0$

89. $\dfrac{12}{x} = \dfrac{1}{2}$

90. $8 + 2x = 42$

91. $(20 - 4) \cdot 4 \div 2$

92. $2 \cdot (5 + 3^2)$

93. answers may vary

94. a. expression

b. equation

c. equation

d. expression

95. answers may vary

96. answers may vary

Name _____

Write each sentence as an equation. Use x to represent any unknown number. See Example 9.

81. One increased by two equals the quotient of nine and three.

82. Four subtracted from eight is equal to two squared.

83. Three is not equal to four divided by two.

84. The difference of sixteen and four is greater than ten.

85. The sum of 5 and a number is 20.

86. Twice a number is 17.

87. Thirteen minus three times a number is 13.

88. Seven subtracted from a number is 0.

89. The quotient of 12 and a number is $\frac{1}{2}$.

90. The sum of 8 and twice a number is 42.

COMBINING CONCEPTS

91. Insert parentheses so that the following expression simplifies to 32.
$$20 - 4 \cdot 4 \div 2$$

92. Insert parentheses so that the following expression simplifies to 28.
$$2 \cdot 5 + 3^2$$

93. In your own words, explain the difference between an expression and an equation.

94. Determine whether each is an expression or an equation.
   a. $3x^2 - 26$
   b. $3x^2 - 26 = 1$
   c. $2x - 5 = 7x - 5$
   d. $9y + x - 8$

95. Why is 8^2 usually read as "eight squared"? (*Hint*: What is the area of the **square** below?)

8 inches

96. Why is 4^3 usually read as "four cubed"? (*Hint*: What is the volume of the **cube** below?)

4 cm

1.3 ADDING REAL NUMBERS

Real numbers can be added, subtracted, multiplied, divided, and raised to powers, just as whole numbers can.

A ADDING REAL NUMBERS

To begin, we will use the number line to help picture the addition of real numbers.

Example 1   Add: $3 + 2$

Solution: We start at 0 on a number line, and draw an arrow representing 3. This arrow is three units long and points to the right since 3 is positive. From the tip of this arrow, we draw another arrow representing 2. The number below the tip of this arrow is the sum, 5.

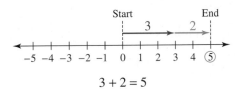

$$3 + 2 = 5$$

Example 2   Add: $-1 + (-2)$

Solution: We start at 0 on a number line, and draw an arrow representing -1. This arrow is one unit long and points to the left since -1 is negative. From the tip of this arrow, we draw another arrow representing -2. The number below the tip of this arrow is the sum, -3.

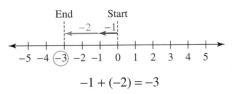

$$-1 + (-2) = -3$$

Thinking of integers as money earned or lost might help make addition more meaningful. Earnings can be thought of as positive numbers. If $1 is earned and later another $3 is earned, the total amount earned is $4. In other words, $1 + 3 = 4$.

On the other hand, losses can be thought of as negative numbers. If $1 is lost and later another $3 is lost, a total of $4 is lost. In other words, $(-1) + (-3) = -4$.

In Examples 1 and 2, we added numbers with the same sign. Adding numbers whose signs are not the same can be pictured on a number line also.

Example 3   Add: $-4 + 6$

Solution:

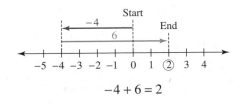

$$-4 + 6 = 2$$

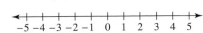

Objectives

 Add real numbers.

 Solve problems that involve addition of real numbers.

 Find the opposite of a number.

SSM   CD-ROM   Video 1.3

Practice Problem 1

Add using a number line: $1 + 5$

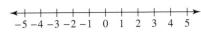

Practice Problem 2

Add using a number line: $-2 + (-4)$

TEACHING TIP

Adding integers may be shown using $+$ and $-$ signs. For example, $3 + (-5)$ can be represented by

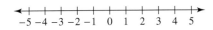

Using the atom analogy, a positive and a negative will neutralize to a zero. In this problem, the three positives are neutralized by 3 of the negatives and 2 negatives remain so the answer is -2.

Practice Problem 3

Add using a number line: $-5 + 8$

Answers

1. 6, **2.** -6, **3.** 3

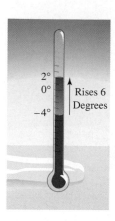

Practice Problem 4

Add using a number line: $5 + (-4)$

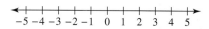

Practice Problem 5

Add without using a number line:
$(-8) + (-5)$

Practice Problem 6

Add without using a number line:
$(-14) + 6$

Practice Problems 7–12

Add without using a number line.

7. $(-17) + (-10)$

8. $(-4) + 12$

9. $1.5 + (-3.2)$

10. $-\dfrac{6}{11} + \left(-\dfrac{3}{11}\right)$

11. $12.8 + (-3.6)$

12. $-\dfrac{4}{5} + \dfrac{2}{3}$

Practice Problem 13

Find each sum.

a. $16 + (-9) + (-9)$

b. $[3 + (-13)] + [-4 + (-7)]$

Answers

4. 1,   **5.** −13,   **6.** −8,   **7.** −27,   **8.** 8,

9. −1.7,   **10.** $-\dfrac{9}{11}$,   **11.** 9.2,   **12.** $-\dfrac{2}{15}$,

13. a. −2,   **b.** −21

Using temperature as an example, if the thermometer registers 4 degrees below 0 degrees and then rises 6 degrees, the new temperature is 2 degrees above 0 degrees. Thus, it is reasonable that $-4 + 6 = 2$.

Example 4   Add: $4 + (-6)$

Solution:

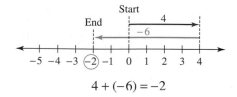

$$4 + (-6) = -2$$

Using a number line each time we add two numbers can be time consuming. Instead, we can notice patterns in the previous examples and write rules for adding real numbers.

> **ADDING REAL NUMBERS**
>
> To add two real numbers
>
> 1. with the *same sign*, add their absolute values. Use their common sign as the sign of the answer.
> 2. with *different signs*, subtract their absolute values. Give the answer the same sign as the number with the larger absolute value.

Example 5   Add without using a number line: $(-7) + (-6)$

Solution: Here, we are adding two numbers with the same sign.

$$(-7) + (-6) = -13$$

↑ ↖ sum of absolute values
same sign

Example 6   Add without using a number line: $(-10) + 4$

Solution: Here, we are adding two numbers with different signs.

$$(-10) + 4 = -6$$

↑ ↖ difference of absolute values
sign of number with larger absolute value, (-10)

Examples   Add without using a number line.

7. $(-8) + (-11) = -19$

8. $(-2) + 10 = 8$

9. $0.2 + (-0.5) = -0.3$

10. $-\dfrac{7}{10} + \left(-\dfrac{1}{10}\right) = -\dfrac{8}{10} = -\dfrac{4}{5}$

11. $11.4 + (-4.7) = 6.7$

12. $-\dfrac{3}{8} + \dfrac{2}{5} = -\dfrac{15}{40} + \dfrac{16}{40} = \dfrac{1}{40}$

Example 13   Find each sum.

a. $3 + (-7) + (-8)$

b. $[7 + (-10)] + [-2 + (-4)]$

Solution: **a.** Perform the additions from left to right.

$$3 + (-7) + (-8) = -4 + (-8)$$

Adding numbers with different signs

$$= -12$$

Adding numbers with like signs

> **HELPFUL HINT:**
> Don't forget that brackets are grouping symbols. We simplify within them first.

b. Simplify inside the brackets first.

$$[7 + (-10)] + [-2 + (-4)] = [-3] + [-6]$$
$$= -9 \quad \text{Add.} \quad \blacksquare$$

B SOLVING PROBLEMS THAT INVOLVE ADDITION

Positive and negative numbers are used in everyday life. Stock market returns show gains and losses as positive and negative numbers. Temperatures in cold climates often dip into the negative range, commonly referred to as "below zero" temperatures. Bank statements report deposits and withdrawals as positive and negative numbers.

Example 14   Calculating Gain or Loss

During a three-day period, a share of Lamplighter's International stock recorded the following gains and losses:

Monday	Tuesday	Wednesday
a gain of $2	a loss of $1	a loss of $3

Find the overall gain or loss for the stock for the three days.

Solution: Gains can be represented by positive numbers. Losses can be represented by negative numbers. The overall gain or loss is the sum of the gains and losses.

In words: gain   plus   loss   plus   loss

Translate: 2   +   (-1)   +   (-3)   = -2

The overall loss is $2.   ▬

C FINDING OPPOSITES

To help us subtract real numbers in the next section, we first review what we mean by opposites. To help us, the graphs of 4 and -4 are shown on the number line below.

Notice that the graph of 4 and -4 lie on opposite sides of 0, and each is 4 units away from 0. Such numbers are known as **opposites** or **additive inverses** of each other.

Practice Problem 14

During a four-day period, a share of Walco stock recorded the following gains and losses:

Tuesday	Wednesday
a loss of $2	a loss of $1
Thursday	**Friday**
a gain of $3	a gain of $3

Find the overall gain or loss for the stock for the four days.

TEACHING TIP

Point out to students that in the next chapter they will be using opposites or additive inverses to solve equations.

Answer

14. a gain of $3

OPPOSITE OR ADDITIVE INVERSE

Two numbers that are the same distance from 0 but lie on opposite sides of 0 are called **opposites** or **additive inverses** of each other.

Practice Problems 15–18

Find the opposite of each number.

15. −35

16. 12

17. $-\dfrac{3}{11}$

18. 1.9

Examples   Find the opposite of each number.

15. 10   The opposite of 10 is −10.

16. −3   The opposite of −3 is 3.

17. $\dfrac{1}{2}$ The opposite of $\dfrac{1}{2}$ is $-\dfrac{1}{2}$.

18. −4.5   The opposite of −4.5 is 4.5.

We use the symbol " − " to represent the phrase "the opposite of" or "the additive inverse of." In general, if a is a number, we write the opposite or additive inverse of a as $-a$. We know that the opposite of −3 is 3. Notice that this translates as

the opposite of	−3	is	3
↓	↓	↓	↓
−	(-3)	=	3

This is true in general.

If a is a number, then $-(-a) = a$.

Practice Problem 19

Simplify each expression.

a. $-(-22)$

b. $-\left(-\dfrac{2}{7}\right)$

c. $-(-x)$

d. $-|-14|$

Example 19   Simplify each expression.

a. $-(-10)$ **b.** $-\left(-\dfrac{1}{2}\right)$

c. $-(-2x)$ **d.** $-|-6|$

Solution: **a.** $-(-10) = 10$ **b.** $-\left(-\dfrac{1}{2}\right) = \dfrac{1}{2}$

c. $-(-2x) = 2x$

d. Since $|-6| = 6$, then $-|-6| = -6$.

Let's discover another characteristic about opposites. Notice that the sum of a number and its opposite is 0.

$$10 + (-10) = 0$$
$$-3 + 3 = 0$$
$$\frac{1}{2} + \left(-\frac{1}{2}\right) = 0$$

In general, we can write the following:

The sum of a number a and its opposite $-a$ is 0.
$$a + (-a) = 0$$

Notice that this means that the opposite of 0 is then 0 since $0 + 0 = 0$.

Answers

15. 35, **16.** −12, **17.** $\dfrac{3}{11}$, **18.** −1.9,

19. a. 22, **b.** $\dfrac{2}{7}$, **c.** x, **d.** −14

Name _____ Section _____ Date _____

Exercise Set 1.3

A *Add. See Examples 1 through 13.*

1. $6 + 3$ **2.** $9 + (-12)$ **3.** $-6 + (-8)$ **4.** $-6 + (-14)$

5. $8 + (-7)$ **6.** $6 + (-4)$ **7.** $-14 + 2$ **8.** $-10 + 5$

9. $-2 + (-3)$ **10.** $-7 + (-4)$ **11.** $-9 + (-3)$ **12.** $7 + (-5)$

13. $-7 + 3$ **14.** $-5 + 9$ **15.** $10 + (-3)$ **16.** $8 + (-6)$

17. $5 + (-7)$ **18.** $3 + (-6)$ **19.** $-16 + 16$ **20.** $23 + (-23)$

21. $27 + (-46)$ **22.** $53 + (-37)$ **23.** $-18 + 49$ **24.** $-26 + 14$

25. $-33 + (-14)$ **26.** $-18 + (-26)$ **27.** $6.3 + (-8.4)$ **28.** $9.2 + (-11.4)$

29. $|-8| + (-16)$ **30.** $|-6| + (-61)$ **31.** $117 + (-79)$ **32.** $144 + (-88)$

33. $-9.6 + (-3.5)$ **34.** $-6.7 + (-7.6)$ **35.** $-\dfrac{3}{8} + \dfrac{5}{8}$ **36.** $-\dfrac{5}{12} + \dfrac{7}{12}$

37. $-\dfrac{7}{16} + \dfrac{1}{4}$ **38.** $-\dfrac{5}{9} + \dfrac{1}{3}$ **39.** $-\dfrac{7}{10} + \left(-\dfrac{3}{5}\right)$ **40.** $-\dfrac{5}{6} + \left(-\dfrac{2}{3}\right)$

41. $-15 + 9 + (-2)$ **42.** $-9 + 15 + (-5)$

ANSWERS

1. 9
2. -3
3. -14
4. -20
5. 1
6. 2
7. -12
8. -5
9. -5
10. -11
11. -12
12. 2
13. -4
14. 4
15. 7
16. 2
17. -2
18. -3
19. 0
20. 0
21. -19
22. 16
23. 31
24. -12
25. -47
26. -44
27. -2.1
28. -2.2
29. -8
30. -55
31. 38
32. 56
33. -13.1
34. -14.3
35. $\dfrac{2}{8} = \dfrac{1}{4}$
36. $\dfrac{2}{12} = \dfrac{1}{6}$
37. $-\dfrac{3}{16}$
38. $-\dfrac{2}{9}$
39. $-\dfrac{13}{10}$
40. $-\dfrac{9}{6} = -\dfrac{3}{2}$
41. -8
42. 1

Name _____

43. −21 + (−16) + (−22)   **44.** −18 + (−6) + (−40)

45. −23 + 16 + (−2)   **46.** −14 + (−3) + 11   **47.** |5 + (−10)|

48. |7 + (−17)| **49.** 6 + (−4) + 9   **50.** 8 + (−2) + 7

51. [−17 + (−4)] + [−12 + 15]   **52.** [−2 + (−7)] + [−11 + 22]

53. |9 + (−12)| + |−16| **54.** |43 + (−73)| + |−20|

55. −13 + [5 + (−3) + 4]   **56.** −30 + [1 + (−6) + 8]

57. Explain why adding a negative number to another negative number always gives a negative sum.

58. When a positive and a negative number are added, sometimes the sum is positive, sometimes it is zero, and sometimes it is negative. Explain why this happens.

B *Solve each of the following. See Example 14.*

59. The low temperature in Anoka, Minnesota, was −15° last night. During the day it rose only 9°. Find the high temperature for the day.

60. On January 2, 1943, the temperature was −4° at 7:30 a.m. in Spearfish, South Dakota. Incredibly, it got 49° warmer in the next 2 minutes. To what temperature did it rise by 7:32?

61. The lowest elevation on Earth is −1296 feet (that is, 1296 feet below sea level) at the Dead Sea. If you are standing 658 feet above the Dead Sea, what is your elevation? (*Source*: Microsoft Encarta)

62. The lowest point in Africa is −502 feet at Lake Assal in Djibouti. If you are standing 658 feet above Lake Assal, what is your elevation? (*Source*: Microsoft Encarta)

Name _____

63. In checking the stock market results, Alexis discovers her stock posted changes of $-1\frac{5}{8}$ points and $-2\frac{1}{2}$ points over the last two days. What is the combined change?

64. Yesterday your stock posted a change of $-1\frac{1}{4}$ points, but today it showed a gain of $+\frac{7}{8}$ point. Find the overall change for the two days.

65. In golf, scores that are under par for the entire round are shown as negative scores; positive scores are shown for scores that are over par, and 0 is par. During the 1997 Buick Open, Tiger Woods had scores of 0, -4, -2, and -4. What was his total overall score? (*Source*: *USA Today*, 8/11/97)

66. During the same Buick Open (see Exercise 67), the winner, Vijay Singh, had scores of -5, $+1$, -5, and -6. What was his total overall score? (*Source*: *USA Today*, 8/11/97)

67. A negative net profit results when a company spends more money than it brings in. The Monterey Pasta Company had net profits of $-\$3.0$ million, $-\$22.0$ million, and $-\$8.2$ million in 1994, 1995, and 1996, respectively. What was the total net profit for these three years? (*Source*: Monterey Pasta)

68. Maytag had net profits of $-\$315.4$ million, $\$51.3$ million, and $\$147.9$ million in 1992, 1993, and 1994, respectively. What was the total net profit for these three years? (*Source*: Maytag)

C *Find each additive inverse or opposite. See Examples 15 through 18.*

69. 6

70. 4

71. -2

72. -8

73. 0

74. $-\frac{1}{4}$

75. $\left|-6\right|$

76. $\left|-11\right|$

77. In your own words, explain how to find the opposite of a number.

78. In your own words, explain why 0 is the only number that is its own opposite.

63. $-4\frac{1}{8}$ points

64. $-\frac{3}{8}$ points

65. -10

66. -15

67. $-\$33.2$ million

68. $-\$116.2$ million

69. -6

70. -4

71. 2

72. 8

73. 0

74. $\frac{1}{4}$

75. -6

76. -11

77. answers may vary

78. answers may vary

30

Name _____

Simplify each of the following. See Example 19.

 **79.** $-|-2|$ **80.** $-(-3)$ **81.** $-|0|$

82. $\left|-\frac{2}{3}\right|$ **83.** $-\left|-\frac{2}{3}\right|$  **84.** $-(-7)$

◤ **COMBINING CONCEPTS**

The following bar graph shows the daily low temperatures for a week in Sioux Falls, South Dakota. Use this graph to answer Exercises 85–90.

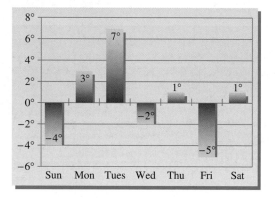

85. On what day of the week was the graphed temperature the highest?

86. On what day of the week was the graphed temperature the lowest?

87. What is the highest temperature shown on the graph?

88. What is the lowest temperature shown on the graph?

89. Find the average daily low temperature for Sunday through Thursday. (*Hint:* To find the average of the five temperatures, find their sum and divide by 5.)

90. Find the average daily low temperature for Tuesday through Thursday.

If a is a positive number and b is a negative number, fill in the blanks with the words positive or negative.

91. $-a$ is a ——— number.

92. $-b$ is a ——— number.

93. $a + a$ is a ——— number.

94. $b + b$ is a ——— number.

1.4 SUBTRACTING REAL NUMBERS

A SUBTRACTING REAL NUMBERS

Now that addition of real numbers has been discussed, we can explore subtraction. We know that $9 - 7 = 2$. Notice that $9 + (-7) = 2$, also. This means that

$$9 - 7 = 9 + (-7)$$

Notice that the *difference* of 9 and 7 is the same as the *sum* of 9 and the opposite of 7. This is how we can subtract real numbers.

> ### SUBTRACTING REAL NUMBERS
>
> If a and b are real numbers, then $a - b = a + (-b)$.

In other words, to find the difference of two numbers, we add the opposite of the number being subtracted.

Example 1 Subtract.

a. $-13 - 4$ **b.** $5 - (-6)$ **c.** $3 - 6$ **d.** $-1 - (-7)$

Solution:

a. $-13 - 4 = -13 + (-4)$ Add -13 to the opposite of 4, which is -4.

$= -17$

b. $5 - (-6) = 5 + (6)$ Add 5 to the opposite of -6, which is 6.

$= 11$

c. $3 - 6 = 3 + (-6)$ Add 3 to the opposite of 6, which is -6.
$= -3$

d. $-1 - (-7) = -1 + (7) = 6$

> ┌ **HELPFUL HINT**
>
> Study the patterns indicated.
>
> No change ↓ ┌─Change to addition.
> ┌─Change to opposite.
>
> $5 - 11 = 5 + (-11) = -6$
> $-3 - 4 = -3 + (-4) = -7$
> $7 - (-1) = 7 + (1) = 8$

Examples Subtract.

2. $5.3 - (-4.6) = 5.3 + (4.6) = 9.9$

3. $-\dfrac{3}{10} - \dfrac{5}{10} = -\dfrac{3}{10} + \left(-\dfrac{5}{10}\right) = -\dfrac{8}{10} = -\dfrac{4}{5}$

4. $-\dfrac{2}{3} - \left(-\dfrac{4}{5}\right) = -\dfrac{2}{3} + \left(\dfrac{4}{5}\right) = -\dfrac{10}{15} + \dfrac{12}{15} = \dfrac{2}{15}$

Objectives

A Subtract real numbers.
B Evaluate algebraic expressions using real numbers.
C Determine whether a number is a solution of a given equation.
D Solve problems that involve subtraction of real numbers.
E Find complementary and supplementary angles.

SSM   CD-ROM   Video
1.4

Practice Problem 1

Subtract.

a. $-20 - 6$

b. $3 - (-5)$

c. $7 - 17$

d. $-4 - (-9)$

Practice Problems 2–4

Subtract.

2. $9.6 - (-5.7)$ **3.** $-\dfrac{4}{9} - \dfrac{2}{9}$

4. $-\dfrac{1}{4} - \left(-\dfrac{2}{5}\right)$

Answers

1. a. -26, **b.** 8, **c.** -10, **d.** 5,

2. 15.3, **3.** $-\dfrac{2}{3}$, **4.** $\dfrac{3}{20}$

Practice Problem 5

Subtract 7 from -11.

Practice Problem 6

Simplify each expression.

a. $-20 - 5 + 12 - (-3)$

b. $5.2 - (-4.4) + (-8.8)$

Practice Problem 7

Simplify each expression.

a. $-9 + [(-4 - 1) - 10]$

b. $5^2 - 20 + [-11 - (-3)]$

TEACHING TIP

Ask students what makes Example 7 easy to follow. Be sure they observe that each step is shown, the equal signs are lined up and that brief comments accompany each step. Encourage them to develop good organizational habits that support their learning of mathematics.

Answers

5. -18,   **6. a.** -10,   **b.** 0.8,

7. a. -24,   **b.** -3

Example 5   Subtract 8 from -4.

Solution: Be careful when interpreting this: The order of numbers in subtraction is important. 8 is to be subtracted **from** -4.

$$-4 - 8 = -4 + (-8) = -12$$

If an expression contains additions and subtractions, just write the subtractions as equivalent additions. Then simplify from left to right.

Example 6   Simplify each expression.

   a. $-14 - 8 + 10 - (-6)$

   b. $1.6 - (-10.3) + (-5.6)$

Solution: a. $-14 - 8 + 10 - (-6) =$

$$-14 + (-8) + 10 + 6 = -6$$

   b. $1.6 - (-10.3) + (-5.6) =$

$$1.6 + 10.3 + (-5.6) = 6.3$$

When an expression contains parentheses and brackets, remember the order of operations. Start with the innermost set of parentheses or brackets and work your way outward.

Example 7   Simplify each expression.

   a. $-3 + [(-2 - 5) - 2]$

   b. $2^3 - 10 + [-6 - (-5)]$

Solution: a. Start with the innermost set of parentheses. Rewrite $-2 - 5$ as an addition.

$$
\begin{aligned}
-3 + [(-2 - 5) - 2] &= -3 + [(-2 + (-5)) - 2] &&\text{Add: } -2 + (-5).\\
&= -3 + [(-7) - 2] &&\text{Write } -7 - 2 \text{ as an addition.}\\
&= -3 + [-7 + (-2)] \\
&= -3 + [-9] &&\text{Add.}\\
&= -12 &&\text{Add.}
\end{aligned}
$$

   b. Start simplifying the expression inside the brackets by writing $-6 - (-5)$ as an addition.

$$
\begin{aligned}
2^3 - 10 + [-6 - (-5)] &= 2^3 - 10 + [-6 + 5]\\
&= 2^3 - 10 + [-1] &&\text{Add.}\\
&= 8 - 10 + (-1) &&\text{Evaluate } 2^3.\\
&= 8 + (-10) + (-1) &&\text{Write } 8 - 10 \text{ as an addition.}\\
&= -2 + (-1) &&\text{Add.}\\
&= -3 &&\text{Add.}
\end{aligned}
$$

B   **Evaluating Algebraic Expressions**

It is important to be able to evaluate expressions for given replacement values. This helps, for example, when checking solutions of equations.

Example 8   Find the value of each expression when $x = 2$ and $y = -5$.

a. $\dfrac{x - y}{12 + x}$ b. $x^2 - y$

Solution: **a.** Replace x with 2 and y with -5. Be sure to put parentheses around -5 to separate signs. Then simplify the resulting expression.

$$\frac{x - y}{12 + x} = \frac{2 - (-5)}{12 + 2} = \frac{2 + 5}{14} = \frac{7}{14} = \frac{1}{2}$$

b. Replace the x with 2 and y with -5 and simplify.

$$x^2 - y = 2^2 - (-5) = 4 - (-5) = 4 + 5 = 9$$

Practice Problem 8

Find the value of each expression when $x = 1$ and $y = -4$.

a. $\dfrac{x - y}{14 + x}$

b. $x^2 - y$

C SOLUTIONS OF EQUATIONS

Recall from Section 1.2 that a solution of an equation is a value for the variable that makes the equation true.

Example 9   Determine whether -4 is a solution of $x - 5 = -9$.

Solution: Replace x with -4 and see if a true statement results.

$$x - 5 = -9 \qquad \text{Original equation}$$
$$-4 - 5 \overset{?}{=} -9 \qquad \text{Replace } x \text{ with } -4.$$
$$-4 + (-5) \overset{?}{=} -9$$
$$-9 = -9 \qquad \text{True.}$$

Thus -4 is a solution of $x - 5 = -9$.

Practice Problem 9

Determine whether -2 is a solution of $-1 + x = 1$.

D SOLVING PROBLEMS THAT INVOLVE SUBTRACTION

Another use of real numbers is in recording altitudes above and below sea level, as shown in the next example.

Example 10   Finding the Difference in Elevations

The lowest point in North America is in Death Valley, at an elevation of 282 feet below sea level. Nearby, Mount Whitney reaches 14,494 feet, the highest point in the United States outside Alaska. How much of a variation in elevation is there between these two extremes? (*Source:* U.S. Geological Survey)

Solution: To find the variation in elevation between the two heights, find the difference of the high point and the low point.

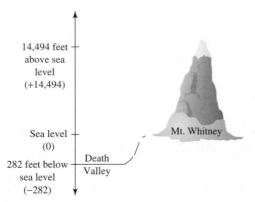

Practice Problem 10

At 6.00 P.M., the temperature at the Winter Olympics was 14°; by morning the temperature dropped to $-23°$. Find the overall change in temperature.

Answers

8. a. $\dfrac{1}{3}$,   **b.** 5,   **9.** -2 is not a solution,
10. $-37°$

In words: | high point | minus | low point |

Translate: $14{,}494$ $-$ (-282) $= 14{,}494 + 282$
$= 14{,}776$ feet

Thus, the variation in elevation is 14,776 feet.   ▬▬▬

E FINDING COMPLEMENTARY AND SUPPLEMENTARY ANGLES

A knowledge of geometric concepts is needed by many professionals, such as doctors, carpenters, electronic technicians, gardeners, machinists, and pilots, just to name a few. With this in mind, we review the geometric concepts of **complementary** and **supplementary angles**.

COMPLEMENTARY AND SUPPLEMENTARY ANGLES

Two angles are **complementary** if their sum is 90°.

Two angles are **supplementary** if their sum is 180°.

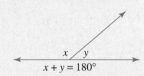

Practice Problem 11

Find each unknown complementary or supplementary angle.

a.

b.

Example 11   Find each unknown complementary or supplementary angle.

a.   **b.**

Solution: **a.** These angles are complementary, so their sum is 90°. This means that x is $90° - 38°$.

$x = 90° - 38° = 52°$

b. These angles are supplementary, so their sum is 180°. This means that y is $180° - 62°$.

$y = 180° - 62° = 118°$ ▬▬▬

Answer
11. a. 102°, **b.** 9°

Name _____ Section _____ Date _____

ANSWERS

1. -10
2. -20
3. -5
4. -3
5. 19
6. 17
7. $\dfrac{1}{6}$
8. $-\dfrac{1}{8}$
9. 2
10. 28
11. -11
12. -12
13. 11
14. 9
15. 5
16. 12
17. 37
18. 48
19. -6.4
20. -2.9
21. -71
22. -87
23. 0
24. 0
25. 4.1
26. -0.8
27. $\dfrac{2}{11}$
28. $-\dfrac{3}{7}$
29. $-\dfrac{22}{24} = -\dfrac{11}{12}$
30. $-\dfrac{78}{80} = -\dfrac{39}{40}$
31. 8.92
32. 5.17
33. sometimes positive and sometimes negative
34. sometimes positive and sometimes negative
35. 13
36. -5

EXERCISE SET 1.4

A *Subtract. See Examples 1 through 4.*

1. $-6 - 4$
2. $-12 - 8$
3. $4 - 9$
4. $8 - 11$

5. $16 - (-3)$
6. $12 - (-5)$
7. $\dfrac{1}{2} - \dfrac{1}{3}$
8. $\dfrac{3}{4} - \dfrac{7}{8}$

9. $-16 - (-18)$
10. $-20 - (-48)$
11. $-6 - 5$
12. $-8 - 4$

13. $7 - (-4)$
14. $3 - (-6)$
15. $-6 - (-11)$
16. $-4 - (-16)$

17. $16 - (-21)$
18. $15 - (-33)$
19. $9.7 - 16.1$
20. $8.3 - 11.2$

21. $-44 - 27$
22. $-36 - 51$
23. $-21 - (-21)$
24. $-17 - (-17)$

25. $-2.6 - (-6.7)$
26. $-6.1 - (-5.3)$
27. $-\dfrac{3}{11} - \left(-\dfrac{5}{11}\right)$

28. $-\dfrac{4}{7} - \left(-\dfrac{1}{7}\right)$
29. $-\dfrac{1}{6} - \dfrac{3}{4}$
30. $-\dfrac{1}{10} - \dfrac{7}{8}$

31. $8.3 - (-0.62)$
32. $4.3 - (-0.87)$

33. If a and b are positive numbers, then is $a - b$ always positive, always negative, or sometimes positive and sometimes negative?

34. If a and b are negative numbers, then is $a - b$ always positive, always negative, or sometimes positive and sometimes negative?

Write each phrase as an expression and simplify. See Example 5.

35. Subtract -5 from 8.
36. Subtract 3 from -2.

37. Subtract -1 from -6. **38.** Subtract 17 from 1.

39. Subtract 8 from 7. ▣ **40.** Subtract 9 from -4.

41. Decrease -8 by 15. **42.** Decrease 11 by -14.

Simplify each expression. (Remember the order of operations.) See Examples 6 and 7.

43. $-10 - (-8) + (-4) - 20$ **44.** $-16 - (-3) + (-11) - 14$

45. $5 - 9 + (-4) - 8 - 8$ **46.** $7 - 12 + (-5) - 2 + (-2)$

47. $-6 - (2 - 11)$ **48.** $-9 - (3 - 8)$ **49.** $3^3 - 8 \cdot 9$

50. $2^3 - 6 \cdot 3$ **51.** $2 - 3(8 - 6)$ **52.** $4 - 6(7 - 3)$

▣ **53.** $(3 - 6) + 4^2$ **54.** $(2 - 3) + 5^2$

55. $-2 + \left[(8 - 11) - (-2 - 9) \right]$ ▣ **56.** $-5 + \left[(4 - 15) - (-6) - 8 \right]$

57. $|-3| + 2^2 + \left[-4 - (-6) \right]$ **58.** $|-2| + 6^2 + (-3 - 8)$

B *Evaluate each expression when $x = -5$, $y = 4$, and $t = 10$. See Example 8.*

59. $x - y$ **60.** $y - x$ **61.** $|x| + 2t - 8y$ **62.** $|x + t - 7y|$

63. $\dfrac{9 - x}{y + 6}$ **64.** $\dfrac{15 - x}{y + 2}$ ▣ **65.** $y^2 - x$ **66.** $t^2 - x$

Answers in left margin:

37. -5
38. -16
39. -1
40. -13
41. -23
42. 25
43. -26
44. -38
45. -24
46. -14
47. 3
48. -4
49. -45
50. -10
51. -4
52. -20
53. 13
54. 24
55. 6
56. -18
57. 9
58. 27
59. -9
60. 9
61. -7
62. 23
63. $\dfrac{7}{5}$
64. $\dfrac{10}{3}$
65. 21
66. 105

67. $\dfrac{|x - (-10)|}{2t}$

68. $\dfrac{|5y - x|}{6t}$

C *Decide whether the given number is a solution of the given equation. See Example 9.*

69. $x - 9 = 5$; -4

70. $x - 10 = -7$; 3

71. $-x + 6 = -x - 1$; -2

72. $-x - 6 = -x - 1$; -10

73. $-x - 13 = -15$; 2

74. $4 = 1 - x$; 5

D *Solve. See Example 10.*

75. Within 24 hours in 1916, the temperature in Browning, Montana, fell from $44°$ to $-56°$. How large a drop in temperature was this?

76. Much of New Orleans is just barely above sea level. If George descends 12 feet from an elevation of 5 feet above sea level, what is his new elevation?

77. In a series of plays, the San Francisco 49ers gain 2 yards, lose 5 yards, and then lose another 20 yards. What is their total gain or loss of yardage?

78. In some card games, it is possible to have a negative score. Lavonne Schultz currently has a score of 15 points. She then loses 24 points. What is her new score?

79. Aristotle died in the year -322 (or 322 B.C.). When was he born, if he was 62 years old when he died?

80. Augustus Caesar died in A.D. 14 in his 77th year. When was he born?

81. A commercial jet liner hits an air pocket and drops 250 feet. After climbing 120 feet, it drops another 178 feet. What is its overall vertical change?

82. Tyson Industries stock posted a loss of $1\frac{5}{8}$ points yesterday. If it drops another $\frac{3}{4}$ points today, find its overall change for the two days.

67. $\dfrac{1}{4}$

68. $\dfrac{5}{12}$

69. not a solution

70. solution

71. not a solution

72. not a solution

73. solution

74. not a solution

75. $100°$

76. 7 feet below sea level

77. lost 23 yards

78. -9

79. 384 B.C.

80. 63 B.C.

81. -308 ft.

82. $-2\frac{3}{8}$ points

83. 22,965 feet

84. 25,191 feet

85. 130°

86. 40°

87. 30°

88. 75°

38

83. The highest point in South America is Mount Aconcagua, Argentina, at an elevation of 22,834 feet. The lowest point is Valdes Peninsula, Argentina, at 131 feet below sea level. How much higher is Mount Aconcagua than Valdes Peninsula? (*Source*: National Geographic Society)

84. The lowest altitude in Antarctica is the Bentley Subglacial Trench at 8327 feet below sea level. The highest altitude is Vinson Massif at an elevation of 16,864 feet above sea level. What is the difference between these altitudes? (*Source*: National Geographic Society)

E *Find each unknown complementary or supplementary angle. See Example 11.*

85.

86.

87.

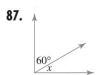

88.

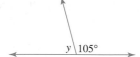

Name _____

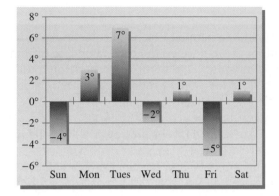

☜ **COMBINING CONCEPTS**

The following bar graph is from an earlier section and shows the daily low temperatures for a week in Sioux Falls, South Dakota. Use this graph to answer Exercises 89–91.

89. Record the daily _increases_ and _decreases_ in the low temperatures from the previous day.

Day	Daily Increase or Decrease
Monday	
Tuesday	
Wednesday	
Thursday	
Friday	
Saturday	

90. Which day of the week had the greatest increase in temperature?

91. Which day of the week had the greatest decrease in temperature?

If a is a positive number and b is a negative number, determine whether each statement is true or false.

92. $a - b$ is always a positive number.

93. $b - a$ is always a negative number.

94. $|b| - |a|$ is always a positive number.

95. $|b - a|$ is always a positive number.

Without calculating, determine whether each answer is positive or negative. Then use a calculator to find the exact difference.

🖩 **96.** $56{,}875 - 87{,}262$

🖩 **97.** $4.362 - 7.0086$

90. Monday

91. Wednesday

92. true

93. true

94. false

95. true

96. negative, −30,387

97. negative, −2.6466

Focus On Study Skills

STUDY TIPS

Have you wondered what you can do to be successful in your algebra course? If so, that may well be your first step to success in algebra! Here are some tips on how to use this text and how to study mathematics in general.

Using this Text

1. Each example in the section has a parallel Practice Problem. As you read a section, try each Practice Problem after you've finished the corresponding example. This "learn-by-doing" approach will help you grasp ideas before you move on to other concepts.
2. The main section of exercises in an exercise set are referenced by an objective, such as **A** or **B** and also an example(s). Use this referencing in case you have trouble completing an assignment from the exercise set.
3. If you need extra help in a particular section, check at the beginning of the section to see what videotapes and software are available.
4. Integrated Reviews in each chapter offer you a chance to practice—in one place—the many concepts that you have learned separately over several sections.
5. There are many opportunities at the end of each chapter to help you understand the concepts of the chapter.

   Chapter Highlights contain chapter summaries with examples.
   Chapter Reviews contain review problems organized by section.
   Chapter Tests are sample tests to help you prepare for an exam.
   Cumulative Reviews are reviews consisting of material from the beginning of the book to the end of the particular chapter.

General Tips

1. *Choose to attend all class periods.* If possible, sit near the front of the classroom. This way, you will see and hear the presentation better. It may also be easier for you to participate in classroom activities.
2. *Do your homework.* You've probably heard the phrase "practice makes perfect" in relation to music and sports. It also applies to mathematics. You will find that the more time you spend solving mathematics problems, the easier the process becomes. Be sure to block out enough time to complete your assignments.
3. *Check your work.* Review the steps you made while working a problem. Learn to check your answers in the original problems. You can also compare your answers to the answers to selected exercises listed in the back of the book. If you have made a mistake, figure out what went wrong. Then correct your mistake.
4. *Learn from your mistakes.* Everyone, even your instructors, makes mistakes. You can use your mistakes to become a better math student. The key is finding and understanding your mistakes. Was your mistake a careless mistake? If so, you can try to work more slowly and make a conscious effort to carefully check your work. Did you make a mistake because you don't understand a concept? If so, take the time to review the concept or ask questions to better understand the concept.
5. *Know how to get help if you need it.* It's OK to ask for help. In fact, it's a good idea to ask for help whenever there is something that you don't understand. Make sure you know when your instructor has office hours and how to find his or her office. Find out if math tutoring services are available on your campus. Check out the hours, location, and requirements of the tutoring service. You might also want to find another student in your class that you can call to discuss your assignment.
6. *Organize your class materials,* including homework assignments, graded quizzes and tests, and notes from your class or lab. All of these items will make valuable references throughout your course and as you study for upcoming tests and your final exam. Make sure you can locate any of these materials when you need them.

1.5 MULTIPLYING REAL NUMBERS

A MULTIPLYING REAL NUMBERS

Multiplication of real numbers is similar to multiplication of whole numbers. We just need to determine when the answer is positive, when it is negative, and when it is zero. To discover sign patterns for multiplication, recall that multiplication is repeated addition. For example, 3(2) means that 2 is added to itself three times, or

$$3(2) = 2 + 2 + 2 = 6$$

Also,

$$3(-2) = (-2) + (-2) + (-2) = -6$$

Since $3(-2) = -6$, this suggests that the product of a positive number and a negative number is a negative number.

What about the product of two negative numbers? To find out, consider the following pattern.

Factor decreases by 1 each time.

$$-3 \cdot 2 = -6$$
$$-3 \cdot 1 = -3 \quad \text{Product increases by 3 each time}$$
$$-3 \cdot 0 = 0$$
$$-3 \cdot -1 = 3$$
$$-3 \cdot -2 = 6$$

This suggests that the product of two negative numbers is a positive number. Our results are given below.

MULTIPLYING REAL NUMBERS

1. The product of two numbers with the same sign is a positive number.
2. The product of two numbers with different signs is a negative number.

Examples   Multiply.

1. $-6(4) = -24$
2. $2(-10) = -20$
3. $-5(-10) = 50$
4. $-\frac{2}{3} \cdot \frac{4}{7} = -\frac{2 \cdot 4}{3 \cdot 7} = -\frac{8}{21}$
5. $5(-1.7) = -8.5$
6. $-18(-3) = 54$

We already know that the product of 0 and any whole number is 0. This is true of all real numbers.

PRODUCTS INVOLVING ZERO

If b is a real number, then $b \cdot 0 = 0$. Also $0 \cdot b = 0$.

Objectives

A Multiply real numbers.
B Evaluate algebraic expressions using real numbers.
C Determine whether a number is a solution of a given equation.

SSM   CD-ROM   Video 1.5

TEACHING TIP

Consider beginning this lesson with the following activity: Students will find and graph the location of a train at a certain time given its velocity (speed and direction). Positive velocities indicate speeds (miles per hour) in the eastward direction, and negative velocities indicate speeds (mph) in the westward direction. Positive times are in the future, and negative times are in the past (i.e., −3 means 3 hours ago). Assume that all trains are currently at the train depot. Interpret each situation, and find how far the train was/will be from the depot. Graph its position on the number line.

Train	Time (hours)	Velocity (mph)	Miles from depot (+ → east; − → west)
A	2	70	
B	5	−45	
C	−4	60	
D	−3	−55	

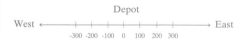
Depot
West ←————————→ East
-300 -200 -100 0 100 200 300

Practice Problems 1–6

Multiply.

1. $-8(3)$ 2. $5(-30)$
3. $-4(-12)$ 4. $-\frac{5}{6} \cdot \frac{1}{4}$
5. $6(-2.3)$ 6. $-15(-2)$

Answers

1. −24, **2.** −150, **3.** 48, **4.** $-\frac{5}{24}$,
5. −13.8, **6.** 30

Practice Problem 7

Use the order of operations and simplify each expression.

a. $5(0)(-3)$ b. $(-1)(-6)(-7)$
c. $(-2)(4)(-8)$ d. $(-2)^2$
e. $-3(-9) - 4(-4)$

HELPFUL HINT

You may have noticed from the example that if we multiply:

▲ an *even* number of negative numbers, the product is *positive*.
▲ an *odd* number of negative numbers, the product is *negative*.

Practice Problem 8

Evaluate each expression when $x = -1$ and $y = -5$.

a. $3x - y$ b. $x^2 - y^3$

Practice Problem 9

Determine whether -10 is a solution of $3x + 4 = -26$.

Answers

7. a. 0, **b.** −42, **c.** 64, **d.** 4, **e.** 43,

8. a. 2 **b.** 126, **9.** −10 is a solution.

Example 7   Use the order of operations and simplify each expression.

a. $7(0)(-6)$ **b.** $(-2)(-3)(-4)$ **c.** $(-1)(5)(-9)$
d. $(-2)^3$ **e.** $(-4)(-11) - 5(-2)$

Solution: **a.** By the order of operations, we multiply from left to right. Notice that because one of the factors is 0, the product is 0.

$$7(0)(-6) = 0(-6) = 0$$

b. Multiply two factors at a time, from left to right.

$$(-2)(-3)(-4) = (6)(-4) \quad \text{Multiply } (-2)(-3).$$
$$= -24$$

c. Multiply from left to right.

$$(-1)(5)(-9) = (-5)(-9) \quad \text{Multiply } (-1)(5).$$
$$= 45$$

d. The exponent 3 means 3 factors of the base -2.

$$(-2)^3 = (-2)(-2)(-2)$$
$$= -8 \quad \text{Multiply.}$$

e. Follow the order of operations.

$$(-4)(-11) - 5(-2) = 44 - (-10) \quad \text{Find the products.}$$
$$= 44 + 10 \quad \text{Add 44 to the opposite of } -10.$$
$$= 54 \quad \text{Add.}$$

B   EVALUATING ALGEBRAIC EXPRESSIONS

Now that we know how to multiply positive and negative numbers, we continue to practice evaluating algebraic expressions.

Example 8   Evaluate each expression when $x = -2$ and $y = -4$.

a. $5x - y$ **b.** $x^3 - y^2$

Solution: **a.** Replace x with -2 and y with -4 and simplify.

$$5x - y = 5(-2) - (-4) = -10 - (-4) = -10 + 4 = -6$$

b. Replace x with -2 and y with -4.

$$x^3 - y^2 = (-2)^3 - (-4)^2 \quad \text{Substitute the given values for the variables.}$$
$$= -8 - (16) \quad \text{Evaluate } (-2)^3 \text{ and } (-4)^2.$$
$$= -8 + (-16) \quad \text{Write as a sum.}$$
$$= -24 \quad \text{Add.}$$

C   SOLUTIONS OF EQUATIONS

To prepare for solving equations, we continue to check possible solutions for an equation.

Example 9   Determine whether -3 is a solution of $-7x + 2 = 23$.

Solution: Replace x with -3 and see if a true statement results.

$$-7x + 2 = 23 \quad \text{Original equation}$$
$$-7(-3) + 2 \stackrel{?}{=} 23 \quad \text{Replace } x \text{ with } -3.$$
$$21 + 2 \stackrel{?}{=} 23 \quad \text{Multiply.}$$
$$23 = 23 \quad \text{True.}$$

Thus, -3 is a solution of $-7x + 2 = 23$.

Exercise Set 1.5

A *Multiply. See Examples 1 through 6.*

▣ **1.** $-6(4)$ **2.** $-8(5)$ ▣ **3.** $2(-1)$ **4.** $7(-4)$

▣ **5.** $-5(-10)$ **6.** $-6(-11)$ **7.** $-3\cdot4$ **8.** $-2\cdot8$

9. $-6(-7)$ **10.** $-6(-9)$ **11.** $2(-9)$ **12.** $3(-5)$

13. $-\dfrac{1}{2}\left(-\dfrac{3}{5}\right)$ **14.** $-\dfrac{1}{8}\left(-\dfrac{1}{3}\right)$ **15.** $-\dfrac{3}{4}\left(-\dfrac{8}{9}\right)$ **16.** $-\dfrac{5}{6}\left(-\dfrac{3}{10}\right)$

17. $5(-1.4)$ **18.** $6(-2.5)$ **19.** $-0.2(-0.7)$ **20.** $-0.5(-0.3)$

21. $-10(80)$ **22.** $-20(60)$ **23.** $4(-7)$ **24.** $5(-9)$

25. $(-5)(-5)$ **26.** $(-7)(-7)$ ▣ **27.** $\dfrac{2}{3}\left(-\dfrac{4}{9}\right)$ **28.** $\dfrac{2}{7}\left(-\dfrac{2}{11}\right)$

29. $-11(11)$ **30.** $-12(12)$ **31.** $-\dfrac{20}{25}\left(\dfrac{5}{16}\right)$ **32.** $-\dfrac{25}{36}\left(\dfrac{6}{15}\right)$

33. $-2.1(-0.4)$ **34.** $-1.3(-0.6)$

Simplify. See Example 7.
35. $(-1)(2)(-3)(-5)$ **36.** $(-2)(-3)(-4)(-2)$

ANSWERS

1. -24
2. -40
3. -2
4. -28
5. 50
6. 66
7. -12
8. -16
9. 42
10. 54
11. -18
12. -15
13. $\dfrac{3}{10}$
14. $\dfrac{1}{24}$
15. $\dfrac{24}{36}=\dfrac{2}{3}$
16. $\dfrac{15}{60}=\dfrac{1}{4}$
17. -7
18. -15
19. 0.14
20. 0.15
21. -800
22. -1200
23. -28
24. -45
25. 25
26. 49
27. $-\dfrac{8}{27}$
28. $-\dfrac{4}{77}$
29. -121
30. -144
31. $-\dfrac{100}{400}=-\dfrac{1}{4}$
32. $-\dfrac{150}{540}=-\dfrac{5}{18}$
33. 0.84
34. 0.78
35. -30
36. 48

37. 90	
38. 60	**37.** $(2)(-1)(-3)(5)(3)$ **38.** $(3)(-5)(-2)(-1)(-2)$
39. 16	
40. −27	
41. −36	🔊 **39.** $(-4)^2$ 🔊 **40.** $(-3)^3$ **41.** $(-6)(3)(-2)(-1)$
42. 12	
43. −125	**42.** $(-3)(-2)(-1)(-2)$ **43.** $(-5)^3$ **44.** $(-2)^5$
44. −32	
45. −16	
46. −36	**45.** -4^2 **46.** -6^2 **47.** $-3(2-8)$
47. 18	
48. 24	**48.** $-4(3-9)$ **49.** $6(3-8)$ **50.** $4(8-11)$
49. −30	
50. −12	
51. −24	**51.** $-3[(2-8)-(-6-8)]$ **52.** $-2[(3-5)-(2-9)]$
52. −10	
53. $\dfrac{9}{16}$	**53.** $\left(-\dfrac{3}{4}\right)^2$ **54.** $\left(-\dfrac{2}{7}\right)^2$
54. $\dfrac{4}{49}$	
55. true	*State whether each statement is true or false.*
56. true	**55.** The product of three negative integers is negative. **56.** The product of three positive integers is positive.
57. false	
58. true	**57.** The product of four negative integers is negative. **58.** The product of four positive integers is positive.
59. −21	
60. −35	**B** *Evaluate each expression when $x = -5$ and $y = -3$. See Example 8.*
61. 41	**59.** $3x + 2y$ **60.** $4x + 5y$ 🔊 **61.** $2x^2 - y^2$ **62.** $x^2 - 2y^2$
62. 7	
63. −134	**63.** $x^3 + 3y$ **64.** $y^3 + 3x$
64. −42	
65. solution	**C** *Decide whether the given number is a solution of the given equation. See Example 9.*
66. not a solution	**65.** $-5x = -35;$ 7 **66.** $2x = x - 1;$ -4

44

Name _____

67. $-3x - 5 = -20$; 5

68. $2x + 4 = x + 8$; -4

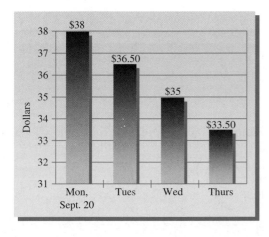 **69.** $9x + 1 = 14$; -1

70. $5x = -60$; -12

71. $3x - 20 = -5$; 5

72. $6x = 3x + 5$; -6

73. $17 - 4x = x + 27$; -2

74. $3x - 1 = 2x + 6$; -2

◤ COMBINING CONCEPTS

If q is a negative number, r is a negative number, and t is a positive number, determine whether each expression simplifies to a positive or negative number. If it is not possible to determine, so state.

75. $q \cdot r \cdot t$

76. $q^2 \cdot r \cdot t$

77. $q + t$

78. $t + r$

79. $t(q + r)$

80. $r(q - t)$

81. The following graph shows Trader's stock consistently decreasing in value by $1.50 per share per day. If this trend continues, when will the stock be worth $20 per share?

82. If a and b are any real numbers, is the statement $a \cdot b = b \cdot a$ always true? Why or why not?

83. If a and b are any real numbers, is the statement $a - b = b - a$ always true? Why or why not?

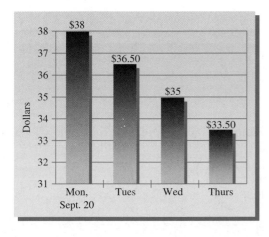

67. solution

68. not a solution

69. not a solution

70. solution

71. solution

72. not a solution

73. solution

74. not a solution

75. positive

76. negative

77. can't determine

78. can't determine

79. negative

80. positive

81. Sat., Oct. 2

82. yes; answers may vary

83. no, answers may vary

Focus On History

Numbers have a long history. The numbers we are accustomed to using probably originated in India in the 3rd century and were later adapted by Arabic cultures. Many other ancient civilizations developed their own unique number systems.

Roman Numerals

I	V	X	L	C	D	M
1	5	10	50	100	500	1000

If numerals decrease in value from left to right, the values are added. If a smaller numeral appears to the left of a larger numeral, the smaller value is subtracted. For example:

$$XVII = 10 + 5 + 1 + 1 = 17 \quad but \quad XLIV = 50 - 10 + 5 - 1 = 44$$

Chinese Numerals

一	二	三	四	五	六	七	八	九	十	百	千	萬	億
1	2	3	4	5	6	7	8	9	10	100	1000	10,000	100,000

Numerals are written vertically. If a digit representing 2–9 appears before a digit representing 10, 100, 1000, 10,000, or 100,000, multiplication is indicated.

(7×1000)

(3×100)

(8×10)

5   is $7000 + 300 + 80 + 5 = 7385$

Egyptian Hieroglyphic Numerals

I	∩	𐏓			
1	10	100	1000	10,000	100,000

The Egyptian system is also multiplicative. For example, 3 ∩ hieroglyphs represents 3×10.

$$= 4 \times 1 + 2 \times 10 + 5 \times 100 + 2 \times 1000 + 1 \times 10,000$$
$$= 12,524$$

GROUP ACTIVITIES

- ▲ Write several numbers using each of the Roman, Chinese, and Egyptian hieroglyphic systems. Trade your numbers with another student in your group to translate into our numerals. Check one another's work.
- ▲ Research the number system of another ancient culture (such as Babylonian, Mayan Indian, or Ionic Greek). Write the numbers 712, 4690, 5113, and 208 using that system. Demonstrate the system to the rest of your group.

1.6 DIVIDING REAL NUMBERS

A FINDING RECIPROCALS

Addition and subtraction are related. Every difference of two numbers $a - b$ can be written as the sum $a + (-b)$. Multiplication and division are related also. For example, the quotient $6 \div 3$ can be written as the product $6 \cdot \frac{1}{3}$. Recall that the pair of numbers 3 and $\frac{1}{3}$ has a special relationship. Their product is 1 and they are called **reciprocals** or **multiplicative inverses** of each other.

> **RECIPROCAL OR MULTIPLICATIVE INVERSE**
>
> Two numbers whose product is 1 are called **reciprocals** or **multiplicative inverses** of each other.

Examples   Find the reciprocal of each number.

1. 22   The reciprocal of 22 is $\frac{1}{22}$ since $22 \cdot \frac{1}{22} = 1$.

2. $\frac{3}{16}$ The reciprocal of $\frac{3}{16}$ is $\frac{16}{3}$ since $\frac{3}{16} \cdot \frac{16}{3} = 1$.

3. -10 The reciprocal of -10 is $-\frac{1}{10}$ since $-10 \cdot -\frac{1}{10} = 1$.

4. $-\frac{9}{13}$ The reciprocal of $-\frac{9}{13}$ is $-\frac{13}{9}$ since $-\frac{9}{13} \cdot -\frac{13}{9} = 1$.

5. 1.7   The reciprocal of 1.7 is $\frac{1}{1.7}$ since $1.7 \cdot \frac{1}{1.7} = 1$.

Does the number 0 have a reciprocal? If it does, it is a number n such that $0 \cdot n = 1$. Notice that this can never be true since $0 \cdot n = 0$. This means that 0 has no reciprocal.

> **QUOTIENTS INVOLVING ZERO**
>
> The number 0 does not have a reciprocal.

B DIVIDING REAL NUMBERS

We may now write a quotient as an equivalent product.

> **QUOTIENT OF TWO REAL NUMBERS**
>
> If a and b are real numbers and b is not 0, then
> $$a \div b = \frac{a}{b} = a \cdot \frac{1}{b}$$

In other words, the quotient of two real numbers is the product of the first number and the multiplicative inverse or reciprocal of the second number.

Example 6   Use the definition of the quotient of two numbers to find each quotient.

a. $-18 \div 3$ **b.** $\frac{-14}{-2}$ **c.** $\frac{20}{-4}$

Objectives

A Find the reciprocal of a real number.
B Divide real numbers.
C Evaluate algebraic expressions using real numbers.
D Determine whether a number is a solution of a given equation.

SSM   CD-ROM   Video 1.6

Practice Problems 1–5

Find the reciprocal of each number.

1. 13   2. $\frac{7}{15}$ 3. -5
4. $-\frac{8}{11}$ 5. 7.9

TEACHING TIP

Point out to students that finding the reciprocals of numbers will be useful in the next chapter when they will use the reciprocal of the coefficient to solve an equation.

Practice Problem 6

Use the definition of the quotient of two numbers to find each quotient.

a. $-12 \div 4$ b. $\frac{-20}{-10}$
c. $\frac{36}{-4}$

Answers

1. $\frac{1}{13}$, 2. $\frac{15}{7}$, 3. $-\frac{1}{5}$, 4. $-\frac{11}{8}$, 5. $\frac{1}{7.9}$,
6. a. -3, b. 2, c. -9

TEACHING TIP

Consider using the following activity: Suppose students are exchanging a $10 bill for smaller units of currency. Ask them how many smaller units they would get in return if the exchange is all in (a) $5 bills, (b) $1 bills, (c) $0.50 coins, (d) $0.25 coins, (e) $0.10 coins, (f) $0.05 coins, and (g) $0.01 coins. Ask students to write each answer in the form

$$\frac{\$10.00}{\text{smaller unit of currency}} = \frac{\text{number of smaller units of currency}}{}$$

What happens to the number of smaller units of currency as the value of the currency unit becomes increasingly smaller? How many units would be received if the change were given in all half-cent units?

Practice Problems 7–10

Divide.

7. $\dfrac{-25}{5}$ 8. $\dfrac{-48}{-6}$

9. $\dfrac{50}{-2}$ 10. $\dfrac{-72}{0.2}$

Practice Problems 11–12

Divide.

11. $-\dfrac{5}{9} \div \dfrac{2}{3}$

12. $-\dfrac{2}{7} \div \left(-\dfrac{1}{5}\right)$

Solution: **a.** $-18 \div 3 = -18 \cdot \dfrac{1}{3} = -6$

b. $\dfrac{-14}{-2} = -14 \cdot -\dfrac{1}{2} = 7$

c. $\dfrac{20}{-4} = 20 \cdot -\dfrac{1}{4} = -5$ ▬

Since the quotient $a \div b$ can be written as the product $a \cdot \dfrac{1}{b}$, it follows that sign patterns for dividing two real numbers are the same as sign patterns for multiplying two real numbers.

> ### DIVIDING REAL NUMBERS
>
> 1. The quotient of two numbers with the same sign is a positive number.
> 2. The quotient of two numbers with different signs is a negative number.

Examples   Divide.

7. $\dfrac{-30}{-10} = 3$ Same sign, so the quotient is positive.

8. $\dfrac{-100}{5} = -20$

9. $\dfrac{20}{-2} = -10$ Unlike signs, so the quotient is negative.

10. $\dfrac{42}{-0.6} = -70$ $0.6\overline{\smash{\big)}42.0}$ $\dfrac{70.}{}$

In the examples above, we divided mentally or by long division. When we divide by a fraction, it is usually easier to multiply by its reciprocal.

Examples   Divide.

11. $\dfrac{2}{3} \div \left(-\dfrac{5}{4}\right) = \dfrac{2}{3} \cdot \left(-\dfrac{4}{5}\right) = -\dfrac{8}{15}$

12. $-\dfrac{1}{6} \div \left(-\dfrac{2}{3}\right) = -\dfrac{1}{6} \cdot \left(-\dfrac{3}{2}\right) = \dfrac{3}{12} = \dfrac{1}{4}$ ▬

Our definition of the quotient of two real numbers does not allow for division by 0 because 0 does not have a reciprocal. How then do we interpret $\dfrac{3}{0}$? We say that an expression such as this one is undefined. Can we divide 0 by a number other than 0? Yes; for example,

$$\frac{0}{3} = 0 \cdot \frac{1}{3} = 0$$

> ### DIVISION INVOLVING ZERO
>
> Division by 0 is undefined. For example, $\dfrac{-5}{0}$ is undefined.
>
> 0 divided by a nonzero number is 0. For example, $\dfrac{0}{-5} = 0$.

Answers

7. -5,   **8.** 8,   **9.** -25,   **10.** -360,   **11.** $-\dfrac{5}{6}$,

12. $\dfrac{10}{7}$

Examples   Perform each indicated operation.

13. $\frac{1}{0}$ is undefined.   **14.** $\frac{0}{-3} = 0$

15. $\frac{0(-8)}{2} = \frac{0}{2} = 0$

Notice that $\frac{12}{-2} = -6$, $-\frac{12}{2} = -6$, and $\frac{-12}{2} = -6$. This means that

$$\frac{12}{-2} = -\frac{12}{2} = \frac{-12}{2}$$

In other words, a single negative sign in a fraction can be written in the denominator, in the numerator, or in front of the fraction without changing the value of the fraction.

In general, if a and b are real numbers, $b \neq 0$, $\frac{a}{-b} = \frac{-a}{b} = -\frac{a}{b}$.

Examples combining basic arithmetic operations along with the principles of the order of operations help us to review these concepts.

Example 16   Simplify each expression.

a. $\frac{(-12)(-3) + 4}{-7 - (-2)}$

b. $\frac{2(-3)^2 - 20}{-5 + 4}$

Solution: **a.** First, simplify the numerator and denominator separately, then divide.

$$\frac{(-12)(-3) + 4}{-7 - (-2)} = \frac{36 + 4}{-7 + 2}$$
$$= \frac{40}{-5}$$
$$= -8 \quad \text{Divide.}$$

b. Simplify the numerator and denominator separately, then divide.

$$\frac{2(-3)^2 - 20}{-5 + 4} = \frac{2 \cdot 9 - 20}{-5 + 4} = \frac{18 - 20}{-5 + 4} = \frac{-2}{-1} = 2$$

C   Evaluating Algebraic Expressions

Using what we have learned about dividing real numbers, we continue to practice evaluating algebraic expressions.

Example 17   Evaluate $\frac{3x}{2y}$ when $x = -2$ and $y = -4$.

Solution: Replace x with -2 and y with -4 and simplify.

$$\frac{3x}{2y} = \frac{3(-2)}{2(-4)} = \frac{-6}{-8} = \frac{3}{4}$$

Practice Problem 18

Determine whether -8 is a solution of $\frac{x}{4} - 3 = x + 3$.

D SOLUTIONS OF EQUATIONS

We use our skills in dividing real numbers to check possible solutions for an equation.

Example 18   Determine whether -10 is a solution of $\frac{-20}{x} + 5 = x$.

Solution:

$$\frac{-20}{x} + 5 = x \qquad \text{Original equation}$$

$$\frac{-20}{-10} + 5 \stackrel{?}{=} -10 \qquad \text{Replace } x \text{ with } -10.$$

$$2 + 5 \stackrel{?}{=} -10 \qquad \text{Divide.}$$

$$7 = -10 \qquad \textbf{False.}$$

Since we have a false statement, -10 is *not* a solution of the equation.

CALCULATOR EXPLORATIONS
ENTERING NEGATIVE NUMBERS ON A SCIENTIFIC CALCULATOR

To enter a negative number on a scientific calculator, find a key marked $\boxed{+ / -}$. (On some calculators, this key is marked $\boxed{\text{CHS}}$ for "change sign.") To enter -8, for example, press the keys $\boxed{8}\ \boxed{+ / -}$. The display will read $\boxed{-8}$.

ENTERING NEGATIVE NUMBERS ON A GRAPHING CALCULATOR

To enter a negative number on a graphing calculator, find a key marked $\boxed{(-)}$. Do not confuse this key with the key $\boxed{-}$, which is used for subtraction. To enter -8, for example, press the keys $\boxed{(-)}\ \boxed{8}$. The display will read $\boxed{-8}$.

OPERATIONS WITH REAL NUMBERS

To evaluate $-2(7 - 9) - 20$ on a calculator, press the keys

$\boxed{2}\ \boxed{+ / -}\ \boxed{\times}\ \boxed{(}\ \boxed{7}\ \boxed{-}\ \boxed{9}\ \boxed{)}\ \boxed{-}\ \boxed{2}\ \boxed{0}\ \boxed{=}$, or

$\boxed{(-)}\ \boxed{2}\ \boxed{(}\ \boxed{7}\ \boxed{-}\ \boxed{9}\ \boxed{)}\ \boxed{-}\ \boxed{2}\ \boxed{0}\ \boxed{\text{ENTER}}$.

The display will read $\boxed{-16}$ or $\boxed{\begin{array}{r}-2(7-9)-20\\-16\end{array}}$

Use a calculator to simplify each expression.

1. $-38(26 - 27)$ 38
2. $-59(-8) + 1726$ 2198
3. $134 + 25(68 - 91)$ -441
4. $45(32) - 8(218)$ -304
5. $\dfrac{-50(294)}{175 - 265}$ $163\frac{1}{3}$
6. $\dfrac{-444 - 444.8}{-181 - 324}$ 1.76
7. $9^5 - 4550$ 54,499
8. $5^8 - 6259$ 384,366
9. $(-125)^2$ (Be careful.)   15,625
10. -125^2 (Be careful.) $-15,625$

Answer

18. -8 is a solution.

EXERCISE SET 1.6

A *Find each reciprocal. See Examples 1 through 5.*

1. 9

2. 100

3. $\dfrac{2}{3}$

4. $\dfrac{1}{7}$

5. -14

6. -8

7. $-\dfrac{3}{11}$

8. $-\dfrac{6}{13}$

9. 0.2

10. 1.5

11. $\dfrac{1}{-6.3}$

12. $\dfrac{1}{-8.9}$

13. Find any real numbers that are their own reciprocal.

14. Explain why 0 has no reciprocal.

B *Divide. See Examples 6 through 15.*

15. $\dfrac{18}{-2}$

16. $\dfrac{20}{-10}$

17. $\dfrac{-16}{-4}$

18. $\dfrac{-18}{-6}$

19. $\dfrac{-48}{12}$

20. $\dfrac{-60}{5}$

21. $\dfrac{0}{-4}$

22. $\dfrac{0}{-9}$

23. $-\dfrac{15}{3}$

24. $-\dfrac{24}{8}$

25. $\dfrac{5}{0}$

26. $\dfrac{3}{0}$

27. $\dfrac{-12}{-4}$

28. $\dfrac{-45}{-9}$

29. $\dfrac{30}{-2}$

30. $\dfrac{14}{-2}$

31. $\dfrac{6}{7} \div \left(-\dfrac{1}{3}\right)$

32. $\dfrac{4}{5} \div \left(-\dfrac{1}{2}\right)$

33. $-\dfrac{5}{9} \div \left(-\dfrac{3}{4}\right)$

34. $-\dfrac{1}{10} \div \left(-\dfrac{8}{11}\right)$

35. $-\dfrac{4}{9} \div \dfrac{4}{9}$

36. $-\dfrac{5}{12} \div \dfrac{5}{12}$

ANSWERS

1. $\dfrac{1}{9}$
2. $\dfrac{1}{100}$
3. $\dfrac{3}{2}$
4. 7
5. $-\dfrac{1}{14}$
6. $-\dfrac{1}{8}$
7. $-\dfrac{11}{3}$
8. $-\dfrac{13}{6}$
9. $\dfrac{1}{0.2}$
10. $\dfrac{1}{1.5}$
11. -6.3
12. -8.9
13. $1, -1$
14. answers may vary
15. -9
16. -2
17. 4
18. 3
19. -4
20. -12
21. 0
22. 0
23. -5
24. -3
25. undefined
26. undefined
27. 3
28. 5
29. -15
30. -7
31. $-\dfrac{18}{7}$
32. $-\dfrac{8}{5}$
33. $\dfrac{20}{27}$
34. $\dfrac{11}{80}$
35. -1
36. -1

37. $-\dfrac{20}{24} = -\dfrac{5}{6}$

38. $-\dfrac{15}{12} = -\dfrac{5}{4}$

39. -40

40. -34.4

41. 160

42. 70

43. $-\dfrac{9}{2}$

44. $-\dfrac{9}{2}$

45. -4

46. 5

47. 16

48. $\dfrac{13}{5}$

49. -3

50. -2

51. $-\dfrac{16}{7}$

52. 8

53. 2

54. 1

55. $\dfrac{6}{5}$

56. $\dfrac{7}{11}$

57. -5

58. -4

59. $\dfrac{3}{2}$

60. 2

61. 3

62. 2

63. -1

64. undefined

65. $\dfrac{8}{9}$

66. $-\dfrac{1}{3}$

52

Name _____

37. $-\dfrac{5}{8} \div \dfrac{3}{4}$

38. $-\dfrac{5}{6} \div \dfrac{2}{3}$

39. $-48 \div 1.2$

40. $-86 \div 2.5$

▨ **41.** $-3.2 \div -0.02$

42. $-4.9 \div -0.07$

Simplify. See Example 16.

43. $\dfrac{-9(-3)}{-6}$

44. $\dfrac{-6(-3)}{-4}$

45. $\dfrac{12}{9-12}$

46. $\dfrac{-15}{1-4}$

47. $\dfrac{-6^2 + 4}{-2}$

48. $\dfrac{3^2 + 4}{5}$

49. $\dfrac{8 + (-4)^2}{4 - 12}$

50. $\dfrac{6 + (-2)^2}{4 - 9}$

51. $\dfrac{22 + (3)(-2)}{-5 - 2}$

52. $\dfrac{-20 + (-4)(3)}{1 - 5}$

53. $\dfrac{-3 - 5^2}{2(-7)}$

54. $\dfrac{-2 - 4^2}{3(-6)}$

▨ **55.** $\dfrac{6 - 2(-3)}{4 - 3(-2)}$

56. $\dfrac{8 - 3(-2)}{2 - 5(-4)}$

57. $\dfrac{-3 - 2(-9)}{-15 - 3(-4)}$

58. $\dfrac{-4 - 8(-2)}{-9 - 2(-3)}$

59. $\dfrac{|5 - 9| + |10 - 15|}{|2(-3)|}$

60. $\dfrac{|-3 + 6| + |-2 + 7|}{|-2 \cdot 2|}$

C *Evaluate each expression when $x = -5$ and $y = -3$. See Example 17.*

61. $\dfrac{2x - 5}{y - 2}$

62. $\dfrac{2y - 12}{x - 4}$

▨ **63.** $\dfrac{6 - y}{x - 4}$

64. $\dfrac{4 - 2x}{y + 3}$

65. $\dfrac{x + y}{3y}$

66. $\dfrac{y - x}{2x}$

D *Decide whether the given number is a solution of the given equation. See Example 18.*

67. $\dfrac{-10}{x} = -5;\quad 2$

68. $\dfrac{-14}{x} = 2;\quad -7$

69. $\dfrac{x}{5} + 2 = -1;\quad 15$

70. $\dfrac{x}{6} - 3 = 5;\quad 48$

71. $\dfrac{x+4}{5} = -6;\quad -30$

72. $\dfrac{x-3}{7} = -2;\quad -11$

COMBINING CONCEPTS

Write each as an algebraic expression. Then simplify the expression.

73. 7 subtracted from the quotient of 0 and 5

74. Twice the sum of -3 and -4

75. -1 added to the product of -8 and -5

76. The difference of -9 and the product of -4 and -6

77. The quotient of -8 and -20

78. The quotient of -9 and -30

If a is a positive number and b is a negative number, determine whether each expression simplifies to a positive number or a negative number.

79. $\dfrac{a}{b}$

80. $\dfrac{b}{a}$

81. $\dfrac{b+b}{a+a}$

82. $\dfrac{-a}{-b}$

Internet Excursions

Go to http://www.prenhall.com/martin-gay
A major stock market in the United States is the National Association of Securities Dealers Automated Quotations (NASDAQ). The given World Wide Web address will provide you with access to the NASDAQ site, or a related site, where you can look up current price information for any stock traded on the NASDAQ exchange. You can also obtain a graph of the closing share price for the past 6 months. By clicking on this graph, you can view a table of the closing prices by date.

67. solution

68. solution

69. not a solution

70. solution

71. not a solution

72. solution

73. $\dfrac{0}{5} - 7 = -7$

74. $2(-3 + -4) = -14$

75. $-8(-5) + (-1) = 39$

76. $-9 - (-4)(-6) = -33$

77. $\dfrac{-8}{-20} = \dfrac{2}{5}$

78. $\dfrac{-9}{-30} = \dfrac{3}{10}$

79. negative

80. negative

81. negative

82. negative

Name _____

83. Look up current stock prices for Intel Corporation (ticker symbol: INTC).
  a. Obtain a graph of the closing share price for the past 6 months. Describe any trends you see.

  b. Complete the following table. (You will need to click on the graph to find the closing price one month ago.)

Date of previous day Previous day's close Date 1 month ago Close 1 month ago	

  c. What is the difference between the previous day's closing price and the closing price one month ago?

  d. If you had bought 100 shares of Intel stock one month ago and sold all your shares at the previous day's closing price, how much money would you have gained or lost?

84. Look up current stock prices for Boston Market, Inc. (ticker symbol: BOST).
  a. Obtain a graph of the closing share price for the past 6 months. Describe any trends you see.

  b. Complete the following table. (You will need to click on the graph to find the closing price one month ago.)

Date of previous day Previous day's close Date 1 month ago Close 1 month ago	

  c. What is the difference between the previous day's closing price and the closing price one month ago?

  d. If you had bought 100 shares of Boston Market stock one month ago and sold all your shares at the previous day's closing price, how much money would you have gained or lost?

INTEGRATED REVIEW—OPERATIONS ON REAL NUMBERS

Perform each indicated operation and simplify.

1. $5(-7)$

2. $-3(-10)$

3. $\dfrac{-20}{-4}$

4. $\dfrac{30}{-6}$

5. $7-(-3)$

6. $-8-10$

7. $-14-(-12)$

8. $-3-(-1)$

9. $-\dfrac{1}{2}\left(-\dfrac{3}{4}\right)$

10. $-\dfrac{2}{7}\left(\dfrac{11}{12}\right)$

11. $\dfrac{-12}{0.2}$

12. $\dfrac{-3.8}{-2}$

13. $-19+(-23)$

14. $18+(-25)$

15. $-15+17$

16. $-2+(-37)$

17. $(-8)^2$

18. -9^2

19. -3^3

20. $(-2)^4$

21. $(2)(-8)(-3)$

22. $3(-2)(5)$

23. $-6(2)-5(2)-4$

ANSWERS

1. -35

2. 30

3. 5

4. -5

5. 10

6. -18

7. -2

8. -2

9. $\dfrac{3}{8}$

10. $-\dfrac{11}{42}$

11. -60

12. 1.9

13. -42

14. -7

15. 2

16. -39

17. 64

18. -81

19. -27

20. 16

21. 48

22. -30

23. -26

24. 6

25. 4

26. -3

27. 2

28. 16

29. 0

30. $-\frac{32}{15}$

24. $(7 - 10)(4 - 6)$

25. $2(19 - 17)^3 - 3(7 - 9)^2$

26. $3(10 - 9)^2 - 6(20 - 19)^3$

27. $\dfrac{19 - 25}{3(-1)}$

28. $\dfrac{8(-4)}{-2}$

29. $\dfrac{-2(3 - 6) - 6(10 - 9)}{-6 - (-5)}$

30. $\dfrac{-4(8 - 10)^3}{-2 - 1 - 12}$

1.7 PROPERTIES OF REAL NUMBERS

A USING THE COMMUTATIVE AND ASSOCIATIVE PROPERTIES

In this section we give names to properties of real numbers with which we are already familiar. Throughout this section, the variables a, b, and c represent real numbers.

We know that order does not matter when adding numbers. For example, we know that $7 + 5$ is the same as $5 + 7$. This property is given a special name—the **commutative property of addition**. We also know that order does not matter when multiplying numbers. For example, we know that $-5(6) = 6(-5)$. This property means that multiplication is commutative also and is called the **commutative property of multiplication**.

COMMUTATIVE PROPERTIES	
Addition:	$a + b = b + a$
Multiplication:	$a \cdot b = b \cdot a$

These properties state that the *order* in which any two real numbers are added or multiplied does not change their sum or product. For example, if we let $a = 3$ and $b = 5$, then the commutative properties guarantee that

$$3 + 5 = 5 + 3 \qquad \text{and} \qquad 3 \cdot 5 = 5 \cdot 3$$

> **HELPFUL HINT**
>
> Is subtraction also commutative? Try an example. Is $3 - 2 = 2 - 3$?
> **No!** The left side of this statement equals 1; the right side equals -1. There is no commutative property of subtraction. Similarly, there is no commutative property for division. For example, $10 \div 2$ does not equal $2 \div 10$.

Example 1   Use a commutative property to complete each statement.

a. $x + 5 =$ _____   **b.** $3 \cdot x =$ _____

Solution: **a.** $x + 5 = 5 + x$ By the commutative property of addition
b. $3 \cdot x = x \cdot 3$ By the commutative property of multiplication

TRY THE CONCEPT CHECK IN THE MARGIN.

Let's now discuss grouping numbers. We know that when we add three numbers, the way in which they are grouped or associated does not change their sum. For example, we know that $2 + (3 + 4) = 2 + 7 = 9$. This result is the same if we group the numbers differently. In other words, $(2 + 3) + 4 = 5 + 4 = 9$, also. Thus, $2 + (3 + 4) = (2 + 3) + 4$. This property is called the **associative property of addition**.

We also know that changing the grouping of numbers when multiplying does not change their product. For example, $2 \cdot (3 \cdot 4) = (2 \cdot 3) \cdot 4$ (check it). This is the **associative property of multiplication**.

Objectives

A Use the commutative and associative properties.

B Use the distributive property.

C Use the identity and inverse properties.

SSM   CD-ROM   Video 1.7

TEACHING TIP

Consider the following activity: Together make a list of situations in which the commutative property holds. Then make another list of situations in which the commutative property does not hold. Which list was easier to generate?

✓ CONCEPT CHECK

Which of the following pairs of actions are commutative?

a. "taking a test" and "studying for the test"

b. "putting on your left shoe" and "putting on your right shoe"

c. "putting on your shoes" and "putting on your socks"

d. "reading the sports section" and "reading the comics section"

Practice Problem 1

Use a commutative property to complete each statement.

a. $7 \cdot y =$ _____
b. $4 + x =$ _____

Answers

1. a. $y \cdot 7$, **b.** $x + 4$
✓ Concept Check: b, d

ASSOCIATIVE PROPERTIES

Addition:	$(a + b) + c = a + (b + c)$
Multiplication:	$(a \cdot b) \cdot c = a \cdot (b \cdot c)$

These properties state that the way in which three numbers are *grouped* does not change their sum or their product.

Example 2   Use an associative property to complete each statement.

a. $5 + (4 + 6) =$ _____
b. $(-1 \cdot 2) \cdot 5 =$ _____

Solution: **a.** $5 + (4 + 6) = (5 + 4) + 6$ By the associative property of addition

b. $(-1 \cdot 2) \cdot 5 = -1 \cdot (2 \cdot 5)$ By the associative property of multiplication

Practice Problem 2

Use an associative property to complete each statement.

a. $5 \cdot (-3 \cdot 6) =$ _____

b. $(-2 + 7) + 3 =$ _____

⌐HELPFUL HINT

Remember the difference between the commutative properties and the associative properties. The commutative properties have to do with the *order* of numbers and the associative properties have to do with the *grouping* of numbers.

Examples   Determine whether each statement is true by an associative property or a commutative property.

3. $(7 + 10) + 4 = (10 + 7) + 4$ Since the order of two numbers was changed and their grouping was not, this is true by the commutative property of addition.

4. $2 \cdot (3 \cdot 1) = (2 \cdot 3) \cdot 1$ Since the grouping of the numbers was changed and their order was not, this is true by the associative property of multiplication.

Practice Problems 3–4

Determine whether each statement is true by an associative property or a commutative property.

3. $5 \cdot (4 \cdot 7) = 5 \cdot (7 \cdot 4)$

4. $-2 + (4 + 9) = (-2 + 4) + 9$

Let's now illustrate how these properties can help us simplify expressions.

Examples   Simplify each expression.

5. $10 + (x + 12) = 10 + (12 + x)$ By the commutative property of addition
$= (10 + 12) + x$ By the associative property of addition
$= 22 + x$ Add.

6. $-3(7x) = (-3 \cdot 7)x$ By the associative property of multiplication
$= -21x$ Multiply.

Practice Problems 5–6

Simplify each expression.

5. $(-3 + x) + 17$

6. $4(5x)$

B USING THE DISTRIBUTIVE PROPERTY

The **distributive property of multiplication over addition** is used repeatedly throughout algebra. It is useful because it allows us to write a product as a sum or a sum as a product.

We know that $7(2 + 4) = 7(6) = 42$. Compare that with $7(2) + 7(4) = 14 + 28 = 42$. Since both original expressions equal 42, they must equal each other, or

$$7(2 + 4) = 7(2) + 7(4)$$

This is an example of the distributive property. The product on the left side of the equal sign is equal to the sum on the right side. We can think of the 7 as being distributed to each number inside the parentheses.

TEACHING TIP

You may want to show students that the distributive property can be used to quickly multiply numbers in their head. The product $12 \cdot 104$ can be rewritten as
$12(100 + 4) = 12 \cdot 100 + 12 \cdot 4$
$= 1200 + 48$
$= 1248$
The product $7 \cdot 98$ can be rewritten as
$7(100 - 2) = 7 \cdot 100 - 7 \cdot 2$
$= 700 - 14$
$= 686$

DISTRIBUTIVE PROPERTY OF MULTIPLICATION OVER ADDITION

$$a(b + c) = ab + ac$$

Answers

2. a. $(5 \cdot -3) \cdot 6$, **b.** $-2 + (7 + 3)$, **3.** commutative, **4.** associative, **5.** $14 + x$, **6.** $20x$

Since multiplication is commutative, this property can also be written as

$$\overparen{(b + c)a} = ba + ca$$

The distributive property can also be extended to more than two numbers inside the parentheses. For example,

$$3(x + y + z) = 3(x) + 3(y) + 3(z)$$
$$= 3x + 3y + 3z$$

Since we define subtraction in terms of addition, the distributive property is also true for subtraction. For example,

$$\overparen{2(x - y)} = 2(x) - 2(y)$$
$$= 2x - 2y$$

Examples

Use the distributive property to write each expression without parentheses. Then simplify the result.

7. $2(x + y) = 2(x) + 2(y)$
$= 2x + 2y$

8. $-5(-3 + 2z) = -5(-3) + (-5)(2z)$
$= 15 - 10z$

9. $5(x + 3y - z) = 5(x) + 5(3y) - 5(z)$
$= 5x + 15y - 5z$

10. $-1(2 - y) = (-1)(2) - (-1)(y)$
$= -2 + y$

11. $-(3 + x - w) = -1(3 + x - w)$
$= (-1)(3) + (-1)(x) - (-1)(w)$
$= -3 - x + w$

12. $4(3x + 7) + 10 = 4(3x) + 4(7) + 10$ *Apply the distributive property.*
$= 12x + 28 + 10$ *Multiply.*
$= 12x + 38$ *Add.*   ▬▬▬

> **HELPFUL HINT**
>
> Notice in Example 11 that $-(3 + x - w)$ is first rewritten as $-1(3 + x - w)$.

The distributive property can also be used to write a sum as a product.

Examples

Use the distributive property to write each sum as a product.

13. $8 \cdot 2 + 8 \cdot x = 8(2 + x)$
14. $7s + 7t = 7(s + t)$ ▬▬▬

C   USING THE IDENTITY AND INVERSE PROPERTIES

Next, we look at the **identity properties**.

The number 0 is called the identity for addition because when 0 is added to any real number, the result is the same real number. In other words, the *identity* of the real number is not changed.

The number 1 is called the identity for multiplication because when a real number is multiplied by 1, the result is the same real number. In other words, the *identity* of the real number is not changed.

> ### IDENTITIES FOR ADDITION AND MULTIPLICATION
>
> 0 is the identity element for addition.
> $$a + 0 = a \quad \text{and} \quad 0 + a = a$$
> 1 is the identity element for multiplication.
> $$a \cdot 1 = a \quad \text{and} \quad 1 \cdot a = a$$

TEACHING TIP

It may be helpful to illustrate the distributive property for students.

Let the tile ⬛ *x* represent *x* and the tile ⬛ represent 1.

$3(x + 2) = 3x + 3 \cdot 2$ can be illustrated as

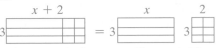

Three sets   is the same   Three + Three
of x + 2   as   sets of   sets of
                               x   2

Practice Problems 7–12

Use the distributive property to write each expression without parentheses. Then simplify the result.

7. $5(x + y)$

8. $-3(2 + 7x)$

9. $4(x + 6y - 2z)$

10. $-1(3 - a)$

11. $-(8 + a - b)$

12. $9(2x + 4) + 9$

Practice Problems 13–14

Use the distributive property to write each sum as a product.

13. $9 \cdot 3 + 9 \cdot y$

14. $4x + 4y$

Notice that 0 is the *only* number that can be added to any real number with the result that the sum is the same real number. Also, 1 is the *only* number that can be multiplied by any real number with the result that the product is the same real number.

Additive inverses or **opposites** were introduced in Section 1.3. Two numbers are called additive inverses or opposites if their sum is 0. The additive inverse or opposite of 6 is -6 because $6 + (-6) = 0$. The additive inverse or opposite of -5 is 5 because $-5 + 5 = 0$.

Reciprocals or **multiplicative inverses** were introduced in Section R.2. Two nonzero numbers are called reciprocals or multiplicative inverses if their product is 1. The reciprocal or multiplicative inverse of $\frac{2}{3}$ is $\frac{3}{2}$ because $\frac{2}{3} \cdot \frac{3}{2} = 1$. Likewise, the reciprocal of -5 is $-\frac{1}{5}$ because $-5\left(-\frac{1}{5}\right) = 1$.

ADDITIVE OR MULTIPLICATIVE INVERSES

The numbers a and $-a$ are additive inverses or opposites of each other because their sum is 0; that is,

$$a + (-a) = 0$$

The numbers b and $\frac{1}{b}$ (for $b \neq 0$) are reciprocals or multiplicative inverses of each other because their product is 1; that is,

$$b \cdot \frac{1}{b} = 1$$

✓ CONCEPT CHECK

Which of the following is the

a. opposite of $-\dfrac{3}{10}$, and which is the

b. reciprocal of $-\dfrac{3}{10}$?

$$1, \quad -\frac{10}{3}, \quad \frac{3}{10}, \quad 0, \quad \frac{10}{3}, \quad -\frac{3}{10}$$

TRY THE CONCEPT CHECK IN THE MARGIN.

Practice Problems 15–21

Name the property illustrated by each true statement.

15. $5 + (-5) = 0$

16. $12 + y = y + 12$

17. $-4 \cdot (6 \cdot x) = (-4 \cdot 6) \cdot x$

18. $6 + (z + 2) = 6 + (2 + z)$

19. $3\left(\dfrac{1}{3}\right) = 1$

20. $(x + 0) + 23 = x + 23$

21. $(7 \cdot y) \cdot 10 = y \cdot (7 \cdot 10)$

Examples   Name the property illustrated by each true statement.

15. $3 \cdot y = y \cdot 3$ — Commutative property of multiplication (order changed)

16. $(x + 7) + 9 = x + (7 + 9)$ — Associative property of addition (grouping changed)

17. $(b + 0) + 3 = b + 3$ — Identity element for addition

18. $2 \cdot (z \cdot 5) = 2 \cdot (5 \cdot z)$ — Commutative property of multiplication (order changed)

19. $-2 \cdot \left(-\dfrac{1}{2}\right) = 1$ — Multiplicative inverse property

20. $-2 + 2 = 0$ — Additive inverse property

21. $-6 \cdot (y \cdot 2) = (-6 \cdot 2) \cdot y$ — Commutative and associative properties of multiplication (order and grouping changed)

Answers

✓ **Concept Check:**   a. $\dfrac{3}{10}$,   b. $-\dfrac{10}{3}$

15. additive inverse property,   **16.** commutative property of addition,   **17.** associative property of multiplication,   **18.** commutative property of addition,   **19.** multiplicative inverse property,   **20.** identity element for addition,   **21.** commutative and associative properties of multiplication

Name _____ **Section** _____ **Date** _____

ANSWERS

1. $16 + x$

2. $y + 4$

3. $y \cdot (-4)$

4. $x \cdot (-2)$

5. yx

6. ba

7. $13 + 2x$

8. $3y + 19$

9. $x \cdot (yz)$

10. $(3x) \cdot y$

11. $(2 + a) + b$

12. $y + (4 + z)$

13. $4a \cdot (b)$

14. $-3 \cdot (yz)$

15. $a + (b + c)$

16. $(6 + r) + s$

17. $17 + b$

18. $r + 14$

19. $24y$

20. $84x$

21. y

22. z

23. $26 + a$

24. $11 + x$

25. $-72x$

26. $-36y$

27. s

28. r

EXERCISE SET 1.7

A *Use a commutative property to complete each statement. See Examples 1 and 3.*

1. $x + 16 =$ _____

2. $4 + y =$ _____

3. $-4 \cdot y =$ _____

4. $-2 \cdot x =$ _____

5. $xy =$ _____

6. $ab =$ _____

7. $2x + 13 =$ _____

8. $19 + 3y =$ _____

Use an associative property to complete each statement. See Examples 2 and 4.

9. $(xy) \cdot z =$ _____

10. $3 \cdot (x \cdot y) =$ _____

11. $2 + (a + b) =$ _____

12. $(y + 4) + z =$ _____

13. $4 \cdot (ab) =$ _____

14. $(-3y) \cdot z =$ _____

15. $(a + b) + c =$ _____

16. $6 + (r + s) =$ _____

Use the commutative and associative properties to simplify each expression. See Examples 5 and 6.

17. $8 + (9 + b)$

18. $(r + 3) + 11$

19. $4(6y)$

20. $2(42x)$

21. $\frac{1}{5}(5y)$

22. $\frac{1}{8}(8z)$

23. $(13 + a) + 13$

24. $7 + (x + 4)$

25. $-9(8x)$

26. $-3(12y)$

27. $\frac{3}{4}\left(\frac{4}{3}s\right)$

28. $\frac{2}{7}\left(\frac{7}{2}r\right)$

29. answers may vary

30. answers may vary

31. $4x + 4y$

32. $7a + 7b$

33. $9x - 54$

34. $11y - 44$

35. $6x + 10$

36. $35 + 40y$

37. $28x - 21$

38. $24x - 3$

39. $18 + 3x$

40. $2x + 10$

41. $-2y + 2z$

42. $-3z + 3y$

43. $-21y - 35$

44. $-10r - 55$

45. $5x + 20m + 10$

46. $24y + 8z - 48$

47. $-4 + 8m - 4n$

48. $-16 - 8p - 20$

49. $-5x - 2$

50. $-9r - 5$

51. $-r + 3 + 7p$

52. $-q + 2 - 6r$

53. $3x + 4$

54. $x - \dfrac{1}{2}$

55. $-x + 3y$

56. $-2a + 5b$

57. $6r + 8$

58. $40s + 20$

59. $-36x - 70$

60. $-55x - 23$

61. $-16x - 25$

62. $-12x - 7$

Name _____

29. Write an example that shows that division is not commutative.

30. Write an example that shows that subtraction is not commutative.

B *Use the distributive property to write each expression without parentheses. Then simplify the result. See Examples 7 through 12.*

31. $4(x + y)$ **32.** $7(a + b)$ **33.** $9(x - 6)$ **34.** $11(y - 4)$

35. $2(3x + 5)$ **36.** $5(7 + 8y)$ **37.** $7(4x - 3)$ **38.** $3(8x - 1)$

39. $3(6 + x)$ **40.** $2(x + 5)$ **41.** $-2(y - z)$ **42.** $-3(z - y)$

43. $-7(3y + 5)$ **44.** $-5(2r + 11)$ **45.** $5(x + 4m + 2)$

46. $8(3y + z - 6)$ **47.** $-4(1 - 2m + n)$ **48.** $-4(4 + 2p + 5)$

49. $-(5x + 2)$ **50.** $-(9r + 5)$ **51.** $-(r - 3 - 7p)$

52. $-(q - 2 + 6r)$ **53.** $\dfrac{1}{2}(6x + 8)$ **54.** $\dfrac{1}{4}(4x - 2)$

55. $-\dfrac{1}{3}(3x - 9y)$ **56.** $-\dfrac{1}{5}(10a - 25b)$ **57.** $3(2r + 5) - 7$

58. $10(4s + 6) - 40$ **59.** $-9(4x + 8) + 2$ **60.** $-11(5x + 3) + 10$

61. $-4(4x + 5) - 5$ **62.** $-6(2x + 1) - 1$

Name _____

Use the distributive property to write each sum as a product. See Examples 13 and 14.

63. $4 \cdot 1 + 4 \cdot y$ **64.** $14 \cdot z + 14 \cdot 5$ **65.** $11x + 11y$

66. $9a + 9b$ **67.** $(-1) \cdot 5 + (-1) \cdot x$ **68.** $(-3)a + (-3)y$

69. $30a + 30b$ **70.** $25x + 25y$

C *Find the additive inverse or opposite of each of the following numbers. See Example 20.*

71. 16 **72.** 14 **73.** -8 **74.** -3

75. $-(-1.2)$ **76.** $-(7.9)$ **77.** $-|-2|$ **78.** $-|-9|$

Find each multiplicative inverse or reciprocal. See Example 19.

79. $\dfrac{2}{3}$ **80.** $\dfrac{3}{4}$ **81.** $-\dfrac{5}{6}$ **82.** $-\dfrac{7}{8}$

83. $3\dfrac{5}{6}$ **84.** $2\dfrac{3}{5}$ **85.** -2 **86.** -5

Name the properties illustrated by each true statement. See Examples 15 through 21.

87. $3 \cdot 5 = 5 \cdot 3$ **88.** $4(3 + 8) = 4 \cdot 3 + 4 \cdot 8$

89. $2 + (x + 5) = (2 + x) + 5$ **90.** $(x + 9) + 3 = (9 + x) + 3$

91. $9(3 + 7) = 9 \cdot 3 + 9 \cdot 7$ **92.** $1 \cdot 9 = 9$

93. $(4 \cdot y) \cdot 9 = 4 \cdot (y \cdot 9)$ **94.** $6 \cdot \dfrac{1}{6} = 1$

63. $4(1 + y)$

64. $14(z + 5)$

65. $11(x + y)$

66. $9(a + b)$

67. $-1(5 + x)$

68. $-3(a + y)$

69. $30(a + b)$

70. $25(x + y)$

71. -16

72. -14

73. 8

74. 3

75. -1.2

76. 7.9

77. 2

78. 9

79. $\dfrac{3}{2}$

80. $\dfrac{4}{3}$

81. $-\dfrac{6}{5}$

82. $-\dfrac{8}{7}$

83. $\dfrac{6}{23}$

84. $\dfrac{5}{13}$

85. $-\dfrac{1}{2}$

86. $-\dfrac{1}{5}$

87. commutative property of multiplication

88. distributive property

89. associative property of addition

90. commutative property of addition

91. distributive property

92. identity property of multiplication

93. associative property of multiplication

94. multiplicative inverse property

95. identity property of addition

95. $0 + 6 = 6$

96. $(a + 9) + 6 = a + (9 + 6)$

96. associative property of addition

97. $-4(y + 7) = -4 \cdot y + (-4) \cdot 7$

98. $(11 + r) + 8 = (r + 11) + 8$

97. distributive property

98. commutative property of addition

99. $-4 \cdot (8 \cdot 3) = (8 \cdot -4) \cdot 3$

100. $r + 0 = r$

99. commutative and associative properties of multiplication

◤ COMBINING CONCEPTS

100. identity property of addition

Fill in the table with the opposite (additive inverse), the reciprocal (multiplicative inverse), or the expression. Assume that the value of each expression is not 0.

	Expression	Opposite	Reciprocal
101.	8	-8	$\dfrac{1}{8}$
102.	$-\dfrac{2}{3}$	$\dfrac{2}{3}$	$-\dfrac{3}{2}$
103.	x	$-x$	$\dfrac{1}{x}$
104.	$4y$	$-4y$	$\dfrac{1}{4y}$
105.	$2x$	$-2x$	$\dfrac{1}{2x}$
106.	$-7x$	$7x$	$-\dfrac{1}{7x}$

107. a. commutative property of addition

b. commutative property of addition

c. associative property of addition

Name the property illustrated by each step.

107. **a.** $\triangle + (\square + \bigcirc) = (\square + \bigcirc) + \triangle$

b. $ = (\bigcirc + \square) + \triangle$

c. $ = \bigcirc + (\square + \triangle)$

108. **a.** $(x + y) + z = x + (y + z)$

b. $ = (y + z) + x$

c. $ = (z + y) + x$

108. a. associative property of addition

b. commutative property of addition

c. commutative property of addition

109. Explain why 0 is called the identity element for addition.

110. Explain why 1 is called the identity element for multiplication.

109. answers may vary

110. answers may vary

1.8 READING GRAPHS

In today's world, where the exchange of information must be fast and entertaining, graphs are becoming increasingly popular. They provide a quick way of making comparisons, drawing conclusions, and approximating quantities.

A READING BAR GRAPHS

A **bar graph** consists of a series of bars arranged vertically or horizontally. The bar graph in Example 1 shows a comparison of the rates charged by selected electricity companies. The names of the companies are listed horizontally and a bar is shown for each company. Corresponding to the height of the bar for each company is a number along a vertical axis. These vertical numbers are cents charged for each kilowatt-hour of electricity used.

Example 1   The following bar graph shows the cents charged per kilowatt-hour for selected electricity companies.

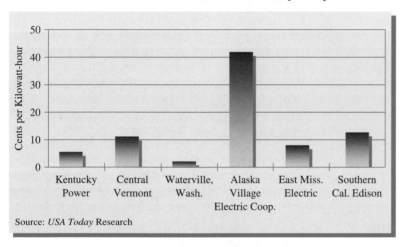

Source: *USA Today* Research

  a. Which company charges the highest rate?
  b. Which company charges the lowest rate?
  c. Approximate the electricity rate charged by the first four companies listed.
  d. Approximate the difference in the rates charged by the companies in parts (a) and (b).

Solution: **a.** The tallest bar corresponds to the company that charges the highest rate. Alaska Village Electric Cooperative charges the highest rate.
  b. The shortest bar corresponds to the company that charges the lowest rate. Waterville, Washington charges the lowest rate.
  c. To approximate the rate charged by Kentucky Power, we go to the top of the bar that corresponds to this company. From the top of the bar, we move horizontally to the left until the vertical axis is reached.

Objectives

A Read bar graphs.
B Read line graphs.

SSM   CD-ROM   Video
                     1.8

Practice Problem 1

Use the bar graph from Example 1 to answer the following.

a. Approximate the rate charged by East Mississippi Electric.

b. Approximate the rate charged by Southern California Edison.

c. Find the difference in rates charged by Southern California Edison and East Mississippi Electric.

Answers

1.a. 8¢ per kilowatt-hour, **b.** 12¢ per kilowatt-hour, **c.** 4¢

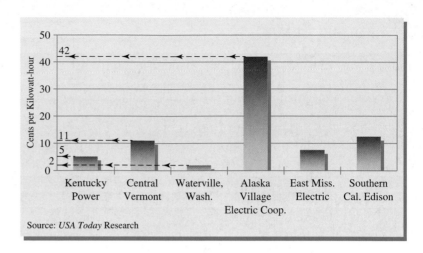

Source: *USA Today* Research

TEACHING TIP

It may be useful to have students record the information from the bar graph in a table and then compare and contrast the two descriptions of the data.

The height of the bar is approximately halfway between the 0 and 10 marks. We therefore conclude that

Kentucky Power charges approximately 5¢ per kilowatt-hour.
Central Vermont charges approximately 11¢ per kilowatt-hour.
Waterville, Washington charges approximately 2¢ per kilowatt-hour.
Alaska Village Electric charges approximately 42¢ per kilowatt-hour.

d. The difference in rates for Alaska Village Electric Cooperative and Waterville, Washington is approximately 42¢ − 2¢ or 40¢.

Practice Problem 2

Use the graph from Example 2 to answer the following.

a. How much money did the film *Snow White and the Seven Dwarfs* generate?

b. How much more money did the film *The Jungle Book* make than the film *Bambi*?

Example 2   The following bar graph shows Disney's top eight animated films and the amount of money they generated at theaters.

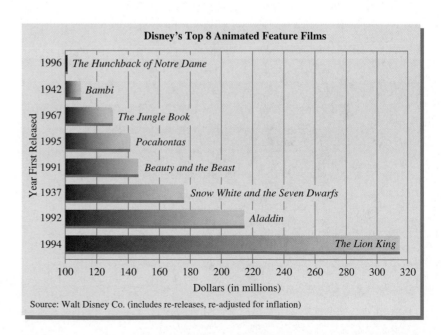

Source: Walt Disney Co. (includes re-releases, re-adjusted for inflation)

Answers

2. a. 175 million,   **b.** 20 million

a. Find the film shown that generated the most income for Disney and approximate the income.

b. How much more money did the film *Aladdin* make than the film *Beauty and the Beast*?

Solution: **a.** Since these bars are arranged horizontally, we look for the longest bar, which is the bar representing the film *The Lion King*. To approximate the income from this film, we move from the right edge of this bar vertically downward to the dollars axis. This film generated approximately 315 million dollars, or $315,000,000, the most income for Disney.

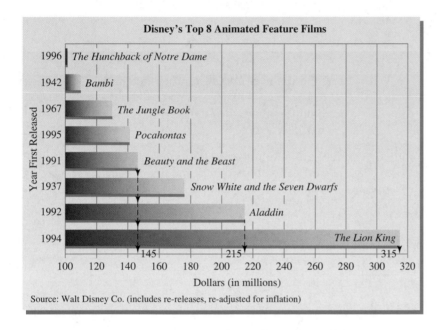

b. *Aladdin* generated approximately 215 million dollars. *Beauty and the Beast* generated approximately 145 million dollars. To find how much more money *Aladdin* generated than *Beauty and the Beast*, we subtract $215 - 145 = 70$ million dollars, or $70,000,000.

B Reading Line Graphs

A **line graph** consists of a series of points connected by a line. The graph in Example 3 is a line graph.

Example 3   The line graph below shows the relationship between the distance driven in a 14-foot U-Haul truck in one day and the total cost of renting this truck for that day. Notice that the horizontal axis is labeled Distance and the vertical axis is labeled Total Cost.

Practice Problem 3

Use the graph from Example 3 to answer the following.

a. Find the total cost of renting the truck if 50 miles are driven.

b. Find the total number of miles driven if the total cost of renting is $100.

Answers

3. a. $50, **b.** 180 miles

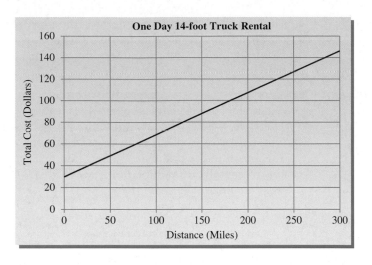

a. Find the total cost of renting the truck if 100 miles are driven.

b. Find the number of miles driven if the total cost of renting is $140.

Solution: **a.** Find the number 100 on the horizontal scale and move vertically upward until the line is reached. From this point on the line, we move horizontally to the left until the vertical scale is reached. We find that the total cost of renting the truck if 100 miles are driven is approximately $70.

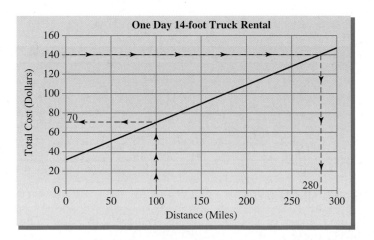

b. We find the number 140 on the vertical scale and move horizontally to the right until the line is reached. From this point on the line, we move vertically downward until the horizontal scale is reached. We find that the truck is driven approximately 280 miles. ▬

From the previous example, we can see that graphing provides a quick way to approximate quantities. In Chapter 6 we show how we can use equations to find exact answers to the questions posed in Example 3. The next graph is another example of a line graph. It is also sometimes called a **broken line graph**.

Example 4 The line graph shows the relationship between time spent smoking a cigarette and pulse rate. Time is recorded along the horizontal axis in minutes, with 0 minutes being the moment a smoker lights a cigarette. Pulse is recorded along the vertical axis in heartbeats per minute.

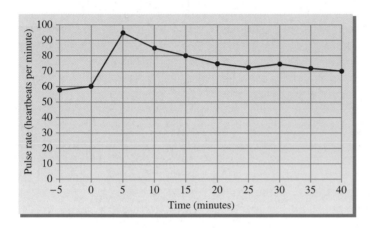

a. What is the pulse rate 15 minutes after lighting a cigarette?
b. When is the pulse rate the lowest?
c. When does the pulse rate show the greatest change?

Solution: **a.** We locate the number 15 along the time axis and move vertically upward until the line is reached. From this point on the line, we move horizontally to the left until the pulse rate axis is reached. Reading the number of beats per minute, we find that the pulse rate is 80 beats per minute 15 minutes after lighting a cigarette.

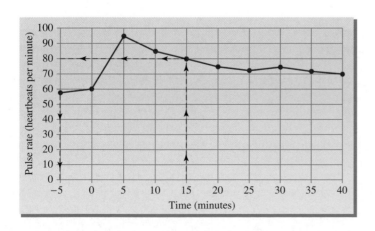

Practice Problem 4

Use the graph from Example 4 to answer the following.

a. What is the pulse rate 40 minutes after lighting a cigarette?

b. What is the pulse rate when the cigarette is being lit?

c. When is the pulse rate the highest?

TEACHING TIP

Have students compare and contrast bar graphs and line graphs.

Answers

4. a. 70, **b.** 60, **c.** 5 min. after lighting

b. We find the lowest point of the line graph, which represents the lowest pulse rate. From this point, we move vertically downward to the time axis. We find that the pulse rate is the lowest at −5 minutes, which means 5 minutes *before* lighting a cigarette.

c. The pulse rate shows the greatest change during the 5 minutes between 0 and 5. Notice that the line graph is *steepest* between 0 and 5 minutes. ▬▬▬▬

Exercise Set 1.8

A *The following bar graph shows the number of teenagers expected to use the Internet for the years shown. Use this graph to answer Exercises 1–4. See Example 1.*

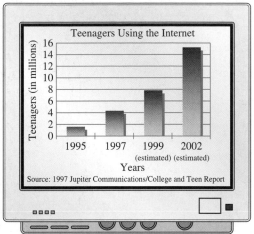

1. Approximate the number of teenagers expected to use the Internet in 1999.

2. Approximate the number of teenagers who use the Internet in 1995.

3. What year shows the greatest *increase* in number of teenagers using the Internet?

4. How many more teenagers are expected to use the Internet in 2002 than in 1999?

The following bar graph shows the amounts of money used by major pro sports for advertising in 1996. Use this graph to answer Exercises 5–10. See Example 2.

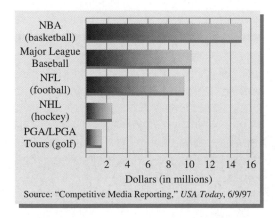

5. Which major pro sports used the least amount of money for advertising?

6. Which major pro sports used the greatest amount of money for advertising?

7. Which major pro sports spent over $10,000,000 in advertising?

8. Which major pro sports spent under $5,000,000 in advertising?

Answers

1. approx. 7.8 million

2. approx. 1.6 million

3. 2002

4. approx. 7.4 million

5. PGA/LPGA tours

6. NBA

7. Major League Baseball, NBA

8. NHL, PGA/LPGA tours

71

9. approx. 15 million

10. approx. 2.5 million

11. Cleveland Indians

12. Philadelphia Flyers

13. tied

14. approx. 9 years

15. approx. 3 years

16. approx. 6 years

17. approx. 142 million

18. approx. 100 million

19. Snow White and the Seven Dwarfs

20. The Lion King

Name _____

9. Estimate the amount of money spent by the NBA for advertising.

10. Estimate the amount of money spent by the NHL for advertising.

The following bar graph shows the team in each sport that has gone the longest time without being in a playoff. Use this graph to answer Exercises 11–16. See Examples 1 and 2.

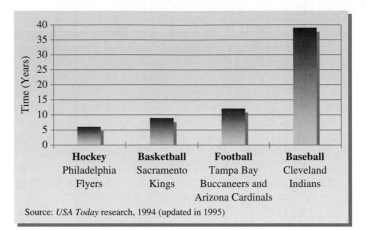

Source: *USA Today* research, 1994 (updated in 1995)

11. Which team for the sports shown has gone the longest time without being in a playoff?

12. Which hockey team has gone the longest without being in a playoff?

13. Why are two football teams listed?

14. Approximate the greatest number of years that a basketball team has gone without being in the playoffs.

15. How many more years has a football team gone without being in the play-offs than a basketball team?

16. How many more years has a football team gone without being in the play-offs than a hockey team?

Use the bar graph in Example 2 to answer Exercises 17–22.

17. Approximate the income generated by the film *Pocahontas*.

18. Approximate the income generated by the film *The Hunchback of Notre Dame*.

19. Before 1990, which Disney film generated the most income?

20. After 1990, which Disney film generated the most income?

Name ______________________________

21. Why do you think that the Disney film *The Little Mermaid* is not shown on this graph?

22. How much less money did the film *The Hunchback of Notre Dame* generate than *Pocahontas*?

B *Many fires are deliberately set. An increasing number of those arrested for arson are juveniles (age 17 and under). The following line graph shows the percent of deliberately set fires started by juveniles. Use this graph to answer Exercises 23–30. See Example 3 and 4.*

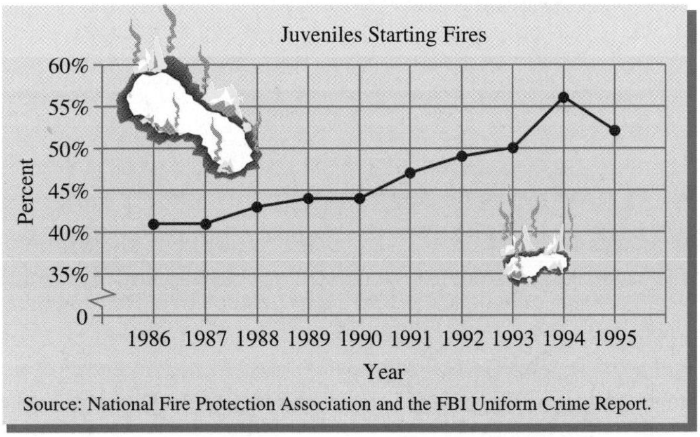

Juveniles Starting Fires
60%
55%
50%
45%
40%
35%
0
Percent
1986 1987 1988 1989 1990 1991 1992 1993 1994 1995
Year
Source: National Fire Protection Association and the FBI Uniform Crime Report.

23. What year shows the highest percent of arson fires started by juveniles?

24. What year since 1990 shows a decrease in the percent of fires started by juveniles?

25. Name two consecutive years where the percent remained the same.

26. What year(s) shows the lowest percent of arson fires started by juveniles?

27. Estimate the percent of arson fires started by juveniles in 1995.

28. Estimate the percent of arson fires started by juveniles in 1992.

29. What year shows the greatest increase in the percent of fires started by juveniles?

30. What trend do you notice from this graph?

Use the line graph in Example 4 to answer Exercises 31–34.

31. Approximate the pulse rate 5 minutes before lighting a cigarette.

32. Approximate the pulse rate 10 minutes after lighting a cigarette.

33. Find the difference in pulse rate between 5 minutes before and 10 minutes after lighting a cigarette.

34. When is the pulse rate less than 60 heartbeats per minute?

21. answers may vary
22. approx. 42 million
23. 1994
24. 1995
25. 1986, 1987 or 1989, 1990
26. 1986 or 1987
27. approx. 52%
28. approx. 49%
29. 1994
30. answers may vary
31. approx. 59 beats per minute
32. approx. 85 beats per minute
33. approx. 26 beats per minute
34. 5 min. before lighting

Name _____

The line graph below shows the average cost of newsprint per metric ton since 1984. Use this graph to answer Exercises 35–38.

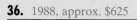

The Bouncing Cost of Newsprint
Average price per metric ton since 1984.

$599

Prices can vary by region. Prices based on East Coast averages

Source: Pulp and Paper Week

35. During what year was the cost of newsprint less than $500 per metric ton?

36. What year shown had the highest cost of newsprint? Estimate this cost.

37. What year shown had the lowest cost of newsprint? Estimate this cost.

38. Estimate the cost of newsprint in 1986.

◆ **COMBINING CONCEPTS**

Geographic locations can be described by a gridwork of lines called latitudes and longitudes, as shown below. For example, the location of Houston, Texas, can be described by latitude 30° north and longitude 95° west. Use the map shown to answer Exercises 39–42.

39. Using latitude and longitude, describe the location of New Orleans, Louisiana.

40. Using latitude and longitude, describe the location of Denver, Colorado.

41. Use an atlas and describe the location of your hometown.

42. Give another name for 0° latitude.

CHAPTER 1 ACTIVITY
ANALYZING WEALTH

This activity may be completed by working in groups or individually.

Tom Stanley and Bill Danko are experts on millionaires. While doing research for their book, *The Millionaire Next Door*, they found that over 3 million households in the United States have net worths* of over $1,000,000. Most of these millionaires are regular people who have made their fortunes not overnight in risky business deals, but over the years through steady saving and wise investing. Stanley and Danko say that almost anyone can become a millionaire—the key is to live below your means. They give the following rule of thumb to test your progress in becoming a millionaire.

> Your net worth should be 10% of your annual income, multiplied by your age, and then doubled.
> (*Source*: *Parade Magazine*, June 22, 1997)

Stanley and Danko caution people to not worry if their net worths don't meet this goal. With some discipline, they can still catch up.

*Net worth is the total value of all your assets (such as property, bank accounts, and investments) minus what you owe (such as a car loan, mortgage, or education loan).

1. Suppose you are a financial planner. A 30-year-old client would like to eventually become a millionaire. He makes $36,000 each year. What should his net worth be?
$2(0.10 \cdot 36,000)(30) = \$216,000$

2. Let N represent net worth. Choose variables to represent annual income and age. Write an algebraic equation that represents Stanley and Danko's rule of thumb for becoming a millionaire. x = annual income; y = age; N = net worth; $2(0.10x)(y) = N$

3. Simplify the equation you wrote in Question 2.
$0.20xy = N$

4. Suppose that a 42-year-old client hopes to reitre as a millionaire. She earns $44,000 each year and has a net worth of $350,000. Using the equation you wrote in Questions 2 and 3, is she on track? What is the difference between her net worth and the goal? $0.20(44,000)(42) = \$369,600$
$369,600 - 350,000 = \$19,600$

5. Suppose that a 35-year-old client earns $40,000 each year. To become a millionaire, what should his net worth be at this age? According to Stanley and Danko's rule of thumb, if he continues to earn $40,000 during the next year, how much additional net worth will he need to still be on track at age 36?
$0.20(40,000)(35) = \$280,000$,
$0.20(40,000)(36) = \$288,000$,
$288,000 - 280,000 = \$8,000$

6. (Optional) Using your own age and annual income, compute the net worth goal for becoming a millionaire in your own situation.
answers may vary

TEACHING TIP

Encourage students to use their notebook along with the Chapter Highlights as a study guide for the test.

CHAPTER 1 HIGHLIGHTS

DEFINITIONS AND CONCEPTS	EXAMPLES

SECTION 1.1   SYMBOLS AND SETS OF NUMBERS

A **set** is a collection of objects, called **elements**, enclosed in braces.

$\{a, c, e\}$

Natural numbers: $\{1, 2, 3, 4, \ldots\}$
Whole numbers: $\{0, 1, 2, 3, 4, \ldots\}$
Integers: $\{\ldots, -3, -2, -1, 0, 1, 2, 3, \ldots\}$
Rational numbers: {real numbers that can be expressed as a quotient of integers}
Irrational numbers: {real numbers that cannot be expressed as a quotient of integers}
Real numbers: {all numbers that correspond to a point on the number line}

Given the set $\{-3.4, \sqrt{3}, 0, \frac{2}{3}, 5, -4\}$ list the numbers that belong to the set of
Natural numbers   5
Whole numbers   0, 5
Integers   −4, 0, 5
Rational numbers $-3.4, 0, \frac{2}{3}, 5, -4$
Irrational numbers $\sqrt{3}$
Real numbers $-3.4, \sqrt{3}, 0, \frac{2}{3}, 5, -4$

A line used to picture numbers is called a **number line**.

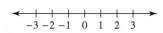

The **absolute value** of a real number a denoted by $|a|$ is the distance between a and 0 on the number line.

$|5| = 5 \quad |0| = 0 \quad |-2| = 2$

SYMBOLS:
= is equal to
≠ is not equal to
> is greater than
< is less than
≤ is less than or equal to
≥ is greater than or equal to

$-7 = -7$
$3 \neq -3$
$4 > 1$
$1 < 4$
$6 \leq 6$
$18 \geq -\dfrac{1}{3}$

ORDER PROPERTY FOR REAL NUMBERS

For any two real numbers a and b, a is less than b if a is to the left of b on the number line.

$-3 < 0 \qquad 0 > -3 \qquad 0 < 2.5 \qquad 2.5 > 0$

SECTION 1.2   INTRODUCTION TO VARIABLE EXPRESSIONS AND EQUATIONS

The expression a^n is an **exponential expression**. The number a is called the **base**; it is the repeated factor. The number n is called the **exponent**; it is the number of times that the base is a factor.

$4^3 = 4 \cdot 4 \cdot 4 = 64$
$7^2 = 7 \cdot 7 = 49$

SECTION 1.2   (CONTINUED)

ORDER OF OPERATIONS

Simplify expressions in the following order. If grouping symbols are present, simplify expressions within those first, starting with the innermost set. Also, simplify the numerator and the denominator of a fraction separately.

1. Simplify exponential expressions.
2. Multiply or divide in order from left to right.
3. Add or subtract in order from left to right.

$$\frac{8^2 + 5(7 - 3)}{3 \cdot 7} = \frac{8^2 + 5(4)}{21}$$
$$= \frac{64 + 5(4)}{21}$$
$$= \frac{64 + 20}{21}$$
$$= \frac{84}{21}$$
$$= 4$$

A symbol used to represent a number is called a **variable**.

Examples of variables are

$$q, \quad x, \quad z$$

An **algebraic expression** is a collection of numbers, variables, operation symbols, and grouping symbols.

Examples of algebraic expressions are

$$5x, \quad 2(y - 6), \quad \frac{q^2 - 3q + 1}{6}$$

To **evaluate an algebraic expression** containing a variable, substitute a given number for the variable and simplify.

Evaluate $x^2 - y^2$ when $x = 5$ and $y = 3$.

$$x^2 - y^2 = (5)^2 - 3^2$$
$$= 25 - 9$$
$$= 16$$

A mathematical statement that two expressions are equal is called an **equation**.

Equations:

$$3x - 9 = 20$$
$$A = \pi r^2$$

A **solution** of an equation is a value for the variable that makes the equation a true statement.

Determine whether 4 is a solution of $5x + 7 = 27$.

$$5x + 7 = 27$$
$$5(4) + 7 \stackrel{?}{=} 27$$
$$20 + 7 \stackrel{?}{=} 27$$
$$27 = 27 \quad \text{True.}$$

4 is a solution.

SECTION 1.3   ADDING REAL NUMBERS

TO ADD TWO NUMBERS WITH THE SAME SIGN

1. Add their absolute values.
2. Use their common sign as the sign of the sum.

TO ADD TWO NUMBERS WITH DIFFERENT SIGNS

1. Subtract their absolute values.
2. Use the sign of the number whose absolute value is larger as the sign of the sum.

Add.

$$10 + 7 = 17$$
$$-3 + (-8) = -11$$

$$-25 + 5 = -20$$
$$14 + (-9) = 5$$

SECTION 1.3 (CONTINUED)	
Two numbers that are the same distance from 0 but lie on opposite sides of 0 are called **opposites** or **additive inverses**. The opposite of a number a is denoted by $-a$.	The opposite of -7 is 7. The opposite of 123 is -123.

SECTION 1.4 SUBTRACTING REAL NUMBERS	
To subtract two numbers a and b, add the first number a to the opposite of the second number b. $$a - b = a + (-b)$$	Subtract. $$3 - (-44) = 3 + 44 = 47$$ $$-5 - 22 = -5 + (-22) = -27$$ $$-30 - (-30) = -30 + 30 = 0$$

SECTION 1.5 MULTIPLYING REAL NUMBERS	
MULTIPLYING REAL NUMBERS The product of two numbers with the same sign is a positive number. The product of two numbers with different signs is a negative number. PRODUCTS INVOLVING ZERO The product of 0 and any number is 0. $$b \cdot 0 = 0 \text{ and } 0 \cdot b = 0$$	Multiply. $$7 \cdot 8 = 56 \qquad -7 \cdot (-8) = 56$$ $$-2 \cdot 4 = -8 \qquad 2 \cdot (-4) = -8$$ $$-4 \cdot 0 = 0 \qquad 0 \cdot \left(-\frac{3}{4}\right) = 0$$

SECTION 1.6 DIVIDING REAL NUMBERS	
QUOTIENT OF TWO REAL NUMBERS $$\frac{a}{b} = a \cdot \frac{1}{b}$$ DIVIDING REAL NUMBERS The quotient of two numbers with the same sign is a positive number. The quotient of two numbers with different signs is a negative number.	Divide. $$\frac{42}{2} = 42 \cdot \frac{1}{2} = 21$$ $$\frac{90}{10} = 9 \qquad \frac{-90}{-10} = 9$$ $$\frac{42}{-6} = -7 \qquad \frac{-42}{6} = -7$$

SECTION 1.6   (CONTINUED)

QUOTIENTS INVOLVING ZERO

The quotient of a nonzero number and 0 is undefined.

$$\frac{b}{0} \text{ is undefined.}$$

The quotient of 0 and any nonzero number is 0.

$$\frac{0}{b} = 0$$

$$\frac{-85}{0} \text{ is undefined.}$$

$$\frac{0}{18} = 0 \qquad \frac{0}{-47} = 0$$

SECTION 1.7   PROPERTIES OF REAL NUMBERS

COMMUTATIVE PROPERTIES

Addition: $a + b = b + a$
Multiplication: $a \cdot b = b \cdot a$

$$3 + (-7) = -7 + 3$$
$$-8 \cdot 5 = 5 \cdot (-8)$$

ASSOCIATIVE PROPERTIES

Addition: $(a + b) + c = a + (b + c)$
Multiplication: $(a \cdot b) \cdot c = a \cdot (b \cdot c)$

$$(5 + 10) + 20 = 5 + (10 + 20)$$
$$(-3 \cdot 2) \cdot 11 = -3 \cdot (2 \cdot 11)$$

Two numbers whose product is 1 are called **multiplicative inverses** or **reciprocals**. The reciprocal of a nonzero number a is $\frac{1}{a}$ because $a \cdot \frac{1}{a} = 1$.

The reciprocal of 3 is $\frac{1}{3}$.

The reciprocal of $-\frac{2}{5}$ is $-\frac{5}{2}$.

DISTRIBUTIVE PROPERTY

$$a(b + c) = a \cdot b + a \cdot c$$

$$5(6 + 10) = 5 \cdot 6 + 5 \cdot 10$$
$$-2(3 + x) = -2 \cdot 3 + (-2)(x)$$

IDENTITIES

$a + 0 = a \qquad 0 + a = a$
$a \cdot 1 = a \qquad 1 \cdot a = a$

$5 + 0 = 5 \qquad 0 + (-2) = -2$
$-14 \cdot 1 = -14 \qquad 1 \cdot 27 = 27$

INVERSES

Additive or opposite: $a + (-a) = 0$

Multiplicative or reciprocal: $b \cdot \frac{1}{b} = 1$

$$7 + (-7) = 0$$

$$3 \cdot \frac{1}{3} = 1$$

SECTION 1.8   READING GRAPHS

To find the value on the vertical axis representing a location on a graph, move horizontally from the location on the graph until the vertical axis is reached. To find the value on the horizontal axis representing a location on a graph, move vertically from the location on the graph until the horizontal axis is reached.

This broken line graph shows the average public classroom teachers' salaries for the school year ending in the years shown.

Estimate the average public teacher's salary for the school year ending in 1989.   The average salary is approximately $29,500.

Find the earliest year that the average salary rose above $32,000.   The year was 1991.

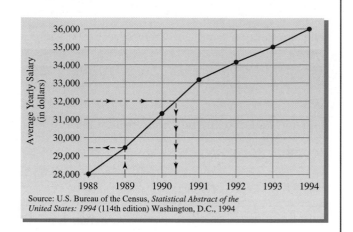

Source: U.S. Bureau of the Census, *Statistical Abstract of the United States: 1994* (114th edition) Washington, D.C., 1994

CHAPTER 1 REVIEW

(1.1) *Insert* $<$, $>$, *or* $=$ *in the appropriate space to make each statement true.*

1. $8 \quad < \quad 10$ **2.** $7 \quad > \quad 2$ **3.** $-4 \quad > \quad -5$ **4.** $\dfrac{12}{2} \quad > \quad -8$

5. $|-7| \quad < \quad |-8|$ **6.** $|-9| \quad > \quad -9$ **7.** $-|-1| \quad = \quad -1$

8. $|-14| \quad = \quad -(-14)$ **9.** $1.2 \quad > \quad 1.02$ **10.** $-\dfrac{3}{2} \quad < \quad -\dfrac{3}{4}$

Translate each statement into symbols.

11. Four is greater than or equal to negative three.
$4 \geq -3$

12. Six is not equal to five. $6 \neq 5$

13. 0.03 is less than 0.3. $0.03 < 0.3$

14. Lions and hyenas were featured in the Disney film *The Lion King*. For short distances, lions can run at a rate of 50 miles per hour whereas hyenas can run at a rate of 40 miles per hour. Write an inequality statement comparing the numbers 50 and 40.
$50 > 40$

Given the following sets of numbers, list the numbers in each set that also belong to the set of:

a. Natural numbers   **b.** Whole numbers
c. Integers   **d.** Rational numbers
e. Irrational numbers   **f.** Real numbers

15. $\left\{-6, 0, 1, 1\frac{1}{2}, 3, \pi, 9.62\right\}$ a. $1, 3$ b. $0, 1, 3$
c. $-6, 0, 1, 3$ d. $-6, 0, 1, 1\frac{1}{2}, 3, 9.62$
e. π f. all numbers in set

16. $\left\{-3, -1.6, 2, 5, \frac{11}{2}, 15.1, \sqrt{5}, 2\pi\right\}$ a. $2, 5$ b. $2, 5$
c. $-3, 2, 5$ d. $-3, -1.6, 2, 5, \frac{11}{2}, 15.1$ e. $\sqrt{5}, 2\pi$
f. all numbers in set

The following chart shows the gains and losses in dollars of Density Oil and Gas stock for a particular week. Use this chart to answer Exercises 17–18.

Day	Gain or Loss (in dollars)
Monday	$+1$
Tuesday	-2
Wednesday	$+5$
Thursday	$+1$
Friday	-4

17. Which day showed the greatest loss?   Friday

18. Which day showed the greatest gain?   Wednesday

Name _____

(1.2) *Choose the correct answer for each statement.*

19. The expression $6 \cdot 3^2 + 2 \cdot 8$ simplifies to
    a. -52 b. 440 c. 70 d. 64 c

20. The expression $68 - 5 \cdot 2^3$ simplifies to
    a. -232 b. 28 c. 38 d. 504 b

Simplify each expression.

21. $3(1 + 2 \cdot 5) + 4$ 37

22. $8 + 3(2 \cdot 6 - 1)$ 41

23. $\dfrac{4 + |6 - 2| + 8^2}{4 + 6 \cdot 4}$ $\dfrac{18}{7}$

24. $5[3(2 + 5) - 5]$ -80

Translate each word statement to symbols.

25. The difference of twenty and twelve is equal to the product of two and four. $20 - 12 = 2 \cdot 4$

26. The quotient of nine and two is greater than negative five. $\dfrac{9}{2} > -5$

Evaluate each expression when $x = 6$, $y = 2$, and $z = 8$.

27. $2x + 3y$ 18

28. $x(y + 2z)$ 108

29. $\dfrac{x}{y} + \dfrac{z}{2y}$ 5

30. $x^2 - 3y^2$ 24

31. The expression $180 - a - b$ represents the measure of the unknown angle of the given triangle. Replace a with 37 and b with 80 to find the measure of the unknown angle. $63°$

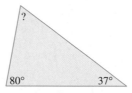

Decide whether the given number is a solution to the given equation.

32. $7x - 3 = 18$; 3   solution

33. $3x^2 + 4 = x - 1$; 1   not a solution

(1.3) *Find each additive inverse or opposite.*

34. -9 9

35. $\dfrac{2}{3}$ $-\dfrac{2}{3}$

36. $|-2|$ -2

37. $-|-7|$ 7

Add.

38. $-15 + 4$ -11

39. $-6 + (-11)$ -17

40. $\dfrac{1}{16} + \left(-\dfrac{1}{4}\right)$ $-\dfrac{3}{16}$

41. $-8 + |-3|$ -5

42. $-4.6 + (-9.3)$ -13.9

43. $-2.8 + 6.7$ 3.9

Name _____

(1.4) *Perform each indicated operation.*

44. $6 - 20$ -14

45. $-3.1 - 8.4$ -11.5

46. $-6 - (-11)$ 5

47. $4 - 15$ -11

48. $-21 - 16 + 3(8 - 2)$ -19

49. $\dfrac{11 - (-9) + 6(8 - 2)}{2 + 3 \cdot 4}$ 4

Suppose $x = 3$, $y = -6$, and $z = -9$. Choose the correct evaluation for each expression.

50. $2x^2 - y + z$ evaluates to
a. 15   b. 3   c. 27   d. -3 a

51. $\dfrac{y - 4x}{2x}$ evaluates to
a. 3   b. 1   c. -1 d. -3 d

52. At the beginning of the week the price of Density Oil and Gas stock from Exercises 17 and 18 is $50 per share. Find the price of a share of stock at the end of the week. $51

Find each multiplicative inverse or reciprocal.

53. -6 $-\dfrac{1}{6}$

54. $\dfrac{3}{5}$ $\dfrac{5}{3}$

(1.5) and (1.6) *Simplify each expression.*

55. $6(-8)$ -48

56. $(-2)(-14)$ 28

57. $\dfrac{-18}{-6}$ 3

58. $\dfrac{42}{-3}$ -14

59. $-3(-6)(-2)$ -36

60. $(-4)(-3)(0)(-6)$ 0

61. $\dfrac{4 \cdot (-3) + (-8)}{2 + (-2)}$ undefined

62. $\dfrac{3(-2)^2 - 5}{-14}$ $-\dfrac{1}{2}$

(1.7) *Name the property illustrated in each equation.*

63. $-6 + 5 = 5 + (-6)$ commutative property of addition

64. $6 \cdot 1 = 6$ multiplicative identity property

65. $3(8 - 5) = 3 \cdot 8 + 3 \cdot (-5)$ distributive property

66. $4 + (-4) = 0$ additive inverse property

67. $2 + (3 + 9) = (2 + 3) + 9$ associative property of addition

68. $2 \cdot 8 = 8 \cdot 2$ commutative property of multiplication

69. $6(8 + 5) = 6 \cdot 8 + 6 \cdot 5$ distributive property

70. $(3 \cdot 8) \cdot 4 = 3 \cdot (8 \cdot 4)$ associative property of multiplication

71. $4 \cdot \dfrac{1}{4} = 1$ multiplicative inverse property

72. $8 + 0 = 8$ additive identity property

73. $4(8 + 3) = 4(3 + 8)$ commutative property of addition

(1.8) *Use the graph below showing Disney's consumer products revenues to answer Exercises 74–77.*

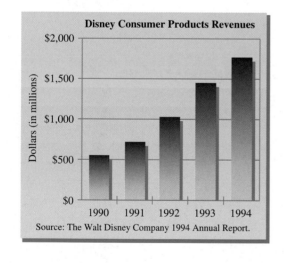

Disney Consumer Products Revenues

Source: The Walt Disney Company 1994 Annual Report.

74. Approximate Disney's consumer products revenue in 1994. $1800 million

75. Approximate the increase in consumer products revenue in 1992. $400 million

76. What year shows the greatest revenue?   1994

77. What trend is shown by this graph?   Revenue is increasing.

Name _____ **Section** _____ **Date** _____

CHAPTER 1 TEST

Translate each statement into symbols.

1. The absolute value of negative seven is greater than five.

2. The sum of nine and five is greater than or equal to four.

Simplify each expression.

3. $-13 + 8$

4. $-13 - (-2)$

5. $6 \cdot 3 - 8 \cdot 4$

6. $(13)(-3)$

7. $(-6)(-2)$

8. $\dfrac{|-16|}{-8}$

9. $\dfrac{-8}{0}$

10. $\dfrac{|-6| + 2}{5 - 6}$

11. $\dfrac{1}{2} - \dfrac{5}{6}$

12. $-1\dfrac{1}{8} + 5\dfrac{3}{4}$

13. $-\dfrac{3}{5} + \dfrac{15}{8}$

14. $3(-4)^2 - 80$

15. $6[5 + 2(3 - 8) - 3]$

16. $\dfrac{-12 + 3 \cdot 8}{4}$

17. $\dfrac{(-2)(0)(-3)}{-6}$

Insert $<$, $>$, *or* $=$ *in the appropriate space to make each statement true.*

18. -3 ___ -7

19. 4 ___ -8

20. $|-3|$ ___ 2

21. $|-2|$ ___ $-1 - (-3)$

22. Given
$\left\{-5, -1, \dfrac{1}{4}, 0, 1, 7, 11.6, \sqrt{7}, 3\pi\right\}$, list the numbers in this set that also belong to the set of:

a. Natural numbers

b. Whole numbers

c. Integers

d. Rational numbers

e. Irrational numbers

f. Real numbers

ANSWERS

1. $|-7| > 5$

2. $(9 + 5) \geq 4$

3. -5

4. -11

5. -14

6. -39

7. 12

8. -2

9. undefined

10. -8

11. $-\dfrac{1}{3}$

12. $4\dfrac{5}{8}$

13. $\dfrac{51}{40}$

14. -32

15. -48

16. 3

17. 0

18. $>$

19. $>$

20. $>$

21. $=$

22. a. $1, 7$

b. $0, 1, 7$

c. $-5, -1, 0, 1, 7$

d. $-5, -1, \frac{1}{4}, 0, 1, 7, 11.6$

e. $\sqrt{7}, 3\pi$

f. $-5, -1\frac{1}{4}, 0, 1, 7, 11.6, \sqrt{7}, 3\pi$

23. 40

24. 12

25. 22

26. −1

27. associative property of addition

28. commutative property of multiplication

29. distributive property

30. multiplicative inverse

31. 9

32. −3

33. second down

34. yes

35. 17°

36. loss of $420

Evaluate each expression when $x = 6$, $y = -2$, and $z = -3$.

23. $x^2 + y^2$ **24.** $x + yz$ **25.** $2 + 3x - y$ **26.** $\dfrac{y + z - 1}{x}$

Identify the property illustrated by each expression.

27. $8 + (9 + 3) = (8 + 9) + 3$ **28.** $6 \cdot 8 = 8 \cdot 6$

29. $-6(2 + 4) = -6 \cdot 2 + (-6) \cdot 4$ **30.** $\dfrac{1}{6}(6) = 1$

31. Find the opposite of -9.   **32.** Find the reciprocal of $-\dfrac{1}{3}$.

The New Orleans Saints were 22 yards from the goal when the series of gains and losses shown in the chart occurred. Use this chart to answer Questions 33–34.

Gains and Losses (in yards)	
First down	5
Second down	−10
Third down	−2
Fourth down	29

33. During which down did the greatest loss of yardage occur?

34. Was a touchdown scored?

35. The temperature at the Winter Olympics was a frigid 14° below zero in the morning, but by noon it had risen 31°. What was the temperature at noon?

36. Jean Avarez decided to sell 280 shares of stock, which decreased in value by $1.50 per share yesterday. How much money did she lose?

Intel is a semiconductor manufacturer that makes almost one-third of the world's computer chips. (You may have seen the slogan "Intel Inside" in commercials on television.) The line graph to the right shows Intel's net revenues in billions of dollars. Use this graph to answer Questions 37–40.

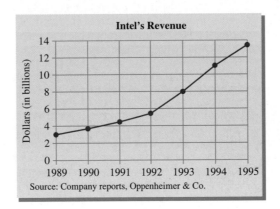

Intel's Revenue

Source: Company reports, Oppenheimer & Co.

37. Estimate Intel's revenue in 1993.

38. Estimate Intel's revenue in 1989.

39. Find the increase in Intel's revenue from 1993 to 1995.

40. What year shows the greatest increase in revenue?

37. $8 billion

38. $3 billion

39. $5.5 billion

40. 1994

Equations, Inequalities, and Problem Solving

In this chapter, we solve equations and inequalities. Once we know how to solve equations and inequalities, we may solve problems. Of course, problem solving is an integral topic in algebra and its discussion is continued throughout this text.

A glacier is a giant mass of rocks and ice that flows downhill like a river. Alaska alone has an estimated 100,000 glaciers. They form high in the mountains where snow does not melt. After years of accumulated snowfall, the weight of the compacted ice causes it to slide downhill, and a glacier is formed. In Example 1 on page 101, we will find the time it takes for ice at the beginning of a glacier to reach its face, or ending wall.

Glaciers account for approximately 77 percent of the world's fresh water. They also contain records of past environments between their layers of snow. Scientists study glaciers to learn about past global climates and then use this information to predict future environmental changes.

(*Source: Blue Ice in Motion—The Story of Alaska's Glaciers*, by Sally D. Wiley, published in 1990 by The Alaska Natural History Association and Judith Ann Rose)

Name _____ **Section** _____ **Date** _____

CHAPTER PRETEST

Simplify.

1. $3c - 4 + 6c - 9$

2. $-5(2y - 3) - 7y + 1$

Solve.

3. $3 - x = -12$

4. $12 - (5 - 4b) = 9 + 3b$

5. $\frac{2}{3}m = -8$

6. $-7 - 3y = 17 + 5y$

7. $3(1 - 4x) + 2(5x) = 9$

8. $0.20x + 0.15(60) = 0.75(18)$

9. $2(x - 1) = 2x + 5$

10. Three times the sum of a number and -2 is the same as 2 more than the number. Find the number.

11. Find two consecutive even integers such that three times the smaller is 16 more than twice the larger.

12. Substitute the given values into the given formula and solve for the unknown variable.

$$V = \frac{1}{3} Ah; V = 60, h = 4$$

13. If the area of a right-triangularly shaped sign is 18 square feet and its height is 4 feet, find the base of the sign.

14. Solve the given formula for the specified variable.

$$2x + y = 8 \quad \text{for} \quad y$$

15. What number is 22% of 90?

16. Write the ratio "4 quarts to 5 gallons" in fractional notation in lowest terms.

17. Solve the following proportion.

$$\frac{3x}{8} = \frac{9}{7}$$

Solve the inequality. Graph the solutions.

18. $-4 + x \leq 2$

19. $-\frac{3}{2} y > 6$

20. $-5x + 3 \leq 4(x - 6)$

2.1 SIMPLIFYING EXPRESSIONS

As we explore in this section, we will see that an expression such as $3x + 2x$ is not as simple as possible. This is because—even without replacing x by a value—we can perform the indicated addition.

A IDENTIFYING TERMS, LIKE TERMS, AND UNLIKE TERMS

Before we practice simplifying expressions, some new language is presented. A **term** is a number or the product of a number and variables raised to powers.

Terms

$$-y, \quad 2x^3, \quad -5, \quad 3xz^2, \quad \frac{2}{y}, \quad 0.8z$$

The **numerical coefficient** of a term is the numerical factor. The numerical coefficient of $3x$ is 3. Recall that $3x$ means $3 \cdot x$.

Term	Numerical Coefficient
$3x$	3
$\dfrac{y^3}{5}$	$\dfrac{1}{5}$ since $\dfrac{y^3}{5}$ means $\dfrac{1}{5} \cdot y^3$
$-0.7ab^3c^5$	-0.7
z	1
$-y$	-1
-5	-5

> **HELPFUL HINT**
>
> The term $-y$ means $-1y$ and thus has a numerical coefficient of -1. The term z means $1z$ and thus has a numerical coefficient of 1.

Example 1   Identify the numerical coefficient in each term.

    a. $-3y$ **b.** $22z^4$ **c.** y **d.** $-x$ **e.** $\dfrac{x}{7}$

Solution: **a.** The numerical coefficient of $-3y$ is -3.
    b. The numerical coefficient of $22z^4$ is 22.
    c. The numerical coefficient of y is 1, since y is $1y$.
    d. The numerical coefficient of $-x$ is -1, since $-x$ is $-1x$.
    e. The numerical coefficient of $\dfrac{x}{7}$ is $\dfrac{1}{7}$, since $\dfrac{x}{7}$ is $\dfrac{1}{7} \cdot x$.

Terms with the same variables raised to exactly the same powers are called **like terms**. Terms that aren't like terms are called **unlike terms**.

Like Terms	Unlike Terms	
$3x, 2x$	$5x, 5x^2$	Why? Same variable x, but different powers x and x^2
$-6x^2y, 2x^2y, 4x^2y$	$7y, 3z, 8x^2$	Why? Different variables
$2ab^2c^3, ac^3b^2$	$6abc^3, 6ab^2$	Why? Different variables and different powers

Objectives

A Identify terms, like terms, and unlike terms.
B Combine like terms.
C Simplify expressions containing parentheses.
D Write word phrases as algebraic expressions.

SSM   CD-ROM   Video 2.1

Practice Problem 1

Identify the numerical coefficient in each term.

a. $-4x$ b. $15y^3$ c. x d. $-y$ e. $\dfrac{z}{4}$

TEACHING TIP

It may be helpful to point out that the reason students must be able to identify like terms is to simplify expressions. This process is very similar to counting coins. People usually first sort the coins by type and then count how many of each type they have. Counting coins compares to collecting and combining like terms.

Answers

1. a. -4, **b.** 15, **c.** 1, **d.** -1, **e.** $\dfrac{1}{4}$

┌───┐
HELPFUL HINT

In like terms, each variable and its exponent must match exactly, but these factors don't need to be in the same order.

   $2x^2y$ and $3yx^2$ are like terms.
└───┘

Practice Problem 2

Determine whether the terms are like or unlike.

a. $7x, -6x$ b. $3x^2y^2, -x^2y^2, 4x^2y^2$

c. $-5ab, 3ba$

Example 2   Determine whether the terms are like or unlike.

   a. $2x, 3x^2$ b. $4x^2y, x^2y, -2x^2y$ c. $-2yz, -3zy$
   d. $-x^4, x^4$

Solution: a. Unlike terms, since the exponents on x are not the same.
   b. Like terms, since each variable and its exponent match.
   c. Like terms, since $zy = yz$ by the commutative property.
   d. Like terms.
   ▬▬▬

TEACHING TIP

It may be helpful to give students a visual representation of some terms. This problem illustrates why like terms are combined and unlike terms are not combined.

Use the tiles, ,
$2x^2 + 3x + x^2 + 1 + x + 2 = 3x^2 + 4x + 3$

B COMBINING LIKE TERMS

An algebraic expression containing the sum or difference of like terms can be simplified by applying the distributive property. For example, by the distributive property, we rewrite the sum of the like terms $3x + 2x$ as

   $$3x + 2x = (3 + 2)x = 5x$$

Also,

   $$-y^2 + 5y^2 = (-1 + 5)y^2 = 4y^2$$

Simplifying the sum or difference of like terms is called **combining like terms**.

Practice Problem 3

Simplify each expression by combining like terms.

a. $9y - 4y$ b. $11x^2 + x^2$

c. $5y - 3x + 6x$

Example 3   Simplify each expression by combining like terms.

   a. $7x - 3x$ b. $10y^2 + y^2$ c. $8x^2 + 2x - 3x$

Solution: a. $7x - 3x = (7 - 3)x = 4x$
   b. $10y^2 + y^2 = (10 + 1)y^2 = 11y^2$
   c. $8x^2 + 2x - 3x = 8x^2 + (2 - 3)x = 8x^2 - x$
   ▬▬▬

Practice Problems 4–7

Simplify each expression by combining like terms.

4. $7y + 2y + 6 + 10$

5. $-2x + 4 + x - 11$

6. $3z - 3z^2$

7. $8.9y + 4.2y - 3$

Examples   Simplify each expression by combining like terms.

4. $2x + 3x + 5 + 2 = (2 + 3)x + (5 + 2)$
                    $= 5x + 7$
5. $-5a - 3 + a + 2 = -5a + 1a + (-3 + 2)$
                    $= (-5 + 1)a + (-3 + 2)$
                    $= -4a - 1$
6. $4y - 3y^2$ These two terms cannot be combined because they are unlike terms.
7. $2.3x + 5x - 6 = (2.3 + 5)x - 6$
                 $= 7.3x - 6$
   ▬▬▬

The examples above suggest the following.

Answers
2. **a.** like, **b.** like, **c.** like,
3. **a.** $5y$, **b.** $12x^2$, **c.** $5y + 3x$,
4. $9y + 16$, 5. $-x - 7$, 6. $3z - 3z^2$,
7. $13.1y - 3$

COMBINING LIKE TERMS

To **combine like terms**, combine the numerical coefficients and multiply the result by the common variable factors.

C SIMPLIFYING EXPRESSIONS CONTAINING PARENTHESES

In simplifying expressions we make frequent use of the distributive property to remove parentheses.

Examples   Find each product by using the distributive property to remove parentheses.

8. $5(x + 2) = 5(x) + 5(2)$ Apply the distributive property.
$\quad = 5x + 10$ Multiply.

9. $-2(y + 0.3z - 1) = -2(y) + (-2)(0.3z)$ Apply the distributive property.
$\quad\quad + (-2)(-1)$
$\quad = -2y - 0.6z + 2$ Multiply.

10. $-(x + y - 2z + 6) = -1(x + y - 2z + 6)$ Distribute -1 over each term.
$\quad = -1(x) - 1(y) - 1(-2z)$
$\quad - 1(6)$
$\quad = -x - y + 2z - 6$

Practice Problems 8–10

Find each product by using the distributive property to remove parentheses.

8. $3(y + 6)$

9. $-4(x + 0.2y - 3)$

10. $-(3x + 2y + z - 1)$

HELPFUL HINT

If a "−" sign precedes parentheses, the sign of each term inside the parentheses is changed when the distributive property is applied to remove parentheses.

Examples:
$-(2x + 1) = -2x - 1$
$-(x - 2y) = -x + 2y$
$-(-5x + y - z) = 5x - y + z$
$-(-3x - 4y - 1) = 3x + 4y + 1$

When simplifying an expression containing parentheses, we often use the distributive property first to remove parentheses and then again to combine any like terms.

Examples   Simplify each expression.

11. $3(2x - 5) + 1 = 6x - 15 + 1$ Apply the distributive property.
$\quad = 6x - 14$ Combine like terms.

12. $8 - (7x + 2) + 3x = 8 - 7x - 2 + 3x$ Apply the distributive property.
$\quad = -7x + 3x + 8 - 2$
$\quad = -4x + 6$ Combine like terms.

Practice Problems 11–13

Simplify each expression.

11. $4(x - 6) + 20$

12. $5 - (3x + 9)$

13. $-3(7x + 1) - (4x - 2)$

Answers

8. $3y + 18$, 9. $-4x - 0.8y + 12$, 10. $-3x - 2y - z + 1$, 11. $4x - 4$, 12. $-3x - 4$, 13. $-25x - 1$

13. $-2(4x + 7) - (3x - 1) = -8x - 14 - 3x + 1$ Apply the distributive property.
$$= -11x - 13$$ Combine like terms.

Practice Problem 14

Subtract $9x - 10$ from $4x - 3$.

Example 14   Subtract $4x - 2$ from $2x - 3$.

Solution: We first note that "subtract $4x - 2$ **from** $2x - 3$" translates to $(2x - 3) - (4x - 2)$. Next, we simplify the algebraic expression.

$(2x - 3) - (4x - 2) = 2x - 3 - 4x + 2$ Apply the distributive property.
$$= -2x - 1$$ Combine like terms.

D WRITING ALGEBRAIC EXPRESSIONS

To prepare for problem solving, we next practice writing word phrases as algebraic expressions.

Practice Problems 15–17

Write each phrase as an algebraic expression and simplify if possible. Let x represent the unknown number.

15. Three times a number, *subtracted from* 10

16. The sum of a number and 2, divided by 5

17. Three times a number, added to the sum of twice a number and 6

Examples   Write each phrase as an algebraic expression and simplify if possible. Let x represent the unknown number.

15. Twice a number, plus 6

$2x$ $+ 6$

This expression cannot be simplified.

16. The difference of a number and 4, divided by 7

$(x - 4)$ $\div$ 7

This expression cannot be simplified.

17. Five plus the sum of a number and 1

5   + $(x + 1)$

Next, we simplify this expression.
$$5 + (x + 1) = 5 + x + 1$$
$$= 6 + x$$

TEACHING TIP

After doing the examples, have students write each of the following phrases as an algebraic expression:
a. Four, added to three times a number.
b. Three times a number, added to four.
c. Four, subtracted from three times a number.
d. Three times a number, subtracted from four.
Then ask students the following questions:
Are the algebraic expressions of **a.** and **b.** equivalent? (Yes)
Why or why not? (Addition is commutative)
Are the algebraic expressions of **c.** and **d.** equivalent? (No)
Why or why not? (Subtraction is not commutative.)

Answers

14. $-5x + 7$, **15.** $10 - 3x$, **16.** $\dfrac{(x + 2)}{5}$, **17.** $5x + 6$

Name _____ Section _____ Date _____

MENTAL MATH

A *Identify the numerical coefficient of each term. See Example 1.*

1. $-7y$ 2. $3x$ 3. x

4. $-y$ 5. $17x^2y$ 6. $1.2xyz$

Indicate whether each list of terms are like or unlike. See Example 2.

7. $5y, -y$ 8. $-2x^2y, 6xy$ 9. $2z, 3z^2$

10. $ab^2, -7ab^2$ 11. $8wz, \frac{1}{7}zw$ 12. $7.4p^3q^2, 6.2p^3q^2r$

EXERCISE SET 2.1

B *Simplify each expression by combining any like terms. See Examples 3 through 7.*

1. $7y + 8y$ 2. $3x + 2x$ 3. $8w - w + 6w$

4. $c - 7c + 2c$ 5. $3b - 5 - 10b - 4$ 6. $6g + 5 - 3g - 7$

7. $m - 4m + 2m - 6$ 8. $a + 3a - 2 - 7a$ 9. $5g - 3 - 5 - 5g$

10. $8p + 4 - 8p - 15$ 11. $6.2x - 4 + x - 1.2$ 12. $7.9y - 0.7 - y + 0.2$

13. $2k - k - 6$ 14. $7c - 8 - c$

15. $-9x + 4x + 18 - 10x$ 16. $5y - 14 + 7y - 20y$

17. $6x - 5x + x - 3 + 2x$ 18. $8h + 13h - 6 + 7h - h$

19. $7x^2 + 8x^2 - 10x^2$ 20. $8x^3 + x^3 - 11x^3$

21. $3.4m - 4 - 3.4m - 7$ 22. $2.8w - 0.9 - 0.5 - 2.8w$

23. $6x + 0.5 - 4.3x - 0.4x + 3$ 24. $0.4y - 6.7 + y - 0.3 - 2.6y$

95

Name _____

C *Simplify each expression. Use the distributive property to remove any parentheses. See Examples 8 through 10.*

25. $5(y + 4)$ **26.** $7(r + 3)$ **27.** $-2(x + 2)$

28. $-4(y + 6)$ **29.** $-5(2x - 3y + 6)$ **30.** $-2(4x - 3z - 1)$

31. $-(3x - 2y + 1)$ **32.** $-(y + 5z - 7)$

Remove parentheses and simplify each expression. See Examples 11 through 13.

33. $7(d - 3) + 10$ **34.** $9(z + 7) - 15$ **35.** $-4(3y - 4) + 12y$

36. $-3(2x + 5) - 6x$ **37.** $3(2x - 5) - 5(x - 4)$ **38.** $2(6x - 1) - (x - 7)$

39. $-2(3x - 4) + 7x - 6$ **40.** $8y - 2 - 3(y + 4)$ **41.** $5k - (3k - 10)$

42. $-11c - (4 - 2c)$ **43.** $(3x + 4) - (6x - 1)$ **44.** $(8 - 5y) - (4 + 3y)$

45. $5(x + 2) - (3x - 4)$ **46.** $4(2x - 3) - 2(x + 1)$

47. $0.5(m + 2) + 0.4m$ **48.** $0.2(k + 8) - 0.1k$

49. In your own words, explain how to combine like terms.

50. Do like terms contain the same numerical coefficients? Explain your answer.

Write each phrase as an algebraic expression. Simplify if possible. See Example 14.
51. Add $6x + 7$ to $4x - 10$

52. Add $3y - 5$ to $y + 16$

53. Subtract $7x + 1$ from $3x - 8$

54. Subtract $4x - 7$ from $12 + x$

55. Subtract $5m - 6$ from $m - 9$

56. Subtract $m - 3$ from $2m - 6$

D *Write each phrase as an algebraic expression and simplify if possible. Let x represent the unknown number. See Examples 15 through 17.*

57. Twice a number, decreased by four

58. The difference of a number and two, divided by five

59. Three-fourths of a number, increased by twelve

60. Eight more than triple a number

61. The sum of 5 times a number and -2, added to 7 times the number

62. The sum of 3 times a number and 10, **subtracted from** 9 times the number

63. Eight times the sum of a number and six

64. Five, subtracted from four times a number

65. Double a number minus the sum of the number and ten

66. Half a number minus the product of the number and eight

REVIEW AND PREVIEW

Evaluate each expression for the given values. See Section 1.5.
67. If $x = -1$ and $y = 3$, find $y - x^2$

68. If $g = 0$ and $h = -4$, find $gh - h^2$

69. If $a = 2$ and $b = -5$, find $a - b^2$

70. If $x = -3$, find $x^3 - x^2 + 4$

51. $10x - 3$

52. $4y + 11$

53. $-4x - 9$

54. $-3x + 19$

55. $-4m - 3$

56. $m - 3$

57. $2x - 4$

58. $\dfrac{(x - 2)}{5}$

59. $\dfrac{3}{4}x + 12$

60. $3x + 8$

61. $12x - 2$

62. $6x - 10$

63. $8(x + 6)$

64. $4x - 5$

65. $x - 10$

66. $-7.5x$

67. 2

68. -16

69. -23

70. -32

71. −25

71. If $y = -5$ and $z = 0$, find $yz - y^2$

72. If $x = -2$, find $x^3 - x^2 - x$

72. −10

COMBINING CONCEPTS

Given the following information, determine whether each scale is balanced or not.

73. balanced

1 cone balances 1 cube   1 cylinder balances 2 cubes

73.    **74.**

74. not balanced

75. balanced

75. 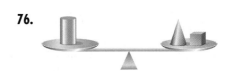   **76.**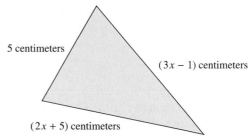

76. balanced

Write each algebraic expression described.

77. Recall that the perimeter of a figure is the total distance around the figure. Given the following rectangle, express the perimeter as an algebraic expression containing the variable x.

77. $(18x - 2)$ ft

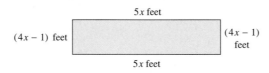

$5x$ feet

$(4x - 1)$ feet $(4x - 1)$ feet

$5x$ feet

78. Given the following triangle, express its perimeter as an algebraic expression containing the variable x.

5 centimeters $(3x - 1)$ centimeters

$(2x + 5)$ centimeters

78. $(5x + 9)$ cm

79. To convert from feet to inches, we multiply by 12. For example, the number of inches in 2 feet is $12 \cdot 2$ inches. If one board has a length of $(x + 2)$ *feet* and a second board has a length of $(3x - 1)$ *inches*, express their total length in inches as an algebraic expression.

79. $(15x + 23)$ in.

80. The value of 7 nickels is $5 \cdot 7$ cents. Likewise, the value of x nickels is $5x$ cents. If the money box in a drink machine contains x *nickels*, $3x$ *dimes*, and $(30x - 1)$ *quarters*, express their total value in cents as an algebraic expression.

80. $(785x - 25)$¢

2.2   THE ADDITION PROPERTY OF EQUALITY

A   USING THE ADDITION PROPERTY

Recall from Section 1.2 that an equation is a statement that two expressions have the same value. Also, a value of the variable that makes an equation a true statement is called a solution or root of the equation. The process of finding the solution of an equation is called **solving** the equation for the variable. In this section, we concentrate on solving **linear equations** in one variable.

LINEAR EQUATION IN ONE VARIABLE

A linear equation in one variable can be written in the form
$$Ax + B = C$$
where A, B, and C are real numbers and $A \neq 0$.

Evaluating a linear equation for a given value of the variable, as we did in Section 1.2, can tell us whether that value is a solution. But we can't rely on evaluating an equation as our method of solving it—with what value would we start?

Instead, to solve a linear equation in x, we write a series of simpler equations, all *equivalent* to the original equation, so that the final equation has the form

$$x = \text{number} \qquad \text{or} \qquad \text{number} = x$$

Equivalent equations are equations that have the same solution. This means that the "number" above is the solution to the original equation.

The first property of equality that helps us write simpler equivalent equations is the **addition property of equality**.

ADDITION PROPERTY OF EQUALITY

If a, b, and c are real numbers, then
$$a = b \qquad \text{and} \qquad a + c = b + c$$
are equivalent equations.

This property guarantees that adding the same number to both sides of an equation does not change the solution of the equation. Since subtraction is defined in terms of addition, we may also **subtract the same number from both sides** without changing the solution.

A good way to picture a true equation is as a balanced scale. Since it is balanced, each side of the scale weighs the same amount.

$x - 2$ 5

Objectives

A   Use the addition property of equality to solve linear equations.

B   Simplify an equation and then use the addition property of equality.

C   Write word phrases as algebraic expressions.

SSM   CD-ROM   Video 2.2

If the same weight is added to or subtracted from each side, the scale remains balanced.

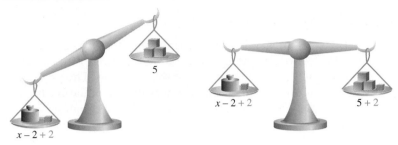

$$x - 2 + 2$$

$$x - 2 + 2 \qquad 5 + 2$$

We use the addition property of equality to write equivalent equations until the variable is alone (by itself on one side of the equation) and the equation looks like "x = number" or "number = x."

TRY THE CONCEPT CHECK IN THE MARGIN.

✓ **CONCEPT CHECK**

Use the addition property to fill in the blank so that the middle equation simplifies to the last equation.

$$x - 5 = 3$$
$$x - 5 + \underline{} = 3 + \underline{}$$
$$x = 8$$

Practice Problem 1

Solve $x - 5 = 8$ for x.

Example 1   Solve $x - 7 = 10$ for x.

Solution: To solve for x, we first get x alone on one side of the equation. To do this, we add 7 to both sides of the equation.

$$x - 7 = 10$$
$$x - 7 + 7 = 10 + 7 \qquad \text{Add 7 to both sides.}$$
$$x = 17 \qquad \text{Simplify.}$$

The solution of the equation $x = 17$ is obviously 17. Since we are writing equivalent equations, the solution of the equation $x - 7 = 10$ is also 17.

Check: To check, replace x with 17 in the original equation.

$$x - 7 = 10 \qquad \text{Original equation.}$$
$$17 - 7 \stackrel{?}{=} 10 \qquad \text{Replace } x \text{ with 17.}$$
$$10 = 10 \qquad \text{True.}$$

Since the statement is true, 17 is the solution.   ■

Practice Problem 2

Solve: $y + 1.7 = 0.3$

Example 2   Solve: $y + 0.6 = -1.0$

Solution: To solve for y, we subtract 0.6 from both sides of the equation.

$$y + 0.6 = -1.0$$
$$y + 0.6 - 0.6 = -1.0 - 0.6 \qquad \text{Subtract 0.6 from both sides.}$$
$$y = -1.6 \qquad \text{Combine like terms.}$$

Check:
$$y + 0.6 = -1.0 \qquad \text{Original equation.}$$
$$-1.6 + 0.6 \stackrel{?}{=} -1.0 \qquad \text{Replace } y \text{ with } -1.6.$$
$$-1.0 = -1.0 \qquad \text{True.}$$

The solution is -1.6.   ■

Answers

✓ Concept Check: 5

1. $x = 13$,   **2.** $y = -1.4$

Example 3   Solve: $\frac{1}{2} = x - \frac{3}{4}$

Solution: To get x alone, we add $\frac{3}{4}$ to both sides.

$$\frac{1}{2} = x - \frac{3}{4}$$

$$\frac{1}{2} + \frac{3}{4} = x - \frac{3}{4} + \frac{3}{4} \qquad \text{Add } \frac{3}{4} \text{ to both sides.}$$

$$\frac{1}{2} \cdot \frac{2}{2} + \frac{3}{4} = x \qquad \text{The LCD is 4.}$$

$$\frac{2}{4} + \frac{3}{4} = x \qquad \text{Add the fractions.}$$

$$\frac{5}{4} = x$$

Check: $\frac{1}{2} = x - \frac{3}{4}$ Original equation.

$$\frac{1}{2} \overset{?}{=} \frac{5}{4} - \frac{3}{4} \qquad \text{Replace } x \text{ with } \frac{5}{4}.$$

$$\frac{1}{2} \overset{?}{=} \frac{2}{4} \qquad \text{Subtract.}$$

$$\frac{1}{2} = \frac{1}{2} \qquad \text{True.}$$

The solution is $\frac{5}{4}$.

Practice Problem 3

Solve: $\frac{7}{8} = y - \frac{1}{3}$

HELPFUL HINT

We may solve an equation so that the variable is alone on *either* side of the equation. For example, $\frac{5}{4} = x$ is equivalent to $x = \frac{5}{4}$.

Example 4   Solve: $5t - 5 = 6t$

Solution: To solve for t, we first want all terms containing t on one side of the equation. To do this, we subtract $5t$ from both sides of the equation.

$$5t - 5 = 6t$$

$$5t - 5 - 5t = 6t - 5t \qquad \text{Subtract } 5t \text{ from both sides.}$$

$$-5 = t \qquad \text{Combine like terms.}$$

Check: $5t - 5 = 6t$ Original equation.

$$5(-5) - 5 \overset{?}{=} 6(-5) \qquad \text{Replace } t \text{ with } -5.$$

$$-25 - 5 \overset{?}{=} -30$$

$$-30 = -30 \qquad \text{True.}$$

The solution is -5.

Practice Problem 4

Solve: $3x + 10 = 4x$

B SIMPLIFYING EQUATIONS

Many times, it is best to simplify one or both sides of an equation before applying the addition property of equality.

Practice Problem 5

Solve: $10w + 3 - 4w + 4$
$$= -2w + 3 + 7w$$

Example 5   Solve: $2x + 3x - 5 + 7 = 10x + 3 - 6x - 4$

Solution: First we simplify both sides of the equation.

$$2x + 3x - 5 + 7 = 10x + 3 - 6x - 4$$
$$5x + 2 = 4x - 1 \qquad \text{Combine like terms on each side of the equation.}$$

Next, we want all terms with a variable on one side of the equation and all numbers on the other side.

$$5x + 2 - 4x = 4x - 1 - 4x \quad \text{Subtract } 4x \text{ from both sides.}$$
$$x + 2 = -1 \qquad \text{Combine like terms.}$$
$$x + 2 - 2 = -1 - 2 \quad \begin{array}{l}\text{Subtract 2 from both sides to}\\ \text{get } x \text{ alone.}\end{array}$$
$$x = -3 \qquad \text{Combine like terms.}$$

Check: $2x + 3x - 5 + 7 = 10x + 3 - 6x - 4$ Original equation.

$$2(-3) + 3(-3) - 5 + 7 \stackrel{?}{=} 10(-3) + 3 - 6(-3) - 4$$

Replace x with -3.

$$-6 - 9 - 5 + 7 \stackrel{?}{=} -30 + 3 + 18 - 4 \quad \text{Multiply.}$$
$$-13 = -13 \qquad \text{True.}$$

The solution is -3. ▬▬▬

If an equation contains parentheses, we use the distributive property to remove them, as before. Then we combine any like terms.

Practice Problem 6

Solve: $3(2w - 5) - (5w + 1) = -3$

Example 6   Solve: $6(2a - 1) - (11a + 6) = 7$

Solution:
$$6(2a - 1) - 1(11a + 6) = 7$$
$$6(2a) + 6(-1) - 1(11a) - 1(6) = 7 \quad \begin{array}{l}\text{Apply the distrib-}\\ \text{utive property.}\end{array}$$
$$12a - 6 - 11a - 6 = 7 \quad \text{Multiply.}$$
$$a - 12 = 7 \quad \text{Combine like terms.}$$
$$a - 12 + 12 = 7 + 12 \quad \begin{array}{l}\text{Add 12 to}\\ \text{both sides.}\end{array}$$
$$a = 19 \quad \text{Simplify.}$$

Check: Check by replacing a with 19 in the original equation. ▬▬▬

Practice Problem 7

Solve: $12 - y = 9$

Example 7   Solve: $3 - x = 7$

Solution: First we subtract 3 from both sides.

$$3 - x = 7$$
$$3 - x - 3 = 7 - 3 \quad \text{Subtract 3 from both sides.}$$
$$-x = 4 \quad \text{Simplify.}$$

Answers

5. $w = -4$, **6.** $w = 13$, **7.** $y = 3$

We have not yet solved for x since x is not alone. However, this equation does say that the opposite of x is 4. If the opposite of x is 4, then x is the opposite of 4, or $x = -4$.

If $-x = 4$,

then $x = -4$.

Check: $3 - x = 7$ Original equation.

$3 - (-4) \stackrel{?}{=} 7$ Replace x with -4.

$3 + 4 \stackrel{?}{=} 7$ Add.

$7 = 7$ True.

The solution is -4.

TEACHING TIP

After solving Example 7, you may want to point out that there is more than one way to solve this problem.

$3 - x = 7$
$3 - x + x = 7 + x$
$3 = 7 + x$
$3 - 7 = 7 + x - 7$
$-4 = x$

C WRITING ALGEBRAIC EXPRESSIONS

In this section, we continue to practice writing algebraic expressions.

Example 8   **a.** The sum of two numbers is 8. If one number is 3, find the other number.

b. The sum of two numbers is 8. If one number is x, write an expression representing the other number.

Solution: **a.** If the sum of two numbers is 8 and one number is 3, we find the other number by subtracting 3 from 8. The other number is $8 - 3$, or 5.

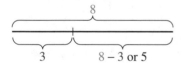

b. If the sum of two numbers is 8 and one number is x, we find the other number by subtracting x from 8. The other number is represented by $8 - x$.

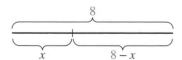

Practice Problem 8

The sum of two numbers is 11. If one number is x, write an expression representing the other number.

Answer

8. $11 - x$

Focus On History

THE GOLDEN RECTANGLE IN ART

The golden rectangle is a rectangle whose length is approximately 1.6 times its width. The early Greeks thought that a rectangle with these dimensions was the most pleasing to the eye. Examples of the golden rectangle are found in many ancient, as well as modern, works of art. For example, the Parthenon in Athens, Greece, shows the golden rectangle in many aspects of its design. Modern-era artists, including Piet Mondrian (1872–1944) and Georges Seurat (1859–1891), also frequently used the proportions of a golden rectangle in their paintings.

Mondrian

Composition with Gray and Light Brown 1918, Oil on canvas, 80.2 x 49.9 cm (31 9/16 x 19 5/8 in); Museum of Fine Arts, Houston, Texas

To test whether a rectangle is a golden rectangle, divide the rectangle's length by its width. If the result is approximately 1.6, we can consider the rectangle to be a golden rectangle. For instance, consider Mondrian's *Composition with Gray and Light Brown*, which was painted on an 80.2 × 49.9 cm canvas. Because $\frac{80.2}{49.9} \approx 1.6$, the dimensions of the canvas form a golden rectangle. In what other ways are golden rectangles connected with this painting?

Examples of golden rectangles can be found in the designs of many everyday objects. Visual artists, from architects to product and package designers, use the golden rectangle shape in such things as the face of a building, the floor of a room, the front of a food package, the front cover of a book, and even the shape of a credit card.

GROUP ACTIVITY

Find an example of a golden rectangle in a building or an everyday object. Use a ruler to measure its dimensions and verify that the length is approximately 1.6 times the width.

Name _____ Section _____ Date _____

MENTAL MATH ANSWERS
1. $x = 2$
2. $x = 3$
3. $n = 12$
4. $z = 18$
5. $b = 17$
6. $d = 21$

MENTAL MATH

Solve each equation mentally. See Examples 1 and 2.

1. $x + 4 = 6$
2. $x + 7 = 10$
3. $n + 18 = 30$
4. $z + 22 = 40$
5. $b - 11 = 6$
6. $d - 16 = 5$

EXERCISE SET 2.2

A *Solve each equation. Check each solution. See Examples 1 through 4.*

1. $x + 7 = 10$
2. $x + 14 = 25$
3. $x - 2 = -4$
4. $y - 9 = 1$

5. $3 + x = -11$
6. $8 + z = -8$
7. $r - 8.6 = -8.1$
8. $t - 9.2 = -6.8$

9. $\frac{1}{3} + f = \frac{3}{4}$
10. $c + \frac{1}{6} = \frac{3}{8}$
11. $5b - 0.7 = 6b$
12. $9x + 5.5 = 10x$

13. $7x - 3 = 6x$
14. $18x - 9 = 19x$

15. In your own words, explain what is meant by the solution of an equation.

16. In your own words, explain how to check a solution of an equation.

B *Solve each equation. Don't forget to first simplify each side of the equation, if possible. Check each solution. See Examples 5 through 7.*

17. $7x + 2x = 8x - 3$
18. $3n + 2n = 7 + 4n$

19. $2y + 10 = 5y - 4y$
20. $4x - 4 = 10x - 7x$

21. $3x - 6 = 2x + 5$
22. $7y + 2 = 6y + 2$

23. $5x - 6 = 6x - 5$
24. $2x + 7 = x - 10$

25. $8y + 2 - 6y = 3 + y - 10$
26. $4p - 11 - p = 2 + 2p - 20$

ANSWERS
1. $x = 3$
2. $x = 11$
3. $x = -2$
4. $y = 10$
5. $x = -14$
6. $z = -16$
7. $r = 0.5$
8. $t = 2.4$
9. $f = \frac{5}{12}$
10. $c = \frac{5}{24}$
11. $b = -0.7$
12. $x = 5.5$
13. $x = 3$
14. $x = -9$
15. answers may vary
16. answers may vary
17. $x = -3$
18. $n = 7$
19. $y = -10$
20. $x = 4$
21. $x = 11$
22. $y = 0$
23. $x = -1$
24. $x = -17$
25. $y = -9$
26. $p = -7$

106

Name _____

27. $13x - 9 + 2x - 5 = 12x - 1 + 2x$

28. $15x + 20 - 10x - 9 = 25x + 8 - 21x - 7$

29. $-6.5 - 4x - 1.6 - 3x = -6x + 9.8$ **30.** $-1.4 - 7x - 3.6 - 2x = -8x + 4.4$

31. $\dfrac{3}{8}x - \dfrac{1}{6} = -\dfrac{5}{8}x - \dfrac{2}{3}$ **32.** $\dfrac{2}{5}x - \dfrac{1}{12} = -\dfrac{3}{5}x - \dfrac{3}{4}$

33. $2(x - 4) = x + 3$ **34.** $3(y + 7) = 2y - 5$

35. $7(6 + w) = 6(2 + w)$ **36.** $6(5 + c) = 5(c - 4)$

37. $10 - (2x - 4) = 7 - 3x$ **38.** $15 - (6 - 7k) = 2 + 6k$

39. $-5(n - 2) = 8 - 4n$ **40.** $-4(z - 3) = 2 - 3z$

41. $-3(x - 4) = -4x$ **42.** $-2(x - 1) = -3x$

43. $3(n - 5) - (6 - 2n) = 4n$ **44.** $5(3 + z) - (8z + 9) = -4z$

45. $-2(x + 6) + 3(2x - 5) = 3(x - 4) + 10$

46. $-5(x + 1) + 4(2x - 3) = 2(x + 2) - 8$

Name _____

C *Write each algebraic expression described. See Example 8.*

47. Two numbers have a sum of 20. If one number is p, express the other number in terms of p.

48. Two numbers have a sum of 13. If one number is y, express the other number in terms of y.

49. A 10-foot board is cut into two pieces. If one piece is x feet long, express the other length in terms of x.

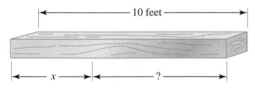

50. A 5-foot piece of string is cut into two pieces. If one piece is x feet long, express the other length in terms of x.

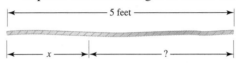

51. Two angles are *supplementary* if their sum is 180°. If one angle measures $x°$, express the measure of its supplement in terms of x.

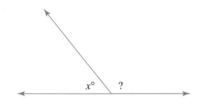

52. Two angles are *complementary* if their sum is 90°. If one angle measures $x°$, express the measure of its complement in terms of x.

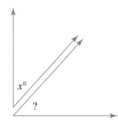

53. In a recent U.S. Senate race in Maine, Susan M. Collins received 30,898 more votes than Joseph E. Brennan. If Joseph received n votes, how many did Susan receive? (*Source*: Voter News Service)

54. The Verrazano-Narrows Bridge in New York City is the longest suspension bridge in North America. The Golden Gate Bridge in San Francisco is 60 feet shorter than the Verrazano-Narrows Bridge. If the length of the Verrazano-Narrows Bridge is m feet, express the length of the Golden Gate Bridge as an algebraic expression in m. (*Source*: Survey of State Highway Engineers, 1996)

55. $\frac{8}{5}$

56. $\frac{6}{7}$

57. $\frac{1}{2}$

58. $\frac{1}{5}$

59. -9

60. $-\frac{5}{3}$

61. x

62. y

63. y

64. r

65. x

66. x

67. $x = -145.478$

68. $x = 12.66$

69. $(173 - 3x)°$

70. $(360 - 9x)°$

Name _____

REVIEW AND PREVIEW

Find each multiplicative inverse or reciprocal. See Section 1.7.

55. $\frac{5}{8}$ **56.** $\frac{7}{6}$ **57.** 2   **58.** 5   **59.** $-\frac{1}{9}$ **60.** $-\frac{3}{5}$

Perform each indicated operation and simplify. See Sections 1.5 and 1.6.

61. $\frac{3x}{3}$ **62.** $\frac{-2y}{-2}$ **63.** $-5\left(-\frac{1}{5}y\right)$

64. $7\left(\frac{1}{7}r\right)$ **65.** $\frac{3}{5}\left(\frac{5}{3}x\right)$ **66.** $\frac{9}{2}\left(\frac{2}{9}x\right)$

◤ COMBINING CONCEPTS

Use a calculator to determine the solution of each equation.

67. $36.766 + x = -108.712$ **68.** $-85.325 = x - 97.985$

Solve.

69. The sum of the angles of a triangle is 180°. If one angle of a triangle measures $x°$ and a second angle measures $(2x + 7)°$, express the measure of the third angle in terms of x. Simplify the expression.

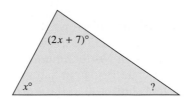

70. A quadrilateral is a four-sided figure (like the one shown in the figure) whose angle sum is 360°. If one angle measures $x°$, a second angle measures $3x°$, and a third angle measures $5x°$, express the measure of the fourth angle in terms of x. Simplify the expression.

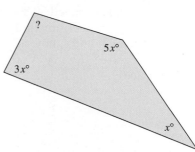

2.3 THE MULTIPLICATION PROPERTY OF EQUALITY

A USING THE MULTIPLICATION PROPERTY

As useful as the addition property of equality is, it cannot help us solve every type of linear equation in one variable. For example, adding or subtracting a value on both sides of the equation does not help solve

$$\frac{5}{2}x = 15$$

Instead, we apply another important property of equality, the **multiplication property of equality**.

MULTIPLICATION PROPERTY OF EQUALITY

If a, b, and c are real numbers and $c \neq 0$, then

$$a = b \quad \text{and} \quad ac = bc$$

are equivalent equations.

This property guarantees that multiplying both sides of an equation by the same nonzero number does not change the solution of the equation. Since division is defined in terms of multiplication, we may also **divide both sides of the equation by the same nonzero number** without changing the solution.

Example 1   Solve: $\frac{5}{2}x = 15$

Solution: To get x alone, we multiply both sides of the equation by the reciprocal of $\frac{5}{2}$, which is $\frac{2}{5}$.

$$\frac{5}{2}x = 15$$

$$\frac{2}{5} \cdot \left(\frac{5}{2}x\right) = \frac{2}{5} \cdot 15 \quad \text{Multiply both sides by } \frac{2}{5}.$$

$$\left(\frac{2}{5} \cdot \frac{5}{2}\right)x = \frac{2}{5} \cdot 15 \quad \text{Apply the associative property.}$$

$$1x = 6 \quad \text{Simplify.}$$

or

$$x = 6$$

Check: Replace x with 6 in the original equation.

$$\frac{5}{2}x = 15 \quad \text{Original equation.}$$

$$\frac{5}{2}(6) \stackrel{?}{=} 15 \quad \text{Replace } x \text{ with 6.}$$

$$15 = 15 \quad \text{True.}$$

The solution is 6.

Objectives

A Use the multiplication property of equality to solve linear equations.

B Use both the addition and multiplication properties of equality to solve linear equations.

C Write word phrases as algebraic expressions.

SSM   CD-ROM   Video 2.3

Practice Problem 1

Solve: $\frac{3}{7}x = 9$

Answer

1. $x = 21$

In the equation $\frac{5}{2}x = 15$, $\frac{5}{2}$ is the coefficient of x. When the coefficient of x is a *fraction*, we will get x alone by multiplying by the reciprocal. When the coefficient of x is an integer or a decimal, it is usually more convenient to divide both sides by the coefficient. (Dividing by a number is, of course, the same as multiplying by the reciprocal of the number.)

Practice Problem 2

Solve: $7x = 42$

Example 2   Solve: $5x = 30$

Solution: To get x alone, we divide both sides of the equation by 5, the coefficient of x.

$$5x = 30$$
$$\frac{5x}{5} = \frac{30}{5} \quad \text{Divide both sides by 5.}$$
$$1 \cdot x = 6 \quad \text{Simplify.}$$
$$x = 6$$

Check:
$$5x = 30 \quad \text{Original equation.}$$
$$5 \cdot 6 \stackrel{?}{=} 30 \quad \text{Replace } x \text{ with 6.}$$
$$30 = 30 \quad \text{True.}$$

The solution is 6.   ▬▬▬

Practice Problem 3

Solve: $-4x = 52$

TEACHING TIP

Before beginning Example 4, try asking students: If one-seventh of a restaurant bill is $20.00, how much is the entire bill? ($140.00) Discuss how students get their answer. Recall that 7 and $\frac{1}{7}$ are reciprocals.

Example 3   Solve: $-3x = 33$

Solution: Recall that $-3x$ means $-3 \cdot x$. To get x alone, we divide both sides by the coefficient of x, that is, -3.

$$-3x = 33$$
$$\frac{-3x}{-3} = \frac{33}{-3} \quad \text{Divide both sides by } -3.$$
$$1x = -11 \quad \text{Simplify.}$$
$$x = -11$$

Check:
$$-3x = 33 \quad \text{Original equation.}$$
$$-3(-11) \stackrel{?}{=} 33 \quad \text{Replace } x \text{ with } -11.$$
$$33 = 33 \quad \text{True.}$$

The solution is -11.   ▬▬▬

Practice Problem 4

Solve: $\frac{y}{5} = 13$

Example 4   Solve: $\frac{y}{7} = 20$

Solution: Recall that $\frac{y}{7} = \frac{1}{7}y$. To get y alone, we multiply both sides of the equation by 7, the reciprocal of $\frac{1}{7}$.

$$\frac{y}{7} = 20$$
$$\frac{1}{7}y = 20$$
$$7 \cdot \frac{1}{7}y = 7 \cdot 20 \quad \text{Multiply both sides by 7.}$$
$$1y = 140 \quad \text{Simplify.}$$
$$y = 140$$

Check: $\dfrac{y}{7} = 20$ Original equation.

$\dfrac{140}{7} \overset{?}{=} 20$ Replace y with 140.

$20 = 20$ True.

The solution is 140.

Example 5   Solve: $3.1x = 4.96$

Solution: $3.1x = 4.96$

$\dfrac{3.1x}{3.1} = \dfrac{4.96}{3.1}$ Divide both sides by 3.1.

$1x = 1.6$ Simplify.

$x = 1.6$

Check: Check by replacing x with 1.6 in the original equation. The solution is 1.6.

Practice Problem 5

Solve: $2.6x = 13.52$

Example 6   Solve: $-\dfrac{2}{3}x = -\dfrac{5}{2}$

Solution: To get x alone, we multiply both sides of the equation by $-\dfrac{3}{2}$, the reciprocal of the coefficient of x.

$-\dfrac{2}{3}x = -\dfrac{5}{2}$

$-\dfrac{3}{2} \cdot -\dfrac{2}{3}x = -\dfrac{3}{2} \cdot -\dfrac{5}{2}$ Multiply both sides by $-\dfrac{3}{2}$, the reciprocal of $-\dfrac{2}{3}$.

$x = \dfrac{15}{4}$ Simplify.

Check: Check by replacing x with $\dfrac{15}{4}$ in the original equation. The solution is $\dfrac{15}{4}$.

Practice Problem 6

Solve: $-\dfrac{5}{6}y = -\dfrac{3}{5}$

B Using Both the Addition and Multiplication Properties

We are now ready to combine the skills learned in the last section with the skills learned from this section to solve equations by applying more than one property.

Example 7   Solve: $-z - 4 = 6$

Solution: First, to get $-z$, the term containing the variable alone, we add 4 to both sides of the equation.

$-z - 4 + 4 = 6 + 4$ Add 4 to both sides.

$-z = 10$ Simplify.

Next, recall that $-z$ means $-1 \cdot z$. Thus to get z alone, we either multiply or divide both sides of the equation by -1. In this example, we divide.

$-z = 10$

$\dfrac{-z}{-1} = \dfrac{10}{-1}$ Divide both sides by the coefficient -1.

$1z = -10$ Simplify.

$z = -10$

Practice Problem 7

Solve: $-x + 7 = -12$

Answers

5. $x = 5.2$, **6.** $y = \dfrac{18}{25}$, **7.** $x = 19$

Check:

$$-z - 4 = 6 \quad \text{Original equation.}$$
$$-(-10) - 4 \overset{?}{=} 6 \quad \text{Replace z with } -10.$$
$$10 - 4 \overset{?}{=} 6$$
$$6 = 6 \quad \text{True.}$$

The solution is -10.

Practice Problem 8

Solve: $-7x + 2x + 3 - 20 = -2$

Example 8   Solve: $a + a - 10 + 7 = -13$

Solution: First, we simplify both sides of the equation by combining like terms.

$$a + a - 10 + 7 = -13$$
$$2a - 3 = -13 \quad \text{Combine like terms.}$$
$$2a - 3 + 3 = -13 + 3 \quad \text{Add 3 to both sides.}$$
$$2a = -10 \quad \text{Simplify.}$$
$$\frac{2a}{2} = \frac{-10}{2} \quad \text{Divide both sides by 2.}$$
$$a = -5 \quad \text{Simplify.}$$

Check: To check, replace a with -5 in the original equation. The solution is -5.

C WRITING ALGEBRAIC EXPRESSIONS

We continue to sharpen our problem-solving skills by writing algebraic expressions.

Practice Problem 9

If x is the first of two consecutive integers, express the sum of the first and the second integer in terms of x. Simplify if possible.

Example 9   Writing an Expression for Consecutive Integers

If x is the first of three consecutive integers, express the sum of the three integers in terms of x. Simplify if possible.

Solution: An example of three consecutive integers is

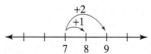

The second consecutive integer is always 1 more than the first, and the third consecutive integer is 2 more than the first. If x is the first of three consecutive integers, the three consecutive integers are

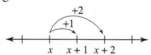

Their sum is

In words: | first integer | $+$ | second integer | $+$ | third integer |

Translate: $x \quad + \quad (x + 1) \quad + \quad (x + 2)$

which simplifies to $3x + 3$.

Study these examples of consecutive even and odd integers.

Even integers:

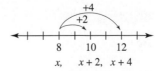

Odd integers:

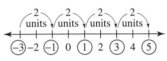

$$x, \quad x+2, \quad x+4$$

HELPFUL HINT

If x is an odd integer, then $x + 2$ is the next odd integer. This 2 simply means that odd integers are always 2 units from each other.

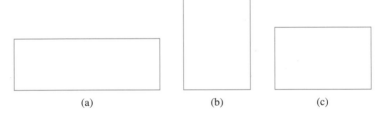

Focus On The Real World

SURVEYS

Recall that the golden rectangle is a rectangle whose length is approximately 1.6 times its width. It is thought that for about 75% of adults, a rectangle in the shape of the golden rectangle is most pleasing to the eye.

(a)   (b)   (c)

GROUP ACTIVITIES

1. Measure the dimensions of each of the three rectangles shown above and decide which one best approximates the shape of the golden rectangle.
2. Using the three rectangles shown above, conduct a survey asking students which rectangle they prefer. (To avoid bias, take care not to reveal which rectangle is the golden rectangle.) Tally your results and find the percent of survey respondents who preferred each rectangle. Do your results agree with the percent suggested above?

Name _____ Section _____ Date _____

MENTAL MATH

Solve each equation mentally. See Examples 2 and 3.

1. $3a = 27$ **2.** $9c = 54$ **3.** $5b = 10$
4. $7t = 14$ **5.** $6x = -30$ **6.** $8r = -64$

EXERCISE SET 2.3

A *Solve each equation. Check each solution. See Examples 1 through 6.*

📼 **1.** $-5x = 20$ **2.** $7x = 49$ **3.** $3x = 0$ **4.** $2x = 0$

5. $-x = -12$ **6.** $-y = 8$ 📼 **7.** $\frac{2}{3}x = -8$ **8.** $\frac{3}{4}n = -15$

9. $\frac{1}{6}d = \frac{1}{2}$ **10.** $\frac{1}{8}v = \frac{1}{4}$ **11.** $\frac{a}{2} = 1$ **12.** $\frac{d}{15} = 2$

13. $\frac{k}{-7} = 0$ **14.** $\frac{f}{-5} = 0$ **15.** $1.7x = 10.71$ **16.** $8.5y = 18.7$

17. $42 = 7x$ **18.** $81 = 3x$ **19.** $4.4 = -0.8x$ **20.** $6.3 = -0.6x$

21. $-\frac{3}{7}p = -2$ **22.** $-\frac{4}{5}r = -5$ **23.** $-\frac{4}{3}x = 12$ **24.** $-\frac{10}{3}x = 30$

B *Solve each equation. Check each solution. See Examples 7 and 8.*

25. $2x - 4 = 16$ **26.** $3x - 1 = 26$ **27.** $-x + 2 = 22$

28. $-x + 4 = -24$ 📼 **29.** $6a + 3 = 3$ **30.** $8t + 5 = 5$

31. $6x + 10 = -20$ **32.** $-10y + 15 = 5$ **33.** $5 - 0.3k = 5$

34. $2 + 0.4p = 2$ **35.** $-2x + \frac{1}{2} = \frac{7}{2}$ **36.** $-3n - \frac{1}{3} = \frac{8}{3}$

37. $\frac{x}{3} + 2 = -5$ **38.** $\frac{b}{4} - 1 = -7$ **39.** $10 = 2x - 1$

40. $j = \dfrac{16}{3}$

41. $z = 1$

42. $a = 2$

43. $x = -\dfrac{1}{4}$

44. $x = -\dfrac{7}{10}$

45. $x = -30$

46. $x = -50$

47. $z = \dfrac{9}{10}$

48. $t = \dfrac{13}{3}$

49. $2x + 2$

50. $4x + 12$

51. $2x + 2$

52. $x + 21$

53. $7x - 12$

54. $-8y - 3$

55. $12z + 44$

56. $-a - 3$

57. 1

58. $15z - 49$

59. $x = 2$

60. $y = -1.23$

61. answers may vary

62. answers may vary

63. answers may vary

64. answers may vary

116

Name _____

40. $12 = 3j - 4$

41. $6z - 8 - z + 3 = 0$

42. $4a + 1 + a - 11 = 0$

43. $10 - 3x - 6 - 9x = 7$

44. $12x + 30 + 8x - 6 = 10$

45. $1 = 0.4x - 0.6x - 5$

46. $19 = 0.4x - 0.9x - 6$

47. $z - 5z = 7z - 9 - z$

48. $t - 6t = -13 + t - 3t$

C *Write each algebraic expression described. Simplify if possible. See Example 9.*

49. If x represents the first of two consecutive odd integers, express the sum of the two integers in terms of x.

50. If x is the first of four consecutive even integers, write their sum as an algebraic expression in x.

51. If x is the first of three consecutive integers, express the sum of the first integer and the third integer as an algebraic expression containing the variable x.

52. If x is the first of two consecutive integers, express the sum of 20 and the second consecutive integer as an algebraic expression containing the variable x.

REVIEW AND PREVIEW

Simplify each expression. See Section 2.1.

53. $5x + 2(x - 6)$

54. $-7y + 2y - 3(y + 1)$

55. $6(2z + 4) + 20$

56. $-(3a - 3) + 2a - 6$

57. $-(x - 1) + x$

58. $8(z - 6) + 7z - 1$

COMBINING CONCEPTS

Solve.

59. $0.07x - 5.06 = -4.92$

60. $0.06y + 2.63 = 2.5562$

61. The equation $3x + 6 = 2x + 10 + x - 4$ is true for all real numbers. Substitute a few real numbers for x to see that this is so and then try solving the equation.

62. The equation $6x + 2 - 2x = 4x + 1$ has no solution. Try solving this equation for x and see what happens.

63. From the results of Exercises 61 and 62, when do you think an equation has all real numbers as its solutions?

64. From the results of Exercises 61 and 62, when do you think an equation has no solution?

2.4   FURTHER SOLVING LINEAR EQUATIONS

A   SOLVING LINEAR EQUATIONS

We now combine our knowledge from the previous sections into a general strategy for solving linear equations. One new piece in this strategy is a suggestion to "clear an equation of fractions" as a first step. Doing so makes the equation more manageable, since working with integers is more convenient than working with fractions.

TO SOLVE LINEAR EQUATIONS IN ONE VARIABLE

Step 1.   Multiply on both sides to clear the equation of fractions if they occur.

Step 2.   Use the distributive property to remove parentheses if they occur.

Step 3.   Simplify each side of the equation by combining like terms.

Step 4.   Get all variable terms on one side and all numbers on the other side by using the addition property of equality.

Step 5.   Get the variable alone by using the multiplication property of equality.

Step 6.   Check the solution by substituting it into the original equation.

Example 1   Solve: $4(2x - 3) + 7 = 3x + 5$

Solution: There are no fractions, so we begin with Step 2.

$$4(2x - 3) + 7 = 3x + 5$$

Step 2. $8x - 12 + 7 = 3x + 5$ Apply the distributive property.

Step 3. $\qquad 8x - 5 = 3x + 5$ Combine like terms.

Step 4. Get all variable terms on the same side of the equation by subtracting $3x$ from both sides; then adding 5 to both sides.

$$8x - 5 - 3x = 3x + 5 - 3x \quad \text{Subtract } 3x \text{ from both}$$
$$5x - 5 = 5 \qquad\qquad \text{sides.}$$
$$\text{Simplify.}$$
$$5x - 5 + 5 = 5 + 5 \qquad \text{Add 5 to both sides.}$$
$$5x = 10 \qquad\qquad \text{Simplify.}$$

Step 5. Use the multiplication property of equality to get x alone.

$$\frac{5x}{5} = \frac{10}{5} \quad \text{Divide both sides by 5.}$$

$$x = 2 \qquad \text{Simplify.}$$

Step 6. Check.

$$4(2x - 3) + 7 = 3x + 5 \qquad \text{Original equation}$$
$$4[2(2) - 3] + 7 \stackrel{?}{=} 3(2) + 5 \qquad \text{Replace } x \text{ with 2.}$$
$$4(4 - 3) + 7 \stackrel{?}{=} 6 + 5$$
$$4(1) + 7 \stackrel{?}{=} 11$$
$$4 + 7 \stackrel{?}{=} 11$$
$$11 = 11 \qquad \text{True.}$$

The solution is 2.

Objectives

A Apply the general strategy for solving a linear equation.

B Solve equations containing fractions or decimals.

C Recognize identities and equations with no solution.

SSM   CD-ROM   Video 2.4

TEACHING TIP

Remind students that even though the equations in this section may be more complicated than previous equations, the goal is still to get an equation that looks like "variable = number" or "number = variable."

Practice Problem 1

Solve: $5(3x - 1) + 2 = 12x + 6$

Answer

1. $x = 3$

> **HELPFUL HINT**
>
> When checking solutions, use the original written equation.

Practice Problem 2

Practice Problem 2

Solve: $9(5 - x) = -3x$

Example 2   Solve: $8(2 - t) = -5t$

Solution: First, we apply the distributive property.

$$\overset{\frown}{8(2 - t)} = -5t$$

Step 2. $16 - 8t = -5t$ Use the distributive property.

Step 4. $16 - 8t + 8t = -5t + 8t$ Add 8t to both sides.

$16 = 3t$ Combine like terms.

Step 5. $\dfrac{16}{3} = \dfrac{3t}{3}$ Divide both sides by 3.

$\dfrac{16}{3} = t$ Simplify.

Step 6.   Check.

$8(2 - t) = -5t$ Original equation

$8\left(2 - \dfrac{16}{3}\right) \overset{?}{=} -5\left(\dfrac{16}{3}\right)$ Replace t with $\dfrac{16}{3}$.

$8\left(\dfrac{6}{3} - \dfrac{16}{3}\right) \overset{?}{=} -\dfrac{80}{3}$ The LCD is 3.

$8\left(-\dfrac{10}{3}\right) \overset{?}{=} -\dfrac{80}{3}$ Subtract fractions.

$-\dfrac{80}{3} \overset{?}{=} -\dfrac{80}{3}$ True.

The solution is $\dfrac{16}{3}$.

B   SOLVING EQUATIONS CONTAINING FRACTIONS OR DECIMALS

If an equation contains fractions, we can clear the equation of fractions by multiplying both sides by the LCD of all denominators. By doing this, we avoid working with time-consuming fractions.

Practice Problem 3

Solve: $\dfrac{5}{2}x - 1 = \dfrac{3}{2}x - 4$

Example 3   Solve: $\dfrac{x}{2} - 1 = \dfrac{2}{3}x - 3$

Solution: We begin by clearing fractions. To do this, we multiply both sides of the equation by the LCD of 2 and 3, which is 6.

$$\dfrac{x}{2} - 1 = \dfrac{2}{3}x - 3$$

Step 1. $6\left(\dfrac{x}{2} - 1\right) = 6\left(\dfrac{2}{3}x - 3\right)$ Multiply both sides by the LCD, 6.

Answers

2. $x = \dfrac{15}{2}$,   **3.** $x = -3$

<table>
<tr><td>

HELPFUL HINT

Don't forget to multiply *each* term by the LCD.

</td><td>

Step 2. $6\left(\dfrac{x}{2}\right) - 6(1) = 6\left(\dfrac{2}{3}x\right) - 6(3)$ Apply the distributive property.

$3x - 6 = 4x - 18$ Simplify.

</td></tr>
</table>

There are no longer grouping symbols and no like terms on either side of the equation, so we continue with Step 4.

$$3x - 6 = 4x - 18$$

Step 4. $3x - 6 - 3x = 4x - 18 - 3x$ Subtract $3x$ from both sides.

$-6 = x - 18$ Simplify.

$-6 + 18 = x - 18 + 18$ Add 18 to both sides.

$12 = x$ Simplify.

Step 5. The variable is now alone, so there is no need to apply the multiplication property of equality.

Step 6. Check.

$$\frac{x}{2} - 1 = \frac{2}{3}x - 3 \qquad \text{Original equation}$$

$$\frac{12}{2} - 1 \stackrel{?}{=} \frac{2}{3} \cdot 12 - 3 \qquad \text{Replace } x \text{ with 12.}$$

$$6 - 1 \stackrel{?}{=} 8 - 3 \qquad \text{Simplify.}$$

$$5 = 5 \qquad \text{True.}$$

The solution is 12.

Example 4   Solve: $\dfrac{2(a + 3)}{3} = 6a + 2$

Solution: We clear the equation of fractions first.

$$\frac{2(a + 3)}{3} = 6a + 2$$

Step 1. $3 \cdot \dfrac{2(a + 3)}{3} = 3(6a + 2)$ Clear the fraction by multiplying both sides by the LCD, 3.

Step 2. Next, we use the distributive property and remove parentheses.

$2a + 6 = 18a + 6$ Apply the distributive property.

Step 4. $2a + 6 - 6 = 18a + 6 - 6$ Subtract 6 from both sides.

$2a = 18a$

$2a - 18a = 18a - 18a$ Subtract $18a$ from both sides.

$-16a = 0$

Step 5. $\dfrac{-16a}{-16} = \dfrac{0}{-16}$ Divide both sides by -16.

$a = 0$ Write the fraction in simplest form.

Step 6. To check, replace a with 0 in the original equation. The solution is 0.

Practice Problem 4

Solve: $\dfrac{3(x - 2)}{5} = 3x + 6$

Answer

4. $x = -3$

Practice Problem 5

Solve:
$$0.06x - 0.10(x - 2) = -0.02(8)$$

When solving a problem about money, you may need to solve an equation containing decimals. If you choose, you may multiply to clear the equation of decimals.

Example 5   Solve: $0.25x + 0.10(x - 3) = 0.05(22)$

Solution: First we clear this equation of decimals by multiplying both sides of the equation by 100. Recall that multiplying a decimal number by 100 has the effect of moving the decimal point 2 places to the right.

$$0.25x + 0.10(x - 3) = 0.05(22)$$

Step 1. $0.25x + 0.10(x - 3) = 0.05(22)$ Multiply both sides by 100.

$$25x + 10(x - 3) = 5(22)$$

Step 2. $25x + 10x - 30 = 110$ Apply the distributive property.

Step 3. $35x - 30 = 110$ Combine like terms.

Step 4. $35x - 30 + 30 = 110 + 30$ Add 30 to both sides.

$$35x = 140$$ Combine like terms.

Step 5. $\dfrac{35x}{35} = \dfrac{140}{35}$ Divide both sides by 35.

$$x = 4$$

Step 6. To check, replace x with 4 in the original equation. The solution is 4.

C   RECOGNIZING IDENTITIES AND EQUATIONS WITH NO SOLUTION

So far, each equation that we have solved has had a single solution. However, not every equation in one variable has a single solution. Some equations have no solution, while others have an infinite number of solutions. For example,

$$x + 5 = x + 7$$

has no solution since no matter which **real number** we replace x with, the equation is false.

$$\text{real number} + 5 = \text{same real number} + 7 \quad \text{FALSE}$$

On the other hand,

$$x + 6 = x + 6$$

has infinitely many solutions since x can be replaced by any real number and the equation is always true.

$$\text{real number} + 6 = \text{same real number} + 6 \quad \text{TRUE}$$

The equation $x + 6 = x + 6$ is called an **identity**. The next few examples illustrate special equations like these.

Answer

5. $x = 9$

Example 6   Solve: $-2(x - 5) + 10 = -3(x + 2) + x$

Solution:
$$-2(x - 5) + 10 = -3(x + 2) + x$$
$$-2x + 10 + 10 = -3x - 6 + x \quad \text{Apply the distributive property on both sides.}$$
$$-2x + 20 = -2x - 6 \quad \text{Combine like terms.}$$
$$-2x + 20 + 2x = -2x - 6 + 2x \quad \text{Add } 2x \text{ to both sides.}$$
$$20 = -6 \quad \text{Combine like terms.}$$

The final equation contains no variable terms, and there is no value for x that makes $20 = -6$ a true equation. We conclude that there is **no solution** to this equation.

Example 7   Solve: $3(x - 4) = 3x - 12$

Solution:
$$3(x - 4) = 3x - 12$$
$$3x - 12 = 3x - 12 \quad \text{Apply the distributive property.}$$

The left side of the equation is now identical to the right side. Every real number may be substituted for x and a true statement will result. We arrive at the same conclusion if we continue.

$$3x - 12 = 3x - 12$$
$$3x - 12 + 12 = 3x - 12 + 12 \quad \text{Add 12 to both sides.}$$
$$3x = 3x \quad \text{Combine like terms.}$$
$$3x - 3x = 3x - 3x \quad \text{Subtract } 3x \text{ from both sides.}$$
$$0 = 0$$

Again, one side of the equation is identical to the other side. Thus, $3(x - 4) = 3x - 12$ is an **identity** and **every real number** is a solution.

TRY THE CONCEPT CHECK IN THE MARGIN.

CALCULATOR EXPLORATIONS
CHECKING EQUATIONS

We can use a calculator to check possible solutions of equations. To do this, replace the variable by the possible solution and evaluate both sides of the equation separately.

Equation: $3x - 4 = 2(x + 6)$ Solution: $x = 16$

$3x - 4 = 2(x + 6)$ Original equation

$3(16) - 4 \stackrel{?}{=} 2(16 + 6)$ Replace x with 16.

Now evaluate each side with your calculator.

Evaluate left side: [3] [×] [16] [−] [4] [=]   Display: [44]
or
[ENTER]

Evaluate right side: [2] [(] [16] [+] [6] [)] [=] Display: [44]
or
[ENTER]

Since the left side equals the right side, the equation checks.

Use a calculator to check the possible solutions to each equation.

1. $2x = 48 + 6x$; $x = -12$ solution

2. $-3x - 7 = 3x - 1$; $x = -1$ solution

3. $5x - 2.6 = 2(x + 0.8)$; $x = 4.4$ not a solution

4. $-1.6x - 3.9 = -6.9x - 25.6$; $x = 5$ not a solution

5. $\dfrac{564x}{4} = 200x - 11(649)$; $x = 121$ solution

6. $20(x - 39) = 5x - 432$; $x = 23.2$ solution

Name _____ **Section** _____ **Date** _____

EXERCISE SET 2.4

A *Solve each equation. See Examples 1 and 2.*

1. $-4y + 10 = -2(3y + 1)$

2. $-3x + 1 = -2(x - 2)$

3. $9x - 8 = 10 + 15x$

4. $15x - 5 = 7 + 12x$

5. $-2(3x - 4) = 2x$

6. $-(5x - 10) = 5x$

7. $4(2n - 1) = (6n + 4) + 1$

8. $4(4y + 2) = 2(1 + 6y) + 8$

9. $5(2x - 1) - 2(3x) = 1$

10. $3(2 - 5x) + 4(6x) = 12$

11. $6(x - 3) + 10 = -8$

12. $-4(2 + n) + 9 = 1$

13. $8 - 2(a - 1) = 7 + a$

14. $5 - 6(2 + b) = b - 14$

15. $4x + 3 = 2x + 11$

16. $6y - 8 = 3y + 7$

17. $-2y - 10 = 5y + 18$

18. $7n + 5 = 10n - 10$

19. $-3(t - 5) + 2t = 5t - 4$

20. $-(4a - 7) - 5a = 10 + a$

ANSWERS

1. $y = -6$

2. $x = -3$

3. $x = -3$

4. $x = 4$

5. $x = 1$

6. $x = 1$

7. $n = \dfrac{9}{2}$

8. $y = \dfrac{1}{2}$

9. $x = \dfrac{3}{2}$

10. $x = \dfrac{2}{3}$

11. $x = 0$

12. $n = 0$

13. $a = 1$

14. $b = 1$

15. $x = 4$

16. $y = 5$

17. $y = -4$

18. $n = 5$

19. $t = \dfrac{19}{6}$

20. $a = -\dfrac{3}{10}$

21. $x = 2$

22. $x = 0$

23. $x = -5$

24. $x = -2$

25. $x = 10$

26. $x = -15$

27. $z = 18$

28. $w = 20$

29. $x = 3$

30. $x = -1$

31. $x = 13$

32. $x = -11$

33. $x = 50$

34. $x = 20$

35. $y = 0.2$

36. $z = 500$

37. $x = 1$

38. $y = -3$

39. $x = \dfrac{7}{3}$

40. $x = \dfrac{5}{6}$

Name _____

B *Solve each equation. See Examples 3 through 5.*

21. $\dfrac{3}{4}x - \dfrac{1}{2} = 1$

22. $\dfrac{2}{3}x + \dfrac{5}{3} = \dfrac{5}{3}$

23. $x + \dfrac{5}{4} = \dfrac{3}{4}x$

24. $\dfrac{7}{8}x + \dfrac{1}{4} = \dfrac{3}{4}x$

25. $\dfrac{x}{2} - 1 = \dfrac{x}{5} + 2$

26. $\dfrac{x}{5} - 2 = \dfrac{x}{3}$

27. $\dfrac{6(3 - z)}{5} = -z$

28. $\dfrac{4(5 - w)}{3} = -w$

29. $0.06 - 0.01(x + 1) = -0.02(2 - x)$

30. $-0.01(5x + 4) = 0.04 - 0.01(x + 4)$

31. $\dfrac{3(x - 5)}{2} = \dfrac{2(x + 5)}{3}$

32. $\dfrac{5(x - 1)}{4} = \dfrac{3(x + 1)}{2}$

33. $0.50x + 0.15(70) = 0.25(142)$

34. $0.40x + 0.06(30) = 0.20(49)$

35. $0.12(y - 6) + 0.06y = 0.08y - 0.07(10)$

36. $0.60(z - 300) + 0.05z = 0.70z - 0.41(500)$

37. $\dfrac{2(x + 1)}{4} = 3x - 2$

38. $\dfrac{3(y + 3)}{5} = 2y + 6$

39. $x + \dfrac{7}{6} = 2x - \dfrac{7}{6}$

40. $\dfrac{5}{2}x - 1 = x + \dfrac{1}{4}$

Name _____

41. Explain the difference between simplifying an expression and solving an equation.

42. When solving an equation, if an equivalent equation is $0 = 5$, what can we conclude? If an equivalent equation is $-2 = -2$, what can we conclude?

C *Solve each equation. See Examples 6 and 7.*

43. $5x - 5 = 2(x + 1) + 3x - 7$

44. $3(2x - 1) + 5 = 6x + 2$

45. $\dfrac{x}{4} + 1 = \dfrac{x}{4}$

46. $\dfrac{x}{3} - 2 = \dfrac{x}{3}$

47. $3x - 7 = 3(x + 1)$

48. $2(x - 5) = 2x + 10$

49. $2(x + 3) - 5 = 5x - 3(1 + x)$

50. $4(2 + x) + 1 = 7x - 3(x - 2)$

51. On your own, construct an equation for which every real number is a solution.

52. On your own, construct an equation that has no solution.

REVIEW AND PREVIEW

Write each algebraic expression described. See Section 2.1.

53. A plot of land is in the shape of a triangle. If one side is x meters, a second side is $(2x - 3)$ meters, and a third side is $(3x - 5)$ meters, express the perimeter of the lot as a simplified expression in x.

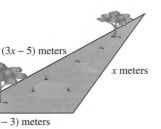

(3x − 5) meters

x meters

(2x − 3) meters

54. A portion of a board has length x feet. The other part has length $(7x - 9)$ feet. Express the total length of the board as a simplified expression in x.

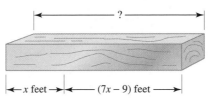

├← x feet →├← (7x − 9) feet →┤

55. $-8 - x$

56. $3x$

57. $-3 + 2x$

58. $8 - 2x$

59. $9(x + 20)$

60. $\dfrac{-12}{x - 3}$

61. $x = 15.3$

62. $x = -128$

63. $x = -0.2$

64. $x = 3.05$

65. $x = 4$ cm;
$2x = 8$ cm

66. $x = 6$ meters;
$2x + 1 = 13$ meters;
$3x - 2 = 16$ meters

Name _____

Write each phrase as an algebraic expression. Use x for the unknown number.

55. A number subtracted from -8

56. Three times a number

57. The sum of -3 and twice a number

58. The difference of 8 and twice a number

59. The product of 9 and the sum of a number and 20

60. The quotient of -12 and the difference of a number and 3

◤ COMBINING CONCEPTS

Solve.

▤ **61.** $1000(7x - 10) = 50(412 + 100x)$

▤ **62.** $1000(x + 40) = 100(16 + 7x)$

▤ **63.** $0.035x + 5.112 = 0.010x + 5.107$

▤ **64.** $0.127x - 2.685 = 0.027x - 2.38$

65. The perimeter of a geometric figure is the sum of the lengths of its sides. If the perimeter of the following pentagon (five-sided figure) is 28 centimeters, find the length of each side.

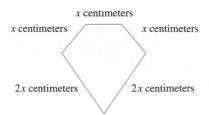

x centimeters
x centimeters
x centimeters
$2x$ centimeters
$2x$ centimeters

66. The perimeter of the following triangle is 35 meters. Find the length of each side.

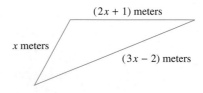

$(2x + 1)$ meters
x meters
$(3x - 2)$ meters

Name _____ **Section** _____ **Date** _____

INTEGRATED REVIEW—SOLVING LINEAR EQUATIONS

Solve:

1. $x - 10 = -4$ **2.** $y + 14 = -3$ **3.** $9y = 108$

4. $-3x = 78$ **5.** $-6x + 7 = 25$ **6.** $5y - 42 = -47$

7. $\frac{2}{3}x = 9$ **8.** $\frac{4}{5}z = 10$ **9.** $\frac{r}{-4} = -2$

10. $\frac{y}{-8} = 8$ **11.** $6 - 2x + 8 = 10$ **12.** $-5 - 6y + 6 = 19$

13. $2x - 7 = 6x - 27$ **14.** $3 + 8y = 3y - 2$

15. $-3a + 6 + 5a = 7a - 8a$ **16.** $4b - 8 - b = 10b - 3b$

17. $-\frac{2}{3}x = \frac{5}{9}$ **18.** $-\frac{3}{8}y = -\frac{1}{16}$

1. $x = 6$

2. $y = -17$

3. $y = 12$

4. $x = -26$

5. $x = -3$

6. $y = -1$

7. $x = 13.5$

8. $z = 12.5$

9. $r = 8$

10. $y = -64$

11. $x = 2$

12. $y = -3$

13. $x = 5$

14. $y = -1$

15. $a = -2$

16. $b = -2$

17. $x = -\frac{5}{6}$

18. $y = \frac{1}{6}$

19. $n = 1$

20. $m = 6$

21. $c = 4$

22. $t = 1$

23. $z = \dfrac{9}{5}$

24. $w = -\dfrac{6}{5}$

25. all real numbers

26. all real numbers

27. $t = 0$

28. $m = -1.6$

19. $10 = -6n + 16$

20. $-5 = -2m + 7$

21. $3(5c - 1) - 2 = 13c + 3$

22. $4(3t + 4) - 20 = 3 + 5t$

23. $\dfrac{2(z + 3)}{3} = 5 - z$

24. $\dfrac{3(w + 2)}{4} = 2w + 3$

25. $-2(2x - 5) = -3x + 7 - x + 3$

26. $-4(5x - 2) = -12x + 4 - 8x + 4$

27. $0.02(6t - 3) = 0.04(t - 2) + 0.02$

28. $0.03(m + 7) = 0.02(5 - m) + 0.03$

2.5   AN INTRODUCTION TO PROBLEM SOLVING

In the preceding sections, we practiced translating phrases into expressions and sentences into equations as well as solving linear equations. We are now ready to put our skills to practical use. To begin, we present a general strategy for problem solving.

Objective

A Translate a problem to an equation, then use the equation to solve the problem.

SSM   CD-ROM   Video 2.5

GENERAL STRATEGY FOR PROBLEM SOLVING

1. UNDERSTAND the problem. During this step, become comfortable with the problem. Some ways of doing this are:

  Read and reread the problem.
  Choose a variable to represent the unknown.
  Construct a drawing.
  Propose a solution and check. Pay careful attention to how you check your proposed solution. This will help when writing an equation to model the problem.

2. TRANSLATE the problem into an equation.
3. SOLVE the equation.
4. INTERPRET the results: *Check* the proposed solution in the stated problem and *state* your conclusion.

A   TRANSLATING AND SOLVING PROBLEMS

Much of problem solving involves a direct translation from a sentence to an equation.

Example 1   Finding an Unknown Number

Twice the sum of a number and 4 is the same as four times the number decreased by 12. Find the number.

Solution: **1.** UNDERSTAND. Read and reread the problem. If we let

$$x = \text{ the unknown number, then}$$

"the sum of a number and 4" translates to "$x + 4$" and

"four times the number" translates to "$4x$"

2. TRANSLATE.

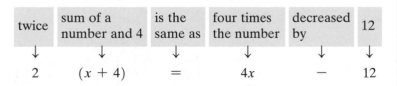

twice	sum of a number and 4	is the same as	four times the number	decreased by	12
↓	↓	↓	↓	↓	↓
2	$(x + 4)$	=	$4x$	−	12

Practice Problem 1

Three times the difference of a number and 5 is the same as twice the number decreased by 3. Find the number.

Answer

1. The number is 12.

3. SOLVE.

$$2(x + 4) = 4x - 12$$
$$2x + 8 = 4x - 12 \qquad \text{Apply the distributive property.}$$
$$2x + 8 - 4x = 4x - 12 - 4x \qquad \text{Subtract } 4x \text{ from both sides.}$$
$$-2x + 8 = -12$$
$$-2x + 8 - 8 = -12 - 8 \qquad \text{Subtract 8 from both sides.}$$
$$-2x = -20$$
$$\frac{2x}{-2} = \frac{-20}{-2} \qquad \text{Divide both sides by } -2.$$
$$x = 10$$

4. INTERPRET.

Check: Check this solution in the problem as it was originally stated. To do so, replace "number" with 10. Twice the sum of "10" and 4 is 28, which is the same as 4 times "10" decreased by 12.

State: The number is 10.   ▬▬▬

Practice Problem 2

An 18-foot wire is to be cut so that the longer piece is 5 times longer than the shorter piece. Find the length of each piece.

TEACHING TIP

You may want to show students how to use a chart to help them set up the correct equation for applications. For Example 2, have students propose a few lengths for the short piece and complete the chart for each proposal. After students see a pattern, have them use x for the short piece and complete the chart.

Length of Short Piece	Length of Long Piece	Sum of Lengths
3	$4 \cdot 3$	$3 + 4 \cdot 3 = 3 + 12 = 15$
x	$4x$	$x + 4x = 10$

Example 2   Finding the Length of a Board

A 10-foot board is to be cut into two pieces so that the longer piece is 4 times the shorter. Find the length of each piece.

Solution: **1.** UNDERSTAND the problem. To do so, read and re-read the problem. You may also want to propose a solution. For example, if 3 feet represents the length of the shorter piece, then $4(3) = 12$ feet is the length of the longer piece, since it is 4 times the length of the shorter piece. This guess gives a total board length of 3 feet $+$ 12 feet $=$ 15 feet, too long. However, the purpose of proposing a solution is not to guess correctly, but to help better understand the problem and how to model it.

In general, if we let

x = length of shorter piece, then

$4x$ = length of longer piece

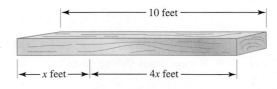

2. TRANSLATE the problem. First, we write the equation in words.

length of shorter piece	added to	length of longer piece	equals	total length of board
↓	↓	↓	↓	↓
x	$+$	$4x$	$=$	10

Answer

2. shorter piece $=$ 3 ft; longer piece $=$ 15 ft

3. SOLVE.

$$x + 4x = 10$$

$$5x = 10 \qquad \text{Combine like terms.}$$

$$\frac{5x}{5} = \frac{10}{5} \qquad \text{Divide both sides by 5.}$$

$$x = 2$$

4. INTERPRET.

Check: Check the solution in the stated problem. If the shorter piece of board is 2 feet, the longer piece is $4 \cdot (2 \text{ feet}) = 8 \text{ feet}$ and the sum of the two pieces is $2 \text{ feet} + 8 \text{ feet} = 10 \text{ feet}$.

State: The shorter piece of board is 2 feet and the longer piece of board is 8 feet.

> **HELPFUL HINT:**
>
> Make sure that units are included in your answer, if appropriate.

Example 3   Finding the Number of Republican and Democratic Senators

In a recent year, Congress had 8 more Republican senators than Democratic. If the total number of senators is 100, how many senators of each party were there?

Solution:

1. UNDERSTAND the problem. Read and reread the problem. Let's suppose that there are 40 Democratic senators. Since there are 8 more Republicans than Democrats, there must be $40 + 8 = 48$ Republicans. The total number of Democrats and Republicans is then $40 + 48 = 88$. This is incorrect since the total should be 100, but we now have a better understanding of the problem.

In general, if we let

$$x = \text{number of Democrats, then}$$

$$x + 8 = \text{number of Republicans}$$

2. TRANSLATE the problem. First, we write the equation in words.

number of Democrats	added to	number of Republicans	equals	100
↓	↓	↓	↓	↓
x	$+$	$(x + 8)$	$=$	100

Practice Problem 3

In a recent year, the total number of Democrats and Republicans in the U.S. House of Representatives was 433. There were 39 more Republicans than Democrats. Find the number of representatives from each party.

Answer

3. Democrats = 197 representatives; Republicans = 236 representatives

3. SOLVE.

$$x + (x + 8) = 100$$

$$2x + 8 = 100 \quad \text{Combine like terms.}$$

$$2x + 8 - 8 = 100 - 8 \quad \text{Subtract 8 from both sides.}$$

$$2x = 92$$

$$\frac{2x}{2} = \frac{92}{2} \quad \text{Divide both sides by 2.}$$

$$x = 46$$

4. INTERPRET.

Check: If there are 46 Democratic senators, then there are $46 + 8 = 54$ Republican senators. The total number of senators is then $46 + 54 = 100$. The results check.

State: There were 46 Democratic and 54 Republican senators.

Practice Problem 4

Enterprise Car Rental charges a daily rate of $34 plus $0.20 per mile. Suppose that you rent a car for a day and your bill (before taxes) is $104. How many miles did you drive?

Example 4   ## Calculating Cellular Phone Usage

A local cellular phone company charges Elaine Chapoton $50 per month and $0.36 per minute of phone use in her usage category. If Elaine was charged $99.68 for a month's cellular phone use, determine the number of whole minutes of phone use.

Solution: **1. UNDERSTAND. Read and reread the problem.** Let's propose that Elaine uses the phone for 70 minutes. Pay careful attention as to how we calculate her bill. For 70 minutes of use, Elaine's phone bill will be $50 plus $0.36 per minute of use. This is $50 + 0.36(70) = 75.20, less than $99.68. We now understand the problem and know that the number of minutes is greater than 70.

If we let

$$x = \text{number of minutes, then}$$
$$0.36x = \text{charge per minute of phone use}$$

2. TRANSLATE.

$50	added to	minute charge	is equal to	$99.68
↓	↓	↓	↓	↓
50	+	0.36x	=	99.68

3. SOLVE.

$$50 + 0.36x = 99.68$$
$$50 + 0.36x - 50 = 99.68 - 50 \quad \text{Subtract 50 from both sides.}$$
$$0.36x = 49.68 \quad \text{Simplify.}$$
$$\frac{0.36x}{0.36} = \frac{49.68}{0.36} \quad \text{Divide both sides by 0.36.}$$
$$x = 138 \quad \text{Simplify.}$$

4. INTERPRET.

Check: If Elaine spends 138 minutes on her cellular phone, her bill is $50 + $0.36(138) = $99.68.

State: Elaine spent 138 minutes on her cellular phone this month.

Example 5   Finding Angle Measures

If the two walls of the Vietnam Veterans Memorial in Washington D.C. were connected, an isosceles triangle would be formed. The measure of the third angle is 97.5° more than the measure of either of the other two equal angles. Find the measure of the third angle. (*Source:* National Park Service)

Solution: **1.** UNDERSTAND. Read and reread the problem. We then draw a diagram (recall that an isosceles triangle has two angles with the same measure) and let

x = degree measure of one angle

x = degree measure of the second equal angle

$x + 97.5$ = degree measure of the third angle

Practice Problem 5

The measure of the second angle of a triangle is twice the measure of the smallest angle. The measure of the third angle of the triangle is three times the measure of the smallest angle. Find the measures of the angles.

Answer

5. smallest = 30°;   second = 60°;   third = 90°

2. TRANSLATE. Recall that the sum of the measures of the angles of a triangle equals 180.

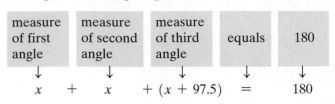

measure of first angle		measure of second angle		measure of third angle		equals		180
↓		↓		↓		↓		↓
x	$+$	x	$+$	$(x + 97.5)$	$=$			180

3. SOLVE.

$$x + x + (x + 97.5) = 180$$
$$3x + 97.5 = 180 \qquad \text{Combine like terms.}$$
$$3x + 97.5 - 97.5 = 180 - 97.5 \qquad \text{Subtract 97.5 from both sides.}$$
$$3x = 82.5$$
$$\frac{3x}{3} = \frac{82.5}{3} \qquad \text{Divide both sides by 3.}$$
$$x = 27.5$$

4. INTERPRET.

Check: If $x = 27.5$, then the measure of the third angle is $x + 97.5 = 125$. The sum of the angles is then $27.5 + 27.5 + 125 = 180$, the correct sum.

State: The third angle measures 125°.*

(*This is rounded to the nearest whole degree. The two walls actually meet at an angle of 125 degrees 12 minutes.)

Name _____ Section _____ Date _____

EXERCISE SET 2.5

A *Solve. See Example 1.*

1. The sum of twice a number and $\frac{1}{5}$ is equal to the difference between three times the number and $\frac{4}{5}$. Find the number.

2. The sum of four times a number and $\frac{2}{3}$ is equal to the difference of five times the number and $\frac{5}{6}$. Find the number.

3. Twice the difference of a number and 8 is equal to three times the sum of the number and 3. Find the number.

4. Five times the sum of a number and -1 is the same as 6 times the number. Find the number.

5. The product of twice a number and three is the same as the difference of five times the number and $\frac{3}{4}$. Find the number.

6. If the difference of a number and four is doubled, the result is $\frac{1}{4}$ less than the number. Find the number.

7. If the sum of a number and five is tripled, the result is one less than twice the number. Find the number.

8. Twice the sum of a number and six equals three times the sum of the number and four. Find the number.

Solve. See Examples 2 through 5.

9. The governor of New York makes twice as much money as the governor of Nebraska. If the total of their salaries is $195,000, find the salary of each.

10. In the 1996 Summer Olympics, the United States Team won 14 more gold medals than the German Team. If the total number of gold medals for both is 64, find the number of gold medals that each team won. (*Source: World Almanac*)

11. A 40-inch board is to be cut into three pieces so that the second piece is twice as long as the first piece and the third piece is 5 times as long as the first piece. If x represents the length of the first piece, find the lengths of all three pieces.

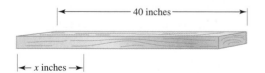

12. A 21-foot beam is to be divided so that the longer piece is 1 foot more than 3 times the shorter piece. If x represents the length of the shorter piece, find the lengths of both pieces.

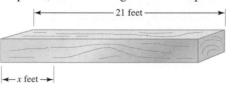

13. 172 miles

14. 12 hours

15. 1st angle: 37.5°;
2nd angle: 37.5°;
3rd angle: 105°

16. 1st angle: 65°;
2nd angle: 115°

17. Little: 46,895 votes;
Brown: 63,845 votes

18. Merritt: 94,107 votes;
Sandlin: 103,924
votes

Name _____

13. A car rental agency advertised renting a Buick Century for $24.95 per day and $0.29 per mile. If you rent this car for 2 days, how many whole miles can you drive on a $100 budget?

14. A plumber gave an estimate for the renovation of a kitchen. Her hourly pay is $27 per hour and the plumber's parts will cost $80. If her total estimate is $404, how many hours does she expect this job to take?

15. The flag of Equatorial Guinea contains an isosceles triangle. (Recall that an isosceles triangle contains two angles with the same measure.) If the measure of the third angle of the triangle is 30° more than twice the measure of either of the other two angles, find the measure of each angle of the triangle. (*Hint*: Recall that the sum of the measures of the angles of a triangle is 180°.)

16. The flag of Brazil contains a parallelogram. One angle of the parallelogram is 15° less than twice the measure of the angle next to it. Find the measure of each angle of the parallelogram. (*Hint*: Recall that opposite angles of a parallelogram have the same measure and that the sum of the measures of the angles is 360°.)

17. In a recent election in Florida for a seat in the United States House of Representatives, Corrine Brown received 16,950 more votes than Marc Little. If the total number of votes was 110,740, find the number of votes for each candidate.

18. In a recent election in Texas for a seat in the United States House of Representatives, Max Sandlin received 9817 more votes than opponent Ed Merritt. If the total number of votes was 198,031, find the number of votes for each candidate. (*Source*: Voter News Service)

19. Two angles are supplementary if their sum is 180°. One angle measures three times the measure of a smaller angle. If *x* represents the measure of the smaller angle and these two angles are supplementary, find the measure of each angle.

20. Two angles are complementary if their sum is 90°. Given the measures of the complementary angles shown, find the measure of each angle.

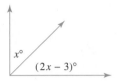

$x°$
$(2x - 3)°$

21. A 17-foot piece of string is cut into two pieces so that one piece is 2 feet longer than twice the shorter piece. If the shorter piece is *x* feet long, find the lengths of both pieces.

22. A woman's $15,000 estate is to be divided so that her husband receives twice as much as her son. If *x* represents the amount of money that her son receives, find the amount of money that her husband receives and the amount of money that her son receives.

23. On December 7, 1995, a probe launched from the robot explorer called Galileo entered the atmosphere of Jupiter at 100,000 miles per hour. The diameter of the probe is 19 inches less than twice its height. If the sum of the height and the diameter is 83 inches, find each dimension.

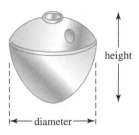

height

diameter

24. Over the past few years the satellite Voyager II has passed by the planets Saturn, Uranus, and Neptune, continually updating information about these planets, including the number of moons for each. Uranus is now believed to have 7 more moons than Neptune. Also, Saturn is now believed to have 2 more than twice the number of moons of Neptune. If the total number of moons for these planets is 41, find the number of moons for each planet. (*Source*: National Space Science Data Center)

25. Midway _____

26. 37 _____

27. 145 _____

28. 192 _____

29. $\frac{1}{2}(x - 1) = 37$ _____

30. $5(-x) = x + 60$ _____

31. $\dfrac{3(x + 2)}{5} = 0$ _____

32. $50 - (x + 9) = 0$ _____

33. 34 _____

34. 154 _____

35. 225π _____

36. 30 _____

138

The graph below is called a three-dimensional bar graph. It shows the five most common names of cities, towns, or villages in the United States. Use this graph to answer Exercises 25–28.

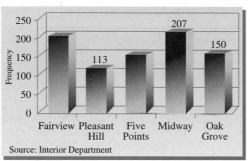

Source: Interior Department

25. What is the most popular name of a city, town, or village in the United States?

26. How many more cities, towns, or villages are named Oak Grove than are named Pleasant Hill?

27. Let *x* represent "the number of towns, cities, or villages named Five Points" and use the information given to determine the unknown number. "The number of towns, cities, or villages named Five Points" added to 55 is equal to twice "the number of towns, cities, or villages named Five Points" minus 90. Check your answer by noticing the height of the bar representing Five Points. Is your answer reasonable?

28. Let *x* represent "the number of towns, cities, or villages named Fairview" and use the information given to determine the unknown number. Three times "the number of towns, cities, or villages named Fairview" added to 24 is equal to 168 subtracted from 4 times "the number of towns, cities, or villages named Fairview." Check your answer by noticing the height of the bar representing Fairview. Is your answer reasonable?

REVIEW AND PREVIEW

Translate each sentence into an equation. See Sections 1.2 and 2.5.

29. Half of the difference of a number and one is thirty-seven.

30. Five times the opposite of a number is the number plus sixty.

31. If three times the sum of a number and 2 is divided by 5, the quotient is 0.

32. If the sum of a number and 9 is subtracted from 50, the result is 0.

Evaluate each expression for the given values. See Section 1.2.

33. $2W + 2L$; $W = 7$ and $L = 10$

34. $\frac{1}{2}Bh$; $B = 14$ and $h = 22$

35. πr^2; $r = 15$

36. $r \cdot t$ $r = 15$ and $t = 2$

COMBINING CONCEPTS

Solve. Recall that examples of consecutive integers are 7, 8, 9, 10, and so on. They can be represented by x, x + 1, x + 2, and so on.
↑
integer

Examples of consecutive even *integers are 20, 22, 24, 26, and so on. They can be represented by x, x + 2, x + 4, and so on.*
↑
even integer

Examples of consecutive odd *integers are −19, −17, −15, −13, and so on. They can be represented by x, x + 2, x + 4, and so on.*
↑
odd integer

37. Find two consecutive odd integers such that twice the larger is 15 more than three times the smaller.

38. Find three consecutive even integers whose sum is negative 114.

39. On June 11, 1997, Luc Longely's slam-dunk for the Chicago Bulls with 6.2 seconds remaining won Game 5 of the NBA Finals. The opposing team was the Utah Jazz. The two teams' final scores for the game were two consecutive even integers whose sum was 178. (*Source*: The National Basketball Associaton) Find each final score. (The Bulls went on to win the 1997 championship in Game 6, their fifth NBA Championship in seven years.)

40. On June 20, 1994, John Paxson sank a 3-point shot with 3.9 seconds left to give the Chicago Bulls their third straight National Basketball Associaton championship. The opposing team was the Phoenix Suns. If the final score of the game was 2 consecutive integers whose sum is 197, find each final score.

41. The sum of three consecutive integers is 13 more than twice the smallest integer. Find the integers.

42. The sum of two consecutive integers is 31. What are the numbers?

37. −11, −9

38. −40, −38, −36

39. Bulls: 90; Jazz: 88

40. Suns: 98; Bulls: 99

41. 10, 11, 12

42. 15, 16

139

43. To make an international telephone call, you need the code for the country you are calling. The codes for Mali Republic, Côte d'Ivoire, and Niger are three consecutive odd integers whose sum is 675. Find the code for each country. (*Source*: NYNEX)

44. The measures of the angles of a triangle are 3 consecutive even integers. Find the measure of each angle.

45. Determine whether there are two consecutive odd integers such that 7 times the first exceeds 5 times the second by 54.

46. Give an example of how you recently solved a problem using mathematics.

47. In your own words, explain why a solution of a word problem should be checked using the original wording of the problem and not the equation written from the wording.

Mali Republic: 223;
Côte d'Ivoire: 225;
43. Niger: 227
_____

44. 58°, 60°, 62°
_____

45. There are none.
_____

46. answers may vary
_____

47. answers may vary

140

2.6   FORMULAS AND PROBLEM SOLVING

A   USING FORMULAS TO SOLVE PROBLEMS

A **formula** describes a known relationship among quantities. Many formulas are given as equations. For example, the formula

$$d = r \cdot t$$

stands for the relationship

distance = rate · time

Let's look at one way that we can use this formula.

If we know we traveled a distance of 100 miles at a rate of 40 miles per hour, we can replace the variables d and r in the formula $d = rt$ and find our time, t.

$d = rt$ Formula.

$100 = 40t$ Replace d with 100 and r with 40.

To solve for t, we divide both sides of the equation by 40.

$\dfrac{100}{40} = \dfrac{40t}{40}$ Divide both sides by 40.

$\dfrac{5}{2} = t$ Simplify.

The time traveled is $\frac{5}{2}$ hours, or $2\frac{1}{2}$ hours.

In this section, we solve problems that can be modeled by known formulas. We use the same problem-solving strategy that was introduced in the previous section.

Example 1   Finding Time Given Rate and Distance

A glacier is a giant mass of rocks and ice that flows downhill like a river. Portage Glacier in Alaska is about 6 miles, or 31,680 *feet*, long and moves 400 *feet* per year. Icebergs are created when the front end of the glacier flows into Portage Lake. How long does it take for ice at the head (beginning) of the glacier to reach the lake?

Objectives

A Use formulas to solve problems.
B Solve a formula or equation for one of its variables.

SSM   CD-ROM   Video
2.6

TEACHING TIP

Remind students to pay attention to the units when working with formulas and to express the answer with the appropriate units.

Practice Problem 1

A family is planning their vacation to Disney World. They will drive from a small town outside New Orleans, Louisiana, to Orlando, Florida, a distance of 700 miles. They plan to average a rate of 55 miles per hour. How long will this trip take?

Answer

1. approximately $12\frac{8}{11}$ hours

Solution: **1.** UNDERSTAND. Read and reread the problem. The appropriate formula needed to solve this problem is the distance formula, $d = rt$. To become familiar with this formula, let's find the distance that ice traveling at a rate of 400 feet per year travels in 100 years. To do so, we let time t be 100 years and rate r be the given 400 feet per year, and substitute these values into the formula $d = rt$. We then have that distance $d = 400(100) = 40,000$ feet. Since we are interested in finding how long it takes ice to travel 31,680 feet, we now know that it is less than 100 years.

Since we are using the formula $d = rt$, we let

$t =$ the time in years for ice to reach the lake

$r =$ rate or speed of ice

$d =$ distance from beginning of glacier to lake

2. TRANSLATE. To translate to an equation, we use the formula $d = rt$ and let distance $d = 31,680$ feet and rate $r = 400$ feet per year.

$$d = r \cdot t$$
$$31,680 = 400 \cdot t \qquad \text{Let } d = 31,680 \text{ and } r = 400.$$

3. SOLVE. Solve the equation for t. To solve for t, divide both sides by 400.

$$\frac{31,680}{400} = \frac{400 \cdot t}{400} \qquad \text{Divide both sides by 400.}$$
$$79.2 = t \qquad \text{Simplify.}$$

4. INTERPRET.

Check: To check, substitute 79.2 for t and 400 for r in the distance formula and check to see that the distance is 31,680 feet.

State: It takes 79.2 years for the ice at the head of Portage Glacier to reach the lake.

HELPFUL HINT

Don't forget to include units, if appropriate.

Practice Problem 2

A wood deck is being built behind a house. The width of the deck must be 18 feet because of the shape of the house. If there is 450 square feet of decking material, find the length of the deck.

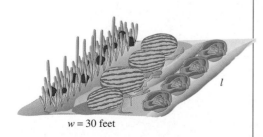

$w = 30$ feet

Example 2   Calculating the Length of a Garden

Charles Pecot can afford enough fencing to enclose a rectangular garden with a perimeter of 140 feet. If the width of his garden is to be 30 feet, find the length.

Solution: **1.** UNDERSTAND. Read and reread the problem. The formula needed to solve this problem is the formula for the perimeter of a rectangle, $P = 2l + 2w$. Before continuing, let's become familar with this formula.

$l =$ the length of the rectangular garden

$w =$ the width of the rectangular garden

$P =$ perimeter of the garden

2. TRANSLATE. To translate to an equation, we use the formula $P = 2l + 2w$ and let perimeter $P = 140$ feet and width $w = 30$ feet.

$$P = 2l + 2w \qquad \text{Let } P = 140 \text{ and } w = 30.$$
$$140 = 2l + 2(30)$$

3. SOLVE.

$$140 = 2l + 2(30)$$
$$140 = 2l + 60 \qquad \text{Multiply } 2(30).$$
$$140 - 60 = 2l + 60 - 60 \qquad \text{Subtract 60 from both sides.}$$
$$80 = 2l \qquad \text{Combine like terms.}$$
$$40 = l \qquad \text{Divide both sides by 2.}$$

4. INTERPRET.

Check: Substitute 40 for l and 30 for w in the perimeter formula and check to see that the perimeter is 140 feet.

State: The length of the rectangular garden is 40 feet. ▬▬▬

B SOLVING A FORMULA FOR A VARIABLE

We say that the formula

$$d = rt$$

is solved for d because d is alone on one side of the equation and the other side contains no d's. Suppose that we have a large number of problems to solve where we are given distance d and rate r and asked to find time t. In this case, it may be easier to first solve the formula $d = rt$ for t. To solve for t, we divide both sides of the equation by r.

$$d = rt$$
$$\frac{d}{r} = \frac{rt}{r} \qquad \text{Divide both sides by } r.$$
$$\frac{d}{r} = t \qquad \text{Simplify.}$$

To solve a formula or an equation for a specified variable, we use the same steps as for solving a linear equation. These steps are listed next.

TO SOLVE EQUATIONS FOR A SPECIFIED VARIABLE

Step 1. Multiply on both sides to clear the equation of fractions if they occur.

Step 2. Use the distributive property to remove parentheses if they occur.

Step 3. Simplify each side of the equation by combining like terms.

Step 4. Get all terms containing the specified variable on one side and all other terms on the other side by using the addition property of equality.

Step 5. Get the specified variable alone by using the multiplication property of equality.

Practice Problem 3

Solve $C = 2\pi r$ for r. (This formula is used to find the circumference C of a circle given its radius r.)

Practice Problem 4

Solve $P = 2l + 2w$ for w.

Practice Problem 5

Solve $A = \dfrac{a + b}{2}$ for b.

Example 3   Solve $V = lwh$ for l.

Solution: This formula is used to find the volume of a box. To solve for l, we divide both sides by wh.

$$V = lwh$$

$$\frac{V}{wh} = \frac{lwh}{wh} \quad \text{Divide both sides by } wh.$$

$$\frac{V}{wh} = l \quad \text{Simplify.}$$

Since we have l alone on one side of the equation, we have solved for l in terms of V, w, and h. Remember that it does not matter on which side of the equation we get the variable alone. ▬▬▬

Example 4   Solve $y = mx + b$ for x.

Solution: First we get mx alone by subtracting b from both sides.

$$y = mx + b$$

$$y - b = mx + b - b \quad \text{Subtract } b \text{ from both sides.}$$

$$y - b = mx \quad \text{Combine like terms.}$$

Next we solve for x by dividing both sides by m.

$$\frac{y - b}{m} = \frac{mx}{m}$$

$$\frac{y - b}{m} = x \quad \text{Simplify.} \quad ▬▬▬$$

Example 5   Solve $A = \dfrac{bh}{2}$ for h.

Solution: First let's clear the equation of fractions by multiplying both sides by 2.

$$A = \frac{bh}{2}$$

$$2 \cdot A = 2\left(\frac{bh}{2}\right) \quad \text{Multiply both sides by 2 to clear fractions.}$$

$$2A = bh$$

$$\frac{2A}{b} = \frac{bh}{b} \quad \text{Divide both sides by } b \text{ to get } h \text{ alone.}$$

$$\frac{2A}{b} = h \quad \text{Simplify.} \quad ▬▬▬$$

Answers

3. $r = \dfrac{C}{2\pi}$,   **4.** $w = \dfrac{P - 2l}{2}$,

5. $b = 2A - a$

Name _____ Section _____ Date _____

EXERCISE SET 2.6

A *Substitute the given values into each given formula and solve for the unknown variable. See Examples 1 and 2.*

1. $A = bh$; $A = 45, b = 15$
   (Area of a parallelogram)

2. $d = rt$; $d = 195, t = 3$
   (Distance formula)

3. $S = 4lw + 2wh$; $S = 102, l = 7, w = 3$
   (Surface area of a special rectangular box)

4. $V = lwh$; $l = 14, w = 8, h = 3$
   (Volume of a rectangular box)

5. $A = \frac{1}{2}(B + b)h$; $A = 180, B = 11, b = 7$
   (Area of a trapezoid)

6. $A = \frac{1}{2}(B + h)h$; $A = 60, B = 7, b = 3$
   (Area of a trapezoid)

7. $P = a + b + c$; $P = 30, a = 8, b = 10$
   (Perimeter of a triangle)

8. $V = \frac{1}{3}Ah$; $V = 45, h = 5$
   (Volume of a pyramid)

9. $C = 2\pi r$; $C = 15.7$ (use the approximation 3.14 for π)
   (Circumference of a circle)

10. $A = \pi r^2$; $r = 4.5$ (use the approximation 3.14 for π)
    (Area of a circle)

11. $I = PRT$; $I = 3750, P = 25,000, R = 0.05$
    (Simple interest formula)

12. $I = PRT$; $I = 1,056,000, R = 0.055, T = 6$
    (Simple interest formula)

13. $V = \frac{1}{3}\pi r^2 h$; $V = 565.2, r = 6$ (use the approximation 3.14 for π)
    (Volume of a cone)

14. $V = \frac{4}{3}\pi r^3$; $r = 3$ (use the approximation 3.14 for π)
    (Volume of a sphere)

Solve. See Examples 1 and 2.

15. The world's largest sign for Coca-Cola is located in Arica, Chile. The rectangular sign has a length of 400 feet and has an area of 52,400 square feet. Find the width of the sign. (*Source: Fabulous Facts about Coca-Cola, Atlanta, GA*)

16. The length of a rectangular garden is 6 meters. If 21 meters of fencing are required to fence the garden, find its width.

6 meters

2. $r = 65$

3. $h = 3$

4. $V = 336$

5. $h = 20$

6. $h = 12$

7. $c = 12$

8. $A = 27$

9. $r = 2.5$

10. $A = 63.585$

11. $T = 3$

12. $P = 3,200,000$

13. $h = 15$

14. $V = 113.04$

15. 131 ft

16. 4.5 meters

17. 2000 mph

18. 7.5 hours

19. −10°C

20. 23°F

21. 96 piranhas

22. 75 goldfish

23. 2.25 hours

24. 7:20 a.m.

25. 6.25 hours

26. $35\frac{11}{17}$ mph

 17. The SR-71 is a top secret spy plane. It is capable of traveling from Rochester, New York, to San Francisco, California—a distance of approximately 3000 miles—in $1\frac{1}{2}$ hours. Find the rate of the SR-71.

18. A limousine built in 1968 for the president cost $500,000 and weighed 5.5 tons. This Lincoln Continental Executive could travel at 50 miles per hour with all of its tires shot away. At this rate, how long would it take to travel from Charleston, West Virginia, to Washington, D.C., a distance of 375 miles?

19. Convert Nome, Alaska's 14°F high temperature to Celsius. (*Source*: The World Almanac, 1997)

20. Convert Paris, France's low temperature of −5°C to Fahrenheit. (*Source*: The World Almanac, 1997)

21. Piranha fish require 1.5 cubic feet of water per fish to maintain a healthy environment. Find the maximum number of piranhas you could put in a tank measuring 8 feet by 3 feet by 6 feet.

22. Find how many goldfish you can put in a cylindrical tank whose diameter is 8 meters and whose height is 3 meters if each goldfish needs 2 cubic meters of water.

3 meters

8 meters

23. Find how long it takes Tran Nguyen to drive 135 miles on I-10 if he merges onto I-10 at 10 a.m. and drives non-stop with his cruise control set on 60 mph.

24. Beaumont, Texas, is about 150 miles from Toledo Bend. If Leo Miller leaves Beaumont at 4 a.m. and averages 45 mph, when should he arrive at Toledo Bend?

25. The X-30 is a new "space plane" being developed that will skim the edge of space at 4000 miles per hour. Neglecting altitude, if the circumference of the Earth is approximately 25,000 miles, how long will it take for the X-30 to travel around the Earth?

26. In the United States, the longest hang glider flight was a 303-mile, $8\frac{1}{2}$-hour flight from New Mexico to Kansas. What was the average rate during this flight?

Name _____

27. A lawn is in the shape of a trapezoid with a height of 60 feet and bases of 70 feet and 130 feet. How many bags of fertilizer must be purchased to cover the lawn if each bag covers 4000 square feet?

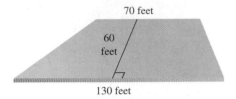

70 feet

60 feet

130 feet

28. If the area of a right-triangularly shaped sail is 20 square feet and its base is 5 feet, find the height of the sail.

?

5 feet

The Dante II is a spider-like robot that is used to map the depths of an active Alaskan volcano.

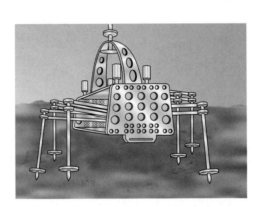

29. The dimensions of the Dante II are 10 feet long by 8 feet wide by 10 feet high. Find the volume of the smallest box needed to store this robot.

30. The Dante II traveled 600 feet into an active Alaskan volcano in $3\frac{1}{3}$ hours. Find the traveling rate of Dante II in feet per minute. (*Hint*: First convert $3\frac{1}{3}$ hours to minutes.)

31. one 16 in pizza

32. 25,120 miles

33. −109.3°F

34. 27,760°C

35. 500 sec. or $8\frac{1}{3}$ min

36. 1.3 sec

37. 33,493,333,333 cubic miles

31. Maria's Pizza sells one 16-inch cheese pizza or two 10-inch cheese pizzas for $9.99. Determine which size gives more pizza.

← 16 inches → | 10 inches | 10 inches

32. Find how much rope is needed to wrap around the Earth at the equator, if the radius of the Earth is 4000 miles. (*Hint*: Use 3.14 for π and the formula for circumference.)

33. Dry ice is a name given to solidified carbon dioxide. At −78.5°C it changes directly from a solid to a gas. Convert this temperature to degrees Fahrenheit.

34. Lightning bolts can reach a temperature of 50,000°F. Convert this temperature to degrees Celsius.

35. The distance from the sun to the Earth is approximately 93,000,000 miles. If light travels at a rate of 186,000 miles per second, how long does it take light from the sun to reach us?

36. Light travels at a rate of 186,000 miles per second. If our moon is 238,860 miles from the Earth, how long does it take light from the moon to reach us? (Round to the nearest tenth of a second.)

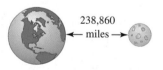

238,860
← miles →

37. On July 16, 1994, the Shoemaker-Levy 9 comet collided with Jupiter. The impact of the largest fragment of the comet, a massive chunk of rock and ice, created a fireball with a radius of 2000 miles. Find the volume of this spherical fireball. (Use 3.14 for π. Round to the nearest whole cubic mile.)

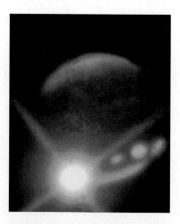

Color image made by the Hubble Space Telescope of the impact site of Fragment G of the comet.

38. The fireball from the largest fragment of the comet (see Exercise 37) immediately collapsed, as it was pulled down by gravity. As it fell, it cooled to approximately −350°F. Convert this temperature to degrees Celsius.

39. Bolts of lightning can travel at 270,000 miles per second. How many times can a lightning bolt travel around the world in one second? (See Exercise 32. Round to the nearest tenth.)

40. A glacier is a giant mass of rocks and ice that flows downhill like a river. Exit Glacier, near Seward, Alaska, moves at a rate of 20 inches a day. Find the distance in feet the glacier moves in a year. (Assume 365 days a year. Round to 2 decimal places.)

41. Flying fish do not *actually* fly, but glide. They have been known to travel a distance of 1300 feet at a rate of 20 miles per hour. How many seconds did it take to travel this distance? (*Hint:* First convert miles per hour to feet per second. Recall that 1 mile = 5280 feet. Round to the nearest tenth of a second.)

42. Stalactites join stalagmites to form columns. A column found at Natural Bridge Caverns near San Antonio, Texas, rises 15 feet and has a *diameter* of only 2 inches. Find the volume of this column in cubic inches. (*Hint:* Use the formula for volume of a cylinder and use 3.14 for π.)

B *Solve each formula for the specified variable. See Examples 3 through 5.*

43. $f = 5gh$ for h

44. $C = 2\pi r$ for r

45. $V = LWH$ for W

46. $T = mnr$ for n

47. $3x + y = 7$ for y

48. $-x + y = 13$ for y

38. $-212\frac{2}{9}°C$

39. 10.8

40. 608.33 ft.

41. 44.3 sec.

42. 565.2 cubic inches

43. $h = \dfrac{f}{5g}$

44. $r = \dfrac{C}{2\pi}$

45. $W = \dfrac{V}{LH}$

46. $n = \dfrac{T}{mr}$

47. $y = 7 - 3x$

48. $y = 13 + x$

49. $R = \dfrac{A - P}{PT}$

50. $T = \dfrac{A - P}{PR}$

51. $A = \dfrac{3V}{h}$

52. $k = \dfrac{4D}{f}$

53. $a = P - b - c$

54. $\begin{array}{l} s_3 = \\ PR - s_1 - s_2 - s_4 \end{array}$

55. $h = \dfrac{S - 2\pi r^2}{2\pi r}$

56. $h = \dfrac{S - 4lw}{2w}$

57. 0.32

58. 0.08

59. 2.00 or 2

60. 0.005

61. 17%

62. 3%

63. 720%

64. 500%

65. $V = G(N - R)$

66. $V = P - \dfrac{F}{B}$

67. multiplies the volume by 8

68. multiplies the area by 4

69. $-40°$

150

49. $A = P + PRT$ for R

50. $A = P + PRT$ for T

51. $V = \dfrac{1}{3}Ah$ for A

52. $D = \dfrac{1}{4}fk$ for k

53. $P = a + b + c$ for a

54. $PR = s_1 + s_2 + s_3 + s_4$ for s_3

55. $S = 2\pi rh + 2\pi r^2$ for h

56. $S = 4lw + 2wh$ for h

REVIEW AND PREVIEW

Write each percent as a decimal. See Section R.3.

57. 32%   **58.** 8%   **59.** 200%   **60.** 0.5%

Write each decimal as a percent. See Section R.3.

61. 0.17   **62.** 0.03   **63.** 7.2   **64.** 5

COMBINING CONCEPTS

Solve.

65. $N = R + \dfrac{V}{G}$ for V
(Urban forestry: tree plantings per year)

66. $B = \dfrac{F}{P - V}$ for V
(Business: break-even point)

67. The formula $V = LWH$ is used to find the volume of a box. If the length of a box is doubled, the width is doubled, and the height is doubled, how does this affect the volume?

68. The formula $A = bh$ is used to find the area of a parallelogram. If the base of a parallelogram is doubled and its height is doubled, how does this affect the area?

69. Find the temperature at which the Celsius measurement and Fahrenheit measurement are the same number.

2.7 PERCENT, RATIO, AND PROPORTION

A SOLVING PERCENT EQUATIONS

Much of today's statistics is given in terms of percent: a basketball player's free throw percent, current interest rates, stock market trends, and nutrition labeling, just to name a few. In this section, we first explore percent, percent equations, and applications involving percents. See Section R.3 if a further review of percents is needed.

Example 1   The number 63 is what percent of 72?

Solution: **1.** UNDERSTAND. Read and reread the problem. Next, let's suppose that the percent is 80%. To check, we find 80% of 72.

$$80\% \text{ of } 72 = 0.80(72) = 57.6$$

Close, but not 63. At this point, though, we have a better understanding of the problem, we know the correct answer is close to and greater than 80%, and we know how to check our proposed solution later.

Let $x =$ the unknown percent.

2. TRANSLATE. Recall that "is" means "equals" and "of" signifies multiplying. Let's translate the sentence directly.

the number 63	is	what percent	of	72
↓	↓	↓	↓	↓
63	=	x	·	72

3. SOLVE.

$$63 = 72x$$
$$0.875 = x \quad \text{Divide both sides by 72.}$$
$$87.5\% = x \quad \text{Write as a percent.}$$

4. INTERPRET.

Check: Verify that 87.5% of 72 is 63.

State: The number 63 is 87.5% of 72.

Example 2   The number 120 is 15% of what number?

Solution: **1.** UNDERSTAND. Read and reread the problem.

Let $x =$ the unknown number.

2. TRANSLATE.

the number 120	is	15%	of	what number
↓	↓	↓	↓	↓
120	=	15%	·	x

3. SOLVE.

$$120 = 0.15x \quad \text{Write 15\% as 0.15.}$$
$$800 = x \quad \text{Divide both sides by 0.15.}$$

4. INTERPRET.

Objectives

A Solve percent equations.
B Solve problems involving percents.
C Write ratios as fractions.
D Solve proportions.
E Solve problems modeled by proportions.

SSM   CD-ROM   Video
2.7

Practice Problem 1

The number 22 is what percent of 40?

TEACHING TIPS

It may be helpful to point out to students that they can also solve these percent problems using

$$part = percent \cdot total$$

For example, in a class of 125 students, 56% of the students, or 70 students, are female. Here,
125 students is the *total*,
56% is the *percent*, and
70 students is the *part*.

Practice Problem 2

The number 150 is 40% of what number?

Answers

1. 55%,   **2.** 375

Check: Check the proposed solution by finding 15% of 800 and verifying that the result is 120.

State: Thus, 120 is 15% of 800.

B SOLVING PROBLEMS INVOLVING PERCENT

As mentioned earlier, percents are often used in statistics. Recall that the graph below is called a circle graph or a pie chart. The circle or pie represents a whole, or 100%. Each circle is divided into sectors (shaped like pieces of a pie) that represent various parts of the whole 100%.

Example 3   The circle graph below shows how much money homeowners in the United States spend annually on maintaining their homes. Use this graph to answer the questions below.

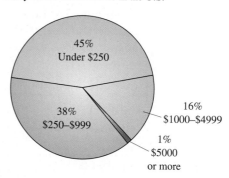

Yearly Home Maintenance in the U.S.

a. What percent of homeowners spend under $250 on yearly home maintenance?
b. What percent of homeowners spend less than $1000 per year on home maintenance?
c. How many of the 22,000 homeowners in a town called Fairview might we expect to spend under $250 a year on home maintenance?

Solution: **a.** From the circle graph, we see that 45% of homeowners spend under $250 per year on home maintenance.
b. From the circle graph, we know that 45% of homeowners spend under $250 per year and 38% of homeowners spend $250–$999 per year, so that the sum 45% + 38% or 83% of homeowners spend less than $1000 per year.
c. Since 45% of homeowners spend under $250 per year on maintenance, we find 45% of 22,000.

$$45\% \text{ of } 22{,}000 = 0.45(22{,}000)$$
$$= 9900$$

We might then expect that 9900 homeowners in Fairview spend under $250 per year on home maintenance.

C WRITING RATIOS AS FRACTIONS

A **ratio** is the quotient of two numbers or two quantities. For example, a percent can be thought of as a ratio, since it is the quotient of a number and 100.

Practice Problem 3

Use the circle graph and part (c) of Example 3 to answer each question.

a. What percent of homeowners spend $250–$4999 on yearly home maintenance?

b. What percent of homeowners spend $250 or more on yearly home maintenance?

c. How many of the homeowners in the town of Fairview might we expect to spend from $250–$999 per year on home maintenance.

Answers

3. a. 54%,   **b.** 55%,   **c.** 8360

$$53\% = \frac{53}{100} \quad \text{or} \quad \text{the ratio of 53 to 100}$$

RATIO

The ratio of a number a to a number b is their quotient. Ways of writing ratios are

$$a \text{ to } b, \quad a : b, \quad \frac{a}{b}$$

Example 4   Write a ratio for each phrase. Use fractional notation.

a. The ratio of 2 parts salt to 5 parts water
b. The ratio of 18 inches to 2 feet

Solution: **a.** The ratio of 2 parts salt to 5 parts water is $\frac{2}{5}$.
b. First we convert to the same unit of measurement. For example,

$$2 \text{ feet} = 2 \cdot 12 \text{ inches} = 24 \text{ inches}$$

The ratio of 18 inches to 2 feet is then

$$\frac{18}{24}, \text{ or } \frac{3}{4} \text{ in lowest terms.}$$

Practice Problem 4

Write a ratio for each phrase. Use fractional notation.

a. The ratio of 3 parts oil to 7 parts gasoline

b. The ratio of 40 minutes to 3 hours

D SOLVING PROPORTIONS

Ratios can be used to form proportions. A **proportion** is a mathematical statement that two ratios are equal.

For example, the equation

$$\frac{1}{2} = \frac{4}{8}$$

is a proportion that says that the ratios $\frac{1}{2}$ and $\frac{4}{8}$ are equal.

Notice that a proportion contains four numbers. If any three numbers are known, we can solve and find the fourth number. One way to do so is to use cross products. To understand cross products, let's start with the proportion

$$\frac{a}{b} = \frac{c}{d}$$

and multiply both sides by the LCD, bd.

$$\frac{a}{b} = \frac{c}{d}$$

$$bd\left(\frac{a}{b}\right) = bd\left(\frac{c}{d}\right) \quad \text{Multiply both sides by the LCD, } bd.$$

$$\underbrace{ad} = \underbrace{bc} \quad \text{Simplify.}$$

cross product   cross product

Notice why ad and bc are called cross products.

CROSS PRODUCTS

If $\frac{a}{b} = \frac{c}{d}$, then $ad = bc$.

Practice Problem 5

Solve for x: $\frac{3}{8} = \frac{63}{x}$

Example 5 Solve for x: $\frac{45}{x} = \frac{5}{7}$

Solution: To solve, we set cross products equal.

$45 \cdot 7 = x \cdot 5$ Set cross products equal.

$315 = 5x$ Multiply.

$\frac{315}{5} = \frac{5x}{5}$ Divide both sides by 5.

$63 = x$ Simplify.

Check: To check, substitute 63 for x in the original proportion. The solution is 63.

Practice Problem 6

Solve for x: $\frac{2x+1}{7} = \frac{x-3}{5}$

Example 6 Solve for x: $\frac{x-5}{3} = \frac{x+2}{5}$

Solution:

$5(x-5) = 3(x+2)$ Set cross products equal.

$5x - 25 = 3x + 6$ Multiply.

$5x = 3x + 31$ Add 25 to both sides.

$2x = 31$ Subtract $3x$ from both sides.

$\frac{2x}{2} = \frac{31}{2}$ Divide both sides by 2.

$x = \frac{31}{2}$

Check: Verify that $\frac{31}{2}$ is the solution.

Answers

5. $x = 168$,   6. $x = -\frac{26}{3}$

TRY THE CONCEPT CHECK IN THE MARGIN.

E SOLVING PROBLEMS MODELED BY PROPORTIONS

Proportions can be used to model and solve many real-life problems. When using proportions in this way, it is important to judge whether the solution is reasonable. Doing so helps us to decide if the proportion has been formed correctly, We use the same problem-solving strategy that was introduced in Section 2.5.

Example 7   Calculating Cost with a Proportion

Three boxes of 3.5-inch high-density diskettes cost $37.47. How much should 5 boxes cost?

Solution:

1. UNDERSTAND. Read and reread the problem. We know that the cost of 5 boxes is more than the cost of 3 boxes, or $37.47, and less than the cost of 6 boxes, which is double the cost of 3 boxes, or 2($37.47) = $74.94. Let's suppose that 5 boxes cost $60.00. To check, we see if 3 boxes is to 5 boxes as the *price* of 3 boxes is to the *price* of 5 boxes. In other words, we see if

$$\frac{3 \text{ boxes}}{5 \text{ boxes}} = \frac{\text{price of 3 boxes}}{\text{price of 5 boxes}}$$

or

$$\frac{3}{5} \diagup \frac{37.47}{60.00}$$

$3(60.00) = 5(37.47)$ Set cross products equal.

or

$180.00 = 187.35$ Not a true statement.

Thus, $60 is not correct but we now have a better understanding of the problem.

Let x = price of 5 boxes of diskettes.

2. TRANSLATE.

$$\frac{3 \text{ boxes}}{5 \text{ boxes}} = \frac{\text{price of 3 boxes}}{\text{price of 5 boxes}}$$

$$\frac{3}{5} = \frac{37.47}{x}$$

3. SOLVE.

$$\frac{3}{5} \diagup \frac{37.47}{x}$$

$3x = 5(37.47)$ Set cross products equal.

$3x = 187.35$

$x = 62.45$ Divide both sides by 3.

4. INTERPRET.

✓ **CONCEPT CHECK**

For which of the following equations can we immediately use cross products to solve for x?

a. $\dfrac{2 - x}{5} = \dfrac{1 + x}{3}$

b. $\dfrac{2}{5} - x = \dfrac{1 + x}{3}$

Practice Problem 7

To estimate the number of people in Jackson, population 50,000, who have no health insurance, 250 people were polled. Of those polled, 39 had no insurance. How many people in the city might we expect to be uninsured?

Answers

✓ **Concept Check:** a

7. 7800 people

Check: Verify that 3 boxes is to 5 boxes as $37.47 is to $62.45. Also, notice that our solution is a reasonable one as discussed in Step 1.

State: Five boxes of high-density diskettes cost $62.45. ▬▬

HELPFUL HINT

The proportion $\dfrac{5 \text{ boxes}}{3 \text{ boxes}} = \dfrac{\text{price of 5 boxes}}{\text{price of 3 boxes}}$ could also have been used to solve the problem above. Notice that the cross products are the same.

When shopping for an item offered in many different sizes, it is important to be able to determine the best buy, or the best price per unit. To find the **unit price** of an item, divide the total price of the item by the total number of units.

$$\text{unit price} = \frac{\text{total price}}{\text{number of units}}$$

For example, if a 16-ounce can of green beans is priced at $0.88, its unit price is

$$\text{unit price} = \frac{\$0.88}{16} = \$0.055$$

Practice Problem 8

Which is the better buy for the same brand of toothpaste?

8 ounces for $2.59

10 ounces for $3.11

Example 8   **Finding the Better Buy**

A supermarket offers a 14-ounce box of cereal for $3.79 and an 18-ounce box of the same brand of cereal for $4.99. Which is the better buy?

Solution: To find the better buy, we compare unit prices. The following unit prices were rounded to three decimal places.

Size	Price	Unit Price
14 ounce	$3.79	$\dfrac{\$3.79}{14} \approx \$0.271$
18 ounce	$4.99	$\dfrac{\$4.99}{18} \approx \$0.277$

The 14-ounce box of cereal has the lower unit price so it is the better buy. ▬▬

Answer

8. 10 ounces

Name _____ **Section** _____ **Date** _____

EXERCISE SET 2.7

A *Find each number described. See Examples 1 and 2.*

1. What number is 16% of 70?

2. What number is 88% of 1000?

3. The number 28.6 is what percent of 52?

4. The number 87.2 is what percent of 436?

5. The number 45 is 25% of what number?

6. The number 126 is 35% of what number?

7. Find 23% of 20.

8. Find 140% of 86.

9. The number 40 is 80% of what number?

10. The number 56.25 is 45% of what number?

11. The number 144 is what percent of 480?

12. The number 42 is what percent of 35?

B *Solve. See Examples 1 through 3. Many applications in this exercise group may be solved more efficiently with the use of a calculator.*

13. Dillard's advertised a 25% off sale. If a London Fog coat originally sold for $156, find the decrease and the sale price.

14. Time Saver increased the price of a $0.75 cola by 15%. Find the increase and the new price.

15. At this writing, the women's world record for throwing a disc (like a heavy Frisbee) is held by Anni Kreml of the United States. Her throw was 447.2 feet. The men's record is held by Niclas Bergehamm of Sweden. His throw was 44.8% further than Anni's. Find the length of his throw. (Round to the nearest tenth of a foot.) (*Source*: World Flying Disc Federation)

16. Scoville units are used to measure the hotness of a pepper. An alkaloid, capsaicin, is the ingredient that makes a pepper hot and liquid chromatography measures the amount of capsaicin in parts per million. The jalapeno measures around 5000 Scoville units, while the hottest pepper, the habanero, measures around 3000% of the measure of the jalapeno. Find the measure of the habanero pepper.

ANSWERS
1. 11.2

2. 880

3. 55%

4. 20%

5. 180

6. 360

7. 4.6

8. 120.4

9. 50

10. 125

11. 30%

12. 120%

13. $39 decrease; $117 sale price

14. $0.11 increase; $0.86 new price

15. 647.5 ft.

16. 150,000 Scoville units

157

Name _____

The graph shows the percent of people in a survey who have used various over-the-counter drugs in a twelve-month period. Use this graph to answer Exercises 17–22. See Example 3.

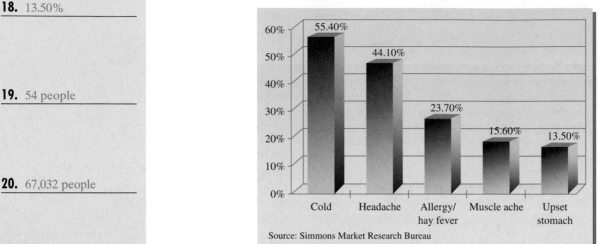

Source: Simmons Market Research Bureau

17. What percent of those surveyed used over-the-counter drugs to combat the common cold?

18. What percent of those surveyed used over-the-counter drugs to combat an upset stomach?

19. If 230 people were surveyed, how many of these used over-the-counter drugs for allergies?

20. The city of Chattanooga, Tenn., has a population of approximately 152,000. How many of these people would you expect to have used over-the-counter drugs for relief of a headache?

21. Do the percents shown in the graph have a sum of 100%? Why or why not?

22. Survey your algebra class and find what percent of the class has used over-the-counter drugs for each of the categories listed. Draw a bar graph of the results.

23. A recent study showed that 26% of men have dozed off at their place of work. If you currently employ 121 men, how many of these men might you expect to have dozed off at work? (*Source*: Better Sleep Council)

24. A recent study showed that women and girls spend 41% of the household clothing budget. If a family spent $2000 last year on clothing, how much might have been spent on clothing for women and girls? (*Source*: The Interep Radio Store)

Name _____

Fill in the percent column in each table. See Examples 1 through 3.

25.

THE WALT DISNEY COMPANY 1996 OPERATING INCOME		
	Millions of Dollars	Percent of Total (Round to nearest whole percent)
Creative Content	1596	$\frac{1596}{3333} \approx 48\%$
Theme Parks	990	
Broadcasting	747	
TOTALS	3333	

Source: Hoover's Company Profiles

26.

NIKE, INC. 1996 SALES		
	Millions of Dollars	Percent of Total (Round to nearest whole percent)
U.S. Footwear	2773	$\frac{2773}{6471} \approx 43\%$
International Footwear	1682	
U.S. Apparel	843	
International Apparel	651	
Other Brands	522	
TOTALS	6471	

Source: NIKE, Inc.

C *Write each ratio in fractional notation in lowest terms. See Example 4.*

27. 2 megabytes to 15 megabytes

28. 18 disks to 41 disks

29. 10 inches to 12 inches

30. 15 miles to 40 miles

31. 5 quarts to 3 gallons

32. 8 inches to 3 feet

33. 4 nickels to 2 dollars

34. 12 quarters to 2 dollars

35. 175 centimeters to 5 meters

36. 90 centimeters to 4 meters

37. 190 minutes to 3 hours

38. 60 hours to 2 days

39. Suppose someone tells you that the ratio of 11 inches to 2 feet is $\frac{11}{2}$. How do you correct that person and explain the error?

40. Write a ratio that can be written in fractional notation as $\frac{3}{2}$.

25. 48%, 30%, 22%, 100%

26. 43%, 26%, 13%, 10%, 8%, 100%

27. $\frac{2}{15}$

28. $\frac{18}{41}$

29. $\frac{5}{6}$

30. $\frac{3}{8}$

31. $\frac{5}{12}$

32. $\frac{2}{9}$

33. $\frac{1}{10}$

34. $\frac{3}{2}$

35. $\frac{7}{20}$

36. $\frac{9}{40}$

37. $\frac{19}{18}$

38. $\frac{5}{4}$

39. answers may vary

40. answers may vary

41. $x = 4$

42. $x = \dfrac{16}{3}$

43. $x = \dfrac{50}{9}$

44. $x = \dfrac{3}{4}$

45. $x = \dfrac{21}{4}$

46. $a = \dfrac{15}{2}$

47. $x = 7$

48. $y = 40$

49. $x = -3$

50. $x = -\dfrac{7}{2}$

51. $x = \dfrac{14}{9}$

52. $x = \dfrac{103}{6}$

53. $x = 5$

54. $x = -4$

55. 123 lb

56. 3441 lb

57. 165 calories

58. $15\frac{1}{2}$ ft

59. 3833 women

60. 360 square ft

61. 9 gal

62. 337 yds/game

Name _____

D *Solve each proportion. See Examples 5 and 6.*

41. $\dfrac{2}{3} = \dfrac{x}{6}$ **42.** $\dfrac{x}{2} = \dfrac{16}{6}$ **43.** $\dfrac{x}{10} = \dfrac{5}{9}$ **44.** $\dfrac{9}{4x} = \dfrac{6}{2}$

45. $\dfrac{4x}{6} = \dfrac{7}{2}$ **46.** $\dfrac{a}{5} = \dfrac{3}{2}$ **47.** $\dfrac{x - 3}{x} = \dfrac{4}{7}$ **48.** $\dfrac{y}{y - 16} = \dfrac{5}{3}$

49. $\dfrac{x + 1}{2x + 3} = \dfrac{2}{3}$ **50.** $\dfrac{x + 1}{x + 2} = \dfrac{5}{3}$ **51.** $\dfrac{9}{5} = \dfrac{12}{3x + 2}$ **52.** $\dfrac{6}{11} = \dfrac{27}{3x - 2}$

53. $\dfrac{3}{x + 1} = \dfrac{5}{2x}$ **54.** $\dfrac{7}{x - 3} = \dfrac{8}{2x}$

E *Solve. See Example 7.*

55. The ratio of the weight of an object on Earth to the weight of the same object on Pluto is 100 to 3. If an elephant weighs 4100 pounds on Earth, find the elephant's weight on Pluto.

56. If a 170-pound person weighs approximately 65 pounds on Mars, how much does a 9000-pound satellite weigh?

57. There are 110 calories per 28.4 grams of Crispy Rice cereal. Find how many calories are in 42.6 grams of this cereal.

58. On an architect's blueprint, 1 inch corresponds to 4 feet. Find the length of a wall represented by a line that is $3\frac{7}{8}$ inches long on the blueprint.

59. A recent headline read, "Women Earn Bigger Check in 1 of Every 6 Couples." If there are 23,000 couples in a nearby metropolitan area, how many women would you expect to earn bigger paychecks?

60. A human factors expert recommends that there be at least 9 square feet of floor space in a college classroom for every student in the class. Find the minimum floor space that 40 students need.

61. To mix weed killer with water correctly, it is necessary to mix 8 teaspoons of weed killer with 2 gallons of water. Find how many gallons of water are needed to mix with the entire box if it contains 36 teaspoons of weed killer.

62. Ken Hall, a tailback, holds the high school sports record for total yards rushed in a season. In 1953, he rushed for 4045 total yards in 12 games. Find his average rushing yards per game.

Name _____

Given the following prices charged for various sizes of an item, find the best buy. See Example 8.

📱 63. Laundry detergent
    110 ounces for $5.79
    240 ounces for $13.99

📱 64. Jelly
    10 ounces for $1.14
    15 ounces for $1.69

📱 65. Tuna (in cans)
    6 ounces for $0.69
    8 ounces for $0.90
    16 ounces for $1.89

📱 66. Picante sauce
    10 ounces for $0.99
    16 ounces for $1.69
    30 ounces for $3.29

REVIEW AND PREVIEW

Place $<$, $>$, *or* $=$ *in the appropriate space to make each a true statement. See Sections 1.1, 1.5, and 1.6.*

67. -5 -7

68. $\dfrac{12}{3}$ 2^2

69. $|-5|$ $-(-5)$

70. -3^3 $(-3)^3$

71. $(-3)^2$ -3^2

72. $|-2|$ $-|-2|$

COMBINING CONCEPTS

Standardized nutrition labels like the one to the right have been displayed on food items since 1994. The percent column on the right shows the percent of daily values based on a 2000-calorie diet shown at the bottom of the label. For example, a serving of this food contains 4 grams of total fat, where the recommended daily fat based on a 2000-calorie diet is 65 grams of fat. This means that $\frac{4}{65}$ or approximately 6% (as shown) of your daily recommended fat is taken in by eating a serving of this food.

Use this nutrition label to answer Exercises 73–75.

Nutrition Facts

Serving Size 18 crackers (31g)
Servings Per Container About 9

Amount Per Serving

Calories 130   Calories from Fat 35

	% Daily Value*
Total Fat 4g	**6%**
Saturated Fat 0.5g	**3%**
Polyunsaturated Fat 0g	
Monounsaturated Fat 1.5g	
Cholesterol 0mg	**0%**
Sodium 230mg	x
Total Carbohydrate 23g	y
Dietary Fiber 2g	**8%**
Sugars 3g	
Protein 2g	

Vitamin A 0%	•	Vitamin C 0%
Calcium 2%	•	Iron 6%

*Percent Daily Values are based on a 2,000 calorie diet. Your daily values may be higher or lower depending on your calorie needs.

		Calories:	2,000	2,500
Total Fat	Less than		65g	80g
Sat Fat	Less than		20g	25g
Cholesterol	Less than		300mg	300mg
Sodium	Less than		2400mg	2400mg
Total Carbohydrate			300g	375g
Dietary Fiber			25g	30g

73. 9.6%

73. Based on a 2000-calorie diet, what percent of daily values of sodium is contained in a serving of this food? In other words, find x in the label on the previous page. (Round to the nearest tenth of a percent.)

75. Notice on the nutrition label that one serving of this food contains 130 calories and 35 of these calories are from fat. Find the percent of calories from fat. (Round to the nearest tenth of a percent.) It is recommended that no more than 30% of calorie intake come from fat. Does this food satisfy this recommendation?

74. 7.7%

74. Based on a 2000-calorie diet, what percent of daily values of total carbohydrate is contained in a serving of this food? In other words, find y in the label on the previous page. (Round to the nearest tenth of a percent.)

75. 26.9%; yes

Use the nutrition label below to answer Exercises 76–77.

NUTRITIONAL INFORMATION PER SERVING
Serving Size: 9.8 oz.   **Servings Per Container: 1**

Calories. 280	Polyunsaturated Fat. 1g	
Protein. 12g	Saturated Fat. 3g	
Carbohydrate. 45g	Cholesterol 20mg	
Fat. 6g	Sodium. 520mg	
Percent of Calories from Fat. . . . . ?	Potassium. 220mg	

76. 19.3%

76. If fat contains approximately 9 calories per gram, find the percent of calories from fat in one serving of this food. (Round to the nearest tenth of a percent.)

77. If protein contains approximately 4 calories per gram, find the percent of calories from protein from one serving of this food. (Round to the nearest tenth of a percent.)

77. 17.1%

78. Find a food that contains more than 30% of its calories per serving from fat. Analyze the nutrition label and verify that the percents shown are correct.

78. answers may vary

2.8   SOLVING LINEAR INEQUALITIES

In Chapter 1, we reviewed these inequality symbols and their meanings:

$<$ means "is less than" $\leq$ means "is less than or equal to"
$>$ means "is greater than" $\geq$ means "is greater than or equal to"

An **inequality** is a statement that contains one of the symbols above.

Equations	Inequalities
$x = 3$	$x \leq 3$
$5n - 6 = 14$	$5n - 6 > 14$
$12 = 7 - 3y$	$12 \leq 7 - 3y$
$\dfrac{x}{4} - 6 = 1$	$\dfrac{x}{4} - 6 > 1$

A GRAPHING INEQUALITIES ON A NUMBER LINE

Recall that the single solution to the equation $x = 3$ is 3. The solutions of the inequality $x \leq 3$ include 3 and *all real numbers* less than 3. Because we can't list all numbers less than 3, we show instead a picture of the solutions by graphing them.

To graph $x \leq 3$, we shade the numbers to the left of 3 since they are less than 3. Then we place a closed circle on the point representing 3. The closed circle indicates that 3 *is* a solution: 3 *is* less than or equal to 3.

To graph $x < 3$, we shade the numbers to the left of 3. Then we place an open circle on the point representing 3. The open circle indicates that 3 *is not* a solution: 3 *is not* less than 3.

Example 1   Graph: $x \geq -1$

Solution: We place a closed circle at -1 since the inequality symbol is $\geq$ and -1 is greater than or equal to -1. Then we shade to the right of -1.

Example 2   Graph: $-1 > x$

Solution: Recall from Chapter 1 that $-1 > x$ means the same as $x < -1$, shown below.

B USING THE ADDITION PROPERTY

When solutions of a linear inequality are not immediately obvious, they are found through a process similar to the one used to solve a linear equation. Our goal is to get the variable alone on one side of the inequality. We use properties of inequality similar to properties of equality.

Objectives

A Graph inequalities on a number line.

B Use the addition property of inequality to solve inequalities.

C Use the multiplication property of inequality to solve inequalities.

D Use both properties to solve inequalities.

E Solve problems modeled by inequalities.

SSM   CD-ROM   Video 2.8

TEACHING TIP

Show students how they can easily check the shading on their graph. Substituting any point from the shaded region into the inequality should give a true statement. Substituting any point which has not been shaded should give a false statement.

Practice Problem 1

Graph: $x \geq -2$

Practice Problem 2

Graph: $5 > x$

Answers

1.
2.

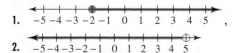

> **ADDITION PROPERTY OF INEQUALITY**
>
> If a, b, and c are real numbers, then
>
> $$a < b \quad \text{and} \quad a + c < b + c$$
>
> are equivalent inequalities.

This property also holds true for subtracting values, since subtraction is defined in terms of addition. In other words, adding or subtracting the same quantity from both sides of an inequality does not change the solutions of the inequality.

Practice Problem 3

Solve $x - 6 \geq -11$. Graph the solutions.

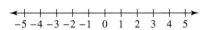

Example 3   Solve $x + 4 \leq -6$. Graph the solutions.

Solution: To solve for x, subtract 4 from both sides of the inequality.

$$
\begin{array}{ll}
x + 4 \leq -6 & \text{Original inequality} \\
x + 4 - 4 \leq -6 - 4 & \text{Subtract 4 from both sides.} \\
x \leq -10 & \text{Simplify.}
\end{array}
$$

The graph of the solutions is shown below.

$$\overset{\longleftarrow \; | \; | \; \bullet \; | \; | \; | \; | \longrightarrow}{\quad -12 \; -11 \; -10 \; -9 \; -8 \; -7 \; -6}$$

HELPFUL HINT

Notice that any number less than or equal to -10 is a solution to $x \leq -10$. For example, solutions include

$$-10, \quad -200, \quad -11\tfrac{1}{2}, \quad -7\pi, \quad -\sqrt{130}, \quad -50.3$$

C USING THE MULTIPLICATION PROPERTY

An important difference between linear equations and linear inequalities is shown when we multiply or divide both sides of an inequality by a nonzero real number. For example, start with the true statement $6 < 8$ and multiply both sides by 2. As we see below, the resulting inequality is also true.

$$
\begin{array}{ll}
6 < 8 & \text{True.} \\
2(6) < 2(8) & \text{Multiply both sides by 2.} \\
12 < 16 & \text{True.}
\end{array}
$$

But if we start with the same true statement $6 < 8$ and multiply both sides by -2, the resulting inequality is not a true statement.

$$
\begin{array}{ll}
6 < 8 & \text{True.} \\
-2(6) < -2(8) & \text{Multiply both sides by } -2. \\
-12 < -16 & \text{False.}
\end{array}
$$

Answer

3. $x \geq -5$

$$\overset{\longleftarrow \; \bullet \; | \; | \; | \; | \; | \; | \; | \; | \; | \longrightarrow}{\;\; -5 \; -4 \; -3 \; -2 \; -1 \;\; 0 \;\; 1 \;\; 2 \;\; 3 \;\; 4 \;\; 5}$$

Notice, however, that if we reverse the direction of the inequality symbol, the resulting inequality is true.

$-12 < -16$ False.

$-12 > -16$ True.

This demonstrates the multiplication property of inequality.

MULTIPLICATION PROPERTY OF INEQUALITY

1. If a, b, and c are real numbers, and c is **positive**, then
   $$a < b \quad \text{and} \quad ac < bc$$
   are equivalent inequalities.

2. If a, b, and c are real numbers, and c is **negative**, then
   $$a < b \quad \text{and} \quad ac > bc$$
   are equivalent inequalities.

Because division is defined in terms of multiplication, this property also holds true when dividing both sides of an inequality by a nonzero number: If we multiply or divide both sides of an inequality by a negative number, **the direction of the inequality sign must be reversed for the inequalities to remain equivalent**.

TRY THE CONCEPT CHECK IN THE MARGIN.

Example 4   Solve $-2x \leq -4$. Graph the solutions.

Solution: Remember to reverse the direction of the inequality symbol when dividing by a negative number.

$$-2x \leq -4$$

$$\frac{-2x}{-2} \geq \frac{-4}{-2} \quad \text{Divide both sides by } -2 \text{ and reverse the inequality sign.}$$

HELPFUL HINT
Don't forget to reverse the direction of the inequality sign.

$x \geq 2$ Simplify.

The graph of the solutions is shown.

Example 5   Solve $2x < -4$. Graph the solutions.

Solution: $2x < -4$

$$\frac{2x}{2} < \frac{-4}{2} \quad \text{Divide both sides by 2. Do not reverse the inequality sign.}$$

HELPFUL HINT
Do not reverse the inequality sign.

$x < -2$ Simplify.

✓ **CONCEPT CHECK**

Fill in the blank with $<$, $>$, $\leq$, or $\geq$.

a. Since $-8 < -4$,
   then $3(-8) \quad \_\_\_ \quad 3(-4)$.

b. Since $5 \geq -2$, then $\dfrac{5}{-7} \quad \_\_\_ \quad \dfrac{-2}{-7}$.

c. If $a < b$, then $2a \quad \_\_\_ \quad 2b$.

d. If $a \geq b$, then $\dfrac{a}{-3} \quad \_\_\_ \quad \dfrac{b}{-3}$.

Practice Problem 4

Solve $-3x \leq 12$. Graph the solutions.

Practice Problem 5

Solve $5x > -20$. Graph the solutions.

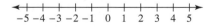

Answers

✓ **Concept Check:**

a. $<$,   **b.** $\leq$,   **c.** $<$,   **d.** $\leq$

4. $x \geq -4$,

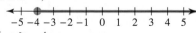

5. $x > -4$

The graph of the solutions is shown.

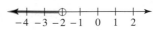

D USING BOTH PROPERTIES OF INEQUALITY

The following steps may be helpful when solving inequalities. Notice that these steps are similar to the ones given in Section 2.4 for solving equations.

> **TO SOLVE INEQUALITIES IN ONE VARIABLE**
>
> **Step 1.** Multiply on both sides to clear the inequality of fractions if they occur.
>
> **Step 2.** Use the distributive property to remove parentheses if they occur.
>
> **Step 3.** Simplify each side of the inequality by combining like terms.
>
> **Step 4.** Get all variable terms on one side and all numbers on the other side by using the addition property of inequality.
>
> **Step 5.** Get the variable alone by using the multiplication property of inequality.

HELPFUL HINT

Don't forget that if both sides of an inequality are multiplied or divided by a negative number, the direction of the inequality sign must be reversed.

Practice Problem 6

Solve $-3x + 11 \leq -13$. Graph the solutions.

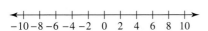

Example 6   Solve $-4x + 7 \geq -9$. Graph the solutions.

Solution:
$$-4x + 7 \geq -9$$
$$-4x + 7 - 7 \geq -9 - 7 \qquad \text{Subtract 7 from both sides.}$$
$$-4x \geq -16 \qquad \text{Simplify.}$$
$$\frac{-4x}{-4} \leq \frac{-16}{-4} \qquad \text{Divide both sides by } -4 \text{ and reverse the direction of the inequality sign.}$$
$$x \leq 4 \qquad \text{Simplify.}$$

The graph of the solutions is shown.

Practice Problem 7

Solve $-6x - 3 > -4(x + 1)$. Graph the solutions.

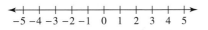

Answers

6. $x \geq 8$,

7. $x < \frac{1}{2}$

Example 7   Solve $-5x + 7 < 2(x - 3)$. Graph the solutions.

Solution:
$$-5x + 7 < 2(x - 3)$$
$$-5x + 7 < 2x - 6 \qquad \text{Apply the distributive property.}$$
$$-5x + 7 - 2x < 2x - 6 - 2x \qquad \text{Subtract } 2x \text{ from both sides.}$$
$$-7x + 7 < -6 \qquad \text{Combine like terms.}$$
$$-7x + 7 - 7 < -6 - 7 \qquad \text{Subtract 7 from both sides.}$$
$$-7x < -13 \qquad \text{Combine like terms.}$$

$$\frac{-7x}{-7} > \frac{-13}{-7}$$

Divide both sides by −7 and reverse the direction of the inequality sign.

$$x > \frac{13}{7}$$

Simplify.

The graph of the solutions is shown.

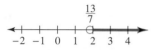

Example 8

Solve: $2(x - 3) - 5 \le 3(x + 2) - 18$

Solution: $2(x - 3) - 5 \le 3(x + 2) - 18$

$2x - 6 - 5 \le 3x + 6 - 18$ Apply the distributive property.

$2x - 11 \le 3x - 12$ Combine like terms.

$-x - 11 \le -12$ Subtract $3x$ from both sides.

$-x \le -1$ Add 11 to both sides.

$\frac{-x}{-1} \ge \frac{-1}{-1}$ Divide both sides by −1 and reverse the direction of the inequality sign.

$x \ge 1$ Simplify.

Practice Problem 8

Solve: $3(x + 5) - 1 \ge 5(x - 1) + 7$

E SOLVING PROBLEMS MODELED BY INEQUALITIES

Problems containing words such as "at least," "at most," "between," "no more than," and "no less than" usually indicate that an inequality should be solved instead of an equation. In solving applications involving linear inequalities, we use the same procedure we used to solve applications involving linear equations.

Example 9   Budgeting for a Wedding

Marie Chase and Jonathan Edwards are having their wedding reception at the Gallery reception hall. They may spend at most $1000 for the reception. If the reception hall charges a $100 cleanup fee plus $14 per person, find the greatest number of people that they can invite and still stay within their budget.

Solution: **1.** UNDERSTAND. Read and reread the problem. Suppose that 50 people attend the reception. The cost is then $100 + $14(50) = $100 + $700 = $800.

Let $x =$ the number of people who attend the reception.

2. TRANSLATE.

Practice Problem 9

Alex earns $600 per month plus 4% of all his sales. Find the minimum sales that will allow Alex to earn at least $3000 per month.

cleanup fee	+	cost per person	must be less than or equal to	$1000
↓		↓	↓	↓
100	+	14x	≤	1000

Answers

8. $x \le 6$,   **9.** $60,000

3. SOLVE.

$$100 + 14x \le 1000$$

$$14x \le 900 \qquad \text{Subtract 100 from both sides.}$$

$$x \le 64\frac{2}{7} \qquad \text{Divide both sides by 14.}$$

4. INTERPRET.

Check: Since x represents the number of people, we round down to the nearest whole, or 64. Notice that if 64 people attend, the cost is $100 + \$14(64) = \$996$. If 65 people attend, the cost is $100 + \$14(65) = \$1010$, which is more than the given $1000.

State: Marie Chase and Jonathan Edwards can invite at most 64 people to the reception. ▬▬▬

Focus On Mathematical Connections

SEQUENCES

The ratio of the lengths of the sides of the golden rectangle is approximately 1.6. This value is also known as the golden ratio. The actual value of the golden ratio is $\dfrac{1 + \sqrt{5}}{2} \approx 1.618033989\ldots,$ which is an irrational number. Interestingly enough, there is a very simple sequence, or ordered list of numbers, that can be used to approximate the golden ratio. The sequence is called the Fibonacci sequence, and it is very easy to construct. Start with the first two terms of the sequence, the numbers 1 and 1. The next term in the sequence is the sum of the two previous terms, and so on. For example,

		$1+1=$	$1+2=$	$2+3=$	$3+5=$	$5+8=$
1,	1,	2,	3,	5,	8,	13,

	$8+13=$	$13+21=$	$21+34=$
	21,	34,	55, . . .

Now, let's look at the ratios of a Fibonacci sequence term to the previous one:

$$\frac{1}{1} = 1, \quad \frac{2}{1} = 2, \quad \frac{3}{2} = 1.5, \quad \frac{5}{3} \approx 1.66667, \quad \frac{8}{5} = 1.6, \quad \frac{13}{8} = 1.625, \quad \frac{21}{13} \approx 1.615, \quad \frac{34}{21} \approx 1.619,$$

Notice that each successive ratio is a bit closer than the one before it to the value of the golden ratio, $1.618033989\ldots.$

CRITICAL THINKING

Generate a few additional terms of the Fibonacci sequence and extend the list of ratios of terms. How much closer to the value of the golden ratio can you get?

Name _____ Section _____ Date _____

MENTAL MATH ANSWERS

1. $x > 2$
2. $x < 5$
3. $x \geq 8$
4. $x \leq 7$
5. -5
6. $|-6|$
7. 4.1
8. -4

MENTAL MATH

Solve each inequality.

1. $5x > 10$ 2. $4x < 20$ 3. $2x \geq 16$ 4. $9x \leq 63$

Decide which number listed is not a solution to each given inequality.

5. $x \geq -3$; $-3, 0, -5, \pi$

6. $x < 6$; $-6, |-6|, 0, -3.2$

7. $x < 4.01$; $4, -4.01, 4.1, -4.1$

8. $x \geq -3$; $-4, -3, -2, -(-2)$

EXERCISE SET 2.8

A *Graph each on a number line. See Examples 1 and 2.*

 1. $x \leq -1$

2. $y < 0$

3. $x > \dfrac{1}{2}$
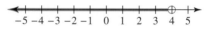

4. $z \geq -\dfrac{2}{3}$

5. $y < 4$
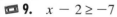

6. $x > 3$
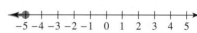

7. $-2 \leq m$
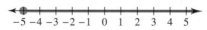

8. $-5 \geq x$
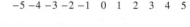

B *Solve each inequality. Graph the solutions. See Example 3.*

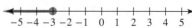 9. $x - 2 \geq -7$

10. $x + 4 \leq 1$

11. $-9 + y < 0$
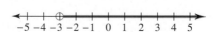

12. $-3 + m > 5$
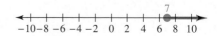

13. $3x - 5 > 2x - 8$

14. $3 - 7x \geq 10 - 8x$

ANSWERS

1. see number line
2. see number line
3   see number line
4. see number line
5. see number line
6. see number line
7. see number line
8. see number line
9. $x \geq -5$
10. $x \leq -3$
11. $y < 9$
12. $m > 8$
13. $x > -3$
14. $x \geq 7$

15. $x \le 1$

16. $x < -3$

17. $x < -3$

18. $x > -3$

19. $x \ge -2$

20. $x > -4$

21. $x < 0$

22. $y \le 0$

23. $y \ge -\dfrac{8}{3}$

24. $x \le -\dfrac{48}{5}$

25. $y > 3$

26. $x < 8$

27. when multiplying or dividing by a negative number

28. no; answers may vary

29. $x > -3$

30. $x \le 2$

31. $x \ge -\dfrac{2}{3}$

32. $x > \dfrac{4}{11}$

33. $x \le -2$

34. $x < -2$

35. $x \le -8$

36. $x \le \dfrac{4}{3}$

170

---

15. $4x - 1 \le 5x - 2x$

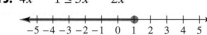

16. $7x + 3 < 9x - 3x$

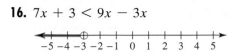

C *Solve each inequality. Graph the solutions. See Examples 4 and 5.*

17. $2x < -6$

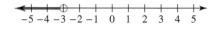

18. $3x > -9$

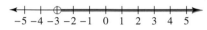

19. $-8x \le 16$

20. $-5x < 20$

21. $-x > 0$

22. $-y \ge 0$

23. $\dfrac{3}{4}y \ge -2$

24. $\dfrac{5}{6}x \le -8$

25. $-0.6y < -1.8$

26. $-0.3x > -2.4$

27. When solving an inequality, explain when the direction of an inequality symbol must be reversed.

28. If both sides of the inequality $-3x < -30$ are divided by 3, do you reverse the direction of the inequality symbol? Why or why not?

D *Solve each inequality. See Examples 6 through 8.*

29. $3x - 7 < 6x + 2$

30. $2x - 1 \ge 4x - 5$

31. $5x - 7x \le x + 2$

32. $4 - x < 8x + 2x$

33. $-6x + 2 \ge 2(5 - x)$

34. $-7x + 4 > 3(4 - x)$

35. $4(3x - 1) \le 5(2x - 4)$

36. $3(5x - 4) \le 4(3x - 2)$

37. $3(x + 2) - 6 > -2(x - 3) + 14$

38. $7(x - 2) + x \leq -4(5 - x) - 12$

39. $-2(x - 4) - 3x < -(4x + 1) + 2x$

40. $-5(1 - x) + x \leq -(6 - 2x) + 6$

41. $\frac{1}{2}(x - 5) < \frac{1}{3}(2x - 1)$

42. $\frac{1}{4}(x + 4) < \frac{1}{5}(2x + 3)$

43. $-5x + 4 \leq -4(x - 1)$

44. $-6x + 2 < -3(x + 4)$

E *Solve the following. See Example 9.*

45. Six more than twice a number is greater than negative fourteen. Find all numbers that make this statement true.

46. Five times a number increased by one is less than or equal to ten. Find all such numbers.

47. The perimeter of a rectangle is to be no greater than 100 centimeters and the width must be 15 centimeters. Find the maximum length of the rectangle.

48. One side of a triangle is four times as long as another side, and the third side is 12 inches long. If the perimeter can be no longer than 87 inches, find the maximum lengths of the other two sides.

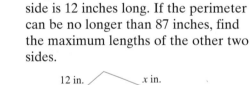

49. Ben Holladay bowled 146 and 201 in his first two games. What must he bowl in his third game to have an average of at least 180? (*Hint*: The average of a list of numbers is their sum divided by the number of numbers in the list.)

50. On an NBA team the two forwards measure 6'8" and 6'6" and the two guards measure 6'0" and 5'9" tall. How tall a center should they hire if they wish to have a starting team average height of at least 6'5"?

37. $x > 4$

38. $x \leq -\frac{9}{2}$

39. $x > 3$

40. $x \leq \frac{5}{4}$

41. $x > -13$

42. $x > \frac{8}{3}$

43. $x \geq 0$

44. $x > \frac{14}{3}$

45. $x > -10$

46. $x \leq \frac{9}{5}$

47. 35 cm

48. 15 in.; 60 in.

49. 193

50. $x \geq 7'2''$

51. final exam ≥ 78.5

51. Eric Daly has scores of 75, 83, and 85 on his history tests. Use an inequality to find the scores he can make on his final exam to receive a B in the class. The final exam counts as two tests, and a B is received if the final course average is greater than or equal to 80.

52. Maria Lipco has scores of 85, 95, and 92 on her algebra tests. Use an inequality to find the scores she can make on her final exam to receive an A in the course. The final exam counts as three tests, and an A is received if the final course average is greater than or equal to 90.

52. final exam ≥ 89.33

53. 8

REVIEW AND PREVIEW

Evaluate each expression. See Section 1.3.

53. $(2)^3$

54. $(3)^3$

55. $(1)^{12}$

54. 27

56. 0^5

57. $\left(\dfrac{4}{7}\right)^2$

58. $\left(\dfrac{2}{3}\right)^3$

55. 1

This broken line graph shows the enrollment of people (members) in a Health Maintenance Organization (HMO). The height of each dot corresponds to the number of members (in millions). Use this graph to answer Exercises 59–62.

56. 0

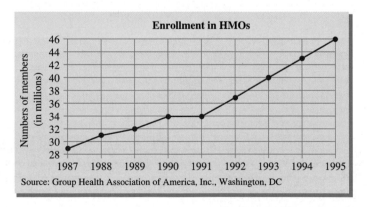

Source: Group Health Association of America, Inc., Washington, DC

57. $\dfrac{16}{49}$

58. $\dfrac{8}{27}$

59. How many people were enrolled in Health Maintenance Organizations in 1992?

60. How many people were enrolled in Health Maintenance Organizations in 1995?

61. Which year shows the greatest increase in number of members?

62. In what year were there 31,000,000 members of HMOs?

COMBINING CONCEPTS

Internet Excursions

Go to http://www.prenhall.com/martin-gay

The National Climatic Data Center (NCDC) is the world's largest active archive of weather data. The given World Wide Web address will provide you with access to NCDC's Climate Visualization Web site, or a related site, where you can find data on precipitation, temperature, and drought conditions for any state. Most states are subdivided into geographical divisions. Data from 1895 through the present is available for many states. For instance, you can graph data for all divisions of a given state and year for a parameter such as temperature.

59. 37 million _____

60. 46 million _____

61. 1992, 1993, 1994, or 1995 _____

62. 1988 _____

Choose a year. Create a graph of the average monthly temperatures for that year for all divisions of the state in which your college or university is located. Use the information on the Web site to answer the following exercises.

63. Fill in the following blanks. In the year _____, the lowest average monthly temperature of ____°F occurred in Division _____ of the state of _____. Let F represent the average monthly temperature in Fahrenheit degrees. Write an inequality that describes the minimum average monthly temperature in Fahrenheit degrees for this year. Then use the relationship $F = \frac{9}{5}C + 32$ to convert this inequality to one that describes the minimum average monthly temperature in degrees Celsius.

64. Fill in the following blanks. In the year _____, the highest average monthly temperature of _____°F occurred in Division _____ of the state _____. Let F represent the average monthly temperature in Fahrenheit degrees. Write an inequality that describes the maximum average monthly temperature in Fahrenheit degrees for this year. Then use the relationship $F = \frac{9}{5}C + 32$ to convert this inequality to one that describes the maximum average monthly temperature in degrees Celsius.

CHAPTER 2 ACTIVITY
INVESTIGATING AVERAGES

MATERIALS:
▲ small rubber ball or crumpled paper ball
▲ bucket or waste can

This activity may be completed by working in groups or individually.

1. Try shooting the ball into the bucket or waste can 5 times. Record your results below.

Shots Made   **Shots Missed**

2. Find your shooting percent for the 5 shots (that is, the percent of the shots you actually made out of the number you tried). answers may vary

3. Suppose you are going to try an additional 5 shots. How many of the next 5 shots will you have to make to have a 50% shooting percent for all 10 shots? An 80% shooting percent?
answers may vary

4. Did you solve an equation in Question 3? If so, explain what you did. If not, explain how you could use an equation to find the answers.
answers may vary

5. Now suppose you are going to try an additional 22 shots. How many of the next 22 shots will you have to make to have at least a 50% shooting percent for all 27 shots? At least a 70% shooting percent. answers may vary

6. Choose one of the sports played at your college that is currently in season. How many regular-season games are scheduled? What is the team's current percentage of games won? answers may vary

7. Suppose the team has a goal of finishing the season with a winning percentage better than 110% of their current wins. At least how many of the remaining games must they win to achieve their goal? answers may vary

CHAPTER 2 HIGHLIGHTS

DEFINITIONS AND CONCEPTS	EXAMPLES

SECTION 2.1 SIMPLIFYING EXPRESSIONS

The **numerical coefficient** of a **term** is its numerical factor.

TERM	NUMERICAL COEFFICIENT
$-7y$	-7
x	1
$\frac{1}{5}a^2b$	$\frac{1}{5}$

Terms with the same variables raised to exactly the same powers are **like terms**.

LIKE TERMS	UNLIKE TERMS
$12x, -x$	$3y, 3y^2$
$-2xy, 5yx$	$7a^2b, -2ab^2$

To combine like terms, add the numerical coefficients and multiply the result by the common variable factor.

$$9y + 3y = 12y$$
$$-4z^2 + 5z^2 - 6z^2 = -5z^2$$

To remove parentheses, apply the distributive property.

$$-4(x + 7) + 10(3x - 1)$$
$$= -4x - 28 + 30x - 10$$
$$= 26x - 38$$

SECTION 2.2 THE ADDITION PROPERTY OF EQUALITY

A **linear equation in one variable** can be written in the form $Ax + B = C$ where A, B, and C are real numbers and $A \neq 0$.

$$-3x + 7 = 2$$
$$3(x - 1) = -8(x + 5) + 4$$

Equivalent equations are equations that have the same solution.

$x - 7 = 10$ and $x = 17$ are equivalent equations.

ADDITION PROPERTY OF EQUALITY

Adding the same number to or subtracting the same number from both sides of an equation does not change its solution.

$$y + 9 = 3$$
$$y + 9 - 9 = 3 - 9$$
$$y = -6$$

SECTION 2.3 THE MULTIPLICATION PROPERTY OF EQUALITY

MULTIPLICATION PROPERTY OF EQUALITY

Multiplying both sides or dividing both sides of an equation by the same nonzero number does not change its solution.

$$\frac{2}{3}a = 18$$
$$\frac{3}{2}\left(\frac{2}{3}a\right) = \frac{3}{2}(18)$$
$$a = 27$$

SECTION 2.4 FURTHER SOLVING LINEAR EQUATIONS

TO SOLVE LINEAR EQUATIONS

1. Clear the equation of fractions.

Solve: $\dfrac{5(-2x + 9)}{6} + 3 = \dfrac{1}{2}$

1. $6 \cdot \dfrac{5(-2x + 9)}{6} + 6 \cdot 3 = 6 \cdot \dfrac{1}{2}$

2. Remove any grouping symbols such as parentheses.

2. $5(-2x + 9) + 18 = 3$ Apply the distributive property.
$-10x + 45 + 18 = 3$

3. Simplify each side by combining like terms.

3. $-10x + 63 = 3$ Combine like terms.

SECTION 2.4 (CONTINUED)	
4. Get all variable terms on one side and all numbers on the other side by using the addition property of equality. **5.** Get the variable alone by using the multiplication property of equality. **6.** Check the solution by substituting it into the original equation.	**4.** $-10x + 63 - 63 = 3 - 63$ Subtract 63. $\qquad -10x = -60$ **5.** $\dfrac{-10x}{-10} = \dfrac{-60}{-10}$ Divide by -10. $\qquad\quad x = 6$

SECTION 2.5 AN INTRODUCTION TO PROBLEM SOLVING	

PROBLEM-SOLVING STEPS

The height of the Hudson volcano in Chili is twice the height of the Kiska volcano in the Aleutian Islands. If the sum of their heights is 12,870 feet, find the height of each.

1. UNDERSTAND the problem.

1. Read and reread the problem. Guess a solution and check your guess.

Let x be the height of the Kiska volcano. Then $2x$ is the height of the Hudson volcano.

$$x \;\text{I} \qquad 2x \;\text{I}$$
$$\text{Kiska} \qquad \text{Hudson}$$

2. TRANSLATE the problem.

2.

height of Kiska	added to	height of Hudson	is	12,870
↓	↓	↓	↓	↓
x	$+$	$2x$	$=$	12,870

3. SOLVE the equation.

3. $x + 2x = 12{,}870$
$\quad\;\; 3x = 12{,}870$
$\qquad x = 4290$

4. INTERPRET the results.

4. *Check*: If x is 4290 then $2x$ is $2(4290)$ or 8580. Their sum is $4290 + 8580$ or 12,870, the required amount.

State: The Kiska volcano is 4290 feet high and the Hudson volcano is 8580 feet high.

SECTION 2.6 FORMULAS AND PROBLEM SOLVING	

An equation that describes a known relationship among quantities is called a **formula**.

To solve a formula for a specified variable, use the same steps as for solving a linear equation. Treat the specified variable as the only variable of the equation.

$A = lw$ (area of a rectangle)
$I = PRT$ (simple interest)

Solve $P = 2l + 2w$ for l.

$$P = 2l + 2w$$
$$P - 2w = 2l + 2w - 2w \qquad \text{Subtract } 2w.$$
$$P - 2w = 2l$$
$$\frac{P - 2w}{2} = \frac{2l}{2} \qquad\qquad\quad \text{Divide by 2.}$$
$$\frac{P - 2w}{2} = l \qquad\qquad\quad\;\; \text{Simplify.}$$

SECTION 2.7   PERCENT, RATIO, AND PROPORTION

Use the same problem-solving steps to solve a problem containing percents.

1. UNDERSTAND.

2. TRANSLATE.

3. SOLVE.

4. INTERPRET.

A **ratio** is the quotient of two numbers or two quantities.

The ratio of *a* to *b* can also be written as

$$\frac{a}{b} \quad \text{or} \quad a{:}b$$

A **proportion** is a mathematical statement that two ratios are equal.

In the proportion $\frac{a}{b} = \frac{c}{d}$, the products ad and bc are called **cross products**.

If $\frac{a}{b} = \frac{c}{d}$, then $ad = bc$.

32% of what number is 36.8?

1. Read and reread. Propose a solution and check. Let $x =$ the unknown number.

2.

32%	of	what number	is	36.8
↓	↓	↓	↓	↓
32%	·	x	=	36.8

3. Solve $32\% \cdot x = 36.8$

$0.32x = 36.8$

$\dfrac{0.32x}{0.32} = \dfrac{36.8}{0.32}$ Divide by 0.32.

$x = 115$ Simplify.

4. 32% of 115 is 36.8.

Write the ratio of 5 hours to 1 day using fractional notation.

$$\frac{5\text{ hours}}{1\text{ day}} = \frac{5\text{ hours}}{24\text{ hours}} = \frac{5}{24}$$

$$\frac{2}{3} = \frac{8}{12} \qquad \frac{x}{7} = \frac{15}{35}$$

$$\frac{2}{3} \diagdown \frac{8}{12} \longrightarrow 3 \cdot 8 \text{ or } 24$$
$$\longrightarrow 2 \cdot 12 \text{ or } 24$$

Solve: $\dfrac{3}{4} = \dfrac{x}{x-1}$

$$\frac{3}{4} \diagup \frac{x}{x-1}$$

$3(x-1) = 4x$ Set cross products equal.

$3x - 3 = 4x$

$-3 = x$

SECTION 2.8   SOLVING LINEAR INEQUALITIES

Properties of inequalities are similar to properties of equations. Don't forget that if you multiply or divide both sides of an inequality by the same *negative* number, you must reverse the direction of the inequality symbol.

TO SOLVE LINEAR INEQUALITIES

1. Clear the inequality of fractions.
2. Remove grouping symbols.
3. Simplify each side by combining like terms.
4. Write all variable terms on one side and all numbers on the other side using the addition property of inequality.
5. Get the variable alone by using the multiplication property of inequality.

$-2x \le 4$

$\dfrac{-2x}{-2} \ge \dfrac{4}{-2}$ Divide by −2, reverse the inequality symbol.

$x \ge -2$

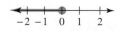

Solve: $3(x+2) \le -2 + 8$

1. $3(x+2) \le -2 + 8$ No fractions to clear.
2. $3x + 6 \le -2 + 8$ Apply the distributive property.
3. $3x + 6 \le 6$ Combine like terms.
4. $3x + 6 - 6 \le 6 - 6$ Subtract 6.

$3x \le 0$

5. $\dfrac{3x}{3} \le \dfrac{0}{3}$ Divide by 3.

$x \le 0$

Name _____ Section _____ Date _____

Chapter 2 Review

(2.1) *Simplify each expression.*

1. $5x - x + 2x$ $6x$

2. $0.2z - 4.6z - 7.4z$ $-11.8z$

3. $\frac{1}{2}x + 3 + \frac{7}{2}x - 5$ $4x - 2$

4. $\frac{4}{5}y + 1 + \frac{6}{5}y + 2$ $2y + 3$

5. $2(n - 4) + n - 10$
$3n - 18$

6. $3(w + 2) - (12 - w)$
$4w - 6$

7. Subtract $7x - 2$ from $x + 5$ $-6x + 7$

8. Subtract $1.4y - 3$ from $y - 0.7$ $-0.4y + 2.3$

Write each phrase as an algebraic expression. Simplify if possible.

9. Three times a number decreased by 7 $3x - 7$

10. Twice the sum of a number and 2.8 added to 3 times the number $5x + 5.6$

(2.2) *Solve each equation.*

11. $8x + 4 = 9x$ $x = 4$

12. $5y - 3 = 6y$ $y = -3$

13. $\frac{2}{7}x + \frac{5}{7}x = 6$ $x = 6$

14. $3x - 5 = 4x + 1$ $x = -6$

15. $2x - 6 = x - 6$ $x = 0$

16. $4(x + 3) = 3(1 + x)$
$x = -9$

17. $6(3 + n) = 5(n - 1)$ $n = -23$

18. $5(2 + x) - 3(3x + 2) = -5(x - 6) + 2$
$x = 28$

Choose the correct algebraic expression.

19. The sum of two numbers is 10. If one number is x, express the other number in terms of x.   b
   a. $x - 10$
   b. $10 - x$
   c. $10 + x$
   d. $10x$

20. Mandy is 5 inches taller than Melissa. If x inches represents the height of Mandy, express Melissa's height in terms of x.   a
   a. $x - 5$
   b. $5 - x$
   c. $5 + x$
   d. $5x$

21. If one angle measures $(x + 5)°$, express the measure of its supplement in terms of x.   c
   a. $185 + x$
   b. $95 + x$
   c. $175 - x$
   d. $x - 170$

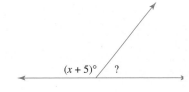

(2.3) *Solve each equation.*

22. $\frac{3}{4}x = -9$ $x = -12$

23. $\frac{x}{6} = \frac{2}{3}$ $x = 4$

24. $-5x = 0$ $x = 0$

25. $-y = 7$ $y = -7$

26. $0.2x = 0.15$ $x = 0.75$

27. $\frac{-x}{3} = 1$ $x = -3$

28. $-3x + 1 = 19$ $x = -6$

29. $5x + 25 = 20$ $x = -1$

30. $5x - 6 + x = 4x$ $x = 3$

31. $-y + 4y = -y$ $y = 0$

32. $-5x + \frac{3}{7} = \frac{10}{7}$ $x = -\frac{1}{5}$

33. Write the sum of three consecutive integers as an expression in x. Let x be the first even integer.
$3x + 3$

(2.4) *Solve each equation.*

34. $\frac{5}{3}x + 4 = \frac{2}{3}x$ $x = -4$

35. $-(5x + 1) = -7x + 3$
$x = 2$

36. $-4(2x + 1) = -5x + 5$
$x = -3$

37. $-6(2x - 5) = -3(9 + 4x)$
no solution

38. $3(8y - 1) = 6(5 + 4y)$
no solution

39. $\frac{3(2 - z)}{5} = z$ $z = \frac{3}{4}$

40. $\frac{4(n + 2)}{5} = -n$ $n = -\frac{8}{9}$

41. $0.5(2n - 3) - 0.1 = 0.4(6 + 2n)$ $n = 20$

42. $-9 - 5a = 3(6a - 1)$ $a = -\frac{6}{23}$

43. $\frac{5(c + 1)}{6} = 2c - 3$ $c = \frac{23}{7}$

44. $\frac{2(8 - a)}{3} = 4 - 4a$ $a = -\frac{2}{5}$

45. $200(70x - 3560) = -179(150x - 19{,}300)$
$x = 102$

46. $1.72y - 0.04y = 0.42$ $y = 0.25$

Name _____

(2.5) *Solve each of the following.*

47. The height of the Eiffel Tower is 68 feet more than three times a side of its square base. If the sum of these two dimensions is 1380 feet, find the height of the Eiffel Tower.

height: 1052 ft

48. A 12-foot board is to be divided into two pieces so that one piece is twice as long as the other. If x represents the length of the shorter piece, find the length of each piece.

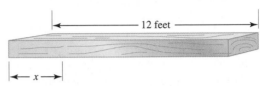

short piece: 4 ft; long piece: 8ft

49. One area code used in Ohio is 34 more than three times another area code used in Ohio. If the sum of these area codes is 1262, find the two area codes.

1st area code: 307; 2nd area code: 955

50. Find three consecutive integers whose sum is negative 114. $-39, -38, -37$

51. The quotient of a number and 3 is the same as the difference of the number and two. Find the number.   3

52. Double the sum of a number and 6 is the opposite of the number. Find the number. -4

(2.6) *Substitute the given values into the given formulas and solve for the unknown variable.*

53. $P = 2l + 2w$; $P = 46, l = 14$ $w = 9$

54. $V = lwh$; $V = 192, l = 8, w = 6$ $h = 4$

Solve each equation for the indicated variable.

55. $y = mx + b$ for m

$m = \dfrac{y - b}{x}$

56. $r = vst - 5$ for s $s = \dfrac{r + 5}{vt}$

57. $2y - 5x = 7$ for x

$x = \dfrac{2y - 7}{5}$

58. $3x - 6y = -2$ for y

$y = \dfrac{2 + 3x}{6}$

59. $C = \pi D$ for π $\pi = \dfrac{C}{D}$

60. $C = 2\pi r$ for π $\pi = \dfrac{C}{2r}$

61. A swimming pool holds 900 cubic meters of water. If its length is 20 meters and its height is 3 meters, find its width.   15 meters

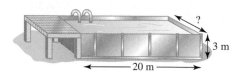

62. The highest temperature on record in Rome, Italy, is 104°F. Convert this temperature to degrees Celsius.   40°C

63. A charity 10K race is given annually to benefit a local hospice organization. How long will it take to run/walk a 10K race (10 kilometers or 10,000 meters) if your average pace is 125 **meters** per minute?   1 hour and 20 min

(2.7) *Find each of the following.*

64. The number 9 is what percent of 45?   20%

65. The number 59.5 is what percent of 85?   70%

66. The number 137.5 is 125% of what number?   110

67. The number 768 is 60% of what number?   1280

68. The state of Mississippi has the highest phoneless rate in the United States, 12.6% of households. If a city in Mississippi has 50,000 households, how many of these would you expect to be phoneless?
6300 households

The graph below shows how business travelers relax when in their hotel rooms. Use this graph to answer Exercises 69–72.

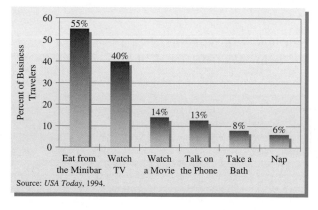

69. What percent of business travelers surveyed relax by taking a nap?   6%

70. What is the most popular way to relax according to the survey?   Eat from the Minibar

71. If a hotel in New York currently has 300 business travelers, how many might you expect to relax by watching TV?   120 travelers

72. Do the percents in the graph above have a sum of 100%? Why or why not?   no; answers may vary

Name _____

Write each phrase as a ratio in fractional notation.

73. 20 cents to 1 dollar $\dfrac{1}{5}$

74. four parts red to six parts white $\dfrac{2}{3}$

Solve each proportion.

75. $\dfrac{x}{2} = \dfrac{12}{4}$
$x = 6$

76. $\dfrac{20}{1} = \dfrac{x}{25}$
$x = 500$

77. $\dfrac{32}{100} = \dfrac{100}{x}$
$x = 312.5$

78. $\dfrac{20}{2} = \dfrac{c}{5}$
$c = 50$

79. $\dfrac{2}{x-1} = \dfrac{3}{x+3}$
$x = 9$

80. $\dfrac{4}{y-3} = \dfrac{2}{y-3}$
no solution

81. $\dfrac{y+2}{y} = \dfrac{5}{3}$
$y = 3$

82. $\dfrac{x-3}{3x+2} = \dfrac{2}{6}$
no solution

Given the following prices charged for various sizes of an item, find the best buy.

83. Shampoo
   10 ounces for $1.29
   16 ounces for $2.15   10 oz. for $1.29

84. Frozen green beans
   8 ounces for $0.89
   15 ounces for $1.63
   20 ounces for $2.36   15 oz. for $1.63

Solve.

85. A machine can process 300 parts in 20 minutes. Find how many parts can be processed in 45 minutes.   675 parts

86. As his consulting fee, Mr. Visconti charges $90.00 per day. Find how much he charges for 3 hours of consulting. Assume an 8-hour work day. $33.75

87. One fund raiser can address 100 letters in 35 minutes. Find how many he can address in 55 minutes.
157 letters

(2.8) *Graph on a number line.*

88. $x \le -2$

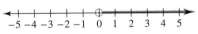

89. $x > 0$

Solve each inequality.

90. $x - 5 \le -4$ $x \le 1$

91. $x + 7 > 2$ $x > -5$

92. $-2x \ge -20$ $x \le 10$

93. $-3x > 12$ $x < -4$

94. $5x - 7 > 8x + 5$ $x < -4$

95. $x + 4 \geq 6x - 16$ $x \leq 4$

96. $\frac{2}{3}y > 6$ $y > 9$

97. $-0.5y \leq 7.5$ $y \geq -15$

98. $-2(x - 5) > 2(3x - 2)$ $x < \frac{7}{4}$

99. $4(2x - 5) \leq 5x - 1$ $x \leq \frac{19}{3}$

100. Tina earns $175 per week plus a 5% commission on all her sales. Find the minimum amount of sales to ensure that she earns at least $300 per week.
at least $2500

101. Ellen shot rounds of 76, 82, and 79 golfing. What must she shoot on her next round so that her average will be below 80?   score must be less than 83

Name _____ Section _____ Date _____

CHAPTER 2 TEST

Simplify each expression.

1. $2y - 6 - y - 4$

2. $2.7x + 6.1 + 3.2x - 4.9$

3. $4(x - 2) - 3(2x - 6)$

4. $-5(y + 1) + 2(3 - 5y)$

Solve each equation.

5. $-\dfrac{4}{5}x = 4$

6. $4(n - 5) = -(4 - 2n)$

7. $5y - 7 + y = -(y + 3y)$

8. $4z + 1 - z = 1 + z$

9. $\dfrac{2(x + 6)}{3} = x - 5$

10. $\dfrac{4(y - 1)}{5} = 2y + 3$

11. $\dfrac{1}{2} - x + \dfrac{3}{2} = x - 4$

12. $\dfrac{5}{y + 1} = \dfrac{4}{y + 2}$

13. $\dfrac{1}{3}(y + 3) = 4y$

14. $-0.3(x - 4) + x = 0.5(3 - x)$

15. $-4(a + 1) - 3a = -7(2a - 3)$

Solve each application.

16. A number increased by two-thirds of the number is 35. Find the number.

17. A gallon of water seal covers 200 square feet. How many gallons are needed to paint two coats of water seal on a deck that measures 20 feet by 35 feet?

20 feet

35 feet

18. Decide which is the best buy in crackers.
   6 ounces for $1.19
   10 ounces for $2.15
   16 ounces for $3.25

19. In a sample of 85 fluorescent bulbs, 3 were found to be defective. At this rate, how many defective bulbs should be found in 510 bulbs?

1. $y - 10$

2. $5.9x + 1.2$

3. $-2x + 10$

4. $-15y + 1$

5. $x = -5$

6. $n = 8$

7. $y = \dfrac{7}{10}$

8. $z = 0$

9. $x = 27$

10. $y = -\dfrac{19}{6}$

11. $x = 3$

12. $y = -6$

13. $y = \dfrac{3}{11}$

14. $x = 0.25$

15. $a = \dfrac{25}{7}$

16. 21

17. 7 gal

18. 6 oz. for $1.19

19. 18 bulbs

20. Find the value of x if
$y = -14$, $m = -2$, and $b = -2$ in
the formula $y = mx + b$.

Solve each equation for the indicated variable.

21. $V = \pi r^2 h$ for h

22. $3x - 4y = 10$ for y

Solve each inequality. Graph the solutions.

23. $3x - 5 > 7x + 3$

$\overset{\longleftarrow}{\underset{-5\ -4\ -3\ -2\ -1\ \ \ 0\ \ \ 1\ \ \ 2\ \ \ 3\ \ \ 4\ \ \ 5}{\vert\ \ \vert\ \ \vert\ \ \oplus\ \vert\ \ \vert\ \ \vert\ \ \vert\ \ \vert\ \ \vert\ \ \vert}}$

24. $x + 6 > 4x - 6$

$\overset{\longleftarrow}{\underset{-5\ -4\ -3\ -2\ -1\ \ \ 0\ \ \ 1\ \ \ 2\ \ \ 3\ \ \ 4\ \ \ 5}{\vert\ \ \vert\ \ \vert\ \ \vert\ \ \vert\ \ \vert\ \ \vert\ \ \vert\ \ \vert\ \ \oplus\ \vert}}$

Solve each inequality.

25. $-0.3x \geq 2.4$

26. $-5(x - 1) + 6 \leq -3(x + 4) + 1$

27. $\dfrac{2(5x + 1)}{3} > 2$

The following graph shows the source of income for charities. Use this graph to answer Exercises 28–29. (Source: American Association of Fund-Raising Counsel, Inc.)

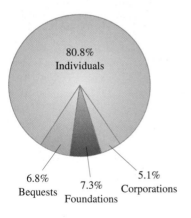

80.8%
Individuals

6.8%
Bequests

7.3%
Foundations

5.1%
Corporations

28. What percent of charity income comes from individuals?

29. In 1995, the total annual income for charities was $143.84 billion. Find the amount that came from corporations.

30. The number 72 is what percent of 180?

31. New York State has more public libraries than any other state. It has 520 more public libraries than Indiana does. If the total number of public libraries for these states is 996, find the number of public libraries in New York and the number in Indiana. (*Source: The World Almanac and Book of Facts*, 1997)

---

20. $x = 6$

21. $h = \dfrac{V}{\pi r^2}$

22. $y = \dfrac{3x - 10}{4}$

23. $x < -2$

24. $x < 4$

25. $x \leq -8$

26. $x \geq 11$

27. $x > \dfrac{2}{5}$

28. 80.8%

29. 7.33584 billion

30. 40%

31. New York: 758
Indiana: 238

Name	Section	Date

CUMULATIVE REVIEW

Determine whether each statement is true or false.

1. $8 \geq 8$ **2.** $8 \leq 8$ **3.** $23 \leq 0$ **4.** $0 \leq 23$

5. Insert $<$, $>$, or $=$ in the appropriate space to make each statement true.
  a. $|0|$ 2 **d.** $|5|$ $|6|$

  b. $|-5|$ 5 **e.** $|-7|$ $|6|$

  c. $|-3|$ $|-2|$

6. Simplify the expression $\dfrac{3 + |4 - 3| + 2^2}{6 - 3}$.

Add without using number lines.

7. $(-8) + (-11)$ **8.** $(-2) + 10$ **9.** $0.2 + (-0.5)$

10. Simplify each expression.
  a. $-3 + [(-2 - 5) - 2]$ **b.** $2^3 - 10 + [-6 - (-5)]$

11. Use order of operations and simplify each expression.
  a. $7 \cdot 0(-6)$ **d.** $(-2)^3$

  b. $(-2)(-3)(-4)$ **e.** $-4(-11) - 5(-2)$

  c. $(-1)(5)(-9)$

12. Use the definition of the quotient of two numbers to find each quotient.
  a. $-18 \div 3$ **b.** $\dfrac{-14}{-2}$ **c.** $\dfrac{20}{-4}$

Use the distributive property to write each expression without parentheses. Then simplify the result.

13. $-5(-3 + 2z)$ **14.** $4(3x + 7) + 10$

a. $70

b. 280 miles

a. Unlike

b. Like

c. Like

d. Like

188

Name _____

15. The line graph shows the relationship between two sets of measurements: the distance driven in a 14-foot U-Haul truck in one day and the total cost of renting this truck for that day. Notice that the horizontal axis is labeled Distance and the vertical axis is labeled Total Cost.

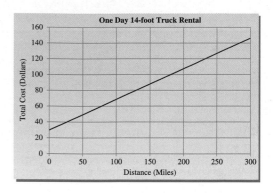

a. Find the total cost of renting the truck if 100 miles are driven.

b. Find the number of miles driven if the total cost of renting is $140.

16. Tell whether the terms are like or unlike.
   a. $2x, 3x^2$ **b.** $4x^2y, x^2y, -2x^2y$ **c.** $-2yz, -3zy$ **d.** $-x^4, x^4$

17. Subtract $4x - 2$ from $2x - 3$.

18. Solve $x - 7 = 10$ for x.

19. Solve: $-z - 4 = 6$

20. Solve: $\dfrac{2(a + 3)}{3} = 6a + 2$

21. In a recent year, Congress had 8 more Republican senators than Democratic. If the total number of senators is 100, how many senators of each party were there?

22. A glacier is a giant mass of rocks and ice that flows downhill like a river. Portage Glacier in Alaska is about 6 miles, or 31,680 feet, long and moves 400 feet per year. Icebergs are created when the front end of the glacier flows into Portage Lake. How long does it take for ice at the head (beginning) of the glacier to reach the lake?

23. The number 63 is what percent of 72?

24. Solve for x: $\dfrac{45}{x} = \dfrac{5}{7}$

25. Graph $-1 > x$.

$$\overset{\oplus}{\underset{-5\ -4\ -3\ -2\ -1\ \ 0\ \ 1\ \ 2\ \ 3\ \ 4\ \ 5}{\longleftrightarrow}}$$

26. Solve $2(x - 3) - 5 \leq 3(x + 2) - 18$.

Exponents and Polynomials

Recall from Chapter 1 that an exponent is a shorthand notation for repeated factors. This chapter explores additional concepts about exponents and exponential expressions. An especially useful type of exponential expression is a polynomial. Polynomials model many real-world phenomena. Our goal in this chapter is to become proficient with operations on polynomials.

One way exponents are used is to express very large or very small numbers in scientific notation. For instance, in Exercise 88 on page 220, you will find an estimate of the U.S. population written in scientific notation.

The U.S. Bureau of the Census conducts a census every 10 years to count U.S. citizens. However, government agencies and private businesses need more up-to-date population figures between censuses. From its research, the Census Bureau devised a way to estimate the U.S. population based on a few simple observations. There is a birth in the U.S. every 8 seconds and there is a death in the U.S. every 14 seconds. The estimate also takes into account the arrival of immigrants (one approximately every 42 seconds) and U.S. citizens returning from abroad (one approximately every 4108 seconds). These facts along with the most recent census figures can be used to estimate the current population. Visit the Bureau of the Census Web site, at http://www.census.gov/cgi-bin/popclock.

Name _____ Section _____ Date _____

CHAPTER PRETEST

1. Evaluate $\left(-\dfrac{3}{4}\right)^2$.

Simplify.

2. $\left(4y^6\right)\left(2y^7\right)$

3. $\dfrac{a^9 b^{16}}{a^{12} b^5}$

4. $4^0 + 2x^0$

5. $\left(-\dfrac{1}{6}\right)^{-3}$

6. $\left(\dfrac{m^{-2}n}{m^6 n^{-8}}\right)^{-2}$

7. $12x^2 + 3x - 5 - 8x^2 + 7x$

8. Express in scientific notation:
    0.000000814

9. Find the degree of the following polynomial.

    $8x - 4x^5 + 6x^3 + 10$

10. Find the value of the given polynomial when $x = -1$.

    $-3x^3 + 2x^2 - 4$

Perform the indicated operations.

11. $\left(4x^2 - 3x + 9\right) + \left(6x^2 + 3x - 8\right)$

12. $\left(6y^2 - 4\right) - \left(-3y^2 + 5y - 1\right)$

13. $\left(2a^2 + 3ab - 7b^2\right) - \left(3a^2 + 3ab + 9b^2\right)$

14. $\left(-\dfrac{2}{7}n^6\right)\left(\dfrac{21}{16}n^3\right)$

15. $-2t^2\left(3t^5 + 4t^3 - 8\right)$

16. $\left(2y - 1\right)\left(5y + 6\right)$

17. $\left(7a - 5\right)^2$

18. $\left(4b + 9\right)\left(4b - 9\right)$

19. $\dfrac{16p^4 - 8p^3 + 20p^2}{4p}$

20. $\dfrac{5x^2 - 28x - 12}{x - 6}$

3.1 EXPONENTS

A EVALUATING EXPONENTIAL EXPRESSIONS

In this section, we continue our work with integer exponents. As we reviewed in Section 1.2, for example,

$$2 \cdot 2 \cdot 2 \cdot 2 \cdot 2 = 2^5$$

The exponent 5 tells us how many times that 2 is a factor. The expression 2^5 is called an **exponential expression**. It is also called the fifth **power** of 2, or we can say that 2 is **raised** to the fifth power.

$$5^6 = \underbrace{5 \cdot 5 \cdot 5 \cdot 5 \cdot 5 \cdot 5}_{\text{6 factors; each factor is 5}} \quad \text{and} \quad (-3)^4 = \underbrace{(-3) \cdot (-3) \cdot (-3) \cdot (-3)}_{\text{4 factors; each factor is } -3}$$

The base of an exponential expression is the repeated factor. The exponent is the number of times that the base is used as a factor.

$$a^n = \underbrace{a \cdot a \cdot a \dots a}_{\substack{\uparrow \\ \text{base} \quad n \text{ factors of } a}} \quad \overset{\text{exponent or power}}{}$$

Examples   Evaluate (find the value of) each expression.

1. $2^3 = 2 \cdot 2 \cdot 2 = 8$

2. $3^1 = 3$. To raise 3 to the first power means to use 3 as a factor only once. When no exponent is shown, the exponent is assumed to be 1.

3. $(-4)^2 = (-4)(-4) = 16$

4. $-4^2 = -(4 \cdot 4) = -16$

5. $\left(\dfrac{1}{2}\right)^4 = \dfrac{1}{2} \cdot \dfrac{1}{2} \cdot \dfrac{1}{2} \cdot \dfrac{1}{2} = \dfrac{1}{16}$

6. $4 \cdot 3^2 = 4 \cdot 9 = 36$

Notice how similar -4^2 is to $(-4)^2$ in the examples above. The difference between the two is the parentheses. In $(-4)^2$, the parentheses tell us that the base, or the repeated factor, is -4. In -4^2, only 4 is the base.

⌐HELPFUL HINT

Be careful when identifying the base of an exponential expression. Pay close attention to the use of parentheses.

$$(-3)^2 \qquad\qquad -3^2 \qquad\qquad 2 \cdot 3^2$$
The base is -3.   The base is 3.   The base is 3.
$$(-3)^2 = (-3)(-3) = 9 \quad -3^2 = -(3 \cdot 3) = -9 \quad 2 \cdot 3^2 = 2 \cdot 3 \cdot 3 = 18$$

An exponent has the same meaning whether the base is a number or a variable. If x is a real number and n is a positive integer, then x^n is the product of n factors, each of which is x.

$$x^n = \underbrace{x \cdot x \cdot x \cdot x \cdot x \dots x}_{n \text{ factors; each factor is } x}$$

Objectives

A Evaluate exponential expressions.

B Use the product rule for exponents.

C Use the power rule for exponents.

D Use the power rules for products and quotients.

E Use the quotient rule for exponents, and define a number raised to the 0 power.

F Decide which rule(s) to use to simplify an expression.

SSM   CD-ROM   Video 3.1

Practice Problems 1–6

Evaluate (find the value of) each expression.

1. 3^4 2. 7^1

3. $(-2)^3$ 4. -2^3

5. $\left(\dfrac{2}{3}\right)^2$ 6. $5 \cdot 6^2$

TEACHING TIP

After the examples, ask students to write each of the following expressions as an exponential expression.

	Answers
1. 27	3^3
2. 7	7^1
3. 36	6^2 or $(-6)^2$
4. -36	-6^2
5. $\dfrac{1}{81}$	$\left(\dfrac{1}{9}\right)^2$
6. 75	$3 \cdot 5^2$

TEACHING TIP

Before doing Example 7, have students fill out the following table.

x	2	-2
x^2	$(2)^2 = 4$	$(-2)^2 = 4$
x^3	$(2)^3 = 8$	$(-2)^3 = -8$
x^4	$(2)^4 = 16$	$(-2)^4 = 16$
x^5	$(2)^5 = 32$	$(-2)^5 = -32$

Answers

1. 81, **2.** 7, **3.** -8, **4.** -8, **5.** $\dfrac{4}{9}$, **6.** 180

Practice Problem 7

Evaluate each expression for the given value of x.

a. $3x^2$ when x is 4

b. $\dfrac{x^4}{-8}$ when x is -2

Example 7

Evaluate each expression for the given value of x.

a. $2x^3$ when x is 5

b. $\dfrac{9}{x^2}$ when x is -3

Solution:

a. When x is 5, $2x^3 = 2 \cdot 5^3$

$$= 2 \cdot (5 \cdot 5 \cdot 5)$$
$$= 2 \cdot 125$$
$$= 250$$

b. When x is -3, $\dfrac{9}{x^2} = \dfrac{9}{-3^2}$

$$= \dfrac{9}{(-3)(-3)}$$
$$= \dfrac{9}{9} = 1$$

B USING THE PRODUCT RULE

Exponential expressions can be multiplied, divided, added, subtracted, and themselves raised to powers. By our definition of an exponent,

$$5^4 \cdot 5^3 = \underbrace{(5 \cdot 5 \cdot 5 \cdot 5)}_{4 \text{ factors of } 5} \cdot \underbrace{(5 \cdot 5 \cdot 5)}_{3 \text{ factors of } 5}$$

$$= \underbrace{5 \cdot 5 \cdot 5 \cdot 5 \cdot 5 \cdot 5 \cdot 5}_{7 \text{ factors of } 5}$$

$$= 5^7$$

Also,

$$x^2 \cdot x^3 = (x \cdot x) \cdot (x \cdot x \cdot x)$$
$$= x \cdot x \cdot x \cdot x \cdot x$$
$$= x^5$$

In both cases, notice that the result is exactly the same if the exponents are added.

$$5^4 \cdot 5^3 = 5^{4+3} = 5^7 \quad \text{and} \quad x^2 \cdot x^3 = x^{2+3} = x^5$$

This suggests the following rule.

> **PRODUCT RULE FOR EXPONENTS**
>
> If m and n are positive integers and a is a real number, then
> $$a^m \cdot a^n = a^{m+n}$$
> For example:
> $$3^5 \cdot 3^7 = 3^{5+7} = 3^{12}$$

In other words, to multiply two exponential expressions with the **same base**, we keep the base and add the exponents. We call this *simplifying* the exponential expression.

Practice Problems 8–12

Use the product rule to simplify each expression.

8. $7^3 \cdot 7^2$ 9. $x^4 \cdot x^9$

10. $r^5 \cdot r$ 11. $s^6 \cdot s^2 \cdot s^3$

12. $(-3)^9 \cdot (-3)$

Examples

Use the product rule to simplify each expression.

8. $4^2 \cdot 4^5 = 4^{2+5} = 4^7$

9. $x^2 \cdot x^5 = x^{2+5} = x^7$

10. $y^3 \cdot y = y^3 \cdot y^1$
$$= y^{3+1}$$
$$= y^4$$

> **HELPFUL HINT**
>
> Don't forget that if no exponent is written, it is assumed to be 1.

Answers

7. a. 48,   **b.** -2,   **8.** 7^5,   **9.** x^{13},   **10.** r^6,
11. s^{11},   **12.** $(-3)^{10}$

11. $y^3 \cdot y^2 \cdot y^7 = y^{3+2+7} = y^{12}$

12. $(-5)^7 \cdot (-5)^8 = (-5)^{7+8} = (-5)^{15}$ ▬▬▬

Example 13   Use the product rule to simplify $(2x^2)(-3x^5)$.

Solution: Recall that $2x^2$ means $2 \cdot x^2$ and $-3x^5$ means $-3 \cdot x^5$.

$(2x^2)(-3x^5) = 2 \cdot x^2 \cdot -3 \cdot x^5$ Remove parentheses.

$= 2 \cdot -3 \cdot x^2 \cdot x^5$ Group factors with common bases.

$= -6x^7$ Simplify.

HELPFUL HINT

These examples will remind you of the difference between adding and multiplying terms.

Addition
$5x^3 + 3x^3 = (5 + 3)x^3 = 8x^3$
$7x + 4x^2 = 7x + 4x^2$

Multiplication
$(5x^3)(3x^3) = 5 \cdot 3 \cdot x^3 \cdot x^3 = 15x^{3+3} = 15x^6$
$(7x)(4x^2) = 7 \cdot 4 \cdot x \cdot x^2 = 28x^{1+2} = 28x^3$

C USING THE POWER RULE

Exponential expressions can themselves be raised to powers. Let's try to discover a rule that simplifies an expression like $(x^2)^3$. By the definition of a^n,

$(x^2)^3 = (x^2)(x^2)(x^2)$ $(x^2)^3$ means 3 factors of (x^2).

which can be simplified by the product rule for exponents.

$(x^2)^3 = (x^2)(x^2)(x^2) = x^{2+2+2} = x^6$

Notice that the result is exactly the same if we multiply the exponents.

$(x^2)^3 = x^{2 \cdot 3} = x^6$

The following rule states this result.

POWER RULE FOR EXPONENTS

If m and n are positive integers and a is a real number, then
$(a^m)^n = a^{mn}$

For example:
$(7^2)^5 = 7^{2 \cdot 5} = 7^{10}$

In other words, to raise an exponential expression to a power, we keep the base and multiply the exponents.

Examples   Use the power rule to simplify each expression.

14. $(5^3)^6 = 5^{3 \cdot 6} = 5^{18}$

15. $(y^8)^2 = y^{8 \cdot 2} = y^{16}$

Practice Problem 13

Use the product rule to simplify $(6x^3)(-2x^9)$.

TEACHING TIP

Consider beginning this section by having students discover the power rule using a numerical expression.
$8^4 = (2^3)^4$
$= 2^3 \cdot 2^3 \cdot 2^3 \cdot 2^3$
$= 2^{3+3+3+3}$
$= 2^{12}$

Optional: Now have students use a calculator to evaluate both 8^4 and 2^{12}.

Practice Problems 14–15

Use the power rule to simplify each expression.

14. $(9^4)^{10}$ 15. $(z^6)^3$

Answers
13. $-12x^{12}$, **14.** 9^{40}, **15.** z^{18}

HELPFUL HINT

Take a moment to make sure that you understand when to apply the product rule and when to apply the power rule.

Product Rule → Add Exponents
$x^5 \cdot x^7 = x^{5+7} = x^{12}$
$y^6 \cdot y^2 = y^{6+2} = y^8$

Power Rule → Multiply Exponents
$(x^5)^7 = x^{5 \cdot 7} = x^{35}$
$(y^6)^2 = y^{6 \cdot 2} = y^{12}$

TEACHING TIP

It may be helpful to emphasize to students that this rule does not apply when a power is taken of a sum. In other words, $(a + b)^{11} \neq a^{11} + b^{11}$. For example,

$(3 + 4)^2 \neq 3^2 + 4^2$

$7^2 \neq 9 + 16$

$49 \neq 25$

Practice Problems 16–18

Simplify each expression.

16. $(xy)^7$ 17. $(3y)^4$

18. $(-2p^4q^2r)^3$

POWER OF A QUOTIENT RULE

If n is a positive integer and a and c are real numbers, then

$$\left(\frac{a}{c}\right)^n = \frac{a^n}{c^n}, \quad c \neq 0$$

For example:

$$\left(\frac{y}{7}\right)^3 = \frac{y^3}{7^3}$$

Answers

16. x^7y^7,   **17.** $81y^4$,   **18.** $-8p^{12}q^6r^3$

D   USING THE POWER RULES FOR PRODUCTS AND QUOTIENTS

When the base of an exponential expression is a product, the definition of a^n still applies. To simplify $(xy)^3$, for example,

$(xy)^3 = (xy)(xy)(xy)$ $(xy)^3$ means 3 factors of (xy).

$= x \cdot x \cdot x \cdot y \cdot y \cdot y$ Group factors with common bases.

$= x^3y^3$ Simplify.

Notice that to simplify the expression $(xy)^3$, we raise each factor within the parentheses to a power of 3.

$(xy)^3 = x^3y^3$

In general, we have the following rule.

POWER OF A PRODUCT RULE

If n is a positive integer and a and b are real numbers, then

$$(ab)^n = a^nb^n$$

For example:

$$(3x)^5 = 3^5x^5$$

In other words, to raise a product to a power, we raise each factor to the power.

Examples   Simplify each expression.

16. $(st)^4 = s^4 \cdot t^4 = s^4t^4$ Use the power of a product rule.
17. $(2a)^3 = 2^3 \cdot a^3 = 8a^3$ Use the power of a product rule.
18. $(-5x^2y^3z)^2 = (-5)^2 \cdot (x^2)^2 \cdot (y^3)^2 \cdot (z^1)^2$ Use the power of a product rule.

     $= 25x^4y^6z^2$ Use the power rule for exponents.

Let's see what happens when we raise a quotient to a power. To simplify $\left(\frac{x}{y}\right)^3$, for example,

$$\left(\frac{x}{y}\right)^3 = \left(\frac{x}{y}\right)\left(\frac{x}{y}\right)\left(\frac{x}{y}\right)$$ $\left(\frac{x}{y}\right)^3$ means 3 factors of $\left(\frac{x}{y}\right)$.

$$= \frac{x \cdot x \cdot x}{y \cdot y \cdot y}$$ Multiply fractions.

$$= \frac{x^3}{y^3}$$ Simplify.

Notice that to simplify the expression, $\left(\frac{x}{y}\right)^3$, we raise both the numerator and the denominator to a power of 3.

$$\left(\frac{x}{y}\right)^3 = \frac{x^3}{y^3}$$

In general, we have the rule shown in the margin.

In other words, to raise a quotient to a power, we raise both the numerator and the denominator to the power.

Examples   Simplify each expression.

19. $\left(\dfrac{m}{n}\right)^7 = \dfrac{m^7}{n^7}, \quad n \neq 0$ Use the power of a quotient rule.

20. $\left(\dfrac{2x^4}{3y^5}\right)^4 = \dfrac{2^4 \cdot (x^4)^4}{3^4 \cdot (y^5)^4}$ Use the power of a quotient rule.

$= \dfrac{16x^{16}}{81y^{20}}, \quad y \neq 0$ Use the power rule for exponents.

Practice Problems 19–20

Simplify each expression.

19. $\left(\dfrac{r}{s}\right)^6$ **20.** $\left(\dfrac{5x^6}{9y^3}\right)^2$

E   USING THE QUOTIENT RULE AND DEFINING THE ZERO EXPONENT

Another pattern for simplifying exponential expressions involves quotients.

$$\dfrac{x^5}{x^3} = \dfrac{x \cdot x \cdot x \cdot x \cdot x}{x \cdot x \cdot x}$$
$$= \dfrac{x \cdot x \cdot x \cdot x \cdot x}{x \cdot x \cdot x}$$
$$= x \cdot x$$
$$= x^2$$

Notice that the result is exactly the same if we subtract exponents of the common bases.

$$\dfrac{x^5}{x^3} = x^{5-3} = x^2$$

The following rule states this result in a general way.

QUOTIENT RULE FOR EXPONENTS

If m and n are positive integers and a is a real number, then

$$\dfrac{a^m}{a^n} = a^{m-n}, \quad a \neq 0$$

For example:

$$\dfrac{x^6}{x^2} = x^{6-2} = x^4, \quad x \neq 0$$

In other words, to divide one exponential expression by another with a common base, we keep the base and subtract the exponents.

TEACHING TIP

Emphasize to students that the quotient rule does not apply when the numerator and denominator are not in factored form. For example,

$$\dfrac{2^3 + 3^4}{2^2 + 3^1} \neq 2^{3-2} + 3^{4-1}$$
$$\dfrac{8 + 81}{4 + 3} \neq 2^1 + 3^3$$
$$\dfrac{89}{7} \neq 2 + 27$$
$$12\dfrac{5}{7} \neq 29$$

Examples   Simplify each quotient.

21. $\dfrac{x^5}{x^2} = x^{5-2} = x^3$ Use the quotient rule.

22. $\dfrac{4^7}{4^3} = 4^{7-3} = 4^4 = 256$ Use the quotient rule.

23. $\dfrac{(-3)^5}{(-3)^2} = (-3)^3 = -27$ Use the quotient rule.

24. $\dfrac{2x^5y^2}{xy} = 2 \cdot \dfrac{x^5}{x^1} \cdot \dfrac{y^2}{y^1}$

$= 2 \cdot (x^{5-1}) \cdot (y^{2-1})$ Use the quotient rule.

$= 2x^4y^1 \quad \text{or} \quad 2x^4y$

Practice Problems 21–24

Simplify each quotient.

21. $\dfrac{y^7}{y^3}$ **22.** $\dfrac{5^9}{5^6}$

23. $\dfrac{(-2)^{14}}{(-2)^{10}}$ **24.** $\dfrac{7a^4b^{11}}{ab}$

Answers

19. $\dfrac{r^6}{s^6}$, $s \neq 0$, **20.** $\dfrac{25x^{12}}{81y^6}$, $y \neq 0$, **21.** y^4, **22.** 125, **23.** 16, **24.** $7a^3b^{10}$

Let's now give meaning to an expression such as x^0. To do so, we will simplify $\frac{x^3}{x^3}$ in two ways and compare the results.

$$\frac{x^3}{x^3} = x^{3-3} = x^0 \qquad \text{Apply the quotient rule.}$$

$$\frac{x^3}{x^3} = \frac{x \cdot x \cdot x}{x \cdot x \cdot x} = 1 \qquad \text{Apply the fundamental principle for fractions.}$$

Since $\frac{x^3}{x^3} = x^0$ and $\frac{x^3}{x^3} = 1$, we define that $x^0 = 1$ as long as x is not 0.

ZERO EXPONENT

$a^0 = 1$, as long as a is not 0.

For example: $5^0 = 1$.

In other words, a base raised to the 0 power is 1, as long as the base is not 0.

Examples   Simplify each expression.

25. $3^0 = 1$
26. $(5x^3y^2)^0 = 1$
27. $(-4)^0 = 1$
28. $-4^0 = -1 \cdot 4^0 = -1 \cdot 1 = -1$ ▬▬▬

TRY THE CONCEPT CHECK IN THE MARGIN.

F   **DECIDING WHICH RULE TO USE**

Let's practice deciding which rule to use to simplify.

Example 29   Simplify each expression.

   a. $x^7 \cdot x^4$
   b. $\left(\dfrac{1}{2}\right)^4$
   c. $(9y^5)^2$

Solution: **a.** Here, we have a product, so we use the product rule to simplify.

$$x^7 \cdot x^4 = x^{7+4} = x^{11}$$

   b. This is a quotient raised to a power, so we use the power of a quotient rule.

$$\left(\frac{1}{2}\right)^4 = \frac{1^4}{2^4} = \frac{1}{16}$$

   c. This is a product raised to a power, so we use the power of a product rule.

$$(9y^5)^2 = 9^2(y^5)^2 = 81y^{10} \qquad ▬▬▬$$

Practice Problems 25–28

Simplify each expression.

25. 8^0 26. $(2r^2s)^0$

27. $(-5)^0$ 28. -5^0

✓ CONCEPT CHECK

To simplify each expression, tell whether you would *add* the exponents, *subtract* the exponents, *multiply* the exponents, or *divide* the exponents, or *none of these*.

a. $\left(x^{63}\right)^{21}$ b. $\dfrac{y^{15}}{y^3}$

c. $z^{16} + z^8$ d. $w^{45} \cdot w^9$

Practice Problem 29

Simplify each expression.

a. $\dfrac{x^7}{x^4}$ b. $(3y^4)^4$ c. $\left(\dfrac{x}{4}\right)^3$

Answers

25. 1,   **26.** 1,   **27.** 1,   **28.** -1,   **29. a.** x^3,
b. $81y^{16}$,   **c.** $\dfrac{x^3}{64}$

✓ Concept Check

a. multiply,   **b.** subtract,   **c.** none of these,
d. add

Name _____ **Section** _____ **Date** _____

MENTAL MATH

State the bases and the exponents for each expression.

1. 3^2 **2.** 5^4 **3.** $(-3)^6$ **4.** -3^7

5. -4^2 **6.** $(-4)^3$ **7.** $5 \cdot 3^4$ **8.** $9 \cdot 7^6$

9. $5x^2$ **10.** $(5x)^2$

EXERCISE SET 3.1

A Evaluate each expression. See Examples 1 through 6.

1. 7^2 **2.** -3^2 **3.** $(-5)^1$ **4.** $(-3)^2$

5. -2^4 **6.** -4^3 **7.** $(-2)^4$ **8.** $(-4)^3$

9. $\left(\dfrac{1}{3}\right)^3$ **10.** $\left(-\dfrac{1}{9}\right)^2$ **11.** $7 \cdot 2^4$ **12.** $9 \cdot 1^2$

13. Explain why $(-5)^4 = 625$, while $-5^4 = -625$.

14. Explain why $5 \cdot 4^2 = 80$, while $(5 \cdot 4)^2 = 400$.

Evaluate each expression with the given replacement values. See Example 7.

15. x^2 when $x = -2$ **16.** x^3 when $x = -2$

17. $5x^3$ when $x = 3$ **18.** $4x^2$ when $x = -1$

19. $2xy^2$ when $x = 3$ and $y = 5$ **20.** $-4x^2y^3$ when $x = 2$ and $y = -1$

21. $\dfrac{2z^4}{5}$ when $z = -2$ **22.** $\dfrac{10}{3y^3}$ when $y = 5$

---

23. x^7

24. y^3

25. $(-3)^{12}$

26. $(-5)^{13}$

27. $15y^5$

28. $4z^5$

29. $-24z^0$

30. $-12x^{15}$

31. $20x^5$ sq. ft

32. $18y^{17}$ sq. m

33. x^{36}

34. y^{35}

35. p^7q^7

36. a^6b^6

37. $8a^{15}$

38. $16x^{12}$

39. $\dfrac{m^9}{n^9}$

40. $\dfrac{x^2y^2}{49}$

41. $x^{10}y^{15}$

42. $a^{28}b^7$

43. $\dfrac{4x^2z^2}{y^{10}}$

44. $-\dfrac{y^{12}}{27z^9}$

45. $64z^{10}$ sq. decimeters

46. $25y^2\pi$ sq. cm

198

Name _____

B *Use the product rule to simplify each expression. Write the results using exponents. See Examples 8 through 13.*

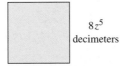 **23.** $x^2 \cdot x^5$ **24.** $y^2 \cdot y$ **25.** $(-3)^3 \cdot (-3)^9$

26. $(-5)^7 \cdot (-5)^6$ **27.** $(5y^4)(3y)$ **28.** $(-2z^3)(-2z^2)$

29. $(4z^{10})(-6z^7)(z^3)$ **30.** $(12x^5)(-x^6)(x^4)$

31. The following rectangle has width $4x^2$ feet and length $5x^3$ feet. Find its area.

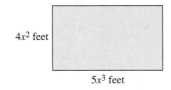
$4x^2$ feet

$5x^3$ feet

32. The following parallelogram has base length $9y^7$ meters and height $2y^{10}$ meters. Find its area.

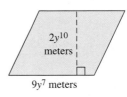
$2y^{10}$ meters

$9y^7$ meters

C *Use the power rule and the power of a product or quotient rule to simplify each expression. See Examples 14 through 20.*

33. $(x^9)^4$ **34.** $(y^7)^5$ **35.** $(pq)^7$ **36.** $(ab)^6$

37. $(2a^5)^3$ **38.** $(4x^6)^2$ **39.** $\left(\dfrac{m}{n}\right)^9$ **40.** $\left(\dfrac{xy}{7}\right)^2$

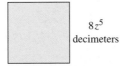 **41.** $(x^2y^3)^5$ **42.** $(a^4b)^7$ **43.** $\left(\dfrac{-2xz}{y^5}\right)^2$ **44.** $\left(\dfrac{y^4}{-3z^3}\right)^3$

45. The square shown has sides of length $8z^5$ decimeters. Find its area.

$8z^5$ decimeters

46. Given the following circle with radius $5y$ centimeters, find its area. Do not approximate π.

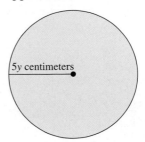
$5y$ centimeters

Name _____

47. The following vault is in the shape of a cube. If each side is $3y^4$ feet, find its volume.

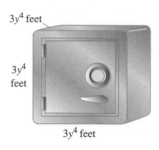

3y⁴ feet

3y⁴ feet

3y⁴ feet

48. The silo shown is in the shape of a cylinder. If its radius is $4x$ meters and its height is $5x^3$ meters, find its volume. Do not approximate π.

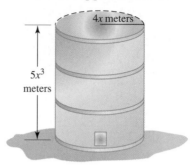

4x meters

5x³ meters

E *Use the quotient rule and simplify each expression. See Examples 21 through 24.*

 49. $\dfrac{x^3}{x}$

50. $\dfrac{y^{10}}{y^9}$

51. $\dfrac{(-2)^5}{(-2)^3}$

52. $\dfrac{(-5)^{14}}{(-5)^{11}}$

53. $\dfrac{p^7 q^{20}}{pq^{15}}$

54. $\dfrac{x^8 y^6}{xy^5}$

55. $\dfrac{7x^2 y^6}{14x^2 y^3}$

56. $\dfrac{9a^4 b^7}{3ab^2}$

Simplify each expression. See Examples 25 through 28.

57. $(2x)^0$

58. $-4x^0$

59. $-2x^0$

60. $(4y)^0$

61. $5^0 + y^0$

62. $-3^0 + 4^0$

63. In your own words, explain why $5^0 = 1$.

64. In your own words, explain when $(-3)^n$ is positive and when it is negative.

F *Simplify each expression. See Example 29.*

65. -5^2

66. $(-5)^2$

67. $\left(\dfrac{1}{4}\right)^3$

68. $\left(\dfrac{2}{3}\right)^3$

69. $\dfrac{z^{12}}{z^4}$

70. $\dfrac{b^4}{b}$

71. $(9xy)^2$

72. $(2ab)^5$

47. $27y^{12}$ cubic ft

48. $80x^5\pi$ cubic m

49. x^2

50. y

51. 4

52. -125

53. $p^6 q^5$

54. $x^7 y$

55. $\dfrac{y^3}{2}$

56. $3a^3 b^5$

57. 1

58. -4

59. -2

60. 1

61. 2

62. 0

63. answers may vary

64. answers may vary

65. -25

66. 25

67. $\dfrac{1}{64}$

68. $\dfrac{8}{27}$

69. z^8

70. b^3

71. $81x^2 y^2$

72. $32a^5 b^5$

73. 1

74. 1

75. 40

76. 48

77. b^6

78. y^5

79. a^9

80. x^{26}

81. $-16x^7$

82. $-15y^5$

83. $64a^3$

84. $16a^4b^4$

85. $36x^2y^2z^6$

86. $-27x^3y^6a^9b^3$

87. $\dfrac{y^{15}}{8x^{12}}$

88. $\dfrac{a^4b^4}{81y^4z^4}$

89. $3x$

90. $5x^6$

91. $2x^2y$

92. x^7y^6

93. -2

94. -3

95. 5

96. 15

97. -7

98. 6

99. 343 cubic meters

100. 150 sq. meters

Name _____

73. $(6b)^0$ **74.** $(5ab)^0$ **75.** $2^3 + 2^5$ **76.** $7^2 - 7^0$

77. b^4b^2 **78.** y^4y^1 **79.** $a^2a^3a^4$ **80.** $x^2x^{15}x^9$

▣ **81.** $(2x^3)(-8x^4)$ **82.** $(3y^4)(-5y)$ **83.** $(4a)^3$ **84.** $(2ab)^4$

85. $(-6xyz^3)^2$ **86.** $(-3xy^2a^3b)^3$ ▣ **87.** $\left(\dfrac{3y^5}{6x^4}\right)^3$ **88.** $\left(\dfrac{2ab}{6yz}\right)^4$

89. $\dfrac{3x^5}{x^4}$ **90.** $\dfrac{5x^9}{x^3}$ **91.** $\dfrac{2x^3y^2z}{xyz}$ **92.** $\dfrac{x^{12}y^{13}}{x^5y^7}$

Review and Preview

Subtract.

93. $5 - 7$ **94.** $9 - 12$ **95.** $3 - (-2)$ **96.** $5 - (-10)$

97. $-11 - (-4)$ **98.** $-15 - (-21)$

◤ Combining Concepts

99. The formula $V = x^3$ can be used to find the volume V of a cube with side length x. Find the volume of a cube with side length 7 meters. (Volume is measured in cubic units.)

x

100. The formula $S = 6x^2$ can be used to find the surface area S of a cube with side length x. Find the surface area of the cube with side length 5 meters. (Surface area is measured in square units.)

101. To find the amount of water that a swimming pool in the shape of a cube can hold, do we use the formula for volume of the cube or surface area of the cube? (See Exercises 99 and 100.)

102. To find the amount of material needed to cover an ottoman in the shape of a cube, do we use the formula for volume of the cube or surface area of the cube? (See Exercises 99 and 100.)

Simplify each expression. Assume that variables represent positive integers.

103. $x^{5a}x^{4a}$

104. $b^{9a}b^{4a}$

105. $(a^b)^5$

106. $(2a^{4b})^4$

107. $\dfrac{x^{9a}}{x^{4a}}$

108. $\dfrac{y^{15b}}{y^{6b}}$

109. Suppose you borrow money for 6 months. If the interest rate is compounded monthly, the formula $A = P\left(1 + \dfrac{r}{12}\right)^6$ gives the total amount to be repaid at the end of 6 months. For a loan of $P = \$1000$ and interest rate of 9% ($r = 0.09$), how much will you need to pay off the loan?

110. On January 31, 1996, the Federal Reserve discount rate was set at 5% (*Source*: Federal Reserve Board). The discount rate is the interest rate at which banks can borrow money from the Federal Reserve System. Suppose a bank needs to borrow money from the Federal Reserve System for 3 months. If the interest is compounded monthly, the formula $A = P\left(1 + \dfrac{r}{12}\right)^3$ gives the total amount to be repaid at the end of 3 months. For a loan of $P = \$100,000$ and interest rate of $r = 0.05$, how much will the bank repay to the Federal Reserve at the end of 3 months?

101. volume

102. surface area

103. x^{9a}

104. b^{13a}

105. a^{5b}

106. $16a^{16b}$

107. x^{5a}

108. y^{9b}

109. $1045.85

110. $101,255.22

Focus On Study Skills

Studying for a Math Exam

Remember that one of the best ways to start preparing for an exam is to keep current with your assignments as they are made. Make an effort to clear up any confusion on topics as you cover them.

Begin reviewing for your exam a few days in advance. This way, if you find a topic that you still don't understand, you'll have plenty of time to ask your instructor, another student in your class, or a math tutor for help. Don't wait until the last minute to "cram" for an exam.

▲ Reread your notes and carefully review the Chapter Highlights at the end of each chapter to be covered.
▲ Work the exercises from the appropriate Chapter Review.
▲ Pay special attention to any new terminology or definitions in the chapter. Be sure you can state the meanings of definitions in your own words.
▲ Find a quiet place to take the Chapter Test found at the end of the chapter to be covered. This gives you a chance to practice taking the real exam, so try the Chapter Test without referring to your notes or looking up anything in your book. Give yourself the same amount of time to take the Chapter Test as you will have to take the exam for which you are preparing. If your exam covers more than one chapter, you should try taking the Chapter Tests for each chapter covered. You may also find working through the Cumulative Reviews helpful when preparing for a multi-chapter test.
▲ When you have finished taking the Chapter Test, check your answers in the back of the book. Redo any of the problems you missed. Then spend extra time solving similar problems.
▲ If you tend to get anxious while taking an exam, try to visualize yourself taking the exam in advance. Picture yourself being calm, clearheaded, and successful. Picture yourself remembering formulas and definitions with no trouble. When you are well prepared for an exam, a lot of nervousness can be avoided through positive thinking.
▲ Get lots of rest the night before the exam. It's hard to show how well you know the material if your brain is foggy from lack of sleep.
▲ Give yourself enough time so that you will arrive early for the exam. This way, if you run into difficulties on the way, you should still arrive on time.

3.2   Negative Exponents and Scientific Notation

A Simplifying Expressions Containing Negative Exponents

Our work with exponential expressions so far has been limited to exponents that are positive integers or 0. Here we expand to give meaning to an expression like x^{-3}.

Suppose that we wish to simplify the expression $\dfrac{x^2}{x^5}$. If we use the quotient rule for exponents, we subtract exponents:

$$\frac{x^2}{x^5} = x^{2-5} = x^{-3}, \quad x \neq 0$$

But what does x^{-3} mean? Let's simplify $\dfrac{x^2}{x^5}$ using the definition of a^n.

$$\frac{x^2}{x^5} = \frac{x \cdot x}{x \cdot x \cdot x \cdot x \cdot x}$$

$$= \frac{x \cdot x}{x \cdot x \cdot x \cdot x \cdot x} \qquad \text{Divide numerator and denominator by common factors by applying the fundamental principle for fractions.}$$

$$= \frac{1}{x^3}$$

If the quotient rule is to hold true for negative exponents, then x^{-3} must equal $\dfrac{1}{x^3}$.

From this example, we state the definition for negative exponents.

Negative Exponents

If a is a real number other than 0 and n is an integer, then

$$a^{-n} = \frac{1}{a^n}$$

For example,

$$x^{-3} = \frac{1}{x^3}$$

In other words, another way to write a^{-n} is to take its reciprocal and change the sign of its exponent.

Examples

Simplify by writing each expression with positive exponents only.

 1. $3^{-2} = \dfrac{1}{3^2} = \dfrac{1}{9}$ Use the definition of negative exponent.

 2. $2x^{-3} = 2 \cdot \dfrac{1}{x^3} = \dfrac{2}{x^3}$ Use the definition of negative exponent.

> **HELPFUL HINT**
>
> Don't forget that since there are no parentheses, only x is the base for the exponent -3.

3. $2^{-1} + 4^{-1} = \dfrac{1}{2} + \dfrac{1}{4} = \dfrac{2}{4} + \dfrac{1}{4} = \dfrac{3}{4}$

4. $(-2)^{-4} = \dfrac{1}{(-2)^4} = \dfrac{1}{(-2)(-2)(-2)(-2)} = \dfrac{1}{16}$

Objectives

A Simplify expressions containing negative exponents.

B Use the rules and definitions for exponents to simplify exponential expressions.

C Write numbers in scientific notation.

D Convert numbers in scientific notation to standard form.

SSM   CD-ROM   Video 3.2

TEACHING TIP

Verbalize the meaning of a negative exponent. For instance 3^{-2} is two factors of the reciprocal of 3 and $2x^{-3}$ is 2 times three factors of the reciprocal of x.

Practice Problems 1–4

Simplify by writing each expression with positive exponents only.

1. 5^{-3} 2. $7x^{-4}$

3. $5^{-1} + 3^{-1}$ 4. $(-3)^{-4}$

Answers

1. $\dfrac{1}{125}$, 2. $\dfrac{7}{x^4}$, 3. $\dfrac{8}{15}$, 4. $\dfrac{1}{81}$

HELPFUL HINT

A negative exponent *does not affect* the sign of its base.
Remember: Another way to write a^{-n} is to take its reciprocal and change the sign of its exponent, $a^{-n} = \dfrac{1}{a^n}$. For example,

$$x^{-2} = \frac{1}{x^2}, \qquad\qquad 2^{-3} = \frac{1}{2^3} \ \text{ or } \ \frac{1}{8}$$

$$\frac{1}{y^{-4}} = \frac{1}{\frac{1}{y^4}} = y^4, \qquad \frac{1}{5^{-2}} = 5^2 \ \text{ or } \ 25$$

Practice Problems 5–8

Simplify each expression. Write each result using positive exponents only.

5. $\left(\dfrac{6}{7}\right)^{-2}$ 6. $\dfrac{x}{x^{-4}}$

7. $\dfrac{y^{-9}}{z^{-5}}$ 8. $\dfrac{y^{-4}}{y^6}$

TEACHING TIP

Consider asking students if they see another approach to Example 5. For example,

$$\left(\frac{2}{3}\right)^{-3} = \left(\frac{3}{2}\right)^{3}$$
$$= \frac{3^3}{2^3}$$
$$= \frac{27}{8}$$

Examples

Simplify each expression. Write each result using positive exponents only.

5. $\left(\dfrac{2}{3}\right)^{-3} = \dfrac{2^{-3}}{3^{-3}} = \dfrac{3^3}{2^3} = \dfrac{27}{8}$ Use the negative exponent rule.

6. $\dfrac{y}{y^{-2}} = \dfrac{y^1}{y^{-2}} = y^{1-(-2)} = y^3$ Use the quotient rule.

7. $\dfrac{p^{-4}}{q^{-9}} = \dfrac{q^9}{p^4}$ Use the negative exponent rule.

8. $\dfrac{x^{-5}}{x^7} = x^{-5-7} = x^{-12} = \dfrac{1}{x^{12}}$

B   SIMPLIFYING EXPONENTIAL EXPRESSIONS

All the previously stated rules for exponents apply for negative exponents also. Here is a summary of the rules and definitions for exponents.

SUMMARY OF EXPONENT RULES

If m and n are integers and a, b, and c are real numbers, then:

Product rule for exponents:	$a^m \cdot a^n = a^{m+n}$
Power rule for exponents:	$(a^m)^n = a^{m \cdot n}$
Power of a product:	$(ab)^n = a^n b^n$
Power of a quotient:	$\left(\dfrac{a}{c}\right)^n = \dfrac{a^n}{c^n}, \quad c \neq 0$
Quotient rule for exponents:	$\dfrac{a^m}{a^n} = a^{m-n}, \quad a \neq 0$
Zero exponent:	$a^0 = 1, \quad a \neq 0$
Negative exponent:	$a^{-n} = \dfrac{1}{a^n}, \quad a \neq 0$

Examples

Simplify each expression. Write each result using positive exponents only.

9. $\dfrac{(x^3)^4 x}{x^7} = \dfrac{x^{12} \cdot x}{x^7} = \dfrac{x^{12+1}}{x^7} = \dfrac{x^{13}}{x^7} = x^{13-7} = x^6$ Use the power rule.

10. $\left(\dfrac{3a^2}{b}\right)^{-3} = \dfrac{3^{-3}(a^2)^{-3}}{b^{-3}}$ Raise each factor in the numerator and the denominator to the -3 power.

$= \dfrac{3^{-3}a^{-6}}{b^{-3}}$ Use the power rule.

$= \dfrac{b^3}{3^3a^6}$ Use the negative exponent rule.

$= \dfrac{b^3}{27a^6}$ Write 3^3 as 27.

11. $(y^{-3}z^6)^{-6} = (y^{-3})^{-6}(z^6)^{-6}$ Raise each factor to the -6 power.

$= y^{18}z^{-36} = \dfrac{y^{18}}{z^{36}}$

Raise each factor in the numerator to the fifth power.

12. $\dfrac{(2x)^5}{x^3} = \dfrac{\overbrace{2^5 \cdot x^5}}{x^3} = 2^5 \cdot x^{5-3} = 32x^2$

13. $\dfrac{x^{-7}}{(x^4)^3} = \dfrac{x^{-7}}{x^{12}} = x^{-7-12} = x^{-19} = \dfrac{1}{x^{19}}$

14. $(5y^3)^{-2} = 5^{-2}(y^3)^{-2}$ Raise each factor to the -2 power.

$= 5^{-2}y^{-6} = \dfrac{1}{5^2y^6} = \dfrac{1}{25y^6}$

C WRITING NUMBERS IN SCIENTIFIC NOTATION

Both very large and very small numbers frequently occur in many fields of science. For example, the distance between the sun and the planet Pluto is approximately 5,906,000,000 kilometers, and the mass of a proton is approximately 0.000000000000000000000000165 gram. It can be tedious to write these numbers in this standard decimal notation, so **scientific notation** is used as a convenient shorthand for expressing very large and very small numbers.

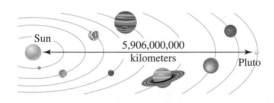

Sun   5,906,000,000 kilometers   Pluto

SCIENTIFIC NOTATION

A positive number is written in scientific notation if it is written as the product of a number a, where $1 \le a < 10$, and an integer power r of 10:

$a \times 10^r$

The following numbers are written in scientific notation. The $\times$ sign for multiplication is used as part of the notation.

2.03×10^2 7.362×10^7 5.906×10^9 (Distance between the sun and Pluto)

1×10^{-3} 8.1×10^{-5} 1.65×10^{-24} (Mass of a proton)

Simplify each expression. Write each result using positive exponents only.

9. $\dfrac{(x^5)^3x}{x^4}$ **10.** $\left(\dfrac{9x^3}{y}\right)^{-2}$

11. $(a^{-4}b^7)^{-5}$ **12.** $\dfrac{(2x)^4}{x^8}$

13. $\dfrac{y^{-10}}{(y^5)^4}$ **14.** $(4a^2)^{-3}$

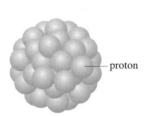

— proton

Mass of proton is approximately
0.000 000 000 000 000 000 000 001 65 gram

TEACHING TIP

Point out that scientific notation is a factored form of a number. For positive numbers, one of the factors is a power of 10 and the other is a number between 1 and 10. It is a useful form for multiplying and dividing very large or very small numbers, as will be seen in Example 17.

Answers

9. x^{12}, **10.** $\dfrac{y^2}{81x^6}$, **11.** $\dfrac{a^{20}}{b^{35}}$, **12.** $\dfrac{16}{x^4}$, **13.** $\dfrac{1}{y^{30}}$,

14. $\dfrac{1}{64a^6}$

The following steps are useful when writing numbers in scientific notation.

> **TO WRITE A NUMBER IN SCIENTIFIC NOTATION**
>
> **Step 1.** Move the decimal point in the original number so that the new number has a value between 1 and 10.
>
> **Step 2.** Count the number of decimal places the decimal point is moved in Step 1. If the decimal point is moved to the left, the count is positive. If the decimal point is moved to the right, the count is negative.
>
> **Step 3.** Multiply the new number in Step 1 by 10 raised to an exponent equal to the count found in Step 2.

Practice Problem 15

Write each number in scientific notation.

a. 420,000   b. 0.00017

c. 9,060,000,000   d. 0.000007

Example 15   Write each number in scientific notation.

a. 367,000,000   **b.** 0.000003
c. 20,520,000,000   **d.** 0.00085

Solution:

a. Step 1. Move the decimal point until the number is between 1 and 10.

367,000,000   8 places

Step 2. The decimal point is moved to the left 8 places, so the count is positive 8.

Step 3. $367,000,000 = 3.67 \times 10^8$.

b. Step 1. Move the decimal point until the number is between 1 and 10.

0.000003   6 places

Step 2. The decimal point is moved 6 places to the right, so the count is -6.

Step 3. $0.000003 = 3.0 \times 10^{-6}$

c. $20,520,000,000 = 2.052 \times 10^{10}$

d. $0.00085 = 8.5 \times 10^{-4}$

D CONVERTING NUMBERS TO STANDARD FORM

A number written in scientific notation can be rewritten in standard form. For example, to write 8.63×10^3 in standard form, recall that $10^3 = 1000$.

$$8.63 \times 10^3 = 8.63(1000) = 8630$$

Notice that the exponent on the 10 is positive 3, and we moved the decimal point 3 places to the right.

To write 7.29×10^{-3} in standard form, recall that $10^{-3} = \frac{1}{10^3} = \frac{1}{1000}$.

$$7.29 \times 10^{-3} = 7.29\left(\frac{1}{1000}\right) = \frac{7.29}{1000} = 0.00729$$

The exponent on the 10 is negative 3, and we moved the decimal to the left 3 places.

In general, **to write a scientific notation number in standard form**, move the decimal point the same number of places as the exponent on 10. If the exponent is positive, move the decimal point to the right; if the exponent is negative, move the decimal point to the left.

Try the Concept Check in the margin.

Example 16   Write each number in standard notation, without exponents.

  a. 1.02×10^5 **b.** 7.358×10^{-3}

  c. 8.4×10^7 **d.** 3.007×10^{-5}

Solution: **a.** Move the decimal point 5 places to the right.

$$1.02 \times 10^5 = 102{,}000.$$

  b. Move the decimal point 3 places to the left.

$$7.358 \times 10^{-3} = 0.007358$$

  c. $8.4 \times 10^7 = 84{,}000{,}000.$ 7 places to the right

  d. $3.007 \times 10^{-5} = 0.00003007$ 5 places to the left

Performing operations on numbers written in scientific notation makes use of the rules and definitions for exponents.

Example 17   Perform each indicated operation. Write each result in standard decimal notation.

  a. $(8 \times 10^{-6})(7 \times 10^3)$

  b. $\dfrac{12 \times 10^2}{6 \times 10^{-3}}$

Solution: **a.** $(8 \times 10^{-6})(7 \times 10^3) = 8 \cdot 7 \cdot 10^{-6} \cdot 10^3$

$$= 56 \times 10^{-3}$$

$$= 0.056$$

  b. $\dfrac{12 \times 10^2}{6 \times 10^{-3}} = \dfrac{12}{6} \times 10^{2-(-3)} = 2 \times 10^5 = 200{,}000$

CALCULATOR EXPLORATIONS
SCIENTIFIC NOTATION

To enter a number written in scientific notation on a scientific calculator, locate the scientific notation key, which may be marked $\boxed{\text{EE}}$ or $\boxed{\text{EXP}}$. To enter 3.1×10^7, press $\boxed{3.1}$ $\boxed{\text{EE}}$ $\boxed{7}$. The display should read $\boxed{3.1 \quad 07}$.

Enter each number written in scientific notation on your calculator.

1. 5.31×10^3 5.31 EE 3

2. -4.8×10^{14} -4.8 EE 14

3. 6.6×10^{-9} 6.6 EE -9

4. -9.9811×10^{-2} -9.9811 EE -2

Multiply each of the following on your calculator. Notice the form of the result.

5. $3,000,000 \times 5,000,000$ 1.5×10^{13}

6. $230,000 \times 1,000$ 2.3×10^8

Multiply each of the following on your calculator. Write the product in scientific notation.

7. $(3.26 \times 10^6)(2.5 \times 10^{13})$ 8.15×10^{19}

8. $(8.76 \times 10^{-4})(1.237 \times 10^9)$ 1.083612×10^6

Name _____ **Section** _____ **Date** _____

MENTAL MATH

State each expression using positive exponents only.

1. $5x^{-2}$

2. $3x^{-3}$

3. $\dfrac{1}{y^{-6}}$

4. $\dfrac{1}{x^{-3}}$

5. $\dfrac{4}{y^{-3}}$

6. $\dfrac{16}{y^{-7}}$

EXERCISE SET 3.2

A Simplify each expression. Write each result using positive exponents only. See Examples 1 through 8.

1. 4^{-3}

2. 6^{-2}

3. $7x^{-3}$

4. $(7x)^{-3}$

5. $\left(-\dfrac{1}{4}\right)^{-3}$

6. $\left(-\dfrac{1}{8}\right)^{-2}$

7. $3^{-1} + 2^{-1}$

8. $4^{-1} + 4^{-2}$

9. $\dfrac{1}{p^{-3}}$

10. $\dfrac{1}{q^{-5}}$

11. $\dfrac{p^{-5}}{q^{-4}}$

12. $\dfrac{r^{-5}}{s^{-2}}$

13. $\dfrac{x^{-2}}{x}$

14. $\dfrac{y}{y^{-3}}$

15. $\dfrac{z^{-4}}{z^{-7}}$

16. $\dfrac{x^{-4}}{x^{-1}}$

17. $2^0 + 3^{-1}$

18. $4^{-2} - 4^{-3}$

19. $(-3)^{-2}$

20. $(-2)^{-6}$

21. $\dfrac{-1}{p^{-4}}$

22. $\dfrac{-1}{y^{-6}}$

23. $-2^0 - 3^0$

24. $5^0 + (-5)^0$

B Simplify each expression. Write each result using positive exponents only. See Examples 9 through 14.

25. $\dfrac{x^2 x^5}{x^3}$

26. $\dfrac{y^4 y^5}{y^6}$

27. $\dfrac{p^2 p}{p^{-1}}$

28. $\dfrac{y^3 y}{y^{-2}}$

MENTAL MATH ANSWERS

1. $\dfrac{5}{x^2}$

2. $\dfrac{3}{x^3}$

3. y^6

4. x^3

5. $4y^3$

6. $16y^7$

ANSWERS

1. $\dfrac{1}{64}$

2. $\dfrac{1}{36}$

3. $\dfrac{7}{x^3}$

4. $\dfrac{1}{343x^3}$

5. -64

6. 64

7. $\dfrac{5}{6}$

8. $\dfrac{5}{16}$

9. p^3

10. q^5

11. $\dfrac{q^4}{p^5}$

12. $\dfrac{s^2}{r^5}$

13. $\dfrac{1}{x^3}$

14. y^4

15. z^3

16. $\dfrac{1}{x^3}$

17. $\dfrac{4}{3}$

18. $\dfrac{3}{64}$

19. $\dfrac{1}{9}$

20. $\dfrac{1}{64}$

21. $-p^4$

22. $-y^6$

23. -2

24. 2

25. x^4

26. y^3

27. p^4

28. y^6

29. m^{11}
30. x^8
31. r^6
32. p^4q^5
33. $\dfrac{1}{x^{15}y^9}$
34. $\dfrac{1}{z^{15}x^{15}}$
35. $\dfrac{1}{x^4}$
36. $\dfrac{1}{y^4}$
37. $\dfrac{1}{a^2}$
38. $\dfrac{1}{x^2}$
39. $4k^3$
40. $\dfrac{9}{r^2}$
41. $3m$
42. $-\dfrac{1}{a}$
43. $-\dfrac{4a^5}{b}$
44. $-\dfrac{y^3}{3}$
45. $-\dfrac{6x}{7y^2}$
46. $\dfrac{8}{5a^3}$
47. $\dfrac{a^{30}}{b^{12}}$
48. $\dfrac{16}{x^{10}}$
49. $\dfrac{1}{x^{10}y^6}$
50. $\dfrac{a^6}{b^9}$
51. $\dfrac{z^2}{4}$
52. $\dfrac{x^{11}}{81}$
53. $\dfrac{1}{32x^5}$
54. $\dfrac{5}{z^2}$
55. $\dfrac{49a^4}{b^6}$
56. $\dfrac{x^3}{216y^2}$
57. $a^{24}b^8$
58. $\dfrac{1}{r^6}$
59. x^9y^{19}
60. $\dfrac{1}{r^7s^9}$
61. $-\dfrac{y^8}{8x^2}$
62. $\dfrac{z^2}{9x^2y^2}$
63. $\dfrac{27}{z^3x^6}$ cubic in.
64. $\dfrac{10}{7x^4}$ sq. m

210

Name _____

29. $\dfrac{(m^5)^4m}{m^{10}}$

30. $\dfrac{(x^2)^8x}{x^9}$

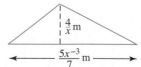 31. $\dfrac{r}{r^{-3}r^{-2}}$

32. $\dfrac{p}{p^{-3}q^{-5}}$

33. $(x^5y^3)^{-3}$

34. $(z^5x^5)^{-3}$

35. $\dfrac{(x^2)^3}{x^{10}}$

36. $\dfrac{(y^4)^2}{y^{12}}$

37. $\dfrac{(a^5)^2}{(a^3)^4}$

38. $\dfrac{(x^2)^5}{(x^4)^3}$

39. $\dfrac{8k^4}{2k}$

40. $\dfrac{27r^4}{3r^6}$

41. $\dfrac{-6m^4}{-2m^3}$

42. $\dfrac{15a^4}{-15a^5}$

43. $\dfrac{-24a^6b}{6ab^2}$

44. $\dfrac{-5x^4y^5}{15x^4y^2}$

45. $\dfrac{6x^2y^3}{-7xy^5}$

46. $\dfrac{-8xa^2b}{-5xa^5b}$

47. $(a^{-5}b^2)^{-6}$

48. $(4^{-1}x^5)^{-2}$

49. $\left(\dfrac{x^{-2}y^4}{x^3y^7}\right)^2$

50. $\left(\dfrac{a^5b}{a^7b^{-2}}\right)^{-3}$

51. $\dfrac{4^2z^{-3}}{4^3z^{-5}}$

52. $\dfrac{3^{-1}x^4}{3^3x^{-7}}$

53. $\dfrac{2^{-3}x^{-4}}{2^2x}$

54. $\dfrac{5^{-1}z^7}{5^{-2}z^9}$

55. $\dfrac{7ab^{-4}}{7^{-1}a^{-3}b^2}$

56. $\dfrac{6^{-5}x^{-1}y^2}{6^{-2}x^{-4}y^4}$

57. $\left(\dfrac{a^{-5}b}{ab^3}\right)^{-4}$

58. $\left(\dfrac{r^{-2}s^{-3}}{r^{-4}s^{-3}}\right)^{-3}$

59. $\dfrac{(xy^3)^5}{(xy)^{-4}}$

60. $\dfrac{(rs)^{-3}}{(r^2s^3)^2}$

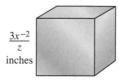 61. $\dfrac{(-2xy^{-3})^{-3}}{(xy^{-1})^{-1}}$

62. $\dfrac{(-3x^2y^2)^{-2}}{(xyz)^{-2}}$

63. Find the volume of the cube.

$\dfrac{3x^{-2}}{z}$ inches

64. Find the area of the triangle.

$\dfrac{4}{x}$ m

$\dfrac{5x^{-3}}{7}$ m

Name _____

C *Write each number in scientific notation. See Example 15.*

65. 78,000   **66.** 9,300,000,000   **67.** 0.00000167   **68.** 0.00000017

 **69.** 0.00635   **70.** 0.00194    **71.** 1,160,000   **72.** 700,000

73. The temperature at the interior of the Earth is 20,000,000 degrees Celsius. Write 20,000,000 in scientific notation.

74. The half-life of a carbon isotope is 5000 years. Write 5000 in scientific notation.

75. The distance between the Earth and the sun is 93,000,000 miles. Write 93,000,000 in scientific notation.

76. The population of the world is 5,887,000,000. Write 5,887,000,000 in scientific notation. (*Source:* U.S. Bureau of the Census)

77. On March 23, 1997, Comet Hale-Bopp passed its closest to earth. It was 120,000,000 miles away. (*Source: World Almanac and Book of Facts*, 1997) Write this number in scientific notation.

78. During fiscal year 1997, Wal-Mart Stores, Inc., conducted $104,859,000,000 in sales. (*Source: Hoover's Company Profiles*, 1997) Write this number in scientific notation.

65. 7.8×10^4

66. 9.3×10^9

67. 1.67×10^{-6}

68. 1.7×10^{-7}

69. 6.35×10^{-3}

70. 1.94×10^{-3}

71. 1.16×10^6

72. 7.0×10^5

73. 2.0×10^7

74. 5.0×10^3

75. 9.3×10^7

76. 5.887×10^9

77. 1.2×10^8

78. 1.04859×10^{11}

79. 0.0000000008673

80. 0.0009056

81. 0.033

82. 0.0000048

83. 20,320

84. 90,700,000,000

85. 6,250,000,000,000,000,000

86. 0.0000000000000000000000017

87. 9,460,000,000,000

88. 268,000,000

89. 0.000036

90. 5

91. 0.0000000000000000028

92. 0.2

93. 0.0000005

94. 500,000

95. 200,000

96. 0.000002

97. 1.512×10^{10} cu. ft

98. 9.46×10^{16} km

Name _____

D *Write each number in standard notation. See Example 16.*

79. 8.673×10^{-10} **80.** 9.056×10^{-4} **81.** 3.3×10^{-2}

82. 4.8×10^{-6} **83.** 2.032×10^{4} **84.** 9.07×10^{10}

85. One coulomb of electricity is 6.25×10^{18} electrons. Write this number in standard notation.

86. The mass of a hydrogen atom is 1.7×10^{-24} grams. Write this number in standard notation.

87. The distance light travels in 1 year is 9.460×10^{12} kilometers. Write this number in standard notation.

88. The population of the United States is 2.68×10^{8}. Write this number in standard notation. (*Source:* U.S. Bureau of the Census)

Evaluate each expression using exponential rules. Write each result in standard notation. See Example 17.

89. $(1.2 \times 10^{-3})(3 \times 10^{-2})$

90. $(2.5 \times 10^{6})(2 \times 10^{-6})$

91. $(4 \times 10^{-10})(7 \times 10^{-9})$

92. $(5 \times 10^{6})(4 \times 10^{-8})$

93. $\dfrac{8 \times 10^{-1}}{16 \times 10^{5}}$

94. $\dfrac{25 \times 10^{-4}}{5 \times 10^{-9}}$

95. $\dfrac{1.4 \times 10^{-2}}{7 \times 10^{-8}}$

96. $\dfrac{0.4 \times 10^{5}}{0.2 \times 10^{11}}$

97. The average amount of water flowing past the mouth of the Amazon River is 4.2×10^{6} cubic feet per second. How much water flows past in an hour? (*Hint:* 1 hour equals 3600 seconds.) Write the result in scientific notation.

98. A beam of light travels 9.460×10^{12} kilometers per year. How far does light travel in 10,000 years? Write the result in scientific notation.

Name _____

REVIEW AND PREVIEW

REVIEW AND PREVIEW

Simplify each expression by combining any like terms. See Section 2.1.

99. $3x - 5x + 7$ **100.** $7w + w - 2w$ **101.** $y - 10 + y$

102. $-6z + 20 - 3z$ **103.** $7x + 2 - 8x - 6$ **104.** $10y - 14 - y - 14$

COMBINING CONCEPTS

Simplify each expression. Write each result in standard notation.

105. $(2.63 \times 10^{12})(-1.5 \times 10^{-10})$ **106.** $(6.785 \times 10^{-4})(4.68 \times 10^{10})$

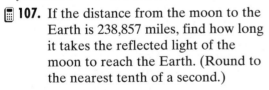

Light travels at a rate of 1.86×10^5 miles per second. Use this information and the distance formula $d = r \cdot t$ to answer Exercises 107 and 108.

107. If the distance from the moon to the Earth is 238,857 miles, find how long it takes the reflected light of the moon to reach the Earth. (Round to the nearest tenth of a second.)

108. If the distance from the sun to the Earth is 93,000,000 miles, find how long it takes the light of the sun to reach the Earth. (Round to the nearest tenth of a second.)

Simplify each expression. Assume that variables represent positive integers.

109. $a^{-4m} \cdot a^{5m}$ **110.** $(x^{-3s})^3$ **111.** $(3y^{2z})^3$ **112.** $a^{4m+1} \cdot a^4$

---

99. $-2x + 7$ _____

100. $6w$ _____

101. $2y - 10$ _____

102. $-9z + 20$ _____

103. $-x - 4$ _____

104. $9y - 28$ _____

105. -394.5 _____

106. $31,753,800$ _____

107. 1.3 sec. _____

108. 500 sec. _____

109. a^m _____

110. $\dfrac{1}{x^{9s}}$ _____

111. $27y^{6z}$ _____

112. a^{4m+5} _____

213

Name _____

113. It was stated earlier that for an integer n,

$$x^{-n} = \frac{1}{x^n}, \quad x \neq 0$$

Explain why x may not equal 0.

114. Determine whether each statement is true or false.

a. $5^{-1} < 5^{-2}$

b. $\left(\frac{1}{5}\right)^{-1} < \left(\frac{1}{5}\right)^{-2}$

c. $a^{-1} < a^{-2}$ for all nonzero numbers.

Internet Excursions

Go to http://www.prenhall.com/martin-gay
The Bureau of the Public Debt is part of the U.S. Department of the Treasury. The Bureau of the Public Debt borrows the money needed to run the federal government and keeps track of the debt. The given World Wide Web address will provide you with access to the Bureau of the Public Debt's Web Site, or a related site, where you can find the current size of the U.S. public debt to the penny. This site is updated daily. It also lists the amount of the public debt for the past month, as well as for selected dates in prior months and years.

115. Find the size of the public debt that is listed most recently on the Bureau of the Public Debt's Website. Use the information on the Website to record the debt amount and its date. Then write the debt amount in scientific notation, rounded to the nearest hundredth.

Debt amount: _____
Date: _____

Scientific notation (rounded to nearest hundredth): _____

116. Look up the size of the public debt at the end of the month six months ago. Use the information on the Website to record the debt amount and its date. Then write the debt amount in scientific notation, rounded to the nearest hundredth.

Debt amount: _____
Date: _____

Scientific notation (rounded to nearest hundredth): _____

3.3 INTRODUCTION TO POLYNOMIALS

A DEFINING TERM AND COEFFICIENT

In this section, we introduce a special algebraic expression called a polynomial. Let's first review some definitions presented in Section 2.1.

Recall that a term is a number or the product of a number and variables raised to powers. The terms of the expression $4x^2 + 3x$ are $4x^2$ and $3x$. The terms of the expression $9x^4 - 7x - 1$ are $9x^4$, $-7x$, and -1.

Expression	Terms
$4x^2 + 3x$	$4x^2, 3x$
$9x^4 - 7x - 1$	$9x^4, -7x, -1$
$7y^3$	$7y^3$

The **numerical coefficient** of a term, or simply the **coefficient**, is the numerical factor of each term. If no numerical factor appears in the term, then the coefficient is understood to be 1. If the term is a number only, it is called a **constant** term or simply a **constant**.

Term	Coefficient
x^5	1
$3x^2$	3
$-4x$	-4
$-x^2y$	-1
3 (constant)	3

Objectives

A Define term and coefficient of a term.

B Define polynomial, monomial, binomial, trinomial, and degree.

C Evaluate polynomials for given replacement values.

D Simplify a polynomial by combining like terms.

E Simplify a polynomial in several variables.

SSM   CD-ROM   Video 3.3

Example 1 Complete the table for the expression $7x^5 - 8x^4 + x^2 - 3x + 5$.

Term	Coefficient
x^2	
	-8
$-3x$	
	7
5	

Solution: The completed table is

Term	Coefficient
x^2	1
$-8x^4$	-8
$-3x$	-3
$7x^5$	7
5	5

Practice Problem 1

Complete the table for the expression $-6x^6 + 4x^5 + 7x^3 - 9x^2 - 1$.

Term	Coefficient
$7x^3$	
	-9
$-6x^6$	
	4
-1	

Answers

1. term: $-9x^2$; $4x^5$,   coefficient: 7; -6; -1

TEACHING TIP

Before discussing the various types of polynomials, consider using the following activity: Ask students to list words having the prefixes *mono-*, *bi-*, *tri-*, and *poly-*. Allow students to discuss the meanings of the prefixes based on the words on their lists. Then ask students to apply the meanings of the prefixes to guess the definitions of *monomial*, *binomial*, *trinomial*, and *polynomial*. Have students share their responses.

TEACHING TIP

Point out that identifying the degree of a polynomial will be useful when they solve polynomial equations.

B   DEFINING POLYNOMIAL, MONOMIAL, BINOMIAL, TRINOMIAL, AND DEGREE

Now we are ready to define what we mean by a polynomial.

> **POLYNOMIAL**
>
> A **polynomial in x** is a finite sum of terms of the form ax^n, where a is a real number and n is a whole number.

For example,

$$x^5 - 3x^3 + 2x^2 - 5x + 1$$

is a polynomial in x. Notice that this polynomial is written in **descending powers** of x because the powers of x decrease from left to right. (Recall that the term 1 can be thought of as $1x^0$.)

On the other hand,

$$x^{-5} + 2x - 3$$

is **not** a polynomial because one of its terms contains a variable with an exponent, -5, that is not a whole number.

> **TYPES OF POLYNOMIALS**
>
> A **monomial** is a polynomial with exactly one term.
> A **binomial** is a polynomial with exactly two terms.
> A **trinomial** is a polynomial with exactly three terms.

The following are examples of monomials, binomials, and trinomials. Each of these examples is also a polynomial.

POLYNOMIALS			
Monomials	Binomials	Trinomials	None of These
ax^2	$x + y$	$x^2 + 4xy + y^2$	$5x^3 - 6x^2 + 3x - 6$
$-3z$	$3p + 2$	$x^5 + 7x^2 - x$	$-y^5 + y^4 - 3y^3 - y^2 + y$
4	$4x^2 - 7$	$-q^4 + q^3 - 2q$	$x^6 + x^4 - x^3 + 1$

Each term of a polynomial has a degree. The **degree** of a term in one variable is the exponent on the variable.

Example 2   Identify the degree of each term of the trinomial $12x^4 - 7x + 3$.

Solution: The term $12x^4$ has degree 4.
The term $-7x$ has degree 1 since $-7x$ is $-7x^1$.
The term 3 has degree 0 since 3 is $3x^0$.

Each polynomial also has a degree.

> **DEGREE OF A POLYNOMIAL**
>
> The **degree of a polynomial** is the greatest degree of any term of the polynomial.

Example 3   Find the degree of each polynomial and tell whether the polynomial is a monomial, binomial, trinomial, or none of these.

a. $-2t^2 + 3t + 6$ **b.** $15x - 10$ **c.** $7x + 3x^3 + 2x^2 - 1$

Practice Problem 2

Identify the degree of each term of the trinomial $-15x^3 + 2x^2 - 5$.

Practice Problem 3

Find the degree of each polynomial and tell whether the polynomial is a monomial, binomial, trinomial, or none of these.

a. $-6x + 14$

b. $9x - 3x^6 + 5x^4 + 2$

c. $10x^2 - 6x - 6$

Answers

2. 3; 2; 0,   **3. a.** binomial, 1,
b. none of these, 6,   **c.** trinomial, 2

Solution:

a. The degree of the trinomial $-2t^2 + 3t + 6$ is 2, the greatest degree of any of its terms.

b. The degree of the binomial $15x - 10$ or $15x^1 - 10$ is 1.

c. The degree of the polynomial $7x + 3x^3 + 2x^2 - 1$ is 3.

C EVALUATING POLYNOMIALS

Polynomials have different values depending on replacement values for the variables. When we find the value of a polynomial for a given replacement value, we are evaluating the polynomial for that value.

Example 4   Evaluate each polynomial when $x = -2$.
    a. $-5x + 6$
    b. $3x^2 - 2x + 1$

Solution: **a.** $-5x + 6 = -5(-2) + 6$ Replace x with -2.
                      $= 10 + 6$
                      $= 16$

    b. $3x^2 - 2x + 1 = 3(-2)^2 - 2(-2) + 1$ Replace x with
                        $= 3(4) + 4 + 1$ -2.
                        $= 12 + 4 + 1$
                        $= 17$

Many physical phenomena can be modeled by polynomials.

Example 5   Finding Free-Fall Time

The CN Tower in Toronto, Ontario, is 1821 feet tall and is the world's tallest self-supporting structure. An object is dropped from the top of this building. Neglecting air resistance, the height of the object at time t seconds is given by the polynomial $-16t^2 + 1821$. Find the height of the object when $t = 1$ second and when $t = 10$ seconds. (*Source*: World Almanac)

Solution: To find each height, we evaluate the polynomial when $t = 1$ and when $t = 10$.

$-16t^2 + 1821 = -16(1)^2 + 1821$ Replace t with 1.
                $= -16(1) + 1821$
                $= -16 + 1821$
                $= 1805$

The height of the object at 1 second is 1805 feet.

$-16t^2 + 1821 = -16(10)^2 + 1821$ Replace t with 10.
                $= -16(100) + 1821$
                $= -1600 + 1821$
                $= 221$

The height of the object at 10 seconds is 221 feet.

Practice Problem 4

Evaluate each polynomial when $x = -1$.

a. $-2x + 10$

b. $6x^2 + 11x - 20$

Practice Problem 5

From Example 5, find the height of the object when $t = 3$ seconds and when $t = 7$ seconds.

TEACHING TIP

Encourage students to think about how the given equation relates to the problem. Here are some questions you could pose about Example 5 to get the students thinking. What does 1821 represent in the polynomial? What polynomial would you use if the building was only 950 feet tall? Do you think this polynomial will give a good estimate of the height of the object for all values of t? (No, once the object hits the ground the polynomial does not apply.)

Answers

4. a. 12,   **b.** -25,   **5.** 1677 feet; 1037 feet

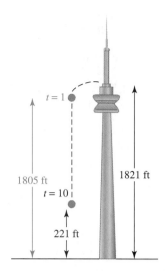

D SIMPLIFYING POLYNOMIALS BY COMBINING LIKE TERMS

Polynomials with like terms can be simplified by combining like terms. Recall that like terms are terms that contain exactly the same variables raised to exactly the same powers.

Like Terms	Unlike Terms
$5x^2, -7x^2$	$3x, 3y$
$y, 2y$	$-2x^2, -5x$
$\frac{1}{2}a^2b, -a^2b$	$6st^2, 4s^2t$

Only like terms can be combined. We combine like terms by applying the distributive property.

Examples   Simplify each polynomial by combining any like terms.

6. $-3x + 7x = (-3 + 7)x = 4x$

7. $11x^2 + 5 + 2x^2 - 7 = 11x^2 + 2x^2 + 5 - 7$
$$= 13x^2 - 2$$

8. $9x^3 + x^3 = 9x^3 + 1x^3$ Write x^3 as $1x^3$.
$$= 10x^3$$

9. $5x^2 + 6x - 9x - 3 = 5x^2 - 3x - 3$ Combine like terms $6x$ and $-9x$.

10. $\frac{2}{5}x^4 + \frac{2}{3}x^3 - x^2 + \frac{1}{10}x^4 - \frac{1}{6}x^3$

$$= \left(\frac{2}{5} + \frac{1}{10}\right)x^4 + \left(\frac{2}{3} - \frac{1}{6}\right)x^3 - x^2$$

$$= \left(\frac{4}{10} + \frac{1}{10}\right)x^4 + \left(\frac{4}{6} - \frac{1}{6}\right)x^3 - x^2$$

$$= \frac{5}{10}x^4 + \frac{3}{6}x^3 - x^2$$

$$= \frac{1}{2}x^4 + \frac{1}{2}x^3 - x^2$$

Example 11

Write a polynomial that describes the total area of the squares and rectangles shown on the next page. Then simplify the polynomial.

Practice Problems 6–10

Simplify each polynomial by combining any like terms.

6. $-6y + 8y$

7. $14y^2 + 3 - 10y^2 - 9$

8. $7x^3 + x^3$

9. $23x^2 - 6x - x - 15$

10. $\frac{2}{7}x^3 - \frac{1}{4}x + 2 - \frac{1}{2}x^3 + \frac{3}{8}x$

Answers

6. $2y$,   **7.** $4y^2 - 6$,   **8.** $8x^3$,

9. $23x^2 - 7x - 15$,   **10.** $-\frac{3}{14}x^3 + \frac{1}{8}x + 2$

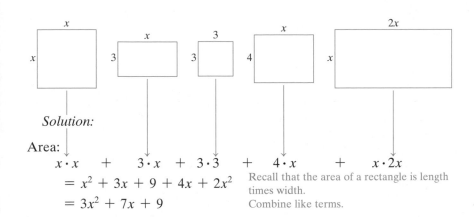

Solution:

Area:
$$x \cdot x \quad + \quad 3 \cdot x \quad + \quad 3 \cdot 3 \quad + \quad 4 \cdot x \quad + \quad x \cdot 2x$$
$$= x^2 + 3x + 9 + 4x + 2x^2$$
$$= 3x^2 + 7x + 9$$

Recall that the area of a rectangle is length times width.
Combine like terms.

E SIMPLIFYING POLYNOMIALS CONTAINING SEVERAL VARIABLES

A polynomial may contain more than one variable, such as

$$5x + 3xy^2 - 6x^2y^2 + x^2y - 2y + 1$$

We call this expression a polynomial in several variables.

The **degree of a term** with more than one variable is the sum of the exponents on the variables. The **degree of the polynomial** in several variables is still the greatest degree of the terms of the polynomial.

Example 12   Identify the degrees of the terms and the degree of the polynomial $5x + 3xy^2 - 6x^2y^2 + x^2y - 2y + 1$.

Solution: To organize our work, we use a table.

Terms of Polynomial	Degree of Term	Degree of Polynomial
$5x$	1	
$3xy^2$	1 + 2 or 3	
$-6x^2y^2$	2 + 2 or 4	4 (highest degree)
x^2y	2 + 1 or 3	
$-2y$	1	
1	0	

To simplify a polynomial containing several variables, we combine any like terms.

Examples   Simplify each polynomial by combining any like terms.

13. $3xy - 5y^2 + 7xy - 9x^2 = (3 + 7)xy - 5y^2 - 9x^2$
$$= 10xy - 5y^2 - 9x^2$$

HELPFUL HINT

This term can be written as $10xy$ or $10yx$

14. $9a^2b - 6a^2 + 5b^2 + a^2b - 11a^2 + 2b^2$
$$= 10a^2b - 17a^2 + 7b^2$$

Practice Problem 11

Write a polynomial that describes the total area of the squares and rectangles shown below. Then simplify the polynomial.

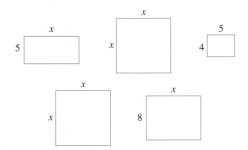

Practice Problem 12

Identify the degrees of the terms and the degree of the polynomial $-2x^3y^2 + 4 - 8xy + 3x^3y + 5xy^2$.

Practice Problems 13–14

Simplify each polynomial by combining any like terms.

13. $11ab - 6a^2 - ba + 8b^2$

14. $7x^2y^2 + 2y^2 - 4y^2x^2 + x^2 - y^2 + 5x^2$

Answers

11. $2x^2 + 13x + 20$, **12.** 5, 0, 2, 4, 3; 5
13. $10ab - 6a^2 + 8b^2$, **14.** $3x^2y^2 + y^2 + 6x^2$

Focus On History

EXPONENTIAL NOTATION

The French mathematician and philosopher René Descartes (1596–1650) is generally credited with devising the system of exponents that we use in math today. His book *La Géométrie* was the first to show successive powers of an unknown quantity x as x, xx, x^3, x^4, x^5, and so on. No one knows why Descartes preferred to write xx instead of x^2. However, the use of xx for the square of the quantity x continued to be popular. Those who used the notation defended it by saying that xx takes up no more space when written than x^2 does.

Before Descartes popularized the use of exponents to indicate powers, other less convenient methods were used. Some mathematicians preferred to write out the Latin words *quadratus* and *cubus* whenever they wanted to indicate that a quantity was to be raised to the second power or the third power. Other mathematicians used the abbreviations of *quadratus* and *cubus*, Q and C, to indicate second and third powers of a quantity.

EXERCISE SET 3.3

A *Complete each table for each polynomial. See Example 1.*

1. $x^2 - 3x + 5$

Term	Coefficient
x^2	1
$-3x$	-3
5	5

2. $2x^3 - x + 4$

Term	Coefficient
$2x^3$	2
$-x$	-1
4	4

3. $-5x^4 + 3.2x^2 + x - 5$

Term	Coefficient
$-5x^4$	-5
$3.2x^2$	3.2
x	1
-5	-5

4. $9.7x^7 - 3x^5 + x^3 - \dfrac{1}{4}x^2$

Term	Coefficient
$9.7x^7$	9.7
$-3x^5$	-3
x^3	1
$-\dfrac{1}{4}x^2$	$-\dfrac{1}{4}$

B *Find the degree of each polynomial and determine whether it is a monomial, binomial, trinomial, or none of these. See Examples 2 and 3.*

5. $x + 2$

6. $-6y + y^2 + 4$

▄▶**7.** $9m^3 - 5m^2 + 4m - 8$

8. $5a^2 + 3a^3 - 4a^4$

9. $12x^4 - x^2 - 12x^2$

10. $7r^2 + 2r - 3r^5$

11. $3z - 5$

12. $5y + 2$

📄 **13.** Describe how to find the degree of a term.

📄 **14.** Describe how to find the degree of a polynomial.

📄 **15.** Explain why xyz is a monomial while $x + y + z$ is a trinomial.

📄 **16.** Explain why the degree of the term $5y^3$ is 3 and the degree of the polynomial $2y + y + 2y$ is 1.

1. see table

2. see table

3. see table

4. see table

5. 1; binomial

6. 2; trinomial

7. 3; none of these

8. 4; trinomial

9. 4; binomial

10. 5; trinomial

11. 1; binomial

12. 1; binomial

13. answers may vary

14. answers may vary

15. answers may vary

16. answers may vary

Answer column (left):

17. (a) 6; (b) 5
18. (a) −10; (b) −12
19. (a) −2; (b) 4
20. (a) −4; (b) −3
21. (a) −15; (b) −16
22. (a) −6; (b) −1
23. 184 ft
24. 600 ft
25. 595.84 ft
26. 362.56 ft
27. 11.7 million
28. 15.1 visits
29. $23x^2$
30. $14x^3$
31. $12x^2 - y$
32. $3k^3 + 11$
33. $7s$
34. $6y$
35. $-1.1y^2 + 4.8$
36. $0.7y^2 - 0.4y$

222

Name _____

C *Evaluate each polynomial when* (a) $x = 0$ *and* (b) $x = -1$. *See Examples 4 and 5.*

17. $x + 6$ 18. $2x - 10$ 19. $x^2 - 5x - 2$ 20. $x^2 - 4$

21. $x^3 - 15$ 22. $-2x^3 + 3x^2 - 6$

A rocket is fired upward from the ground with an initial velocity of 200 feet per second. Neglecting air resistance, the height of the rocket at any time t can be described by the polynomial $-16t^2 + 200t$. *Find the height of the rocket at the time given in Exercises 23–26. See Example 5.*

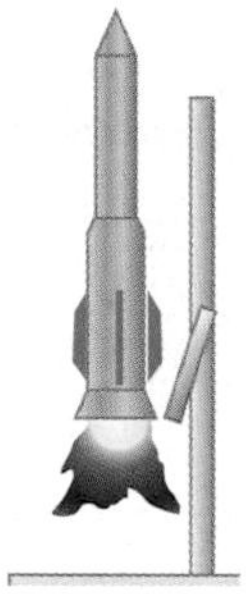

23. $t = 1$ second
24. $t = 5$ seconds
25. $t = 7.6$ seconds
26. $t = 10.3$ seconds

27. The number of U.S. airline departures (in millions) x years after 1993 is given by the polynomial $0.15x^2 + 0.15x + 7.2$ million departures for 1993–1995. Use this model to predict how many departures there were in 1998 $(x = 5)$. (*Source*: Based on data from the National Transportation Safety Board)

28. The average number of physician visits per person age 75 years or older in the United States x years after 1992 is given by the polynomial $0.1x^2 + 0.1x + 12.1$ for 1992–1994. Use this model to predict how many visits a person 75 years or older had in 1997 $(x = 5)$. (*Source*: Based on data from the U.S. Department of Health and Human Services)

D *Simplify each expression by combining like terms. See Examples 6 through 10.*

29. $14x^2 + 9x^2$ 30. $18x^3 - 4x^3$ 31. $15x^2 - 3x^2 - y$

32. $12k^3 - 9k^3 + 11$ 33. $8s - 5s + 4s$ 34. $5y + 7y - 6y$

35. $0.1y^2 - 1.2y^2 + 6.7 - 1.9$ 36. $7.6y + 3.2y^2 - 8y - 2.5y^2$

Name _____

Recall that the perimeter of a figure such as the ones shown in Exercises 37–38 is the sum of the lengths of its sides. Write each perimeter as a polynomial. Then simplify the polynomial.

37.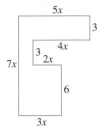

38.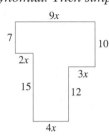

Write a polynomial that describes each total area of the rectangles and squares shown in Exercises 39–40. Then simplify the polynomial. See Example 11.

39.

40.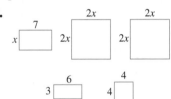

E *Identify the degrees of the terms and the degree of the polynomial. See Example 12.*

41. $9ab - 6a + 5b - 3$

42. $y^4 - 6y^3x + 2x^2y^2 - 5y^2 + 3$

43. $x^3y - 6 + 2x^2y^2 + 5y^3$

44. $2a^2b + 10a^4b - 9ab + 6$

Simplify each polynomial by combining any like terms. See Examples 13 and 14.

45. $3ab - 4a + 6ab - 7a$

46. $-9xy + 7y - xy - 6y$

47. $4x^2 - 6xy + 3y^2 - xy$

48. $3a^2 - 9ab + 4b^2 - 7ab$

49. $5x^2y + 6xy^2 - 5yx^2 + 4 - 9y^2x$

50. $17a^2b - 16ab^2 + 3a^3 + 4ba^3 - b^2a$

51. $14y^3 - 19 - 16a^2b^2$

52. $x^4 - 1$

53. $10x + 19$

54. $-30x + 3$

55. $-x + 5$

56. $2w - 16$

57. answers may vary

58. 12.5 sec

59. $11.1x^2 - 7.97x + 10.76$

60. $-5.42x^2 + 7.75x - 19.61$

Name _____

51. $14y^3 - 9 + 3a^2b^2 - 10 - 19b^2a^2$ **52.** $18x^4 + 2x^3y^3 - 1 - 2y^3x^3 - 17x^4$

REVIEW AND PREVIEW

Simplify each expression. See Section 2.1.

53. $4 + 5(2x + 3)$ **54.** $9 - 6(5x + 1)$

55. $2(x - 5) + 3(5 - x)$ **56.** $-3(w + 7) + 5(w + 1)$

COMBINING CONCEPTS

57. Explain why the height of the rocket in Exercises 23–26 increases and then decreases as time passes.

58. Approximate (to the nearest tenth of a second) how long before the rocket in Exercises 23–26 hits the ground.

Simplify each polynomial by combining like terms.

59. $1.85x^2 - 3.76x + 9.25x^2 + 10.76 - 4.21x$

60. $7.75x + 9.16x^2 - 1.27 - 14.58x^2 - 18.34$

3.4 ADDING AND SUBTRACTING POLYNOMIALS

A ADDING POLYNOMIALS

To add polynomials, we use commutative and associative properties and then combine like terms. To see if you are ready to add polynomials,

TRY THE CONCEPT CHECK IN THE MARGIN.

> **TO ADD POLYNOMIALS**
>
> To add polynomials, combine all like terms.

Examples   Add.

1. $(4x^3 - 6x^2 + 2x + 7) + (5x^2 - 2x)$

$\quad = 4x^3 - 6x^2 + 2x + 7 + 5x^2 - 2x$ Remove parentheses.

$\quad = 4x^3 + (-6x^2 + 5x^2) + (2x - 2x) + 7$ Combine like terms.

$\quad = 4x^3 - x^2 + 7$ Simplify.

2. $(-2x^2 + 5x - 1)$ and $(-2x^2 + x + 3)$ translates to

$\quad (-2x^2 + 5x - 1) + (-2x^2 + x + 3)$

$\quad = -2x^2 + 5x - 1 - 2x^2 + x + 3$ Remove parentheses.

$\quad = (-2x^2 - 2x^2) + (5x + 1x) + (-1 + 3)$ Combine like terms.

$\quad = -4x^2 + 6x + 2$ Simplify.   ▬▬

Polynomials can be added vertically if we line up like terms underneath one another.

Example 3   Add $(7y^3 - 2y^2 + 7)$ and $(6y^2 + 1)$ using a vertical format.

Solution: Vertically line up like terms and add.

$$\begin{array}{r} 7y^3 - 2y^2 + 7 \\ 6y^2 + 1 \\ \hline 7y^3 + 4y^2 + 8 \end{array}$$ ▬▬

B SUBTRACTING POLYNOMIALS

To subtract one polynomial from another, recall the definition of subtraction. To subtract a number, we add its opposite: $a - b = a + (-b)$. To subtract a polynomial, we also add its opposite. Just as $-b$ is the opposite of b, $-(x^2 + 5)$ is the opposite of $(x^2 + 5)$.

Example 4   Subtract: $(5x - 3) - (2x - 11)$

Solution: From the definition of subtraction, we have

$(5x - 3) - (2x - 11) = (5x - 3) + [-(2x - 11)]$ Add the opposite.

$\qquad\qquad\qquad\quad = (5x - 3) + (-2x + 11)$ Apply the distributive property.

$\qquad\qquad\qquad\quad = 3x + 8$ Combine like terms.   ▬▬

Objectives

A Add polynomials.

B Subtract polynomials.

C Add or subtract polynomials in one variable.

D Add or subtract polynomials in several variables.

SSM   CD-ROM   Video
3.4

✓ CONCEPT CHECK

When combining like terms in the expression $5x - 8x^2 - 8x$, which of the following is the proper result?

a. $-11x^2$ b. $-3x - 8x^2$

c. $-11x$ d. $-11x^4$

Practice Problems 1–2

Add.

1. $(3x^5 - 7x^3 + 2x - 1) + (3x^3 - 2x)$

2. $(5x^2 - 2x + 1)$ and $(-6x^2 + x - 1)$

Practice Problem 3

Add: $(9y^2 - 6y + 5)$ and $(4y + 3)$ using a vertical format.

> **TO SUBTRACT POLYNOMIALS**
>
> To subtract two polynomials, change the signs of the terms of the polynomial being subtracted and then add.

Practice Problem 4

Subtract: $(9x + 5) - (4x - 3)$

Answers

1. $3x^5 - 4x^3 - 1$,   **2.** $-x^2 - x$,

3. $9y^2 - 2y + 8$,   **4.** $5x + 8$

✓ Concept Check: b

Practice Problem 5

Subtract: $(4x^3 - 10x^2 + 1) -$
$(-4x^3 + x^2 - 11)$

Example 5   Subtract: $(2x^3 + 8x^2 - 6x) - (2x^3 - x^2 + 1)$

Solution: First, we change the sign of each term of the second polynomial, then we add.

$$(2x^3 + 8x^2 - 6x) - (2x^3 - x^2 + 1) =$$
$$(2x^3 + 8x^2 - 6x) + (-2x^3 + x^2 - 1)$$
$$= 2x^3 - 2x^3 + 8x^2 + x^2 - 6x - 1$$
$$= 9x^2 - 6x - 1 \quad \text{Combine like terms.}$$

Just as polynomials can be added vertically, so can they be subtracted vertically.

Practice Problem 6

Subtract $(6y^2 - 3y + 2)$ from $(2y^2 - 2y + 7)$ using a vertical format.

Example 6   Subtract $(5y^2 + 2y - 6)$ from $(-3y^2 - 2y + 11)$ using a vertical format.

Solution: Arrange the polynomials in a vertical format, lining up like terms.

$$
\begin{array}{r}
-3y^2 - 2y + 11 \\
-(5y^2 + 2y - 6) \\
\end{array}
\qquad
\begin{array}{r}
-3y^2 - 2y + 11 \\
-5y^2 - 2y + 6 \\
\hline
-8y^2 - 4y + 17 \\
\end{array}
$$

> **HELPFUL HINT**
>
> Don't forget to change the sign of each term in the polynomial being subtracted.

C   ADDING AND SUBTRACTING POLYNOMIALS IN ONE VARIABLE

Let's practice adding and subtracting polynomials in one variable.

Example 7   Subtract $(5z - 7)$ from the sum of $(8z + 11)$ and $(9z - 2)$.

Solution: Notice that $(5z - 7)$ is to be subtracted **from** a sum. The translation is

$$[(8z + 11) + (9z - 2)] - (5z - 7)$$
$$= 8z + 11 + 9z - 2 - 5z + 7 \quad \text{Remove grouping symbols.}$$
$$= 8z + 9z - 5z + 11 - 2 + 7 \quad \text{Group like terms.}$$
$$= 12z + 16 \quad \text{Combine like terms.}$$

Practice Problem 7

Subtract $(3x + 1)$ from the sum of $(4x - 3)$ and $(12x - 5)$.

TEACHING TIP

Consider having students also try the following problem.
Subtract the sum of $(3x + 6)$ and $(8x - 5)$ from $(5x + 2)$.
$(5x + 2) - [(3x + 6) + (8x - 5)]$
$= (5x + 2 - (11x + 1) = -6x + 1$

D   ADDING AND SUBTRACTING POLYNOMIALS IN SEVERAL VARIABLES

Now that we know how to add or subtract polynomials in one variable, we can also add and subtract polynomials in several variables.

Practice Problems 8–9

Add or subtract as indicated.

8. $(2a^2 - ab + 6b^2) -$
$(-3a^2 + ab - 7b^2)$

9. $(5x^2y^2 + 3 - 9x^2y + y^2) -$
$(-x^2y^2 + 7 - 8xy^2 + 2y^2)$

Examples   Add or subtract as indicated.

8. $(3x^2 - 6xy + 5y^2) + (-2x^2 + 8xy - y^2)$
$= 3x^2 - 6xy + 5y^2 - 2x^2 + 8xy - y^2$
$= x^2 + 2xy + 4y^2$ Combine like terms.

9. $(9a^2b^2 + 6ab - 3ab^2) - (5b^2a + 2ab - 3 - 9b^2)$ Change the sign of each term of the polynomial being subtracted.
$= 9a^2b^2 + 6ab - 3ab^2 - 5b^2a - 2ab + 3 + 9b^2$

$= 9a^2b^2 + 4ab - 8ab^2 + 3 + 9b^2$ Combine like terms.

Answers

5. $8x^3 - 11x^2 + 12$,   6. $-4y^2 + y + 5$,
7. $13x - 9$,   8. $5a^2 - 2ab + 13b^2$,
9. $6x^2y^2 - 4 - 9x^2y + 8xy^2 - y^2$

EXERCISE SET 3.4

A *Add. See Examples 1 through 3.*

1. $(3x + 7) + (9x + 5)$

2. $(3x^2 + 7) + (3x^2 + 9)$

3. $(-7x + 5) + (-3x^2 + 7x + 5)$

4. $(3x - 8) + (4x^2 - 3x + 3)$

5. $(-5x^2 + 3) + (2x^2 + 1)$

6. $(-y - 2) + (3y + 5)$

7. $(-3y^2 - 4y) + (2y^2 + y - 1)$

8. $(7x^2 + 2x - 9) + (-3x^2 + 5)$

Add using a vertical format. See Example 3.

9.
$$\begin{array}{r} 3t^2 + 4 \\ + 5t^2 - 8 \\ \hline \end{array}$$

10.
$$\begin{array}{r} 7x^3 + 3 \\ + 2x^3 + 1 \\ \hline \end{array}$$

11.
$$\begin{array}{r} 10a^3 - 8a^2 + 9 \\ + \ 5a^3 + 9a^2 + 7 \\ \hline \end{array}$$

12.
$$\begin{array}{r} 2x^3 - 3x^2 + x - 4 \\ + 5x^3 + 2x^2 - 3x + 2 \\ \hline \end{array}$$

B *Subtract. See Examples 4 and 5.*

13. $(2x + 5) - (3x - 9)$

14. $(5x^2 + 4) - (-2y^2 + 4)$

15. $3x - (5x - 9)$

16. $4 - (-y - 4)$

17. $(2x^2 + 3x - 9) - (-4x + 7)$

18. $(-7x^2 + 4x + 7) - (-8x + 2)$

19. $(-7y^2 + 5) - (-8y^2 + 12)$

20. $(4 + 5a) - (-a - 5)$

1. $12x + 12$

2. $6x^2 + 16$

3. $-3x^2 + 10$

4. $4x^2 - 5$

5. $-3x^2 + 4$

6. $2y + 3$

7. $-y^2 - 3y - 1$

8. $4x^2 + 2x - 4$

9. $8t^2 - 4$

10. $9x^3 + 4$

11. $15a^3 + a^2 + 16$

12. $7x^3 - x^2 - 2x - 2$

13. $-x + 14$

14. $5x^2 + 2y^2$

15. $-2x + 9$

16. $y + 8$

17. $2x^2 + 7x - 16$

18. $-7x^2 + 12x + 5$

19. $y^2 - 7$

20. $6a + 9$

21. $2x^2 + 11x$

22. $-15y^2 + 6y - 4$

23. $-2z^2 - 16z + 6$

24. $-4a^2 - 5a + 4$

25. $2u^5 - 10u^2 + 11u - 9$

26. $2x^3 - 2x^2 + 7x + 2$

27. $5x - 9$

28. $4x - 3$

29. $6y + 13$

30. $11y + 7$

31. $-2x^2 + 8x - 1$

32. $3y^2 - 4y - 2$

33. $7x^2 + 14x + 18$

34. $3a^2 - 6a + 11$

35. $3x - 3$

36. $y^2 - 5y + 1$

37. $7x^2 - 4x + 2$

38. $-2x^2 - 11x + 11$

39. $7x^2 - 2x + 2$

40. $2y^2 + y - 10$

Name _____

21. $(5x + 8) - (-2x^2 - 6x + 8)$ **22.** $(-6y^2 + 3y - 4) - (9y^2 - 3y)$

Subtract using a vertical format. See Example 6.

23. $\begin{array}{r} 4z^2 - 8z + 3 \\ -\ (6z^2 + 8z - 3) \\ \hline \end{array}$ **24.** $\begin{array}{r} 7a^2 - 9a + 6 \\ -\ (11a^2 - 4a + 2) \\ \hline \end{array}$

25. $\begin{array}{r} 5u^5 - 4u^2 + 3u - 7 \\ -\ (3u^5 + 6u^2 - 8u + 2) \\ \hline \end{array}$ **26.** $\begin{array}{r} 5x^3 - 4x^2 + 6x - 2 \\ -\ (3x^3 - 2x^2 - x - 4) \\ \hline \end{array}$

C *Add or subtract as indicated. See Example 7.*

27. $(3x + 5) + (2x - 14)$ **28.** $(9x - 1) - (5x + 2)$

29. $(7y + 7) - (y - 6)$ **30.** $(14y + 12) + (-3y - 5)$

31. $(x^2 + 2x + 1) - (3x^2 - 6x + 2)$ **32.** $(5y^2 - 3y - 1) - (2y^2 + y + 1)$

33. $(3x^2 + 5x - 8) + (5x^2 + 9x + 12) - (x^2 - 14)$

34. $(-a^2 + 1) - (a^2 - 3) + (5a^2 - 6a + 7)$

Perform each indicated operation. See Examples 2, 6, and 7.

35. Subtract $4x$ from $7x - 3$.   **36.** Subtract y from $y^2 - 4y + 1$.

37. Add $(4x^2 - 6x + 1)$ and $(3x^2 + 2x + 1)$.   **38.** Add $(-3x^2 - 5x + 2)$ and $(x^2 - 6x + 9)$.

39. Subtract $(5x + 7)$ from $(7x^2 + 3x + 9)$.   **40.** Subtract $(5y^2 + 8y + 2)$ from $(7y^2 + 9y - 8)$.

Name _____

41. Subtract $(4y^2 - 6y - 3)$ from the sum of $(8y^2 + 7)$ and $(6y + 9)$.

42. Subtract $(4x^2 - 2x + 2)$ from the sum of $(x^2 + 7x + 1)$ and $(7x + 5)$.

D *Add or subtract as indicated. See Examples 8 and 9.*

43. $(9a + 6b - 5) + (-11a - 7b + 6)$

44. $(3x - 2 + 6y) + (7x - 2 - y)$

45. $(4x^2 + y^2 + 3) - (x^2 + y^2 - 2)$

46. $(7a^2 - 3b^2 + 10) - (-2a^2 + b^2 - 12)$

47. $(x^2 + 2xy - y^2) + (5x^2 - 4xy + 20y^2)$

48. $(a^2 - ab + 4b^2) + (6a^2 + 8ab - b^2)$

49. $(11r^2s + 16rs - 3 - 2r^2s^2) - (3sr^2 + 5 - 9r^2s^2)$

50. $(3x^2y - 6xy + x^2y^2 - 5) - (11x^2y^2 - 1 + 5yx^2)$

REVIEW AND PREVIEW

Multiply. See Section 3.1.

51. $3x(2x)$

52. $-7x(x)$

53. $(12x^3)(-x^5)$

54. $6r^3(7r^{10})$

55. $10x^2(20xy^2)$

56. $-z^2y(11zy)$

41. $4y^2 + 12y + 19$

42. $-3x^2 + 16x + 4$

43. $-2a - b + 1$

44. $10x - 4 + 5y$

45. $3x^2 + 5$

46. $9a^2 - 4b^2 + 22$

47. $6x^2 - 2xy + 19y^2$

48. $7a^2 + 7ab + 3b^2$

49. $8r^2s + 16rs - 8 + 7r^2s^2$

50. $-2x^2y - 6xy - 10x^2y^2 - 4$

51. $6x^2$

52. $-7x^2$

53. $-12x^8$

54. $42r^{13}$

55. $200x^3y^2$

56. $-11y^2z^3$

57. $(x^2 + 7x + 4)$ ft

58. $(3y^2 + 4y + 11)$ m

59. $(2x^2 - 2x + 2)$ cm

60. $(11x - 9)$ in.

61. $-6.6x^2 - 1.8x - 1.8$

62. $3.7y^4 - 0.7y^3 + 2.2y - 4$

63. $-464.5x^2 + 2388.5x + 40,759$

64. $92x^2 - 106x - 1775$

COMBINING CONCEPTS

57. Given the following triangle, find its perimeter.

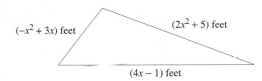

$(-x^2 + 3x)$ feet $(2x^2 + 5)$ feet

$(4x - 1)$ feet

58. A wooden beam is $(4y^2 + 4y + 1)$ meters long. If a piece $(y^2 - 10)$ meters is cut, express the length of the remaining piece of beam as a polynomial in y.

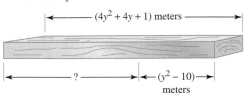

$(4y^2 + 4y + 1)$ meters

? $(y^2 - 10)$
meters

59. Given the following quadrilateral, find its perimeter.

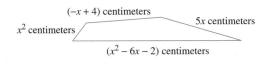

$(-x + 4)$ centimeters
x^2 centimeters $5x$ centimeters
$(x^2 - 6x - 2)$ centimeters

60. A piece of quarter-round molding is $(13x - 7)$ inches long. If a piece $(2x + 2)$ inches is removed, express the length of the remaining piece of molding as a polynomial in x.

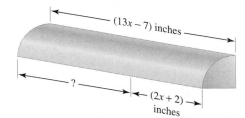

$(13x - 7)$ inches

? $(2x + 2)$
inches

Perform each indicated operation.

61. $\left[(1.2x^2 - 3x + 9.1) - (7.8x^2 - 3.1 + 8)\right] + (1.2x - 6)$

62. $\left[(7.9y^4 - 6.8y^3 + 3.3y) + (6.1y^3 - 5)\right] - (4.2y^4 + 1.1y - 1)$

63. The polynomial
$-250.5x^2 + 1587.5x + 23{,}049$
represents the annual U.S. production of beef (in millions of pounds) during 1993–1995. The polynomial $-214x^2 + 801x + 17{,}710$ represents the combined annual U.S. production of veal, lamb, and pork (in millions of pounds) during 1993–1995. In both polynomials, x represents the number of years after 1993. Find a polynomial for the *total* U.S. meat production in millions of pounds during this period. (*Source:* Based on data from the U.S. Department of Agriculture)

64. The polynomial $61x^2 - 58x + 2719$ represents the total number of traffic fatalities in Florida for 1993–1995. The polynomial $944 + 48x - 31x^2$ represents the total number of alcohol-related traffic fatalities in Florida for 1993–1995. In both polynomials, x represents the number of years after 1993. Find a polynomial for the number of nonalcohol-related traffic fatalities in Florida during this period by subtracting the polynomials. (*Source:* Based on data from the Florida Highway Patrol)

3.5   MULTIPLYING POLYNOMIALS

A   MULTIPLYING MONOMIALS

Recall from Section 3.1 that to multiply two monomials such as $(-5x^3)$ and $(-2x^4)$, we use the associative and commutative properties and regroup. Remember also that to multiply exponential expressions with a common base, we add exponents.

$$(-5x^3)(-2x^4) = (-5)(-2)(x^3 \cdot x^4)$$

Use the commutative and associative properties.

$$= 10x^7$$

Multiply.

Examples   Multiply.

1. $6x \cdot 4x = (6 \cdot 4)(x \cdot x)$ Use the commutative and associative properties.

$$= 24x^2$$ Multiply.

2. $-7x^2 \cdot 2x^5 = (-7 \cdot 2)(x^2 \cdot x^5)$

$$= -14x^7$$

3. $(-12x^5)(-x) = (-12x^5)(-1x)$

$$= (-12)(-1)(x^5 \cdot x)$$

$$= 12x^6$$

B   MULTIPLYING MONOMIALS BY POLYNOMIALS

To multiply a monomial such as $7x$ by a trinomial such as $x^3 + 2x + 5$, we use the distributive property.

Examples   Multiply.

4. $7x(x^2 + 2x + 5) = 7x(x^2) + 7x(2x) + 7x(5)$ Apply the distributive property.

$$= 7x^3 + 14x^2 + 35x$$ Multiply.

5. $5x(2x^3 + 6) = 5x(2x^3) + 5x(6)$ Apply the distributive property.

$$= 10x^4 + 30x$$ Multiply.

6. $-3x^2(5x^2 + 6x - 1)$

$$= (-3x^2)(5x^2) + (-3x^2)(6x) + (-3x^2)(-1)$$ Apply the distributive property.

$$= -15x^4 - 18x^3 + 3x^2$$ Multiply.

C   MULTIPLYING TWO POLYNOMIALS

We also use the distributive property to multiply two binomials.

Example 7   Multiply: $(3x + 2)(2x - 5)$

Solution:

$$(3x + 2)(2x - 5) = 3x(2x - 5) + 2(2x - 5)$$ Use the distributive property.

$$= 3x(2x) + 3x(-5) + 2(2x) + 2(-5)$$

$$= 6x^2 - 15x + 4x - 10$$ Multiply. Combine like terms.

$$= 6x^2 - 11x - 10$$

This idea can be expanded so that we can multiply any two polynomials.

Objectives

A Multiply monomials.

B Multiply a monomial by a polynomial.

C Multiply two polynomials.

D Multiply polynomials vertically.

SSM   CD-ROM   Video 3.5

Practice Problems 1–3

Multiply.

1. $10x \cdot 9x$

2. $8x^3(-11x^7)$

3. $(-5x^4)(-x)$

Practice Problems 4–6

Multiply.

4. $4x(x^2 + 4x + 3)$

5. $8x(7x^4 + 1)$

6. $-2x^3(3x^2 - x + 2)$

TEACHING TIP

Example 4 can be illustrated with an area diagram. Note that the result of the multiplication is written inside the rectangles.

	x^2	$+ 2x +$	5
$7x$	$7x^3$	$14x^2$	$35x$

Practice Problem 7

Multiply: $(4x + 5)(3x - 4)$

Answers

1. $90x^2$,   **2.** $-88x^{10}$,   **3.** $5x^5$,
4. $4x^3 + 16x^2 + 12x$,   **5.** $56x^5 + 8x$,
6. $-6x^5 + 2x^4 - 4x^3$,   **7.** $12x^2 - x - 20$

TEACHING TIP

Example 7 can be illustrated with an area diagram.

	$3x$	$+$	2
$2x$	$6x^2$		$4x$
-5	$-15x$		-10

Practice Problems 8–9

Multiply.

8. $(3x - 2y)^2$

9. $(x + 3)(2x^2 - 5x + 4)$

TEACHING TIP

Before doing Example 10, you may wish to review vertical multiplication using the following example.

$$\begin{array}{r} 134 \\ \times\ 25 \\ \hline 670 \\ 268 \\ \hline 3350 \end{array}$$

After doing Example 10, have students compare and contrast the numerical example with multiplying polynomials vertically.

Practice Problem 10

Multiply vertically:
$(3y^2 + 1)(y^2 - 4y + 5)$

Practice Problem 11

Find the product of $(4x^2 - x - 1)$ and $(3x^2 + 6x - 2)$ using a vertical format.

To Multiply Two Polynomials

Multiply each term of the first polynomial by each term of the second polynomial, and then combine like terms.

Examples   Multiply.

8. $(2x - y)^2$

$= (2x - y)(2x - y)$

$= 2x(2x) + 2x(-y) + (-y)(2x) + (-y)(-y)$

$= 4x^2 - 2xy - 2xy + y^2$ Multiply.

$= 4x^2 - 4xy + y^2$ Combine like terms.

9. $(t + 2)(3t^2 - 4t + 2)$

$= t(3t^2) + t(-4t) + t(2) + 2(3t^2) + 2(-4t) + 2(2)$

$= 3t^3 - 4t^2 + 2t + 6t^2 - 8t + 4$

$= 3t^3 + 2t^2 - 6t + 4$ Combine like terms.

D MULTIPLYING POLYNOMIALS VERTICALLY

Another convenient method for multiplying polynomials is to multiply vertically, similar to how we multiply real numbers. This method is shown in the next examples.

Example 10   Multiply vertically: $(2y^2 + 5)(y^2 - 3y + 4)$

Solution:

$$\begin{array}{r} y^2 - 3y + 4 \\ 2y^2 + 5 \\ \hline 5y^2 - 15y + 20 \\ 2y^4 - 6y^3 + 8y^2 \\ \hline 2y^4 - 6y^3 + 13y^2 - 15y + 20 \end{array}$$

Multiply $y^2 - 3y + 4$ by 5

Multiply $y^2 - 3y + 4$ by $2y^2$

Combine like terms.

Example 11   Find the product of $(2x^2 - 3x + 4)$ and $(x^2 + 5x - 2)$ using a vertical format.

Solution: First, we arrange the polynomials in a vertical format. Then we multiply each term of the second polynomial by each term of the first polynomial.

$$\begin{array}{r} 2x^2 - 3x + 4 \\ x^2 + 5x - 2 \\ \hline -4x^2 + 6x - 8 \\ 10x^3 - 15x^2 + 20x \\ 2x^4 - 3x^3 + 4x^2 \\ \hline 2x^4 + 7x^3 - 15x^2 + 26x - 8 \end{array}$$

Multiply $2x^2 - 3x + 4$ by -2.

Multiply $2x^2 - 3x + 4$ by $5x$.

Multiply $2x^2 - 3x + 4$ by x^2.

Combine like terms.

Answers

8. $9x^2 - 12xy + 4y^2$,

9. $2x^3 + x^2 - 11x + 12$,

10. $3y^4 - 12y^3 + 16y^2 - 4y + 5$,

11. $12x^4 + 21x^3 - 17x^2 - 4x + 2$

Name _____ Section _____ Date _____

MENTAL MATH

Find each product.

1. $x^3 \cdot x^5$ 2. $x^2 \cdot x^6$ 3. $y^4 \cdot y$
4. $y^9 \cdot y$ 5. $x^7 \cdot x^7$ 6. $x^{11} \cdot x^{11}$

EXERCISE SET 3.5

A Multiply. See Examples 1 through 3.

1. $8x^2 \cdot 3x$ ▣ 2. $6x \cdot 3x^2$ 3. $(-3.1x^3)(4x^9)$

4. $(-5.2x^4)(3x^4)$ 5. $(-x^3)(-x)$ 6. $(-x^6)(-x)$

7. $\left(-\frac{1}{3}y^2\right)\left(\frac{2}{5}y\right)$ 8. $\left(-\frac{3}{4}y^7\right)\left(\frac{1}{7}y^4\right)$

9. $(2x)(-3x^2)(4x^5)$ 10. $(x)(5x^4)(-6x^7)$

B Multiply. See Examples 4 through 6.

11. $3x(2x + 5)$ 12. $2x(6x + 3)$ ▣ 13. $7x(x^2 + 2x - 1)$

14. $5y(y^2 + y - 10)$ 15. $-2a(a + 4)$ 16. $-3a(2a + 7)$

17. $3x(2x^2 - 3x + 4)$ 18. $4x(5x^2 - 6x - 10)$ 19. $3a(a^2 + 2)$

20. $x^3(x + 12)$ 21. $-2a^2(3a^2 - 2a + 3)$ 22. $-4b^2(3b^3 - 12b^2 - 6)$

23. $3x^2y(2x^3 - x^2y^2 + 8y^3)$ 24. $4xy^2(7x^3 + 3x^2y^2 - 9y^3)$

MENTAL MATH ANSWERS
1. x^8
2. x^8
3. y^5
4. y^{10}
5. x^{14}
6. x^{22}

ANSWERS
1. $24x^3$
2. $18x^3$
3. $-12.4x^{12}$
4. $-15.6x^8$
5. x^4
6. x^7
7. $-\frac{2}{15}y^3$
8. $-\frac{3}{28}y^{11}$
9. $-24x^8$
10. $-30x^{12}$
11. $6x^2 + 15x$
12. $12x^2 + 6x$
13. $7x^3 + 14x^2 - 7x$
14. $5y^3 + 5y^2 - 50y$
15. $-2a^2 - 8a$
16. $-6a^2 - 21a$
17. $6x^3 - 9x^2 + 12x$
18. $20x^3 - 24x^2 - 40x$
19. $3a^3 + 6a$
20. $x^4 + 12x^3$
21. $-6a^4 + 4a^3 - 6a^2$
22. $-12b^5 + 48b^4 + 24b^2$
23. $6x^5y - 3x^4y^3 + 24x^2y^4$
24. $28x^4y^2 + 12x^3y^4 - 36xy^5$

25. $x^2 + 3x$

26. $x + 2x^2$; $x(1 + 2x)$

27. $x^2 + 7x + 12$

28. $x^2 + 11x + 18$

29. $a^2 + 5a - 14$

30. $y^2 + y - 110$

31. $x^2 + \frac{1}{3}x - \frac{2}{9}$

32. $x^2 + \frac{1}{5}x - \frac{6}{25}$

33. $12x^2 + 25x + 7$

34. $30x^2 + 22x + 4$

35. $12x^2 - 29x + 15$

36. $16x^2 - 38x + 12$

37. $1 - 7a + 12a^2$

38. $6 - 7a + 2a^2$

39. $4y^2 - 16y + 16$

40. $36x^2 - 84x + 49$

41. $x^3 - 5x^2 + 13x - 14$

42. $x^3 + 8x^2 + 7x - 24$

43. $x^4 + 5x^3 - 3x^2 - 11x + 20$

44. $a^4 - a^3 - 6a^2 + 7a + 14$

45. $10a^3 - 27a^2 + 26a - 12$

46. $-3b^3 - 14b^2 - 13b + 6$

47. $49x^2y^2 - 14xy^2 + y^2$

48. $x^4 - 8x^2 + 16$

234

Name _____

25. The area of the larger rectangle below is $x(x + 3)$. Find another expression for this area by finding the sum of the areas of the smaller rectangles.

26. Write an expression for the area of the larger rectangle below in two different ways.

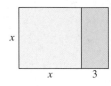

Multiply. See Examples 7 through 9.

27. $(x + 4)(x + 3)$

28. $(x + 2)(x + 9)$

29. $(a + 7)(a - 2)$

30. $(y - 10)(y + 11)$

31. $\left(x + \frac{2}{3}\right)\left(x - \frac{1}{3}\right)$

32. $\left(x + \frac{3}{5}\right)\left(x - \frac{2}{5}\right)$

33. $(3x^2 + 1)(4x^2 + 7)$

34. $(5x^2 + 2)(6x^2 + 2)$

35. $(4x - 3)(3x - 5)$

36. $(8x - 3)(2x - 4)$

37. $(1 - 3a)(1 - 4a)$

38. $(3 - 2a)(2 - a)$

39. $(2y - 4)^2$

40. $(6x - 7)^2$

41. $(x - 2)(x^2 - 3x + 7)$

42. $(x + 3)(x^2 + 5x - 8)$

43. $(x + 5)(x^3 - 3x + 4)$

44. $(a + 2)(a^3 - 3a^2 + 7)$

45. $(2a - 3)(5a^2 - 6a + 4)$

46. $(3 + b)(2 - 5b - 3b^2)$

47. $(7xy - y)^2$

48. $(x^2 - 4)^2$

Name _____

49. The area of the figure below is $(x + 2)(x + 3)$. Find another expression for this area by finding the sum of the areas of the smaller rectangles.

	x	3
x		
2		

50. Write an expression for the area of the figure below in two different ways.

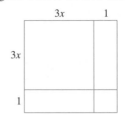

D *Multiply vertically. See Examples 10 and 11.*

51. $(2x - 11)(6x + 1)$

52. $(4x - 7)(5x + 1)$

53. $(x + 3)(2x^2 + 4x - 1)$

54. $(4x - 5)(8x^2 + 2x - 4)$

55. $(x^2 + 5x - 7)(x^2 - 7x - 9)$

56. $(3x^2 - x + 2)(x^2 + 2x + 1)$

REVIEW AND PREVIEW

Perform each indicated operation. See Section 3.1.

57. $(5x)^2$

58. $(4p)^2$

59. $(-3y^3)^2$

60. $(-7m^2)^2$

*For income tax purposes, Rob Calcutta, the owner of Copy Services, uses a method called **straight-line depreciation** to show the depreciated (or decreased) value of a copy machine he recently purchased. Rob assumes that he can use the machine for 7 years. The graph below shows the depreciated (or decreased) value of the machine over the years. Use this graph to answer Exercises 61–66. See Section 1.8.*

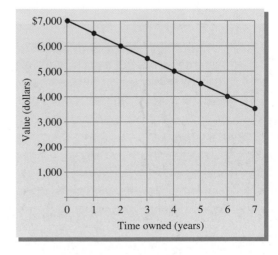

61. What was the purchase price of the copy machine? (*Hint:* This is when time owned is 0 years.)

62. What is the depreciated value of the machine in 7 years?

49. $x^2 + 5x + 6$

50. $(3x + 1)(3x + 1)$; $9x^2 + 6x + 1$

51. $12x^2 - 64x - 11$

52. $20x^2 - 31x - 7$

53. $2x^3 + 10x^2 + 11x - 3$

54. $32x^3 - 32x^2 - 26x + 20$

55. $x^4 - 2x^3 - 51x^2 + 4x + 63$

56. $3x^4 + 5x^3 + 3x^2 + 3x + 2$

57. $25x^2$

58. $16p^2$

59. $9y^6$

60. $49m^4$

61. $7000

62. $3500

235

63. $500

64. $500

65. answers may vary

66. answers may vary

67. $\left(4x^2 - 25\right)$ sq. yds

68. $\left(x^2 + 8x + 16\right)$ sq. ft

69. $\left(6x^2 - 4x\right)$ sq. in.

70. $\left(y^3 - 3y^2 + 3y - 1\right)$ cu. m

71. a. $6x + 12$; answers may vary

b. $9x^2 + 36x + 35$; answers may vary

72. a. 25; 13

b. 324; 164; no; answers may vary

73. a. $a^2 - b^2$

b. $4x^2 - 9y^2$

c. $16x^2 - 49$; answers may vary

Name _____

63. What loss in value occurred during the first year?

64. What loss in value occurred during the second year?

65. Why do you think this method of depreciating is called straight-line depreciation?

66. Why is the line tilted downward?

COMBINING CONCEPTS

Express as the product of polynomials. Then multiply.

67. Find the area of the rectangle.

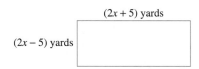

(2x + 5) yards

(2x – 5) yards

68. Find the area of the square field.

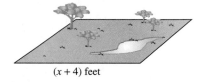

(x + 4) feet

69. Find the area of the triangle.

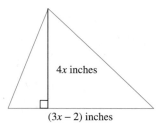

4x inches

(3x – 2) inches

70. Find the volume of the cube-shaped glass block.

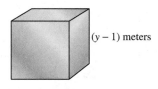

(y – 1) meters

71. Perform each indicated operation. Explain the difference between the two expressions.
   a. $(3x + 5) + (3x + 7)$

   b. $(3x + 5)(3x + 7)$

72. Evaluate each of the following.
   a. $(2 + 3)^2$; $2^2 + 3^2$

   b. $(8 + 10)^2$; $8^2 + 10^2$

   Does $(a + b)^2 = a^2 + b^2$ no matter what the values of a and b are? Why or why not?

73. Multiply each of the following polynomials.
   a. $(a + b)(a - b)$

   c. $(4x + 7)(4x - 7)$

   b. $(2x + 3y)(2x - 3y)$

   Can you make a general statement about all products of the form $(x + y)(x - y)$?

236

3.6 SPECIAL PRODUCTS

A USING THE FOIL METHOD

In this section, we multiply binomials using special products. First, we introduce a special order for multiplying binomials called the FOIL order or method. We demonstrate by multiplying $(3x + 1)$ by $(2x + 5)$.

THE FOIL METHOD

F stands for the
product of the **First** terms. $(3x + 1)(2x + 5)$
$$(3x)(2x) = 6x^2 \quad \text{F}$$

O stands for the
product of the **Outer** terms. $(3x + 1)(2x + 5)$
$$(3x)(5) \;\; = 15x \quad \text{O}$$

I stands for the
product of the **Inner** terms. $(3x + 1)(2x + 5)$
$$(1)(2x) \;\; = 2x \quad \text{I}$$

L stands for the
product of the **Last** terms. $(3x + 1)(2x + 5)$
$$(1)(5) \;\; = 5 \quad \text{L}$$

$$
\begin{array}{cccc}
& \text{F} & \text{O} & \text{I} & \text{L}
\end{array}
$$
$$(3x + 1)(2x + 5) = 6x^2 + 15x + 2x + 5$$
$$= 6x^2 + 17x + 5 \qquad \text{Combine like terms.}$$

Let's practice multiplying binomials using the FOIL method.

Example 1   Multiply: $(x - 3)(x + 4)$

Solution:

$$
(x - 3)(x + 4) = \overset{\text{F}}{(x)(x)} + \overset{\text{O}}{(x)(4)} + \overset{\text{I}}{(-3)(x)} + \overset{\text{L}}{(-3)(4)}
$$

$$= x^2 + 4x - 3x - 12$$
$$= x^2 + x - 12 \qquad \text{Combine like terms.}$$

Example 2   Multiply: $(5x - 7)(x - 2)$

Solution:

$$
(5x - 7)(x - 2) = \overset{\text{F}}{5x(x)} + \overset{\text{O}}{5x(-2)} + \overset{\text{I}}{(-7)(x)} + \overset{\text{L}}{(-7)(-2)}
$$

$$= 5x^2 - 10x - 7x + 14$$
$$= 5x^2 - 17x + 14 \qquad \text{Combine like terms.}$$

Objectives

A Multiply two binomials using the FOIL method.

B Square a binomial.

C Multiply the sum and difference of two terms.

D Use special products to multiply binomials.

SSM   CD-ROM   Video
3.6

TEACHING TIP

Point out that the special products in this section are shortcuts for multiplying binomials. They can all be worked out by the method in the previous section in which each term in the first binomial is multiplied by every term in the second binomial.

Practice Problem 1

Multiply: $(x + 7)(x - 5)$

TEACHING TIP

After doing Example 1, it may be helpful to point out to students that just as 6 and 4 are factors of 24 because $6 \cdot 4 = 24$, $x - 3$ and $x + 4$ are factors of $x^2 + x - 12$ because $(x - 3)(x + 4) = x^2 + x - 12$.

Practice Problem 2

Multiply: $(6x - 1)(x - 4)$

Answers

1. $x^2 + 2x - 35$,   **2.** $6x^2 - 25x + 4$

Practice Problem 3

Multiply: $(2y^2 + 3)(y - 4)$

TEACHING TIP

Consider having students discover patterns for squaring binomials themselves. Before doing Example 4, you may want to have students multiply $(4x + 3)^2$. Then have them multiply $(4x - 3)^2$. Ask if they notice any relationship between the problem and its solution.

Practice Problem 4

Multiply: $(2x + 9)^2$

Example 3   Multiply: $(y^2 + 6)(2y - 1)$

Solution: $\quad$ $(y^2 + 6)(2y - 1) = \overset{\text{F}}{2y^3} - \overset{\text{O}}{1y^2} + \overset{\text{I}}{12y} - \overset{\text{L}}{6}$

Notice in this example that there are no like terms that can be combined, so the product is $2y^3 - y^2 + 12y - 6$.

B   SQUARING BINOMIALS

An expression such as $(3y + 1)^2$ is called the square of a binomial. Since $(3y + 1)^2 = (3y + 1)(3y + 1)$, we can use the FOIL method to find this product.

Example 4   Multiply: $(3y + 1)^2$

Solution: $\quad (3y + 1)^2 = (3y + 1)(3y + 1)$

$$= \overset{\text{F}}{(3y)(3y)} + \overset{\text{O}}{(3y)(1)} + \overset{\text{I}}{1(3y)} + \overset{\text{L}}{1(1)}$$
$$= 9y^2 + 3y + 3y + 1$$
$$= 9y^2 + 6y + 1$$

Notice the pattern that appears in Example 4.

$$(3y + 1)^2 = 9y^2 + 6y + 1$$

$9y^2$ is the first term of the binomial squared. $(3y)^2 = 9y^2$.

$6y$ is 2 times the product of both terms of the binomial. $(2)(3y)(1) = 6y$.

1 is the second term of the binomial squared. $(1)^2 = 1$.

This pattern leads to the following, which can be used when squaring a binomial. We call these **special products**.

> **SQUARING A BINOMIAL**
>
> A binomial squared is equal to the square of the first term plus or minus twice the product of both terms plus the square of the second term.
>
> $$(a + b)^2 = a^2 + 2ab + b^2$$
> $$(a - b)^2 = a^2 - 2ab + b^2$$

This product can be visualized geometrically.

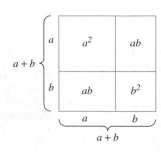

The area of the large square is side · side.

Area $= (a + b)(a + b) = (a + b)^2$

The area of the large square is also the sum of the areas of the smaller rectangles.

Area $= a^2 + ab + ab + b^2 = a^2 + 2ab + b^2$

Thus, $(a + b)^2 = a^2 + 2ab + b^2$.

Answers

3. $2y^3 - 8y^2 + 3y - 12$,   **4.** $4x^2 + 36x + 81$

Examples Use a special product to square each binomial.

first term squared	plus or minus	twice the product of the terms	plus	second term squared

$$
\begin{aligned}
&\textbf{5.} \quad (t + 2)^2 = & t^2 & + & 2(t)(2) & + & 2^2 = t^2 + 4t + 4 \\
&\textbf{6.} \quad (p - q)^2 = & p^2 & - & 2(p)(q) & + & q^2 = p^2 - 2pq + q^2 \\
&\textbf{7.} \quad (2x + 5)^2 = (2x)^2 & & + & 2(2x)(5) & + & 5^2 = 4x^2 + 20x + 25 \\
&\textbf{8.} \quad (x^2 - 7y)^2 = (x^2)^2 & & - & 2(x^2)(7y) & + & (7y)^2 = x^4 - 14x^2y + 49y^2
\end{aligned}
$$

┌ **HELPFUL HINT**

Notice that

$$(a + b)^2 \neq a^2 + b^2 \qquad \text{The middle term } 2ab \text{ is missing.}$$
$$(a + b)^2 = (a + b)(a + b) = a^2 + 2ab + b^2$$

Likewise,

$$(a - b)^2 \neq a^2 - b^2$$
$$(a - b)^2 = (a - b)(a - b) = a^2 - 2ab + b^2$$

└

C ▪ **MULTIPLYING THE SUM AND DIFFERENCE OF TWO TERMS**

Another special product is the product of the sum and difference of the same two terms, such as $(x + y)(x - y)$. Finding this product by the FOIL method, we see a pattern emerge.

$$
\overset{\text{F O I L}}{(x + y)(x - y)} = x^2 - xy + xy - y^2
$$

$$= x^2 - y^2$$

Notice that the middle two terms subtract out. This is because the **Outer** product is the opposite of the **Inner** product. Only the **difference of squares** remains.

┌───┐
MULTIPLYING THE SUM AND DIFFERENCE OF TWO TERMS

The product of the sum and difference of two terms is the square of the first term minus the square of the second term.
$$(a + b)(a - b) = a^2 - b^2$$
└───┘

Examples Use a special product to multiply.

first term squared	minus	second term squared

$$
\begin{aligned}
&\textbf{9.} \quad (x + 4)(x - 4) = x^2 & - & 4^2 = x^2 - 16 \\
&\textbf{10.} \quad (6t + 7)(6t - 7) = (6t)^2 & - & 7^2 = 36t^2 - 49
\end{aligned}
$$

Practice Problems 5–8

Use a special product to square each binomial.

5. $(y + 3)^2$ 6. $(r - s)^2$

7. $(6x + 5)^2$ 8. $(x^2 - 3y)^2$

TEACHING TIP

It may be helpful for students to draw a square diagram of $(a + b)(a - b)$.
$$(a + b)(a - b) = a^2 + ab - ab - b^2$$
$$= a^2 - b^2$$

	a	b
a	a^2	ab
$-b$	$-ab$	$-b^2$

Practice Problems 9–13

Use a special product to multiply.

9. $(x + 7)(x - 7)$

10. $(4y + 5)(4y - 5)$

11. $\left(x - \dfrac{1}{3}\right)\left(x + \dfrac{1}{3}\right)$

12. $(3a - b)(3a + b)$

13. $(2x^2 - 6y)(2x^2 + 6y)$

Answers

5. $y^2 + 6y + 9$, 6. $r^2 - 2rs + s^2$,
7. $36x^2 + 60x + 25$, 8. $x^4 - 6x^2y + 9y^2$,
9. $x^2 - 49$, 10. $16y^2 - 25$, 11. $x^2 - \dfrac{1}{9}$,
12. $9a^2 - b^2$, 13. $4x^4 - 36y^2$

TEACHING TIP

Point out to students that the ability to recognize the patterns they discovered in this section will help them factor binomials in the next chapter.

11. $\left(x - \dfrac{1}{4} \right)\left(x + \dfrac{1}{4} \right) = x^2 - \left(\dfrac{1}{4} \right)^2 = x^2 - \dfrac{1}{16}$

12. $(2p - q)(2p + q) = (2p)^2 - q^2 = 4p^2 - q^2$

13. $(3x^2 - 5y)(3x^2 + 5y) = (3x^2)^2 - (5y)^2 = 9x^4 - 25y^2$ ▬▬▬

TRY THE CONCEPT CHECK IN THE MARGIN.

✓ CONCEPT CHECK

Match each expression on the left to the equivalent expression or expressions in the list on the right.

$(a + b)^2$ a. $(a + b)(a + b)$
$(a + b)(a - b)$ b. $a^2 - b^2$
             c. $a^2 + b^2$
             d. $a^2 - 2ab + b^2$
             e. $a^2 + 2ab + b^2$

D USING SPECIAL PRODUCTS

Let's now practice using our special products on a variety of multiplications. This practice will help us recognize when to apply what special product.

Practice Problems 14–16

Use a special product to multiply.

14. $(7x - 1)^2$

15. $(5y + 3)(2y - 5)$

16. $(2a - 1)(2a + 1)$

Examples   Use a special product to multiply.

14. $(x - 9)(x + 9)$ This is the sum and difference of the same two terms.

$\quad = x^2 - 9^2 = x^2 - 81$

15. $(3y + 2)^2$ This is a binomial squared.

$\quad = (3y)^2 + 2(3y)(2) + 2^2$

$\quad = 9y^2 + 12y + 4$

16. $(6a + 1)(a - 7)$ No special product applies.

            F   O   I   L   Use the FOIL method.

$\quad = 6a \cdot a + 6a(-7) + 1 \cdot a + 1(-7)$

$\quad = 6a^2 - 42a + a - 7$

$\quad = 6a^2 - 41a - 7$ ▬▬▬

TEACHING TIP

You may want to check that students know the difference between the following phrases by having them write an example of each.
▲ The sum of squares
▲ The sum squared
▲ The difference squared
▲ The difference of squares
▲ The product of the sum and difference of two terms

Answers

14. $49x^2 - 14x + 1$,   15. $10y^2 - 19y - 15$,
16. $4a^2 - 1$

✓ **Concept Check:** a or e, b

Name _____ **Section** _____ **Date** _____

EXERCISE SET 3.6

A *Multiply using the FOIL method. See Examples 1 through 3.*

1. $(x + 3)(x + 4)$ **2.** $(x + 5)(x + 1)$ **3.** $(x - 5)(x + 10)$

4. $(y - 12)(y + 4)$ **5.** $(5x - 6)(x + 2)$ **6.** $(3y - 5)(2y - 7)$

7. $(y - 6)(4y - 1)$ **8.** $(2x - 9)(x - 11)$ **9.** $(2x + 5)(3x - 1)$

10. $(6x + 2)(x - 2)$ **11.** $(y^2 + 7)(6y + 4)$ **12.** $(y^2 + 3)(5y + 6)$

13. $\left(x - \dfrac{1}{3}\right)\left(x + \dfrac{2}{3}\right)$ **14.** $\left(x - \dfrac{2}{5}\right)\left(x + \dfrac{1}{5}\right)$ **15.** $(4 - 3a)(2 - 5a)$

16. $(3 - 2a)(6 - 5a)$ **17.** $(x + 5y)(2x - y)$ **18.** $(x + 4y)(3x - y)$

B *Multiply. See Examples 4 through 8.*

19. $(x + 2)^2$ **20.** $(x + 7)^2$ **21.** $(2x - 1)^2$ **22.** $(7x - 3)^2$

23. $(3a - 5)^2$ **24.** $(5a + 2)^2$ **25.** $(x^2 + 5)^2$ **26.** $(x^2 + 3)^2$

27. $\left(y - \dfrac{2}{7}\right)^2$ **28.** $\left(y - \dfrac{3}{4}\right)^2$ **29.** $(2a - 3)^2$ **30.** $(5b - 4)^2$

31. $(5x + 9)^2$ **32.** $(6s + 2)^2$ **33.** $(3x - 7y)^2$ **34.** $(4s - 2y)^2$

35. $(4m + 5n)^2$ **36.** $(3n + 5m)^2$

37. answers may vary

38. answers may vary

39. $a^2 - 49$

40. $b^2 - 9$

41. $x^2 - 36$

42. $x^2 - 64$

43. $9x^2 - 1$

44. $16x^2 - 25$

45. $x^4 - 25$

46. $a^4 - 36$

47. $4y^4 - 1$

48. $9x^4 - 1$

49. $16 - 49x^2$

50. $64 - 49x^2$

51. $9x^2 - \dfrac{1}{4}$

52. $100x^2 - \dfrac{4}{49}$

53. $81x^2 - y^2$

54. $4x^2 - y^2$

55. $4m^2 - 25n^2$

56. $25m^2 - 16n^2$

57. $a^2 + 9a + 20$

58. $a^2 + 12a + 35$

59. $a^2 - 14a + 49$

60. $b^2 - 4b + 4$

61. $12a^2 - a - 1$

62. $36a^2 + 72a + 35$

63. $x^2 - 4$

64. $x^2 - 100$

65. $9a^2 + 6a + 1$

242

Name _____

37. Using your own words, explain how to square a binomial such as $(a + b)^2$.

38. Explain how to find the product of two binomials using the FOIL method.

C *Multiply. See Examples 9 through 13.*

39. $(a - 7)(a + 7)$

40. $(b + 3)(b - 3)$

41. $(x + 6)(x - 6)$

42. $(x - 8)(x + 8)$

43. $(3x - 1)(3x + 1)$

44. $(4x - 5)(4x + 5)$

45. $(x^2 + 5)(x^2 - 5)$

46. $(a^2 + 6)(a^2 - 6)$

47. $(2y^2 - 1)(2y^2 + 1)$

48. $(3x^2 + 1)(3x^2 - 1)$

49. $(4 - 7x)(4 + 7x)$

50. $(8 - 7x)(8 + 7x)$

51. $\left(3x - \dfrac{1}{2}\right)\left(3x + \dfrac{1}{2}\right)$

52. $\left(10x + \dfrac{2}{7}\right)\left(10x - \dfrac{2}{7}\right)$

53. $(9x + y)(9x - y)$

54. $(2x - y)(2x + y)$

55. $(2m + 5n)(2m - 5n)$

56. $(5m + 4n)(5m - 4n)$

D *Multiply. See Examples 14 through 16.*

57. $(a + 5)(a + 4)$

58. $(a + 5)(a + 7)$

59. $(a - 7)^2$

60. $(b - 2)^2$

61. $(4a + 1)(3a - 1)$

62. $(6a + 7)(6a + 5)$

63. $(x + 2)(x - 2)$

64. $(x - 10)(x + 10)$

65. $(3a + 1)^2$

66. $(4a + 2)^2$

67. $(x + y)(4x - y)$

68. $(3x + 2)(4x - 2)$

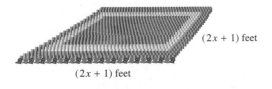 **69.** $\left(a - \dfrac{1}{2}y\right)\left(a + \dfrac{1}{2}y\right)$

70. $\left(\dfrac{a}{2} + 4y\right)\left(\dfrac{a}{2} - 4y\right)$

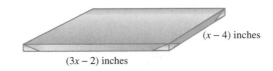 **71.** $(3b + 7)(2b - 5)$

72. $(3y - 13)(y - 3)$

73. $(x^2 + 10)(x^2 - 10)$

74. $(x^2 + 8)(x^2 - 8)$

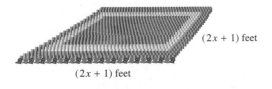 **75.** $(4x + 5)(4x - 5)$

76. $(3x + 5)(3x - 5)$

77. $(5x - 6y)^2$

78. $(4x - 9y)^2$

79. $(2r - 3s)(2r + 3s)$

80. $(6r - 2x)(6r + 2x)$

Review and Preview

Simplify each expression. See Section 3.1.

81. $\dfrac{50b^{10}}{70b^5}$

82. $\dfrac{60y^6}{80y^2}$

83. $\dfrac{8a^{17}b^5}{-4a^7b^{10}}$

84. $\dfrac{-6a^8y}{3a^4y}$

85. $\dfrac{2x^4y^{12}}{3x^4y^4}$

86. $\dfrac{-48ab^6}{32ab^3}$

Combining Concepts

Express each as a product of polynomials in x. Then multiply and simplify.

87. Find the area of the square rug if its side is $(2x + 1)$ feet.

88. Find the area of the rectangular canvas if its length is $(3x - 2)$ inches and its width is $(x - 4)$ inches.

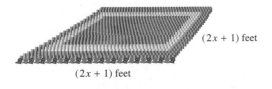

$(2x + 1)$ feet

$(2x + 1)$ feet

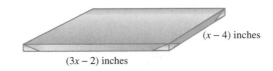

$(x - 4)$ inches

$(3x - 2)$ inches

66. $16a^2 + 16a + 4$

67. $x^2 + 3xy - y^2$

68. $12x^2 + 2x - 4$

69. $a^2 - \dfrac{1}{4}y^2$

70. $\dfrac{a^2}{4} - 16y^2$

71. $6b^2 - b - 35$

72. $3y^2 - 22y + 39$

73. $x^4 - 100$

74. $x^4 - 64$

75. $16x^2 - 25$

76. $9x^2 - 25$

77. $25x^2 - 60xy + 36y^2$

78. $16x^2 - 72xy + 81y^2$

79. $4r^2 - 9s^2$

80. $36r^2 - 4x^2$

81. $\dfrac{5b^5}{7}$

82. $\dfrac{3y^4}{4}$

83. $-\dfrac{2a^{10}}{b^5}$

84. $-2a^4$

85. $\dfrac{2y^8}{3}$

86. $-\dfrac{3b^3}{2}$

87. $(4x^2 + 4x + 1)$ sq. ft

88. $(3x^2 - 14x + 8)$ sq. in.

Name _____

89. Find the area of the shaded region.

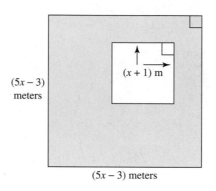

90. Find the area of the shaded region.

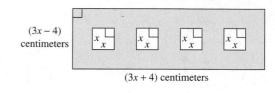

(3x + 4) centimeters

Name _____ Section _____ Date _____

INTEGRATED REVIEW—OPERATIONS ON POLYNOMIALS

Add, subtract, or multiply as indicated.

1. $(5x^2)(7x^3)$

2. $(4y^2)(8y^7)$

3. $(x - 5)(2x + 1)$

4. $(3x - 2)(x + 5)$

5. $(x - 5) + (2x + 1)$

6. $(3x - 2) + (x + 5)$

7. $(4y - 3)(4y + 3)$

8. $(7x - 1)(7x + 1)$

9. $(7x^2 - 2x + 3) - (5x^2 + 9)$

10. $(10x^2 + 7x - 9) - (4x^2 - 6x + 2)$

11. $(x + 4)^2$

12. $(y - 9)^2$

13. $(x - 3)(x^2 + 5x - 1)$

14. $(x + 1)(x^2 - 3x - 2)$

Focus On The Real World

SPACE EXPLORATION

From scientific observations on Earth, we know that Saturn, the second largest planet in our solar system, is a giant ball of gas surrounded by rings and orbited by 19 moons. We also know that Saturn has a diameter of 120,000 kilometers and a mass of 569,000,000,000,000,000,000,000,000 kilograms. But what is Saturn like below its outer layer of clouds? What are Saturn's rings made of? What is the surface of Saturn's largest moon, Titan, like? Could life ever be supported on Titan?

NASA is hoping to answer these questions and more about the sixth planet from the sun in our solar system with its $3,400,000,000 Cassini mission. The Cassini spacecraft is scheduled to arrive in orbit around Saturn in June 2004. The goal of this mission is to study Saturn, its rings, and its moons. The Cassini spacecraft will also launch the Huygens probe to study Titan.

The Cassini mission began on October 15, 1997, with the launch of a mighty Titan IV booster rocket. The entire launch vehicle including rocket fuel weighed more than 2,000,000 pounds before launch. The Titan IV flung Cassini into space at a speed of 14,400 kilometers per hour. To take advantage of something called *gravity assist*, Cassini is taking a roundabout path to Saturn past Venus (twice), Earth, and Jupiter. Altogether, Cassini will travel

(continued)

1. $35x^5$

2. $32y^9$

3. $2x^2 - 9x - 5$

4. $3x^2 + 13x - 10$

5. $3x - 4$

6. $4x + 3$

7. $16y^2 - 9$

8. $49x^2 - 1$

9. $2x^2 - 2x - 6$

10. $6x^2 + 13x - 11$

11. $x^2 + 8x + 16$

12. $y^2 - 18y + 81$

13. $x^3 + 2x^2 - 16x + 3$

14. $x^3 - 2x^2 - 5x - 2$

3,540,000,000 kilometers before reaching Saturn, a planet that is only 1,430,000,000 kilometers from the sun.

Once Cassini reaches Saturn, it will begin collecting all kinds of data about the planet and its moons. Over the course of Cassini's mission, it will collect over 2,000,000,000,000 bits of scientific information, or about the same amount of data in 800 sets of the *Encyclopedia Britannica*. About once per day, Cassini will use its 4-meter antenna to transmit the latest data that it has collected back to Earth at a frequency of 8,400,000,000 cycles per second. For comparison, the FM band on a radio is centered around 100,000,000 cycles per second. It will take from 70 to 90 minutes for Cassini's transmissions to reach Earth and by then the signals are very weak. The power of the signal transmitted by the spacecraft is 20 watts but, even with the huge antennas used on Earth, only 0.0000000000000001 watt can be received. (*Source*: based on data from National Aeronautics and Space Administration)

CRITICAL THINKING

1. Make a list of the numbers (other than those in dates) used in the article. Rewrite each number in scientific notation.
2. What are the advantages of scientific notation?
3. What are the disadvantages of scientific notation?
4. In your opinion, how large or small should a number be to make using scientific notation worthwhile?

3.7   DIVIDING POLYNOMIALS

A   DIVIDING BY A MONOMIAL

To divide a polynomial by a monomial, recall addition of fractions. Fractions that have a common denominator are added by adding the numerators:

$$\frac{a}{c} + \frac{b}{c} = \frac{a + b}{c}$$

If we read this equation from right to left and let a, b, and c be monomials, $c \neq 0$, we have the following:

TO DIVIDE A POLYNOMIAL BY A MONOMIAL

Divide each term of the polynomial by the monomial.

$$\frac{a + b}{c} = \frac{a}{c} + \frac{b}{c}, \quad c \neq 0$$

Throughout this section, we assume that denominators are not 0.

Example 1   Divide: $6m^2 + 2m$ by $2m$

Solution: We begin by writing the quotient in fraction form. Then we divide each term of the polynomial $6m^2 + 2m$ by the monomial $2m$.

$$\frac{6m^2 + 2m}{2m} = \frac{6m^2}{2m} + \frac{2m}{2m}$$

$$= 3m + 1 \qquad \text{Simplify.}$$

Check: To check, we multiply.

$$2m(3m + 1) = 2m(3m) + 2m(1) = 6m^2 + 2m$$

The quotient $3m + 1$ checks.   ▬▬▬

TRY THE CONCEPT CHECK IN THE MARGIN.

Example 2   Divide: $\dfrac{9x^5 - 12x^2 + 3x}{3x^2}$

Solution:

$$\frac{9x^5 - 12x^2 + 3x}{3x^2} = \frac{9x^5}{3x^2} - \frac{12x^2}{3x^2} + \frac{3x}{3x^2} \quad \text{Divide each term by } 3x^2.$$

$$= 3x^3 - 4 + \frac{1}{x} \qquad \text{Simplify.}$$

Notice that the quotient is not a polynomial because of the term $\dfrac{1}{x}$. This expression is called a rational expression— we will study rational expressions further in Chapter 5. Although the quotient of two polynomials is not always a polynomial, we may still check by multiplying.

Check: $3x^2\left(3x^3 - 4 + \dfrac{1}{x}\right) = 3x^2(3x^3) - 3x^2(4) + 3x^2\left(\dfrac{1}{x}\right)$

$$= 9x^5 - 12x^2 + 3x \qquad \text{▬▬▬}$$

Objectives

A   Divide a polynomial by a monomial.

B   Use long division to divide a polynomial by a polynomial other than a monomial.

SSM   CD-ROM   Video 3.7

TEACHING TIP

It may be helpful to students to verify that the answer to Example 1 is true for $m = 1$, 2, 3. Then ask if it would be true for all values. Have them verify that it is not true for $m = 0$.

Practice Problem 1

Divide: $25x^3 + 5x^2$ by $5x^2$

✓ CONCEPT CHECK

In which of the following is $\dfrac{x + 5}{5}$ simplified correctly?

a. $\dfrac{x}{5} + 1$ b. x c. $x + 1$

Practice Problem 2

Divide: $\dfrac{30x^7 + 10x^2 - 5x}{5x^2}$

Answers

1. $5x + 1$,   **2.** $6x^5 + 2 - \dfrac{1}{x}$

✓ Concept Check: a

Practice Problem 3

Divide: $\dfrac{12x^3y^3 - 18xy + 6y}{3xy}$

Example 3

Divide: $\dfrac{8x^2y^2 - 16xy + 2x}{4xy}$

Solution:

$$\frac{8x^2y^2 - 16xy + 2x}{4xy} = \frac{8x^2y^2}{4xy} - \frac{16xy}{4xy} + \frac{2x}{4xy} \qquad \text{Divide each term by } 4xy.$$

$$= 2xy - 4 + \frac{1}{2y} \qquad \text{Simplify.}$$

Check:

$$4xy\left(2xy - 4 + \frac{1}{2y}\right) = 4xy(2xy) - 4xy(4) + 4xy\left(\frac{1}{2y}\right)$$

$$= 8x^2y^2 - 16xy + 2x$$

B DIVIDING BY A POLYNOMIAL OTHER THAN A MONOMIAL

To divide a polynomial by a polynomial other than a monomial, we use a process known as long division. Polynomial long division is similar to number long division, so we review long division by dividing 13 into 3660.

$$
\begin{array}{r}
281 \\
13\overline{)3660} \\
\underline{26\downarrow} \quad\;\; 2 \cdot 13 = 26 \\
106 \quad\;\; \text{Subtract and bring down the next digit in the dividend.} \\
\underline{104\downarrow} \quad\;\; 8 \cdot 13 = 104 \\
20 \quad\;\; \text{Subtract and bring down the next digit in the dividend.} \\
\underline{13} \quad\;\; 1 \cdot 13 = 13 \\
7 \quad\;\; \text{Subtract. There are no more digits to bring down, so the remainder is 7.}
\end{array}
$$

The quotient is 281 R 7, which can be written as $281\dfrac{7}{13}$ $\begin{array}{l}\leftarrow \text{remainder} \\ \leftarrow \text{divisor}\end{array}$.

Recall that division can be checked by multiplication. To check a division problem such as this one, we see that

$$13 \cdot 281 + 7 = 3660$$

Now we demonstrate long division of polynomials.

TEACHING TIP

In Example 4, you may want to point out that a term of the quotient is placed over the like term in the dividend.

Practice Problem 4

Divide: $x^2 + 12x + 35$ by $x + 5$

Example 4

Divide $x^2 + 7x + 12$ by $x + 3$ using long division.

Solution:

To subtract, change the signs of these terms and add.

$$
\begin{array}{r}
x \\
x + 3\overline{)x^2 + 7x + 12} \\
\underline{x^2 + 3x} \\
4x + 12
\end{array}
$$

How many times does x divide x^2? $\dfrac{x^2}{x} = x$.

Multiply: $x(x + 3)$.

Subtract and bring down the next term.

Now we repeat this process.

$$
\begin{array}{r}
x + 4 \\
x + 3\overline{)x^2 + 7x + 12} \\
\underline{x^2 + 3x} \\
4x + 12 \\
\underline{4x + 12} \\
0
\end{array}
$$

To subtract, change the signs of these terms and add.

How many times does x divide $4x$? $\dfrac{4x}{x} = 4$.

Multiply: $4(x + 3)$.

Subtract. The remainder is 0.

The quotient is $x + 4$.

Answers

3. $4x^2y^2 - 6 + \dfrac{2}{x}$, **4.** $x + 7$

Check: We check by multiplying.

$$\boxed{\text{divisor}} \cdot \boxed{\text{quotient}} + \boxed{\text{remainder}} = \boxed{\text{dividend}}$$

or $\downarrow$ $\downarrow$ $\downarrow$ $\downarrow$

$$(x + 3) \cdot (x + 4) + \qquad 0 \qquad = x^2 + 7x + 12$$

The quotient checks.   ▬▬▬

Example 5   Divide $6x^2 + 10x - 5$ by $3x - 1$ using long division.

Solution:

$$
\begin{array}{r}
2x + 4 \\
3x - 1 \overline{)6x^2 + 10x - 5} \\
\underline{6x^2 \mp 2x} \\
12x - 5 \\
\underline{12x \mp 4} \\
-1
\end{array}
$$

$\dfrac{6x^2}{3x} = 2x$, so $2x$ is a term of the quotient.

$2x(3x - 1)$.

Subtract and bring down the next term.

$\dfrac{12x}{3x} = 4, \; 4(3x - 1)$

Subtract. The remainder is -1.

Thus $\left(6x^2 + 10x - 5\right)$ divided by $\left(3x - 1\right)$ is $\left(2x + 4\right)$ with a remainder of -1. This can be written as

$$\frac{6x^2 + 10x - 5}{3x - 1} = 2x + 4 + \frac{-1}{3x - 1} \quad \begin{array}{l} \leftarrow \text{remainder} \\ \leftarrow \text{divisor} \end{array}$$

Check: To check, we multiply $\left(3x - 1\right)\left(2x + 4\right)$. Then we add the remainder, -1, to this product.

$$(3x - 1)(2x + 4) + (-1) = (6x^2 + 12x - 2x - 4) - 1$$
$$= 6x^2 + 10x - 5$$

The quotient checks.   ▬▬▬

Notice that the division process is continued until the degree of the remainder polynomial is less than the degree of the divisor polynomial.

Example 6   Divide: $\dfrac{4x^2 + 7 + 8x^3}{2x + 3}$

Solution: Before we begin the division process, we rewrite $4x^2 + 7 + 8x^3$ as $8x^3 + 4x^2 + 0x + 7$. Notice that we have written the polynomial in descending order and have represented the missing x term by $0x$.

$$
\begin{array}{r}
4x^2 - 4x + 6 \\
2x + 3 \overline{)8x^3 + 4x^2 + 0x + 7} \\
\underline{8x^3 \mp 12x^2} \\
-8x^2 + 0x \\
\underline{\pm 8x^2 \mp 12x} \\
12x + 7 \\
\underline{12x \mp 18} \\
-11
\end{array}
$$

Remainder.

Thus, $\dfrac{4x^2 + 7 + 8x^3}{2x + 3} = 4x^2 - 4x + 6 + \dfrac{-11}{2x + 3}$.   ▬▬▬

Practice Problem 5

Divide: $6x^2 + 7x - 5$ by $2x - 1$

Practice Problem 6

Divide: $\dfrac{5 - x + 9x^3}{3x + 2}$

Answers

5. $3x + 5$,   **6.** $3x^2 - 2x + 1 + \dfrac{3}{3x + 2}$

Focus On History

NEGATIVE EXPONENTS

Negative exponents were the invention of the English mathematician John Wallis (1616–1703). His book *Arithmetica Infinitorum*, published in 1656, begins with proofs of the laws of exponents. He extended these to cover negative exponents as well and showed that $x^0 = 1$, $x^{-1} = \dfrac{1}{x}$, $x^{-2} = \dfrac{1}{x^2}$, and so on. He also showed that, in general, x^{-n} represents the reciprocal of x^n.

Not long after Wallis published his *Arithmetica Infinitorum*, Sir Issac Newton (1642–1727) adopted Wallis's definition and use of negative exponents. Newton's widely circulated mathematical and scientific writings helped the use of negative exponents become universally accepted.

Name _____ Section _____ Date _____

MENTAL MATH

Simplify each expression.

1. $\dfrac{a^6}{a^4}$

2. $\dfrac{y^2}{y}$

3. $\dfrac{a^3}{a}$

4. $\dfrac{p^8}{p^3}$

5. $\dfrac{k^5}{k^2}$

6. $\dfrac{k^7}{k^5}$

EXERCISE SET 3.7

A *Perform each division. See Examples 1 through 3.*

1. $\dfrac{20x^2 + 5x + 9}{5}$

2. $\dfrac{8x^3 - 4x^2 + 6x + 2}{2}$

3. $\dfrac{12x^4 + 3x^2}{x}$

4. $\dfrac{15x^2 - 9x^5}{x}$

5. $\dfrac{15p^3 + 18p^2}{3p}$

6. $\dfrac{14m^2 - 27m^3}{7m}$

7. $\dfrac{-9x^4 + 18x^5}{6x^5}$

8. $\dfrac{6x^5 + 3x^4}{3x^4}$

▭ **9.** $\dfrac{-9x^5 + 3x^4 - 12}{3x^3}$

10. $\dfrac{6a^2 - 4a + 12}{-2a^2}$

11. $\dfrac{4x^4 - 6x^3 + 7}{-4x^4}$

12. $\dfrac{-12a^3 + 36a - 15}{3a}$

13. $\dfrac{a^2b^2 - ab^3}{ab}$

14. $\dfrac{m^3n^2 - mn^4}{mn}$

15. $\dfrac{2x^2y + 8x^2y^2 - xy^2}{2xy}$

16. $\dfrac{11x^3y^3 - 33xy + x^2y^2}{11xy}$

17. $x + 1$

18. $x + 2$

19. $2x + 3$

20. $3x + 2$

21. $2x + 1 + \dfrac{7}{x - 4}$

22. $3x + 2 - \dfrac{2}{x - 1}$

23. $4x + 9$

24. $6w - 2$

25. $\dfrac{3a^2 - 3a + 1 + \dfrac{2}{3a + 2}}{}$

26. $2x^2 + 3x - 4$

27. $\dfrac{2b^2 + b + 2 - \dfrac{12}{b + 4}}{}$

28. $\dfrac{2x^2 - x - 1 + \dfrac{6}{x + 2}}{}$

29. $4x + 3 - \dfrac{2}{2x + 1}$

30. $x + 5 + \dfrac{3}{3x + 2}$

31. $\dfrac{2x^2 + 6x - 5 - \dfrac{2}{x - 2}}{}$

32. $\dfrac{4x^2 - x - 5 + \dfrac{5}{x + 3}}{}$

33. $x^2 + 3x + 9$

34. $x^2 - 4x + 16$

35. $-3x + 6 - \dfrac{11}{x + 2}$

36. $-5x + 15 - \dfrac{38}{x + 3}$

252

Name _____

B *Find each quotient using long division. See Examples 4 and 5.*

17. $\dfrac{x^2 + 4x + 3}{x + 3}$

18. $\dfrac{x^2 + 7x + 10}{x + 5}$

19. $\dfrac{2x^2 + 13x + 15}{x + 5}$

20. $\dfrac{3x^2 + 8x + 4}{x + 2}$

21. $\dfrac{2x^2 - 7x + 3}{x - 4}$

22. $\dfrac{3x^2 - x - 4}{x - 1}$

23. $\dfrac{8x^2 + 6x - 27}{2x - 3}$

24. $\dfrac{18w^2 + 18w - 8}{3w + 4}$

25. $\dfrac{9a^3 - 3a^2 - 3a + 4}{3a + 2}$

26. $\dfrac{4x^3 + 12x^2 + x - 12}{2x + 3}$

27. $\dfrac{2b^3 + 9b^2 + 6b - 4}{b + 4}$

28. $\dfrac{2x^3 + 3x^2 - 3x + 4}{x + 2}$

29. $\dfrac{8x^2 + 10x + 1}{2x + 1}$

30. $\dfrac{3x^2 + 17x + 7}{3x + 2}$

31. $\dfrac{2x^3 + 2x^2 - 17x + 8}{x - 2}$

32. $\dfrac{4x^3 + 11x^2 - 8x - 10}{x + 3}$

Find each quotient using long division. Don't forget to write the polynomials in descending order and fill in any missing terms. See Example 6.

33. $\dfrac{x^3 - 27}{x - 3}$

34. $\dfrac{x^3 + 64}{x + 4}$

35. $\dfrac{1 - 3x^2}{x + 2}$

36. $\dfrac{7 - 5x^2}{x + 3}$

Name _____

37. $\dfrac{-4b + 4b^2 - 5}{2b - 1}$ **38.** $\dfrac{-3y + 2y^2 - 15}{2y + 5}$

REVIEW AND PREVIEW

Fill in each blank. See Sections 3.1 and 3.2.

39. $12 = 4 \cdot$ _____   **40.** $12 = 2 \cdot$ _____   **41.** $20 = -5 \cdot$ _____

42. $20 = -4 \cdot$ _____   **43.** $9x^2 = 3x \cdot$ _____   **44.** $9x^2 = 9x \cdot$ _____

45. $36x^2 = 4x \cdot$ _____   **46.** $36x^2 = 2x \cdot$ _____

COMBINING CONCEPTS

Divide.

47. $\dfrac{x^5 + x^2}{x^2 + x}$ **48.** $\dfrac{x^6 - x^4}{x^3 + 1}$

Solve.

49. The perimeter of a square is $(12x^3 + 4x - 16)$ feet. Find the length of its side.

Perimeter is
$(12x^3 + 4x - 16)$
feet

50. The volume of the swimming pool shown is $(36x^5 - 12x^3 + 6x^2)$ cubic feet. If its height is $2x$ feet and its width is $3x$ feet, find its length.

3x feet

2x feet

51. The area of the following parallelogram is $(10x^2 + 31x + 15)$ square meters. If its base is $(5x + 3)$ meters, find its height.

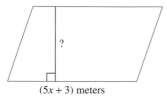

(5x + 3) meters

52. The area of the top of the Ping-Pong table is $(49x^2 + 70x - 200)$ square inches. If its length is $(7x + 20)$ inches, find its width.

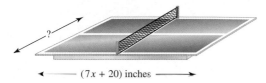

← (7x + 20) inches →

53. Explain how to check a polynomial long division result when the remainder is 0.

54. Explain how to check a polynomial long division result when the remainder is not 0.

CHAPTER 3 ACTIVITY
MODELING WITH POLYNOMIALS

MATERIALS:
▲ Calculator

This activity may be completed by working in groups or individually.

The polynomial model $5.31x^2 - 1.93x + 126.46$ billion dollars represents the total amount donated to charities in the United States by all sources from 1993–1995. The polynomial model $4.65x^2 - 2.25x + 102.13$ billion dollars represents the amount donated to charities in the United States by individuals from 1993–1995. In both models, x is the number of years after 1993. The other major sources of charitable contributions include corporations, foundations, and bequests. (*Source*: Based on data from the American Association of Fund-Raising Counsel, Inc.)

1. Use the polynomials to complete the following table showing the amounts of charitable giving over the period 1993–1995 by evaluating each polynomial at the given values of x. Then subtract each value in the fourth column from the corresponding value in the third column. Record the result in the last column, titled "Difference." What do you think these values represent?

Year	x	Total Charitable Giving (billions of dollars)	Charitable Giving by Individuals (billions of dollars)	Difference
1993	0	126.46	102.13	24.33
1994	1	129.84	104.53	25.31
1995	2	143.84	116.23	27.61

2. Use the polynomial models to find a new polynomial model representing the amount of charitable giving from other sources, including corporations, foundations, and bequests. Then use your new polynomial model to complete the accompanying table. $0.66x^2 + 0.32x + 24.33$

Year	x	Charitable Giving by Other Sources (billions of dollars)
1993	0	24.33
1994	1	25.31
1995	2	27.61

3. Compare the values in the last column of the table in Question 1 to the values in the last column of the table in Question 2. What do you notice? What can you conclude?   answers may vary

4. Make a bar graph of the data in the table in Question 2. What trend do you notice?
answers may vary

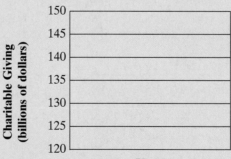

CHAPTER 3 HIGHLIGHTS

DEFINITIONS AND CONCEPTS	EXAMPLES

SECTION 3.1   EXPONENTS

a^n means the product of n factors, each of which is a.	$3^2 = 3 \cdot 3 = 9$ $(-5)^3 = (-5)(-5)(-5) = -125$ $\left(\dfrac{1}{2}\right)^4 = \dfrac{1}{2} \cdot \dfrac{1}{2} \cdot \dfrac{1}{2} \cdot \dfrac{1}{2} = \dfrac{1}{16}$
If m and n are integers and no denominators are 0, **Product Rule:** $a^m \cdot a^n = a^{m+n}$ **Power Rule:** $(a^m)^n = a^{mn}$ **Power of a Product Rule:** $(ab)^n = a^n b^n$ **Power of a Quotient Rule:** $\left(\dfrac{a}{b}\right)^n = \dfrac{a^n}{b^n}$ **Quotient Rule:** $\dfrac{a^m}{a^n} = a^{m-n}$ **Zero Exponent:** $a^0 = 1, a \neq 0$	$x^2 \cdot x^7 = x^{2+7} = x^9$ $(5^3)^8 = 5^{3 \cdot 8} = 5^{24}$ $(7y)^4 = 7^4 y^4$ $\left(\dfrac{x}{8}\right)^3 = \dfrac{x^3}{8^3}$ $\dfrac{x^9}{x^4} = x^{9-4} = x^5$ $5^0 = 1, x^0 = 1, x \neq 0$

SECTION 3.2   NEGATIVE EXPONENTS AND SCIENTIFIC NOTATION

If $a \neq 0$ and n is an integer, $a^{-n} = \dfrac{1}{a^n}$	$3^{-2} = \dfrac{1}{3^2} = \dfrac{1}{9}; \; 5x^{-2} = \dfrac{5}{x^2}$ Simplify: $\left(\dfrac{x^{-2}y}{x^5}\right)^{-2} = \dfrac{x^4 y^{-2}}{x^{-10}}$ $\qquad\qquad = x^{4-(-10)}y^{-2}$ $\qquad\qquad = \dfrac{x^{14}}{y^2}$
A positive number is written in scientific notation if it is written as the product of a number a, $1 \leq a < 10$, and an integer power r of 10. $a \times 10^r$	$12000 = 1.2 \times 10^4$ $0.00000568 = 5.68 \times 10^{-6}$

SECTION 3.3   INTRODUCTION TO POLYNOMIALS

A **term** is a number or the product of a number and variables raised to powers.	$-5x, \; 7a^2b, \; \dfrac{1}{4}y^4, \; 0.2$
The **numerical coefficient** or **coefficient** of a term is its numerical factor.	TERM COEFFICIENT $7x^2$ 7 y 1 $-a^2b$ -1
A **polynomial** is a finite sum of terms of the form ax^n where a is a real number and n is a whole number.	$5x^3 - 6x^2 + 3x - 6$ (Polynomial)
A **monomial** is a polynomial with exactly 1 term.	$\dfrac{5}{6}y^3$ (Monomial)
A **binomial** is a polynomial with exactly 2 terms.	$-0.2a^2b - 5b^2$ (Binomial)

SECTION 3.3 CONTINUED	
A **trinomial** is a polynomial with exactly 3 terms.	$3x^2 - 2x + 1$ (Trinomial)
	POLYNOMIAL DEGREE
The **degree of a polynomial** is the greatest degree of any term of the polynomial.	$5x^2 - 3x + 2$ 2 $7y + 8y^2z^3 - 12$ $2 + 3 = 5$

SECTION 3.4 ADDING AND SUBTRACTING POLYNOMIALS	
To add polynomials, combine like terms.	Add. $(7x^2 - 3x + 2) + (-5x - 6)$ $\quad = 7x^2 - 3x + 2 - 5x - 6$ $\quad = 7x^2 - 8x - 4$
To subtract two polynomials, change the signs of the terms of the second polynomial, then add.	Subtract. $(17y^2 - 2y + 1) - (-3y^3 + 5y - 6)$ $\quad = (17y^2 - 2y + 1) + (3y^3 - 5y + 6)$ $\quad = 17y^2 - 2y + 1 + 3y^3 - 5y + 6$ $\quad = 3y^3 + 17y^2 - 7y + 7$

SECTION 3.5 MULTIPLYING POLYNOMIALS	
To multiply two polynomials, multiply each term of one polynomial by each term of the other polynomial, and then combine like terms.	Multiply. $(2x + 1)(5x^2 - 6x + 2)$ $= 2x(5x^2 - 6x + 2) + 1(5x^2 - 6x + 2)$ $= 10x^3 - 12x^2 + 4x + 5x^2 - 6x + 2$ $= 10x^3 - 7x^2 - 2x + 2$

SECTION 3.6 SPECIAL PRODUCTS	
The **FOIL method** may be used when multiplying two binomials.	Multiply: $(5x - 3)(2x + 3)$ $\quad\quad\text{First}\quad\quad\text{Last}$ $(5x - 3)(2x + 3)$ $\quad\text{Outer}\quad\text{Inner}$ $= \overset{F}{(5x)(2x)} + \overset{O}{(5x)(3)} + \overset{I}{(-3)(2x)} + \overset{L}{(-3)(3)}$ $= 10x^2 + 15x - 6x - 9$ $= 10x^2 + 9x - 9$
Squaring a Binomial $(a + b)^2 = a^2 + 2ab + b^2$ $(a - b)^2 = a^2 - 2ab + b^2$	Square each binomial. $(x + 5)^2 = x^2 + 2(x)(5) + 5^2$ $\quad\quad\quad = x^2 + 10x + 25$ $(3x - 2y)^2 = (3x)^2 - 2(3x)(2y) + (2y)^2$ $\quad\quad\quad\quad = 9x^2 - 12xy + 4y^2$
Multiplying the Sum and Difference of Two Terms $(a + b)(a - b) = a^2 - b^2$	Multiply. $(6y + 5)(6y - 5) = (6y)^2 - 5^2$ $\quad\quad\quad\quad\quad = 36y^2 - 25$

SECTION 3.7   DIVIDING POLYNOMIALS

To divide a polynomial by a monomial: $$\frac{a+b}{c} = \frac{a}{c} + \frac{b}{c}, c \neq 0$$	Divide. $$\frac{15x^5 - 10x^3 + 5x^2 - 2x}{5x^2}$$ $$= \frac{15x^5}{5x^2} - \frac{10x^3}{5x^2} + \frac{5x^2}{5x^2} - \frac{2x}{5x^2}$$ $$= 3x^3 - 2x + 1 - \frac{2}{5x}$$
To divide a polynomial by a polynomial other than a monomial, use long division.	$$5x - 1 + \frac{-4}{2x+3} \text{ or}$$ $$\begin{array}{r} 5x - 1 \phantom{+ \frac{4}{2x+3}} \\ 2x+3\overline{)10x^2 + 13x - 7} \\ \underline{10x^2 + 15x} \\ -2x - 7 \\ \underline{-2x - 3} \\ -4 \end{array} \qquad 5x - 1 - \frac{4}{2x+3}$$

Focus On Study Skills

TAKING A MATH EXAM

When it's time to take your exam, remember these hints:

▲ Make sure you have all the tools you will need to take the exam, including an extra pencil and eraser, paper (if needed), and calculator (if allowed).

▲ Try to relax. Taking a few deep breaths, inhaling and then exhaling slowly before you begin, might help.

▲ Are there any special formulas or definitions that you'll need to remember during the exam? As soon as you get your exam, write these down at the top, bottom, or on the back of your paper.

▲ Scan the entire test to get an idea of what questions are being asked.

▲ Start with the questions that are easiest for you. This will help build your confidence. Then return to the harder ones.

▲ Read all directions carefully. Make sure that your final result answers the question being asked.

▲ Show all of your work. Try to work neatly.

▲ Don't spend too much time on a single problem. If you get stuck, try moving on to other problems so you can increase your chances of finishing the test. If you have time, you can return to the problem giving you trouble.

▲ Before turning in your exam, check your work carefully if time allows. Be on the lookout for careless mistakes.

CHAPTER 3 REVIEW

(3.1) *State the base and the exponent for each expression.*

1. 3^2 base: 3; exponent: 2

2. $(-5)^4$ base: -5; exponent: 4

3. -5^4 base: 5; exponent: 4

4. x^6 base: x; exponent: 6

Evaluate each expression.

5. 8^3 512

6. $(-6)^2$ 36

7. -6^2 -36

8. $-4^3 - 4^0$ -65

9. $(3b)^0$ 1

10. $\dfrac{8b}{8b}$ 1

Simplify each expression.

11. $y^2 \cdot y^7$ y^9

12. $x^9 \cdot x^5$ x^{14}

13. $(2x^5)(-3x^6)$ $-6x^{11}$

14. $(-5y^3)(4y^4)$ $-20y^7$

15. $(x^4)^2$ x^8

16. $(y^3)^5$ y^{15}

17. $(3y^6)^4$ $81y^{24}$

18. $(2x^3)^3$ $8x^9$

19. $\dfrac{x^9}{x^4}$ x^5

20. $\dfrac{z^{12}}{z^5}$ z^7

21. $\dfrac{a^5b^4}{ab}$ a^4b^3

22. $\dfrac{x^4y^6}{xy}$ x^3y^5

23. $\dfrac{12xy^6}{3x^4y^{10}}$ $\dfrac{4}{x^3y^4}$

24. $\dfrac{2x^7y^8}{8xy^2}$ $\dfrac{x^6y^6}{4}$

25. $5a^7(2a^4)^3$ $40a^{19}$

26. $(2x)^2(9x)$ $36x^3$

27. $(-5a)^0 + 7^0 + 8^0$ 3

28. $8x^0 + 9^0$ 9

Name _____

Choose the correct answer for each statement.

29. $\left(\dfrac{3x^4}{4y}\right)^3$ simplifies to   b

  a. $\dfrac{27x^{64}}{64y^3}$

  b. $\dfrac{27x^{12}}{64y^3}$

  c. $\dfrac{9x^{12}}{12y^3}$

  d. $\dfrac{3x^{12}}{4y^3}$

30. $\left(\dfrac{5a^6}{b^3}\right)^2$ simplifies to   c

  a. $\dfrac{10a^{12}}{b^6}$

  b. $\dfrac{25a^{36}}{b^9}$

  c. $\dfrac{25a^{12}}{b^6}$

  d. $25a^{12}b^6$

(3.2) *Simplify each expression.*

31. 7^{-2} $\dfrac{1}{49}$

32. -7^{-2} $-\dfrac{1}{49}$

33. $2x^{-4}$ $\dfrac{2}{x^4}$

34. $(2x)^{-4}$ $\dfrac{1}{16x^4}$

35. $\left(\dfrac{1}{5}\right)^{-3}$ 125

36. $\left(\dfrac{-2}{3}\right)^{-2}$ $\dfrac{9}{4}$

37. $2^0 + 2^{-4}$ $\dfrac{17}{16}$

38. $6^{-1} - 7^{-1}$ $\dfrac{1}{42}$

Simplify each expression. Assume that variables in an exponent represent positive integers only. Write each answer using positive exponents only.

39. $\dfrac{x^5}{x^{-3}}$ x^8

40. $\dfrac{z^4}{z^{-4}}$ z^8

41. $\dfrac{r^{-3}}{r^{-4}}$ r

42. $\dfrac{y^{-2}}{y^{-5}}$ y^3

43. $\left(\dfrac{bc^{-2}}{bc^{-3}}\right)^4$ c^4

44. $\left(\dfrac{x^{-3}y^{-4}}{x^{-2}y^{-5}}\right)^{-3}$ $\dfrac{x^3}{y^3}$

45. $\dfrac{x^{-4}y^{-6}}{x^2y^7}$ $\dfrac{1}{x^6y^{13}}$

46. $\dfrac{a^5b^{-5}}{a^{-5}b^5}$ $\dfrac{a^{10}}{b^{10}}$

Write each number in scientific notation.

47. 0.00027 2.7×10^{-4}

48. 0.8868 8.868×10^{-1}

49. 80,800,000 8.08×10^7

50. −868,000 -8.68×10^5

51. The population of California is 31,880,000. Write this number in scientific notation. (*Source*: Federal–State Cooperative Program for Population Estimates) 3.188×10^7

52. The radius of the earth is 4000 miles. Write 4000 in scientific notation. 4.0×10^3

Write each number in standard form.

53. 8.67×10^5 867,000

54. 3.86×10^{-3} 0.00386

55. 8.6×10^{-4} 0.00086

56. 8.936×10^5 893,600

57. The number of photons of light emitted by a 100-watt bulb every second is 1×10^{20}. Write 1×10^{20} in standard notation.   100,000,000,000,000,000,000

58. The real mass of all the galaxies in the constellation of Virgo is 3×10^{-25}. Write 3×10^{-25} in standard notation.   0.0000000000000000000000003

Simplify. Express each result in standard form.

59. $(8 \times 10^4)(2 \times 10^{-7})$ 0.016

60. $\dfrac{8 \times 10^4}{2 \times 10^{-7}}$ 400,000,000,000

(3.3) *Find the degree of each polynomial.*

61. $y^5 + 7x - 8x^4$ 5

62. $9y^2 + 30y + 25$ 2

63. $-14x^2y - 28x^2y^3 - 42x^2y^2$ 5

64. $6x^2y^2z^2 + 5x^2y^3 - 12xyz$ 6

65. The surface area of a box with a square base and a height of 5 units is given by the polynomial $2x^2 + 20x$. Fill in the table below by evaluating $2x^2 + 20x$ for the given values of x.

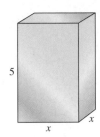

x	1	3	5.1	10
$2x^2 + 20x$	22	78	154.02	400

Combine like terms in each expression.

66. $7a^2 - 4a^2 - a^2$ $2a^2$

67. $9y + y - 14y$ $-4y$

68. $6a^2 + 4a + 9a^2$ $15a^2 + 4a$

69. $21x^2 + 3x + x^2 + 6$ $22x^2 + 3x + 6$

70. $4a^2b - 3b^2 - 8q^2 - 10a^2b + 7q^2$
$-6a^2b - 3b^2 - a^2$

71. $2s^{14} + 3s^{13} + 12s^{12} - s^{10}$
cannot be combined

(3.4) *Add or subtract as indicated.*

72. $\left(3x^2 + 2x + 6\right) + \left(5x^2 + x\right)$ $8x^2 + 3x + 6$

73. $\left(2x^5 + 3x^4 + 4x^3 + 5x^2\right) + \left(4x^2 + 7x + 6\right)$
$2x^5 + 3x^4 + 4x^3 + 9x^2 + 7x + 6$

74. $\left(-5y^2 + 3\right) - \left(2y^2 + 4\right)$ $-7y^2 - 1$

75. $\left(2m^7 + 3x^4 + 7m^6\right) - \left(8m^7 + 4m^2 + 6x^4\right)$
$-6m^7 - 3x^4 + 7m^6 - 4m^2$

76. $\left(3x^2 - 7xy + 7y^2\right) - \left(4x^2 - xy + 9y^2\right)$
$-x^2 - 6xy - 2y^2$

77. Add $\left(-9x^2 + 6x + 2\right)$ and $\left(4x^2 - x - 1\right)$
$-5x^2 + 5x + 1$

78. Subtract $\left(4x^2 + 8x - 7\right)$ from the sum of $\left(x^2 + 7x + 9\right)$ and $\left(x^2 + 4\right)$. $-2x^2 - x + 20$

(3.5) *Multiply each expression.*

79. $6(x + 5)$ $6x + 30$

80. $9(x - 7)$ $9x - 63$

81. $4(2a + 7)$ $8a + 28$

82. $9(6a - 3)$ $54a - 27$

83. $-7x(x^2 + 5)$ $-7x^3 - 35x$

84. $-8y(4y^2 - 6)$ $-32y^3 + 48y$

85. $-2(x^3 - 9x^2 + x)$ $-2x^3 + 18x^2 - 2x$

86. $-3a(a^2b + ab + b^2)$ $-3a^3b - 3a^2b - 3ab^2$

87. $(3a^3 - 4a + 1)(-2a)$ $-6a^4 + 8a^2 - 2a$

88. $(6b^3 - 4b + 2)(7b)$ $42b^4 - 28b^2 + 14b$

89. $(2x + 2)(x - 7)$ $2x^2 - 12x - 14$

90. $(2x - 5)(3x + 2)$ $6x^2 - 11x - 10$

91. $(4a - 1)(a + 7)$ $4a^2 + 27a - 7$

92. $(6a - 1)(7a + 3)$ $42a^2 + 11a - 3$

93. $(x + 7)(x^3 + 4x - 5)$
$x^4 + 7x^3 + 4x^2 + 23x - 35$

94. $(x + 2)(x^5 + x + 1)$
$x^6 + 2x^5 + x^2 + 3x + 2$

95. $(x^2 + 2x + 4)(x^2 + 2x - 4)$
$x^4 + 4x^3 + 4x^2 - 16$

96. $(x^3 + 4x + 4)(x^3 + 4x - 4)$
$x^6 + 8x^4 + 16x^2 - 16$

97. $(x + 7)^3$ $x^3 + 21x^2 + 147x + 343$

98. $(2x - 5)^3$ $8x^3 - 60x^2 + 150x - 125$

(3.6) _Use special products to multiply each of the following._
99. $(x + 7)^2$ $x^2 + 14x + 49$

100. $(x - 5)^2$ $x^2 - 10x + 25$

101. $(3x - 7)^2$ $9x^2 - 42x + 49$

102. $(4x + 2)^2$ $16x^2 + 16x + 4$

103. $(5x - 9)^2$ $25x^2 - 90x + 81$

104. $(5x + 1)(5x - 1)$ $25x^2 - 1$

105. $(7x + 4)(7x - 4)$ $49x^2 - 16$

106. $(a + 2b)(a - 2b)$ $a^2 - 4b^2$

107. $(2x - 6)(2x + 6)$ $4x^2 - 36$

108. $(4a^2 - 2b)(4a^2 + 2b)$ $16a^4 - 4b^2$

(3.7) *Perform each division.*

109. $\dfrac{x^2 + 21x + 49}{7x^2}$ $\dfrac{1}{7} + \dfrac{3}{x} + \dfrac{7}{x^2}$

110. $\dfrac{5a^3b - 15ab^2 + 20ab}{-5ab}$ $-a^2 + 3b - 4$

111. $(a^2 - a + 4) \div (a - 2)$ $a + 1 + \dfrac{6}{a - 2}$

112. $(4x^2 + 20x + 7) \div (x + 5)$ $4x + \dfrac{7}{x + 5}$

113. $\dfrac{a^3 + a^2 + 2a + 6}{a - 2}$ $a^2 + 3a + 8 + \dfrac{22}{a - 2}$

114. $\dfrac{9b^3 - 18b^2 + 8b - 1}{3b - 2}$ $3b^2 - 4b - \dfrac{1}{3b - 2}$

115. $\dfrac{4x^4 - 4x^3 + x^2 + 4x - 3}{2x - 1}$

$2x^3 - x^2 + 2 - \dfrac{1}{2x - 1}$

116. $\dfrac{-10x^2 - x^3 - 21x + 18}{x - 6}$

$-x^2 - 16x - 117 - \dfrac{684}{x - 6}$

Name _____ Section _____ Date _____

Chapter 3 Test

Evaluate each expression.

1. 2^5 **2.** $(-3)^4$ **3.** -3^4 **4.** 4^{-3}

Simplify each exponential expression.

5. $(3x^2)(-5x^9)$ **6.** $\dfrac{y^7}{y^2}$ **7.** $\dfrac{r^{-8}}{r^{-3}}$

Simplify each expression. Write the result using only positive exponents.

8. $\left(\dfrac{x^2 y^3}{x^3 y^{-4}}\right)^2$ **9.** $\dfrac{6^2 x^{-4} y^{-1}}{6^3 x^{-3} y^7}$

Express each number in scientific notation.

10. 563,000   **11.** 0.0000863

Write each number in standard form.

12. 1.5×10^{-3} **13.** 6.23×10^4

14. Simplify. Write the answer in standard form.

$(1.2 \times 10^5)(3 \times 10^{-7})$

15. Find the degree of the following polynomial.

$4xy^2 + 7xyz + 9x^3yz$

16. Simplify by combining like terms.

$5x^2 + 4x - 7x^2 + 11 + 8x$

Perform each indicated operation.

17. $(8x^3 + 7x^2 + 4x - 7) + (8x^3 - 7x - 6)$

18.
$$\begin{aligned}5x^3 + \ x^2 + 5x - 2 \\ -\ (8x^3 - 4x^2 + \ x - 7)\end{aligned}$$

19. Subtract $(4x + 2)$ from the sum of $(8x^2 + 7x + 5)$ and $(x^3 - 8)$.

20. Multiply: $(3x + 7)(x^2 + 5x + 2)$

Answers

1. 32

2. 81

3. -81

4. $\dfrac{1}{64}$

5. $-15x^{11}$

6. y^5

7. $\dfrac{1}{r^5}$

8. $\dfrac{y^{14}}{x^2}$

9. $\dfrac{1}{6xy^8}$

10. 5.63×10^5

11. 8.63×10^{-5}

12. 0.0015

13. 62,300

14. 0.036

15. 5

16. $-2x^2 + 12x + 11$

17. $16x^3 + 7x^2 - 3x - 13$

18. $-3x^3 + 5x^2 + 4x + 5$

19. $x^3 + 8x^2 + 3x - 5$

20. $3x^3 + 22x^2 + 41x + 14$

21. $2x^5 - 5x^4 + 12x^3 - 8x^2 + 4x + 7$

21. Multiply $x^3 - x^2 + x + 1$ by $2x^2 - 3x + 7$ using a vertical format.

22. Use the FOIL method to multiply $(x + 7)(3x - 5)$.

22. $3x^2 + 16x - 35$

Use special products to multiply each of the following.

23. $(3x - 7)(3x + 7)$

24. $(4x - 2)^2$

23. $9x^2 - 49$

24. $16x^2 - 16x + 4$

25. $(8x + 3)^2$

26. $(x^2 - 9b)(x^2 + 9b)$

25. $64x^2 + 48x + 9$

26. $x^4 - 81b^2$

27. The height of the Bank of China in Hong Kong is 1001 feet. Neglecting air resistance, the height of an object dropped from this building at time t seconds is given by the polynomial $-16t^2 + 1001$. Find the height of the object at the given times below. (*Source*: The World Almanac, 1997)

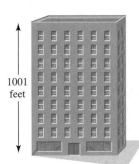

1001 feet

27. see table

28. $\dfrac{x}{2y} + \dfrac{1}{4} - \dfrac{7}{8y}$

t	0 seconds	1 second	3 seconds	5 seconds
$-16t^2 + 1001$	1001 ft	985 ft	857 ft	601 ft

29. $x + 2$

Divide.

28. $\dfrac{4x^2 + 2xy - 7x}{8xy}$

29. $(x^2 + 7x + 10) \div (x + 5)$

30. $\dfrac{27x^3 - 8}{3x + 2}$

30. $9x^2 - 6x + 4 - \dfrac{16}{3x + 2}$

266

Name _____ **Section** _____ **Date** _____

CUMULATIVE REVIEW

1. Given the set
$\left\{ -2, 0, \frac{1}{4}, 112, -3, 11, \sqrt{2} \right\}$, list the
numbers in this set that belong to the set of:
   a. Natural numbers

   b. Whole Numbers

   c. Integers

   d. Rational numbers

   e. Irrational numbers

   f. Real numbers

3. Simplify: $\dfrac{3}{2} \cdot \dfrac{1}{2} - \dfrac{1}{2}$

5. Add: $11.4 + (-4.7)$

2. Evaluate (find the value of) the following:
   a. 3^2

   b. 5^3

   c. 2^4

   d. 7^1

   e. $\left(\dfrac{3}{7} \right)^2$

4. Write an algebraic expression that represents each phrase. Let the variable x represent the unknown number.
   a. The sum of a number and 3

   b. The product of 3 and a number

   c. Twice a number

   d. 10 decreased by a number

   e. 5 times a number increased by 7

6. If $x = 2$ and $y = -5$, find the value of each expression.
   a. $\dfrac{x - y}{12 + x}$

   b. $x^2 - y$

Name _____

Divide.

7. $\dfrac{-30}{-10}$

8. $\dfrac{42}{-0.6}$

Find each product by using the distributive property to remove parentheses.

9. $5(x + 2)$

10. $-2(y + 0.3z - 1)$

11. $-(x + y - 2z + 6)$

12. Solve: $6(2a - 1) - (11a + 6) = 7$

13. Solve: $\dfrac{y}{7} = 20$

14. Solve: $0.25x + 0.10(x - 3) = 0.05(22)$

15. Twice the sum of a number and 4 is the same as four times the number decreased by 12. Find the number.

16. Charles Pecot can afford enough fencing to enclose a rectangular garden with a perimeter of 140 feet. If the width of his garden is to be 30 feet, find the length.

17. The number 120 is 15% of what number?

18. Solve: $-4x + 7 \ge -9$. Graph the solutions.

19. Simplify each expression.

a. $x^7 \cdot x^4$

b. $\left(\dfrac{1}{2}\right)^4$

c. $(9y^5)^2$

Simplify the following expressions. Write each result using positive exponents only.

20. $\left(\dfrac{3a^2}{b}\right)^{-3}$

21. $(5y^3)^{-2}$

Simplify each polynomial by combining any like terms.

22. $9x^3 + x^3$

23. $5x^2 + 6x - 9x - 3$

24. Multiply: $7x(x^2 + 2x + 5)$

25. Divide: $\dfrac{9x^5 - 12x^2 + 3x}{3x^2}$

Factoring Polynomials

In Chapter 3, we learned how to multiply polynomials. Now we will deal with an operation that is the reverse process of multiplying—factoring. Factoring is an important algebraic skill because it allows us to write a sum as a product. As we will see in Sections 4.6 and 4.7, factoring can be used to solve equations other than linear equations. In Chapter 5, we will also use factoring to simplify and perform arithmetic operations on rational expressions.

The America's Cup sailing competition is held roughly every three years. This international yacht race dates from 1851. The sailboats were originally built from wood and later from aluminum. Today's International America's Cup Class sailboats are made of carbon fiber. This strong but lightweight material reduced the boats' weight by 50% and made them much faster. In Exercise 44 on page 346, a quadratic equation is factored to find the height of the sail and length of the boom of an America's Cup sailboat.

CHAPTER 4 PRETEST

Factor each polynomial completely. If a polynomial cannot be factored, write "prime."

1. $2x^3y - 6x^2y^2$

2. $xy + 6x - 4y - 24$

3. $a^2 + 8a + 12$

4. $m^2 + 4m - 3$

5. $3x^3 - 18x^2 + 15x$

6. $2x^2 + 5x - 12$

7. $14x^2 + 63x + 70$

8. $24b^2 - 25b + 6$

9. $15y^2 + 38y + 7$

10. $x^2 + 24x + 144$

11. $4x^2 - 12xy + 9y^2$

12. $a^2 - 49b^2$

13. $1 - 64t^2$

14. $25b^2 + 4$

15. Fill in the blank so that $x^2 + $ _____ $x + 81$ is a perfect square trinomial.

Solve each equation.

16. $(x - 12)(x + 5) = 0$

17. $y^2 - 13y = 0$

18. $2m^3 - 2m^2 - 24m = 0$

19. The length of a rectangle is 7 inches more than its width. Its area is 120 square inches. Find the dimensions of the rectangle.

20. The sum of a number and its square is 240. Find the number.

1. $2x^2y(x - 3y)$; (4.1B)

2. $(x - 4)(y + 6)$; (4.1C)

3. $(a + 6)(a + 2)$; (4.2A)

4. prime; (4.2A)

5. $3x(x - 1)(x - 5)$; (4.2B)

6. $(2x - 3)(x + 4)$; (4.3A)

7. $7(2x + 5)(x + 2)$; (4.3B)

8. $(3b - 2)(8b - 3)$; (4.4A)

9. $(5y + 1)(3y + 7)$; (4.4A)

10. $(x + 12)^2$; (4.5B)

11. $(2x - 3y)^2$; (4.5B)

12. $(a - 7b)(a + 7b)$; (4.5C)

13. $(1 - 8t)(1 + 8t)$; (4.5C)

14. prime; (4.5C)

15. 18; (4.5A)

16. $x = 12, x = -5$; (4.6A)

17. $y = 0, y = 13$; (4.6A)

18. $m = 0, m = 4, m = -3$; (4.6B)

19. 8 in. $\times$ 15 in.; (4.7A)

20. -16 or 15; (4.7A)

4.1 THE GREATEST COMMON FACTOR

In the product $2 \cdot 3 = 6$, 2 and 3 are called **factors** of 6 and $2 \cdot 3$ is a **factored form** of 6. This is true of polynomials also. Since $(x + 2)(x + 3) = x^2 + 5x + 6$, then $(x + 2)$ and $(x + 3)$ are factors of $x^2 + 5x + 6$, and $(x + 2)(x + 3)$ is a factored form of the polynomial.

> The process of writing a polynomial as a product is called **factoring** the polynomial.

Study the examples below and look for a pattern.

TRY THE CONCEPT CHECK IN THE MARGIN.

Multiplying: $5(\overbrace{x^2 + 3}) = 5x^2 + 15$ $2x(\overbrace{x - 7}) = 2x^2 - 14x$

Factoring: $5x^2 + 15 = 5(\overbrace{x^2 + 3})$ $2x^2 - 14x = 2x(\overbrace{x - 7})$

Do you see that factoring is the reverse process of multiplying?

$$x^2 + 5x + 6 \underset{\text{multiplying}}{\overset{\text{factoring}}{=}} (x + 2)(x + 3)$$

A FINDING THE GREATEST COMMON FACTOR

The first step in factoring a polynomial is to see whether the terms of the polynomial have a common factor. If there is one, we can write the polynomial as a product by **factoring out** the common factor. We will usually factor out the *greatest* common factor (GCF).

The **greatest common factor (GCF) of a list of terms** is the product of the GCF of the numerical coefficients and the GCF of the variable factors.

$$20x^2y^2 = 2 \cdot 2 \cdot 5 \cdot x \cdot x \cdot y \cdot y$$
$$6xy^3 = 2 \cdot 3 \cdot x \cdot y \cdot y \cdot y$$
$$\text{GCF} = 2 \cdot x \cdot y \cdot y = 2xy^2$$

Notice that the GCF of a list of variables raised to powers is the variables raised to the *smallest* exponent in the list. For example,

the GCF of x^2, x^5, and x^3 is x^2,

because x^2 is the largest *common* factor.

Example 1   Find the greatest common factor of each list of terms.

    a. $6x^2$, $10x^3$, and $-8x$
    b. $-18y^2$, $-63y^3$, and $27y^4$
    c. a^3b^2, a^5b, and a^6b^2

Solution: **a.** $\left.\begin{array}{l} 6x^2 = 2 \cdot 3 \cdot x^2 \\ 10x^3 = 2 \cdot 5 \cdot x^3 \\ -8x = -1 \cdot 2 \cdot 2 \cdot 2 \cdot x^1 \end{array}\right\} \rightarrow$ The GCF of x^2, x^3, and x is x.

                $\text{GCF} = 2 \cdot x^1$ or $2x$

    b. $\left.\begin{array}{l} -18y^2 = -1 \cdot 2 \cdot 3 \cdot 3 \cdot y^2 \\ -63y^3 = -1 \cdot 3 \cdot 3 \cdot 7 \cdot y^3 \\ 27y^4 = 3 \cdot 3 \cdot 3 \cdot y^4 \end{array}\right\} \rightarrow$ The GCF of y^2, y^3, and y^4 is y^2.

                $\text{GCF} = 3 \cdot 3 \cdot y^2$ or $9y^2$

Objectives

A Find the greatest common factor of a list of terms.

B Factor out the greatest common factor from the terms of a polynomial.

C Factor by grouping.

   SSM   CD-ROM   Video
                         4.1

✓ CONCEPT CHECK

Multiply: $2(x - 4)$
What do you think the result of factoring $2x - 8$ would be? Why?

TEACHING TIP

You may want to begin this chapter by pointing out that students have the tools to solve linear equations such as $8x = x - 12$, but have not learned the tools to solve equations involving higher order polynomials such as $x^2 + 8x = x - 12$. The factoring skills students learn in this chapter will give them the tools needed to begin solving these more complicated kinds of equations.

TEACHING TIP

This might be a good time to remind students that no matter what number they substitute for x in the equation $x^2 + 5x + 6 = (x + 2)(x + 3)$, it will always be a true statement. Ask students to verify this with several numerical values.

Practice Problem 1

Find the greatest common factor of each list of terms.

a. $6x^2$, $9x^4$, and $-12x^5$

b. $-16y$, $-20y^6$, and $40y^4$

c. a^5b^4, ab^3, and a^3b^2

Answers

1. a. $3x^2$, **b.** $4y$, **c.** ab^2

✓ Concept Check

$2x - 8$; The result would be $2(x - 4)$ because factoring is the reverse process of multiplying.

c. The GCF of a^3, a^5, and a^6 is a^3.
The GCF of b^2, b, and b^2 is b. Thus,
the GCF of a^3b^2, a^5b, and a^6b^2 is a^3b.

B   FACTORING OUT THE GREATEST COMMON FACTOR

To factor a polynomial such as $8x + 14$, we first see whether the terms have a greatest common factor other than 1. In this case, they do: The GCF of $8x$ and 14 is 2.

We factor out 2 from each term by writing each term as a product of 2 and the term's remaining factors.

$$8x + 14 = 2 \cdot 4x + 2 \cdot 7$$

Using the distributive property, we can write

$$8x + 14 = 2 \cdot 4x + 2 \cdot 7$$
$$= 2(4x + 7)$$

Thus, a factored form of $8x + 14$ is $2(4x + 7)$. We can check by multiplying: $2(4x + 7) = 2 \cdot 4x + 2 \cdot 7 = 8x + 14$.

TRY THE CONCEPT CHECK IN THE MARGIN.

✓ **CONCEPT CHECK**

Which of the following is/are factored form(s) of $6t + 18$?

a. 6

b. $6 \cdot t + 6 \cdot 3$

c. $6(t + 3)$

d. $3(t + 6)$

HELPFUL HINT

A factored form of $8x + 14$ is *not*
$$2 \cdot 4x + 2 \cdot 7$$
Although the *terms* have been factored (written as a product), the *polynomial* $8x + 14$ has not been factored. A factored form of $8x + 14$ is the *product* $2(4x + 7)$.

Practice Problem 2

Factor each polynomial by factoring out the greatest common factor (GCF).

a. $10y + 25$

b. $x^4 - x^9$

Example 2   Factor each polynomial by factoring out the greatest common factor (GCF).

    a. $6t + 18$ **b.** $y^5 - y^7$

Solution: **a.** The GCF of terms $6t$ and 18 is 6. Thus,

$$6t + 18 = 6 \cdot t + 6 \cdot 3$$
$$= 6(t + 3) \quad \text{Apply the distributive property.}$$

We can check our work by multiplying 6 and $(t + 3)$.

$6(t + 3) = 6 \cdot t + 6 \cdot 3 = 6t + 18$, the original polynomial.

b. The GCF of y^5 and y^7 is y^5. Thus,

$$y^5 - y^7 = (y^5)1 - (y^5)y^2$$
$$= y^5(1 - y^2)$$

HELPFUL HINT   Don't forget the 1.

Example 3   Factor: $-9a^5 + 18a^2 - 3a$

Solution:

$$-9a^5 + 18a^2 - 3a = (3a)(-3a^4) + (3a)(6a) + (3a)(-1)$$
$$= 3a(-3a^4 + 6a - 1)$$

> **HELPFUL HINT**   Don't forget the -1.

In Example 3, we could have chosen to factor out $-3a$ instead of $3a$. If we factor out $-3a$, we have

$$-9a^5 + 18a^2 - 3a = (-3a)(3a^4) + (-3a)(-6a) + (-3a)(1)$$
$$= -3a(3a^4 - 6a + 1)$$

> **HELPFUL HINT**   Notice the changes in signs when factoring out $-3a$.

Examples   Factor.

4. $6a^4 - 12a = 6a(a^3 - 2)$

5. $\frac{3}{7}x^4 + \frac{1}{7}x^3 - \frac{5}{7}x^2 = \frac{1}{7}x^2(3x^2 + x - 5)$

6. $15p^2q^4 + 20p^3q^5 + 5p^3q^3 = 5p^2q^3(3q + 4pq^2 + p)$

Example 7   Factor: $5(x + 3) + y(x + 3)$

Solution: The binomial $(x + 3)$ is present in both terms and is the greatest common factor. We use the distributive property to factor out $(x + 3)$.

$$5(x + 3) + y(x + 3) = (x + 3)(5 + y)$$

C FACTORING BY GROUPING

Once the GCF is factored out, we can often continue to factor the polynomial, using a variety of techniques. We discuss here a technique called **factoring by grouping**. This technique can be used to factor some polynomials with four terms.

Example 8   Factor $xy + 2x + 3y + 6$ by grouping.

Solution: The GCF of the first two terms is x, and the GCF of the last two terms is 3.

$$xy + 2x + 3y + 6 = (xy + 2x) + (3y + 6)$$
$$= x(y + 2) + 3(y + 2)$$

> **HELPFUL HINT**
> Notice that this is *not* a factored form of the original polynomial. It is a sum, not a product.

Next we factor out the common binomial factor, $(y + 2)$.

$$x(y + 2) + 3(y + 2) = (y + 2)(x + 3)$$

Check: Multiply $(y + 2)$ by $(x + 3)$.

$$(y + 2)(x + 3) = xy + 2x + 3y + 6,$$

the original polynomial.

Thus, the factored form of $xy + 2x + 3y + 6$ is the product $(y + 2)(x + 3)$

Practice Problems 9–10

Factor by grouping.

9. $28x^3 - 7x^2 + 12x - 3$

10. $2xy + 5y^2 - 4x - 10y$

Examples   Factor by grouping.

9. $15x^3 - 10x^2 + 6x - 4$
$= (15x^3 - 10x^2) + (6x - 4)$
$= 5x^2(3x - 2) + 2(3x - 2)$ Factor each group.
$= (3x - 2)(5x^2 + 2)$ Factor out the common factor, $(3x - 2)$.

10. $3x^2 + 4xy - 3x - 4y$
$= (3x^2 + 4xy) + (-3x - 4y)$
$= x(3x + 4y) - 1(3x + 4y)$ Factor each group. A -1 is factored from the second pair of terms so that there is a common factor, $(3x + 4y)$.
$= (3x + 4y)(x - 1)$ Factor out the common factor, $(3x + 4y)$.

Practice Problems 11–13

Factor by grouping.

11. $4x^3 + x - 20x^2 - 5$

12. $2x - 2 + x^3 - 3x^2$

13. $3xy - 4 + x - 12y$

Examples   Factor by grouping.

11. $3x^3 - 2x - 9x^2 + 6$ Factor each group. A -3 is factored from the second pair of terms so that there is a common factor, $(3x^2 - 2)$.
$= x(3x^2 - 2) - 3(3x^2 - 2)$
$= (3x^2 - 2)(x - 3)$ Factor out the common factor, $(3x^2 - 2)$.

12. $5x - 10 + x^3 - x^2 = 5(x - 2) + x^2(x - 1)$

There is no common binomial factor that can now be factored out. No matter how we rearrange the terms, no grouping will lead to a common factor. Thus, this polynomial is not factorable by grouping.

13. $3xy + 2 - 3x - 2y$

Notice that the first two terms have no common factor other than 1. However, if we rearrange these terms, a grouping emerges that does lead to a common factor.

$3xy + 2 - 3x - 2y$
$= (3xy - 3x) + (-2y + 2)$
$= 3x(y - 1) - 2(y - 1)$ Factor -2 from the second group.
$= (y - 1)(3x - 2)$ Factor out the common factor, $(y - 1)$.

HELPFUL HINT

Throughout this chapter, we will be factoring polynomials. Even when the instructions do not so state, it is always a good idea to check your answers by multiplying.

Answers

9. $(4x - 1)(7x^2 + 3)$, **10.** $(2x + 5y)(y - 2)$,
11. $(4x^2 + 1)(x - 5)$, **12.** Can't be factored,
13. $(3y + 1)(x - 4)$

Name _____ **Section** _____ **Date** _____

MENTAL MATH

Find the GCF of each pair of integers.

1. 2, 16
2. 3, 18
3. 6, 15

4. 20, 15
5. 14, 35
6. 27, 36

EXERCISE SET 4.1

A *Find the GCF for each list. See Example 1.*

1. y^2, y^4, y^7
2. x^3, x^2, x^3
3. $x^{10}y^2, xy^2, x^3y^3$

4. p^7q, p^8q^2, p^9q^3
5. $8x, 4$
6. $9y, y$

7. $12y^4, 20y^3$
8. $32x, 18x^2$
9. $-10x^2, 15x^3$

10. $-21x^3, 14x$
11. $12x^3, -6x^4, 3x^5$
12. $15y^2, 5y^7, -20y^3$

13. $-18x^2y, 9x^3y^3, 36x^3y$
14. $7x, -21x^2y^2, 14xy$

B *Factor out the GCF from each polynomial. See Examples 2 through 7.*

15. $3a + 6$
16. $18a + 12$
17. $30x - 15$
18. $42x - 7$

19. $x^3 + 5x^2$
20. $y^5 - 6y^4$
21. $6y^4 - 2y$
22. $5x^2 + 10x^6$

23. $32xy - 18x^2$
24. $10xy - 15x^2$
25. $4x - 8y + 4$
26. $7x + 21y - 7$

27. $6x^3 - 9x^2 + 12x$
28. $12x^3 + 16x^2 - 8x$

29. $\underline{a^2b^2(a^5b^4 - a + b^3 - 1)}$

30. $\underline{x^3y^3(x^6y^3 + y^2 - x + 1)}$

31. $\underline{5xy(x^2 - 3x + 2)}$

32. $\underline{7xy(2x^2 + x - 1)}$

33. $\underline{4(2x^5 + 4x^4 - 5x^3 + 3)}$

34. $\underline{3(3y^6 - 9y^4 + 6y^2 + 2)}$

35. $\underline{\frac{1}{3}x(x^3 + 2x^2 - 4x + 1)}$

36. $\underline{\frac{1}{5}y(2y^6 - 4y^4 + 3y - 2)}$

37. $\underline{(x + 2)(y + 3)}$

38. $\underline{(y + 4)(z + 3)}$

39. $\underline{(x + 2)(8 - y)}$

40. $\underline{(y^2 + 1)(x - 3)}$

41. answers may vary

42. answers may vary

43. $\underline{(x^2 + 5)(x + 2)}$

44. $\underline{(x^2 + 3)(x + 4)}$

45. $\underline{(x + 3)(5 + y)}$

46. $\underline{(x + 1)(y + 2)}$

47. $\underline{(2x^2 + 5)(3x - 2)}$

48. $\underline{(4x^2 + 3)(4x - 7)}$

49. $\underline{(y - 4)(2 + x)}$

50. $\underline{(x - 7)(6 + y)}$

29. $a^7b^6 - a^3b^2 + a^2b^5 - a^2b^2$

30. $x^9y^6 + x^3y^5 - x^4y^3 + x^3y^3$

31. $5x^3y - 15x^2y + 10xy$

32. $14x^3y + 7x^2y - 7xy$

33. $8x^5 + 16x^4 - 20x^3 + 12$

34. $9y^6 - 27y^4 + 18y^2 + 6$

35. $\frac{1}{3}x^4 + \frac{2}{3}x^3 - \frac{4}{3}x^5 + \frac{1}{3}x$

36. $\frac{2}{5}y^7 - \frac{4}{5}y^5 + \frac{3}{5}y^2 - \frac{2}{5}y$

37. $y(x + 2) + 3(x + 2)$

38. $z(y + 4) + 3(y + 4)$

39. $8(x + 2) - y(x + 2)$

40. $x(y^2 + 1) - 3(y^2 + 1)$

41. Construct a binomial whose greatest common factor is $5a^3$. (*Hint:* Multiply $5a^3$ by a binomial whose terms contain no common factor other than 1. $5a^3(\square + \square)$.)

42. Construct a trinomial whose greatest common factor is $2x^2$. See the hint for Exercise 41.

C *Factor each four-term polynomial by grouping. See Examples 8 through 13.*

43. $x^3 + 2x^2 + 5x + 10$

44. $x^3 + 4x^2 + 3x + 12$

45. $5x + 15 + xy + 3y$

46. $xy + y + 2x + 2$

47. $6x^3 - 4x^2 + 15x - 10$

48. $16x^3 - 28x^2 + 12x - 21$

49. $2y - 8 + xy - 4x$

50. $6x - 42 + xy - 7y$

51. $2x^3 + x^2 + 8x + 4$

52. $2x^3 - x^2 - 10x + 5$

53. $4x^2 - 8xy - 3x + 6y$

54. $5xy - 15x - 6y + 18$

55. Explain how you can tell whether a polynomial is written in factored form.

56. Construct a four-term polynomial that can be factored by grouping.

REVIEW AND PREVIEW

Multiply. See Section 3.5.

57. $(x + 2)(x + 5)$

58. $(y + 3)(y + 6)$

59. $(b + 1)(b - 4)$

60. $(x - 5)(x + 10)$

Fill in the chart by finding two numbers that have the given product and sum. The first row is filled in for you.

Two Numbers	Their Product	Their Sum
4, 7	28	11
61.	12	8
62.	20	9
63.	8	−9
64.	16	−10
65.	−10	3
66.	−9	0
67.	−24	−5
68.	−36	−5

COMBINING CONCEPTS

Factor out the GCF from each polynomial. Then factor by grouping.

69. $12x^2y - 42x^2 - 4y + 14$

70. $90 + 15y^2 - 18x - 3xy^2$

51. $(2x + 1)(x^2 + 4)$

52. $(2x - 1)(x^2 - 5)$

53. $(x - 2y)(4x - 3)$

54. $(y - 3)(5x - 6)$

55. answers may vary

56. answers may vary

57. $x^2 + 7x + 10$

58. $y^2 + 9y + 18$

59. $b^2 - 3b - 4$

60. $x^2 + 5x - 50$

61. 2, 6

62. 4, 5

63. −1, −8

64. −2, −8

65. −2, 5

66. −3, 3

67. −8, 3

68. −9, 4

69. $2(3x^2 - 1)(2y - 7)$

70. $3(6 + y^2)(5 - x)$

71. $12x^3 - 2x;$
$2x(6x^2 - 1)$

72. $4x^2 - \pi x^2;$
$x^2(4 - \pi)$

73. $(n^3 - 6)$ units

74. $(x^3 - 1)$ units

75. a. 6 million

b. 12 million

c. $3(x^2 - 7x + 14)$

76. a. 2320 (in thousands)

b. 2476 (in thousands)

$2(9x^2 - 19x + 1170)$
c.

Write an expression for the area of each shaded region. Then write the expression as a factored polynomial.

71.

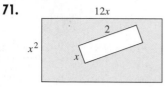

72.

Write an expression for the length of each rectangle. (Hint: Factor the area binomial and recall that Area = width · length.)

73.

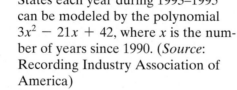

Area is
$(4n^4 - 24n)$
square
units

← 4n → units

74.

Area is
$(5x^5 - 5x^2)$
square units

$5x^2$ units

← ? →

75. The number (in millions) of CD singles sold annually in the United States each year during 1993–1995 can be modeled by the polynomial $3x^2 - 21x + 42$, where x is the number of years since 1990. (*Source:* Recording Industry Association of America)

a. Find the number of CD singles sold in 1994. To do so, let $x = 4$ and evaluate $3x^2 - 21x + 42$.

b. Find the number of CD singles sold in 1995.

c. Factor the polynomial $3x^2 - 21x + 42$.

76. The number (in thousands) of students who graduated from U.S. high schools each year during 1990–1995 can be modeled by $18x^2 - 38x + 2340$, where x is the number of years since 1990. (*Source:* U.S. Bureau of the Census)

a. Find the number of students who graduated from U.S. high schools in 1991. To do so, let $x = 1$ and evaluate $18x^2 - 38x + 2340$.

b. Find the number of students who graduated from U.S. high schools in 1994.

c. Factor the polynomial $18x^2 - 38x + 2340$.

4.2   FACTORING TRINOMIALS OF THE FORM $x^2 + bx + c$

A FACTORING TRINOMIALS OF THE FORM $x^2 + bx + c$

In this section, we factor trinomials of the form $x^2 + bx + c$, such as

$$x^2 + 4x + 3, \quad x^2 - 8x + 15, \quad x^2 + 4x - 12, \quad r^2 - r - 42$$

Notice that for these trinomials, the coefficient of the squared variable is 1.

Recall that factoring means to write as a product and that factoring and multiplying are reverse processes. Using the FOIL method of multiplying binomials, we have that

$$
\begin{array}{c}
\quad\quad\;\; \overset{\text{F}}{}\;\; \overset{\text{O}}{}\;\; \overset{\text{I}}{}\;\; \overset{\text{L}}{} \\
(x + 3)(x + 1) = x^2 + 1x + 3x + 3 \\
= x^2 + 4x + 3
\end{array}
$$

Thus, a factored form of $x^2 + 4x + 3$ is $(x + 3)(x + 1)$.

Notice that the product of the first terms of the binomials is $x \cdot x = x^2$, the first term of the trinomial. Also, the product of the last two terms of the binomials is $3 \cdot 1 = 3$, the third term of the trinomial. The sum of these same terms is $3 + 1 = 4$, the coefficient of the middle, x, term of the trinomial.

The product of these numbers is 3.

$$x^2 + 4x + 3 = (x + 3)(x + 1)$$

The sum of these numbers is 4.

Many trinomials, such as the one above, factor into two binomials. To factor $x^2 + 7x + 10$, let's assume that it factors into two binomials and begin by writing two pairs of parentheses. The first term of the trinomial is x^2, so we use x and x as the first terms of the binomial factors.

$$x^2 + 7x + 10 = (x + \square)(x + \square)$$

To determine the last term of each binomial factor, we look for two integers whose product is 10 and whose sum is 7. The integers are 2 and 5. Thus,

$$x^2 + 7x + 10 = (x + 2)(x + 5)$$

To see if we have factored correctly, we multiply.

$$
\begin{array}{ll}
(x + 2)(x + 5) = x^2 + 5x + 2x + 10 & \\
\quad\quad\quad\quad\;\; = x^2 + 7x + 10 & \text{Combine like terms.}
\end{array}
$$

HELPFUL HINT

Since multiplication is commutative, the factored form of $x^2 + 7x + 10$ can be written as either $(x + 2)(x + 5)$ or $(x + 5)(x + 2)$.

Objectives

A Factor trinomials of the form $x^2 + bx + c$.

B Factor out the greatest common factor and then factor a trinomial of the form $x^2 + bx + c$.

SSM   CD-ROM   Video
4.2

TEACHING TIP

Encourage students to look at the problems in this section as brainteasers to solve. Students should look for two numbers such that the constant term of the polynomial is their product and the x-coefficient is their sum.

> **TO FACTOR A TRINOMIAL OF THE FORM $x^2 + bx + c$**
>
> The product of these numbers is c.
>
> $$x^2 + bx + c = (x + \Box)(x + \Box)$$
>
> The sum of these numbers is b.

Practice Problem 1

Factor: $x^2 + 9x + 20$

TEACHING TIP

For Examples 1 and 2, you may want to use an area model to help students visualize what they are finding.

	Example 1			Example 2	
	x	a		x	a
x	x^2	ax	x	x^2	ax
b	bx	ab	b	bx	ab

$$ax + bx = 7x \qquad ax + bx = -8x$$
$$ab = 12 \qquad\qquad ab = 15$$

Practice Problem 2

Factor each trinomial.

a. $x^2 - 13x + 22$

b. $x^2 - 27x + 50$

TEACHING TIP

Example 3 may be a good opportunity to challenge students. They know that a pair of factors with opposite signs is needed. Since the x-term coefficient is positive, what does this tell us about the factors? Lead them to discover that in this case the absolute value of the positive factor will be larger than the absolute value of the negative factor. Point out that realizing this fact will cut their factor search in half.

Practice Problem 3

Factor: $x^2 + 5x - 36$

Answers

1. $(x + 4)(x + 5)$, **2. a.** $(x - 2)(x - 11)$,
b. $(x - 2)(x - 25)$, **3.** $(x + 9)(x - 4)$

Example 1   Factor: $x^2 + 7x + 12$

Solution: We begin by writing the first terms of the binomial factors.

$$(x + \Box)(x + \Box)$$

Next we look for two numbers whose product is 12 and whose sum is 7. Since our numbers must have a positive product and a positive sum, we look at pairs of positive factors of 12 only.

Factors of 12	Sum of Factors
1, 12	13
2, 6	8
3, 4	7

Correct sum, so the numbers are 3 and 4.

$$x^2 + 7x + 12 = (x + 3)(x + 4)$$

Check: Multiply $(x + 3)$ by $(x + 4)$.

——

Example 2   Factor: $x^2 - 8x + 15$

Solution: Again, we begin by writing the first terms of the binomials.

$$(x + \Box)(x + \Box)$$

Now we look for two numbers whose product is 15 and whose sum is -8. Since our numbers must have a positive product and a negative sum, we look at pairs of negative factors of 15 only.

Factors of 15	Sum of Factors
$-1, -15$	-16
$-3, -5$	-8

Correct sum, so the numbers are -3 and -5.

$$x^2 - 8x + 15 = (x - 3)(x - 5)$$

——

Example 3   Factor: $x^2 + 4x - 12$

Solution: $x^2 + 4x - 12 = (x + \Box)(x + \Box)$

We look for two numbers whose product is -12 and whose sum is 4. Since our numbers must have a negative product, we look at pairs of factors with opposite signs.

Factors of -12	Sum of Factors
$-1, 12$	11
$1, -12$	-11
$-2, 6$	4
$2, -6$	-4
$-3, 4$	1
$3, -4$	-1

Correct sum, so the numbers are -2 and 6.

$$x^2 + 4x - 12 = (x - 2)(x + 6)$$

Example 4   Factor: $r^2 - r - 42$

Solution: Because the variable in this trinomial is r, the first term of each binomial factor is r.

$$r^2 - r - 42 = (r + \square)(r + \square)$$

Now we look for two numbers whose product is -42 and whose sum is -1, the numerical coefficient of r. The numbers are 6 and -7. Therefore,

$$r^2 - r - 42 = (r + 6)(r - 7)$$

Example 5   Factor: $a^2 + 2a + 10$

Solution: Look for two numbers whose product is 10 and whose sum is 2. Neither 1 and 10 nor 2 and 5 give the required sum, 2. We conclude that $a^2 + 2a + 10$ is not factorable with integers. A polynomial such as $a^2 + 2a + 10$ is called a **prime polynomial**.

Example 6   Factor: $x^2 + 5xy + 6y^2$

Solution: $x^2 + 5xy + 6y^2 = (x + \square)(x + \square)$

Recall that the middle term $5xy$ is the same as $5yx$. Notice that $5y$ is the "coefficient" of x. We then look for two terms whose product is $6y^2$ and whose sum is $5y$. The terms are $2y$ and $3y$ because $2y \cdot 3y = 6y^2$ and $2y + 3y = 5y$. Therefore,

$$x^2 + 5xy + 6y^2 = (x + 2y)(x + 3y)$$

Example 7   Factor: $x^4 + 5x^2 + 6$

Solution: As usual, we begin by writing the first terms of the binomials. Since the greatest power of x in this polynomial is x^4, we write

$$(x^2 + \square)(x^2 + \square) \quad \text{since } x^2 \cdot x^2 = x^4$$

Now we look for two factors of 6 whose sum is 5. The numbers are 2 and 3. Thus,

$$x^4 + 5x^2 + 6 = (x^2 + 2)(x^2 + 3)$$

Practice Problem 4

Factor each trinomial.

a. $q^2 - 3q - 40$

b. $y^2 + 2y - 48$

Practice Problem 5

Factor: $x^2 + 6x + 15$

Practice Problem 6

Factor each trinomial.

a. $x^2 + 6xy + 8y^2$

b. $a^2 - 13ab + 30b^2$

Practice Problem 7

Factor: $x^4 + 8x^2 + 12$

Answers

4. a. $(q - 8)(q + 5)$, **b.** $(y + 8)(y - 6)$,
5. prime polynomial, **6. a.** $(x + 2y)(x + 4y)$,
b. $(a - 3b)(a - 10b)$, **7.** $(x^2 + 6)(x^2 + 2)$

The following sign patterns may be useful when factoring trinomials.

HELPFUL HINT—SIGN PATTERNS

A positive constant in a trinomial tells us to look for two numbers with the same sign. The sign of the coefficient of the middle term tells us whether the signs are both positive or both negative.

both positive → same sign →

$$x^2 + 10x + 16 = (x + 2)(x + 8)$$

both negative → same sign →

$$x^2 - 10x + 16 = (x - 2)(x - 8)$$

A negative constant in a trinomial tells us to look for two numbers with opposite signs.

opposite signs ↓   opposite signs ↓

$$x^2 + 6x - 16 = (x + 8)(x - 2) \qquad x^2 - 6x - 16 = (x - 8)(x + 2)$$

B FACTORING OUT THE GREATEST COMMON FACTOR

Remember that the first step in factoring any polynomial is to factor out the greatest common factor (if there is one other than 1 or −1).

Practice Problem 8

Factor each trinomial.

a. $x^3 + 3x^2 - 4x$

b. $4x^2 - 24x + 36$

Example 8   Factor: $3m^2 - 24m - 60$

Solution: First we factor out the greatest common factor, 3, from each term.

$$3m^2 - 24m - 60 = 3(m^2 - 8m - 20)$$

Now we factor $m^2 - 8m - 20$ by looking for two factors of −20 whose sum is −8. The factors are −10 and 2. Therefore, the complete factored form is

$$3m^2 - 24m - 60 = 3(m + 2)(m - 10)$$

HELPFUL HINT

Remember to write the common factor 3 as part of the factored form.

Answers

8. a. $x(x + 4)(x - 1)$, **b.** $4(x - 3)(x - 3)$

Name _____ **Section** _____ **Date** _____

MENTAL MATH

Complete each factored form.

1. $x^2 + 9x + 20 = (x + 4)(x \quad)$

2. $x^2 + 12x + 35 = (x + 5)(x \quad)$

3. $x^2 - 7x + 12 = (x - 4)(x \quad)$

4. $x^2 - 13x + 22 = (x - 2)(x \quad)$

5. $x^2 + 4x + 4 = (x + 2)(x \quad)$

6. $x^2 + 10x + 24 = (x + 6)(x \quad)$

EXERCISE SET 4.2

A Factor each trinomial completely. If a polynomial can't be factored, write "prime." See Examples 1 through 7.

▣ 1. $x^2 + 7x + 6$

2. $x^2 + 6x + 8$

3. $x^2 - 10x + 9$

4. $x^2 - 6x + 9$

▣ 5. $x^2 - 3x - 18$

6. $x^2 - x - 30$

7. $x^2 + 3x - 70$

8. $x^2 + 4x - 32$

9. $x^2 + 5x + 2$

10. $x^2 - 7x + 5$

11. $x^2 + 8xy + 15y^2$

12. $x^2 + 6xy + 8y^2$

13. $a^4 - 2a^2 - 15$

14. $y^4 - 3y^2 - 70$

◲ 15. Write a polynomial that factors as $(x - 3)(x + 8)$.

◲ 16. To factor $x^2 + 13x + 42$, think of two numbers whose _____ is 42 and whose _____ is 13.

Complete each sentence in your own words.

◲ 17. If $x^2 + bx + c$ is factorable and c is negative, then the signs of the last-term factors of the binomials are opposite because . . . .

◲ 18. If $x^2 + bx + c$ is factorable and c is positive, then the signs of the last-term factors of the binomials are the same because . . . .

19. $2(z + 8)(z + 2)$

20. $3(x + 7)(x + 3)$

21. $2x(x - 5)(x - 4)$

22. $x(x - 8)(x + 7)$

23. $(x - 4y)(x + y)$

24. $(x - 11y)(x + 7y)$

25. $(x + 12)(x + 3)$

26. $(x + 4)(x + 15)$

27. $(x - 2)(x + 1)$

28. $(x - 7)(x + 2)$

29. $(r - 12)(r - 4)$

30. $(r - 7)(r - 3)$

31. $(x + 2y)(x - y)$

32. $(x - 3y)(x + 2y)$

33. $3(x + 5)(x - 2)$

34. $4(x - 4)(x + 3)$

35. $3(x - 18)(x - 2)$

36. $2(x - 7)(x - 5)$

37. $(x - 24)(x + 6)$

38. $(x + 7)(x - 6)$

39. prime

40. prime

41. $(x - 5)(x - 3)$

42. $(x - 7)(x - 2)$

43. $6x(x + 4)(x + 5)$

44. $3x(x + 7)(x - 6)$

45. $4y(x^2 + x - 3)$

284

B *Factor each trinomial completely. See Examples 1 through 8.*

19. $2z^2 + 20z + 32$ **20.** $3x^2 + 30x + 63$ **21.** $2x^3 - 18x^2 + 40x$

22. $x^3 - x^2 - 56x$ **23.** $x^2 - 3xy - 4y^2$ **24.** $x^2 - 4xy - 77y^2$

25. $x^2 + 15x + 36$ **26.** $x^2 + 19x + 60$ **27.** $x^2 - x - 2$

28. $x^2 - 5x - 14$ **29.** $r^2 - 16r + 48$ **30.** $r^2 - 10r + 21$

31. $x^2 + xy - 2y^2$ **32.** $x^2 - xy - 6y^2$ **33.** $3x^2 + 9x - 30$

34. $4x^2 - 4x - 48$ **35.** $3x^2 - 60x + 108$ **36.** $2x^2 - 24x + 70$

37. $x^2 - 18x - 144$ **38.** $x^2 + x - 42$ **39.** $r^2 - 3r + 6$

40. $x^2 + 4x - 10$ **41.** $x^2 - 8x + 15$ **42.** $x^2 - 9x + 14$

43. $6x^3 + 54x^2 + 120x$ **44.** $3x^3 + 3x^2 - 126x$ **45.** $4x^2y + 4xy - 12y$

Name _____

46. $3x^2y - 9xy + 45y$ **47.** $x^2 - 4x - 21$ **48.** $x^2 - 4x - 32$

49. $x^2 + 7xy + 10y^2$ **50.** $x^2 - 3xy - 4y^2$ **51.** $64 + 24t + 2t^2$

52. $50 + 20t + 2t^2$ **53.** $x^3 - 2x^2 - 24x$ **54.** $x^3 - 3x^2 - 28x$

55. $2t^5 - 14t^4 + 24t^3$ **56.** $3x^6 + 30x^5 + 72x^4$

57. $5x^3y - 25x^2y^2 - 120xy^3$ **58.** $3x^2 - 6xy - 72y^2$

REVIEW AND PREVIEW

Multiply. See Section 3.5.

59. $(2x + 1)(x + 5)$ **60.** $(3x + 2)(x + 4)$ **61.** $(5y - 4)(3y - 1)$

62. $(4z - 7)(7z - 1)$ **63.** $(a + 3)(9a - 4)$ **64.** $(y - 5)(6y + 5)$

46. $3y(x^2 - 3x + 15)$

47. $(x - 7)(x + 3)$

48. $(x - 8)(x + 4)$

49. $(x + 5y)(x + 2y)$

50. $(x - 4y)(x + y)$

51. $2(t + 8)(t + 4)$

52. $2(t + 5)(t + 5)$

53. $x(x - 6)(x + 4)$

54. $x(x - 7)(x + 4)$

55. $2t^3(t - 4)(t - 3)$

56. $3x^4(x + 6)(x + 4)$

57. $5xy(x - 8y)(x + 3y)$

58. $3(x - 6y)(x + 4y)$

59. $2x^2 + 11x + 5$

60. $3x^2 + 14x + 8$

61. $15y^2 - 17y + 4$

62. $28z^2 - 53z + 7$

63. $9a^2 + 23a - 12$

64. $6y^2 - 25y - 25$

285

65. $2x^2 + 28x + 66$

$2(x + 3)(x + 11)$

66. $4x^3 + 24x^2 + 32x$

$4x(x + 4)(x + 2)$

$(x + 1)(y - 5)(y + 3)$
67.

$(x + 1)(z - 10)(z + 7)$
68.

69. $3; 4$

70. $15; 28; 39; 48; 55; 60;$
$63; 64$

71. $8; 16$

72. $9; 12; 21$

73. $(x^n + 2)(x^n + 3)$

74. $(x^n + 10)(x^n - 2)$

 COMBINING CONCEPTS

Write the perimeter of each figure as a simplified polynomial. Then factor the polynomial.

65.

$4x + 33$

$x^2 + 10x$

66.

$12x^2$

$2x^3 + 16x$

Factor each trinomial completely.

67. $y^2(x + 1) - 2y(x + 1) - 15(x + 1)$ **68.** $z^2(x + 1) - 3z(x + 1) - 70(x + 1)$

Find a positive value of c so that each trinomial is factorable.

69. $y^2 - 4y + c$ **70.** $n^2 - 16n + c$

Find a positive value of b so that each trinomial is factorable.

71. $x^2 + bx + 15$ **72.** $y^2 + by + 20$

Factor each trinomial. (Hint: Notice that $x^{2n} + 4x^n + 3$ factors as $(x^n + 1)(x^n + 3)$.)

73. $x^{2n} + 5x^n + 6$ **74.** $x^{2n} + 8x^n - 20$

4.3   FACTORING TRINOMIALS OF THE FORM $ax^2 + bx + c$

A FACTORING TRINOMIALS OF THE FORM $ax^2 + bx + c$

In this section, we factor trinomials of the form $ax^2 + bx + c$, such as

$$3x^2 + 11x + 6, \qquad 8x^2 - 22x + 5, \qquad 2x^2 + 13x - 7$$

Notice that the coefficient of the squared variable in these trinomials is a number other than 1. We will factor these trinomials using a trial-and-check method based on our work in the last section.

To begin, let's review the relationship between the numerical coefficients of the trinomial and the numerical coefficients of its factored form. For example, since $(2x + 1)(x + 6) = 2x^2 + 13x + 6$, the factored form of $2x^2 + 13x + 6$ is

$$2x^2 + 13x + 6 = (2x + 1)(x + 6)$$

Notice that $2x$ and x are factors of $2x^2$, the first term of the trinomial. Also, 6 and 1 are factors of 6, the last term of the trinomial, as shown:

$$\overset{\displaystyle \overset{2x \cdot x}{\frown}}{2x^2 + 13x + 6 = (2x + 1)(x + 6)}$$
$$\underset{1 \cdot 6}{\smile}$$

Also notice that $13x$, the middle term, is the sum of the following products:

$$2x^2 + 13x + 6 = (2x + \underline{1})(x + 6)$$
$$\begin{array}{c} 1x \\ +\ 12x \\ \hline 13x \end{array} \quad \text{Middle term}$$

Let's use this pattern to factor $5x^2 + 7x + 2$. First, we find factors of $5x^2$. Since all numerical coefficients in this trinomial are positive, we will use factors with positive numerical coefficients only. Thus, the factors of $5x^2$ are $5x$ and x. Let's try these factors as first terms of the binomials. Thus far, we have

$$5x^2 + 7x + 2 = (5x + \square)(x + \square)$$

Next, we need to find positive factors of 2. Positive factors of 2 are 1 and 2. Now we try possible combinations of these factors as second terms of the binomials until we obtain a middle term of $7x$.

$$(5x + \underline{1})(x + 2) = 5x^2 + 11x + 2$$
$$\begin{array}{c} 1x \\ +\ 10x \\ \hline 11x \end{array} \longrightarrow \textbf{Incorrect} \text{ middle term}$$

Let's try switching factors 2 and 1.

$$(5x + \underline{2})(x + 1) = 5x^2 + 7x + 2$$
$$\begin{array}{c} 2x \\ +\ 5x \\ \hline 7x \end{array} \longrightarrow \textbf{Correct} \text{ middle term}$$

Thus the factored form of $5x^2 + 7x + 2$ is $(5x + 2)(x + 1)$. To check, we multiply $(5x + 2)$ and $(x + 1)$. The product is $5x^2 + 7x + 2$.

Objectives

A Factor trinomials of the form $ax^2 + bx + c$, where $a \neq 1$.

B Factor out the GCF before factoring a trinomial of the form $ax^2 + bx + c$.

SSM   CD-ROM   Video
4.3

TEACHING TIP

Point out that when the coefficient of the squared term is prime and the constant term is prime, there are only 4 possible ways to factor the trinomial. Have students create the 4 trinomials which can be factored when 7 is the coefficient of the squared term and 3 is the constant:

$(7x + 3)(x + 1) = 7x^2 + 10x + 3$
$(7x + 1)(x + 3) = 7x^2 + 22x + 3$
$(7x - 3)(x - 1) = 7x^2 - 10x + 3$
$(7x - 1)(x - 3) = 7x^2 - 22x + 3$

Practice Problem 1

Factor each trinomial.

a. $4x^2 + 12x + 5$

b. $5x^2 + 27x + 10$

✓ **CONCEPT CHECK**

Do the terms of $3x^2 + 29x + 18$ have a common factor? Without multiplying, decide which of the following factored forms could not be a factored form of $3x^2 + 29x + 18$.

a. $(3x + 18)(x + 1)$

b. $(3x + 2)(x + 9)$

c. $(3x + 6)(x + 3)$

d. $(3x + 9)(x + 2)$

Practice Problem 2

Factor each trinomial.

a. $6x^2 - 5x + 1$

b. $2x^2 - 11x + 12$

Answers

1. a. $(2x + 5)(2x + 1)$, **b.** $(5x + 2)(x + 5)$,
2. a. $(3x - 1)(2x - 1)$, **b.** $(2x - 3)(x - 4)$

✓ Concept Check: no; a, c, d

Example 1

Factor: $3x^2 + 11x + 6$

Solution: Since all numerical coefficients are positive, we use factors with positive numerical coefficients. We first find factors of $3x^2$.

Factors of $3x^2$: $3x^2 = 3x \cdot x$

If factorable, the trinomial will be of the form

$3x^2 + 11x + 6 = (3x + \square)(x + \square)$

Next we factor 6.

Factors of 6: $6 = 1 \cdot 6$, $6 = 2 \cdot 3$

Now we try combinations of factors of 6 until a middle term of $11x$ is obtained. Let's try 1 and 6 first.

$$(3x + 1)(x + 6) = 3x^2 + 19x + 6$$

$$\underbrace{\begin{array}{c} 1x \\ + 18x \\ \hline 19x \end{array}}_{} \longrightarrow \textbf{Incorrect} \text{ middle term}$$

Now let's next try 6 and 1.

$$(3x + 6)(x + 1)$$

Before multiplying, notice that the terms of the factor $3x + 6$ have a common factor of 3. The terms of the original trinomial $3x^2 + 11x + 6$ have no common factor other than 1, so the terms of the factored form of $3x^2 + 11x + 6$ can contain no common factor other than 1. This means that $(3x + 6)(x + 1)$ is not a factored form.

Next let's try 2 and 3 as last terms.

$$(3x + 2)(x + 3) = 3x^2 + 11x + 6$$

$$\underbrace{\begin{array}{c} 2x \\ + 9x \\ \hline 11x \end{array}}_{} \longrightarrow \textbf{Correct} \text{ middle term}$$

Thus the factored form of $3x^2 + 11x + 6$ is $(3x + 2)(x + 3)$.

HELPFUL HINT

If the terms of a trinomial have no common factor (other than 1), then the terms of neither of its binomial factors will contain a common factor (other than 1).

TRY THE CONCEPT CHECK IN THE MARGIN.

Example 2

Factor: $8x^2 - 22x + 5$

Solution: Factors of $8x^2$: $8x^2 = 8x \cdot x$, $8x^2 = 4x \cdot 2x$

We'll try $8x$ and x.

$8x^2 - 22x + 5 = (8x + \square)(x + \square)$

Since the middle term, $-22x$, has a negative numerical coefficient, we factor 5 into negative factors.

Factors of 5: $5 = -1 \cdot -5$

Let's try -1 and -5.

$$(8x - 1)(x - 5) = 8x^2 - 41x + 5$$

$$\underbrace{\begin{aligned} -1x \\ + (-40x) \\ \hline -41x \end{aligned}} \longrightarrow \textbf{Incorrect } \text{middle term}$$

Now let's try -5 and -1.

$$(8x - 5)(x - 1) = 8x^2 - 13x + 5$$

$$\underbrace{\begin{aligned} -5x \\ + (-8x) \\ \hline -13x \end{aligned}} \longrightarrow \textbf{Incorrect } \text{middle term}$$

Don't give up yet! We can still try other factors of $8x^2$. Let's try $4x$ and $2x$ with -1 and -5.

$$(4x - 1)(2x - 5) = 8x^2 - 22x + 5$$

$$\underbrace{\begin{aligned} -2x \\ + (-20x) \\ \hline -22x \end{aligned}} \longrightarrow \textbf{Correct } \text{middle term}$$

The factored form of $8x^2 - 22x + 5$ is $(4x - 1)(2x - 5)$.

TEACHING TIP

Encourage students to analyze a problem before trying to solve it. Ask them to list some key characteristics about the trinomial in Example 3 before factoring it. (Key characteristics: Squared term is prime, constant term is prime and negative, x-term is positive.)

Example 3

Solution: Factor: $2x^2 + 13x - 7$

Factors of $2x^2$: $2x^2 = 2x \cdot x$

Factors of -7: $-7 = -1 \cdot 7,$ $-7 = 1 \cdot -7$

We try possible combinations of these factors:

$(2x + 1)(x - 7) = 2x^2 - 13x - 7$ **Incorrect** middle term
$(2x - 1)(x + 7) = 2x^2 + 13x - 7$ **Correct** middle term

The factored form of $2x^2 + 13x - 7$ is $(2x - 1)(x + 7)$.

Practice Problem 3

Factor each trinomial.

a. $35x^2 + 4x - 4$

b. $4x^2 + 3x - 7$

Example 4

Solution: Factor: $10x^2 - 13xy - 3y^2$

Factors of $10x^2$: $10x^2 = 10x \cdot x,$ $10x^2 = 2x \cdot 5x$
Factors of $-3y^2$: $-3y^2 = -3y \cdot y,$
$-3y^2 = 3y \cdot -y$

We try some combinations of these factors:

$$(10x - 3y)(x + y) = 10x^2 + 7xy - 3y^2$$
$$(x + 3y)(10x - y) = 10x^2 + 29xy - 3y^2$$
$$(5x + 3y)(2x - y) = 10x^2 + xy - 3y^2$$
$$(2x - 3y)(5x + y) = 10x^2 - 13xy - 3y^2$$

Correct middle term

Practice Problem 4

Factor each trinomial.

a. $14x^2 - 3xy - 2y^2$

b. $12a^2 - 16ab - 3b^2$

Answers

3. a. $(5x + 2)(7x - 2)$,   b. $(4x + 7)(x - 1)$,
4. a. $(7x + 2y)(2x - y)$,
b. $(6a + b)(2a - 3b)$

The factored form of $10x^2 - 13xy - 3y^2$ is $(2x - 3y)(5x + y)$.

B FACTORING OUT THE GREATEST COMMON FACTOR

Don't forget that the first step in factoring any polynomial is to look for a common factor to factor out.

Practice Problem 5

Factor each trinomial.

a. $3x^3 + 17x^2 + 10x$

b. $6xy^2 + 33xy - 18x$

Example 5   Factor: $24x^4 + 40x^3 + 6x^2$

Solution: Notice that all three terms have a common factor of $2x^2$. Thus we factor out $2x^2$ first.

$$24x^4 + 40x^3 + 6x^2 = 2x^2(12x^2 + 20x + 3)$$

Next we factor $12x^2 + 20x + 3$.

Factors of $12x^2$: $12x^2 = 4x \cdot 3x$, $12x^2 = 12x \cdot x$, $12x^2 = 6x \cdot 2x$

Since all terms in the trinomial have positive numerical coefficients, we factor 3 using positive factors only.

Factors of 3: $3 = 1 \cdot 3$

We try some combinations of the factors.

$$2x^2(4x + 3)(3x + 1) = 2x^2(12x^2 + 13x + 3)$$
$$2x^2(12x + 1)(x + 3) = 2x^2(12x^2 + 37x + 3)$$
$$2x^2(2x + 3)(6x + 1) = 2x^2(12x^2 + 20x + 3)$$

                                                                   Correct middle term

The factored form of $24x^4 + 40x^3 + 6x^2$ is $2x^2(2x + 3)(6x + 1)$.

HELPFUL HINT

Don't forget to include the common factor in the factored form.

TEACHING TIP

Have students write down their own strategy for factoring trinomials of the form $ax^2 + bx + c$.

Answers

5. a. $x(3x + 2)(x + 5)$,
b. $3x(2y - 1)(y + 6)$

Name _____ **Section** _____ **Date** _____

Exercise Set 4.3

A *Complete each factored form.*

1. $5x^2 + 22x + 8 = (5x + 2)(\quad)$

2. $2y^2 + 15y + 25 = (2y + 5)(\quad)$

3. $50x^2 + 15x - 2 = (5x + 2)(\quad)$

4. $6y^2 + 11y - 10 = (2y + 5)(\quad)$

5. $20x^2 - 7x - 6 = (5x + 2)(\quad)$

6. $8y^2 - 2y - 55 = (2y + 5)(\quad)$

Factor each trinomial completely. See Examples 1 through 4.

7. $2x^2 + 13x + 15$

8. $3x^2 + 8x + 4$

9. $8y^2 - 17y + 9$

10. $21x^2 - 41x + 10$

11. $2x^2 - 9x - 5$

12. $36r^2 - 5r - 24$

13. $20r^2 + 27r - 8$

14. $3x^2 + 20x - 63$

15. $10x^2 + 17x + 3$

16. $2x^2 + 7x + 5$

17. $3x^2 + x - 2$

18. $8y^2 + y - 9$

19. $6x^2 - 13xy + 5y^2$

20. $8x^2 - 14xy + 3y^2$

21. $15x^2 - 16x - 15$

22. $25x^2 - 5x - 6$

23. $x^2 - 9x + 20$

24. $x^2 - 7x + 12$

ANSWERS

1. $x + 4$

2. $y + 5$

3. $10x - 1$

4. $3y - 2$

5. $4x - 3$

6. $4y - 11$

7. $(2x + 3)(x + 5)$

8. $(3x + 2)(x + 2)$

9. $(y - 1)(8y - 9)$

10. $(7x - 2)(3x - 5)$

11. $(2x + 1)(x - 5)$

12. $(9r - 8)(4r + 3)$

13. $(4r - 1)(5r + 8)$

14. $(3x - 7)(x + 9)$

15. $(5x + 1)(2x + 3)$

16. $(2x + 5)(x + 1)$

17. $(3x - 2)(x + 1)$

18. $(y - 1)(8y + 9)$

19. $(3x - 5y)(2x - y)$

20. $(4x - y)(2x - 3y)$

21. $(3x - 5)(5x + 3)$

22. $(5x + 2)(5x - 3)$

23. $(x - 4)(x - 5)$

24. $(x - 3)(x - 4)$

25. $(2x + 11)(x - 9)$

26. $(2x - 9)(x + 8)$

27. $(7t + 1)(t - 4)$

28. $(4t - 7)(t + 1)$

29. $(3a + b)(a + 3b)$

30. $(2a + b)(a + 5b)$

31. $(7x + 1)(7x - 2)$

32. $(3x - 2)(x + 4)$

33. $(6x - 7)(3x + 2)$

34. $(7a - 6)(6a - 1)$

35. $x(3x + 2)(4x + 1)$

36. $a(4a + 1)(2a + 3)$

37. $3(7x + 5)(x - 3)$

38. $2(3x - 5)(2x + 1)$

39. $(3x + 4)(4x - 3)$

40. $(5x + 3)(3x - 5)$

41. $2y^2(3x - 10)(x + 3)$

42. $2y(4x - 7)(x + 6)$

43. $(2x - 7)(2x + 3)$

44. $(3x + 2)(2x - 5)$

45. $3(x^2 - 14x + 21)$

46. $5(x^2 - 15x + 12)$

47. $(4x + 9)(2x - 3)$

48. $(9a + 8b)(6a - b)$

49. $x(4x + 3)(x - 3)$

292

Name _____

25. $2x^2 - 7x - 99$ **26.** $2x^2 + 7x - 72$ **27.** $-27t + 7t^2 - 4$

28. $4t^2 - 7 - 3t$ **29.** $3a^2 + 10ab + 3b^2$ **30.** $2a^2 + 11ab + 5b^2$

31. $49x^2 - 7x - 2$ **32.** $3x^2 + 10x - 8$ **33.** $18x^2 - 9x - 14$

34. $42a^2 - 43a + 6$

B *Factor each trinomial completely. See Examples 1 through 5.*

35. $12x^3 + 11x^2 + 2x$ **36.** $8a^3 + 14a^2 + 3a$ **37.** $21x^2 - 48x - 45$

38. $12x^2 - 14x - 10$ **39.** $12x^2 + 7x - 12$ **40.** $15x^2 - 16x - 15$

41. $6x^2y^2 - 2xy^2 - 60y^2$ **42.** $8x^2y + 34xy - 84y$ **43.** $4x^2 - 8x - 21$

44. $6x^2 - 11x - 10$ **45.** $3x^2 - 42x + 63$ **46.** $5x^2 - 75x + 60$

47. $8x^2 + 6x - 27$ **48.** $54a^2 + 39ab - 8b^2$ **49.** $4x^3 - 9x^2 - 9x$

Name _____

50. $6x^3 - 31x^2 + 5x$

51. $24x^2 - 58x + 9$

52. $36x^2 + 55x - 14$

53. $40a^2b + 9ab - 9b$

54. $24y^2x + 7yx - 5x$

55. $15x^4 + 19x^2 + 6$

56. $6x^3 - 28x^2 + 16x$

57. $6y^3 - 8y^2 - 30y$

58. $12x^3 - 34x^2 + 24x$

59. $10x^3 + 25x^2y - 15xy^2$

60. $42x^4 - 99x^3y - 15x^2y^2$

REVIEW AND PREVIEW

As of June 1995, approximately 40% of U.S. households had a personal computer. The following graph shows the percent of households having a computer grouped according to household income. Use this graph to answer Exercises 61–64. See Section 1.8.

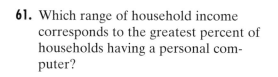

70%
60%
50%
40%
30%
20%
10%
0%

Less than 25,000– 35,000– 45,000– 60,000
25,000 34,000 44,000 59,000 and above

Income (dollars)

Source: Link Resources. AP *Times Picayune*, June 18, 1995

61. Which range of household income corresponds to the greatest percent of households having a personal computer?

62. Which range of household income corresponds to the greatest *increase* in percent of households having a personal computer?

50. $x(x - 5)(6x - 1)$

51. $(4x - 9)(6x - 1)$

52. $(4x + 7)(9x - 2)$

53. $b(8a - 3)(5a + 3)$

54. $x(8y + 5)(3y - 1)$

55. $(3x^2 + 2)(5x^2 + 3)$

56. $2x(3x - 2)(x - 4)$

57. $2y(3y + 5)(y - 3)$

58. $2x(3x - 4)(2x - 3)$

59. $5x(2x - y)(x + 3y)$

60. $3x^2(2x - 5y)(7x + y)$

61. $60,000 and above

62. $45,000–$59,000

63. answers may vary

63. Describe any trend you notice from this graph.

64. Why don't the percents shown in the graph add to 100%?

64. answers may vary

65. $\dfrac{(y-1)^2}{(4x^2+10x+25)}$

▷ **COMBINING CONCEPTS**

Factor each trinomial completely.

65. $4x^2(y-1)^2 + 10x(y-1)^2 + 25(y-1)^2$

66. $\dfrac{(a+3)^3}{(3x-25)(x-1)}$

66. $3x^2(a+3)^3 - 28x(a+3)^3 + 25(a+3)^3$

67. $-3xy^2(4x-5)(x+1)$

67. $-12x^3y^2 + 3x^2y^2 + 15xy^2$ (*Hint*: Begin by factoring out $-3xy^2$.)

68. 2; 14

Find a positive value of b so that each trinomial is factorable.

68. $3x^2 + bx - 5$

69. $2z^2 + bz - 7$

69. 5; 13

Find a positive value of c so that each trinomial is factorable.

70. $5x^2 + 7x + c$

71. $3x^2 - 8x + c$

70. 2

71. 4; 5

4.4 FACTORING TRINOMIALS OF THE FORM $ax^2 + bx + c$ BY GROUPING

A USING THE GROUPING METHOD

There is an alternative method that can be used to factor trinomials of the form $ax^2 + bx + c$, $a \neq 1$. This method is called the **grouping method** because it uses factoring by grouping as we learned in Section 4.1.

To see how this method works, let's multiply the following:

$$(2x + 1)(3x + 5) = 6x^2 + 10x + 3x + 5$$

$$10 \cdot 3 = 30$$
$$6 \cdot 5 = 30$$

$$= 6x^2 + 13x + 5$$

Notice that the product of the coefficients of the first and last terms is $6 \cdot 5 = 30$. This is the same as the product of the coefficients of the two middle terms, $10 \cdot 3 = 30$.

Let's use this pattern to write $2x^2 + 11x + 12$ as a four-term polynomial. We will then factor the polynomial by grouping.

$$2x^2 + 11x + 12 \qquad \text{Find two numbers whose product is } 2 \cdot 12 = 24$$
$$= 2x^2 + \Box x + \Box x + 12 \qquad \text{and whose sum is 11.}$$

Since we want a positive product and a positive sum, we consider pairs of positive factors of 24 only.

Factors of 24	Sum of Factors
1, 24	25
2, 12	14
3, 8	11

Correct sum

The factors are 3 and 8. Now we use these factors to write the middle term $11x$ as $3x + 8x$ (or $8x + 3x$). We replace $11x$ with $3x + 8x$ in the original trinomial and then we can factor by grouping.

$$2x^2 + 11x + 12 = 2x^2 + 3x + 8x + 12$$
$$= (2x^2 + 3x) + (8x + 12) \qquad \text{Group the terms.}$$
$$= x(2x + 3) + 4(2x + 3) \qquad \text{Factor each group.}$$
$$= (2x + 3)(x + 4) \qquad \text{Factor out } (2x + 3).$$

In general, we have the following procedure.

TO FACTOR TRINOMIALS OF THE FORM $ax^2 + bx + c$ BY GROUPING

Step 1. Factor out a greatest common factor, if there is one other than 1 (or −1).

Step 2. Find two numbers whose product is $a \cdot c$ and whose sum is b.

Step 3. Write the middle term, bx using the factors found in Step 2.

Step 4. Factor by grouping.

Objectives

A Use the grouping method to factor trinomials of the form $ax^2 + bx + c$, $a \neq 1$.

SSM   CD-ROM   Video 4.4

TEACHING TIP

Consider beginning this lesson with an example reminding students of the grouping method they learned in Section 4.1. For instance, factor $2y^3 + 5y^2 - 4y - 10$ and $3x^2 - x + 6x - 2$.

Practice Problem 1

Practice Problem 1

Factor each trinomial by grouping.

a. $3x^2 + 14x + 8$

b. $12x^2 + 19x + 5$

Example 1   Factor $8x^2 - 14x + 5$ by grouping.

Solution:

Step 1. The terms of this trinomial contain no greatest common factor other than 1 (or −1).

Step 2. This trinomial is of the form $ax^2 + bx + c$ with $a = 8$, $b = -14$, and $c = 5$. Find two numbers whose product is $a \cdot c$ or $8 \cdot 5 = 40$, and whose sum is b or -14. The numbers are -4 and -10.

Step 3. Write $-14x$ as $-4x - 10x$ so that

$$8x^2 - 14x + 5 = 8x^2 - 4x - 10x + 5$$

Step 4. Factor by grouping.

$$8x^2 - 4x - 10x + 5 = 4x(2x - 1) - 5(2x - 1)$$
$$= (2x - 1)(4x - 5)$$

Practice Problem 2

Factor each trinomial by grouping.

a. $6x^2y - 7xy - 5y$

b. $30x^2 - 26x + 4$

TEACHING TIP

In Example 2, remind students that if they forget to factor out the GCF in Step 1, they may end up with a factored form that is not factored *completely*.

Example 2   Factor $6x^2 - 2x - 20$ by grouping.

Solution:

Step 1. First factor out the greatest common factor, 2.

$$6x^2 - 2x - 20 = 2(3x^2 - x - 10)$$

Step 2. Next notice that $a = 3$, $b = -1$, and $c = -10$ in this trinomial. Find two numbers whose product is $a \cdot c$ or $3(-10) = -30$ and whose sum is b, -1. The numbers are -6 and 5.

Step 3. $3x^2 - x - 10 = 3x^2 - 6x + 5x - 10$

Step 4.
$$= 3x(x - 2) + 5(x - 2)$$
$$= (x - 2)(3x + 5)$$

The factored form of $6x^2 - 2x - 20 = 2(x - 2)(3x + 5)$.

⌐ Don't forget to include the common factor of 2.

Answers

1. a. $(x + 4)(3x + 2)$, **b.** $(4x + 5)(3x + 1)$,
2. a. $y(2x + 1)(3x - 5)$,
b. $2(5x - 1)(3x - 2)$

Exercise Set 4.4

A *Factor each polynomial by grouping. Notice that Step 3 has already been done in these exercises. See Examples 1 and 2.*

1. $x^2 + 3x + 2x + 6$

2. $x^2 + 5x + 3x + 15$

3. $x^2 - 4x + 7x - 28$

4. $x^2 - 6x + 2x - 12$

5. $y^2 + 8y - 2y - 16$

6. $z^2 + 10z - 7z - 70$

7. $3x^2 + 4x + 12x + 16$

8. $2x^2 + 5x + 14x + 35$

9. $8x^2 - 5x - 24x + 15$

10. $4x^2 - 9x - 32x + 72$

11. $5x^4 - 3x^2 + 25x^2 - 15$

12. $2y^4 - 10y^2 + 7y^2 - 35$

ANSWERS

1. $(x + 3)(x + 2)$

2. $(x + 3)(x + 5)$

3. $(x - 4)(x + 7)$

4. $(x - 6)(x + 2)$

5. $(y + 8)(y - 2)$

6. $(z + 10)(z - 7)$

7. $(3x + 4)(x + 4)$

8. $(2x + 5)(x + 7)$

9. $(8x - 5)(x - 3)$

10. $(4x - 9)(x - 8)$

11. $(5x^2 - 3)(x^2 + 5)$

12. $(2y^2 + 7)(y^2 - 5)$

297

13. a. $9, 2$

b. $9x + 2x$

c. $(2x + 3)(3x + 1)$

14. a. $2, 12$

b. $2x + 12x$

c. $(4x + 1)(2x + 3)$

15. a. $-20, -3$

b. $-20x - 3x$

c. $(5x - 1)(3x - 4)$

16. a. $-10, -3$

b. $-10x - 3x$

c. $(3x - 5)(2x - 1)$

17. $(3y + 2)(7y + 1)$

18. $(3x + 1)(5x + 2)$

19. $(7x - 11)(x + 1)$

20. $(8x - 9)(x + 1)$

21. $(5x - 2)(2x - 1)$

22. $(5x - 3)(6x - 1)$

Factor each trinomial by grouping. Exercises 13–16 are broken into parts to help you get started. See Examples 1 and 2.

13. $6x^2 + 11x + 3$
    a. Find two numbers whose product is $6 \cdot 3 = 18$ and whose sum is 11.

    b. Write $11x$ using the factors from part (a).

    c. Factor by grouping.

14. $8x^2 + 14x + 3$
    a. Find two numbers whose product is $8 \cdot 3 = 24$ and whose sum is 14.

    b. Write $14x$ using the factors from part (a).

    c. Factor by grouping.

15. $15x^2 - 23x + 4$
    a. Find two numbers whose product is $15 \cdot 4 = 60$ and whose sum is -23.

    b. Write $-23x$ using the factors from part (a).

    c. Factor by grouping.

16. $6x^2 - 13x + 5$
    a. Find two numbers whose product is $6 \cdot 5 = 30$ and whose sum is -13.

    b. Write $-13x$ using the factors from part (a).

    c. Factor by grouping.

17. $21y^2 + 17y + 2$ **18.** $15x^2 + 11x + 2$ **19.** $7x^2 - 4x - 11$

20. $8x^2 - x - 9$ **21.** $10x^2 - 9x + 2$ **22.** $30x^2 - 23x + 3$

Name _____

23. $2x^2 - 7x + 5$

24. $2x^2 - 7x + 3$

25. $4x^2 + 12x + 9$

26. $25x^2 + 20x + 4$

27. $4x^2 - 8x - 21$

28. $6x^2 - 11x - 10$

29. $10x^2 - 23x + 12$

30. $21x^2 - 13x + 2$

31. $2x^3 + 13x^2 + 15x$

32. $3x^3 + 8x^2 + 4x$

33. $16y^2 - 34y + 18$

34. $4y^2 - 2y - 12$

35. $6x^2 - 13x + 6$

36. $12x^2 - 25x + 12$

37. $54a^2 - 9a - 30$

38. $30a^2 + 38a - 20$

39. $20a^3 + 37a^2 + 8a$

40. $10a^3 + 17a^2 + 3a$

41. $12x^3 - 27x^2 - 27x$

42. $30x^3 - 155x^2 + 25x$

23. $(2x - 5)(x - 1)$

24. $(2x - 1)(x - 3)$

25. $(2x + 3)(2x + 3)$ or $(2x + 3)^2$

26. $(5x + 2)(5x + 2)$ or $(5x + 2)^2$

27. $(2x + 3)(2x - 7)$

28. $(3x + 2)(2x - 5)$

29. $(5x - 4)(2x - 3)$

30. $(7x - 2)(3x - 1)$

31. $x(2x + 3)(x + 5)$

32. $x(3x + 2)(x + 2)$

33. $2(8y - 9)(y - 1)$

34. $2(2y + 3)(y - 2)$

35. $(2x - 3)(3x - 2)$

36. $(3x - 4)(4x - 3)$

37. $3(3a + 2)(6a - 5)$

38. $2(3a + 5)(5a - 2)$

39. $a(4a + 1)(5a + 8)$

40. $a(2a + 3)(5a + 1)$

41. $3x(4x + 3)(x - 3)$

42. $5x(6x - 1)(x - 5)$

299

43. $x^2 - 4$

44. $y^2 - 25$

45. $y^2 + 8y + 16$

46. $x^2 + 14x + 49$

47. $81z^2 - 25$

48. $64y^2 - 81$

49. $16x^2 - 24x + 9$

50. $4z^2 - 4z + 1$

51. $(x^n + 2)(x^n + 3)$

52. $(x^n + 6)(x^n + 10)$

53. $(3x^n - 5)(x^n + 7)$

54. $(2x^n - 5)(6x^n - 5)$

Name _____

REVIEW AND PREVIEW

Multiply. See Section 3.6.

43. $(x - 2)(x + 2)$

44. $(y - 5)(y + 5)$

45. $(y + 4)(y + 4)$

46. $(x + 7)(x + 7)$

47. $(9z + 5)(9z - 5)$

48. $(8y + 9)(8y - 9)$

49. $(4x - 3)^2$

50. $(2z - 1)^2$

◤ COMBINING CONCEPTS

Factor each polynomial by grouping.

51. $x^{2n} + 2x^n + 3x^n + 6$
(*Hint:* Don't forget that $x^{2n} = x^n \cdot x^n$.)

52. $x^{2n} + 6x^n + 10x^n + 60$

53. $3x^{2n} + 16x^n - 35$

54. $12x^{2n} - 40x^n + 25$

4.5 FACTORING PERFECT SQUARE TRINOMIALS AND THE DIFFERENCE OF TWO SQUARES

A RECOGNIZING PERFECT SQUARE TRINOMIALS

A trinomial that is the square of a binomial is called a **perfect square trinomial**. For example,

$$(x + 3)^2 = (x + 3)(x + 3)$$
$$= x^2 + 6x + 9$$

Thus $x^2 + 6x + 9$ is a perfect square trinomial.

In Chapter 3, we discovered special product formulas for squaring binomials, recognizing that

$$(a + b)^2 = a^2 + 2ab + b^2 \quad \text{and} \quad (a - b)^2 = a^2 - 2ab + b^2$$

Because multiplication and factoring are reverse processes, we can now use these special products to help us factor perfect square trinomials. If we reverse these equations, we have the following.

FACTORING PERFECT SQUARE TRINOMIALS

$$a^2 + 2ab + b^2 = (a + b)^2$$
$$a^2 - 2ab + b^2 = (a - b)^2$$

To use these equations to help us factor, we must first be able to recognize a perfect square trinomial. A trinomial is a perfect square when

1. Two terms, a^2 and b^2, are squares, and
2. another term is $2 \cdot a \cdot b$ or $-2 \cdot a \cdot b$. That is, this term is twice the product of a and b, or its opposite.

Example 1   Decide whether $x^2 + 8x + 16$ is a perfect square trinomial.

Solution: **1.** Two terms, x^2 and 16, are squares ($16 = 4^2$).

   2. Twice the product of x and 4 is the other term of the trinomial.

   $$2 \cdot x \cdot 4 = 8x$$

   Thus, $x^2 + 8x + 16$ is a perfect square trinomial.

Example 2   Decide whether $4x^2 + 10x + 9$ is a perfect square trinomial.

Solution: **1.** Two terms, $4x^2$ and 9, are squares.

   $$4x^2 = (2x)^2 \quad \text{and} \quad 9 = 3^2$$

   2. Twice the product of $2x$ and 3 is not the other term of the trinomial.

   $$2 \cdot 2x \cdot 3 = 12x, \, not \, 10x$$

   The trinomial is *not* a perfect square trinomial. ▬

TEACHING TIP

Remind students that learning math can become easier when their experience allows them to recognize certain patterns which occur over and over again. In algebra we extend the concept of perfect square from the specific ($1^2, 2^2, 3^2, 4^2, 5^2, 6^2, 7^2, \ldots$) to the abstract ($(x + 1)^2, (x + 2)^2, (x + 3)^2, \ldots$). Just as students have come to recognize 1, 4, 9, 16, 25, 36, and 49 as perfect squares, they will come to recognize certain simplified algebraic expressions as perfect squares.
Make an area diagram of Example 1 to show that it is a square because the length = width.

	$x + 4$	
x	x^2	$4x$
$+$ 4	$4x$	16

Practice Problem 1

Decide whether each trinomial is a perfect square trinomial.

a. $x^2 + 12x + 36$

b. $x^2 + 20x + 100$

Practice Problem 2

Decide whether each trinomial is a perfect square trinomial.

a. $9x^2 + 20x + 25$

b. $4x^2 + 8x + 11$

Answers

1. a. yes, **b.** yes, **2. a.** no, **b.** no

Practice Problem 3

Decide whether each trinomial is a perfect square trinomial.

a. $25x^2 - 10x + 1$

b. $9x^2 - 42x + 49$

TEACHING TIP

After students do Example 3, have them find the two perfect square trinomials whose coefficient of the squared term is 4 and whose constant term is 9:
$4x^2 + 12x + 9$ and $4x^2 - 12x + 9$.

Practice Problem 4

Factor: $x^2 + 16x + 64$

Practice Problem 5

Factor: $9r^2 + 24rs + 16s^2$

Practice Problem 6

Factor: $9n^2 - 6n + 1$

TEACHING TIP

After completing Example 6, ask what the sign of the middle term of the trinomial tells them about the factored form. Lead them to realize that it tells whether the trinomial is a sum squared or a difference squared.

Practice Problem 7

Factor: $9x^2 + 15x + 4$

Example 3   Decide whether $9x^2 - 12x + 4$ is a perfect square trinomial.

Solution: **1.** Two terms, $9x^2$ and 4, are squares.

$$9x^2 = (3x)^2 \quad \text{and} \quad 4 = 2^2$$

2. Twice the product of $3x$ and 2 is the opposite of the other term of the trinomial.

$$2 \cdot 3x \cdot 2 = 12x, \text{ the opposite of } -12x$$

Thus, $9x^2 - 12x + 4$ is a perfect square trinomial.

B FACTORING PERFECT SQUARE TRINOMIALS

Now that we can recognize perfect square trinomials, we are ready to factor them.

Example 4   Factor: $x^2 + 12x + 36$

Solution: $x^2 + 12x + 36 = x^2 + 2 \cdot x \cdot 6 + 6^2$ $36 = 6^2$ and $12x = 2 \cdot x \cdot 6$

$$a^2 + 2 \cdot a \cdot b + b^2$$

$$= (x + 6)^2$$

$$(a + b)^2$$

Example 5   Factor: $25x^2 + 20xy + 4y^2$

Solution: $25x^2 + 20xy + 4y^2 = (5x)^2 + 2 \cdot 5x \cdot 2y + (2y)^2$
$$= (5x + 2y)^2$$

Example 6   Factor: $4m^2 - 4m + 1$

Solution: $4m^2 - 4m + 1 = (2m)^2 - 2 \cdot 2m \cdot 1 + 1^2$

$$a^2 - 2 \cdot a \cdot b + b^2$$

$$= (2m - 1)^2$$

$$(a - b)^2$$

Example 7   Factor: $25x^2 + 50x + 9$

Solution: Notice that this trinomial is not a perfect square trinomial.

$$25x^2 = (5x)^2, \ 9 = 3^2$$

but

$$2 \cdot 5x \cdot 3 = 30x$$

and $30x$ is not the middle term $50x$.

Although $25x^2 + 50x + 9$ is not a perfect square trinomial, it is factorable. Using techniques we learned in Section 4.3, we find that

$$25x^2 + 50x + 9 = (5x + 9)(5x + 1)$$

Example 8   Factor: $162x^3 - 144x^2 + 32x$

Solution: Don't forget to first look for a common factor. There is a greatest common factor of $2x$ in this trinomial.

$$162x^3 - 144x^2 + 32x = 2x(81x^2 - 72x + 16)$$
$$= 2x[(9x)^2 - 2 \cdot 9x \cdot 4 + 4^2]$$
$$= 2x(9x - 4)^2$$

C FACTORING THE DIFFERENCE OF TWO SQUARES

In Chapter 3, we discovered another special product, the product of the sum and difference of two terms a and b:

$$(a + b)(a - b) = a^2 - b^2$$

Reversing this equation gives us another factoring pattern, which we use to factor the difference of two squares.

> **FACTORING THE DIFFERENCE OF TWO SQUARES**
>
> $$a^2 - b^2 = (a + b)(a - b)$$

Let's practice using this pattern.

Examples   Factor each binomial.

9. $x^2 - 4 = x^2 - 2^2 = (x + 2)(x - 2)$

$$a^2 - b^2 = (a + b)(a - b)$$

10. $y^2 - 25 = y^2 - 5^2 = (y + 5)(y - 5)$

11. $y^2 - \dfrac{4}{9} = y^2 - \left(\dfrac{2}{3}\right)^2 = \left(y + \dfrac{2}{3}\right)\left(y - \dfrac{2}{3}\right)$

12. $x^2 + 4$

Note that the binomial $x^2 + 4$ is the *sum* of two squares since we can write $x^2 + 4$ as $x^2 + 2^2$. We might try to factor using $(x + 2)(x + 2)$ or $(x - 2)(x - 2)$. But when we multiply to check, we find that neither factoring is correct.

$$(x + 2)(x + 2) = x^2 + 4x + 4$$
$$(x - 2)(x - 2) = x^2 - 4x + 4$$

In both cases, the product is a trinomial, not the required binomial. In fact, $x^2 + 4$ is a prime polynomial.

Factor: $12x^3 - 84x^2 + 147x$

Factor each binomial.

9. $x^2 - 9$ 10. $a^2 - 16$

11. $c^2 - \dfrac{9}{25}$ 12. $s^2 + 9$

TEACHING TIP

Point out to students that each difference of squares can be rewritten with an x term having a coefficient of zero. To factor it as we did in the previous section, find two numbers which multiply to give the constant and add to give zero. Notice that only additive inverses will give zero when added.

Answers

8. $3x(2x - 7)^2$, **9.** $(x - 3)(x + 3)$,

10. $(a - 4)(a + 4)$, **11.** $\left(c - \dfrac{3}{5}\right)\left(c + \dfrac{3}{5}\right)$,

12. prime polynomial

Practice Problems 13–15

Factor each difference of two squares.

13. $9s^2 - 1$
14. $16x^2 - 49y^2$
15. $p^4 - 81$

Examples   Factor each difference of two squares.

13. $4x^2 - 1 = (2x)^2 - 1^2 = (2x + 1)(2x - 1)$

14. $25a^2 - 9b^2 = (5a)^2 - (3b)^2 = (5a + 3b)(5a - 3b)$

15. $y^4 - 16 = (y^2)^2 - 4^2$
$= (y^2 + 4)(y^2 - 4)$ Factor the difference of two squares.
$= (y^2 + 4)(y + 2)(y - 2)$ Factor the difference of two squares.

HELPFUL HINTS

1. Don't forget to first see whether there's a greatest common factor (other than 1 or −1) that can be factored out.
2. Factor completely. In other words, check to see whether any factors can be factored further (as in Example 15).

Practice Problems 16–17

Factor each difference of two squares.

16. $9x^3 - 25x$

17. $48x^4 - 3$

Examples   Factor each difference of two squares.

16. $4x^3 - 49x = x(4x^2 - 49)$ Factor out the common factor x.
$= x[(2x)^2 - 7^2]$
$= x(2x + 7)(2x - 7)$ Factor the difference of two squares.

17. $162x^4 - 2 = 2(81x^4 - 1)$ Factor out the common factor, 2.
$= 2(9x^2 + 1)(9x^2 - 1)$ Factor the difference of two squares.
$= 2(9x^2 + 1)(3x + 1)(3x - 1)$ Factor the difference of two squares.

Answers

13. $(3s - 1)(3s + 1)$,
14. $(4x - 7y)(4x + 7y)$,
15. $(p^2 + 9)(p + 3)(p - 3)$,
16. $x(3x - 5)(3x + 5)$,
17. $3(4x^2 + 1)(2x + 1)(2x - 1)$

CALCULATOR EXPLORATION
GRAPHING

A graphing calculator is a convenient tool for evaluating an expression at a given replacement value. For example, let's evaluate $x^2 - 6x$ when $x = 2$. To do so, store the value 2 in the variable x and then enter and evaluate the algebraic expression.

```
2 → X
              2
X² − 6X
             −8
```

The value of $x^2 - 6x$ when $x = 2$ is -8. You may want to use this method for evaluating expressions as you explore the following.

   We can use a graphing calculator to explore factoring patterns numerically. Use your calculator to evaluate $x^2 - 2x + 1$, $x^2 - 2x - 1$, and $(x - 1)^2$ for each value of x given in the table. What do you observe?

	$x^2 - 2x + 1$	$x^2 - 2x - 1$	$(x - 1)^2$
$x = 5$	16	14	16
$x = -3$	16	14	16
$x = 2.7$	2.89	0.89	2.89
$x = -12.1$	171.61	169.61	171.61
$x = 0$	1	−1	1

Notice in each case that $x^2 - 2x - 1 \neq (x - 1)^2$. Because for each x in the table the value of $x^2 - 2x + 1$ and the value of $(x - 1)^2$ are the same, we might guess that $x^2 - 2x + 1 = (x - 1)^2$. We can verify our guess algebraically with multiplication:

$$(x - 1)(x - 1) = x^2 - x - x + 1 = x^2 - 2x + 1$$

TEACHING TIP

Some graphing calculators have a TABLE feature which allows the user to evaluate an expression for various values. Enter the expressions using the Y = key. Then use the TABLE feature.

Focus On Mathematical Connections

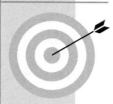

GEOMETRY

Factoring polynomials can be visualized using areas of rectangles. To see this, let's first find the areas of the following squares and rectangles. (Recall that Area = Length · Width)

Area: $x \cdot x = x^2$ square units

Area: $1 \cdot x = x$ square units

Area: $1 \cdot 1 = 1$ square unit

To use these areas to visualize factoring the polynomial $x^2 + 3x + 2$, for example, use the shapes below to form a rectangle. The factored form is found by reading the length and the width of the rectangle as shown below.

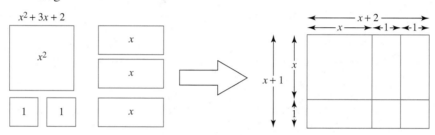

Thus, $x^2 + 3x + 2 = (x + 2)(x + 1)$.

Try using this method to visualize the factored form of each polynomial below.

GROUP ACTIVITY

Work in a group and use files to find the factored form of the polynomials below. (Tiles can be hand made from index cards.)

1. $x^2 + 6x + 5$ **4.** $x^2 + 4x + 3$

2. $x^2 + 5x + 6$ **5.** $x^2 + 6x + 9$

3. $x^2 + 5x + 4$ **6.** $x^2 + 4x + 4$

Name _____ Section _____ Date _____

MENTAL MATH

State each number as a square.

1. 1 **2.** 25 **3.** 81 **4.** 64

5. 9 **6.** 100

State each term as a square.

7. $9x^2$ **8.** $16y^2$ **9.** $25a^2$ **10.** $81b^2$

11. $36p^4$ **12.** $4q^4$

EXERCISE SET 4.5

A *Determine whether each trinomial is a perfect square trinomial. See Examples 1 through 3.*

1. $x^2 + 16x + 64$ **2.** $x^2 + 22x + 121$ **3.** $y^2 + 5y + 25$

4. $y^2 + 4y + 16$ **5.** $m^2 - 2m + 1$ **6.** $p^2 - 4p + 4$

7. $a^2 - 16a + 49$ **8.** $n^2 - 20n + 144$ **9.** $4x^2 + 12xy + 8y^2$

10. $25x^2 + 20xy + 2y^2$ **11.** $25a^2 - 40ab + 16b^2$ **12.** $36a^2 - 12ab + b^2$

13. Fill in the blank so that $x^2 + \_ x + 16$ is a perfect square trinomial.

14. Fill in the blank so that $9x^2 + \_ x + 25$ is a perfect square trinomial.

B *Factor each trinomial completely. See Examples 4 through 8.*

15. $x^2 + 22x + 121$ **16.** $x^2 + 18x + 81$ **17.** $x^2 - 16x + 64$

18. $x^2 - 12x + 36$ **19.** $16a^2 - 24a + 9$ **20.** $25x^2 + 20x + 4$

21. $x^4 + 4x^2 + 4$ **22.** $m^4 + 10m^2 + 25$ **23.** $2n^2 - 28n + 98$

24. $3y^2 - 6y + 3$ **25.** $16y^2 + 40y + 25$ **26.** $9y^2 + 48y + 64$

27. $x^2y^2 - 10xy + 25$ **28.** $4x^2y^2 - 28xy + 49$ **29.** $m^3 + 18m^2 + 81m$

MENTAL MATH ANSWERS

1. 1^2
2. 5^2
3. 9^2
4. 8^2
5. 3^2
6. 10^2
7. $(3x)^2$
8. $(4y)^2$
9. $(5a)^2$
10. $(9b)^2$
11. $(6p^2)^2$
12. $(2q^2)^2$

ANSWERS

1. yes
2. yes
3. no
4. no
5. yes
6. yes
7. no
8. no
9. no
10. no
11. yes
12. yes
13. 8
14. 30
15. $(x + 11)^2$
16. $(x + 9)^2$
17. $(x - 8)^2$
18. $(x - 6)^2$
19. $(4a - 3)^2$
20. $(5x + 2)^2$
21. $(x^2 + 2)^2$
22. $(m^2 + 5)^2$
23. $2(n - 7)^2$
24. $3(y - 1)^2$
25. $(4y + 5)^2$
26. $(3y + 8)^2$
27. $(xy - 5)^2$
28. $(2xy - 7)^2$
29. $m(m + 9)^2$

30. $y(y + 6)^2$
31. prime
32. prime
33. $(3x - 4y)^2$
34. $(5x - 6y)^2$
35. $(x + 7y)^2$
36. $(x + 5y)^2$
37. answers may vary
38. $x^2 + 6xy + 9y^2$
39. $(x - 2)(x + 2)$
40. $(x + 6)(x - 6)$
41. $(9 - p)(9 + p)$
42. $(10 - t)(10 + t)$
43. $(2r - 1)(2r + 1)$
44. $(3t - 1)(3t + 1)$
45. $(3x - 4)(3x + 4)$
46. $(6y - 5)(6y + 5)$
47. prime
48. prime
49. $(-6 + x)(6 + x)$
50. $(-1 + y)(1 + y)$
51. $(m^2 + 1)(m + 1)$ $(m - 1)$
52. $(n^2 + 4)(n + 2)$ $(n - 2)$
53. $(x - 13y)(x + 13y)$
54. $(x - 15y)(x + 15y)$
55. $2(3r - 2)(3r + 2)$
56. $2(4t - 5)(4t + 5)$
57. $x(3y - 2)(3y + 2)$
58. $16x(y - 2)(y + 2)$
59. $25y^2(y - 2)(y + 2)$
60. $xy(y - 3z)(y + 3z)$
61. $xy(x - 2y)(x + 2y)$
62. $4(3x - 4)(3x + 4)$
63. $9(5a - 3b)(5a + 3b)$
64. $9(4 - 3x)(4 + 3x)$
65. $3(2x - 3)(2x + 3)$
66. $(5y - 3)(5y + 3)$
67. $(7a - 4)(7a + 4)$
68. $(11 - 10x)(11 + 10x)$
69. $(13a - 7b)(13a + 7b)$
70. $(xy - 1)(xy + 1)$
71. $(4 - ab)(4 + ab)$
72. $\left(x - \dfrac{1}{2}\right)\left(x + \dfrac{1}{2}\right)$
73. $\left(y - \dfrac{1}{4}\right)\left(y + \dfrac{1}{4}\right)$
74. $\left(7 - \dfrac{3}{5}m\right)\left(7 + \dfrac{3}{5}m\right)$
75. $\left(10 - \dfrac{2}{9}n\right)\left(10 + \dfrac{2}{9}n\right)$

308

Name _____

30. $y^3 + 12y^2 + 36y$ **31.** $1 + 6x^2 + x^4$ **32.** $1 + 16x^2 + x^4$

33. $9x^2 - 24xy + 16y^2$ **34.** $25x^2 - 60xy + 36y^2$

35. $x^2 + 14xy + 49y^2$ **36.** $x^2 + 10xy + 25y^2$

37. Describe a perfect square trinomial.   **38.** Write a perfect square trinomial that factors as $(x + 3y)^2$.

C *Factor each binomial completely. See Examples 9 through 17.*

39. $x^2 - 4$ **40.** $x^2 - 36$ **41.** $81 - p^2$ **42.** $100 - t^2$

43. $4r^2 - 1$ **44.** $9t^2 - 1$ **45.** $9x^2 - 16$ **46.** $36y^2 - 25$

47. $16r^2 + 1$ **48.** $49y^2 + 1$ **49.** $-36 + x^2$ **50.** $-1 + y^2$

51. $m^4 - 1$ **52.** $n^4 - 16$ **53.** $x^2 - 169y^2$ **54.** $x^2 - 225y^2$

55. $18r^2 - 8$ **56.** $32t^2 - 50$ **57.** $9xy^2 - 4x$ **58.** $16xy^2 - 64x$

59. $25y^4 - 100y^2$ **60.** $xy^3 - 9xyz^2$ **61.** $x^3y - 4xy^3$ **62.** $36x^2 - 64$

63. $225a^2 - 81b^2$ **64.** $144 - 81x^2$ **65.** $12x^2 - 27$ **66.** $25y^2 - 9$

67. $49a^2 - 16$ **68.** $121 - 100x^2$ **69.** $169a^2 - 49b^2$ **70.** $x^2y^2 - 1$

71. $16 - a^2b^2$ **72.** $x^2 - \dfrac{1}{4}$ **73.** $y^2 - \dfrac{1}{16}$ **74.** $49 - \dfrac{9}{25}m^2$

75. $100 - \dfrac{4}{81}n^2$

Name _____

76. What binomial multiplied by $(x - 6)$ gives the difference of two squares?

77. What binomial multiplied by $(5 + y)$ gives the difference of two squares?

REVIEW AND PREVIEW

Solve each equation. See Section 2.4.

78. $x - 6 = 0$

79. $y + 5 = 0$

80. $2m + 4 = 0$

81. $3x - 9 = 0$

82. $5z - 1 = 0$

83. $4a + 2 = 0$

Solve each of the following. See Section 2.6.

84. A suitcase has a volume of 960 cubic inches. Find x.

10 inches

x inches

12 inches

85. The sail shown has an area of 25 square feet. Find its height, x.

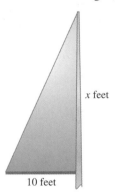

x feet

10 feet

COMBINING CONCEPTS

Factor each expression completely.

86. $(x + 2)^2 - y^2$

87. $(y - 6)^2 - z^2$

88. $a^2(b - 4) - 16(b - 4)$

89. $m^2(n + 8) - 9(n + 8)$

90. $(x^2 + 6x + 9) - 4y^2$ (*Hint:* Factor the trinomial in parentheses first.)

91. $(x^2 + 2x + 1) - 36y^2$

92. $x^{2n} - 100$

93. $x^{2n} - 81$

76. $(x + 6)$

77. $(5 - y)$

78. $x = 6$

79. $y = -5$

80. $m = -2$

81. $x = 3$

82. $z = \dfrac{1}{5}$

83. $a = -\dfrac{1}{2}$

84. 8 in.

85. 5 ft

86. $\begin{array}{l}(x + 2 - y)\\(x + 2 + y)\end{array}$

87. $\begin{array}{l}(y - 6 - z)\\(y - 6 + z)\end{array}$

88. $(a + 4)(a - 4)(b - 4)$

89. $(m - 3)(m + 3)(n + 8)$

90. $\begin{array}{l}(x + 3 - 2y)\\(x + 3 + 2y)\end{array}$

91. $\begin{array}{l}(x + 1 - 6y)\\(x + 1 + 6y)\end{array}$

92. $(x^n + 10)(x^n - 10)$

93. $(x^n + 9)(x^n - 9)$

Name _____

The area of the largest square in the figure is $(a + b)^2$. Use this figure to answer Exercises 94–95.

94. Write the area of the largest square as the sum of the areas of the smaller squares and rectangles.

95. What factoring formula from this section is visually represented by this square?

96. An object is dropped from the top of Pittsburgh's USX Towers, which is 841 feet tall. (*Source: World Almanac* research) The height of the object after t seconds is given by the expression $841 - 16t^2$.
   a. Find the height of the object after 2 seconds.

97. A worker on the top of the Aetna Life Building in San Francisco accidentally drops a bolt. The Aetna Life Building is 529 feet tall. (*Source: World Almanac* research) The height of the bolt after t seconds is given by the expression $529 - 16t^2$.
   a. Find the height of the bolt after 1 second.

   b. Find the height of the object after 5 seconds.

   b. Find the height of the bolt after 4 seconds.

   c. To the nearest whole second, estimate when the object hits the ground.

   c. To the nearest whole second, estimate when the bolt hits the ground.

   d. Factor $841 - 16t^2$.

   d. Factor $529 - 16t^2$.

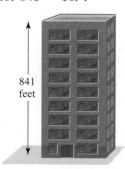

841 feet

Name _____ Section _____ Date _____

INTEGRATED REVIEW—CHOOSING A FACTORING STRATEGY

The following steps may be helpful when factoring polynomials.

TO FACTOR A POLYNOMIAL

Step 1. Are there any common factors? If so, factor out the GCF.

Step 2. How many terms are in the polynomial?
    a. Two terms: Is it the difference of two squares? $a^2 - b^2 = (a - b)(a + b)$
    b. Three terms: Try one of the following.
        i. Perfect square trinomial: $a^2 + 2ab + b^2 = (a + b)^2$
                          $a^2 - 2ab + b^2 = (a - b)^2$
        ii. If not a perfect square trinomial, factor using the methods presented in Sections 4.2 through 4.4.
    c. Four terms: Try factoring by grouping.

Step 3. See if any factors in the factored polynomial can be factored further.

Step 4. Check by multiplying.

Factor each polynomial completely.

1. $x^2 + x - 12$

2. $x^2 - 10x + 16$

3. $x^2 - x - 6$

4. $x^2 + 2x + 1$

5. $x^2 - 6x + 9$

6. $x^2 + x - 2$

7. $x^2 + x - 6$

8. $x^2 + 7x + 12$

9. $x^2 - 7x + 10$

10. $x^2 - x - 30$

11. $2x^2 - 98$

12. $3x^2 - 75$

13. $x^2 + 3x + 5x + 15$

14. $3y - 21 + xy - 7x$

15. $x^2 + 6x - 16$

16. $x^2 - 3x - 28$

17. $4x^3 + 20x^2 - 56x$

18. $6x^3 - 6x^2 - 120x$

19. $12x^2 + 34x + 24$

20. $8a^2 + 6ab - 5b^2$

21. $4a^2 - b^2$

22. $x^2 - 25y^2$

23. $28 - 13x - 6x^2$

24. $20 - 3x - 2x^2$

25. $x^2 - 2x + 4$

26. $a^2 + a - 3$

27. $6y^2 + y - 15$

28. $4x^2 - x - 5$

29. $18x^3 - 63x^2 + 9x$

30. $12a^3 - 24a^2 + 4a$

1. $(x - 3)(x + 4)$

2. $(x - 8)(x - 2)$

3. $(x + 2)(x - 3)$

4. $(x + 1)^2$

5. $(x - 3)^2$

6. $(x + 2)(x - 1)$

7. $(x + 3)(x - 2)$

8. $(x + 3)(x + 4)$

9. $(x - 5)(x - 2)$

10. $(x - 6)(x + 5)$

11. $2(x - 7)(x + 7)$

12. $3(x - 5)(x + 5)$

13. $(x + 3)(x + 5)$

14. $(y - 7)(3 + x)$

15. $(x + 8)(x - 2)$

16. $(x - 7)(x + 4)$

17. $4x(x + 7)(x - 2)$

18. $6x(x - 5)(x + 4)$

19. $2(3x + 4)(2x + 3)$

20. $(2a - b)(4a + 5b)$

21. $(2a - b)(2a + b)$

22. $(x - 5y)(x + 5y)$

23. $(4 - 3x)(7 + 2x)$

24. $(5 - 2x)(4 + x)$

25. prime

26. prime

27. $(3y + 5)(2y - 3)$

28. $(4x - 5)(x + 1)$

29. $9x(2x^2 - 7x + 1)$

30. $4a(3a^2 - 6a + 1)$

31. $(4a - 7)^2$

32. $(5p - 7)^2$

33. $(7 - x)(2 + x)$

34. $(3 + x)(1 - x)$

35. $3x^2y(x + 6)(x - 4)$

36. $2xy(x + 5y)(x - y)$

37. $3xy(4x^2 + 81)$

38. $2xy^2(3x^2 + 4)$

39. $2xy(1 - 6x)(1 + 6x)$

40. $2x(x - 3)(x + 3)$

41. $(x - 2)(x + 2)(x + 6)$

42. $(x - 2)(x - 6)(x + 6)$

43. $2a^2(3a + 5)$

44. $2n(2n - 3)$

45. $(x^2 + 4)(3x - 1)$

46. $(x - 2)(x^2 + 3)$

47. $6(x + 2y)(x + y)$

48. $2(x + 4y)(6x - y)$

49. $(5 + x)(x + y)$

50. $(x - y)(7 + y)$

51. $(7t - 1)(2t - 1)$

52. prime

53. $(3x + 5)(x - 1)$

54. $(7x - 2)(x + 3)$

55. $(1 - 10a)(1 + 2a)$

56. $(1 + 5a)(1 - 12a)$

57. $\dfrac{(x - 3)(x + 3)}{(x - 1)(x + 1)}$

58. $\dfrac{(x - 3)(x + 3)}{(x - 2)(x + 2)}$

59. $(x - 15)(x - 8)$

60. $(y + 16)(y + 6)$

61. answers may vary

62. yes; $9(x^2 + 9y^2)$

31. $16a^2 - 56a + 49$

32. $25p^2 - 70p + 49$

33. $14 + 5x - x^2$

34. $3 - 2x - x^2$

35. $3x^4y + 6x^3y - 72x^2y$

36. $2x^3y + 8x^2y^2 - 10xy^3$

37. $12x^3y + 243xy$

38. $6x^3y^2 + 8xy^2$

39. $2xy - 72x^3y$

40. $2x^3 - 18x$

41. $x^3 + 6x^2 - 4x - 24$

42. $x^3 - 2x^2 - 36x + 72$

43. $6a^3 + 10a^2$

44. $4n^2 - 6n$

45. $3x^3 - x^2 + 12x - 4$

46. $x^3 - 2x^2 + 3x - 6$

47. $6x^2 + 18xy + 12y^2$

48. $12x^2 + 46xy - 8y^2$

49. $5(x + y) + x(x + y)$

50. $7(x - y) + y(x - y)$

51. $14t^2 - 9t + 1$

52. $3t^2 - 5t + 1$

53. $3x^2 + 2x - 5$

54. $7x^2 + 19x - 6$

55. $1 - 8a - 20a^2$

56. $1 - 7a - 60a^2$

57. $x^4 - 10x^2 + 9$

58. $x^4 - 13x^2 + 36$

59. $x^2 - 23x + 120$

60. $y^2 + 22y + 96$

61. Explain why it makes good sense to factor out the GCF first, before using other methods of factoring.

62. The sum of two squares usually does not factor. Is the sum of two squares $9x^2 + 81y^2$ factorable?

4.6 SOLVING QUADRATIC EQUATIONS BY FACTORING

In this section, we introduce a new type of equation—the **quadratic equation**.

> ## QUADRATIC EQUATION
>
> A quadratic equation is one that can be written in the form
> $$ax^2 + bx + c = 0$$
> where a, b, and c are real numbers and $a \neq 0$.

Some examples of quadratic equations are

$$3x^2 + 5x + 6 = 0 \qquad x^2 = 9 \qquad y^2 + y = 1$$

The form $ax^2 + bx + c = 0$ is called the **standard form** of a quadratic equation. The quadratic equation $3x^2 + 5x + 6 = 0$ is the only equation above that is in standard form.

Quadratic equations model many real-life situations. For example, let's suppose an object is dropped from the top of a 256-foot cliff and we want to know how long before the object strikes the ground. The answer to this question is found by solving the quadratic equation $-16t^2 + 256 = 0$. (See Example 1 in Section 4.7.)

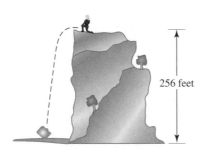

256 feet

A SOLVING QUADRATIC EQUATIONS BY FACTORING

Some quadratic equations can be solved by making use of factoring and the **zero factor property**.

> ## ZERO FACTOR PROPERTY
>
> If a and b are real numbers and if $ab = 0$, then $a = 0$, or $b = 0$.

In other words, if the product of two numbers is 0, then at least one of the numbers must be 0.

Example 1   Solve: $(x - 3)(x + 1) = 0$

Solution: If this equation is to be a true statement, then either the factor $x - 3$ must be 0 or the factor $x + 1$ must be 0. In other words, either

$$x - 3 = 0 \qquad \text{or} \qquad x + 1 = 0$$

If we solve these two linear equations, we have

$$x = 3 \qquad \text{or} \qquad x = -1$$

TEACHING TIP

You may want to begin this lesson with the following activity. Ask students to suppose they are told that the product of two factors is 12. Do they know for certain what either factor is? Why or why not? Now tell students the product of two factors is 0. Do they know for certain what either factor is? Why or why not?

Practice Problem 1

Solve: $(x - 7)(x + 2) = 0$

Answer

1. 7 and -2

Thus, 3 and -1 are both solutions of the equation $(x - 3)(x + 1) = 0$. To check, we replace x with 3 in the original equation. Then we replace x with -1 in the original equation.

Check:

$$(x - 3)(x + 1) = 0$$
$$(3 - 3)(3 + 1) = 0 \quad \text{Replace } x \text{ with 3.}$$
$$0(4) = 0 \quad \text{True.}$$

$$(x - 3)(x + 1) = 0$$
$$(-1 - 3)(-1 + 1) = 0 \quad \text{Replace } x \text{ with } -1.$$
$$(-4)(0) = 0 \quad \text{True.}$$

The solutions are 3 and -1.

HELPFUL HINT

The zero factor property says that *if a product is 0, then a factor is 0.*

If $a \cdot b = 0$, then $a = 0$ or $b = 0$.
If $x(x + 5) = 0$, then $x = 0$ or $x + 5 = 0$.
If $(x + 7)(2x - 3) = 0$, then $x + 7 = 0$ or $2x - 3 = 0$.

Use this property only when the product is 0. For example, if $a \cdot b = 8$, we do not know the value of a or b. The values may be $a = 2, b = 4$ or $a = 8, b = 1$, or any other two numbers whose product is 8.

Practice Problem 2

Solve: $(x - 10)(3x + 1) = 0$

Example 2   Solve: $(x - 5)(2x + 7) = 0$

Solution: The product is 0. By the zero factor property, this is true only when a factor is 0. To solve, we set each factor equal to 0 and solve the resulting linear equations.

$$(x - 5)(2x + 7) = 0$$
$$x - 5 = 0 \quad \text{or} \quad 2x + 7 = 0$$
$$x = 5 \qquad\qquad 2x = -7$$
$$x = -\frac{7}{2}$$

Check: Let $x = 5$.
$$(x - 5)(2x + 7) = 0$$
$$(5 - 5)(2 \cdot 5 + 7) \stackrel{?}{=} 0 \quad \text{Replace } x \text{ with 5.}$$
$$0 \cdot 17 \stackrel{?}{=} 0$$
$$0 = 0 \quad \text{True.}$$

Let $x = -\frac{7}{2}$.
$$(x - 5)(2x + 7) = 0$$
$$\left(-\frac{7}{2} - 5\right)\left(2\left(-\frac{7}{2}\right) + 7\right) \stackrel{?}{=} 0 \quad \text{Replace } x \text{ with } -\frac{7}{2}.$$
$$\left(-\frac{17}{2}\right)(-7 + 7) \stackrel{?}{=} 0$$
$$\left(-\frac{17}{2}\right) \cdot 0 \stackrel{?}{=} 0$$
$$0 = 0 \quad \text{True.}$$

The solutions are 5 and $-\frac{7}{2}$.

Answer

2. 10 and $-\frac{1}{3}$

Example 3   Solve: $x(5x - 2) = 0$

Solution: $x(5x - 2) = 0$

$x = 0$ or $5x - 2 = 0$ Use the zero factor property.

$5x = 2$

$x = \dfrac{2}{5}$

Check these solutions in the original equation. The solutions are 0 and $\dfrac{2}{5}$.

━━━━━

Practice Problem 3

Solve each equation.

a. $y(y + 3) = 0$

b. $x(4x - 3) = 0$

Example 4   Solve: $x^2 - 9x - 22 = 0$

Solution: One side of the equation is 0. However, to use the zero factor property, one side of the equation must be 0 *and* the other side must be written as a product (must be factored). Thus, we must first factor this polynomial.

$x^2 - 9x - 22 = 0$

$(x - 11)(x + 2) = 0$ Factor.

Now we can apply the zero factor property.

$x - 11 = 0$ or $x + 2 = 0$

$x = 11$ $x = -2$

Check:

Let $x = 11$.

$x^2 - 9x - 22 = 0$

$11^2 - 9 \cdot 11 - 22 \overset{?}{=} 0$

$121 - 99 - 22 \overset{?}{=} 0$

$22 - 22 \overset{?}{=} 0$

$0 = 0$ True.

Let $x = -2$.

$x^2 - 9x - 22 = 0$

$(-2)^2 - 9(-2) - 22 \overset{?}{=} 0$

$4 + 18 - 22 \overset{?}{=} 0$

$22 - 22 \overset{?}{=} 0$

$0 = 0$ True.

The solutions are 11 and -2.

━━━━━

Practice Problem 4

Solve: $x^2 - 3x - 18 = 0$

Example 5   Solve: $x^2 - 9x = -20$

Solution: First we rewrite the equation in standard form so that one side is 0. Then we factor the polynomial.

$x^2 - 9x = -20$

$x^2 - 9x + 20 = 0$ Write in standard form by adding 20 to both sides.

$(x - 4)(x - 5) = 0$ Factor.

Next we use the zero factor property and set each factor equal to 0.

$x - 4 = 0$ or $x - 5 = 0$ Set each factor equal to 0.

$x = 4$ $x = 5$ Solve.

Check: Check these solutions in the original equation. The solutions are 4 and 5.

━━━━━

Practice Problem 5

Solve: $x^2 - 14x = -24$

TEACHING TIP

Consider asking your students why we want to rewrite the equation so that one side is equal to zero. What does that accomplish?

Answers

3. a. 0 and -3,   **b.** 0 and $\dfrac{3}{4}$,   **4.** 6 and -3,

5. 12 and 2

The following steps may be used to solve a quadratic equation by factoring.

TO SOLVE QUADRATIC EQUATIONS BY FACTORING

Step 1. Write the equation in standard form so that one side of the equation is 0.

Step 2. Factor the quadratic equation completely.

Step 3. Set each factor containing a variable equal to 0.

Step 4. Solve the resulting equations.

Step 5. Check each solution in the original equation.

Since it is not always possible to factor a quadratic polynomial, not all quadratic equations can be solved by factoring. Other methods of solving quadratic equations are presented in Chapter 9.

Practice Problem 6

Solve each equation.

a. $x(x - 4) = 5$

b. $x(3x + 7) = 6$

Example 6   Solve: $x(2x - 7) = 4$

Solution: First we write the equation in standard form; then we factor.

$$x(2x - 7) = 4$$
$$2x^2 - 7x = 4 \qquad \text{Multiply.}$$
$$2x^2 - 7x - 4 = 0 \qquad \text{Write in standard form.}$$
$$(2x + 1)(x - 4) = 0 \qquad \text{Factor.}$$

$$2x + 1 = 0 \quad \text{or} \quad x - 4 = 0 \qquad \text{Set each factor equal to zero.}$$
$$2x = -1 \qquad\qquad x = 4 \qquad \text{Solve.}$$
$$x = -\frac{1}{2}$$

Check the solutions in the original equation. The solutions are $-\frac{1}{2}$ and 4.

HELPFUL HINT

To solve the equation $x(2x - 7) = 4$, do **not** set each factor equal to 4. Remember that to apply the zero factor property, one side of the equation must be 0 and the other side of the equation must be in factored form.

B   SOLVING EQUATIONS WITH DEGREE GREATER THAN TWO BY FACTORING

Some equations with degree greater than 2 can be solved by factoring and then using the zero factor property.

Practice Problem 7

Solve: $2x^3 - 18x = 0$

Example 7   Solve: $3x^3 - 12x = 0$

Solution: To factor the left side of the equation, we begin by factoring out the greatest common factor, $3x$.

Answers

6. a. 5 and -1,   **b.** $\frac{2}{3}$ and -3,   **7.** 0, 3, and -3

$$3x^3 - 12x = 0$$
$$3x(x^2 - 4) = 0 \quad \text{Factor out the GCF, } 3x.$$
$$3x(x + 2)(x - 2) = 0 \quad \begin{array}{l}\text{Factor } x^2 - 4, \text{ a differ-}\\ \text{ence of two squares.}\end{array}$$

$3x = 0$ or $x + 2 = 0$ or $x - 2 = 0$ Set each factor equal to 0.

$x = 0$ $x = -2$ $x = 2$ Solve.

Thus, the equation $3x^3 - 12x = 0$ has three solutions: 0, -2, and 2.

Check: Replace x with each solution in the original equation.

Let $x = 0$.   Let $x = -2$.   Let $x = 2$.

$3(0)^3 - 12(0) \stackrel{?}{=} 0$ $3(-2)^3 - 12(-2) \stackrel{?}{=} 0$ $3(2)^3 - 12(2) \stackrel{?}{=} 0$

          $0 = 0$ $3(-8) + 24 \stackrel{?}{=} 0$ $3(8) - 24 \stackrel{?}{=} 0$

          True. $0 = 0$ True. $0 = 0$ True.

The solutions are 0, -2, and 2.   ▬▬▬

Example 8   Solve: $(5x - 1)(2x^2 + 15x + 18) = 0$

Solution:

$(5x - 1)(2x^2 + 15x + 18) = 0$

$(5x - 1)(2x + 3)(x + 6) = 0$ Factor the trinomial.

$5x - 1 = 0$ or $2x + 3 = 0$ or $x + 6 = 0$ $\begin{array}{l}\text{Set each factor}\\ \text{equal to 0.}\end{array}$

    $5x = 1$ $2x = -3$ $x = -6$ Solve.

    $x = \dfrac{1}{5}$ $x = -\dfrac{3}{2}$

Check each solution in the original equation. The solutions are $\dfrac{1}{5}$, $-\dfrac{3}{2}$, and -6.   ▬▬▬

Practice Problem 8

Solve: $(x + 3)(3x^2 - 20x - 7) = 0$

TEACHING TIP

Consider challenging students to notice some patterns. Ask them to make a conjecture about the number of solutions that a quadratic equation has, that a cubic equation has, and so on. Also ask students if they know a solution of a quadratic equation, do they know a factor of the equation?

Answer

8. -3, $-\dfrac{1}{3}$, and 7

Focus On Mathematical Connections

NUMBER THEORY

By now, you have realized that being able to write a number as the product of prime numbers is very useful in the process of factoring polynomials. You probably also know at least a few numbers that are prime (such as 2, 3, and 5). But what about the other prime numbers? When we come across a number, how will we know if it is a prime number? Apparently, the ancient Greek mathematician Eratosthenes had a similar question because in the third century B.C. he devised a simple method for identifying primes. The method is called the Sieve of Eratosthenes because it "sifts out" the primes in a list of numbers. The Sieve of Eratosthenes is generally considered to be the most useful for identifying primes less than 1,000,000.

Here's how the sieve works: suppose you want to find the prime numbers in the first n natural numbers. Write the numbers, in order, from 2 to n. We know that 2 is prime, so circle it. Now, cross out each number greater than 2 that is a multiple of 2. Consider the next number in the list that is not crossed out. This number is 3, which we know is prime. Circle 3 and then cross out all multiples of 3 in the remainder of the list. Continue considering each uncircled number in the list. Once you have reached the largest prime number less than or equal to $\sqrt{n}$, and you have eliminated all its multiples in the remainder of the list, you can stop. Circle all the numbers left in the list that have not yet been circled. Now all of the circled numbers in the list are prime numbers. The list below demonstrates the Sieve of Eratosthenes on the numbers 2 through 30. Because $\sqrt{30} \approx 5.477$, we need only check and eliminate the multiples of primes up to and including 5, which is the largest prime less than or equal to the square root of 30.

We can see that the prime numbers less than 30 are 2, 3, 5, 7, 11, 13, 17, 19, 23, and 29.

GROUP ACTIVITY

Work with your group to identify the prime numbers less than 300 using the Sieve of Eratosthenes. What is the largest prime number that you will need to check in this process?

Name _____ Section _____ Date _____

MENTAL MATH

Solve each equation by inspection.

1. $(a - 3)(a - 7) = 0$ **2.** $(a - 5)(a - 2) = 0$ **3.** $(x + 8)(x + 6) = 0$

4. $(x + 2)(x + 3) = 0$ **5.** $(x + 1)(x - 3) = 0$ **6.** $(x - 1)(x + 2) = 0$

EXERCISE SET 4.6

A Solve each equation. See Examples 1 through 3.

1. $(x - 2)(x + 1) = 0$ **2.** $(x + 3)(x + 2) = 0$ **3.** $(x - 6)(x - 7) = 0$

4. $(x + 4)(x - 10) = 0$ **5.** $(x + 9)(x + 17) = 0$ **6.** $(x - 11)(x - 1) = 0$

7. $x(x + 6) = 0$ **8.** $x(x - 7) = 0$ **9.** $3x(x - 8) = 0$

10. $2x(x + 12) = 0$ **11.** $(2x + 3)(4x - 5) = 0$

12. $(3x - 2)(5x + 1) = 0$ **13.** $(2x - 7)(7x + 2) = 0$

14. $(9x + 1)(4x - 3) = 0$ **15.** $\left(x - \dfrac{1}{2}\right)\left(x + \dfrac{1}{3}\right) = 0$

16. $\left(x + \dfrac{2}{9}\right)\left(x - \dfrac{1}{4}\right) = 0$ **17.** $(x + 0.2)(x + 1.5) = 0$

18. $(x + 1.7)(x + 2.3) = 0$

19. Write a quadratic equation that has two solutions, 6 and −1. Leave the polynomial in the equation in factored form.

20. Write a quadratic equation that has two solutions, 0 and −2. Leave the polynomial in the equation in factored form.

1. $a = 3, a = 7$

2. $a = 5, a = 2$

3. $x = -8, x = -6$

4. $x = -2, x = -3$

5. $x = -1, x = 3$

6. $x = 1, x = -2$

ANSWERS

1. $x = 2, x = -1$

2. $x = -3, x = -2$

3. $x = 6, x = 7$

4. $x = -4, x = 10$

5. $x = -9, x = -17$

6. $x = 11, x = 1$

7. $x = 0, x = -6$

8. $x = 0, x = 7$

9. $x = 0, x = 8$

10. $x = 0, x = -12$

11. $x = -\frac{3}{2}, x = \frac{5}{4}$

12. $x = \frac{2}{3}, x = -\frac{1}{5}$

13. $x = \frac{7}{2}, x = -\frac{2}{7}$

14. $x = -\frac{1}{9}, x = \frac{3}{4}$

15. $x = \frac{1}{2}, x = -\frac{1}{3}$

16. $x = -\frac{2}{9}, x = \frac{1}{4}$

17. $x = -0.2, x = -1.5$

18. $x = -1.7, x = -2.3$

19. $(x - 6)(x + 1) = 0$

20. $x(x + 2) = 0$

21. $x = 9, x = 4$

22. $x = -9, x = 7$

23. $x = -4, x = 2$

24. $x = 3, x = 2$

25. $x = 0, x = 7$

26. $x = 0, x = 3$

27. $x = 0, x = -20$

28. $x = 0, x = -15$

29. $x = 4, x = -4$

30. $x = 3, x = -3$

31. $x = 8, x = -4$

32. $x = 8, x = -3$

33. $x = \frac{7}{3}, x = -2$

34. $x = -\frac{1}{4}, x = 3$

35. $x = \frac{8}{3}, x = -9$

36. $x = \frac{3}{4}, x = -\frac{7}{9}$

37. $x = 0, x = \frac{1}{2}, x = -\frac{1}{2}$

38. $y = 0, y = 3, y = -3$

39. $x = \frac{17}{2}$

40. $x = \frac{3}{10}$

41. $x = \frac{3}{4}$

42. $x = -\frac{5}{2}$

43. $y = \frac{1}{2}, y = -\frac{1}{2}$

44. $y = \frac{9}{2}, y = -\frac{9}{2}$

45. $x = -\frac{3}{2}, x = -\frac{1}{2}, x = 3$

46. $x = \frac{9}{2}, x = -9, x = 4$

47. $x = -5, x = 3$

48. $x = -13, x = 2$

49. $x = 0, x = 16$

320

B *Solve each equation. See Examples 4 through 8.*

21. $x^2 - 13x + 36 = 0$ **22.** $x^2 + 2x - 63 = 0$ **23.** $x^2 + 2x - 8 = 0$

24. $x^2 - 5x + 6 = 0$ **25.** $x^2 - 7x = 0$ **26.** $x^2 - 3x = 0$

27. $x^2 + 20x = 0$ **28.** $x^2 + 15x = 0$ **29.** $x^2 = 16$

30. $x^2 = 9$ **31.** $x^2 - 4x = 32$ **32.** $x^2 - 5x = 24$

33. $x(3x - 1) = 14$ **34.** $x(4x - 11) = 3$ **35.** $3x^2 + 19x - 72 = 0$

36. $36x^2 + x - 21 = 0$ **37.** $4x^3 - x = 0$ **38.** $4y^3 - 36y = 0$

39. $4(x - 7) = 6$ **40.** $5(3 - 4x) = 9$

41. $(4x - 3)(16x^2 - 24x + 9) = 0$ **42.** $(2x + 5)(4x^2 - 10x + 25) = 0$

43. $4y^2 - 1 = 0$ **44.** $4y^2 - 81 = 0$

45. $(2x + 3)(2x^2 - 5x - 3) = 0$ **46.** $(2x - 9)(x^2 + 5x - 36) = 0$

47. $x^2 - 15 = -2x$ **48.** $x^2 - 26 = -11x$ **49.** $x^2 - 16x = 0$

Name _____

50. $x^2 + 5x = 0$ **51.** $x^2 - x = 30$ **52.** $x^2 + 13x = -36$

53. $6y^2 - 22y - 40 = 0$ **54.** $3x^2 - 6x - 9 = 0$ **55.** $(y - 2)(y + 3) = 6$

56. $(y - 5)(y - 2) = 28$ **57.** $x^3 - 12x^2 + 32x = 0$ **58.** $x^3 - 14x^2 + 49x = 0$

59. Write a quadratic equation in standard form that has two solutions, 5 and 7.

60. Write an equation that has three solutions, 0, 1, and 2.

REVIEW AND PREVIEW

Perform each indicated operation. Write all results in lowest terms. See Section R.2.

61. $\dfrac{3}{5} + \dfrac{4}{9}$ **62.** $\dfrac{2}{3} + \dfrac{3}{7}$ **63.** $\dfrac{7}{10} - \dfrac{5}{12}$

64. $\dfrac{5}{9} - \dfrac{5}{12}$ **65.** $\dfrac{4}{5} \cdot \dfrac{7}{8}$ **66.** $\dfrac{3}{7} \cdot \dfrac{12}{17}$

COMBINING CONCEPTS

67. Find the error.

$x(x - 2) = 6$

$x = 6$ or $x - 2 = 6$

$\qquad\qquad\qquad x = 8$

68. Find the error.

$(x - 4)(x + 2) = 0$

$x = -4$ or $x = 2$

50. $x = 0, x = -5$

51. $x = -5, x = 6$

52. $x = -9, x = -4$

53. $y = -\frac{4}{3}, y = 5$

54. $x = 3, x = -1$

55. $y = -4, y = 3$

56. $y = 9, y = -2$

57. $x = 0, x = 8, x = 4$

58. $x = 0, x = 7$

59. $x^2 - 12x + 35 = 0$

60. $x^3 - 3x^2 + 2x = 0$

61. $\dfrac{47}{45}$

62. $\dfrac{23}{21}$

63. $\dfrac{17}{60}$

64. $\dfrac{5}{36}$

65. $\dfrac{7}{10}$

66. $\dfrac{36}{119}$

67. didn't write equation in standard form

68. should be $x = 4$ or $x = -2$

321

69. a. see table

b. 5 sec

c. 304 ft

70. a. see table

b. 6.3 sec

c. 156 ft

71. $x = 0, x = \frac{1}{2}$

72. $x = 0, x = -16$

73. $x = 0, x = -15$

74. $x = 0, x = 1$

69. A compass is accidentally thrown upward and out of an air balloon at a height of 300 feet. The height, y, of the compass at time x is given by the equation $y = -16x^2 + 20x + 300$.

300 ft

a. Find the height of the compass at the given times by filling in the table below.

Time, x (in seconds)	0	1	2	3	4	5	6
Height, y (in feet)	300	304	276	216	124	0	−156

b. Use the table to determine when the compass strikes the ground.

c. Use the table to approximate the maximum height of the compass.

70. A rocket is fired upward from the ground with an initial velocity of 100 feet per second. The height, y, of the rocket at any time x is given by the equation $y = -16x^2 + 100x$.

y

a. Find the height of the rocket at the given times by filling in the table below.

Time, x (in seconds)	0	1	2	3	4	5	6	7
Height, y (in feet)	0	84	136	156	144	100	24	−84

b. Use the table to help approximate when the rocket strikes the ground to the nearest tenth of a second.

c. Use the table to approximate the maximum height of the rocket.

Solve each equation.

71. $(x - 3)(3x + 4) = (x + 2)(x - 6)$

72. $(2x - 3)(x + 6) = (x - 9)(x + 2)$

73. $(2x - 3)(x + 8) = (x - 6)(x + 4)$

74. $(x + 6)(x - 6) = (2x - 9)(x + 4)$

4.7 QUADRATIC EQUATIONS AND PROBLEM SOLVING

A SOLVING PROBLEMS MODELED BY QUADRATIC EQUATIONS

Some problems may be modeled by quadratic equations. To solve these problems, we use the same problem-solving steps that were introduced in Section 2.5. When solving these problems, keep in mind that a solution of an equation that models a problem may not be a solution to the problem. For example, a person's age or the length of a rectangle is always a positive number. Thus we discard solutions that do not make sense as solutions of the problem.

Example 1   Finding Free-Fall Time

For a TV commercial, a piece of luggage is dropped from a cliff 256 feet above the ground to show the durability of the luggage. Neglecting air resistance, the height h in feet of the luggage above the ground after t seconds is given by the quadratic equation

$$h = -16t^2 + 256$$

Find how long it takes for the luggage to hit the ground.

Solution: **1.** UNDERSTAND. Read and reread the problem. Then draw a picture of the problem.

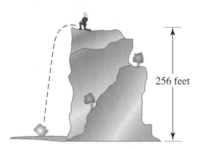

256 feet

The equation $h = -16t^2 + 256$ models the height of the falling luggage at time t. Familiarize yourself with this equation by finding the height of the luggage at $t = 1$ second and $t = 2$ seconds.

When $t = 1$ second, the height of the suitcase is
$h = -16(1)^2 + 256 = 240$ feet.
When $t = 2$ seconds, the height of the suitcase is
$h = -16(2)^2 + 256 = 192$ feet.

2. TRANSLATE. To find how long it takes the luggage to hit the ground, we want to know the value of t for which the height $h = 0$.

$$0 = -16t^2 + 256$$

3. SOLVE. We solve the quadratic equation by factoring.

$$0 = -16t^2 + 256$$
$$0 = -16(t^2 - 16)$$
$$0 = -16(t - 4)(t + 4)$$
$$t - 4 = 0 \quad \text{or} \quad t + 4 = 0$$
$$t = 4 \qquad\qquad t = -4$$

Objectives

A Solve problems that can be modeled by quadratic equations.

SSM   CD-ROM   Video
4.7

Practice Problem 1

An object is dropped from the roof of a 144-foot-tall building. Neglecting air resistance, the height h in feet of the object above ground after t seconds is given by the quadratic equation

$$h = -16t^2 + 144$$

Find how long it takes the object to hit the ground.

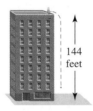

144 feet

Answer

1. 3 seconds

4. INTERPRET. Since the time t cannot be negative, the proposed solution is 4 seconds.

Check: Verify that the height of the luggage when t is 4 seconds is 0.

When $t = 4$ seconds, $h = -16(4)^2 + 256 = -256 + 256 = 0$ feet.

State: The solution checks and the luggage hits the ground 4 seconds after it is dropped.   ▬▬▬

Practice Problem 2

The square of a number minus twice the number is 63. Find the number.

Example 2   Finding a Number

The square of a number plus three times the number is 70. Find the number.

Solution: 1. UNDERSTAND. Read and reread the problem. Suppose that the number is 5. The square of 5 is 5^2 or 25. Three times 5 is 15. Then $25 + 15 = 40$, not 70, so the number must be greater than 5. Remember, the purpose of proposing a number, such as 5, is to better understand the problem. Now that we do, we will let $x =$ the number.

2. TRANSLATE.

the square of a number	plus	three times the number	is	70
↓	↓	↓	↓	↓
x^2	$+$	$3x$	$=$	70

3. SOLVE.

$$x^2 + 3x = 70$$
$$x^2 + 3x - 70 = 0 \qquad \text{Subtract 70 from both sides.}$$
$$(x + 10)(x - 7) = 0 \qquad \text{Factor.}$$
$$x + 10 = 0 \quad \text{or} \quad x - 7 = 0 \qquad \text{Set each factor equal to 0.}$$
$$x = -10 \qquad\qquad x = 7 \qquad \text{Solve.}$$

4. INTERPRET.

Check: The square of -10 is $(-10)^2$, or 100. Three times -10 is $3(-10)$ or -30. Then $100 + (-30) = 70$, the correct sum, so -10 checks.

The square of 7 is 7^2 or 49. Three times 7 is $3(7)$, or 21. Then $49 + 21 = 70$, the correct sum, so 7 checks.

State: There are two numbers. They are -10 and 7.   ▬▬▬

Practice Problem 3

The length of a rectangle is 5 feet more than its width. The area of the rectangle is 176 square feet. Find the length and the width of the rectangle.

Example 3   Finding the Dimensions of a Sail

The height of a triangular sail is 2 meters less than twice the length of the base. If the sail has an area of 30 square meters, find the length of its base and the height.

Solution: 1. UNDERSTAND. Read and reread the problem. Since we are finding the length of the base and the height, we let

$x =$ the length of the base

Answers

2. 9 and -7,   **3.** length $= 16$ ft; width $= 11$ ft

Height = 2x − 2

Base = x

and since the height is 2 meters less than twice the base,

$2x - 2 =$ the height

An illustration is shown to the left.

2. TRANSLATE. We are given that the area of the triangle is 30 square meters, so we use the formula for area of a triangle.

area of triangle	=	$\frac{1}{2}$	·	base	·	height

$$30 = \frac{1}{2} \cdot x \cdot (2x - 2)$$

3. SOLVE. Now we solve the quadratic equation.

$$30 = \frac{1}{2}x(2x - 2)$$

$$30 = x^2 - x \qquad \text{Multiply.}$$

$$x^2 - x - 30 = 0 \qquad \text{Write in standard form.}$$

$$(x - 6)(x + 5) = 0 \qquad \text{Factor.}$$

$$x - 6 = 0 \quad \text{or} \quad x + 5 = 0 \qquad \text{Set each factor equal to 0.}$$

$$x = 6 \qquad\qquad x = -5$$

4. INTERPRET. Since x represents the length of the base, we discard the solution -5. The base of a triangle cannot be negative. The base is then 6 feet and the height is $2(6) - 2 = 10$ feet.

Check: To check this problem, we recall that $\frac{1}{2}$ base · height = area, or

$$\frac{1}{2}(6)(10) = 30 \qquad \text{The required area}$$

State: The base of the triangular sail is 6 meters and the height is 10 meters.

The next example makes use of the **Pythagorean theorem** and consecutive integers. Before we review this theorem, recall that a **right triangle** is a triangle that contains a 90° or right angle. The **hypotenuse** of a right triangle is the side opposite the right angle and is the longest side of the triangle. The **legs** of a right triangle are the other sides of the triangle.

PYTHAGOREAN THEOREM

In a right triangle, the sum of the squares of the lengths of the two legs is equal to the square of the length of the hypotenuse.

$$(\text{leg})^2 + (\text{leg})^2 = (\text{hypotenuse})^2 \qquad \text{or} \qquad a^2 + b^2 = c^2$$

Leg b Hypotenuse c

Leg a

TEACHING TIP

Remind students that although the equation that models an application may have more than one solution, the solution to the application must make sense with the problem. Always check to see that a solution makes sense for the question being asked.

Study the following diagrams for a review of consecutive integers.

Consecutive integers:

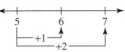

If x is the first integer: x, $x + 1$, $x + 2$

Consecutive even integers:

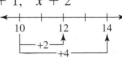

If x is the first even integer: x, $x + 2$, $x + 4$

Consecutive odd integers:

If x is the first odd integer: x, $x + 2$, $x + 4$

Practice Problem 4

Solve.

a. Find two consecutive odd integers whose product is 23 more than their sum.

b. The length of one leg of a right triangle is 7 meters less than the length of the other leg. The length of the hypotenuse is 13 meters. Find the lengths of the legs.

TEACHING TIP

Example 4 may be a good opportunity to point out to students that they are not finished with an application when they have found the values for x. They must answer the question which was asked.

Example 4   Finding the Dimensions of a Triangle

Find the lengths of the sides of a right triangle if the lengths can be expressed as three consecutive even integers.

Solution:

1. UNDERSTAND. Read and reread the problem. Let's suppose that the length of one leg of the right triangle is 4 units. Then the other leg is the next even integer, or 6 units, and the hypotenuse of the triangle is the next even integer, or 8 units. Remember that the hypotenuse is the longest side. Let's see if a triangle with sides of these lengths forms a right triangle. To do this, we check to see whether the Pythagorean theorem holds true.

$$4^2 + 6^2 \stackrel{?}{=} 8^2$$

$$16 + 36 \stackrel{?}{=} 64$$

$$52 = 64 \qquad \text{False.}$$

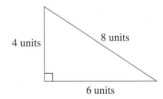

Our proposed numbers do not check, but we now have a better understanding of the problem.

We let x, $x + 2$, and $x + 4$ be three consecutive even integers. Since these integers represent lengths of the sides of a right triangle, we have

$$x = \text{one leg}$$
$$x + 2 = \text{other leg}$$
$$x + 4 = \text{hypotenuse (longest side)}$$

Answers

4. a. 5 and 7 or −5 and −3, **b.** 5 meters, 12 meters

2. TRANSLATE. By the Pythagorean theorem, we have that

$$(\text{hypotenuse})^2 = (\text{leg})^2 + (\text{leg})^2$$
$$(x + 4)^2 = (x)^2 + (x + 2)^2$$

3. SOLVE. Now we solve the equation.

$$(x + 4)^2 = x^2 + (x + 2)^2$$

$x^2 + 8x + 16 = x^2 + x^2 + 4x + 4$ Multiply.

$x^2 + 8x + 16 = 2x^2 + 4x + 4$ Combine like terms.

$x^2 - 4x - 12 = 0$ Write in standard form.

$(x - 6)(x + 2) = 0$ Factor.

$x - 6 = 0$ or $x + 2 = 0$ Set each factor equal to 0.

     $x = 6$ $x = -2$

4. INTERPRET. We discard $x = -2$ since length cannot be negative. If $x = 6$, then $x + 2 = 8$ and $x + 4 = 10$.

Check: Verify that $(\text{hypotenuse})^2 = (\text{leg})^2 + (\text{leg})^2$, or $10^2 = 6^2 + 8^2$, or $100 = 36 + 64$.

State: The sides of the right triangle have lengths 6 units, 8 units, and 10 units.

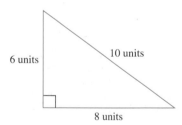

Focus On The Real World

Now that you have discovered a way to identify relatively small prime numbers with the Sieve of Eratosthenes, perhaps you are wondering whether there are very large primes. The answer is yes. Another ancient Greek mathematician, Euclid, proved that there are an infinite number of primes. Thus, there do exist huge numbers that are prime. Researchers call prime numbers with more than 1000 digits, *titanic primes*.

Knowing that very large prime numbers exist and finding them and proving that they are in fact prime are two very different things. David Slowinski, a research scientist at Silicon Graphics Inc.'s Cray Research unit and a discoverer or co-discoverer of seven titanic primes, compares the hunt for large primes with looking for a needle in a haystack. At the time that this book was written, the largest-known prime number, discovered November 13, 1996, had 420,921 digits which would fill up about 13.5 newspaper pages in standard-size type. The following newspaper article from the *San Jose Mercury News* describes how this prime number was found.

San Jose Mercury News
Published: Nov. 23, 1996

"Move Over, Supercomputers"
Net-linked band of off-the shelf PC
users finds largest example of
Mersenne prime number

By Dan Gillmor
Mercury News Computing Editor

In an undertaking previously reserved for powerful supercomputers, a worldwide band of personal computer users has found the largest known example of a special kind of prime number. Combining the Internet's global reach with hundreds of off-the-shelf PCs running Intel microprocessors, they divided an enormous problem into megabyte-sized tasks—and showed again how PCs are moving rapidly up computing's evolutionary scale.

Ten days ago in Paris, a PC operated by 29-year-old programmer Joel Armengaud found what is now the biggest-known Mersenne prime number. Mersenne primes are named after a 17th-century French monk, Father Marin Mersenne, who was fascinated by mathematics.

To test whether the nearly 421,000-digit number was such a prime, Armengaud was running a program written by Florida-based programmer George Woltman. Early this year, Woltman organized what he dubbed "The Great Internet Mersenne Prime Search," a hunt that has attracted more than 750 people from around the globe, some of whom have devoted more than one machine to the task.

A prime number is an integer greater than zero whose divisors are only itself and 1. (The number 2 is prime because it can only be divided evenly by 1 and 2, for example.) Mersenne primes take the form 2 to some power, minus 1—in other words, 2 multiplied by itself a certain number of times with 1 subtracted from the result.

The smallest Mersenne prime is 3, or 2 to the 2nd power $(2 \times 2 = 4)$ minus 1. The next largest "regular" prime number is 5, and the next largest Mersenne prime is 7, or 2 to the 3rd power $(2 \times 2 \times 2 = 8)$ minus 1.

The latest discovery—only the 35th known Mersenne prime—is 2 to

the 1,398,269th power (2 multiplied by itself 1,398,269 times) minus 1. It's 420,921 digits long, or more than 150 pages of single-spaced text.

The previous largest-known Mersenne prime, discovered earlier this year, is 2 to the 1,257,787th power minus 1. It was found by David Slowinski and Paul Gage, computer scientists at Silicon Graphics Inc.'s Cray Research unit, using a Cray supercomputer.

While Armengaud, Woltman and others in the search are in it mainly for fun, finding more efficient ways to crunch huge numbers is more than an abstract exercise. Related techniques have helped people handle other important computing tasks—such as modeling complex weather patterns and exploring for oil. And cryptography, the art of scrambling messages to ensure privacy, relies on the difficulty of performing complex operations on extremely large numbers.

And the way the latest Mersenne prime was found represents another step in computing's evolution.

GROUP ACTIVITIES

1. The following World Wide Web site contains information on the current status of the largest-known prime numbers: http://www.utm.edu/research/primes/largest.html. Visit this site and report on the five largest-known prime numbers. Be sure to include information about the numbers themselves, who found them, when they were found, and how they were found (if possible). How many numbers larger than that found by Armengaud, Woltman, et. al. in November 1996 have been found?

2. Explore some of the links from the Web site listed in Activity 1 to find information on the history of the search for the largest-known primes (look for information on the largest known prime by year). Summarize the trends in the methods used to hunt for titanic primes. When is a prime number with over 1,000,000 digits expected to be discovered?

Name _____ Section _____ Date _____

EXERCISE SET 4.7

A *See Examples 1 through 4 for all exercises. Represent each given condition using a single variable, x.*

1. The length and width of a rectangle whose length is 4 centimeters more than its width

2. The length and width of a rectangle whose length is twice its width

3. Two consecutive odd integers

4. Two consecutive even integers

5. The base and height of a triangle whose height is one more than four times its base

6. The base and height of a trapezoid whose base is three less than five times its height

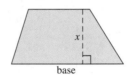

Use the information given to find the dimensions of each figure.

7.

The *area* of the square is 121 square units. Find the length of its sides.

8.

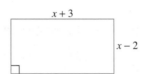

The *area* of the rectangle is 84 square inches. Find its length and width.

9.

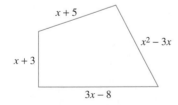

The *perimeter* of the quadrilateral is 120 centimeters. Find the lengths of the sides.

10.

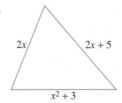

The *perimeter* of the triangle is 85 feet. Find the lengths of its sides.

329

11. base = 16 mi;
height = 6 mi

12. radius = 5 kilo-meters

13. 5 sec

14. 5 sec

15. length = 5 cm;
width = 6 cm

16. length = 16 in.;
width = 7 in.

17. 54 diagonals

18. 90 diagonals

330

11.

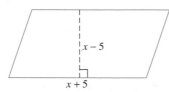

$x - 5$

$x + 5$

The *area* of the parallelogram is 96 square miles. Find its base and height.

12.

x

The *area* of the circle is 25π square kilometers. Find its radius.

Solve.

13. An object is thrown upward from the top of an 80-foot building with an initial velocity of 64 feet per second. The height h of the object after t seconds is given by the quadratic equation $h = -16t^2 + 64t + 80$. When will the object hit the ground?

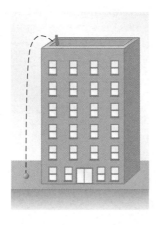

14. A hang glider pilot accidentally drops her compass from the top of a 400-foot cliff. The height h of the compass after t seconds is given by the quadratic equation $h = -16t^2 + 400$. When will the compass hit the ground?

15. The length of a rectangle is 7 centimeters less than twice its width. Its area is 30 square centimeters. Find the dimensions of the rectangle.

16. The length of a rectangle is 9 inches more than its width. Its area is 112 square inches. Find the dimensions of the rectangle.

The equation $D = \frac{1}{2}n(n - 3)$ gives the number of diagonals D for a polygon with n sides. For example, a polygon with 6 sides has $D = \frac{1}{2} \cdot 6(6 - 3)$ or $D = 9$ diagonals. (See if you can count all 9 diagonals. Some are shown in the figure.) Use this equation, $D = \frac{1}{2}n(n - 3)$, for Exercises 17–20.

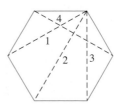

17. Find the number of diagonals for a polygon that has 12 sides.

18. Find the number of diagonals for a polygon that has 15 sides.

Name _____

19. Find the number of sides *n* for a polygon that has 35 diagonals.

20. Find the number of sides *n* for a polygon that has 14 diagonals.

Solve.

21. The sum of a number and its square is 132. Find the number.

22. The sum of a number and its square is 182. Find the number.

23. Two boats travel at a right angle to each other after leaving the same dock at the same time. One hour later the boats are 17 miles apart. If one boat travels 7 miles per hour faster than the other boat, find the rate of each boat.

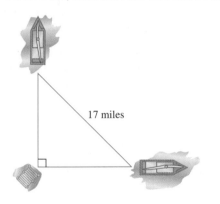

17 miles

24. The side of a square equals the width of a rectangle. The length of the rectangle is 6 meters longer than its width. The sum of the areas of the square and the rectangle is 176 square meters. Find the side of the square.

25. The sum of two numbers is 20, and the sum of their squares is 218. Find the numbers.

26. The sum of two numbers is 25, and the sum of their squares is 325. Find the numbers.

27. If the sides of a square are increased by 3 inches, the area becomes 64 square inches. Find the length of the sides of the original square.

x   *x* + 3

28. If the sides of a square are increased by 5 meters, the area becomes 100 square meters. Find the length of the sides of the original square.

x   *x* + 5

19. 10 sides

20. 7 sides

21. −12 or 11

22. −14 or 13

23. slow boat: 8 mph; fast boat: 15 mph

24. 8 m

25. 13 and 7

26. 10 and 15

27. 5 in.

28. 5 m

331

29. 12 mm, 16 mm, 20 mm

29. One leg of a right triangle is 4 millimeters longer than the smaller leg and the hypotenuse is 8 millimeters longer than the smaller leg. Find the lengths of the sides of the triangle.

30. One leg of a right triangle is 9 centimeters longer than the other leg and the hypotenuse is 45 centimeters. Find the lengths of the legs of the triangle.

30. 27 cm and 36 cm

31. The length of the base of a triangle is twice its height. If the area of the triangle is 100 square kilometers, find the height.

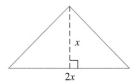

32. The height of a triangle is 2 millimeters less than the base. If the area is 60 square millimeters, find the base.

31. 10 km

32. 12 mm

33. Find the length of the shorter leg of a right triangle if the longer leg is 12 feet more than the shorter leg and the hypotenuse is 12 feet less than twice the shorter leg.

34. Find the length of the shorter leg of a right triangle if the longer leg is 10 miles more than the shorter leg and the hypotenuse is 10 miles less than twice the shorter leg.

33. 36 ft

35. An object is dropped from the top of the 625-foot-tall Waldorf-Astoria Hotel on Park Avenue in New York City. (*Source: World Almanac* research) The height h of the object after t seconds is given by the equation $h = -16t^2 + 625$. Find how many seconds pass before the object reaches the ground.

36. A 6-foot-tall person drops an object from the top of the Westin Peachtree Plaza in Atlanta, Georgia. The Westin building is 723 feet tall. (*Source: World Almanac* research) The height h of the object after t seconds is given by the equation $h = -16t^2 + 729$. Find how many seconds pass before the object reaches the ground.

34. 30 miles

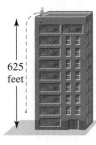

625 feet

723 feet

35. 6.25 seconds

36. 6.75 seconds

Name _____

REVIEW AND PREVIEW

The following double line graph shows a comparison of the number of farms in the United States and the size of the average farm. Use this graph to answer Exercises 37–43. See Section 1.8.

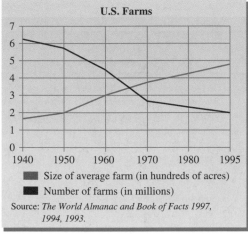

U.S. Farms

■ Size of average farm (in hundreds of acres)
■ Number of farms (in millions)

Source: *The World Almanac and Book of Facts 1997, 1994, 1993.*

37. Approximate the size of the average farm in 1940.

38. Approximate the size of the average farm in 1995.

39. Approximate the number of farms in 1940.

40. Approximate the number of farms in 1995.

41. Approximate the year that the colored lines in this graph intersect.

42. In your own words, explain the meaning of the point of intersection in the graph.

43. Describe the trends shown in this graph and speculate as to why these trends have occurred.

COMBINING CONCEPTS

44. According to the International America's Cup Class (IACC) rule, a sailboat competing in the America's Cup match must have a 110-foot-tall mast and a combined mainsail and jib sail area of 3000 square feet. (*Source*: America's Cup Organizing Committee) A design for an IACC-class sailboat calls for the mainsail to be 60% of the combined sail area. If the height of the triangular mainsail is 28 feet more than twice the length of the boom, find the length of the boom and the height of the mainsail.

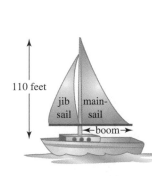

110 feet

jib sail   main-sail   ←boom→

Name _____

45. A rectangular pool is surrounded by a walk 4 meters wide. The pool is 6 meters longer than its width. If the total area is 576 square meters more than the area of the pool, find the dimensions of the pool.

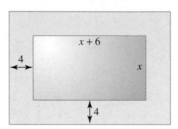

46. A rectangular garden is surrounded by a walk of uniform width. The area of the garden is 180 square yards. If the dimensions of the garden plus the walk are 16 yards by 24 yards, find the width of the walk.

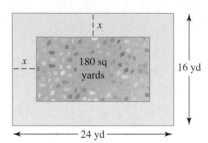

Internet Excursions

Go to http://www.prenhall.com/martin-gay

The United States Postal Service uses five-digit ZIP codes to simplify mail distribution. Each ZIP code corresponds to a unique post office location. If you know a ZIP code, you can look up the city and state associated with it by visiting the above World Wide Web address where you will gain access to the United States Postal Service Web site, or a related site.

47. The first three digits in the ZIP codes of two cities are 601. The remaining digits of the two ZIP codes are consecutive integers whose product is 870. Find the names of the cities and their states.

48. The first three digits in the ZIP codes of two cities are 448. The remaining digits of the two ZIP codes are consecutive even integers whose product is 1848. Find the names of the cities and their states.

CHAPTER 4 ACTIVITY
FACTORING

Choosing Among Building Options: Shaunesa has just had a 10-foot-by-15-foot, in-ground swimming pool installed in her backyard. She has $3000 left from the building project that she would like to spend on surrounding the pool with a patio, equally wide on all sides (see the figure). She has talked to several local suppliers about options for building this patio and must choose among the following.

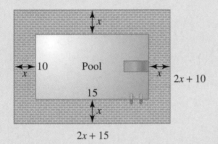

Option	Material	Price
A	Poured cement	$5 per square foot
B	Brick	$7.50 per square foot plus a $30 flat fee for delivering the bricks
C	Outdoor carpeting	$4.50 per square foot plus $10.86 per foot of the pool's perimeter to install an edging

1. Find the area of the swimming pool.
    150 sq. ft

2. Write an algebraic expression for the total area of the region containing both the pool and patio.
    $4x^2 + 50x + 150$

3. Use subtraction to find an algebraic expression for the area of just the patio (not including the pool). $4x^2 + 50x$

4. Find the perimeter of the swimming pool alone.
    50 ft

5. For each patio material option, write an algebraic expression for the total cost of installing the patio based on its area and the given price information.   A: $20x^2 + 250x$;
    B: $30x^2 + 375x + 30$;
    C: $18x^2 + 225x + 543$

6. If Shaunesa plans to spend the entire $3000 she has saved for the patio, how wide would be the patio in option A?   7.5 ft

7. If Shaunesa plans to spend the entire $3000 she has saved for the patio, how wide would be the patio in option B?   5.5 ft

8. If Shaunesa plans to spend the entire $3000 she has saved for the patio, how wide would be the patio in option C?   7 ft

9. Which option should Shaunesa choose? Why? Discuss the pros and cons of each option.
    answers may vary

CHAPTER 4 HIGHLIGHTS

DEFINITIONS AND CONCEPTS	EXAMPLES

SECTION 4.1 THE GREATEST COMMON FACTOR

Factoring is the process of writing an expression as a product.

The GCF of a list of common variables raised to powers is the variable raised to the smallest exponent in the list.

The GCF of a list of terms is the product of all common factors.

Factor: $6 = 2 \cdot 3$
Factor: $x^2 + 5x + 6 = (x + 2)(x + 3)$

The GCF of z^5, z^3, and z^{10} is z^3.

Find the GCF of $8x^2y$, $10x^3y^2$, and $50x^2y^3$.

$$8x^2y = 2 \cdot 2 \cdot 2 \cdot x^2 \cdot y$$
$$10x^3y^2 = 2 \cdot 5 \cdot x^3 \cdot y^2$$
$$50x^2y^3 = 2 \cdot 5 \cdot 5 \cdot x^2 \cdot y^3$$
$$\text{GCF} = 2 \cdot x^2 \cdot y \quad \text{or} \quad 2x^2y$$

TO FACTOR BY GROUPING

Step 1. Group the terms into two groups of two terms.

Step 2. Factor out the GCF from each group.

Step 3. If there is a common binomial factor, factor it out.

Step 4. If not, rearrange the terms and try Steps 1–3 again.

Factor: $10ax + 15a - 6xy - 9y$

Step 1. $(10ax + 15a) + (-6xy - 9y)$
Step 2. $5a(2x + 3) - 3y(2x + 3)$
Step 3. $(2x + 3)(5a - 3y)$

SECTION 4.2 FACTORING TRINOMIALS OF THE FORM $x^2 + bx + c$

The sum of these numbers is b.

$$x^2 + bx + c = (x + \square)(x + \square)$$

The product of these numbers is c.

Factor: $x^2 + 7x + 12$

$3 + 4 = 7 \qquad 3 \cdot 4 = 12$

$$x^2 + 7x + 12 = (x + 3)(x + 4)$$

SECTION 4.3 FACTORING TRINOMIALS OF THE FORM $ax^2 + bx + c$

To factor $ax^2 + bx + c$, try various combinations of factors of ax^2 and c until a middle term of bx is obtained when checking.

Factor: $3x^2 + 14x - 5$

Factors of $3x^2$: $3x$, x

Factors of -5: -1, 5 and 1, -5.

$(3x - 1)(x + 5)$

$$\begin{array}{r} -1x \\ + 15x \\ \hline 14x \end{array}$$ **Correct** middle term

SECTION 4.4 FACTORING TRINOMIALS OF THE FORM $ax^2 + bx + c$ BY GROUPING

TO FACTOR $ax^2 + bx + c$ BY GROUPING

Step 1. Find two numbers whose product is $a \cdot c$ and whose sum is b.

Step 2. Rewrite bx, using the factors found in step 1.

Step 3. Factor by grouping.

Factor: $3x^2 + 14x - 5$

Step 1. Find two numbers whose product is $3 \cdot (-5)$ or -15 and whose sum is 14. They are 15 and -1.

Step 2. $3x^2 + 14x - 5$
$= 3x^2 + 15x - 1x - 5$

Step 3. $= 3x(x + 5) - 1(x + 5)$
$= (x + 5)(3x - 1)$

SECTION 4.5 FACTORING PERFECT SQUARE TRINOMIALS AND THE DIFFERENCE OF TWO SQUARES

A **perfect square trinomial** is a trinomial that is the square of some binomial.

PERFECT SQUARE TRINOMIAL = SQUARE OF BINOMIAL

$$x^2 + 4x + 4 = (x + 2)^2$$
$$25x^2 - 10x + 1 = (5x - 1)^2$$

Factoring Perfect Square Trinomials

$$a^2 + 2ab + b^2 = (a + b)^2$$
$$a^2 - 2ab + b^2 = (a - b)^2$$

Factor:

$x^2 + 6x + 9 =$
$x^2 + 2(x \cdot 3) + 3^2 = (x + 3)^2$
$4x^2 - 12x + 9 =$
$(2x)^2 - 2(2x \cdot 3) + 3^2 = (2x - 3)^2$

Difference of Two Squares

$$a^2 - b^2 = (a + b)(a - b)$$

Factor:

$$x^2 - 9 = x^2 - 3^2 = (x + 3)(x - 3)$$

SECTION 4.6 SOLVING QUADRATIC EQUATIONS BY FACTORING

A **quadratic equation** is an equation that can be written in the form $ax^2 + bx + c = 0$ with a not 0.
The form $ax^2 + bx + c = 0$ is called the **standard form** of a quadratic equation.

Zero Factor Property
If a and b are real numbers and if $ab = 0$, then $a = 0$ or $b = 0$.

TO SOLVE QUADRATIC EQUATIONS BY FACTORING

Step 1. Write the equation in standard form so that one side of the equation is 0.

Step 2. Factor completely.

Step 3. Set each factor containing a variable equal to 0.

Step 4. Solve the resulting equations.

Step 5. Check in the original equation.

Quadratic Equation	Standard Form
$x^2 = 16$	$x^2 - 16 = 0$
$y = -2y^2 + 5$	$2y^2 + y - 5 = 0$

If $(x + 3)(x - 1) = 0$, then $x + 3 = 0$ or $x - 1 = 0$.

Solve: $3x^2 = 13x - 4$

Step 1. $3x^2 - 13x + 4 = 0$

Step 2. $(3x - 1)(x - 4) = 0$

Step 3. $3x - 1 = 0$ or $x - 4 = 0$

Step 4. $3x = 1$ $x = 4$
$x = \dfrac{1}{3}$

Step 5. Check both $\frac{1}{3}$ and 4 in the original equation.

PROBLEM-SOLVING STEPS

A garden is in the shape of a rectangle whose length is two feet more than its width. If the area of the garden is 35 square feet, find its dimensions.

1. UNDERSTAND the problem.

1. Read and reread the problem. Guess a solution and check your guess. Draw a diagram.

Let x be the width of the rectangular garden. Then $x + 2$ is the length.

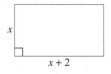

2. TRANSLATE.

2.

length	·	width	=	area
↓		↓		↓
$(x + 2)$	·	x	=	35

3. SOLVE.

3.
$$(x + 2)x = 35$$
$$x^2 + 2x - 35 = 0$$
$$(x - 5)(x + 7) = 0$$
$$x - 5 = 0 \quad \text{or} \quad x + 7 = 0$$
$$x = 5 \qquad\qquad x = -7$$

4. INTERPRET.

4. Discard the solution of -7 since x represents width.

Check: If x is 5 feet then $x + 2 = 5 + 2 = 7$ feet. The area of a rectangle whose width is 5 feet and whose length is 7 feet is (5 feet)(7 feet) or 35 square feet.

State: The garden is 5 feet by 7 feet.

Name _____ Section _____ Date _____

CHAPTER 4 REVIEW

(4.1) *Complete each factoring.*

1. $6x^2 - 15x = 3x(2x - 5)$

2. $4x^5 + 2x - 10x^4 = 2x(2x^4 + 1 - 5x^3)$

Factor out the GCF from each polynomial.

3. $5m + 30$ $5(m + 6)$

4. $20x^3 + 12x^2 + 24x$ $4x(5x^2 + 3x + 6)$

5. $3x(2x + 3) - 5(2x + 3)$ $(2x + 3)(3x - 5)$

6. $5x(x + 1) - (x + 1)$ $(x + 1)(5x - 1)$

Factor each polynomial by grouping.

7. $3x^2 - 3x + 2x - 2$ $(x - 1)(3x + 2)$

8. $6x^2 + 10x - 3x - 5$ $(2x - 1)(3x + 5)$

9. $3a^2 + 9ab + 3b^2 + ab$ $(a + 3b)(3a + b)$

(4.2) *Factor each trinomial.*

10. $x^2 + 6x + 8$ $(x + 4)(x + 2)$

11. $x^2 - 11x + 24$ $(x - 8)(x - 3)$

12. $x^2 + x + 2$ prime

13. $x^2 - 5x - 6$ $(x - 6)(x + 1)$

14. $x^2 + 2x - 8$ $(x + 4)(x - 2)$

15. $x^2 + 4xy - 12y^2$ $(x + 6y)(x - 2y)$

16. $x^2 + 8xy + 15y^2$ $(x + 5y)(x + 3y)$

17. $72 - 18x - 2x^2$ $2(3 - x)(12 + x)$

18. $32 + 12x - 4x^2$ $4(8 + 3x - x^2)$

19. $5y^3 - 50y^2 + 120y$ $5y(y - 6)(y - 4)$

20. To factor $x^2 + 2x - 48$, think of two numbers whose product is _____ and whose sum is _____.
$-48, 2$

21. What is the first step to factoring $3x^2 + 15x + 30$?
factor out the GCF, 3

(4.3) or (4.4) *Factor each trinomial.*

22. $2x^2 + 13x + 6$ $(2x + 1)(x + 6)$

23. $4x^2 + 4x - 3$ $(2x + 3)(2x - 1)$

24. $6x^2 + 5xy - 4y^2$ $(3x + 4y)(2x - y)$

25. $x^2 - x + 2$ prime

26. $2x^2 - 23x - 39$ $(2x + 3)(x - 13)$

27. $18x^2 - 9xy - 20y^2$ $(6x + 5y)(3x - 4y)$

28. $10y^3 + 25y^2 - 60y$ $5y(2y - 3)(y + 4)$

Write the perimeter of each figure as a simplified polynomial. Then factor each polynomial.

29.

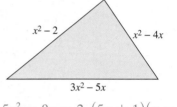

$5x^2 - 9x - 2, (5x + 1)(x - 2)$

30.

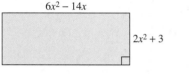

$16x^2 - 28x + 6, 2(4x - 1)(2x - 3)$

(4.5) *Determine whether each polynomial is a perfect square trinomial.*

31. $x^2 + 6x + 9$ yes

32. $x^2 + 8x + 64$ no

33. $9m^2 - 12m + 16$ no

34. $4y^2 - 28y + 49$ yes

Determine whether each binomial is a difference of two squares.

35. $x^2 - 9$ yes

36. $x^2 + 16$ no

37. $4x^2 - 25y^2$ yes

38. $9a^3 - 1$ no

Factor each polynomial completely.

39. $x^2 - 81$ $(x - 9)(x + 9)$

40. $x^2 + 12x + 36$ $(x + 6)^2$

41. $4x^2 - 9$ $(2x - 3)(2x + 3)$

42. $9t^2 - 25s^2$ $(3t - 5s)(3t + 5s)$

43. $16x^2 + y^2$ prime

44. $n^2 - 18n + 81$ $(n - 9)^2$

45. $3r^2 + 36r + 108$ $3(r + 6)^2$

46. $9y^2 - 42y + 49$ $(3y - 7)^2$

47. $5m^8 - 5m^6$ $5m^6(m + 1)(m - 1)$

48. $4x^2 - 28xy + 49y^2$ $(2x - 7y)^2$

49. $3x^2y + 6xy^2 + 3y^3$ $3y(x + y)^2$

50. $16x^4 - 1$ $(2x - 1)(2x + 1)(4x^2 + 1)$

(4.6) *Solve each equation.*

51. $(x + 6)(x - 2) = 0$ $x = -6, x = 2$

52. $3x(x + 1)(7x - 2) = 0$ $x = 0, x = -1, x = \dfrac{2}{7}$

53. $4(5x + 1)(x + 3) = 0$ $x = -\dfrac{1}{5}, x = -3$

54. $x^2 + 8x + 7 = 0$ $x = -7, x = -1$

55. $x^2 - 2x - 24 = 0$ $x = -4, x = 6$

56. $x^2 + 10x = -25$ $x = -5$

57. $x(x - 10) = -16$ $x = 2, x = 8$

58. $(3x - 1)(9x^2 + 3x + 1) = 0$ $x = \dfrac{1}{3}$

59. $56x^2 - 5x - 6 = 0$ $x = -\dfrac{2}{7}, x = \dfrac{3}{8}$

60. $m^2 = 6m$ $m = 0, m = 6$

61. $r^2 = 25$ $r = 5, r = -5$

62. Write a quadratic equation that has the two solutions 4 and 5. $x^2 - 9x + 20 = 0$

(4.7) *Use the given information to choose the correct dimensions.*

63. The perimeter of a rectangle is 24 inches. The length is twice the width. Find the dimensions of the rectangle.   (c)
 a. 5 inches by 7 inches

 b. 5 inches by 10 inches

 c. 4 inches by 8 inches

 d. 2 inches by 10 inches

64. The area of a rectangle is 80 meters. The length is one more than three times the width. Find the dimensions of the rectangle.   (d)
 a. 8 meters by 10 meters

 b. 4 meters by 13 meters

 c. 4 meters by 20 meters

 d. 5 meters by 16 meters

Use the given information to find the dimensions of each figure.

65.

x

The *area* of the square is 81 square units. Find the length of a side.   9 units

66.

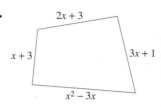

2x + 3

x + 3   3x + 1

x² − 3x

The *perimeter* of the quadrilateral is 47 units. Find the lengths of the sides.   8 units, 13 units, 16 units, 10 units

Solve.

67. A flag for a local organization is in the shape of a rectangle whose length is 15 inches less than twice its width. If the area of the flag is 500 square inches, find its dimensions.   width: 20 in.; length: 25 in.

x

68. The base of a triangular sail is four times its height. If the area of the triangle is 162 square yards, find the base.   36 yd

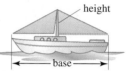

height

base

69. Find two consecutive positive integers whose product is 380.   19 and 20

70. A rocket is fired from the ground with an initial velocity of 440 feet per second. Its height h after t seconds is given by the equation $h = -16t^2 + 440t$.

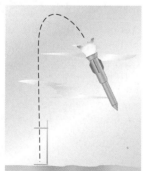

a. Find how many seconds pass before the rocket reaches a height of 2800 feet. Explain why two answers are obtained.   17.5 sec. and 10 sec.; answers may vary

b. Find how many seconds pass before the rocket reaches the ground again.   27.5 sec.

71. An architect's squaring instrument is in the shape of a right triangle. Find the length of the longer leg of the right triangle if the hypotenuse is 8 centimeters longer than the longer leg and the shorter leg is 8 centimeters shorter than the longer leg.   32 cm

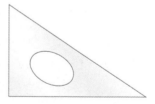

Name _____ Section _____ Date _____

CHAPTER 4 TEST

Factor each polynomial completely. If a polynomial cannot be factored, write "prime."

1. $9x^2 - 3x$

2. $x^2 + 11x + 28$

3. $49 - m^2$

4. $y^2 + 22y + 121$

5. $x^4 - 16$

6. $4(a + 3) - y(a + 3)$

7. $x^2 + 4$

8. $y^2 - 8y - 48$

9. $3a^2 + 3ab - 7a - 7b$

10. $3x^2 - 5x + 2$

11. $180 - 5x^2$

12. $3x^3 - 21x^2 + 30x$

13. $6t^2 - t - 5$

14. $xy^2 - 7y^2 - 4x + 28$

15. $x - x^5$

16. $x^2 + 14xy + 24y^2$

Solve each equation.

17. $(x - 3)(x + 9) = 0$

18. $x^2 + 10x + 24 = 0$

19. $x^2 + 5x = 14$

20. $3x(2x - 3)(3x + 4) = 0$

21. $5t^3 - 45t = 0$

22. $3x^2 = -12x$

23. $t^2 - 2t - 15 = 0$

24. $(x - 1)(3x^2 - x - 2) = 0$

1. $3x(3x - 1)$

2. $(x + 7)(x + 4)$

3. $(7 - m)(7 + m)$

4. $(y + 11)^2$

5. $(x^2 + 4)(x + 2)(x - 2)$

6. $(4 - y)(a + 3)$

7. prime

8. $(y - 12)(y + 4)$

9. $(3a - 7)(a + b)$

10. $(3x - 2)(x - 1)$

11. $5(6 - x)(6 + x)$

12. $3x(x - 5)(x - 2)$

13. $(6t + 5)(t - 1)$

14. $(y - 2)(y + 2)(x - 7)$

15. $x(1 - x)(1 + x)(1 + x^2)$

16. $(x + 12y)(x + 2y)$

17. $x = 3, x = -9$

18. $x = -6, x = -4$

19. $x = -7, x = 2$

20. $x = 0, x = \frac{3}{2}, x = -\frac{4}{3}$

21. $t = 0, t = 3, t = -3$

22. $x = 0, x = -4$

23. $t = -3, t = 5$

24. $x = -\frac{2}{3}, x = 1$

343

25. The *area* of the rectangle is 54 square units. Find the dimensions of the rectangle.

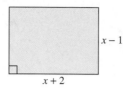

$x - 1$

$x + 2$

Solve.

26. A deck for a home is in the shape of a triangle. The length of the base of the triangle is 9 feet longer than its height. If the area of the triangle is 68 square feet find the length of the base.

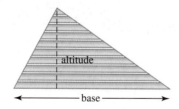

altitude

base

27. The sum of two numbers is 17, and the sum of their squares is 145. Find the numbers.

28. An object is dropped from the top of the Woolworth Building on Broadway in New York City. The height h of the object after t seconds is given by the equation $h = -16t^2 + 784$.

Find how many seconds pass before the object reaches the ground.

29. Find the lengths of the sides of a right triangle if the hypotenuse is 10 centimeters longer than the shorter leg and 5 centimeters longer than the longer leg.

CUMULATIVE REVIEW

1. Translate each sentence into a mathe-matical statement.

   a. Nine is less than or equal to eleven.

   b. Eight is greater than one.

   c. Three is not equal to four.

2. Decide whether 2 is a solution of $3x + 10 = 8x$.

3. Subtract 8 from -4

4. If $x = -2$ and $y = -4$, evaluate each expression.

   a. $5x - y$

   b. $x^3 - y^2$

Simplify each expression by combining like terms.

5. $2x + 3x + 5 + 2$

6. $-5a - 3 + a + 2$

7. $2.3x + 5x - 6$

8. Solve: $-3x = 33$

9. Solve: $3(x - 4) = 3x - 12$

10. Solve for l: $V = lwh$

11. Solve for x: $\dfrac{x - 5}{3} = \dfrac{x + 2}{5}$

Simplify each expression.

12. $(5^3)^6$

13. $(y^8)^2$

Simplify the following expressions. Write each result using positive exponents only.

14. $\dfrac{(x^3)^4 x}{x^7}$

15. $(y^{-3}z^6)^{-6}$

16. $\dfrac{x^{-7}}{(x^4)^3}$

ANSWERS

1. a. $9 \le 11$

   b. $8 > 1$

   c. $3 \ne 4$ (Sec. 1.1, Ex. 7)

2. solution (Sec. 1.2, Ex. 7)

3. -12 (Sec 1.4, Ex. 5)

4. a. -6

   b. -24 (Sec. 1.5, Ex. 8)

5. $5x + 7$ (Sec. 2.1, Ex. 4)

6. $-4a - 1$ (Sec 2.1, Ex. 5)

7. $7.3x - 6$ (Sec. 2.1, Ex. 7)

8. $x = -11$ (Sec. 2.3, Ex. 3)

9. every real number (Sec. 2.4, Ex. 7)

10. $l = \dfrac{V}{wh}$ (Sec. 2.6, Ex. 3)

11. $x = \dfrac{31}{2}$ (Sec. 2.7, Ex. 6)

12. 5^{18} (Sec. 3.1, Ex. 14)

13. y^{16} (Sec. 3.1, Ex. 15)

14. x^6 (Sec. 3.2, Ex. 9)

15. $\dfrac{y^{18}}{z^{36}}$ (Sec. 3.2, Ex. 11)

16. $\dfrac{1}{x^{19}}$ (Sec. 3.2, Ex. 13)

345

Name _____

Simplify each polynomial by combining any like terms.

17. $-3x + 7x$ **18.** $11x^2 + 5 + 2x^2 - 7$ **19.** Multiply: $(2x - y)^2$

Use a special product to square each binomial.

20. $(t + 2)^2$ **21.** $(x^2 - 7y)^2$

22. Divide: $\dfrac{8x^2y^2 - 16xy + 2x}{4xy}$ **23.** Factor: $5(x + 3) + y(x + 3)$

24. Factor: $x^4 + 5x^2 + 6$ **25.** Factor by grouping: $6x^2 - 2x - 20$

Rational Expressions

In this chapter, we expand our knowledge of algebraic expressions to include algebraic fractions, called **rational expressions**. We explore the operations of addition, subtraction, multiplication, and division using principles similar to the principles for number fractions.

Rational expressions can be found in formulas for many real-world situations. In Exercise 70 on page 356 and in Exercise 45 on page 389, you will find a rational expression used in a formula for rainfall intensity. Rainfall intensity describes the depth of rain that falls during a certain time period. According to the National Geographic Society, the rainiest location in the world is Mount Waialeale, Hawaii. Mount Waialeale receives an average of 460 inches of rain per year. Most of the United States only receives an average of 10 to 60 inches of rain per year.

Rainfall also varies in intensity from storm to storm. Most rainfall can be classified on a scale ranging from 0.10 inches per hour (light rain) to 0.30 inches per hour (heavy rain). However, the intensities on this scale are occasionally exceeded by particularly strong storms.

1. $x = -1, x = 10$; (5.1B)

2. $\dfrac{4}{x + 2}$; (5.1C)

3. $10(x + 2)(x + 3)$; (5.3B)

4. 3; (5.2A)

5. $\dfrac{5(x + 5)}{x^3(x - 5)}$; (5.2B)

6. $\dfrac{1}{b - 11}$; (5.3A)

7. $\dfrac{7 - 4x}{x - 1}$; (5.4A)

8. $\dfrac{9}{x - 5}$; (5.4A)

9. $\dfrac{x^2 + 12}{(x + 4)(x - 4)(x - 3)}$; (5.4A)

10. $b = -7$; (5.5A)

11. no solution; (5.5A)

12. $y = -1$; (5.5A)

13. $b = \dfrac{2A}{h}$; (5.5B)

14. $\dfrac{15n^6}{m^3}$; (5.7A)

15. $4a - 1$; (5.7A, B)

16. $x = 5$; (5.6D)

17. 2 or 5; (5.6A)

18. $3\frac{1}{13}$ hr.; (5.6B)

19. 250 mph; (5.6C)

CHAPTER 5 PRETEST

1. Find any real numbers for which the following expression is undefined.

$$\frac{x + 2}{x^2 - 9x - 10}$$

2. Simplify: $\dfrac{4x + 32}{x^2 + 10x + 16}$

3. Find the LCD for the following rational expressions.

$$\frac{1}{5x + 10}, \quad \frac{3}{2x^2 + 10x + 12}$$

Perform the indicated operations and simplify if possible.

4. $\dfrac{y^2 - 8y + 7}{2y - 14} \cdot \dfrac{6y + 18}{y^2 + 2y - 3}$

5. $\dfrac{5x^3}{x^2 - 25} \div \dfrac{x^6}{(x + 5)^2}$

6. $\dfrac{b}{b^2 - 9b - 22} + \dfrac{2}{b^2 - 9b - 22}$

7. $\dfrac{3}{x - 1} - 4$

8. $\dfrac{2}{x - 5} - \dfrac{7}{5 - x}$

9. $\dfrac{x}{x^2 - 16} + \dfrac{3}{x^2 - 7x + 12}$

Solve each equation.

10. $\dfrac{5}{b} + \dfrac{3}{5} = \dfrac{4}{5b}$

11. $9 + \dfrac{7}{d - 7} = \dfrac{d}{d - 7}$

12. $\dfrac{4y + 5}{y^2 + 5y + 6} + \dfrac{3}{y + 3} = \dfrac{2}{y + 2}$

13. Solve the equation for the indicated variable. $\dfrac{2A}{b} = h$; for b.

Simplify each complex fraction.

14. $\dfrac{\dfrac{12m^3}{5n^2}}{\dfrac{4m^6}{25n^8}}$

15. $\dfrac{16 - \dfrac{1}{a^2}}{\dfrac{4}{a} + \dfrac{1}{a^2}}$

16. Given that the two triangles are similar, find x.

17. A number added to the product of 10 and the reciprocal of the number equals 7. Find the number.

18. Sonya can wash the windows in her house in 5 hours. Her daughter completes the same job in 8 hours. Find how long it takes if they work together.

19. Tom flies his airplane 495 miles with a tail wind of 25 miles per hour. Against the wind, he flies only 405 miles. Find the rate of the plane in still air.

5.1   SIMPLIFYING RATIONAL EXPRESSIONS

A   EVALUATING RATIONAL EXPRESSIONS

A rational number is a number that can be written as a quotient of integers. A **rational expression** is also a quotient; it is a quotient of polynomials. Examples are

$$\frac{3y^3}{8}, \quad \frac{-4p}{p^3 + 2p + 1}, \quad \frac{5x^2 - 3x + 2}{3x + 7}$$

RATIONAL EXPRESSION

A rational expression is an expression that can be written in the form

$$\frac{P}{Q}$$

where P and Q are polynomials and $Q \neq 0$.

Rational expressions have different numerical values depending on what value replaces the variable.

Example 1   Find the numerical value of $\frac{x + 4}{2x - 3}$ for each replacement value.

a. $x = 5$ **b.** $x = -2$

Solution: **a.** We replace each x in the expression with 5 and then simplify.

$$\frac{x + 4}{2x - 3} = \frac{5 + 4}{2(5) - 3} = \frac{9}{10 - 3} = \frac{9}{7}$$

b. We replace each x in the expression with -2 and then simplify.

$$\frac{x + 4}{2x - 3} = \frac{-2 + 4}{2(-2) - 3} = \frac{2}{-7} \quad \text{or} \quad -\frac{2}{7}$$ ▬▬▬

In the example above, we wrote $\frac{2}{-7}$ as $-\frac{2}{7}$. For a negative fraction such as $\frac{2}{-7}$, recall from Section 1.6 that

$$\frac{2}{-7} = \frac{-2}{7} = -\frac{2}{7}$$

In general, for any fraction

$$\frac{-a}{b} = \frac{a}{-b} = -\frac{a}{b}, \quad b \neq 0$$

This is also true for rational expressions. For example,

$$\underbrace{\frac{-(x + 2)}{x}}_{\uparrow} = \frac{x + 2}{-x} = -\frac{x + 2}{x}$$

Notice the parentheses.

B   IDENTIFYING WHEN A RATIONAL EXPRESSION IS UNDEFINED

In the preceding box, notice that we wrote $b \neq 0$ for the denominator b. The denominator of a rational expression must not equal 0 since division by

---

Objectives

A Find the value of a rational expression given a replacement number.

B Identify values for which a rational expression is undefined.

C Simplify, or write rational expressions in lowest terms.

SSM   CD-ROM   Video 5.1

Practice Problem 1

Find the value of $\frac{x - 3}{5x + 1}$ for each replacement value.

a. $x = 4$

b. $x = -3$

TEACHING TIP

In Example 1, have students evaluate, with a scientific or graphing calculator, to see if they get the same result. If they do not get the same result, show them how to use parentheses to group the numerator and the denominator.

Answers

1. a. $\frac{1}{21}$,   **b.** $\frac{-6}{-14} = \frac{3}{7}$

0 is not defined. This means we must be careful when replacing the variable in a rational expression by a number. For example, suppose we replace x with 5 in the rational expression $\dfrac{2 + x}{x - 5}$. The expression becomes

$$\frac{2 + x}{x - 5} = \frac{2 + 5}{5 - 5} = \frac{7}{0}$$

But division by 0 is undefined. Therefore, in this expression we can allow x to be any real number *except* 5. **A rational expression is undefined for values that make the denominator 0.**

Practice Problem 2

Are there any values for x for which each rational expression is undefined?

a. $\dfrac{x}{x + 2}$ b. $\dfrac{x - 3}{x^2 + 5x + 4}$

c. $\dfrac{x^2 - 3x + 2}{5}$

Example 2   Are there any values for x for which each expression is undefined?

a. $\dfrac{x}{x - 3}$ b. $\dfrac{x^2 + 2}{x^2 - 3x + 2}$

c. $\dfrac{x^3 - 6x^2 - 10x}{3}$

Solution: To find values for which a rational expression is undefined, we find values that make the denominator 0.

a. The denominator of $\dfrac{x}{x - 3}$ is 0 when $x - 3 = 0$ or when $x = 3$. Thus, when $x = 3$, the expression $\dfrac{x}{x - 3}$ is undefined.

b. We set the denominator equal to 0.

$$x^2 - 3x + 2 = 0$$
$$(x - 2)(x - 1) = 0 \qquad \text{Factor.}$$
$$x - 2 = 0 \quad \text{or} \quad x - 1 = 0 \quad \text{Set each factor equal to 0.}$$
$$x = 2 \qquad\qquad x = 1 \quad \text{Solve.}$$

Thus, when $x = 2$ or $x = 1$, the denominator $x^2 - 3x + 2$ is 0. So the rational expression $\dfrac{x^2 + 2}{x^2 - 3x + 2}$ is undefined when $x = 2$ or when $x = 1$.

c. The denominator of $\dfrac{x^3 - 6x^2 - 10x}{3}$ is never 0, so there are no values of x for which this expression is undefined.

C SIMPLIFYING RATIONAL EXPRESSIONS

A fraction is said to be written in lowest terms or simplest form when the numerator and denominator have no common factors other than 1 (or -1). For example, the fraction $\frac{7}{10}$ is in lowest terms since the numerator and denominator have no common factors other than 1 (or -1).

The process of writing a rational expression in lowest terms or simplest form is called **simplifying** a rational expression. The following **fundamental principle of rational expressions** is used to simplify a rational expression.

FUNDAMENTAL PRINCIPLE OF RATIONAL EXPRESSIONS

If $\dfrac{P}{Q}$ is a rational expression and R is a nonzero polynomial, then

$$\frac{PR}{QR} = \frac{P}{Q}$$

Answers

2. a. $x = -2$,   **b.** $x = -4, x = -1$,   **c.** no

Simplifying a rational expression is similar to simplifying a fraction.

Simplify: $\dfrac{15}{20}$

$$\dfrac{15}{20} = \dfrac{3 \cdot 5}{2 \cdot 2 \cdot 5} \qquad \text{Factor the numerator and the denominator.}$$

$$= \dfrac{3 \cdot 5}{2 \cdot 2 \cdot 5} \qquad \text{Look for common factors.}$$

$$= \dfrac{3}{2 \cdot 2} = \dfrac{3}{4} \qquad \text{Apply the fundamental principle.}$$

Simplify: $\dfrac{x^2 - 9}{x^2 + x - 6}$

$$\dfrac{x^2 - 9}{x^2 + x - 6} = \dfrac{(x - 3)(x + 3)}{(x - 2)(x + 3)} \qquad \text{Factor the numerator and the denominator.}$$

$$= \dfrac{(x - 3)(x + 3)}{(x - 2)(x + 3)} \qquad \text{Look for common factors.}$$

$$= \dfrac{x - 3}{x - 2} \qquad \text{Apply the fundamental principle.}$$

Thus, the rational expression $\dfrac{x^2 - 9}{x^2 + x - 6}$ has the same value as the rational expression $\dfrac{x - 3}{x - 2}$ for all values of x except 2 and -3. (Remember that when x is 2, the denominator of both rational expressions is 0 and when x is -3, the original rational expression has a denominator of 0.)

As we simplify rational expressions, we will assume that the simplified rational expression is equal to the original rational expression for all real numbers except those for which either denominator is 0. The following steps may be used to simplify rational expressions.

TEACHING TIP

It may be useful for students to use the Table feature on a graphing calculator to evaluate $\dfrac{x^2 - 9}{x^2 + x - 6}$ and $\dfrac{x - 3}{x - 2}$ for various values of x to verify that the expressions are the same for all values except -3 and 2. Then ask students which expression is easier to evaluate? Finally, point out that one reason to simplify rational expressions before evaluating them is to simplify your work.

> **TO SIMPLIFY A RATIONAL EXPRESSION**
>
> **Step 1.** Completely factor the numerator and denominator.
>
> **Step 2.** Apply the fundamental principle of rational expressions to divide out common factors.

Example 3   Simplify: $\dfrac{5x - 5}{x^3 - x^2}$

Solution: To begin, we factor the numerator and denominator if possible. Then we apply the fundamental principle.

$$\dfrac{5x - 5}{x^3 - x^2} = \dfrac{5(x - 1)}{x^2(x - 1)} = \dfrac{5}{x^2}$$

Example 4   Simplify: $\dfrac{x^2 + 8x + 7}{x^2 - 4x - 5}$

Solution: We factor the numerator and denominator and then apply the fundamental principle.

$$\dfrac{x^2 + 8x + 7}{x^2 - 4x - 5} = \dfrac{(x + 7)(x + 1)}{(x - 5)(x + 1)} = \dfrac{x + 7}{x - 5}$$

Practice Problem 3

Simplify: $\dfrac{x^4 + x^3}{5x + 5}$

Practice Problem 4

Simplify: $\dfrac{x^2 + 11x + 18}{x^2 + x - 2}$

Answers

3. $\dfrac{x^3}{5}$,   **4.** $\dfrac{x + 9}{x - 1}$

Practice Problem 5

Simplify: $\dfrac{x^2 + 10x + 25}{x^2 + 5x}$

TEACHING TIP

Give students a concrete example which illustrates why $\dfrac{2x}{x}$ can be simplified to 2 and $\dfrac{x+2}{x}$ cannot be simplified to 2. For example, evaluate each expression for $x = 1$ and then $x = 5$. Have students notice that the first expression evaluates to 2 for both values of x but the second expression does not.

Practice Problem 6

Simplify: $\dfrac{x+5}{x^2 - 25}$

Practice Problem 7

Simplify each rational expression.

a. $\dfrac{x+4}{4+x}$ b. $\dfrac{x-4}{4-x}$

✓ Concept Check

Recall that the fundamental principle applies to common factors only. Which of the following are *not* true? Explain why.

a. $\dfrac{3-1}{3+5} = -\dfrac{1}{5}$

b. $\dfrac{2x+10}{2} = x+5$

c. $\dfrac{37}{72} = \dfrac{3}{2}$

d. $\dfrac{2x+3}{2} = x+3$

Answers

5. $\dfrac{x+5}{x}$, 6. $\dfrac{1}{x-5}$, 7. a. 1, b. −1

✓ Concept Check: a, c, d

Example 5   Simplify: $\dfrac{x^2 + 4x + 4}{x^2 + 2x}$

Solution: We factor the numerator and denominator and then apply the fundamental principle.

$$\dfrac{x^2 + 4x + 4}{x^2 + 2x} = \dfrac{(x+2)(x+2)}{x(x+2)} = \dfrac{x+2}{x}$$

HELPFUL HINT

When simplifying a rational expression, the fundamental principle applies to **common *factors*, not common *terms*.**

$\dfrac{x \cdot (x+2)}{x \cdot x} = \dfrac{x+2}{x}$

Common factors. These can be divided out.

$\dfrac{x+2}{x}$

Common terms. Fundamental principle does not apply. This is in simplest form.

TRY THE CONCEPT CHECK IN THE MARGIN.

Example 6   Simplify: $\dfrac{x+9}{x^2 - 81}$

Solution: We factor and then apply the fundamental principle.

$$\dfrac{x+9}{x^2 - 81} = \dfrac{x+9}{(x+9)(x-9)} = \dfrac{1}{x-9}$$

Example 7   Simplify each rational expression.

a. $\dfrac{x+y}{y+x}$ b. $\dfrac{x-y}{y-x}$

Solution: a. The expression $\dfrac{x+y}{y+x}$ can be simplified by using the commutative property of addition to rewrite the denominator $y+x$ as $x+y$.

$$\dfrac{x+y}{y+x} = \dfrac{x+y}{x+y} = 1$$

b. The expression $\dfrac{x-y}{y-x}$ can be simplified by recognizing that $y-x$ and $x-y$ are opposites. In other words, $y-x = -1(x-y)$. We proceed as follows:

$$\dfrac{x-y}{y-x} = \dfrac{1 \cdot (x-y)}{(-1)(x-y)} = \dfrac{1}{-1} = -1$$

MENTAL MATH

Find any real numbers for which each rational expression is undefined. See Example 2.

1. $\dfrac{x + 5}{x}$

2. $\dfrac{x^2 - 5x}{x - 3}$

3. $\dfrac{x^2 + 4x - 2}{x(x - 1)}$

4. $\dfrac{x + 2}{(x - 5)(x - 6)}$

EXERCISE SET 5.1

A *Find the value of the following expressions when $x = 2$, $y = -2$, and $z = -5$. See Example 1.*

1. $\dfrac{x + 5}{x + 2}$

2. $\dfrac{x + 8}{2x + 5}$

3. $\dfrac{y^3}{y^2 - 1}$

4. $\dfrac{z}{z^2 - 5}$

5. $\dfrac{x^2 + 8x + 2}{x^2 - x - 6}$

📼 6. $\dfrac{x + 5}{x^2 + 4x - 8}$

📩 7. The total revenue R from the sale of a popular music compact disc is approximately given by the equation

$$R = \dfrac{150x^2}{x^2 + 3}$$

where x is the number of years since the CD has been released and revenue R is in millions of dollars.

  a. Find the total revenue generated by the end of the first year.

  b. Find the total revenue generated by the end of the second year. Round to the nearest tenth.

  c. Find the total revenue generated in the second year only.

📩 8. For a certain model fax machine, the manufacturing cost C per machine is given by the equation.

$$C = \dfrac{250x + 10,000}{x}$$

where x is the number of fax machines manufactured and cost C is in dollars per machine.

  a. Find the cost per fax machine when manufacturing 100 fax machines.

  b. Find the cost per fax machine when manufacturing 1000 fax machines.

  c. Does the cost per machine decrease or increase when more machines are manufactured? Explain why this is so.

B *Find any real numbers for which each rational expression is undefined. See Example 2.*

9. $\dfrac{7}{2x}$

10. $\dfrac{3}{5x}$

📼 11. $\dfrac{x + 3}{x + 2}$

12. $\dfrac{5x + 1}{x - 3}$

353

13. $x = 4$

14. $x = -3$

15. $x = -2$

$x = 0, x = 1,$
16. $x = -1$

17. none

18. none

19. answers may vary

no; answers may
20. vary

21. $\dfrac{1}{4(x + 2)}$

22. $\dfrac{1}{3x + 2}$

23. $\dfrac{1}{x + 2}$

24. $\dfrac{1}{x - 5}$

25. can't simplify

26. $\dfrac{3}{4}$

27. 1

28. 1

29. -1

30. -1

31. -5

32. $\dfrac{7}{x}$

33. $\dfrac{1}{x - 9}$

34. $\dfrac{1}{x - 3}$

35. $5x + 1$

36. $6x - 1$

37. $\dfrac{1}{x - 2}$

38. $\dfrac{1}{x - 7}$

39. $x + 2$

40. $4x$

41. $\dfrac{x + 5}{x - 5}$

13. $\dfrac{4x^2 + 9}{2x - 8}$

14. $\dfrac{9x^3 + 4x}{15x + 45}$

15. $\dfrac{9x^3 + 4}{15x + 30}$

16. $\dfrac{19x^3 + 2}{x^3 - x}$

17. $\dfrac{x^2 - 5x - 2}{4}$

18. $\dfrac{9y^5 + y^3}{9}$

19. Explain why the denominator of a fraction or a rational expression must not equal 0.

20. Does $\dfrac{(x - 3)(x + 3)}{x - 3}$ have the same value as $x + 3$ for all real numbers? Explain why or why not.

C *Simplify each expression. See Examples 3 through 7.*

21. $\dfrac{2}{8x + 16}$

22. $\dfrac{3}{9x + 6}$

23. $\dfrac{x - 2}{x^2 - 4}$

24. $\dfrac{x + 5}{x^2 - 25}$

25. $\dfrac{2x - 10}{3x - 30}$

26. $\dfrac{3x - 12}{4x - 16}$

27. $\dfrac{x + 7}{7 + x}$

28. $\dfrac{y + 9}{9 + y}$

29. $\dfrac{x - 7}{7 - x}$

30. $\dfrac{y - 9}{9 - y}$

31. $\dfrac{-5a - 5b}{a + b}$

32. $\dfrac{7x + 35}{x^2 + 5x}$

33. $\dfrac{x + 5}{x^2 - 4x - 45}$

34. $\dfrac{x - 3}{x^2 - 6x + 9}$

35. $\dfrac{5x^2 + 11x + 2}{x + 2}$

36. $\dfrac{12x^2 + 4x - 1}{2x + 1}$

37. $\dfrac{x + 7}{x^2 + 5x - 14}$

38. $\dfrac{x - 10}{x^2 - 17x + 70}$

39. $\dfrac{2x^2 + 3x - 2}{2x - 1}$

40. $\dfrac{4x^2 + 24x}{x + 6}$

41. $\dfrac{x^2 + 7x + 10}{x^2 - 3x - 10}$

Name _____

42. $\dfrac{2x^2 + 7x - 4}{x^2 + 3x - 4}$ **43.** $\dfrac{3x^2 + 7x + 2}{3x^2 + 13x + 4}$ **44.** $\dfrac{4x^2 - 4x + 1}{2x^2 + 9x - 5}$

45. $\dfrac{2x^2 - 8}{4x - 8}$ **46.** $\dfrac{5x^2 - 500}{35x + 350}$ **47.** $\dfrac{11x^2 - 22x^3}{6x - 12x^2}$

48. $\dfrac{16r^2 - 4s^2}{4r - 2s}$ **49.** $\dfrac{2 - x}{x - 2}$ **50.** $\dfrac{7 - y}{y - 7}$

51. $\dfrac{x^2 - 1}{x^2 - 2x + 1}$ **52.** $\dfrac{x^2 - 16}{x^2 - 8x + 16}$

53. $\dfrac{m^2 - 6m + 9}{m^2 - 9}$ **54.** $\dfrac{m^2 - 4m + 4}{m^2 + m - 6}$

REVIEW AND PREVIEW

Perform each indicated operation. See Section R.2.

55. $\dfrac{1}{3} \cdot \dfrac{9}{11}$ **56.** $\dfrac{5}{27} \cdot \dfrac{2}{5}$ **57.** $\dfrac{5}{6} \cdot \dfrac{10}{11} \cdot \dfrac{2}{3}$ **58.** $\dfrac{4}{3} \cdot \dfrac{1}{7} \cdot \dfrac{10}{13}$

59. $\dfrac{1}{3} \div \dfrac{1}{4}$ **60.** $\dfrac{7}{8} \div \dfrac{1}{2}$ **61.** $\dfrac{13}{20} \div \dfrac{2}{9}$ **62.** $\dfrac{8}{15} \div \dfrac{5}{8}$

COMBINING CONCEPTS

Simplify each expression. Each exercise contains a four-term polynomial that should be factored by grouping.

63. $\dfrac{x^2 + xy + 2x + 2y}{x + 2}$ **64.** $\dfrac{ab + ac + b^2 + bc}{b + c}$

---

42. $\dfrac{2x - 1}{x - 1}$

43. $\dfrac{x + 2}{x + 4}$

44. $\dfrac{2x - 1}{x + 5}$

45. $\dfrac{x + 2}{2}$

46. $\dfrac{x - 10}{7}$

47. $\dfrac{11x}{6}$

48. $2(2r + s)$

49. -1

50. -1

51. $\dfrac{x + 1}{x - 1}$

52. $\dfrac{x + 4}{x - 4}$

53. $\dfrac{m - 3}{m + 3}$

54. $\dfrac{m - 2}{m + 3}$

55. $\dfrac{3}{11}$

56. $\dfrac{2}{27}$

57. $\dfrac{50}{99}$

58. $\dfrac{40}{273}$

59. $\dfrac{4}{3}$

60. $\dfrac{7}{4}$

61. $\dfrac{117}{40}$

62. $\dfrac{64}{75}$

63. $x + y$

64. $a + b$

355

65. $\dfrac{5-y}{2}$

66. $\dfrac{x+2}{y}$

67. answers may vary

68. answers may vary

69. 400 mg

70. 53.6 millimeters per hour

71. no; $B \approx 24$

72. $C = 78.125$; medium

Name _____

65. $\dfrac{5x + 15 - xy - 3y}{2x + 6}$

66. $\dfrac{xy - 6x + 2y - 12}{y^2 - 6y}$

67. Explain how to write a fraction in lowest terms.

68. Explain how to write a rational expression in lowest terms.

69. The dose of medicine prescribed for a child depends on the child's age A in years and the adult dose D for the medication. Young's Rule is a formula used by pediatricians that gives a child's dose C as

$$C = \dfrac{DA}{A + 12}$$

Suppose that an 8-year-old child needs medication, and the normal adult dose is 1000 mg. What size dose should the child receive?

70. During a storm, water treatment engineers monitor how quickly rain is falling. If too much rain comes too fast, there is a danger of sewers backing up. A formula that gives the rainfall intensity i in millimeters per hour for a certain strength storm in eastern Virginia is

$$i = \dfrac{5840}{t + 29}$$

where t is the duration of the storm in minutes. What rainfall intensity should engineers expect for a storm of this strength in eastern Virginia that lasts for 80 minutes? Round answer to one decimal place.

71. Calculating body-mass index is a way to gauge whether a person should lose weight. Doctors recommend that body-mass index values fall between 19 and 25. The formula for body-mass index B is

$$B = \dfrac{705w}{h^2}$$

where w is weight in pounds and h is height in inches. Should a 148-pound person who is 5 feet 6 inches tall lose weight?

72. Anthropologists and forensic scientists use a measure called the cephalic index to help classify skulls. The cephalic index of a skull with width W and length L from front to back is given by the formula

$$C = \dfrac{100W}{L}$$

A long skull has an index value less than 75, a medium skull has an index value between 75 and 85, and a broad skull has an index value over 85. Find the cephalic index of a skull that is 5 inches wide and 6.4 inches long. Classify the skull.

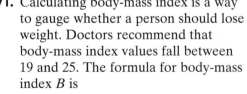

5.2 MULTIPLYING AND DIVIDING RATIONAL EXPRESSIONS

A MULTIPLYING RATIONAL EXPRESSIONS

Just as simplifying rational expressions is similar to simplifying number fractions, multiplying and dividing rational expressions is similar to multiplying and dividing number fractions.

Multiply: $\dfrac{3}{5} \cdot \dfrac{10}{11}$

Multiply: $\dfrac{x-3}{x+5} \cdot \dfrac{2x+10}{x^2-9}$

Multiply numerators and then multiply denominators.

$$\dfrac{3}{5} \cdot \dfrac{10}{11} = \dfrac{3 \cdot 10}{5 \cdot 11}$$

$$\dfrac{x-3}{x+5} \cdot \dfrac{2x+10}{x^2-9} = \dfrac{(x-3) \cdot (2x+10)}{(x+5) \cdot (x^2-9)}$$

Simplify by factoring numerators and denominators.

$$= \dfrac{3 \cdot 2 \cdot 5}{5 \cdot 11}$$

$$= \dfrac{(x-3) \cdot 2(x+5)}{(x+5)(x+3)(x-3)}$$

Apply the fundamental principle.

$$= \dfrac{3 \cdot 2}{11} \quad \text{or} \quad \dfrac{6}{11}$$

$$= \dfrac{2}{x+3}$$

MULTIPLYING RATIONAL EXPRESSIONS

If $\dfrac{P}{Q}$ and $\dfrac{R}{S}$ are rational expressions, then

$$\dfrac{P}{Q} \cdot \dfrac{R}{S} = \dfrac{PR}{QS}$$

To multiply rational expressions, multiply the numerators and then multiply the denominators.

Example 1   Multiply.

  a. $\dfrac{25x}{2} \cdot \dfrac{1}{y^3}$ **b.** $\dfrac{-7x^2}{5y} \cdot \dfrac{3y^5}{14x^2}$

Solution: To multiply rational expressions, we first multiply the numerators and then multiply the denominators of both expressions. Then we write the product in lowest terms.

  a. $\dfrac{25x}{2} \cdot \dfrac{1}{y^3} = \dfrac{25x \cdot 1}{2 \cdot y^3} = \dfrac{25x}{2y^3}$

  The expression $\dfrac{25x}{2y^3}$ is in lowest terms.

  b. $\dfrac{-7x^2}{5y} \cdot \dfrac{3y^5}{14x^2} = \dfrac{-7x^2 \cdot 3y^5}{5y \cdot 14x^2}$ Multiply.

  The expression $\dfrac{-7x^2 \cdot 3y^5}{5y \cdot 14x^2}$ is not in lowest terms, so we factor the numerator and the denominator and apply the fundamental principle.

Objectives

A Multiply rational expressions.
B Divide rational expressions.
C Multiply and divide rational expressions.
D Converting between units of measure.

SSM   CD-ROM   Video
                        5.2

TEACHING TIP

Point out to students that rational expressions are just generalized forms of number fractions. In fact, when the variable(s) in a rational expression are given replacement values, the expression simplifies to a number fraction.

Practice Problem 1

Multiply.

  a. $\dfrac{16y}{3} \cdot \dfrac{1}{x^2}$ **b.** $\dfrac{-5a^3}{3b^3} \cdot \dfrac{2b^2}{15a}$

Answers

1. a. $\dfrac{16y}{3x^2}$, **b.** $-\dfrac{2a^2}{9b}$

$$= \frac{-1 \cdot 7 \cdot 3 \cdot x^2 \cdot y \cdot y^4}{5 \cdot 2 \cdot 7 \cdot x^2 \cdot y}$$

$$= -\frac{3y^4}{10}$$

When multiplying rational expressions, it is usually best to first factor each numerator and denominator. This will help us when we apply the fundamental principle to write the product in lowest terms.

Practice Problem 2

Multiply: $\frac{6x + 6}{7} \cdot \frac{14}{x^2 - 1}$

Example 2   Multiply: $\frac{x^2 + x}{3x} \cdot \frac{6}{5x + 5}$

Solution:

$$\frac{x^2 + x}{3x} \cdot \frac{6}{5x + 5} = \frac{x(x + 1)}{3x} \cdot \frac{2 \cdot 3}{5(x + 1)} \quad \text{Factor numerators and denominators.}$$

$$= \frac{x(x + 1) \cdot 2 \cdot 3}{3x \cdot 5(x + 1)} \quad \text{Multiply.}$$

$$= \frac{2}{5} \quad \text{Apply the fundamental principle.}$$

The following steps may be used to multiply rational expressions.

✓ CONCEPT CHECK

Which of the following is a true statement?

a. $\frac{1}{3} \cdot \frac{1}{2} = \frac{1}{5}$ b. $\frac{2}{x} \cdot \frac{5}{x} = \frac{10}{x}$

c. $\frac{3}{x} \cdot \frac{1}{2} = \frac{3}{2x}$

d. $\frac{x}{7} \cdot \frac{x + 5}{4} = \frac{2x + 5}{28}$

> **TO MULTIPLY RATIONAL EXPRESSIONS**
>
> **Step 1.** Completely factor numerators and denominators.
> **Step 2.** Multiply numerators and multiply denominators.
> **Step 3.** Simplify or write the product in lowest terms by applying the fundamental principle to all common factors.

TRY THE CONCEPT CHECK IN THE MARGIN.

Practice Problem 3

Multiply: $\frac{4x + 8}{7x^2 - 14x} \cdot \frac{3x^2 - 5x - 2}{9x^2 - 1}$

Example 3   Multiply: $\frac{3x + 3}{5x^2 - 5x} \cdot \frac{2x^2 + x - 3}{4x^2 - 9}$

Solution:

$$\frac{3x + 3}{5x^2 - 5x} \cdot \frac{2x^2 + x - 3}{4x^2 - 9} = \frac{3(x + 1)}{5x(x - 1)} \cdot \frac{(2x + 3)(x - 1)}{(2x - 3)(2x + 3)} \quad \text{Factor.}$$

$$= \frac{3(x + 1)(2x + 3)(x - 1)}{5x(x - 1)(2x - 3)(2x + 3)} \quad \text{Multiply.}$$

$$= \frac{3(x + 1)}{5x(2x - 3)} \quad \text{Simplify.}$$

Answers

2. $\frac{12}{x - 1}$, **3.** $\frac{4(x + 2)}{7x(3x - 1)}$

✓ Concept Check: c

B DIVIDING RATIONAL EXPRESSIONS

We can divide by a rational expression in the same way we divide by a number fraction. Recall that to divide by a fraction, we multiply by its reciprocal.

HELPFUL HINT

Don't forget how to find reciprocals. The reciprocal of $\frac{a}{b}$ is $\frac{b}{a}$, $a \neq 0$, $b \neq 0$.

For example, to divide $\frac{3}{2}$ by $\frac{7}{8}$, we multiply $\frac{3}{2}$ by $\frac{8}{7}$.

$$\frac{3}{2} \div \frac{7}{8} = \frac{3}{2} \cdot \frac{8}{7} = \frac{3 \cdot 4 \cdot 2}{2 \cdot 7} = \frac{12}{7}$$

DIVIDING RATIONAL EXPRESSIONS

If $\frac{P}{Q}$ and $\frac{R}{S}$ are rational expressions and $\frac{R}{S}$ is not 0, then

$$\frac{P}{Q} \div \frac{R}{S} = \frac{P}{Q} \cdot \frac{S}{R} = \frac{PS}{QR}$$

To divide two rational expressions, multiply the first rational expression by the reciprocal of the second rational expression.

Example 4   Divide: $\dfrac{3x^3}{40} \div \dfrac{4x^3}{y^2}$

Solution: $\dfrac{3x^3}{40} \div \dfrac{4x^3}{y^2} = \dfrac{3x^3}{40} \cdot \dfrac{y^2}{4x^3}$ Multiply by the reciprocal of $\frac{4x^3}{y^2}$.

$$= \frac{3}{160} \frac{x^3 y^2}{x^3}$$

$$= \frac{3y^2}{160} \qquad \text{Simplify.}$$

Example 5   Divide: $\dfrac{(x-1)(x+2)}{10} \div \dfrac{2x+4}{5}$

Solution:

$$\frac{(x-1)(x+2)}{10} \div \frac{2x+4}{5} = \frac{(x-1)(x+2)}{10} \cdot \frac{5}{2x+4} \qquad \begin{array}{l}\text{Multiply by the}\\\text{reciprocal}\\\text{of } \frac{2x+4}{5}.\end{array}$$

$$= \frac{(x-1)(x+2) \cdot 5}{5 \cdot 2 \cdot 2 \cdot (x+2)} \qquad \begin{array}{l}\text{Factor and}\\\text{multiply.}\end{array}$$

$$= \frac{x-1}{4} \qquad \text{Simplify.}$$

Practice Problem 4

Divide: $\dfrac{7x^2}{6} \div \dfrac{x}{2y}$

Practice Problem 5

Divide: $\dfrac{(2x+3)(x-4)}{6} \div \dfrac{3x-12}{2}$

Answers

4. $\dfrac{7xy}{3}$,   **5.** $\dfrac{2x+3}{9}$

The following may be used to divide by a rational expression.

> **TO DIVIDE BY A RATIONAL EXPRESSION**
>
> Multiply by its reciprocal.

Practice Problem 6

Divide: $\dfrac{10x + 4}{x^2 - 4} \div \dfrac{5x^3 + 2x^2}{x + 2}$

TEACHING TIP

Consider telling students that it is OK to leave their final answer in factored form when working with rational expressions.

Example 6   Divide: $\dfrac{6x + 2}{x^2 - 1} \div \dfrac{3x^2 + x}{x - 1}$

Solution:

$\dfrac{6x + 2}{x^2 - 1} \div \dfrac{3x^2 + x}{x - 1} = \dfrac{6x + 2}{x^2 - 1} \cdot \dfrac{x - 1}{3x^2 + x}$ Multiply by the reciprocal.

$= \dfrac{2(3x + 1)(x - 1)}{(x + 1)(x - 1) \cdot x(3x + 1)}$ Factor and multiply.

$= \dfrac{2}{x(x + 1)}$ Simplify.

Practice Problem 7

Divide: $\dfrac{3x^2 - 10x + 8}{7x - 14} \div \dfrac{9x - 12}{21}$

Example 7   Divide: $\dfrac{2x^2 - 11x + 5}{5x - 25} \div \dfrac{4x - 2}{10}$

Solution:

$\dfrac{2x^2 - 11x + 5}{5x - 25} \div \dfrac{4x - 2}{10} = \dfrac{2x^2 - 11x + 5}{5x - 25} \cdot \dfrac{10}{4x - 2}$ Multiply by the reciprocal.

$= \dfrac{(2x - 1)(x - 5) \cdot 2 \cdot 5}{5(x - 5) \cdot 2(2x - 1)}$ Factor and multiply.

$= \dfrac{1}{1}$ or 1 Simplify.

C   MULTIPLYING AND DIVIDING RATIONAL EXPRESSIONS

Let's make sure that we understand the difference between multiplying and dividing rational expressions.

Rational Expressions	
Multiplication	Multiply the numerators and multiply the denominators.
Division	Multiply by the reciprocal of the divisor.

Practice Problem 8

Multiply or divide as indicated.

a. $\dfrac{x + 3}{x} \cdot \dfrac{7}{x + 3}$

b. $\dfrac{x + 3}{x} \div \dfrac{7}{x + 3}$

Example 8   Multiply or divide as indicated.

a. $\dfrac{x - 4}{5} \cdot \dfrac{x}{x - 4}$ b. $\dfrac{x - 4}{5} \div \dfrac{x}{x - 4}$

Solution: a. $\dfrac{x - 4}{5} \cdot \dfrac{x}{x - 4} = \dfrac{(x - 4) \cdot x}{5 \cdot (x - 4)} = \dfrac{x}{5}$

b. $\dfrac{x - 4}{5} \div \dfrac{x}{x - 4} = \dfrac{x - 4}{5} \cdot \dfrac{x - 4}{x} = \dfrac{(x - 4)^2}{5x}$

Answers

6. $\dfrac{2}{x^2(x - 2)}$,   7. 1,   8. a. $\dfrac{7}{x}$,   b. $\dfrac{(x + 3)^2}{7x}$

D CONVERTING BETWEEN UNITS OF MEASURE

Now that we know how to multiply fractions and rational expressions, we can use this knowledge to help us convert between units of measure. To do so, we will use **unit fractions**. A unit fraction is a fraction that equals 1. For example, since 12 in. = 1 ft, we have the unit fractions

$$\frac{12 \text{ in.}}{1 \text{ ft}} = 1 \quad \text{and} \quad \frac{1 \text{ ft}}{12 \text{ in.}} = 1$$

Example 9   Converting from Square Yards to Square Feet

The largest casino in the world is the Foxwoods Resort Casino in Ledyard, CT. The gaming area for this casino is 21,444 *square yards*. Find the size of the gaming area in *square feet*. (*Source*: *The Guinness Book of Records*, 1996)

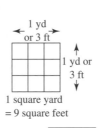

Solution: There are 9 square feet in 1 square yard.

$$21{,}444 \text{ square yards} = 21{,}444 \; \cancel{\text{sq. yd}} \cdot \frac{9 \text{ sq. ft}}{1 \; \cancel{\text{sq. yd}}}$$

$$= 192{,}996 \text{ square feet}$$

$$\approx 193{,}000 \text{ square feet}$$

HELPFUL HINT

When converting among units of measurement, if possible, write the unit fraction so that **the numerator contains units converting to** and **the denominator contains original units**.

Unit fraction

$$48 \text{ ft} = \frac{48 \; \cancel{\text{in.}}}{1} \cdot \frac{1 \text{ ft}}{12 \; \cancel{\text{in.}}} \quad \begin{array}{l} \leftarrow \text{Units converting to} \\ \leftarrow \text{Original units} \end{array}$$

$$= \frac{48}{12} \text{ ft} = 4 \text{ ft}$$

Practice Problem 9

Convert 40,000 square feet to square yards.

Practice Problem 10

Carl Lewis holds the men's record for speed. He has been timed at 39.5 feet per second. Convert this to miles per hour. (*Source: The Guinness Book of Records*, 1996)

Example 10 Converting from Feet per Second to Miles per Hour

Florence Griffith Joyner holds the women's record for speed. She has been timed at 36 feet per second. Convert this to miles per hour. (*Source: The Guinness Book of Records*, 1996)

Solution: Recall that 1 mile = 5280 feet and 1 hour = 3600 seconds $(60 \cdot 60)$.

Unit fractions

$$36 \text{ feet/second} = \frac{36 \text{ feet}}{1 \text{ second}} \cdot \frac{3600 \text{ seconds}}{1 \text{ hour}} \cdot \frac{1 \text{ mile}}{5280 \text{ feet}}$$

$$= \frac{36 \cdot 3600}{5280} \text{ miles/hour}$$

$$\approx 24.5 \text{ miles/hour (rounded to the nearest tenth)}$$

Answer

10. 26.9 miles/hour

Name _____ **Section** _____ **Date** _____

MENTAL MATH

Find each product. See Example 1.

1. $\dfrac{2}{y} \cdot \dfrac{x}{3}$ **2.** $\dfrac{3x}{4} \cdot \dfrac{1}{y}$ **3.** $\dfrac{5}{7} \cdot \dfrac{y^2}{x^2}$

4. $\dfrac{x^5}{11} \cdot \dfrac{4}{z^3}$ **5.** $\dfrac{9}{x} \cdot \dfrac{x}{5}$ **6.** $\dfrac{y}{7} \cdot \dfrac{3}{y}$

EXERCISE SET 5.2

A *Find each product and simplify if possible. See Examples 1 through 3.*

1. $\dfrac{3x}{y^2} \cdot \dfrac{7y}{4x}$ **2.** $\dfrac{9x^2}{y} \cdot \dfrac{4y}{3x^2}$ ▣ **3.** $\dfrac{8x}{2} \cdot \dfrac{x^5}{4x^2}$

4. $\dfrac{6x^2}{10x^3} \cdot \dfrac{5x}{12}$ **5.** $-\dfrac{5a^2b}{30a^2b^2} \cdot b^3$ **6.** $-\dfrac{9x^3y^2}{18xy^5} \cdot y^3$

7. $\dfrac{x}{2x - 14} \cdot \dfrac{x^2 - 7x}{5}$ **8.** $\dfrac{4x - 24}{20x} \cdot \dfrac{5}{x - 6}$ **9.** $\dfrac{6x + 6}{5} \cdot \dfrac{10}{36x + 36}$

10. $\dfrac{x^2 + x}{8} \cdot \dfrac{16}{x + 1}$ **11.** $\dfrac{m^2 - n^2}{m + n} \cdot \dfrac{m}{m^2 - mn}$ **12.** $\dfrac{(m - n)^2}{m + n} \cdot \dfrac{m}{m^2 - mn}$

13. $\dfrac{x^2 - 25}{x^2 - 3x - 10} \cdot \dfrac{x + 2}{x}$ **14.** $\dfrac{a^2 + 6a + 9}{a^2 - 4} \cdot \dfrac{a + 3}{a - 2}$

▢ **15.** Find the area of the rectangle.   ▢ **16.** Find the area of the square.

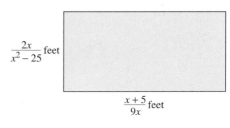

$\dfrac{2x}{x^2 - 25}$ feet

$\dfrac{x + 5}{9x}$ feet

$\dfrac{2x}{5x^2 + 3x}$ meters

364

Name _____

B *Find each quotient and simplify. See Examples 4 through 7.*

17. $\dfrac{5x^7}{2x^5} \div \dfrac{10x}{4x^3}$

18. $\dfrac{9y^4}{6y} \div \dfrac{y^2}{3}$

19. $\dfrac{8x^2}{y^3} \div \dfrac{4x^2y^3}{6}$

20. $\dfrac{7a^2b}{3ab^2} \div \dfrac{21a^2b^2}{14ab}$

21. $\dfrac{(x - 6)(x + 4)}{4x} \div \dfrac{2x - 12}{8x^2}$

22. $\dfrac{(x + 3)^2}{5} \div \dfrac{5x + 15}{25}$

23. $\dfrac{3x^2}{x^2 - 1} \div \dfrac{x^5}{(x + 1)^2}$

24. $\dfrac{x + 1}{(x + 1)(2x + 3)} \div \dfrac{20}{2x + 3}$

25. $\dfrac{m^2 - n^2}{m + n} \div \dfrac{m}{m^2 + nm}$

26. $\dfrac{(m - n)^2}{m + n} \div \dfrac{m^2 - mn}{m}$

27. $\dfrac{x + 2}{7 - x} \div \dfrac{x^2 - 5x + 6}{x^2 - 9x + 14}$

28. $(x - 3) \div \dfrac{x^2 + 3x - 18}{x}$

29. $\dfrac{x^2 + 7x + 10}{x - 1} \div \dfrac{x^2 + 2x - 15}{x - 1}$

30. $\dfrac{a^2 - b^2}{9} \div \dfrac{3b - 3a}{27x^2}$

C *Multiply or divide as indicated. See Example 8.*

31. $\dfrac{5x - 10}{12} \div \dfrac{4x - 8}{8}$

32. $\dfrac{6x + 6}{5} \div \dfrac{3x + 3}{10}$

33. $\dfrac{x^2 + 5x}{8} \cdot \dfrac{9}{3x + 15}$

34. $\dfrac{3x^2 + 12x}{6} \cdot \dfrac{9}{2x + 8}$

35. $\dfrac{7}{6p^2 + q} \div \dfrac{14}{18p^2 + 3q}$

36. $\dfrac{3x + 6}{20} \div \dfrac{4x + 8}{8}$

37. $\dfrac{3x + 4y}{x^2 + 4xy + 4y^2} \cdot \dfrac{x + 2y}{2}$

38. $\dfrac{x^2 - y^2}{3x^2 + 3xy} \cdot \dfrac{3x^2 + 6x}{3x^2 - 2xy - y^2}$

39. $\dfrac{(x + 2)^2}{x - 2} \div \dfrac{x^2 - 4}{2x - 4}$

40. $\dfrac{x^2 - 4}{2y} \div \dfrac{2 - x}{6xy}$

41. $\dfrac{a^2 + 7a + 12}{a^2 + 5a + 6} \cdot \dfrac{a^2 + 8a + 15}{a^2 + 5a + 4}$

42. $\dfrac{b^2 + 2b - 3}{b^2 + b - 2} \cdot \dfrac{b^2 - 4}{b^2 + 6b + 8}$

D *Convert as indicated. See Examples 9 and 10.*

43. 10 square feet = _____ square inches.

44. 1008 square inches = _____ square feet.

45. 50 miles per hour = _____ feet per second (round to the nearest whole).

46. 10 feet per second = _____ miles per hour (round to the nearest tenth).

47. The speed of sound is 5023 feet per second in ocean water whose temperature is 77°F. Convert this speed of sound to miles per hour. Round to the nearest tenth. (*Source: CRC Handbook of Chemistry and Physics*, 65th edition)

48. On October 28, 1996, Craig Breedlove tried unsuccessfully to break the world land speed record. He reached an unofficial speed of 675 mph before losing control of his car, the Spirit of America. Find this speed in feet per second. (*Source: The World Almanac and Book of Facts*, 1997)

49. In 1995, 1 U.S. dollar was equivalent to 94.06 Japanese yen on average. If you had wanted to exchange 1000 Japanese yen to U.S. dollars in 1995, how much would you have received? Round to the nearest cent. (*Source: International Monetary Fund*)

50. In 1995, 1 British pound was equivalent to 1.5785 U.S. dollars on average. If you had wanted to exchange $500 U.S. to British pounds in 1995, how much would you have received? Round to the nearest hundredth. (*Source: International Monetary Fund*)

37. $\dfrac{3x + 4y}{2(x + 2y)}$

38. $\dfrac{x + 2}{3x + y}$

39. $\dfrac{2(x + 2)}{x - 2}$

40. $-3x(x + 2)$

41. $\dfrac{(a + 5)(a + 3)}{(a + 2)(a + 1)}$

42. $\dfrac{(b + 3)(b - 2)}{(b + 2)(b + 4)}$

43. 1440

44. 7

45. 73

46. 6.8

47. 3424.8 mph

48. 990 feet per second

49. $10.63

50. 316.76 pounds

51. 1

52. $\dfrac{3}{5}$

53. $-\dfrac{10}{9}$

54. $-\dfrac{4}{3}$

55. $-\dfrac{1}{5}$

56. $-\dfrac{5}{2}$

57. $\dfrac{x}{2}$

58. $-\dfrac{(x-3)(x+1)}{(2x+3)^2}$

59. $\dfrac{5a(2a+b)(3a-2b)}{b^2(a-b)(a+2b)}$

60. $\dfrac{xy(xy-1)}{12}$

61. answers may vary

62. answers may vary

63. 50 8-ounce cups

64. no

Name _____

REVIEW AND PREVIEW

Perform each indicated operation. See Section R.2.

51. $\dfrac{1}{5} + \dfrac{4}{5}$

52. $\dfrac{3}{15} + \dfrac{6}{15}$

53. $\dfrac{9}{9} - \dfrac{19}{9}$

54. $\dfrac{4}{3} - \dfrac{8}{3}$

55. $\dfrac{6}{5} + \left(\dfrac{1}{5} - \dfrac{8}{5} \right)$

56. $-\dfrac{3}{2} + \left(\dfrac{1}{2} - \dfrac{3}{2} \right)$

COMBINING CONCEPTS

Multiply or divide as indicated.

57. $\left(\dfrac{x^2 - y^2}{x^2 + y^2} \div \dfrac{x^2 - y^2}{3x} \right) \cdot \dfrac{x^2 + y^2}{6}$

58. $\left(\dfrac{x^2 - 9}{x^2 - 1} \cdot \dfrac{x^2 + 2x + 1}{2x^2 + 9x + 9} \right) \div \dfrac{2x + 3}{1 - x}$

59. $\left(\dfrac{2a + b}{b^2} \cdot \dfrac{3a^2 - 2ab}{ab + 2b^2} \right) \div \dfrac{a^2 - 3ab + 2b^2}{5ab - 10b^2}$

60. $\left(\dfrac{x^2 y^2 - xy}{4x - 4y} \div \dfrac{3y - 3x}{8x - 8y} \right) \cdot \dfrac{y - x}{8}$

61. In your own words, explain how you multiply rational expressions.

62. Explain how dividing rational expressions is similar to dividing rational numbers.

63. A coffee urn holds 3.125 gallons of coffee. How many 8-ounce cups of coffee can be dispensed from the urn? (*Hint*: There are 64 fluid ounces in a half gallon.)

64. An environmental technician finds that warm water from an industrial process is being discharged into a nearby pond at a rate of 30 gallons per minute. Plant regulations state that the flow rate should be no more than 0.1 cubic feet per second. Is the flow rate of 30 gallons per minute in violation of the plant regulations? (*Hint*: 1 cubic foot is equivalent to 7.48 gallons.)

5.3   ADDING AND SUBTRACTING RATIONAL EXPRESSIONS WITH THE SAME DENOMINATOR AND LEAST COMMON DENOMINATOR

A   ADDING AND SUBTRACTING RATIONAL EXPRESSIONS WITH THE SAME DENOMINATOR

Like multiplication and division, addition and subtraction of rational expressions is similar to addition and subtraction of rational numbers. In this section, we add and subtract rational expressions with a common denominator.

Add: $\dfrac{6}{5} + \dfrac{2}{5}$ Add: $\dfrac{9}{x+2} + \dfrac{3}{x+2}$

Add the numerators and place the sum over the common denominator.

$\dfrac{6}{5} + \dfrac{2}{5} = \dfrac{6+2}{5}$ $\dfrac{9}{x+2} + \dfrac{3}{x+2} = \dfrac{9+3}{x+2}$

$\qquad = \dfrac{8}{5}$ Simplify. $\qquad\qquad = \dfrac{12}{x+2}$ Simplify.

> ### ADDING AND SUBTRACTING RATIONAL EXPRESSIONS WITH COMMON DENOMINATORS
>
> If $\dfrac{P}{R}$ and $\dfrac{Q}{R}$ are rational expressions, then
>
> $$\dfrac{P}{R} + \dfrac{Q}{R} = \dfrac{P+Q}{R} \qquad \text{and} \qquad \dfrac{P}{R} - \dfrac{Q}{R} = \dfrac{P-Q}{R}$$
>
> To add or subtract rational expressions, add or subtract numerators and place the sum or difference over the common denominator.

Example 1   Add: $\dfrac{5m}{2n} + \dfrac{m}{2n}$

Solution: $\dfrac{5m}{2n} + \dfrac{m}{2n} = \dfrac{5m+m}{2n}$ Add the numerators.

$\qquad\qquad = \dfrac{6m}{2n}$ Simplify the numerator by combining like terms.

$\qquad\qquad = \dfrac{3m}{n}$ Simplify by applying the fundamental principle.

Example 2   Subtract: $\dfrac{2y}{2y-7} - \dfrac{7}{2y-7}$

Solution: $\dfrac{2y}{2y-7} - \dfrac{7}{2y-7} = \dfrac{2y-7}{2y-7}$ Subtract the numerators.

$\qquad\qquad = \dfrac{1}{1}$ or 1 Simplify.

Example 3   Subtract: $\dfrac{3x^2 + 2x}{x-1} - \dfrac{10x - 5}{x-1}$

TEACHING TIP

Before beginning this lesson, consider asking students to explain why the denominators must be the same when adding and subtracting fractions. Be sure they understand that the denominator gives the unit to which the numerator is referring. If all fractions are referring to the same unit, the expressions can be added (subtracted) by adding (subtracting) the numerators and writing the sum (difference) over the common denominator. You may want to illustrate this with a diagram.

TEACHING TIP

Before doing Example 1, ask students to simplify a particular form of this expression. For instance, have them use $m = 4$ and $n = 5$. After they have completed Example 1, have them compare their answer with the simplified form of the sum.

Practice Problem 1

Add: $\dfrac{8x}{3y} + \dfrac{x}{3y}$

Practice Problem 2

Subtract: $\dfrac{3x}{3x-7} - \dfrac{7}{3x-7}$

Practice Problem 3

Subtract: $\dfrac{2x^2 + 5x}{x+2} - \dfrac{4x+6}{x+2}$

Answers

1. $\dfrac{3x}{y}$,   **2.** 1,   **3.** $2x - 3$

Solution:

$$\frac{3x^2 + 2x}{x - 1} - \frac{10x - 5}{x - 1} = \frac{3x^2 + 2x - (10x - 5)}{x - 1} \quad \text{Subtract the numerators. Notice the parentheses.}$$

$$= \frac{3x^2 + 2x - 10x + 5}{x - 1} \quad \text{Use the distributive property.}$$

$$= \frac{3x^2 - 8x + 5}{x - 1} \quad \text{Combine like terms.}$$

$$= \frac{(x - 1)(3x - 5)}{x - 1} \quad \text{Factor.}$$

$$= 3x - 5 \quad \text{Simplify.}$$

HELPFUL HINT

Notice how the numerator $10x - 5$ has been subtracted in Example 3.

This $-$ sign applies to the entire numerator of $10x - 5$. So parentheses are inserted here to indicate this.

$$\frac{3x^2 + 2x}{x - 1} - \frac{10x - 5}{x - 1} = \frac{3x^2 + 2x - (10x - 5)}{x - 1}$$

B   FINDING THE LEAST COMMON DENOMINATOR

Recall from Chapter R that to add and subtract fractions with different denominators, we first find a least common denominator (LCD). Then we write all fractions as equivalent fractions with the LCD.

For example, suppose we add $\frac{8}{3}$ and $\frac{2}{5}$. The LCD of denominators 3 and 5 is 15, since 15 is the smallest number that both 3 and 5 divide into evenly. So we rewrite each fraction so that its denominator is 15. (Notice how we apply the fundamental principle.)

$$\frac{8}{3} + \frac{2}{5} = \frac{8(5)}{3(5)} + \frac{2(3)}{5(3)} = \frac{40}{15} + \frac{6}{15} = \frac{40 + 6}{15} = \frac{46}{15}$$

To add or subtract rational expressions with different denominators, we also first find an LCD and then write all rational expressions as equivalent expressions with the LCD. The **least common denominator (LCD) of a list of rational expressions** is a polynomial of least degree whose factors include all the factors of the denominators in the list.

TO FIND THE LEAST COMMON DENOMINATOR (LCD)

Step 1. Factor each denominator completely.

Step 2. The least common denominator (LCD) is the product of all unique factors found in Step 1, each raised to a power equal to the greatest number of times that the factor appears in any one factored denominator.

Practice Problem 4

Find the LCD for each pair.

a. $\frac{2}{9}, \frac{7}{15}$ b. $\frac{5}{6x^3}, \frac{11}{8x^5}$

Answers

4. a. 45,   **b.** $24x^5$

Example 4   Find the LCD for each pair.

a. $\frac{1}{8}, \frac{3}{22}$ **b.** $\frac{7}{5x}, \frac{6}{15x^2}$

Solution: **a.** We start by finding the prime factorization of each denominator.

$$8 = 2 \cdot 2 \cdot 2 = 2^3 \quad \text{and} \quad 22 = 2 \cdot 11$$

Next we write the product of all the unique factors, each raised to a power equal to the greatest number of times that the factor appears.

The greatest number of times that the factor 2 appears is 3.

The greatest number of times that the factor 11 appears is 1.

$$\text{LCD} = 2^3 \cdot 11^1 = 8 \cdot 11 = 88$$

b. We factor each denominator.

$$5x = 5 \cdot x \quad \text{and} \quad 15x^2 = 3 \cdot 5 \cdot x^2$$

The greatest number of times that the factor 5 appears is 1.

The greatest number of times that the factor 3 appears is 1.

The greatest number of times that the factor x appears is 2.

$$\text{LCD} = 3^1 \cdot 5^1 \cdot x^2 = 15x^2$$

Example 5   Find the LCD of $\dfrac{7x}{x+2}$ and $\dfrac{5x^2}{x-2}$.

Solution: The denominators $x + 2$ and $x - 2$ are completely factored already. The factor $x + 2$ appears once and the factor $x - 2$ appears once.

$$\text{LCD} = (x + 2)(x - 2)$$

Example 6   Find the LCD of $\dfrac{6m^2}{3m+15}$ and $\dfrac{2}{(m+5)^2}$.

Solution: We factor each denominator.

$$3m + 15 = 3(m + 5)$$
$$(m + 5)^2 = (m + 5)^2 \quad \text{This denominator is already factored.}$$

The greatest number of times that the factor 3 appears is 1.

The greatest number of times that the factor $m + 5$ appears *in any one denominator* is 2.

$$\text{LCD} = 3(m + 5)^2$$

TRY THE CONCEPT CHECK IN THE MARGIN.

Example 7   Find the LCD of $\dfrac{t-10}{t^2-t-6}$ and $\dfrac{t+5}{t^2+3t+2}$.

Solution: $t^2 - t - 6 = (t - 3)(t + 2)$
$t^2 + 3t + 2 = (t + 1)(t + 2)$
$\text{LCD} = (t - 3)(t + 2)(t + 1)$

Practice Problem 5

Find the LCD of $\dfrac{3a}{a+5}$ and $\dfrac{7a}{a-5}$.

Practice Problem 6

Find the LCD of $\dfrac{7x^2}{(x-4)^2}$ and $\dfrac{5x}{3x-12}$.

✓ CONCEPT CHECK

Choose the correct LCD of $\dfrac{x}{(x+1)^2}$ and $\dfrac{5}{x+1}$.

a. $x + 1$ b. $(x + 1)^2$

c. $(x + 1)^3$ d. $5x(x + 1)^2$

Practice Problem 7

Find the LCD of $\dfrac{y+5}{y^2+2y-3}$ and $\dfrac{y+4}{y^2-3y+2}$.

Answers

5. $(a + 5)(a - 5)$, **6.** $3(x - 4)^2$,
7. $(y + 3)(y - 2)(y - 1)$
✓ Concept Check: b

Practice Problem 8

Find the LCD of $\dfrac{6}{x-4}$ and $\dfrac{9}{4-x}$.

TEACHING TIP

Before discussing Example 8, have your students find the LCD of

$\dfrac{2}{15}$ and $\dfrac{10}{-15}$

$\dfrac{2}{17-2}$ and $\dfrac{10}{2-17}$

$\dfrac{2}{x-2}$ and $\dfrac{10}{2-x}$ (Example 8)

Practice Problem 9

Write the rational expression as an equivalent rational expression with the given denominator.

$$\dfrac{2x}{5y} = \dfrac{}{20x^2y^2}$$

Practice Problem 10

Write the rational expression as an equivalent rational expression with the given denominator.

$$\dfrac{3}{x^2-25} = \dfrac{}{(x+5)(x-5)(x-3)}$$

Answers

8. $(x-4)$ or $(4-x)$,   **9.** $\dfrac{8x^3y}{20x^2y^2}$,

10. $\dfrac{3x-9}{(x+5)(x-5)(x-3)}$

Example 8   Find the LCD of $\dfrac{2}{x-2}$ and $\dfrac{10}{2-x}$.

Solution: The denominators $x-2$ and $2-x$ are opposites. That is, $2-x = -1(x-2)$. We can use either $x-2$ or $2-x$ as the LCD.

$$\text{LCD} = x-2 \quad \text{or} \quad \text{LCD} = 2-x$$

C WRITING EQUIVALENT RATIONAL EXPRESSIONS

Next we practice writing a rational expression as an equivalent rational expression with a given denominator. To do this, we apply the fundamental principle, which says that $\dfrac{PR}{QR} = \dfrac{P}{Q}$, or equivalently that $\dfrac{P}{Q} = \dfrac{PR}{QR}$. This can be seen by recalling that multiplying an expression by 1 produces an equivalent expression. In other words,

$$\dfrac{P}{Q} = \dfrac{P}{Q} \cdot 1 = \dfrac{P}{Q} \cdot \dfrac{R}{R} = \dfrac{PR}{QR}$$

Example 9   Write the rational expression as an equivalent rational expression with the given denominator.

$$\dfrac{4b}{9a} = \dfrac{}{27a^2b}$$

Solution: We can ask ourselves: "What do we multiply $9a$ by to get $27a^2b$?" The answer is $3ab$, since $9a(3ab) = 27a^2b$. So we multiply the numerator and denominator by $3ab$.

$$\dfrac{4b}{9a} = \dfrac{4b(3ab)}{9a(3ab)} = \dfrac{12ab^2}{27a^2b}$$

Example 10   Write the rational expression as an equivalent rational expression with the given denominator.

$$\dfrac{5}{x^2-4} = \dfrac{}{(x-2)(x+2)(x-4)}$$

Solution: First we factor the denominator x^2-4 as $(x-2)(x+2)$. If we multiply the original denominator $(x-2)(x+2)$ by $x-4$, the result is the new denominator $(x-2)(x+2)(x-4)$. Thus, we multiply the numerator and the denominator by $x-4$.

$$\dfrac{5}{x^2-4} = \dfrac{5}{(x-2)(x+2)} = \dfrac{5(x-4)}{(x-2)(x+2)(x-4)}$$

$$= \dfrac{5x-20}{(x-2)(x+2)(x-4)}$$

Name _____ Section _____ Date _____

MENTAL MATH ANSWERS

1. 1

2. $\dfrac{6}{11}$

3. $\dfrac{7x}{9}$

4. $\dfrac{5y}{8}$

5. $\dfrac{1}{9}$

6. $\dfrac{11}{12}$

7. $\dfrac{17y}{5}$

8. $\dfrac{8x}{7}$

MENTAL MATH

Perform each indicated operation.

1. $\dfrac{2}{3} + \dfrac{1}{3}$
2. $\dfrac{5}{11} + \dfrac{1}{11}$
3. $\dfrac{3x}{9} + \dfrac{4x}{9}$
4. $\dfrac{3y}{8} + \dfrac{2y}{8}$

5. $\dfrac{8}{9} - \dfrac{7}{9}$
6. $\dfrac{14}{12} - \dfrac{3}{12}$
7. $\dfrac{7y}{5} + \dfrac{10y}{5}$
8. $\dfrac{12x}{7} - \dfrac{4x}{7}$

EXERCISE SET 5.3

A *Add or subtract as indicated. Simplify the result if possible. See Examples 1 through 3.*

1. $\dfrac{a}{13} + \dfrac{9}{13}$
2. $\dfrac{x+1}{7} + \dfrac{6}{7}$
3. $\dfrac{4m}{3n} + \dfrac{5m}{3n}$

4. $\dfrac{3p}{2} + \dfrac{11p}{2}$
5. $\dfrac{4m}{m-6} - \dfrac{24}{m-6}$
6. $\dfrac{8y}{y-2} - \dfrac{16}{y-2}$

7. $\dfrac{9}{3+y} + \dfrac{y+1}{3+y}$
8. $\dfrac{9}{y+9} + \dfrac{y}{y+9}$
9. $\dfrac{5x+4}{x-1} - \dfrac{2x+7}{x-1}$

10. $\dfrac{x^2+9x}{x+7} - \dfrac{4x+14}{x+7}$
11. $\dfrac{a}{a^2+2a-15} - \dfrac{3}{a^2+2a-15}$

12. $\dfrac{3y}{y^2+3y-10} - \dfrac{6}{y^2+3y-10}$
13. $\dfrac{2x+3}{x^2-x-30} - \dfrac{x-2}{x^2-x-30}$

14. $\dfrac{3x-1}{x^2+5x-6} - \dfrac{2x-7}{x^2+5x-6}$

ANSWERS

1. $\dfrac{a+9}{13}$

2. $\dfrac{x+7}{7}$

3. $\dfrac{3m}{n}$

4. $7p$

5. 4

6. 8

7. $\dfrac{y+10}{3+y}$

8. 1

9. 3

10. $x-2$

11. $\dfrac{1}{a+5}$

12. $\dfrac{3}{y+5}$

13. $\dfrac{1}{x-6}$

14. $\dfrac{1}{x-1}$

15. $\dfrac{20}{x-2}$ m

16. $\dfrac{2x+15}{x+3}$ in.

17. answers may vary

18. answers may vary

19. $4x^3$

20. $8y^5$

21. $8x(x+2)$

22. $12y(y+3)$

23. $(x+3)(x-2)$

24. $(x-1)(x+5)$

25. $3(x+6)$

26. $4(x+5)$

27. $6(x+1)^2$

28. $4(x-3)^2$

29. $x-8$ or $8-x$

30. $3x-7$ or $7-3x$

31. $(x-1)(x+4)(x+3)$

32. $(x+3)(x+1)(x+7)$

372

☐ **15.** A square has a side of length $\dfrac{5}{x-2}$ meters. Express its perimeter as a rational expression.

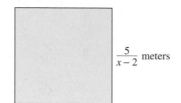
$\dfrac{5}{x-2}$ meters

☐ **16.** A trapezoid has sides of the indicated lengths. Find its perimeter.

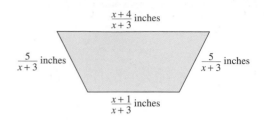
$\dfrac{x+4}{x+3}$ inches
$\dfrac{5}{x+3}$ inches $\dfrac{5}{x+3}$ inches
$\dfrac{x+1}{x+3}$ inches

☐ **17.** In your own words, describe how to add or subtract two rational expressions with the same denominators.

☐ **18.** Explain the similarities between subtracting $\dfrac{3}{8}$ from $\dfrac{7}{8}$ and subtracting $\dfrac{6}{x+3}$ from $\dfrac{9}{x+3}$.

B *Find the LCD for each list of rational expressions. See Examples 4 through 8.*

19. $\dfrac{19}{2x}$, $\dfrac{5}{4x^3}$

20. $\dfrac{17x}{4y^5}$, $\dfrac{2}{8y}$

21. $\dfrac{9}{8x}$, $\dfrac{3}{2x+4}$

22. $\dfrac{1}{6y}$, $\dfrac{3x}{4y+12}$

23. $\dfrac{2}{x+3}$, $\dfrac{5}{x-2}$

24. $\dfrac{-6}{x-1}$, $\dfrac{4}{x+5}$

25. $\dfrac{x}{x+6}$, $\dfrac{10}{3x+18}$

26. $\dfrac{12}{x+5}$, $\dfrac{x}{4x+20}$

▭ **27.** $\dfrac{1}{3x+3}$, $\dfrac{8}{2x^2+4x+2}$

28. $\dfrac{19x+5}{4x-12}$, $\dfrac{3}{2x^2-12x+18}$

29. $\dfrac{5}{x-8}$, $\dfrac{3}{8-x}$

30. $\dfrac{2x+5}{3x-7}$, $\dfrac{5}{7-3x}$

31. $\dfrac{5x+1}{x^2+3x-4}$, $\dfrac{3x}{x^2+2x-3}$

32. $\dfrac{4}{x^2+4x+3}$, $\dfrac{4x-2}{x^2+10x+21}$

Name _____

33. answers may vary

33. Write some instructions to help a friend who is having difficulty finding the LCD of two rational expressions.

34. Explain why the LCD of the rational expressions $\dfrac{7}{x+1}$ and $\dfrac{9x}{(x+1)^2}$ is $(x+1)^2$ and not $(x+1)^3$.

34. answers may vary

35. $\dfrac{6x}{4x^2}$

C *Rewrite each rational expression as an equivalent rational expression with the given denominator. See Examples 9 and 10.*

35. $\dfrac{3}{2x} = \dfrac{}{4x^2}$

36. $\dfrac{3}{9y^5} = \dfrac{}{72y^9}$

37. $\dfrac{6}{3a} = \dfrac{}{12ab^2}$

36. $\dfrac{24y^4}{72y^9}$

37. $\dfrac{24b^2}{12ab^2}$

38. $\dfrac{136ayxz}{32y^3x^2z}$

38. $\dfrac{17a}{4y^2x} = \dfrac{}{32y^3x^2z}$

39. $\dfrac{9}{x+3} = \dfrac{}{2(x+3)}$

40. $\dfrac{4x+1}{3x+6} = \dfrac{}{3y(x+2)}$

39. $\dfrac{18}{2(x+3)}$

40. $\dfrac{4xy+y}{3y(x+2)}$

41. $\dfrac{9a+2}{5a+10} = \dfrac{}{5b(a+2)}$

42. $\dfrac{5+y}{2x^2+10} = \dfrac{}{4(x^2+5)}$

41. $\dfrac{9ba+2b}{5b(a+2)}$

42. $\dfrac{10+2y}{4(x^2+5)}$

43. $\dfrac{x}{x^3+6x^2+8x} = \dfrac{}{x(x+4)(x+2)(x+1)}$

43. $\dfrac{x^2+x}{x(x+4)(x+2)(x+1)}$

44. $\dfrac{5x}{x^2+2x-3} = \dfrac{}{(x-1)(x-5)(x+3)}$

44. $\dfrac{5x^2-25x}{(x-1)(x-5)(x+3)}$

45. $\dfrac{18y-2}{30x^2-60}$

45. $\dfrac{9y-1}{15x^2-30} = \dfrac{}{30x^2-60}$

46. $\dfrac{6}{x^2-9} = \dfrac{}{(x+3)(x-3)(x+2)}$

46. $\dfrac{6x+12}{(x+3)(x-3)(x+2)}$

373

47. $\dfrac{29}{21}$

48. $\dfrac{3}{10}$

49. $-\dfrac{5}{12}$

50. $\dfrac{58}{45}$

51. $\dfrac{7}{30}$

52. $\dfrac{2}{5}$

53. 3 packages hot dogs and 2 packages buns

54. 95,304 Earth days

55. answers may vary

REVIEW AND PREVIEW

Perform each indicated operation. See Section R.2.

47. $\dfrac{2}{3} + \dfrac{5}{7}$

48. $\dfrac{9}{10} - \dfrac{3}{5}$

49. $\dfrac{2}{6} - \dfrac{3}{4}$

50. $\dfrac{11}{15} + \dfrac{5}{9}$

51. $\dfrac{1}{12} + \dfrac{3}{20}$

52. $\dfrac{7}{30} + \dfrac{3}{18}$

◆ COMBINING CONCEPTS

53. You are throwing a barbecue and you want to make sure that you purchase the same number of hot dogs as hot dog buns. Hot dogs come 8 to a package and hot dog buns come 12 to a package. What is the least number of each type of package you should buy?

54. The planet Mercury revolves around the sun in 88 Earth days. It takes Jupiter 4332 Earth days to make one revolution around the sun. (*Source*: National Space Science Data Center) If the two planets are aligned as shown in the figure, how long will it take for them to align again?

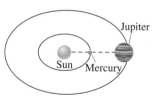

55. An algebra student approaches you with a problem. He's tried to subtract two rational expressions, but his result does not match the book's. Check to see if the student has made an error. If so, correct his work shown below.

$$\dfrac{2x - 6}{x - 5} - \dfrac{x + 4}{x - 5}$$
$$= \dfrac{2x - 6 - x + 4}{x - 5}$$
$$= \dfrac{x - 2}{x - 5}$$

5.4 ADDING AND SUBTRACTING RATIONAL EXPRESSIONS WITH DIFFERENT DENOMINATORS

A ADDING AND SUBTRACTING RATIONAL EXPRESSIONS WITH DIFFERENT DENOMINATORS

In the previous section, we practiced all the skills we need to add and subtract rational expressions with different denominators. The steps are as follows:

> **TO ADD OR SUBTRACT RATIONAL EXPRESSIONS WITH DIFFERENT DENOMINATORS**
>
> **Step 1.** Find the LCD of the rational expressions.
>
> **Step 2.** Rewrite each rational expression as an equivalent expression whose denominator is the LCD found in Step 1.
>
> **Step 3.** Add or subtract numerators and write the sum or difference over the common denominator.
>
> **Step 4.** Simplify or write the rational expression in lowest terms.

Objective

A Add and subtract rational expressions with different denominators.

SSM   CD-ROM   Video 5.4

Example 1   Perform each indicated operation.

a. $\dfrac{a}{4} - \dfrac{2a}{8}$ **b.** $\dfrac{3}{10x^2} + \dfrac{7}{25x}$

Solution: **a.** First, we must find the LCD. Since $4 = 2^2$ and $8 = 2^3$, the LCD $= 2^3 = 8$. Next we write each fraction as an equivalent fraction with the denominator 8, then we subtract.

$$\dfrac{a}{4} - \dfrac{2a}{8} = \dfrac{a(2)}{4(2)} - \dfrac{2a}{8} = \dfrac{2a}{8} - \dfrac{2a}{8} = \dfrac{2a - 2a}{8} = \dfrac{0}{8} = 0$$

b. Since $10x^2 = 2 \cdot 5 \cdot x \cdot x$ and $25x = 5 \cdot 5 \cdot x$, the LCD $= 2 \cdot 5^2 \cdot x^2 = 50x^2$. We write each fraction as an equivalent fraction with a denominator of $50x^2$.

$$\dfrac{3}{10x^2} + \dfrac{7}{25x} = \dfrac{3(5)}{10x^2(5)} + \dfrac{7(2x)}{25x(2x)}$$

$$= \dfrac{15}{50x^2} + \dfrac{14x}{50x^2}$$

$$= \dfrac{15 + 14x}{50x^2} \quad \text{Add numerators. Write the sum over the common denominator.}$$

Example 2   Subtract: $\dfrac{6x}{x^2 - 4} - \dfrac{3}{x + 2}$

Solution: Since $x^2 - 4 = (x + 2)(x - 2)$, the LCD $= (x - 2)(x + 2)$. We write equivalent expressions with the LCD as denominators.

Practice Problem 1

Perform each indicated operation.

a. $\dfrac{y}{5} - \dfrac{3y}{15}$ **b.** $\dfrac{5}{8x} + \dfrac{11}{10x^2}$

TEACHING TIP

In Example 1b, for instance, help students understand that the expression $\dfrac{3}{10x^2} + \dfrac{7}{25x}$ has the same value as $\dfrac{15 + 14x}{50x^2}$, for all numbers except 0. To see this, have students evaluate $\dfrac{3}{10x^2} + \dfrac{7}{25x}$ and the simplified form $\dfrac{15 + 14x}{50x^2}$ for $x = 2$. They should get $\dfrac{43}{200}$ in both cases.

Practice Problem 2

Subtract: $\dfrac{10x}{x^2 - 9} - \dfrac{5}{x + 3}$

Answers

1. a. 0, **b.** $\dfrac{25x + 44}{40x^2}$, **2.** $\dfrac{5}{x - 3}$

$$\frac{6x}{x^2 - 4} - \frac{3}{x + 2} = \frac{6x}{(x - 2)(x + 2)} - \frac{3(x - 2)}{(x + 2)(x - 2)}$$

$$= \frac{6x - 3(x - 2)}{(x + 2)(x - 2)} \quad \text{Subtract numerators. Write the difference over the common denominator.}$$

$$= \frac{6x - 3x + 6}{(x + 2)(x - 2)} \quad \text{Apply the distributive property in the numerator.}$$

$$= \frac{3x + 6}{(x + 2)(x - 2)} \quad \text{Combine like terms in the numerator.}$$

Next we factor the numerator to see if this rational expression can be simplified.

$$= \frac{3(x + 2)}{(x + 2)(x - 2)} \quad \text{Factor.}$$

$$= \frac{3}{x - 2} \quad \text{Apply the fundamental principle to simplify.}$$

Practice Problem 3

Add: $\dfrac{5}{7x} + \dfrac{2}{x + 1}$

Example 3   Add: $\dfrac{2}{3t} + \dfrac{5}{t + 1}$

Solution: The LCD is $3t(t + 1)$. We write each rational expression as an equivalent rational expression with a denominator of $3t(t + 1)$.

$$\frac{2}{3t} + \frac{5}{t + 1} = \frac{2(t + 1)}{3t(t + 1)} + \frac{5(3t)}{(t + 1)(3t)}$$

$$= \frac{2(t + 1) + 5(3t)}{3t(t + 1)} \quad \text{Add numerators. Write the sum over the common denominator.}$$

$$= \frac{2t + 2 + 15t}{3t(t + 1)} \quad \text{Apply the distributive property in the numerator.}$$

$$= \frac{17t + 2}{3t(t + 1)} \quad \text{Combine like terms in the numerator.}$$

Practice Problem 4

Subtract: $\dfrac{10}{x - 6} - \dfrac{15}{6 - x}$

TEACHING TIP

In Example 4, to verify for students that $x - 3$ and $3 - x$ are opposites, replace x with several values and have students notice the results. To verify that $3 - x$ and $-(x - 3)$ are equal, have students replace x with several values and notice the results.

Example 4   Subtract: $\dfrac{7}{x - 3} - \dfrac{9}{3 - x}$

Solution: To find a common denominator, we notice that $x - 3$ and $3 - x$ are opposites. That is, $3 - x = -(x - 3)$. We write the denominator $3 - x$ as $-(x - 3)$ and simplify.

$$\frac{7}{x - 3} - \frac{9}{3 - x} = \frac{7}{x - 3} - \frac{9}{-(x - 3)}$$

$$= \frac{7}{x - 3} - \frac{-9}{x - 3} \quad \text{Apply } \frac{a}{-b} = \frac{-a}{b}.$$

$$= \frac{7 - (-9)}{x - 3} \quad \text{Subtract numerators. Write the difference over the common denominator.}$$

$$= \frac{16}{x - 3}$$

Practice Problem 5

Add: $2 + \dfrac{x}{x + 5}$

Example 5   Add: $1 + \dfrac{m}{m + 1}$

Solution: Recall that 1 is the same as $\dfrac{1}{1}$. The LCD of $\dfrac{1}{1}$ and $\dfrac{m}{m + 1}$ is $m + 1$.

Answers

3. $\dfrac{19x + 5}{7x(x + 1)}$,   **4.** $\dfrac{25}{x - 6}$,   **5.** $\dfrac{3x + 10}{x + 5}$

$$1 + \frac{m}{m+1} = \frac{1}{1} + \frac{m}{m+1}$$ Write 1 as $\frac{1}{1}$.

$$= \frac{1(m+1)}{1(m+1)} + \frac{m}{m+1}$$ Multiply both the numerator and the denominator of $\frac{1}{1}$ by $m+1$.

$$= \frac{m+1+m}{m+1}$$ Add numerators. Write the sum over the common denominator.

$$= \frac{2m+1}{m+1}$$ Combine like terms in the numerator.

Example 6   Subtract: $\dfrac{3}{2x^2 + x} - \dfrac{2x}{6x+3}$

Solution: First, we factor the denominators.

$$\frac{3}{2x^2+x} - \frac{2x}{6x+3} = \frac{3}{x(2x+1)} - \frac{2x}{3(2x+1)}$$

The LCD is $3x(2x+1)$. We write equivalent expressions with denominators of $3x(2x+1)$.

$$= \frac{3(3)}{x(2x+1)(3)} - \frac{2x(x)}{3(2x+1)(x)}$$

$$= \frac{9 - 2x^2}{3x(2x+1)}$$ Subtract numerators. Write the difference over the common denominator.

Practice Problem 6

Subtract: $\dfrac{4}{3x^2 + 2x} - \dfrac{3x}{12x+8}$

Example 7   Add: $\dfrac{2x}{x^2 + 2x + 1} + \dfrac{x}{x^2 - 1}$

Solution: First we factor the denominators.

$$\frac{2x}{x^2 + 2x + 1} + \frac{x}{x^2 - 1} = \frac{2x}{(x+1)(x+1)} + \frac{x}{(x+1)(x-1)}$$

Now we write the rational expressions as equivalent expressions with denominators of $(x+1)(x+1)(x-1)$, the LCD.

$$= \frac{2x(x-1)}{(x+1)(x+1)(x-1)} + \frac{x(x+1)}{(x+1)(x-1)(x+1)}$$

$$= \frac{2x(x-1) + x(x+1)}{(x+1)^2(x-1)}$$ Add numerators. Write the sum over the common denominator.

$$= \frac{2x^2 - 2x + x^2 + x}{(x+1)^2(x-1)}$$ Apply the distributive property in the numerator.

$$= \frac{3x^2 - x}{(x+1)^2(x-1)} \quad \text{or} \quad \frac{x(3x-1)}{(x+1)^2(x-1)}$$

The numerator was factored as a last step to see if the rational expression could be simplified further. Since there are no factors common to the numerator and the denominator, we can't simplify further.

Practice Problem 7

Add: $\dfrac{6x}{x^2 + 4x + 4} + \dfrac{x}{x^2 - 4}$

Answers

6. $\dfrac{16 - 3x^2}{4x(3x+2)}$,   **7.** $\dfrac{x(7x-10)}{(x+2)^2(x-2)}$

Focus On Business and Career

According to U.S. Bureau of Labor Statistics projections, the careers listed below will have the largest job growth into the next century.

Occupation	Employment [Numbers in thousands]		
	1994	2005	Change
1. Cashiers	3,005	3,567	+562
2. Janitors and cleaners, including maids and housekeeping cleaners	3,043	3,602	+559
3. Salespersons, retail	3,842	4,374	+532
4. Waiters and waitresses	1,847	2,326	+479
5. Registered nurses	1,906	2,379	+473
6. General managers and top executives	3,046	3,512	+466
7. Systems analysts	483	928	+445
8. Home health aides	420	848	+428
9. Guards	867	1,282	+415
10. Nursing aides, orderlies, and attendants	1,265	1,652	+387
11. Teachers, secondary	1,340	1,726	+386
12. Marketing and sales worker supervisors	2,293	2,673	+380
13. Teacher's aides and educational associates	932	1,296	+364
14. Receptionists and information clerks	1,019	1,337	+318
15. Truck drivers, light and heavy	2,565	2,837	+271
16. Secretaries, except legal and medical	2,842	3,109	+267
17. Clerical supervisors and managers	1,340	1,600	+261
18. Child care workers	757	1,005	+248
19. Maintenance repairers, general utility	1,273	1,505	+231
20. Teachers, elementary	1,419	1,639	+220

Source: Bureau of Labor Statistics, Office of Employment Projections, November 1995

What do all of these in-demand occupations have in common? They all require a knowledge of math! For some careers like cashiers, salespersons, waiters and waitresses, financial managers, and computer engineers, the ways math is used on the job may be obvious. For other occupations, the use of math may not be quite as obvious. However, tasks common to many jobs like filling in a time sheet, writing up an expense or mileage report, planning a budget, figuring a bill, ordering supplies, completing a packing list, and even making a work schedule all require math.

CRITICAL THINKING

Suppose that your college placement office is planning to publish an occupational handbook on math in popular occupations. Choose one of the occupations from the list above that interests you. Research the occupation. Then write a brief entry for the occupational handbook that describes how a person in that career would use math in his or her job. Include an example if possible.

Exercise Set 5.4

A *Perform each indicated operation. Simplify if possible. See Example 1.*

1. $\dfrac{4}{2x} + \dfrac{9}{3x}$

2. $\dfrac{15}{7a} + \dfrac{8}{6a}$

📼 **3.** $\dfrac{15a}{b} + \dfrac{6b}{5}$

4. $\dfrac{4c}{d} - \dfrac{8x}{5}$

5. $\dfrac{3}{x} + \dfrac{5}{2x^2}$

6. $\dfrac{14}{3x^2} + \dfrac{6}{x}$

7. $\dfrac{6}{x+1} + \dfrac{10}{2x+2}$

8. $\dfrac{8}{x+4} - \dfrac{3}{3x+12}$

9. $\dfrac{15}{2x-4} + \dfrac{x}{x^2-4}$

10. $\dfrac{3}{x+2} - \dfrac{1}{x^2-4}$

11. $\dfrac{3}{4x} + \dfrac{8}{x-2}$

12. $\dfrac{5}{y^2} - \dfrac{y}{2y+1}$

📼 **13.** $\dfrac{6}{x-3} + \dfrac{8}{3-x}$

14. $\dfrac{9}{x-3} + \dfrac{9}{3-x}$

15. $\dfrac{-8}{x^2-1} - \dfrac{7}{1-x^2}$

16. $\dfrac{-9}{25x^2-1} + \dfrac{7}{1-25x^2}$

17. $\dfrac{5}{x} + 2$

18. $\dfrac{7}{x^2} - 5x$

📼 **19.** $\dfrac{5}{x-2} + 6$

20. $\dfrac{6y}{y+5} + 1$

21. $\dfrac{y+2}{y+3} - 2$

22. $\dfrac{7}{2x-3} - 3$

ANSWERS

1. $\dfrac{5}{x}$

2. $\dfrac{73}{21a}$

3. $\dfrac{75a + 6b^2}{5b}$

4. $\dfrac{20c - 8dx}{5d}$

5. $\dfrac{6x + 5}{2x^2}$

6. $\dfrac{14 + 18x}{3x^2}$

7. $\dfrac{11}{x+1}$

8. $\dfrac{7}{x+4}$

9. $\dfrac{17x + 30}{2(x-2)(x+2)}$

10. $\dfrac{3x - 7}{(x-2)(x+2)}$

11. $\dfrac{35x - 6}{4x(x-2)}$

12. $\dfrac{5 + 10y - y^3}{y^2(2y+1)}$

13. $-\dfrac{2}{x-3}$

14. 0

15. $\dfrac{-1}{x^2-1}$

16. $\dfrac{-16}{25x^2-1}$

17. $\dfrac{5 + 2x}{x}$

18. $\dfrac{7 - 5x^3}{x^2}$

19. $\dfrac{6x - 7}{x-2}$

20. $\dfrac{7y + 5}{y+5}$

21. $\dfrac{-y + 4}{y+3}$

22. $\dfrac{16 - 6x}{2x-3}$

23. _answers may vary_

24. _answers may vary_

25. 2

26. 3

27. $3x^3 - 4$

28. $\dfrac{5x + 45x^2}{6}$

29. $\dfrac{x + 2}{(x + 3)^2}$

30. $\dfrac{2(x + 3)}{(x - 2)^2}$

31. $\dfrac{9b - 4}{5b(b - 1)}$

32. $\dfrac{5(y + 2)}{3y(y + 5)}$

33. $\dfrac{2 + m}{m}$

34. $\dfrac{6 - x}{x}$

35. $\dfrac{10}{1 - 2x}$

36. $\dfrac{15}{3n - 4}$

37. $\dfrac{15x - 1}{(x + 1)^2(x - 1)}$

38. $\dfrac{3x}{(x + 1)(x + 5)}$

39. $\dfrac{x^2 - 3x - 2}{(x - 1)^2(x + 1)}$

40. $\dfrac{x^2 - 7x - 10}{(x + 2)(x - 2)^2}$

41. $\dfrac{a + 2}{2(a + 3)}$

42. $\dfrac{x - 2y}{(x - y)(x + y)}$

43. $\dfrac{x - 10}{2(x - 2)}$

44. $\dfrac{3}{2 - a}$

23. In your own words, explain how to add two rational expressions with different denominators.

24. In your own words, explain how to subtract two rational expressions with different denominators.

Perform each indicated operation. Simplify if possible. See Examples 4 through 7.

25. $\dfrac{5x}{x + 2} - \dfrac{3x - 4}{x + 2}$

26. $\dfrac{7x}{x - 3} - \dfrac{4x + 9}{x - 3}$

27. $\dfrac{3x^4}{x} - \dfrac{4x^2}{x^2}$

28. $\dfrac{5x}{6} + \dfrac{15x^2}{2}$

29. $\dfrac{1}{x + 3} - \dfrac{1}{(x + 3)^2}$

30. $\dfrac{5x}{(x - 2)^2} - \dfrac{3}{x - 2}$

31. $\dfrac{4}{5b} + \dfrac{1}{b - 1}$

32. $\dfrac{1}{y + 5} + \dfrac{2}{3y}$

33. $\dfrac{2}{m} + 1$

34. $\dfrac{6}{x} - 1$

35. $\dfrac{6}{1 - 2x} - \dfrac{4}{2x - 1}$

36. $\dfrac{10}{3n - 4} - \dfrac{5}{4 - 3n}$

37. $\dfrac{7}{(x + 1)(x - 1)} + \dfrac{8}{(x + 1)^2}$

38. $\dfrac{5x + 2}{(x + 1)(x + 5)} - \dfrac{2}{x + 5}$

39. $\dfrac{x}{x^2 - 1} - \dfrac{2}{x^2 - 2x + 1}$

40. $\dfrac{x}{x^2 - 4} - \dfrac{5}{x^2 - 4x + 4}$

41. $\dfrac{3a}{2a + 6} - \dfrac{a - 1}{a + 3}$

42. $\dfrac{1}{x + y} - \dfrac{y}{x^2 - y^2}$

43. $\dfrac{5}{2 - x} + \dfrac{x}{2x - 4}$

44. $\dfrac{-1}{a - 2} + \dfrac{4}{4 - 2a}$

Name _____

45. $\dfrac{-7}{y^2 - 3y + 2} - \dfrac{2}{y - 1}$

46. $\dfrac{2}{x^2 + 4x + 4} + \dfrac{1}{x + 2}$

47. $\dfrac{13}{x^2 - 5x + 6} - \dfrac{5}{x - 3}$

48. $\dfrac{27}{y^2 - 81} + \dfrac{3}{2(y + 9)}$

49. $\dfrac{x + 8}{x^2 - 5x - 6} + \dfrac{x + 1}{x^2 - 4x - 5}$

50. $\dfrac{x}{x^2 + 12x + 20} - \dfrac{1}{x^2 + 8x - 20}$

REVIEW AND PREVIEW

Solve each linear or quadratic equation. See Sections 2.4 and 4.6.

51. $3x + 5 = 7$

52. $5x - 1 = 8$

53. $2x^2 - x - 1 = 0$

54. $4x^2 - 9 = 0$

55. $4(x + 6) + 3 = -3$

56. $2(3x + 1) + 15 = -7$

COMBINING CONCEPTS

57. A board of length $\dfrac{3}{x + 4}$ inches was cut into two pieces. If one piece is $\dfrac{1}{x - 4}$ inches, express the length of the other board as a rational expression.

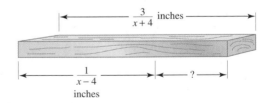

$\dfrac{3}{x+4}$ inches

$\dfrac{1}{x-4}$ inches

?

58. The length of a rectangle is $\dfrac{3}{y - 5}$ feet, while its width is $\dfrac{2}{y}$ feet. Find its perimeter and then find its area.

$\dfrac{3}{y-5}$ feet

$\dfrac{2}{y}$ feet

45. $\dfrac{-3 - 2y}{(y - 2)(y - 1)}$

46. $\dfrac{x + 4}{(x + 2)(x + 2)}$

47. $\dfrac{-5x + 23}{(x - 2)(x - 3)}$

48. $\dfrac{3}{2(y - 9)}$

49. $\dfrac{2x^2 - 2x - 46}{(x + 1)(x - 6)(x - 5)}$

50. $\dfrac{x^2 - 3x - 2}{(x + 10)(x + 2)(x - 2)}$

51. $x = \dfrac{2}{3}$

52. $x = \dfrac{9}{5}$

53. $x = -\dfrac{1}{2}, x = 1$

54. $x = \dfrac{3}{2}, x = -\dfrac{3}{2}$

55. $x = -\dfrac{15}{2}$

56. $x = -4$

57. $\dfrac{2x - 16}{(x - 4)(x + 4)}$ in.

58. $\dfrac{10y - 20}{y(y - 5)}$ ft; $\dfrac{6}{y^2 - 5y}$ sq. ft

59. $\dfrac{11DA - DA^2 - 12D}{24(A + 12)}$

60. answers may vary

61. answers may vary

62. $\dfrac{4x^2 - 15x + 6}{(x - 2)^2(x + 2)(x - 3)}$

63. $\dfrac{-3x^2 + 7x + 2}{(x + 5)(x + 1)(x - 1)}$

64. $\dfrac{-3x^2 + 7x + 55}{(x + 2)(x + 7)(x + 3)}$

65. $\dfrac{73 - 7x}{(x - 4)(x + 1)(x + 5)}$

59. The dose of medicine prescribed for a child depends on the child's age A in years and the adult dose D for the medication. Two expressions that give a child's dose are Young's Rule, $\dfrac{DA}{A + 12}$, and Cowling's Rule, $\dfrac{D(A + 1)}{24}$. Find an expression for the difference in the doses given by these expressions.

60. Explain when the LCD is the product of the denominators.

61. Explain when the LCD is the same as one of the denominators of a rational expression to be added or subtracted.

Perform each indicated operation:

62. $\dfrac{5}{x^2 - 4} + \dfrac{2}{x^2 - 4x + 4} - \dfrac{3}{x^2 - x - 6}$

63. $\dfrac{8}{x^2 + 6x + 5} - \dfrac{3x}{x^2 + 4x - 5} + \dfrac{2}{x^2 - 1}$

64. $\dfrac{9}{x^2 + 9x + 14} - \dfrac{3x}{x^2 + 10x + 21} + \dfrac{4}{x^2 + 5x + 6}$

65. $\dfrac{10}{x^2 - 3x - 4} - \dfrac{8}{x^2 + 6x + 5} - \dfrac{9}{x^2 + x - 20}$

5.5 SOLVING EQUATIONS CONTAINING RATIONAL EXPRESSIONS

A SOLVING EQUATIONS CONTAINING RATIONAL EXPRESSIONS

In Chapter 2, we solved equations containing fractions. In this section, we continue the work we began in Chapter 2 by solving equations containing rational expressions. For example,

$$\frac{x}{5} + \frac{x+2}{9} = 8 \quad \text{and} \quad \frac{x+1}{9x-5} = \frac{2}{3x}$$

are equations containing rational expressions. To solve equations such as these, we use the multiplication property of equality to clear the equation of fractions by multiplying both sides of the equation by the LCD.

Example 1   Solve: $\frac{x}{2} + \frac{8}{3} = \frac{1}{6}$

Solution: The LCD of denominators 2, 3, and 6 is 6, so we multiply both sides of the equation by 6.

$$6\left(\frac{x}{2} + \frac{8}{3}\right) = 6\left(\frac{1}{6}\right)$$

> **HELPFUL HINT**
> Make sure that *each* term is multiplied by the LCD.

$$6\left(\frac{x}{2}\right) + 6\left(\frac{8}{3}\right) = 6\left(\frac{1}{6}\right) \quad \text{Use the distributive property.}$$

$$3 \cdot x + 16 = 1 \quad \text{Multiply and simplify.}$$
$$3x = -15 \quad \text{Subtract 16 from both sides.}$$
$$x = -5 \quad \text{Divide both sides by 3.}$$

Check: To check, we replace x with -5 in the original equation.

$$\frac{-5}{2} + \frac{8}{3} \stackrel{?}{=} \frac{1}{6} \quad \text{Replace } x \text{ with } -5.$$

$$\frac{1}{6} = \frac{1}{6} \quad \text{True.}$$

This number checks, so the solution is -5.

Example 2   Solve: $\frac{t-4}{2} - \frac{t-3}{9} = \frac{5}{18}$

Solution: The LCD of denominators 2, 9, and 18 is 18, so we multiply both sides of the equation by 18.

$$18\left(\frac{t-4}{2} - \frac{t-3}{9}\right) = 18\left(\frac{5}{18}\right)$$

$$18\left(\frac{t-4}{2}\right) - 18\left(\frac{t-3}{9}\right) = 18\left(\frac{5}{18}\right) \quad \text{Use the distributive property.}$$

> **HELPFUL HINT**
> Multiply *each* term by 18.

$$9(t-4) - 2(t-3) = 5 \quad \begin{array}{l}\text{Simplify.}\\ \text{Use the distributive property.}\end{array}$$
$$9t - 36 - 2t + 6 = 5$$
$$7t - 30 = 5 \quad \text{Combine like terms.}$$
$$7t = 35$$
$$t = 5 \quad \text{Solve for } t.$$

Objectives

A Solve equations containing rational expressions.

B Solve equations containing rational expressions for a specified variable.

SSM   CD-ROM   Video
5.5

Practice Problem 1

Solve: $\frac{x}{4} + \frac{4}{5} = \frac{1}{20}$

TEACHING TIP

You may want to begin this lesson with a review of solving linear equations, such as $3x + 16 = 1$.

Practice Problem 2

Solve: $\frac{x+2}{3} - \frac{x-1}{5} = \frac{1}{15}$

Answers
1. $x = -3$,   **2.** $x = -6$

Check:

$$\frac{t-4}{2} - \frac{t-3}{9} = \frac{5}{18}$$

$$\frac{5-4}{2} - \frac{5-3}{9} \stackrel{?}{=} \frac{5}{18} \quad \text{Replace } t \text{ with 5.}$$

$$\frac{1}{2} - \frac{2}{9} \stackrel{?}{=} \frac{5}{18} \quad \text{Simplify.}$$

$$\frac{5}{18} = \frac{5}{18} \quad \text{True.}$$

The solution is 5. ▬▬▬

Recall from Section 5.1 that a rational expression is defined for all real numbers except those that make the denominator of the expression 0. This means that if an equation contains *rational expressions with variables in the denominator*, we must be certain that the proposed solution does not make the denominator 0. If replacing the variable with the proposed solution makes the denominator 0, the rational expression is undefined and this proposed solution must be rejected.

Practice Problem 3

Solve: $2 + \dfrac{6}{x} = x + 7$

Example 3

Solve: $3 - \dfrac{6}{x} = x + 8$

Solution:

In this equation, 0 cannot be a solution because if x is 0, the rational expression $\dfrac{6}{x}$ is undefined. The LCD is x, so we multiply both sides of the equation by x.

$$x\left(3 - \frac{6}{x}\right) = x(x + 8)$$

HELPFUL HINT

Multiply *each* term by x.

$$x(3) - x\left(\frac{6}{x}\right) = x \cdot x + x \cdot 8 \quad \begin{array}{l}\text{Use the distributive}\\\text{property.}\end{array}$$

$$3x - 6 = x^2 + 8x \quad \text{Simplify.}$$

Now we write the quadratic equation in standard form and solve for x.

$$0 = x^2 + 5x + 6$$
$$0 = (x + 3)(x + 2) \quad \text{Factor.}$$
$$x + 3 = 0 \quad \text{or} \quad x + 2 = 0 \quad \begin{array}{l}\text{Set each factor equal}\\\text{to 0 and solve.}\end{array}$$
$$x = -3 \qquad\qquad x = -2$$

Notice that neither -3 nor -2 makes the denominator in the original equation equal to 0.

Check:

To check these solutions, we replace x in the original equation by -3, and then by -2.

If $x = -3$:

$$3 - \frac{6}{x} = x + 8$$

$$3 - \frac{6}{-3} \stackrel{?}{=} -3 + 8$$

$$3 - (-2) \stackrel{?}{=} 5$$

$$5 = 5 \quad \text{True.}$$

If $x = -2$:

$$3 - \frac{6}{x} = x + 8$$

$$3 - \frac{6}{-2} \stackrel{?}{=} -2 + 8$$

$$3 - (-3) \stackrel{?}{=} 6$$

$$6 = 6 \quad \text{True.}$$

Answer

3. $x = -6, x = 1$

Both -3 and -2 are solutions. ▬▬▬

The following steps may be used to solve an equation containing rational expressions.

TO SOLVE AN EQUATION CONTAINING RATIONAL EXPRESSIONS

Step 1. Multiply both sides of the equation by the LCD of all rational expressions in the equation.

Step 2. Remove any grouping symbols and solve the resulting equation.

Step 3. Check the solution in the original equation.

Example 4   Solve: $\dfrac{4x}{x^2 - 25} + \dfrac{2}{x - 5} = \dfrac{1}{x + 5}$

Solution: The denominator $x^2 - 25$ factors as $(x + 5)(x - 5)$. The LCD is then $(x + 5)(x - 5)$, so we multiply both sides of the equation by this LCD.

Multiply by the LCD.

$$(x + 5)(x - 5)\left(\dfrac{4x}{x^2 - 25} + \dfrac{2}{x - 5}\right) = (x + 5)(x - 5)\left(\dfrac{1}{x + 5}\right)$$

$$(x + 5)(x - 5) \cdot \dfrac{4x}{x^2 - 25} + (x + 5)(x - 5) \cdot \dfrac{2}{x - 5}$$ Use the distributive property.

$$= (x + 5)(x - 5) \cdot \dfrac{1}{x + 5}$$

$$4x + 2(x + 5) = x - 5 \quad \text{Simplify.}$$

$$4x + 2x + 10 = x - 5 \quad \text{Use the distributive property.}$$

$$6x + 10 = x - 5 \quad \text{Combine like terms.}$$

$$5x = -15$$

$$x = -3 \qquad \text{Divide both sides by 5.}$$

Check: Check by replacing x with -3 in the original equation. The solution is -3.   ▬▬

Practice Problem 4

Solve: $\dfrac{2}{x + 3} + \dfrac{3}{x - 3} = \dfrac{-2}{x^2 - 9}$

Example 5   Solve: $\dfrac{2x}{x - 4} = \dfrac{8}{x - 4} + 1$

Solution: Multiply both sides by the LCD, $x - 4$.

$$(x - 4)\left(\dfrac{2x}{x - 4}\right) = (x - 4)\left(\dfrac{8}{x - 4} + 1\right)$$ Multiply by the LCD.

$$(x - 4) \cdot \dfrac{2x}{x - 4} = (x - 4) \cdot \dfrac{8}{x - 4} + (x - 4) \cdot 1$$ Use the distributive property.

$$2x = 8 + (x - 4) \qquad \text{Simplify.}$$

$$2x = 4 + x$$

$$x = 4$$

Notice that 4 makes the denominator 0 in the original equation. Therefore, 4 is *not* a solution and this equation has *no solution*.   ▬▬

TRY THE CONCEPT CHECK IN THE MARGIN.

Practice Problem 5

Solve: $\dfrac{5x}{x - 1} = \dfrac{5}{x - 1} + 3$

✓ **CONCEPT CHECK**

When can fractions be cleared by multiplying through by the LCD?

a. When adding or subtracting rational expressions

b. When solving an equation containing rational expressions

c. Both of these

d. Neither of these

⌐**HELPFUL HINT**

As we can see from Example 5, it is important to check the proposed solution(s) in the original equation.

Answers

4. $x = -1$,   **5.** No solution

✓ Concept Check: b

Practice Problem 6

Solve: $x - \dfrac{6}{x+3} = \dfrac{2x}{x+3} + 2$

Example 6   Solve: $x + \dfrac{14}{x-2} = \dfrac{7x}{x-2} + 1$

Solution: Notice the denominators in this equation. We can see that 2 can't be a solution. The LCD is $x - 2$, so we multiply both sides of the equation by $x - 2$.

$$(x-2)\left(x + \frac{14}{x-2}\right) = (x-2)\left(\frac{7x}{x-2} + 1\right)$$

$$(x-2)(x) + (x-2)\left(\frac{14}{x-2}\right) = (x-2)\left(\frac{7x}{x-2}\right) + (x-2)(1)$$

$x^2 - 2x + 14 = 7x + x - 2$	Simplify.
$x^2 - 2x + 14 = 8x - 2$	Combine like terms.
$x^2 - 10x + 16 = 0$	Write the quadratic equation in standard form.
$(x-8)(x-2) = 0$	Factor.
$x - 8 = 0 \quad \text{or} \quad x - 2 = 0$	Set each factor equal to 0.
$x = 8 \qquad\qquad x = 2$	Solve.

As we have already noted, 2 can't be a solution of the original equation. So we need only replace x with 8 in the original equation. We find that 8 is a solution; the only solution is 8.   ━━━

B   SOLVING EQUATIONS FOR A SPECIFIED VARIABLE

The last example in this section is an equation containing several variables, and we are directed to solve for one of the variables. The steps used in the preceding examples can be applied to solve equations for a specified variable as well.

Practice Problem 7

Solve $\dfrac{1}{a} + \dfrac{1}{b} = \dfrac{1}{x}$ for a.

Example 7   Solve $\dfrac{1}{a} + \dfrac{1}{b} = \dfrac{1}{x}$ for x

Solution: (This type of equation often models a work problem, as we shall see in the next section.) The LCD is abx, so we multiply both sides by abx.

$$abx\left(\frac{1}{a} + \frac{1}{b}\right) = abx\left(\frac{1}{x}\right)$$

$$abx\left(\frac{1}{a}\right) + abx\left(\frac{1}{b}\right) = abx \cdot \frac{1}{x}$$

$bx + ax = ab$	Simplify.
$x(b + a) = ab$	Factor out x from each term on the left side.
$\dfrac{x(b+a)}{b+a} = \dfrac{ab}{b+a}$	Divide both sides by $b + a$.
$x = \dfrac{ab}{b+a}$	Simplify.

This equation is now solved for x.   ━━━

Answers

6. $x = 4$,   **7.** $a = \dfrac{bx}{b-x}$

MENTAL MATH

Solve each equation for the variable.

1. $\dfrac{x}{5} = 2$ **2.** $\dfrac{x}{8} = 4$ **3.** $\dfrac{z}{6} = 6$ **4.** $\dfrac{y}{7} = 8$

EXERCISE SET 5.5

A Solve each equation and check each solution. See Examples 1 and 2.

1. $\dfrac{x}{5} + 3 = 9$

2. $\dfrac{x}{5} - 2 = 9$

3. $\dfrac{x}{2} + \dfrac{5x}{4} = \dfrac{x}{12}$

4. $\dfrac{x}{6} + \dfrac{4x}{3} = \dfrac{x}{18}$

5. $2 - \dfrac{8}{x} = 6$

6. $5 + \dfrac{4}{x} = 1$

7. $2 + \dfrac{10}{x} = x + 5$

8. $6 + \dfrac{5}{y} = y - \dfrac{2}{y}$

9. $\dfrac{a}{5} = \dfrac{a-3}{2}$

10. $\dfrac{b}{5} = \dfrac{b+2}{6}$

11. $\dfrac{x-3}{5} + \dfrac{x-2}{2} = \dfrac{1}{2}$

12. $\dfrac{a+5}{4} + \dfrac{a+5}{2} = \dfrac{a}{8}$

Solve each equation and check each answer. See Examples 3 through 6.

13. $\dfrac{2}{y} + \dfrac{1}{2} = \dfrac{5}{2y}$

14. $\dfrac{6}{3y} + \dfrac{3}{y} = 1$

15. $\dfrac{11}{2x} + \dfrac{2}{3} = \dfrac{7}{2x}$

16. $\dfrac{5}{3} - \dfrac{3}{2x} = \dfrac{3}{2}$

17. $2 + \dfrac{3}{a-3} = \dfrac{a}{a-3}$

18. $\dfrac{2y}{y-2} - \dfrac{4}{y-2} = 4$

19. $\dfrac{3}{2a-5} = -1$

20. $\dfrac{6}{4-3x} = -3$

21. $\dfrac{y}{y+4} + \dfrac{4}{y+4} = 3$

22. $\dfrac{5y}{y+1} - \dfrac{3}{y+1} = 4$ **23.** $\dfrac{a}{a-6} = \dfrac{-2}{a-1}$ **24.** $\dfrac{5}{x-6} = \dfrac{x}{x-2}$

25. $\dfrac{2x}{x+2} - 2 = \dfrac{x-8}{x-2}$ **26.** $\dfrac{4y}{y-3} - 3 = \dfrac{3y-1}{y+3}$

27. $\dfrac{4y}{y-4} + 5 = \dfrac{5y}{y-4}$ **28.** $\dfrac{2a}{a+2} - 5 = \dfrac{7a}{a+2}$

29. $\dfrac{2}{x-2} + 1 = \dfrac{x}{x+2}$ **30.** $1 + \dfrac{3}{x+1} = \dfrac{x}{x-1}$

31. $\dfrac{t}{t-4} = \dfrac{t+4}{6}$ **32.** $\dfrac{15}{x+4} = \dfrac{x-4}{x}$

33. $\dfrac{x+1}{3} - \dfrac{x-1}{6} = \dfrac{1}{6}$ **34.** $\dfrac{3x}{5} - \dfrac{x-6}{3} = -\dfrac{2}{5}$

35. $\dfrac{y}{2y+2} + \dfrac{2y-16}{4y+4} = \dfrac{2y-3}{y+1}$ **36.** $\dfrac{1}{x+2} = \dfrac{4}{x^2-4} - \dfrac{1}{x-2}$

37. $\dfrac{4r-4}{r^2+5r-14} + \dfrac{2}{r+7} = \dfrac{1}{r-2}$ **38.** $\dfrac{3}{x+3} = \dfrac{12x+19}{x^2+7x+12} - \dfrac{5}{x+4}$

39. $\dfrac{x+1}{x+3} = \dfrac{x^2-11x}{x^2+x-6} - \dfrac{x-3}{x-2}$ **40.** $\dfrac{2t+3}{t-1} - \dfrac{2}{t+3} = \dfrac{5-6t}{t^2+2t-3}$

---

22. $y = 7$

23. $x = 3, x = -4$

24. $x = 10, x = 1$

25. $x = 6, x = -4$

26. $y = 12, y = -1$

27. $y = 5$

28. $a = -1$

29. $x = 0$

30. $x = 2$

31. $t = 8, t = -2$

32. $x = 16, x = -1$

33. $x = -2$

34. $x = -9$

35. no solution

36. no solution

37. $r = 3$

38. $x = 2$

39. $x = -11, x = 1$

40. $t = -\dfrac{1}{2}, t = -6$

B *Solve each equation for the indicated variable. See Example 7.*

41. $\dfrac{d}{r} = t$ for r (Physics: distance, rate, time formula)

42. $\dfrac{A}{W} = L$ for W (Geometry: area of a rectangle)

43. $T = \dfrac{V}{Q}$ for Q (Water purification: settling time)

44. $T = \dfrac{2U}{B + E}$ for B (Merchandising: stock turnover rate)

45. $i = \dfrac{A}{t + B}$ for t (Hydrology: rainfall intensity)

46. $C = \dfrac{D(A + 1)}{24}$ for A (Medicine: Cowling's Rule for child's dose)

47. $N = R + \dfrac{V}{G}$ for G (Urban forestry: tree plantings per year)

48. $B = \dfrac{705w}{h^2}$ for w (Health: body-mass index)

49. $\dfrac{C}{\pi r} = 2$ for r (Geometry: circumference of a circle)

50. $\dfrac{V}{\pi r^2 h} = 1$ for h (Geometry: volume of a right circular cylinder)

REVIEW AND PREVIEW

Write each phrase as an expression.

51. The reciprocal of x

52. The reciprocal of $x + 1$

53. The reciprocal of x, added to the reciprocal of 2

54. The reciprocal of x, subtracted from the reciprocal of 5

Answer each question.

55. If a tank is filled in 3 hours, what part of the tank is filled in 1 hour?

56. If a strip of beach is cleaned in 4 hours, what part of the beach is cleaned in 1 hour?

41. $r = \dfrac{d}{t}$

42. $W = \dfrac{A}{L}$

43. $Q = \dfrac{V}{T}$

44. $B = \dfrac{2U - TE}{T}$

45. $t = \dfrac{A - Bi}{i}$

46. $A = \dfrac{24C - D}{D}$

47. $G = \dfrac{V}{N - R}$

48. $W = \dfrac{Bh^2}{705}$

49. $r = \dfrac{C}{2\pi}$

50. $h = \dfrac{V}{\pi r^2}$

51. $\dfrac{1}{x}$

52. $\dfrac{1}{x + 1}$

53. $\dfrac{1}{x} + \dfrac{1}{2}$

54. $\dfrac{1}{5} - \dfrac{1}{x}$

55. $\dfrac{1}{3}$

56. $\dfrac{1}{4}$

Name _____

COMBINING CONCEPTS

Recall that two angles are supplementary if the sum of their measures is 180°. Find the measure of each supplementary angle.

57.

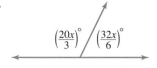

58.

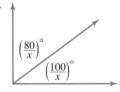

Recall that two angles are complementary if the sum of their measures is 90°. Find the measure of each complementary angle.

59.

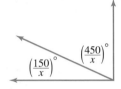

60.

Solve each equation.

61. $\dfrac{4}{a^2 + 4a + 3} + \dfrac{2}{a^2 + a - 6} - \dfrac{3}{a^2 - a - 2} = 0$

62. $\dfrac{-4}{a^2 + 2a - 8} + \dfrac{1}{a^2 + 9a + 20} = \dfrac{-4}{a^2 + 3a - 10}$

Name _____ **Section** _____ **Date** _____

INTEGRATED REVIEW—SUMMARY ON RATIONAL EXPRESSIONS

It is important to know the difference between performing operations on rational expressions and solving an equation containing rational expressions. Study the examples below.

Performing Operations

Adding: $\dfrac{1}{x} + \dfrac{1}{x + 5} = \dfrac{1 \cdot (x + 5)}{x(x + 5)} + \dfrac{1 \cdot x}{x(x + 5)} = \dfrac{x + 5 + x}{x(x + 5)} = \dfrac{2x + 5}{x(x + 5)}$

Subtracting: $\dfrac{3}{x} - \dfrac{5}{x^2 y} = \dfrac{3 \cdot xy}{x \cdot xy} - \dfrac{5}{x^2 y} = \dfrac{3xy - 5}{x^2 y}$

Multiplying: $\dfrac{2}{x} \cdot \dfrac{5}{x - 1} = \dfrac{2 \cdot 5}{x(x - 1)} = \dfrac{10}{x(x - 1)}$

Dividing: $\dfrac{4}{2x + 1} \div \dfrac{x - 3}{x} = \dfrac{4}{2x + 1} \cdot \dfrac{x}{x - 3} = \dfrac{4x}{(2x + 1)(x - 3)}$

Solving an Equation

To solve an equation containing rational expressions, we clear the equation of fractions by multiplying both sides by the LCD.

$$\dfrac{3}{x} - \dfrac{5}{x - 1} = \dfrac{1}{x(x - 1)}$$

$$x(x - 1)\left(\dfrac{3}{x}\right) - x(x - 1)\left(\dfrac{5}{x - 1}\right) = x(x - 1) \cdot \dfrac{1}{x(x - 1)} \quad \text{Multiply both sides by the LCD.}$$

$$3(x - 1) - 5x = 1 \quad \text{Simplify.}$$

$$3x - 3 - 5x = 1 \quad \text{Use the distributive property.}$$

$$-2x - 3 = 1 \quad \text{Combine like terms.}$$

$$-2x = 4 \quad \text{Add 3 to both sides.}$$

$$x = -2 \quad \text{Divide both sides by } -2.$$

Determine whether each of the following is an equation or an expression. If it is an equation, solve it for its variable. If it is an expression, perform the indicated operation.

1. $\dfrac{1}{x} + \dfrac{2}{3}$ **2.** $\dfrac{3}{a} + \dfrac{5}{6}$ **3.** $\dfrac{1}{x} + \dfrac{2}{3} = \dfrac{3}{x}$

4. $\dfrac{3}{a} + \dfrac{5}{6} = 1$ **5.** $\dfrac{2}{x + 1} - \dfrac{1}{x}$ **6.** $\dfrac{4}{x - 3} - \dfrac{1}{x}$

7. $\dfrac{2}{x + 1} - \dfrac{1}{x} = 1$ **8.** $\dfrac{4}{x - 3} - \dfrac{1}{x} = \dfrac{6}{x(x - 3)}$

9. $\dfrac{15x}{x+8} \cdot \dfrac{2x+16}{3x}$

10. $\dfrac{9z+5}{15} \cdot \dfrac{5z}{81z^2-25}$

11. $\dfrac{2x+1}{x-3} + \dfrac{3x+6}{x-3}$

12. $\dfrac{4p-3}{2p+7} + \dfrac{3p+8}{2p+7}$

13. $\dfrac{x+5}{7} = \dfrac{8}{2}$

14. $\dfrac{1}{2} = \dfrac{x+1}{8}$

15. $\dfrac{5a+10}{18} \div \dfrac{a^2-4}{10a}$

16. $\dfrac{9}{x^2-1} \div \dfrac{12}{3x+3}$

17. Explain the difference between solving an equation such as $\dfrac{x}{2} + \dfrac{3}{4} = \dfrac{x}{4}$ for x and performing an operation such as adding $\dfrac{x}{2} + \dfrac{3}{4}$.

18. When solving an equation such as $\dfrac{y}{4} = \dfrac{y}{2} - \dfrac{1}{4}$, we may multiply all terms by 4. When subtracting two rational expressions such as $\dfrac{y}{2} - \dfrac{1}{4}$, we may not. Explain why.

---

9. expression, 10

10. expression; $\dfrac{z}{3(9z-5)}$

11. expression; $\dfrac{5x+7}{x-3}$

12. expression; $\dfrac{7p+5}{2p+7}$

13. equation; $x = 23$

14. equation; $x = 3$

15. expression; $\dfrac{25a}{9(a-2)}$

16. expression; $\dfrac{9}{4(x-1)}$

17. answers may vary

18. answers may vary

5.6 RATIONAL EQUATIONS AND PROBLEM SOLVING

A SOLVING PROBLEMS ABOUT NUMBERS

In this section, we solve problems that can be modeled by equations containing rational expressions. To solve these problems, we use the same problem-solving steps that were first introduced in Section 2.5. In our first example, our goal is to find an unknown number.

Example 1   Finding an Unknown Number

The quotient of a number and 6 minus $\frac{5}{3}$ is the quotient of the number and 2. Find the number.

Solution: 1. UNDERSTAND. Read and reread the problem. Suppose that the unknown number is 2, then we see if the quotient of 2 and 6, or $\frac{2}{6}$, minus $\frac{5}{3}$ is equal to the quotient of 2 and 2, or $\frac{2}{2}$.

$$\frac{2}{6} - \frac{5}{3} = \frac{1}{3} - \frac{5}{3} = -\frac{4}{3}, \text{ not } \frac{2}{2}$$

Don't forget that the purpose of a proposed solution is to better understand the problem.

Let x = the unknown number.

2. TRANSLATE.

In words:

the quotient of x and 6	minus	$\frac{5}{3}$	is	the quotient of x and 2
↓	↓	↓	↓	↓

Translate: $\frac{x}{6}$ $-$ $\frac{5}{3}$ $=$ $\frac{x}{2}$

3. SOLVE. Here, we solve the equation $\frac{x}{6} - \frac{5}{3} = \frac{x}{2}$. We begin by multiplying both sides of the equation by the LCD 6.

$$6\left(\frac{x}{6} - \frac{5}{3}\right) = 6\left(\frac{x}{2}\right)$$

$$6\left(\frac{x}{6}\right) - 6\left(\frac{5}{3}\right) = 6\left(\frac{x}{2}\right) \quad \text{Apply the distributive property.}$$

$$x - 10 = 3x \quad \text{Simplify.}$$

$$-10 = 2x \quad \text{Subtract } x \text{ from both sides.}$$

$$-\frac{10}{2} = \frac{2x}{2} \quad \text{Divide both sides by 2.}$$

$$-5 = x \quad \text{Simplify.}$$

Objectives

A Solve problems about numbers.
B Solve problems about work.
C Solve problems about distance, rate, and time.
D Solve problems about similar triangles.

SSM   CD-ROM   Video 5.6

Practice Problem 1

The quotient of a number and 2 minus $\frac{1}{3}$ is the quotient of the number and 6.

Answer

1. $x = 1$

4. INTERPRET.

Check: To check, we verify that "the quotient of -5 and 6 minus $\frac{5}{3}$ is the quotient of -5 and 2, or $-\frac{5}{6} - \frac{5}{3} = -\frac{5}{2}$."

State: The unknown number is -5.   ▬▬▬

B   SOLVING PROBLEMS ABOUT WORK

The next example is often called a work problem. Work problems usually involve people or machines doing a certain task.

Example 2   Finding Work Rates

Sam Waterton and Frank Schaffer work in a plant that manufactures automobiles. Sam can complete a quality control tour of the plant in 3 hours while his assistant, Frank, needs 7 hours to complete the same job. The regional manager is coming to inspect the plant facilities, so both Sam and Frank are directed to complete a quality control tour together. How long will this take?

Solution: **1.** UNDERSTAND. Read and reread the problem. The key idea here is the relationship between the **time** (hours) it takes to complete the job and the **part of the job** completed in 1 unit of time (hour). For example, if the **time** it takes Sam to complete the job is 3 hours, the **part of the job** he can complete in 1 hour is $\frac{1}{3}$.

Similarly, Frank can complete $\frac{1}{7}$ of the job in 1 hour.

Let $x =$ the **time** in hours it takes Sam and Frank to complete the job together.

Then $\frac{1}{x} =$ the **part of the job** they complete in 1 hour.

	Hours to Complete Total Job	Part of Job Completed in 1 Hour
Sam	3	$\frac{1}{3}$
Frank	7	$\frac{1}{7}$
Together	x	$\frac{1}{x}$

2. TRANSLATE.

In words:

part of job Sam completed in 1 hour	added to	part of job Frank completed in 1 hour	is equal to	part of job they completed together in 1 hour
↓	↓	↓	↓	↓

Translate: $\frac{1}{3}$ $+$ $\frac{1}{7}$ $=$ $\frac{1}{x}$

Practice Problem 2

Andrew and Timothy Larson volunteer at a local recycling plant. Andrew can sort a batch of recyclables in 2 hours alone while his brother Timothy needs 3 hours to complete the same job. If they work together, how long will it take them to sort one batch?

Answer

2. $1\frac{1}{5}$ hours

3. SOLVE. Here, we solve the equation $\frac{1}{3} + \frac{1}{7} = \frac{1}{x}$. We begin by multiplying both sides of the equation by the LCD, $21x$.

$$21x\left(\frac{1}{3}\right) + 21x\left(\frac{1}{7}\right) = 21x\left(\frac{1}{x}\right)$$

$$7x + 3x = 21 \qquad \text{Simplify.}$$

$$10x = 21$$

$$x = \frac{21}{10} \quad \text{or} \quad 2\frac{1}{10} \text{ hours}$$

4. INTERPRET.

Check: Our proposed solution is $2\frac{1}{10}$ hours. This proposed solution is reasonable since $2\frac{1}{10}$ hours is more than half of Sam's time and less than half of Frank's time. Check this solution in the originally *stated* problem.

State: Sam and Frank can complete the quality control tour in $2\frac{1}{10}$ hours.

TEACHING TIP

Have students determine how much of the tour each person does.

Sam: $\frac{21}{10} \cdot \frac{1}{3} = \frac{21}{30} = \frac{7}{10}$

Frank: $\frac{21}{10} \cdot \frac{1}{7} = \frac{21}{70} = \frac{3}{10}$

and verify that the entire tour would be done: $\frac{7}{10} + \frac{3}{10} = \frac{10}{10} = 1$.

C SOLVING PROBLEMS ABOUT DISTANCE, RATE, AND TIME

Next we look at a problem solved by the distance formula.

▪ Example 3   Finding Speeds of Vehicles

A car travels 180 miles in the same time that a truck travels 120 miles. If the car's speed is 20 miles per hour faster than the truck's, find the car's speed and the truck's speed.

Solution: **1.** UNDERSTAND. Read and reread the problem. Suppose that the truck's speed is 45 miles per hour. Then the car's speed is 20 miles per hour more, or 65 miles per hour.

We are given that the car travels 180 miles in the same time that the truck travels 120 miles. To find the time it takes the car to travel 180 miles, remember that since $d = rt$, we know that $\frac{d}{r} = t$.

Car's Time

$$t = \frac{d}{r} = \frac{180}{65} = 2\frac{50}{65} = 2\frac{10}{13} \text{ hours}$$

Truck's Time

$$t = \frac{d}{r} = \frac{120}{45} = 2\frac{30}{45} = 2\frac{2}{3} \text{ hours}$$

Since the times are not the same, our proposed solution is not correct. But we have a better understanding of the problem.

Practice Problem 3

A car travels 280 miles in the same time that a motorcycle travels 240 miles. If the car's speed is 10 miles per hour more than the motorcycle's, find the speed of the car and the speed of the motorcycle.

TEACHING TIP

When solving distance, rate, and time problems, it may be helpful to have students identify equations equivalent to $d = r \cdot t$. For example, solve this equation for r, and then t.

Answer

3. car: 70 mph; motorcycle: 60 mph

Let x = the speed of the truck.

Since the car's speed is 20 miles per hour faster than the truck's, then

$x + 20$ = the speed of the car

Use the formula $d = r \cdot t$ or **d**istance = **r**ate $\cdot$ **t**ime. Prepare a chart to organize the information in the problem.

	Distance	=	Rate	$\cdot$	Time
Truck	120		x		$\dfrac{120}{x} \begin{array}{l}\leftarrow \text{distance} \\ \leftarrow \text{rate}\end{array}$
Car	180		$x + 20$		$\dfrac{180}{x + 20} \begin{array}{l}\leftarrow \text{distance} \\ \leftarrow \text{rate}\end{array}$

> **HELPFUL HINT**
>
> If $d = r \cdot t$, then $t = \dfrac{d}{r}$ or *time* $= \dfrac{distance}{rate}$.

2. **TRANSLATE.** Since the car and the truck traveled the same amount of time, we have that

In words: car's time = truck's time
                    ↓   ↓

Translate: $\dfrac{180}{x + 20}$ = $\dfrac{120}{x}$

3. **SOLVE.** We begin by multiplying both sides of the equation by the LCD, $x(x + 20)$, or cross multiplying.

$$\frac{180}{x + 20} = \frac{120}{x}$$

$180x = 120(x + 20)$

$180x = 120x + 2400$ Use the distributive property.

$60x = 2400$ Subtract $120x$ from both sides.

$x = 40$ Divide both sides by 60.

4. **INTERPRET.** The speed of the truck is 40 miles per hour. The speed of the car must then be $x + 20$ or 60 miles per hour.

Check: Find the time it takes the car to travel 180 miles and the time it takes the truck to travel 120 miles.

Car's Time

$t = \dfrac{d}{r} = \dfrac{180}{60} = 3$ hours

Truck's Time

$t = \dfrac{d}{r} = \dfrac{120}{40} = 3$ hours

Since both travel the same amount of time, the proposed solution is correct.

State: The car's speed is 60 miles per hour and the truck's speed
is 40 miles per hour.

D SOLVING PROBLEMS ABOUT SIMILAR TRIANGLES

Similar triangles have the same shape but not necessarily the same size. In
similar triangles, the measures of corresponding angles are equal, and corresponding sides are in proportion.

If triangle ABC and triangle XYZ shown are similar, then we know that
the measure of angle $A =$ the measure of angle X, the measure of angle
$B =$ the measure of angle Y, and the measure of angle $C =$ the measure
of angle Z. We also know that corresponding sides are in proportion:
$\dfrac{a}{x} = \dfrac{b}{y} = \dfrac{c}{z}$.

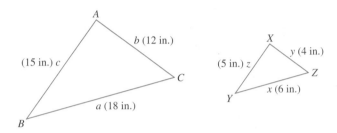

In this section, we will position similar triangles so that they have the same
orientation.

To show that corresponding sides are in proportion for the triangles
above, we write the ratios of the corresponding sides.

$$\frac{a}{x} = \frac{18}{6} = 3 \qquad \frac{b}{y} = \frac{12}{4} = 3 \qquad \frac{c}{z} = \frac{15}{5} = 3$$

Example 4   Finding the Length of a Side of a Triangle

If the following two triangles are similar, find the missing
length x.

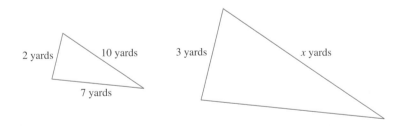

Solution: Since the triangles are similar, their corresponding sides
are in proportion and we have

$$\frac{2}{3} = \frac{10}{x}$$

Practice Problem 4

If the following two triangles are similar, find the missing length x.

TEACHING TIP

Consider ending this section with the following activity: Have students work in
small groups to create their own word
problem based on one of the types in this
section. Then have groups trade their
problems with other groups to solve.

Answer

4. $x = 20$ units

To solve, we multiply both sides by the LCD, $3x$, or cross multiply.

$$2x = 30$$
$$x = 15 \qquad \text{Divide both sides by 2.}$$

The missing length is 15 yards.

EXERCISE SET 5.6

A *Solve. See Example 1.*

1. Three times the reciprocal of a number equals 9 times the reciprocal of 6. Find the number.

2. Twelve divided by the sum of a number and 2 equals the quotient of 4 and the difference of the number and 2. Find the number.

3. If twice a number added to 3 is divided by the number plus 1, the result is three halves. Find the number.

4. A number added to the product of 6 and the reciprocal of the number equals -5. Find the number.

5. Two divided by the difference of a number and 3, minus 4 divided by the number plus 3, equals 8 times the reciprocal of the difference of the number squared and 9. What is the number?

6. If 15 times the reciprocal of a number is added to the ratio of 9 times the number minus 7 and the number plus 2, the result is 9. What is the number?

7. One-fourth equals the quotient of a number and 8. Find the number.

8. Four times a number added to 5 is divided by 6. The result is $\frac{7}{2}$. Find the number.

B *Solve. See Example 2.*

9. Smith Engineering found that an experienced surveyor surveys a roadbed in 4 hours. An apprentice surveyor needs 5 hours to survey the same stretch of road. If the two work together, find how long it takes them to complete the job.

10. An experienced bricklayer constructs a small wall in 3 hours. The apprentice completes the job in 6 hours. Find how long it takes if they work together.

11. In 2 minutes, a conveyor belt moves 300 pounds of recyclable aluminum from the delivery truck to a storage area. A smaller belt moves the same quantity of cans the same distance in 6 minutes. If both belts are used, find how long it takes to move the cans to the storage area.

12. Find how long it takes the conveyor belts described in Exercise 11 to move 1200 pounds of cans. (*Hint*: Think of 1200 pounds as four 300-pound jobs.)

1. 2

2. 4

3. -3

4. -3 or -2

5. 5

6. 3

7. 2

8. 4

9. $2\frac{2}{9}$ hr

10. 2 hr

11. $1\frac{1}{2}$ min

12. 6 min

13. $108.00

14. $2\frac{2}{9}$ days

15. 3 hr

16. $6\frac{2}{3}$ hr

17. 20 hr

18. first pump: 28 min; second pump: 84 min

19. 6 mph

20. time traveled by jet: 3 hr; time traveled by car: 4hr

13. Marcus and Tony work for Lombardo's Pipe and Concrete. Mr. Lombardo is preparing an estimate for a customer. He knows that Marcus lays a slab of concrete in 6 hours. Tony lays the same size slab in 4 hours. If both work on the job and the cost of labor is $45.00 per hour, decide what the labor estimate should be.

14. Mr. Dodson can paint his house by himself in 4 days. His son needs an additional day to complete the job if he works by himself. If they work together, find how long it takes to paint the house.

15. One custodian cleans a suite of offices in 3 hours. When a second worker is asked to join the regular custodian, the job takes only $1\frac{1}{2}$ hours. How long does it take the second worker to do the same job alone?

16. One person proofreads copy for a small newspaper in 4 hours. If a second proofreader is also employed, the job can be done in $2\frac{1}{2}$ hours. How long does it take for the second proofreader to do the same job alone?

17. One pipe fills a storage pond in 20 hours. A second pipe fills the same pond in 15 hours. When a third pipe is added and all three are used to fill the pond, it takes only 6 hours. Find how long it takes the third pipe to do the job.

18. One pump fills a tank 3 times as fast as another pump. If the pumps work together, they fill the tank in 21 minutes. How long does it take for each pump to fill the tank?

C *Solve. See Example 3.*

19. A jogger begins her workout by jogging to the park, a distance of 3 miles. She then jogs home at the same speed but along a different route. This return trip is 9 miles and her time is one hour longer. Complete the accompanying chart and use it to find her jogging speed.

	Distance $=$ Rate $\cdot$ Time		
Trip to park	3		x
Return trip	9		$x + 1$

20. A marketing manager travels 1080 miles in a corporate jet and then an additional 240 miles by car. If the car ride takes one hour longer than the jet ride takes, and if the rate of the jet is 6 times the rate of the car, find the time the manager travels by jet and find the time the manager travels by car.

Name _____

21. A cyclist rode the first 20-mile portion of his workout at a constant speed. For the 16-mile cooldown portion of his workout, he reduced his speed by 2 miles per hour. Each portion of the workout took the same time. Find the cyclist's speed during the first portion and find his speed during the cooldown portion.

22. A tractor-trailer travels 300 miles through the flatland in the same amount of time that it travels 180 miles through mountains. The rate of the tractor-trailer is 20 miles per hour slower in the mountains than in the flatland. Find both the flatland rate and mountain rate.

23. A boat can travel 9 miles upstream in the same amount of time it takes to travel 11 miles downstream. If the current of the river is 3 miles per hour, complete the chart below and use it to find the speed r of the boat in still water.

	Distance	=	Rate	·	Time
Upstream	9		$r - 3$		
Downstream	11		$r + 3$		

24. A pilot flies 630 miles with a tail wind of 35 miles per hour. Against the wind, she flies only 455 miles. Find the rate of the plane in still air.

25. A cyclist rides 16 miles per hour on level ground on a still day. He finds that he rides 48 miles with the wind behind him in the same amount of time that he rides 16 miles into the wind. Find the rate of the wind.

26. The current on a portion of the Mississippi River is 3 miles per hour. A barge can go 6 miles upstream in the same amount of time it takes to go 10 miles downstream. Find the speed of the boat in still water.

27. While road testing a new make of car, the editor of a consumer magazine finds that she can go 10 miles into a 3-mile-per-hour wind in the same amount of time she can go 11 miles with a 3-mile-per-hour wind behind her. Find the speed of the car in still air.

28. A fisherman on Pearl River rows 9 miles downstream in the same amount of time he rows 3 miles upstream. If the current is 6 miles per hour, find how long it takes him to cover the 12 miles.

D *Given that the following pairs of triangles are similar, find each missing length. See Example 4.*

29.
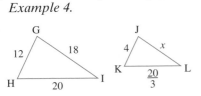

30.
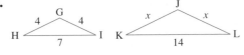

1st portion speed: 10 mph; cooldown
21. speed: 8 mph

flatland rate: 50 mph; mountain rate:
22. 30 mph

23. $r = 30$ mph

rate in still air:
24. 217 mph

25. 8 mph

26. 12 mph

27. 63 mph

28. 1 hour

29. $x = 6$

30. $x = 8$

401

31. *x* = 5

31.
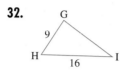
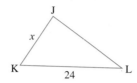

32. *x* = 13.5

32.
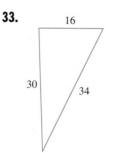

33. *y* = 21.25

33.
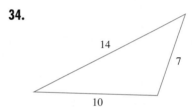

34. *y* = 3.5

34.

35. A toy maker wishes to make a triangular mainsail for a toy sailboat that will be the same shape as a regular-size sailboat's mainsail. Use the diagram to find the missing dimensions.

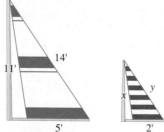

36. A seamstress wishes to make a doll's triangular diaper that will have the same shape as a full-size diaper. Use the diagram to find the missing dimensions of the doll's diaper.

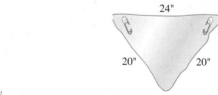

35. *x* = 4.4 ft; *y* = 5.6 ft

36. *x* = 9.6 in.; *y* = 8 in.

Name _____

REVIEW AND PREVIEW

Simplify. Follow the order shown. See Section R.2.

37. $\left.\begin{array}{c}\dfrac{3}{4}+\dfrac{1}{4}\\[2mm]\dfrac{3}{8}+\dfrac{13}{8}\end{array}\right\}$ ←  ① Add. ③ Divide. ② Add.

38. $\left.\begin{array}{c}\dfrac{9}{5}+\dfrac{6}{5}\\[2mm]\dfrac{17}{6}+\dfrac{7}{6}\end{array}\right\}$ ← ① Add. ③ Divide. ② Add.

39. $\left.\begin{array}{c}\dfrac{2}{5}+\dfrac{1}{5}\\[2mm]\dfrac{7}{10}+\dfrac{7}{10}\end{array}\right\}$ ← ① Add. ③ Divide. ② Add.

40. $\left.\begin{array}{c}\dfrac{1}{4}+\dfrac{5}{4}\\[2mm]\dfrac{3}{8}+\dfrac{7}{8}\end{array}\right\}$ ← ① Add. ③ Divide. ② Add.

COMBINING CONCEPTS

41. During the 1997 U.S. 500 race held in Brooklyn, Michigan, Mauricio Gugelmin posted the fastest lap speed but Alex Zanardi won the race. The track is 2 miles long. When traveling at their fastest lap speeds, Zanardi drove 1.952 miles in the same time that it took Gugelmin to complete an entire lap. Gugelmin's fastest lap speed was 5.6 mph faster than Zanardi's fastest lap speed. Find each driver's fastest lap speed. (*Source:* Based on data from Championship Auto Racing Teams, Inc.)

42. A hyena spots a giraffe 0.5 mile away and begins running toward it. The giraffe starts running away from the hyena just as the hyena begins running toward it. A hyena can run at a speed of 40 mph and a giraffe can run at 32 mph. How long will it take for the hyena to overtake the giraffe? (*Source:* Based on data from *Natural History*, March 1974)

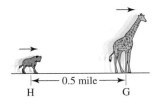

37. $\dfrac{1}{2}$ _____

38. $\dfrac{3}{4}$ _____

39. $\dfrac{3}{7}$ _____

40. $\dfrac{6}{5}$ _____

41. Zanardi's speed: 227.7 mph; Gugelmin's speed: 233.3 mph _____

42. 3.75 minutes _____

Focus On History

EPIGRAM OF DIOPHANTES

One of the great algebraists of ancient times was a man named Diophantus. Little is known of his life other than that he lived and worked in Alexandria. Some historians believe he lived during the first century of the Christian era, about the time of Nero. The only clue to his personal life is the following epigram found in a collection called the Palatine Anthology.

God granted him youth for a sixth of his life and added a twelfth part to this. He clothed his cheeks in down. He lit him the light of wedlock after a seventh part, and five years after his marriage, He granted him a son. Alas, lateborn wretched child. After attaining the measure of half his father's life, cruel fate overtook him, thus leaving Diophantus during the last four years of his life only such consolation as the science of numbers. How old was Diophantus at his death?*

*From *The Nature and Growth of Modern Mathematics*, Edna Kramer, 1970, Fawcett Premier Books, Vol. 1, pages 107–108.

We are looking for Diophantus' age when he died, so let x represent that age. If we sum the parts of his life, we should get the total age.

Parts of his life
$$\begin{cases} \frac{1}{6} \cdot x + \frac{1}{12} \cdot x \text{ is the time of his youth.} \\ \frac{1}{7} \cdot x \text{ is the time between his youth and when he married.} \\ 5 \text{ years is the time between his marriage and the birth of his son.} \\ \frac{1}{2} \cdot x \text{ is the time Diophantus had with his son.} \\ 4 \text{ years is the time between his son's death and his own.} \end{cases}$$

The sum of these parts should equal Diophantus' age when he died.

$$\frac{1}{6} \cdot x + \frac{1}{12} \cdot x + \frac{1}{7} \cdot x + 5 + \frac{1}{2} \cdot x + 4 = x$$

CRITICAL THINKING

1. Solve the epigram.   84 yr
2. How old was Diophantus when his son was born? How old was the son when he died?
   38 yr; 42 yr
3. Solve the following epigram:
   I was four when my mother packed my lunch and sent me off to school. Half my life was spent in school and another sixth was spent on a farm. Alas, hard times befell me. My crops and cattle fared poorly and my land was sold. I returned to school for 3 years and have spent one tenth of my life teaching. How old am I?   30 yr

GROUP ACTIVITY

4. Write an epigram describing your life. Be sure that none of the time periods in your epigram overlap. Exchange epigrams with a partner to solve and check.

Name _____ Section _____ Date _____

EXERCISE SET 5.7

A **B** *Simplify each complex fraction. See Examples 1 through 6.*

1. $\dfrac{\dfrac{1}{2}}{\dfrac{3}{4}}$

2. $\dfrac{\dfrac{1}{8}}{-\dfrac{5}{12}}$

3. $\dfrac{-\dfrac{4x}{9}}{-\dfrac{2x}{3}}$

4. $\dfrac{-\dfrac{6y}{11}}{\dfrac{4y}{9}}$

5. $\dfrac{-\dfrac{5}{12x^2}}{\dfrac{25}{16x^3}}$

6. $\dfrac{-\dfrac{7}{8y}}{\dfrac{21}{4y}}$

7. $\dfrac{\dfrac{1}{3}}{\dfrac{1}{2}-\dfrac{1}{4}}$

8. $\dfrac{\dfrac{7}{10}-\dfrac{3}{5}}{\dfrac{1}{2}}$

9. $\dfrac{2+\dfrac{7}{10}}{1+\dfrac{3}{5}}$

10. $\dfrac{4-\dfrac{11}{12}}{5+\dfrac{1}{4}}$

11. $\dfrac{\dfrac{m}{n}-1}{\dfrac{m}{n}+1}$

12. $\dfrac{\dfrac{x}{2}+2}{\dfrac{x}{2}-2}$

13. $\dfrac{\dfrac{1}{5}-\dfrac{1}{x}}{\dfrac{7}{10}+\dfrac{1}{x^2}}$

14. $\dfrac{\dfrac{1}{y^2}+\dfrac{2}{3}}{\dfrac{1}{y}-\dfrac{5}{6}}$

15. $\dfrac{1+\dfrac{1}{y-2}}{y+\dfrac{1}{y-2}}$

16. $\dfrac{x-\dfrac{1}{2x+1}}{1-\dfrac{x}{2x+1}}$

17. $\dfrac{\dfrac{4y-8}{16}}{\dfrac{6y-12}{4}}$

18. $\dfrac{\dfrac{7y+21}{3}}{\dfrac{3y+9}{8}}$

19. $\dfrac{\dfrac{x}{y}+1}{\dfrac{x}{y}-1}$

20. $\dfrac{\dfrac{3}{5y}+8}{\dfrac{3}{5y}-8}$

ANSWERS

1. $\dfrac{2}{3}$

2. $-\dfrac{3}{10}$

3. $\dfrac{2}{3}$

4. $-\dfrac{27}{22}$

5. $-\dfrac{4x}{15}$

6. $-\dfrac{1}{6}$

7. $\dfrac{4}{3}$

8. $\dfrac{1}{5}$

9. $\dfrac{27}{16}$

10. $\dfrac{37}{63}$

11. $\dfrac{m-n}{m+n}$

12. $\dfrac{x+4}{x-4}$

13. $\dfrac{2x(x-5)}{7x^2+10}$

14. $\dfrac{6+4y^2}{6y-5y^2}$

15. $\dfrac{1}{y-1}$

16. $2x-1$

17. $\dfrac{1}{6}$

18. $\dfrac{56}{9}$

19. $\dfrac{x+y}{x-y}$

20. $\dfrac{3+40y}{3-40y}$

21. $\dfrac{3}{7}$

22. -9

23. $\dfrac{a}{x+b}$

24. $\dfrac{m+2}{2}$

25. $\dfrac{x+8}{2-x}$ or $-\dfrac{x+8}{x-2}$

26. $-\dfrac{5x+50}{5x+22}$

27. $\dfrac{s^2+r^2}{s^2-r^2}$

28. $\dfrac{4+x^2}{4-x^2}$

29. answers may vary

30. answers may vary

410

---

21. $\dfrac{1}{2+\dfrac{1}{3}}$

22. $\dfrac{3}{1-\dfrac{4}{3}}$

23. $\dfrac{\dfrac{ax+ab}{x^2-b^2}}{\dfrac{x+b}{x-b}}$

24. $\dfrac{\dfrac{m+2}{m-2}}{\dfrac{2m+4}{m^2-4}}$

25. $\dfrac{\dfrac{8}{x+4}+2}{\dfrac{12}{x+4}-2}$

26. $\dfrac{\dfrac{25}{x+5}+5}{\dfrac{3}{x+5}-5}$

27. $\dfrac{\dfrac{s}{r}+\dfrac{r}{s}}{\dfrac{s}{r}-\dfrac{r}{s}}$

28. $\dfrac{\dfrac{2}{x}+\dfrac{x}{2}}{\dfrac{2}{x}-\dfrac{x}{2}}$

29. Explain how to simplify a complex fraction using method 1.

30. Explain how to simplify a complex fraction using method 2.

REVIEW AND PREVIEW

Use the bar graph below to answer Exercises 31–34. See Section 1.8.

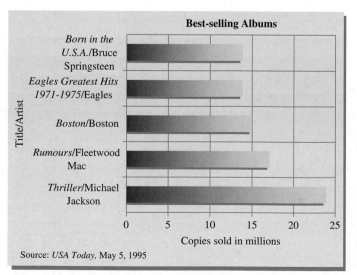

Source: *USA Today*, May 5, 1995

Name _____

31. Which album sold the most copies?

32. Estimate how many more copies the album *Rumours* has sold than the album *Boston*.

33. Which album(s) shown sold the fewest copies?

34. If Michael Jackson made $7 for every *Thriller* album sold, what was his income for this album?

 COMBINING CONCEPTS

To find the average of two numbers, we find their sum and divide by 2. For example, the average of 65 and 81 is found by simplifying $\dfrac{65 + 81}{2}$. *This simplifies to* $\dfrac{146}{2} = 73$.

35. Find the average of $\dfrac{1}{3}$ and $\dfrac{3}{4}$.

36. Write the average of $\dfrac{3}{n}$ and $\dfrac{5}{n^2}$ as a simplified rational expression.

37. In electronics, when two resistors R_1 (read *R* sub 1) and R_2 (read *R* sub 2) are connected in parallel, the total resistance is given by the complex fraction

$$\dfrac{1}{\dfrac{1}{R_1} + \dfrac{1}{R_2}}.$$

Simplify this expression.

Resistance R_1 R_2

38. Astronomers occasionally need to know the day of the week a particular date fell on. The complex fraction

$$\dfrac{J + \dfrac{3}{2}}{7}$$

where *J* is the *Julian day number*, is used to make this calculation. Simplify this expression.

Simplify each of the following. First, write each expression without exponents. Then simplify the complex fraction. The first step has been completed for Exercise 39.

39.

$$\frac{x^{-1} + 2^{-1}}{x^{-2} - 4^{-1}} = \frac{\dfrac{1}{x} + \dfrac{1}{2}}{\dfrac{1}{x^2} - \dfrac{1}{4}}$$

40. $\dfrac{3^{-1} - x^{-1}}{9^{-1} - x^{-2}}$

41. $\dfrac{y^{-2}}{1 - y^{-2}}$

42. $\dfrac{4 + x^{-1}}{3 + x^{-1}}$

CHAPTER 5 ACTIVITY
ANALYZING A TABLE

MATERIALS:
▲ Calculator

This activity may be completed by working in groups or individually.

A person's body-mass index (BMI) can be used as an indicator of whether the person should lose weight. BMI can be calculated with the formula $B = \dfrac{705w}{h^2}$ where w is weight in pounds and h is height in inches. Doctors recommend that body-mass index values fall between 19 and 25.

1. Use the BMI formula to complete the table for the given combinations of height and weight. (*Hint:* You may want to use a spreadsheet software or a calculator that has a table feature to help fill in the table.)

2. Use the table to find the BMI for (a) a person who is 66 inches tall and weighs 160 pounds; (b) a person who is 62 inches tall and weighs 120 pounds; (c) a person who is 70 inches tall and weighs 170 pounds. Do these BMI values indicate that any of these people should try to lose weight? (a) 25.9; (b) 22.01; (c) 24.46; person "a" should lose weight

3. Examine the table. What pattern do you notice as you look across the rows? What pattern do you notice as you look down the columns? BMI decreases; BMI increases

4. Mark the table to show weight and height combinations with corresponding BMI values that fall within the recommended range.

5. Why would having a table like this on hand be beneficial to a doctor? Explain. answers may vary

Height

		60	62	64	66	68	70	72	74
	100	19.58	18.34	17.21	16.19	15.25	14.39	13.60	12.87
	110	21.54	20.17	18.93	17.80	16.77	15.83	14.96	14.16
W	120	23.50	22.01	20.65	19.42	18.30	17.27	16.32	15.45
e	130	25.46	23.84	22.38	21.04	19.82	18.70	17.68	16.74
i	140	27.42	25.68	24.10	22.66	21.35	20.14	19.04	18.02
g	150	29.38	27.51	25.82	24.28	22.87	21.58	20.40	19.31
h	160	31.33	29.34	27.54	25.90	24.39	23.02	21.76	20.60
t	170	33.29	31.18	29.26	27.51	25.92	24.46	23.12	21.89
	180	35.25	33.01	30.98	29.13	27.44	25.90	24.48	23.17
	190	37.21	34.85	32.70	30.75	28.97	27.34	25.84	24.46
	200	39.17	36.68	34.42	32.37	30.49	28.78	27.20	25.75

CHAPTER 5 HIGHLIGHTS

DEFINITIONS AND CONCEPTS	EXAMPLES

SECTION 5.1   SIMPLIFYING RATIONAL EXPRESSIONS

A **rational expression** is an expression that can be written in the form $\dfrac{P}{Q}$, where P and Q are polynomials and Q does not equal 0.

$$\frac{7y^3}{4}, \; \frac{x^2 + 6x + 1}{x - 3}, \; \frac{-5}{s^3 + 8}$$

To find values for which a rational expression is undefined, find values for which the denominator is 0.

Find any values for which the expression
$\dfrac{5y}{y^2 - 4y + 3}$ is undefined.

$$y^2 - 4y + 3 = 0 \quad \text{Set the denominator equal to 0.}$$
$$(y - 3)(y - 1) = 0 \quad \text{Factor.}$$
$$y - 3 = 0 \text{ or } y - 1 = 0 \quad \text{Set each factor equal to 0.}$$
$$y = 3 \qquad y = 1 \quad \text{Solve.}$$

The expression is undefined when y is 3 and when y is 1.

FUNDAMENTAL PRINCIPLE OF RATIONAL EXPRESSIONS

If P, Q, and R are polynomials, and Q and R are not 0, then

$$\frac{PR}{QR} = \frac{P}{Q}$$

By the fundamental principle,

$$\frac{(x - 3)(x + 1)}{x(x + 1)} = \frac{x - 3}{x}$$

as long as $x \neq 0$ and $x \neq -1$.

TO SIMPLIFY A RATIONAL EXPRESSION

Step 1. Factor the numerator and denominator.

Step 2. Apply the fundamental principle to divide out common factors.

Simplify: $\dfrac{4x + 20}{x^2 - 25}$

$$\frac{4x + 20}{x^2 - 25} = \frac{4(x + 5)}{(x + 5)(x - 5)} = \frac{4}{x - 5}$$

SECTION 5.2   MULTIPLYING AND DIVIDING RATIONAL EXPRESSIONS

TO MULTIPLY RATIONAL EXPRESSIONS

Step 1. Factor numerators and denominators.

Step 2. Multiply numerators and multiply denominators.

Step 3. Write the product in lowest terms.

$$\frac{P}{Q} \cdot \frac{R}{S} = \frac{PR}{QS}$$

Multiply: $\dfrac{4x + 4}{2x - 3} \cdot \dfrac{2x^2 + x - 6}{x^2 - 1}$

$$\frac{4x + 4}{2x - 3} \cdot \frac{2x^2 + x - 6}{x^2 - 1}$$
$$= \frac{4(x + 1)}{2x - 3} \cdot \frac{(2x - 3)(x + 2)}{(x + 1)(x - 1)}$$
$$= \frac{4(x + 1)(2x - 3)(x + 2)}{(2x - 3)(x + 1)(x - 1)}$$
$$= \frac{4(x + 2)}{x - 1}$$

Section 5.2 (Continued)	
To divide by a rational expression, multiply by the reciprocal. $$\frac{P}{Q} \div \frac{R}{S} = \frac{P}{Q} \cdot \frac{S}{R} = \frac{PS}{QR}$$	Divide: $\dfrac{15x + 5}{3x^2 - 14x - 5} \div \dfrac{15}{3x - 12}$ $\dfrac{15x + 5}{3x^2 - 14x - 5} \div \dfrac{15}{3x - 12}$ $= \dfrac{5(3x + 1)}{(3x + 1)(x - 5)} \cdot \dfrac{3(x - 4)}{3 \cdot 5}$ $= \dfrac{x - 4}{x - 5}$

Section 5.3   Adding and Subtracting Rational Expressions with the Same Denominator and Least Common Denominator

To add or subtract rational expressions with the same denominator, add or subtract numerators, and place the sum or difference over the common denominator. $$\frac{P}{R} + \frac{Q}{R} = \frac{P + Q}{R}$$ $$\frac{P}{R} - \frac{Q}{R} = \frac{P - Q}{R}$$	Perform each indicated operation. $$\frac{5}{x + 1} + \frac{x}{x + 1} = \frac{5 + x}{x + 1}$$ $\dfrac{2y + 7}{y^2 - 9} - \dfrac{y + 4}{y^2 - 9}$ $= \dfrac{(2y + 7) - (y + 4)}{y^2 - 9}$ $= \dfrac{2y + 7 - y - 4}{y^2 - 9}$ $= \dfrac{y + 3}{(y + 3)(y - 3)}$ $= \dfrac{1}{y - 3}$
To Find the Least Common Denominator (LCD) **Step 1.** Factor the denominators. **Step 2.** The LCD is the product of all unique factors, each raised to a power equal to the greatest number of times that it appears in any one factored denominator.	Find the LCD for $\dfrac{7x}{x^2 + 10x + 25}$ and $\dfrac{11}{3x^2 + 15x}$ $x^2 + 10x + 25 = (x + 5)(x + 5)$ $3x^2 + 15x = 3x(x + 5)$ $\text{LCD} = 3x(x + 5)(x + 5)$ or $3x(x + 5)^2$

Section 5.4   Adding and Subtracting Rational Expressions with Different Denominators

To Add or Subtract Rational Expressions with Different Denominators **Step 1.** Find the LCD.	Perform the indicated operation. $\dfrac{9x + 3}{x^2 - 9} - \dfrac{5}{x - 3}$ $= \dfrac{9x + 3}{(x + 3)(x - 3)} - \dfrac{5}{x - 3}$ LCD is $(x + 3)(x - 3)$.

SECTION 5.4   (CONTINUED)

Step 2. Rewrite each rational expression as an equivalent expression whose denominator is the LCD.

Step 3. Add or subtract numerators and place the sum or difference over the common denominator.

Step 4. Write the result in lowest terms.

$$= \frac{9x + 3}{(x + 3)(x - 3)} - \frac{5(x + 3)}{(x - 3)(x + 3)}$$

$$= \frac{9x + 3 - 5(x + 3)}{(x + 3)(x - 3)}$$

$$= \frac{9x + 3 - 5x - 15}{(x + 3)(x - 3)}$$

$$= \frac{4x - 12}{(x + 3)(x - 3)}$$

$$= \frac{4(x - 3)}{(x + 3)(x - 3)} = \frac{4}{x + 3}$$

SECTION 5.5   SOLVING EQUATIONS CONTAINING RATIONAL EXPRESSIONS

TO SOLVE AN EQUATION CONTAINING RATIONAL EXPRESSIONS

Step 1. Multiply both sides of the equation by the LCD of all rational expressions in the equation.

Step 2. Remove any grouping symbols and solve the resulting equation.

Step 3. Check the solution in the original equation.

Solve: $\dfrac{5x}{x + 2} + 3 = \dfrac{4x - 6}{x + 2}$ The LCD is $x + 2$.

$$(x + 2)\left(\frac{5x}{x + 2} + 3\right) = (x + 2)\left(\frac{4x - 6}{x + 2}\right)$$

$$(x + 2)\left(\frac{5x}{x + 2}\right) + (x + 2)(3)$$

$$= (x + 2)\left(\frac{4x - 6}{x + 2}\right)$$

$$5x + 3x + 6 = 4x - 6$$
$$4x = -12$$
$$x = -3$$

The solution checks; the solution is -3.

SECTION 5.6   RATIONAL EQUATIONS AND PROBLEM SOLVING

PROBLEM-SOLVING STEPS

1. UNDERSTAND. Read and reread the problem.

A small plane and a car leave Kansas City, Missouri, and head for Minneapolis, Minnesota, a distance of 450 miles. The speed of the plane is 3 times the speed of the car, and the plane arrives 6 hours ahead of the car. Find the speed of the car.

$$\text{Let } x = \text{the speed of the car.}$$
$$\text{Then } 3x = \text{ the speed of the plane.}$$

	Distance =	Rate ·	Time
Car	450	x	$\dfrac{450}{x}\left(\dfrac{\text{distance}}{\text{rate}}\right)$
Plane	450	$3x$	$\dfrac{450}{3x}\left(\dfrac{\text{distance}}{\text{rate}}\right)$

2. TRANSLATE.

In words: plane's time + 6 hours = car's time

$$\downarrow \qquad\qquad \downarrow \qquad\quad \downarrow$$

Translate: $\dfrac{450}{3x} \quad + \quad 6 \quad = \quad \dfrac{450}{x}$

3. SOLVE.

$$\frac{450}{3x} + 6 = \frac{450}{x}$$

$$3x\left(\frac{450}{3x}\right) + 3x(6) = 3x\left(\frac{450}{x}\right)$$

$$450 + 18x = 1350$$

$$18x = 900$$

$$x = 50$$

4. INTERPRET.

Check the solution by replacing x with 50 in the original equation. **State** the conclusion: The speed of the car is 50 miles per hour.

SECTION 5.7   SIMPLIFYING COMPLEX FRACTIONS

METHOD 1: TO SIMPLIFY A COMPLEX FRACTION

Step 1. Add or subtract fractions in the numerator and the denominator of the complex fraction.

Step 2. Perform the indicated division.

Step 3. Write the result in lowest terms.

Simplify:

$$\frac{\dfrac{1}{x} + 2}{\dfrac{1}{x} - \dfrac{1}{y}} = \frac{\dfrac{1}{x} + \dfrac{2x}{x}}{\dfrac{y}{xy} - \dfrac{x}{xy}}$$

$$= \frac{\dfrac{1 + 2x}{x}}{\dfrac{y - x}{xy}}$$

$$= \frac{1 + 2x}{x} \cdot \frac{xy}{y - x}$$

$$= \frac{y(1 + 2x)}{y - x}$$

SECTION 5.7   (CONTINUED)

METHOD 2. TO SIMPLIFY A COMPLEX FRACTION

Step 1. Find the LCD of all fractions in the complex fraction.

Step 2. Multiply the numerator and the denominator of the complex fraction by the LCD.

Step 3. Perform the indicated operations and write the result in lowest terms.

$$\frac{\dfrac{1}{x} + 2}{\dfrac{1}{x} - \dfrac{1}{y}} = \frac{xy\left(\dfrac{1}{x} + 2\right)}{xy\left(\dfrac{1}{x} - \dfrac{1}{y}\right)}$$

$$= \frac{xy\left(\dfrac{1}{x}\right) + xy(2)}{xy\left(\dfrac{1}{x}\right) - xy\left(\dfrac{1}{y}\right)}$$

$$= \frac{y + 2xy}{y - x} \quad \text{or} \quad \frac{y(1 + 2x)}{y - x}$$

CHAPTER 5 REVIEW

(5.1) *Find any real number for which each rational expression is undefined.*

1. $\dfrac{x+5}{x^2-4}$ $x=2, x=-2$

2. $\dfrac{5x+9}{4x^2-4x-15}$ $x=\dfrac{5}{2}, x=-\dfrac{3}{2}$

Find the value of each rational expression when $x=5$, $y=7$, and $z=-2$.

3. $\dfrac{2-z}{z+5}$ $\dfrac{4}{3}$

4. $\dfrac{x^2+xy-y^2}{x+y}$ $\dfrac{11}{12}$

Simplify each rational expression.

5. $\dfrac{2x+6}{x^2+3x}$ $\dfrac{2}{x}$

6. $\dfrac{3x-12}{x^2-4x}$ $\dfrac{3}{x}$

7. $\dfrac{x+2}{x^2-3x-10}$ $\dfrac{1}{x-5}$

8. $\dfrac{x+4}{x^2+5x+4}$ $\dfrac{1}{x+1}$

9. $\dfrac{x^3-4x}{x^2+3x+2}$ $\dfrac{x(x-2)}{x+1}$

10. $\dfrac{5x^2-125}{x^2+2x-15}$ $\dfrac{5(x-5)}{x-3}$

11. $\dfrac{x^2-x-6}{x^2-3x-10}$ $\dfrac{x-3}{x-5}$

12. $\dfrac{x^2-2x}{x^2+2x-8}$ $\dfrac{x}{x+4}$

Simplify each expression. First, factor the four-term polynomials by grouping.

13. $\dfrac{x^2+xa+xb+ab}{x^2-xc+bx-bc}$ $\dfrac{x+a}{x-c}$

14. $\dfrac{x^2+5x-2x-10}{x^2-3x-2x+6}$ $\dfrac{x+5}{x-3}$

(5.2) *Perform each indicated operation and simplify.*

15. $\dfrac{15x^3y^2}{z} \cdot \dfrac{z}{5xy^3}$ $\dfrac{3x^2}{y}$

16. $\dfrac{-y^3}{8} \cdot \dfrac{9x^2}{y^3}$ $-\dfrac{9x^2}{8}$

17. $\dfrac{x^2 - 9}{x^2 - 4} \cdot \dfrac{x - 2}{x + 3}$ $\quad \dfrac{x - 3}{x + 2}$

18. $\dfrac{2x + 5}{x - 6} \cdot \dfrac{2x}{-x + 6}$ $\quad \dfrac{-2x(2x + 5)}{(x - 6)^2}$

19. $\dfrac{x^2 - 5x - 24}{x^2 - x - 12} \div \dfrac{x^2 - 10x + 16}{x^2 + x - 6}$ $\quad \dfrac{x + 3}{x - 4}$

20. $\dfrac{4x + 4y}{xy^2} \div \dfrac{3x + 3y}{x^2y}$ $\quad \dfrac{4x}{3y}$

21. $\dfrac{x^2 + x - 42}{x - 3} \cdot \dfrac{(x - 3)^2}{x + 7}$ $\quad (x - 6)(x - 3)$

22. $\dfrac{2a + 2b}{3} \cdot \dfrac{a - b}{a^2 - b^2}$ $\quad \dfrac{2}{3}$

23. $\dfrac{x^2 - 9x + 14}{x^2 - 5x + 6} \cdot \dfrac{x + 2}{x^2 - 5x - 14}$ $\quad \dfrac{1}{x - 3}$

24. $(x - 3) \cdot \dfrac{x}{x^2 + 3x - 18}$ $\quad \dfrac{x}{x + 6}$

25. $\dfrac{2x^2 - 9x + 9}{8x - 12} \div \dfrac{x^2 - 3x}{2x}$ $\quad \dfrac{1}{2}$

26. $\dfrac{x^2 - y^2}{x^2 + xy} \div \dfrac{3x^2 - 2xy - y^2}{3x^2 + 6x}$ $\quad \dfrac{3(x + 2)}{3x + y}$

(5.3) *Perform each indicated operation and simplify.*

27. $\dfrac{x}{x^2 + 9x + 14} + \dfrac{7}{x^2 + 9x + 14}$ $\quad \dfrac{1}{x + 2}$

28. $\dfrac{x}{x^2 + 2x - 15} + \dfrac{5}{x^2 + 2x - 15}$ $\quad \dfrac{1}{x - 3}$

29. $\dfrac{4x - 5}{3x^2} - \dfrac{2x + 5}{3x^2}$ $\quad \dfrac{2x - 10}{3x^2}$

30. $\dfrac{9x + 7}{6x^2} - \dfrac{3x + 4}{6x^2}$ $\quad \dfrac{2x + 1}{2x^2}$

Find the LCD of each pair of rational expressions.

31. $\dfrac{x + 4}{2x}, \dfrac{3}{7x}$ $\quad 14x$

32. $\dfrac{x - 2}{x^2 - 5x - 24}, \dfrac{3}{x^2 + 11x + 24}$
$(x - 8)(x + 8)(x + 3)$

Rewrite each rational expression as an equivalent expression whose denominator is the given polynomial.

33. $\dfrac{5}{7x} = \dfrac{}{14x^3y}$ $\quad \dfrac{10x^2y}{14x^3y}$

34. $\dfrac{9}{4y} = \dfrac{}{16y^3x}$ $\quad \dfrac{36y^2x}{16y^3x}$

35. $\dfrac{x+2}{x^2+11x+18} = \dfrac{}{(x+2)(x-5)(x+9)}$

$\dfrac{x^2-3x-10}{(x+2)(x-5)(x+9)}$

36. $\dfrac{3x-5}{x^2+4x+4} = \dfrac{}{(x+2)^2(x+3)}$

$\dfrac{3x^2+4x-15}{(x+2)^2(x+3)}$

(5.4) *Perform each indicated operation and simplify.*

37. $\dfrac{4}{5x^2} - \dfrac{6}{y}$ $\dfrac{4y-30x^2}{5x^2y}$

38. $\dfrac{2}{x-3} - \dfrac{4}{x-1}$ $\dfrac{-2x+10}{(x-3)(x-1)}$

39. $\dfrac{x+7}{x+3} - \dfrac{x-3}{x+7}$ $\dfrac{14x+58}{(x+3)(x+7)}$

40. $\dfrac{4}{x+3} - 2$ $\dfrac{-2x-2}{x+3}$

41. $\dfrac{3}{x^2+2x-8} + \dfrac{2}{x^2-3x+2}$

$\dfrac{5x+5}{(x+4)(x-2)(x-1)}$

42. $\dfrac{2x-5}{6x+9} - \dfrac{4}{2x^2+3x}$

$\dfrac{x-4}{3x}$

43. $\dfrac{x-1}{x^2-2x+1} - \dfrac{x+1}{x-1}$ $-\dfrac{x}{x-1}$

44. $\dfrac{x-1}{x^2+4x+4} + \dfrac{x-1}{x+2}$ $\dfrac{x^2+2x-3}{(x+2)^2}$

Find the perimeter and the area of each figure.

45. 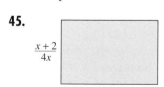 $\dfrac{x+2x+4}{4x}; \dfrac{x+2}{32}$

$\dfrac{x+2}{4x}$

$\dfrac{x}{8}$

46.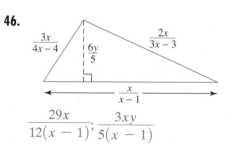

$\dfrac{3x}{4x-4}$ $\dfrac{6y}{5}$ $\dfrac{2x}{3x-3}$

$\dfrac{x}{x-1}$

$\dfrac{29x}{12(x-1)}; \dfrac{3xy}{5(x-1)}$

(5.5) *Solve each equation.*

47. $\dfrac{x+4}{9} = \dfrac{5}{9}$ $x=1$

48. $\dfrac{n}{10} = 9 - \dfrac{n}{5}$ $n=30$

Name _____

49. $\dfrac{5y - 3}{7} = \dfrac{15y - 2}{28}$ $y = 2$

50. $\dfrac{2}{x + 1} - \dfrac{1}{x - 2} = -\dfrac{1}{2}$ $x = 3, x = -4$

51. $\dfrac{1}{a + 3} + \dfrac{1}{a - 3} = -\dfrac{5}{a^2 - 9}$ $a = -\dfrac{5}{2}$

52. $\dfrac{y}{2y + 2} + \dfrac{2y - 16}{4y + 4} = \dfrac{y - 3}{y + 1}$ no solution

53. $\dfrac{4}{x + 3} + \dfrac{8}{x^2 - 9} = 0$ $x = 1$

54. $\dfrac{2}{x - 3} - \dfrac{4}{x + 3} = \dfrac{8}{x^2 - 9}$ $x = 5$

55. $\dfrac{x - 3}{x + 1} - \dfrac{x - 6}{x + 5} = 0$ $x = \dfrac{9}{7}$

56. $x + 5 = \dfrac{6}{x}$ $x = -6, x = 1$

Solve each equation for the indicated variable.

57. $\dfrac{4A}{5b} = x^2$ for b $b = \dfrac{4A}{5x^2}$

58. $\dfrac{x}{7} + \dfrac{y}{8} = 10$ for y $y = \dfrac{560 - 8x}{7}$

(5.6) *Solve each problem.*

59. Five times the reciprocal of a number equals the sum of $\frac{3}{2}$ times the reciprocal of the number and $\frac{7}{6}$. What is the number?   3

60. The reciprocal of a number equals the reciprocal of the difference of 4 and the number. Find the number.   2

61. A car travels 90 miles in the same time that a car traveling 10 miles per hour slower travels 60 miles. Find the speed of each car.   fast car speed: 30 mph, slow car speed: 20 mph

62. The current in a bayou near Lafayette, Louisiana, is 4 miles per hour. A paddle boat travels 48 miles upstream in the same amount of time it takes to travel 72 miles downstream. Find the speed of the boat in still water.   20 mph

63. When Mark and Maria manicure Mr. Stergeon's lawn, it takes them 5 hours. If Mark works alone, it takes 7 hours. Find how long it takes Maria alone. $17\frac{1}{2}$ hr

64. It takes pipe A 20 days to fill a fish pond. Pipe B takes 15 days. Find how long it takes both pipes together to fill the pond. $8\frac{4}{7}$ days

Given that the pairs of triangles are similar, find each missing length x.

65.   $x = 15$ **66.**  $x = 6$

67. $x = 15$ **68.** $x = 60$

(5.7) *Simplify each complex fraction.*

69. $\dfrac{\dfrac{5x}{27}}{\dfrac{10xy}{21}}$ $-\dfrac{7}{18y}$

70. $\dfrac{\dfrac{8x}{x^2 - 9}}{\dfrac{4}{x + 3}}$ $\dfrac{2x}{x - 3}$

71. $\dfrac{\dfrac{3}{5} + \dfrac{2}{7}}{\dfrac{1}{5} + \dfrac{5}{6}}$ $\dfrac{6}{7}$

72. $\dfrac{2 + \dfrac{1}{x^2}}{\dfrac{1}{x} + \dfrac{2}{x^2}}$ $\dfrac{2x^2 + 1}{x + 2}$

73. $\dfrac{3 - \dfrac{1}{y}}{2 - \dfrac{1}{y}}$ $\dfrac{3y - 1}{2y - 1}$

74. $\dfrac{\dfrac{6}{x + 2} + 4}{\dfrac{8}{x + 2} - 4}$ $-\dfrac{7 + 2x}{2x}$

Focus On Business and Career

MORTGAGES

A loan for which the purpose is to buy a house or other property is called a **mortgage**. When you are thinking of getting a mortgage to buy a house, it is helpful to know how much your monthly mortgage payment will be. One way to calculate the monthly payment is to use the formula

$$P = \frac{\dfrac{Ar}{12}}{1 - \dfrac{1}{\left(1 + \dfrac{r}{12}\right)^{12t}}},$$

where A = is the amount of the mortgage, r = annual interest rate (written as a decimal), and t = loan term in years. Try the exercises below.

CRITICAL THINKING

1. The average mortgage rate in the United States in 1996 was 7.93% (Source: National Association of Realtors®). Suppose you had borrowed $80,000 to buy a house in 1996. If your loan term is 30 years, calculate your monthly mortgage payment. $583.11

2. The average mortgage interest rate in the United States in 1989 was 10.11% (Source: National Association of Realtors®). Suppose you had borrowed $71,000 to buy a house in 1989. If your loan term is 20 years, calculate your monthly mortgage payment. $690.35

Another way to calculate a monthly mortgage payment is to use one of the many sites on the World Wide Web that offer an interactive mortgage calculator. For instance, by visiting the given World Wide Web address, you will be able to access the Fleet Bank Web site, or a related site, where you can calculate a monthly mortgage payment by entering the amount to be borrowed, the interest rate as a percent, and the term of the loan in years. Use the site below to solve the given exercises.

Go to http://www.prenhall.com/martin-gay

 Internet Excursions

3. Suppose you would like to borrow $64,000 to buy a house. If the interest rate is 8.2% and you plan to take out a 25-year loan, what will be your monthly mortgage payment? $502.47

4. Suppose you would like to borrow $100,000 to buy a house. If the interest rate is 7.5% and you plan to take out a 20-year loan, what will be your monthly mortgage payment? $805.59

CHAPTER 5 TEST

1. Find any real numbers for which the following expression is undefined.

$$\frac{x + 5}{x^2 + 4x + 3}$$

2. For a certain computer desk, the manufacturing cost C per desk (in dollars) is

$$C = \frac{100x + 3000}{x}$$

where x is the number of desks manufactured.

a. Find the average cost per desk when manufacturing 200 computer desks.

b. Find the average cost per desk when manufacturing 1000 computer desks.

Simplify each rational expression.

3. $\dfrac{3x - 6}{5x - 10}$

4. $\dfrac{x + 10}{x^2 - 100}$

5. $\dfrac{x + 6}{x^2 + 12x + 36}$

6. $\dfrac{7 - x}{x - 7}$

7. $\dfrac{2m^3 - 2m^2 - 12m}{m^2 - 5m + 6}$

8. $\dfrac{y - x}{x^2 - y^2}$

Perform each indicated operation and simplify if possible.

9. $\dfrac{x^2 - 13x + 42}{x^2 + 10x + 21} \div \dfrac{x^2 - 4}{x^2 + x - 6}$

10. $\dfrac{3}{x - 1} \cdot (5x - 5)$

11. $\dfrac{y^2 - 5y + 6}{2y + 4} \cdot \dfrac{y + 2}{2y - 6}$

12. $\dfrac{5}{2x + 5} - \dfrac{6}{2x + 5}$

13. $\dfrac{5a}{a^2 - a - 6} - \dfrac{2}{a - 3}$

14. $\dfrac{6}{x^2 - 1} + \dfrac{3}{x + 1}$

425

Name _____

15. $\dfrac{x^2-9}{x^2-3x} \div \dfrac{x^2+4x+1}{2x+10}$

16. $\dfrac{x+2}{x^2+11x+18} + \dfrac{5}{x^2-3x-10}$

17. $\dfrac{4y}{y^2+6y+5} - \dfrac{3}{y^2+5y+4}$

Solve each equation.

18. $\dfrac{4}{y} - \dfrac{5}{3} = \dfrac{-1}{5}$

19. $\dfrac{5}{y+1} = \dfrac{4}{y+2}$

20. $\dfrac{a}{a-3} = \dfrac{3}{a-3} - \dfrac{3}{2}$

21. $\dfrac{10}{x^2-25} = \dfrac{3}{x+5} + \dfrac{1}{x-5}$

Simplify each complex fraction.

22. $\dfrac{\dfrac{5x^2}{yz^2}}{\dfrac{10x}{z^3}}$

23. $\dfrac{\dfrac{b}{a} - \dfrac{a}{b}}{\dfrac{1}{b} + \dfrac{1}{a}}$

24. $\dfrac{5 - \dfrac{1}{y^2}}{\dfrac{1}{y} + \dfrac{2}{y^2}}$

25. Given that the two triangles are similar, find x.

26. A number plus five times its reciprocal is equal to six. Find the number.

27. A pleasure boat traveling down the Red River takes the same time to go 14 miles upstream as it takes to go 16 miles downstream. If the current of the river is 2 miles per hour, find the speed of the boat in still water.

28. An inlet pipe can fill a tank in 12 hours. A second pipe can fill the tank in 15 hours. If both pipes are used, find how long it takes to fill the tank.

Cumulative Review

1. Write each sentence as an equation. Let x represent the unknown number.

   a. The quotient of 15 and a number is 4.

   b. Three subtracted from 12 is a number.

   c. Four times a number added to 17 is 21.

2. Find the sums.

   a. $3 + (-7) + (-8)$

   b. $[7 + (-10)] + [-2 + (-4)]$

Name the property illustrated by each true statement.

3. $3 \cdot y = y \cdot 3$

4. $(x + 7) + 9 = x + (7 + 9)$

5. Solve: $3 - x = 7$

6. A 10-foot board is to be cut into two pieces so that the longer piece is 4 times longer than the shorter. Find the length of each piece.

7. Solve $y = mx + b$ for x.

8. Solve $x + 4 \le -6$. Graph the solutions.

Simplify each quotient.

9. $\dfrac{x^5}{x^2}$

10. $\dfrac{4^7}{4^3}$

11. $\dfrac{(-3)^5}{(-3)^2}$

12. $\dfrac{2x^5y^2}{xy}$

Simplify by writing each expression with positive exponents only.

13. $2x^{-3}$

14. $(-2)^{-4}$

Multiply.

15. $5x(2x^3 + 6)$

16. $-3x^2(5x^2 + 6x - 1)$

427

17. Divide: $\dfrac{4x^2 + 7 + 8x^3}{2x + 3}$

18. Factor $x^2 + 7x + 12$

19. Factor: $25x^2 + 20xy + 4y^2$

20. Solve: $x^2 - 9x - 22 = 0$

21. Multiply: $\dfrac{x^2 + x}{3x} \cdot \dfrac{6}{5x + 5}$

22. Subtract: $\dfrac{3x^2 + 2x}{x - 1} - \dfrac{10x - 5}{x - 1}$

23. Subtract: $\dfrac{6x}{x^2 - 4} - \dfrac{3}{x + 2}$

24. Solve $\dfrac{t - 4}{2} - \dfrac{t - 3}{9} = \dfrac{5}{18}$

25. Sam Waterton and Frank Schaffer work in a plant that manufactures automobiles. Sam can complete a quality control tour of the plant in 3 hours while his assistant, Frank, needs 7 hours to complete the same job. The regional manager is coming to inspect the plant facilities, so both Sam and Frank are directed to complete a quality control tour together. How long will this take?

26. Simplify: $\dfrac{\dfrac{1}{z} - \dfrac{1}{2}}{\dfrac{1}{3} - \dfrac{z}{6}}$

17. $4x^2 - 4x + 6 + \dfrac{-11}{2x + 3}$ (Sec. 3.7, Ex. 6)

18. $(x + 3)(x + 4)$ (Sec. 4.2, Ex. 1)

19. $(5x + 2y)^2$ (Sec. 4.5, Ex. 5)

20. $x = 11, x = -2$ (Sec. 4.6, Ex. 4)

21. $\dfrac{2}{5}$ (Sec. 5.2, Ex. 2)

22. $3x - 5$ (Sec. 5.3, Ex. 3)

23. $\dfrac{3}{x - 2}$ (Sec. 5.4, Ex. 2)

24. $t = 5$ (Sec. 5.5, Ex. 2)

25. $2\frac{1}{10}$ hr (Sec. 5.6, Ex. 2)

26. $\dfrac{3}{z}$ (Sec. 5.7, Ex. 3)

Graphing Equations and Inequalities

CHAPTER 6

In Chapter 2 we learned to solve and graph the solutions of linear equations and inequalities in one variable on number lines. Now we define and present techniques for solving and graphing linear equations and inequalities in two variables on grids. Two-variable equations lead directly to the concept of *function*, perhaps the most important concept in all mathematics. Functions are introduced in Section 6.6.

In Exercise 36 on page 494, we will use the methods of this chapter to find a linear equation in two variables that represents the crude oil production of OPEC countries. OPEC, the Organization of Petroleum Exporting Countries, was established in 1960. OPEC's goal is to control the price of crude oil worldwide by controlling oil production. For example, if OPEC countries agree to limit the amount of oil they produce, an oil shortage is created and oil prices rise. In 1997, the members of OPEC were Algeria, Indonesia, Iran, Iraq, Kuwait, Libya, Nigeria, Qatar, Saudi Arabia, United Arab Emirates, and Venezuela. OPEC's headquarters is in Vienna, Austria.

Name _____ Section _____ Date _____

CHAPTER 6 PRETEST

1. _____ (6.1A)

2. _____ $(-2, -6)$; (6.1C)

3. _____ (6.2A)

4. _____ (6.3B)

5. _____ (6.7B)

6. _____ $-\dfrac{3}{10}$; (6.4A)

7. _____ $\dfrac{4}{5}$; (6.4B)

8. _____ undefined slope; (6.4C)

9. _____ $x + 3y = -15$; (6.5C)

10. _____ $x + 8y = 0$; (6.5D)

11. _____ $2x - 7y = -98$; (6.5A)

12. _____ domain: $\{-3, 0, 2, 7\}$; range: $\{-1, 6, 8\}$ (6.6A)

13. _____ function; (6.6B)

14. _____ not a function; (6.6B)

15. _____ a. 11   b. 8   c. -22; (6.6D)

1. Plot each ordered pair: $(-4, 3)$, $(0, -2)$, and $(5, 0)$

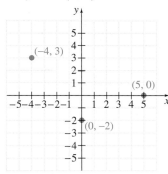

2. Complete the ordered pair solution for the given linear equation.
$8x - 3y = 2$; $(-2, \quad)$

Graph.

3. $3x - y = 6$

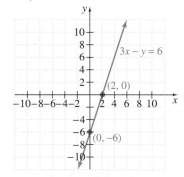

4. The line with x-intercept: -1, y-intercept: 4

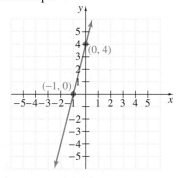

5. $3x + 2y \le 6$

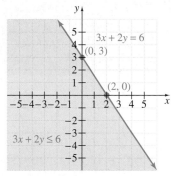

Find the slope of each line.

6. Passes through $(-7, 8)$ and $(3, 5)$

7. $4x - 5y = 20$

8. $x = 10$

Find the equation of each line. Write the equation in the form $Ax + By = C$.

9. Slope $-\dfrac{1}{3}$; passes through $(3, -6)$

10. Passes through the origin and $(-8, 1)$

11. Slope $\dfrac{2}{7}$; y-intercept 14

12. Find the domain and range of the given relation. $\{(-3, 8), (7, -1), (0, 6), (2, -1)\}$

Determine whether each relation is a function.

13. $\{(1, 7), (-8, 7), (6, 3), (9, 2)\}$

14. $\{(0, 4), (1, 3), (2, -5), (1, 10), (-2, -8)\}$

15. Given the function $f(x) = -3x + 8$, find each function value.

   a. $f(-1)$ **b.** $f(0)$ **c.** $f(10)$

6.1 THE RECTANGULAR COORDINATE SYSTEM

In Section 1.8, we learned how to read graphs. Example 4 in Section 1.8 presented the broken line graph below showing the relationship between time spent smoking a cigarette and pulse rate. The horizontal line or axis shows time in minutes and the vertical line or axes shows the pulse rate in heartbeats per minute. Notice in this graph that there are two numbers associated with each point of the graph. For example, we discussed earlier that 15 minutes after "lighting up," the pulse rate is 80 beats per minute. If we agree to write the time first and the pulse rate second, we can say there is a point on the graph corresponding to the **ordered pair** of numbers (15, 80). A few more ordered pairs are shown alongside their corresponding points.

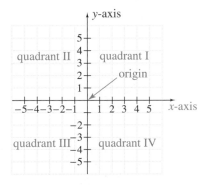

A PLOTTING ORDERED PAIRS OF NUMBERS

In general, we use the idea of ordered pairs to describe the location of a point in a plane (such as a piece of paper). We start with a horizontal and a vertical axis. Each axis is a number line, and for the sake of consistency we construct our axes to intersect at the 0 coordinate of both. This point of intersection is called the **origin**. Notice that these two number lines or axes divide the plane into four regions called **quadrants**. The quadrants are usually numbered with Roman numerals as shown. The axes are not considered to be in any quadrant.

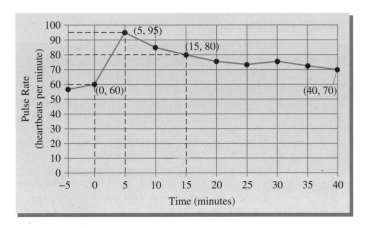

It is helpful to label axes, so we label the horizontal axis the **x-axis** and the vertical axis the **y-axis**. We call the system described above the **rectangular coordinate system**, or the **coordinate plane**. Just as with other graphs shown, we can then describe the locations of points by ordered pairs of numbers. We list the horizontal **x-axis** measurement first and the vertical **y-axis** measurement second.

Objectives

A Plot ordered pairs of numbers on the rectangular coordinate system.

B Graph paired data to create a scatter diagram.

C Find the missing coordinate of an ordered pair solution, given one coordinate of the pair.

SSM   CD-ROM   Video 6.1

TEACHING TIP

Remind students that having the right tools makes learning easier. In this chapter, graph paper and a straight edge will be helpful.

TEACHING TIP

If your class is arranged in rows, begin this lesson by discussing ways to describe the location of desks in the classroom. Point out that you need to define a reference point, that each location needs a row value and a column value and that you need to agree on which value to state first. Then connect these ideas to the origin and ordered pairs. Test students' understanding by asking them to find the location of their desk. Then have groups of students stand up who meet certain criteria.

TEACHING TIP

Point out to students that the x-axis is simply a horizontal number line. Treat every point on that number line as if there is a vertical number line with the positive values above and the negative values below. When we plot a point, we first find its location on the x-axis number line and then find its location on the vertical number line corresponding to that point.

To plot or graph the point corresponding to the ordered pair

$$(a, b)$$

We start at the origin. We then move a units left or right (right if a is positive, left if a is negative). From there, we move b units up or down (up if b is positive, down if b is negative). For example, to plot the point corresponding to the ordered pair $(3, 2)$, we start at the origin, move 3 units right, and from there move 2 units up. (See the figure below.) The x-value, 3, is also called the **x-coordinate** and the y-value, 2, is also called the **y-coordinate**. From now on, we will call the point with coordinates $(3, 2)$ simply the point $(3, 2)$. The point $(-2, 5)$ is graphed below also.

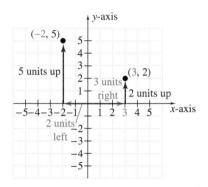

✓ CONCEPT CHECK

Is the graph of the point $(-5, 1)$ in the same location as the graph of the point $(1, -5)$? Explain.

Practice Problem 1

On a single coordinate system, plot each ordered pair. State in which quadrant, if any, each point lies.

a. $(4, 2)$ b. $(-1, -3)$

c. $(2, -2)$ d. $(-5, 1)$

e. $(0, 3)$ f. $(3, 0)$

g. $(0, -4)$ h. $\left(-2\frac{1}{2}, 0\right)$

Answers

1.

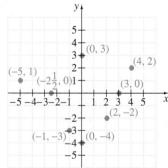

Point $(4, 2)$ lies in quadrant I.
Point $(-5, 1)$ lies in quadrant II.
Point $(-1, -3)$ lies in quadrant III.
Point $(2, -2)$ lies in quadrant IV.

Points $(0, 3)$, $(3, 0)$, $(0, -4)$, and $\left(-2\frac{1}{2}, 0\right)$

lie on axes, so they are not in any quadrant.

✓ **Concept Check:**

The graph of point $(-5, 1)$ lies in quadrant II and the graph of point $(1, -5)$ lies in quadrant IV. They are *not* in the same location.

HELPFUL HINT

Don't forget that **each ordered pair corresponds to exactly one point in the plane and that each point in the plane corresponds to exactly one ordered pair.**

TRY THE CONCEPT CHECK IN THE MARGIN.

Example 1

On a single coordinate system, plot each ordered pair. State in which quadrant, if any, each point lies.

a. $(5, 3)$ b. $(-2, -4)$ c. $(1, -2)$ d. $(-5, 3)$

e. $(0, 0)$ f. $(0, 2)$ g. $(-5, 0)$ h. $\left(0, -5\frac{1}{2}\right)$

Solution:

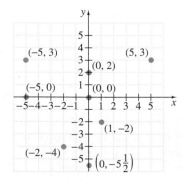

Point $(5, 3)$ lies in quadrant I.
Point $(-5, 3)$ lies in quadrant II.
Point $(-2, -4)$ lies in quadrant III.
Point $(1, -2)$ lies in quadrant IV.
Points $(0, 0)$, $(0, 2)$, $(-5, 0)$,

and $\left(0, -5\frac{1}{2}\right)$ lie on axes, so

they are not in any quadrant.   ▬▬▬

TRY THE CONCEPT CHECK IN THE MARGIN.

From Example 1, notice that the *y*-coordinate of any point on the *x*-axis is 0. For example, the point $(-5, 0)$ lies on the *x*-axis. Also, the *x*-coordinate of any point on the *y*-axis is 0. For example, the point $(0, 2)$ lies on the *y*-axis.

B CREATING SCATTER DIAGRAMS

Data that can be represented as an ordered pair is called **paired data**. Many types of data collected from the real world are paired data. For instance, the annual measurement of a child's height can be written as an ordered pair of the form (year, height in inches) and is paired data. The graph of paired data as points in the rectangular coordinate system is called a **scatter diagram**. Scatter diagrams can be used to look for patterns and trends in paired data.

Example 2   The table gives the annual revenues for Wal-Mart Stores for the years shown. (*Source:* Wal-Mart Stores, Inc.)

Year	Wal-Mart Revenue (in billions of dollars)
1993	56
1994	68
1995	83
1996	95
1997	106

a. Write this paired data as a set of ordered pairs of the form (year, revenue in billions of dollars).
b. Create a scatter diagram of the paired data.
c. What trend in the paired data does the scatter diagram show?

Solution: **a.** The ordered pairs are $(1993, 56)$, $(1994, 68)$, $(1995, 83)$, $(1996, 95)$, and $(1997, 106)$.
b. We begin by plotting the ordered pairs. Because the *x*-coordinate in each ordered pair is a year, we label the *x*-axis "Year" and mark the horizontal axis with the years given. Then we label the *y*-axis or vertical axis "Wal-Mart Revenue (in billions of dollars)." It is convenient to mark the vertical axis in multiples of 20, starting with 0.

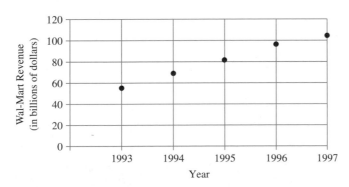

c. The scatter diagram shows that Wal-Mart revenue steadily increased over the years 1993–1997. ▬▬▬

✓ **CONCEPT CHECK**

For each description of a point in the rectangular coordinate system, write an ordered pair that represents it.
a. Point A is located three units to the left of the *y*-axis and five units above the *x*-axis.
b. Point B is located six units below the origin.

Practice Problem 2

The table gives the number of cable TV subscribers (in millions) for the years shown. (*Source: Television and Cable Factbook*, Warren Publishing, Inc., Washington, D.C.)

Year	Cable TV Subscribers (in millions)
1986	38
1988	44
1990	50
1992	53
1994	57
1996	62

a. Write this paired data as a set of ordered pairs of the form (year, number of cable TV subscribers in millions).
b. Create a scatter diagram of the paired data.

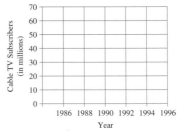

c. What trend in the paired data does the scatter diagram show?

Answers

2. a. $(1986, 38)$, $(1988, 44)$, $(1990, 50)$, $(1992, 53)$, $(1994, 57)$, $(1996, 62)$,
b.

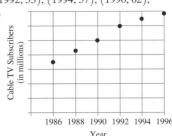

c. The number of cable TV subscribers has steadily increased.
✓ Concept Check: **a.** $(-3, 5)$,   **b.** $(0, -6)$

C COMPLETING ORDERED PAIRS SOLUTIONS

Let's see how we can use ordered pairs to record solutions of equations containing two variables. An equation in one variable such as $x + 1 = 5$ has one solution, which is 4: the number 4 is the value of the variable x that makes the equation true.

An equation in two variables, such as $2x + y = 8$, has solutions consisting of two values, one for x and one for y. For example, $x = 3$ and $y = 2$ is a solution of $2x + y = 8$ because, if x is replaced with 3 and y with 2, we get a true statement.

$$2x + y = 8$$
$$2(3) + 2 \stackrel{?}{=} 8$$
$$8 = 8 \quad \text{True.}$$

The solution $x = 3$ and $y = 2$ can be written as $(3, 2)$, an ordered pair of numbers.

In general, an ordered pair is a **solution** of an equation in two variables if replacing the variables by the values of the ordered pair results in a *true statement*. For example, another ordered pair solution of $2x + y = 8$ is $(5, -2)$. Replacing x with 5 and y with -2 results in a true statement.

$$2x + y = 8$$
$$2(5) + (-2) \stackrel{?}{=} 8 \quad \text{Replace } x \text{ with 5 and } y \text{ with } -2.$$
$$10 - 2 \stackrel{?}{=} 8$$
$$8 = 8 \quad \text{True.}$$

Example 3   Complete each ordered pair so that it is a solution to the equation $3x + y = 12$.

    a. $(0, \ )$ **b.** $(\ , 6)$ **c.** $(-1, \ )$

Solution: **a.** In the ordered pair $(0, \ )$, the x-value is 0. We let $x = 0$ in the equation and solve for y.

$$3x + y = 12$$
$$3(0) + y = 12 \quad \text{Replace } x \text{ with 0.}$$
$$0 + y = 12$$
$$y = 12$$

The completed ordered pair is $(0, 12)$.

    b. In the ordered pair $(\ , 6)$, the y-value is 6. We let $y = 6$ in the equation and solve for x.

$$3x + y = 12$$
$$3x + 6 = 12 \quad \text{Replace } y \text{ with 6.}$$
$$3x = 6 \quad \text{Subtract 6 from both sides.}$$
$$x = 2 \quad \text{Divide both sides by 3.}$$

The ordered pair is $(2, 6)$.

    c. In the ordered pair $(-1, \ )$, the x-value is -1. We let $x = -1$ in the equation and solve for y.

$$3x + y = 12$$
$$3(-1) + y = 12 \quad \text{Replace } x \text{ with } -1.$$
$$-3 + y = 12$$
$$y = 15 \quad \text{Add 3 to both sides.}$$

The ordered pair is $(-1, 15)$.

Practice Problem 3

Complete each ordered pair so that it is a solution to the equation $x + 2y = 8$.

a. $(0, \ )$

b. $(\ , 3)$

c. $(-4, \ )$

Answers

3. a. $(0, 4)$,   **b.** $(2, 3)$,   **c.** $(-4, 6)$

Solutions of equations in two variables can also be recorded in a **table of paired values**, as shown in the next example.

Example 4 Complete the table for the equation $y = 3x$.

x	y
a. −1	
b.	0
c.	−9

Solution: **a.** We replace x with −1 in the equation and solve for y.

$$y = 3x$$
$$y = 3(-1) \quad \text{Let } x = -1.$$
$$y = -3$$

The ordered pair is $(-1, -3)$

b. We replace y with 0 in the equation and solve for x.

$$y = 3x$$
$$0 = 3x \quad \text{Let } y = 0.$$
$$0 = x \quad \text{Divide both sides by 3.}$$

The ordered pair is $(0, 0)$.

c. We replace y with −9 in the equation and solve for x.

$$y = 3x$$
$$-9 = 3x \quad \text{Let } y = -9.$$
$$-3 = x \quad \text{Divide both sides by 3.}$$

The ordered pair is $(-3, -9)$. The completed table is shown to the right.

x	y
−1	−3
0	0
−3	−9

Practice Problem 4

Complete the table for the equation $y = -2x$.

x	y
a. −3	
b.	0
c.	10

Example 5 Complete the table for the equation $y = 3$.

x	y
−2	
0	
−5	

Solution: The equation $y = 3$ is the same as $0x + y = 3$. No matter what value we replace x by, y always equals 3. The completed table is shown to the right.

x	y
−2	3
0	3
−5	3

Practice Problem 5

Complete the table for the equation $x = 5$.

x	y
	−2
	0
	4

Answers

4.

x	y
a. −3	6
b. 0	0
c. −5	10

5.

x	y
5	−2
5	0
5	4

By now, you have noticed that equations in two variables often have more than one solution. We discuss this more in the next section.

A table showing ordered pair solutions may be written vertically, or horizontally as shown in the next example.

Practice Problem 6

A company purchased a fax machine for $400. The business manager of the company predicts that the fax machine will be used for 7 years and the value in dollars y of the machine in x years is $y = -50x + 400$. Complete the table.

x	1	2	3	4	5	6	7
y							

Example 6

A small business purchased a computer for $2000. The business predicts that the computer will be used for 5 years and the value in dollars y of the computer in x years is $y = -300x + 2000$. Complete the table.

x	0	1	2	3	4	5
y						

Solution: To find the value of y when x is 0, we replace x with 0 in the equation. We use this same procedure to find y when x is 1 and when x is 2.

WHEN $x = 0$,

$y = -300x + 2000$
$y = -300 \cdot 0 + 2000$
$y = 0 + 2000$
$y = 2000$

WHEN $x = 1$,

$y = -300x + 2000$
$y = -300 \cdot 1 + 2000$
$y = -300 + 2000$
$y = 1700$

WHEN $x = 2$,

$y = -300x + 2000$
$y = -300 \cdot 2 + 2000$
$y = -600 + 2000$
$y = 1400$

We have the ordered pairs $(0, 2000)$, $(1, 1700)$, and $(2, 1400)$. This means that in 0 years the value of the computer is $2000, in 1 year the value of the computer is $1700, and in 2 years the value is $1400. To complete the table of values, we continue the procedure for $x = 3$, $x = 4$, and $x = 5$.

WHEN $x = 3$,

$y = -300x + 2000$
$y = -300 \cdot 3 + 2000$
$y = -900 + 2000$
$y = 1100$

WHEN $x = 4$,

$y = -300x + 2000$
$y = -300 \cdot 4 + 2000$
$y = -1200 + 2000$
$y = 800$

WHEN $x = 5$,

$y = -300x + 2000$
$y = -300 \cdot 5 + 2000$
$y = -1500 + 2000$
$y = 500$

The completed table is

x	0	1	2	3	4	5
y	2000	1700	1400	1100	800	500

The ordered pair solutions recorded in the completed table for Example 6 are another set of paired data. They are graphed next. Notice that this scatter diagram gives a visual picture of the decrease in value of the computer.

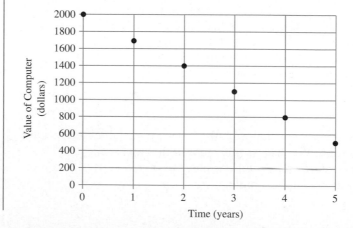

x	y
0	2000
1	1700
2	1400
3	1100
4	800
5	500

Answer

6.

x	1	2	3	4	5	6	7
y	350	300	250	200	150	100	50

Name _____ Section _____ Date _____

MENTAL MATH

Give two ordered pair solutions for each linear equation.

1. $x + y = 10$ **2.** $x + y = 6$

EXERCISE SET 6.1

A Plot each ordered pair. State in which quadrant, if any, each point lies. See Example 1.

1. $(1, 5)$ $(-5, -2)$ $(-3, 0)$
$(0, -1)$ $(2, -4)$ $\left(-1, 4\frac{1}{2}\right)$

2. $(2, 4)$ $(0, 2)$ $(-2, 1)$
$(-3, -3)$ $\left(3\frac{3}{4}, 0\right)$ $(5, -4)$

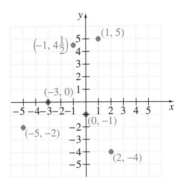

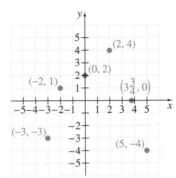

3. When is the graph of the ordered pair (a, b) the same as the graph of the ordered pair (b, a)?

4. In your own words, describe how to plot an ordered pair.

Find the x- and y-coordinates of each labeled point. See Example 1.

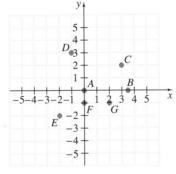

5. A
6. B
7. C
8. D
9. E
10. F
11. G

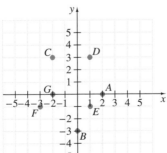

12. A
13. B
14. C
15. D
16. E
17. F
18. G

ANSWERS

$(1, 5)$ is in quadrant I, $\left(-1, 4\frac{1}{2}\right)$ is in quadrant II, $(-5, -2)$ is in quadrant III, $(2, -4)$ is in qudrant IV, $(-3, 0)$ lies on the x-axis, $(0, -1)$ lies on the
1. y-axis

$(2, 4)$ is in quadrant I, $(-2, 1)$ is in quadrant II, $(-3, -3)$ is in quadrant III, $(5, -4)$ is in quadrant IV, $\left(3\frac{3}{4}, 0\right)$ lies on the x-axis, $(0, 2)$
2. lies on the y-axis

3. $a = b$

4. answers may vary

5. $(0, 0)$

6. $\left(3\frac{1}{2}, 0\right)$

7. $(3, 2)$

8. $(-1, 3)$

9. $(-2, -2)$

10. $(0, -1)$

11. $(2, -1)$

12. $(2, 0)$

13. $(0, -3)$

14. $(-2, 3)$

15. $(1, 3)$

16. $(1, -1)$

17. $(-3, -1)$

18. $(-2, 0)$

(1991, 80),
(1992, 79),
(1993, 77),
(1994, 74),
(1995, 77),
(1996, 85),
19. a. (1997, 86)

b. see graph

b. see graph

Name _____

B *Solve. See Example 2.*

19. The table shows the average price of a gallon of gasoline (in cents) for the years shown. (*Source*: Energy Information Administration)

Year	Price of Gasoline (per gallon) (in cents)
1991	80
1992	79
1993	77
1994	74
1995	77
1996	85
1997	86

a. Write each paired data as an ordered pair of the form (year, gasoline price).

b. Create a scatter diagram of the paired data. Be sure to label the axes appropriately.

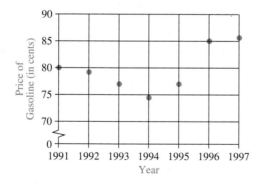

$\dfrac{}{}$ indicates that some numbers on the axis are missing.

20. The table shows the number of regular-season NFL football games won by the winner of the Super Bowl for the years shown. (*Source*: National Football League)

Year	Regular-Season Games Won by Super Bowl Winner
1994	12
1995	13
1996	12
1997	13
1998	12

a. Write each paired data as an ordered pair of the form (year, games won).

b. Create a scatter diagram of the paired data. Be sure to label the axes appropriately.

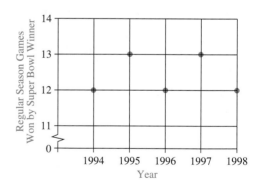

21. The table shows the average monthly mortgage payment made by Americans during the years shown. (*Source:* National Association of REALTORS®)

Year	Average Monthly Mortgage Payment (in dollars)
1994	578
1995	613
1996	654
1997	675
1998	717

a. Write each paired data as an ordered pair of the form (year, mortgage payment).

b. Create a scatter diagram of the paired data. Be sure to label the axes appropriately.

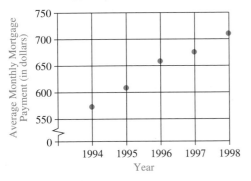

c. What trend in the paired data does the scatter diagram show?

22. The table shows the number of institutions of higher education in the United States for the years shown. (*Source:* U.S. Department of Education)

Year	Number of Institutions of Higher Learning
1909	951
1919	1041
1929	1409
1939	1708
1949	1851
1959	2008
1969	2525
1979	3152
1989	3535
1999	3800

a. Write each paired data as an ordered pair of the form (year, number of institutions).

b. Create a scatter diagram of the paired data. Be sure to label the axes appropriately.

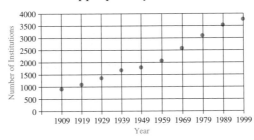

c. What trend in the paired data does the scatter diagram show?

23. Minh, a psychology student, kept a record of how much time she spent studying for each of her 20-point psychology quizzes and her score on each quiz.

Hours Spent Studying	Quiz Score
0.50	10
0.75	12
1.00	15
1.25	16
1.50	18
1.50	19
1.75	19
2.00	20

a. Write each paired data as an ordered pair of the form (hours spent studying, quiz score).

b. Create a scatter diagram of the paired data. Be sure to label the axes appropriately.

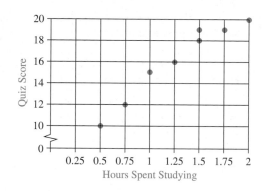

c. What might Minh conclude from the scatter diagram?

24. A local lumberyard uses quantity pricing. The table shows the price per board for different amounts of lumber purchased.

Price per Board (in dollars)	Number of Boards Purchased
8.00	1
7.50	10
6.50	25
5.00	50
2.00	100

a. Write each paired data as an ordered pair of the form (price per board, number of boards purchased).

b. Create a scatter diagram of the paired data. Be sure to label the axes appropriately.

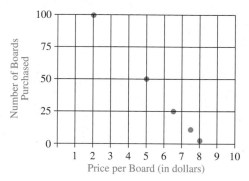

c. What trend in the paired data does the scatter diagram show?

Name _____

C *Complete each ordered pair so that it is a solution of the given linear equation. See Example 3.*

📼 **25.** $x - 4y = 4$; (, -2), (4,)

26. $x - 5y = -1$; (, -2), (4,)

27. $3x + y = 9$; (0,), (, 0)

28. $x + 5y = 15$; (0,), (, 0)

29. $y = -7$; (11,), (, -7)

30. $x = \frac{1}{2}$; (, 0), $\left(\frac{1}{2}, \quad \right)$

Complete the table of ordered pairs for each linear equation. See Examples 4 through 6.

31. $x + 3y = 6$

x	y
0	
	0
	1

32. $2x + y = 4$

x	y
0	
	0
	2

33. $2x - y = 12$

x	y
0	
	-2
3	

34. $-5x + y = 10$

x	y
	0
	5
0	

35. $2x + 7y = 5$

x	y
0	
	0
	1

36. $x - 6y = 3$

x	y
0	
1	
	-1

25. $(-4, -2), (4, 0)$

26. $(-11, -2), (4, 1)$

27. $(0, 9), (3, 0)$

28. $(0, 3), (15, 0)$

29. $(11, -7)$, any x

30. $\left(\frac{1}{2}, 0\right)$, any y

31. $(0, 2), (6, 0), (3, 1)$

32. $(0, 4), (2, 0), (1, 2)$

33. $(0, -12), (5, -2), (3, -6)$

34. $(-2, 0), (-1, 5), (0, 10)$

35. $\left(0, \frac{5}{7}\right), \left(\frac{5}{2}, 0\right), (-1, 1)$

36. $\left(0, -\frac{1}{2}\right), \left(1, -\frac{1}{3}\right), (-3, -1)$

37. $(3, 0), (3, -0.5),$ $\left(3, \dfrac{1}{4}\right)$

Complete the table of ordered pairs for each equation. Then plot the ordered pair solutions.
See Examples 4 through 6.

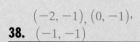 **37.** $x = 3$

x	y
	0
	-0.5
	$\dfrac{1}{4}$

38. $y = -1$

x	y
-2	
0	
	-1

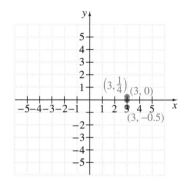

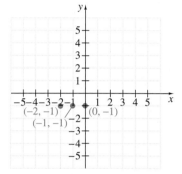

38. $(-2, -1), (0, -1),$ $(-1, -1)$

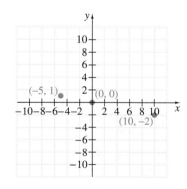 **39.** $x = -5y$

x	y
	0
	1
10	

40. $y = -3x$

x	y
0	
-2	
	9

39. $(0, 0), (-5, 1),$ $(10, -2)$

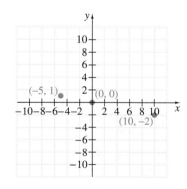

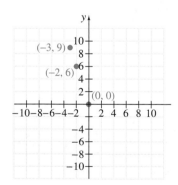

40. $(0, 0), (-2, 6),$ $(-3, 9)$

Name _____

41. a. see table

b. 45 desks

42. a. see table

b. 13 units

43. $y = 5 - x$

44. $y = x - 3$

45. $y = \dfrac{5 - 2x}{4}$

46. $y = \dfrac{7 - 5x}{2}$

47. $y = -2x$

48. $y = -2x$

49. 26 units

50. 40 sq. units

Solve. See Example 6.

41. The cost in dollars y of producing x computer desks is given by
$y = 80x + 5000$.
   a. Complete the table.

x	100	200	300
y	13,000	21,000	29,000

   b. Find the number of computer desks that can be producded for $8600. (*Hint*: Find x when $y = 8600$.)

42. The hourly wage y of an employee at a certain production company is given by $y = 0.25x + 9$ where x is the number of units produced in an hour.
   a. Complete the table.

x	0	1	5	10
y	9	9.25	10.25	11.50

   b. Find the number of units that must be produced each hour to earn an hourly wage of $12.25. (*Hint*: Find x when $y = 12.25$.)

REVIEW AND PREVIEW

Solve each equation for y. See Section 2.4.

43. $x + y = 5$

44. $x - y = 3$

45. $2x + 4y = 5$

46. $5x + 2y = 7$

47. $10x = -5y$

48. $4y = -8x$

COMBINING CONCEPTS

49. Find the perimeter of the rectangle whose vertices are the points with coordinates $(-1, 5)$, $(3, 5)$, $(3, -4)$, and $(-1, -4)$.

50. Find the area of the rectangle whose vertices are the points with coordinates $(5, 2)$, $(5, -6)$, $(0, -6)$, and $(0, 2)$.

Name _____

The scatter diagram below shows Walt Disney Company's annual revenues. The horizontal axis represents the number of years after 1990.

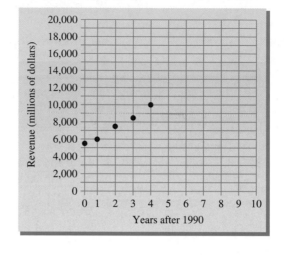

51. Estimate the increase in revenues for years 1, 2, 3, and 4.

52. Use a straight edge or ruler and this scatter diagram to predict Disney's revenue in the year 2000.

53. Discuss any similarities in the graphs of the ordered pair solutions for Exercises 37–40.

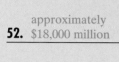 **54.** The percentage y of recorded music sales that were in cassette format from 1991 through 1995 is given by $y = -6.09x + 55.99$. In the equation, x represents the number of years after 1991. (*Source*: Based on data from the Recording Industry Association of America)
a. Complete the table.

x	1	3	5
y	49.9	37.72	25.54

b. Find the year in which approximately 31.6% of recorded music sales were cassettes. (*Hint*: Find x when $y = 31.6$ and round to the nearest whole number.)

55. The population density y of Minnesota (in people per square mile of land) from 1920 through 1990 is given by $y = 0.364x + 21.939$. In the equation, x represents the number of years after 1900. (*Source*: Based on data from the U.S. Bureau of the Census)
a. Complete the table.

x	20	65	90
y	29.219	45.599	54.699

b. Find the year in which the population density was approximately 50 people per square mile. (*Hint*: Find x when $y = 50$ and round to the nearest whole number.)

6.2 GRAPHING LINEAR EQUATIONS

In the previous section, we found that equations in two variables may have more than one solution. For example, both $(2, 2)$ and $(0, 4)$ are solutions of the equation $x + y = 4$. In fact, this equation has an infinite number of solutions. Other solutions include $(-2, 6)$, $(4, 0)$, and $(6, -2)$. Notice the pattern that appears in the graph of these solutions.

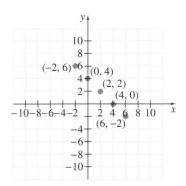

These solutions all appear to lie on the same line, as seen in the second graph. It can be shown that every ordered pair solution of the equation corresponds to a point on this line, and every point on this line corresponds to an ordered pair solution. Thus, we say that this line is the **graph of the equation** $x + y = 4$.

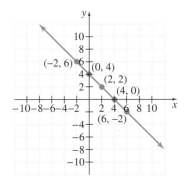

The equation $x + y = 4$ is called a **linear equation in two variables** and **the graph of every linear equation in two variables is a straight line**.

LINEAR EQUATION IN TWO VARIABLES

A linear equation in two variables is an equation that can be written in the form

$$Ax + By = C$$

where A, B, and C are real numbers and A and B are not both 0. The graph of a linear equation in two variables is a straight line.

The form $Ax + By = C$ is called **standard form.** Following are examples of linear equation in two variables.

$$2x + y = 8 \qquad -2x = 7y \qquad y = \frac{1}{3}x + 2 \qquad y = 7$$

Objectives

A Graph a linear equation by finding and plotting ordered pair solutions.

SSM   CD-ROM   Video
6.2

TEACHING TIP

Begin this section by graphing 4 linear equations on 4 sheets of graph paper using sticky dots. Draw axes on each sheet of paper and label with one of the following equations: $x + y = 2, 2x - y = 6, x = 5$, and $y = -4$. Divide the class into 9 groups. Assign each group an integer from -4 to 4 to use as their x-value. Have each group find the y-value for each equation which corresponds to their x-value and plot each point using a sticky dot. When the graphs are completed, ask what similarities and differences they notice in the graphs.

TEACHING TIP

For contrast, you may want to give some examples of equations in two variables that are not linear. For instance,

$$x^2 = 6y + 4$$
$$y = 3^x$$
$$y + 9 = \sqrt{x}$$
$$3x - \frac{1}{y} = 5$$

A GRAPHING LINEAR EQUATIONS

From geometry, we know that a straight line is determined by just two points. Thus, to graph a linear equation in two variables we need to find just two of its infinitely many solutions. Once we do so, we plot the solution points and draw the line connecting the points. Usually, we find a third solution as well, as a check.

Practice Problem 1

Graph the linear equation $x + 3y = 6$.

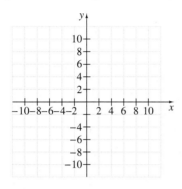

Example 1   Graph the linear equation $2x + y = 5$.

Solution: To graph this equation, we find three ordered pair solutions of $2x + y = 5$. To do this, we choose a value for one variable, x or y, and solve for the other variable. For example, if we let $x = 1$, then $2x + y = 5$ becomes

$$2x + y = 5$$
$$2(1) + y = 5 \quad \text{Replace } x \text{ with 1.}$$
$$2 + y = 5 \quad \text{Multiply.}$$
$$y = 3 \quad \text{Subtract 2 from both sides.}$$

Since $y = 3$ when $x = 1$, the ordered pair $(1, 3)$ is a solution of $2x + y = 5$. Next, we let $x = 0$.

$$2x + y = 5$$
$$2(0) + y = 5 \quad \text{Replace } x \text{ with 0.}$$
$$0 + y = 5$$
$$y = 5$$

The ordered pair $(0, 5)$ is a second solution.

The two solutions found so far allow us to draw the straight line that is the graph of all solutions of $2x + y = 5$. However, we will find a third ordered pair as a check. Let $y = -1$.

$$2x + y = 5$$
$$2x + (-1) = 5 \quad \text{Replace } y \text{ with } -1.$$
$$2x - 1 = 5$$
$$2x = 6 \quad \text{Add 1 to both sides.}$$
$$x = 3 \quad \text{Divide both sides by 2.}$$

The third solution is $(3, -1)$. These three ordered pair solutions are listed in the table and plotted on the coordinate plane. The graph of $2x + y = 5$ is the line through the three points.

x	y
1	3
0	5
3	-1

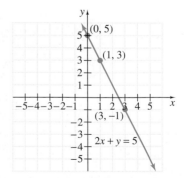

Answer

1.

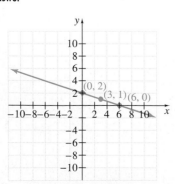

Helpful Hint

All three points should fall on the same straight line. If not, check your ordered pair solutions for a mistake.

TEACHING TIP
Point out the arrowheads on the graph in Example 2. See if the students understand their meaning. Remind them that when they are graphing an equation, they are to graph all the solutions to the equation.

Practice Problem 2

Graph the linear equation $-2x + 4y = 8$.

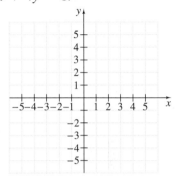

Example 2   Graph the linear equation $-5x + 3y = 15$.

Solution: We find three ordered pair solutions of $-5x + 3y = 15$.

Let x = 0.

$-5x + 3y = 15$
$-5 \cdot 0 + 3y = 15$
$0 + 3y = 15$
$3y = 15$
$y = 5$

Let y = 0.

$-5x + 3y = 15$
$-5x + 3 \cdot 0 = 15$
$-5x + 0 = 15$
$-5x = 15$
$x = -3$

Let x = -2.

$-5x + 3y = 15$
$-5 \cdot -2 + 3y = 15$
$10 + 3y = 15$
$3y = 5$
$y = \dfrac{5}{3}$

The ordered pairs are $(0, 5)$, $(-3, 0)$, and $\left(-2, \dfrac{5}{3}\right)$. The graph of $-5x + 3y = 15$ is the line through the three points.

x	y
0	5
-3	0
-2	$\frac{5}{3} = 1\frac{2}{3}$

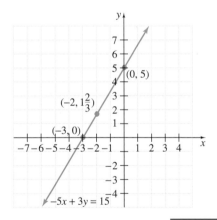

Practice Problem 3

Graph the linear equation $y = 2x$.

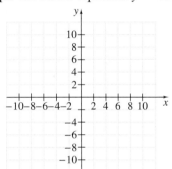

Answers

2.

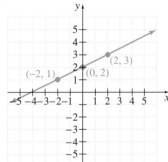

3.

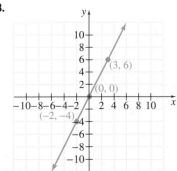

Example 3   Graph the linear equation $y = 3x$.

Solution: We find three ordered pair solutions. Since this equation is solved for y, we'll choose three x values.

If $x = 2$, $y = 3 \cdot 2 = 6$.
If $x = 0$, $y = 3 \cdot 0 = 0$.
If $x = -1$, $y = 3 \cdot -1 = -3$.

Next, we plot the ordered pair solutions and draw a line through the plotted points. The line is the graph of $y = 3x$. Every point on the graph represents an ordered pair solution of the equation and every ordered pair solution is a point on this line.

x	y
2	6
0	0
−1	−3

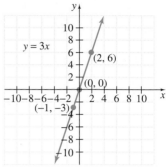

Practice Problem 4

Graph the linear equation $y = -\frac{1}{2}x$.

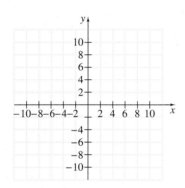

Example 4   Graph the linear equation $y = -\frac{1}{3}x$.

Solution   We find three ordered pair solutions, plot the solutions, and draw a line through the plotted solutions. To avoid fractions, we'll choose x values that are multiples of 3 to substitute into the equation.

If $x = 6$, then $y = -\frac{1}{3} \cdot 6 = -2$.

If $x = 0$, then $y = -\frac{1}{3} \cdot 0 = 0$.

If $x = -3$, then $y = -\frac{1}{3} \cdot -3 = 1$.

x	y
6	−2
0	0
−3	1

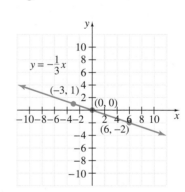

Let's compare the graphs in Examples 3 and 4. The graph of $y = 3x$ tilts upward (as we follow the line from left to right) and the graph of $y = -\frac{1}{3}x$ tilts downward (as we follow the line from left to right). Also notice that both lines go through the origin or that $(0, 0)$ is an ordered pair solution of both equations. We will learn more about the tilt, or slope, of a line in Section 6.4.

Answer

4.

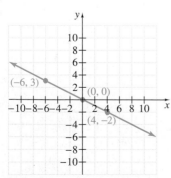

Example 5   Graph the linear equation $y = 3x + 6$ and compare this graph with the graph of $y = 3x$ in Example 3.

Solution   We find three ordered pair solutions, plot the solutions, and draw a line through the plotted solutions. We choose x values and substitute into the equation $y = 3x + 6$.

If $x = -3$, then $y = 3(-3) + 6 = -3$.
If $x = 0$, then $y = 3(0) + 6 = 6$.
If $x = 1$, then $y = 3(1) + 6 = 9$.

x	y
-3	-3
0	6
1	9

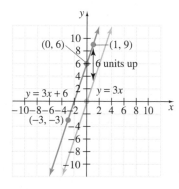

The most startling similarity is that both graphs appear to have the same upward tilt as we move from left to right. Also, the graph of $y = 3x$ crosses the y-axis at the origin, while the graph of $y = 3x + 6$ crosses the y-axis at 6. It appears that the graph of $y = 3x + 6$ is the same as the graph of $y = 3x$ except that the graph of $y = 3x + 6$ is moved 6 units upward.

Example 6   Graph the linear equation $y = -2$.

Solution: Recall from Section 6.1 that the equation $y = -2$ is the same as $0x + y = -2$. No matter what value we replace x with, y is -2.

x	y
0	-2
3	-2
-2	-2

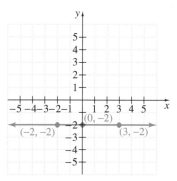

Notice that the graph of $y = -2$ is a horizontal line.

Practice Problem 5

Graph the linear equation $y = 2x + 3$ and compare this graph with the graph of $y = 2x$ in Practice Problem 3.

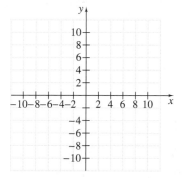

Practice Problem 6

Graph the linear equation $x = 3$.

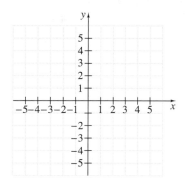

Answers
5.

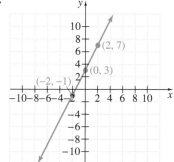

Same as the graph of $y = 2x$ except that the graph of $y = 2x + 3$ is moved 3 units upward.

6.

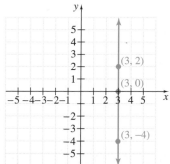

GRAPHING CALCULATOR EXPLORATIONS

In this section, we begin an optional study of graphing calculators and graphing software packages for computers. These graphers use the same point plotting technique that was introduced in this section. The advantage of this graphing technology is, of course, that graphing calculators and computers can find and plot ordered pair solutions much faster than we can. Note, however, that the features described in these boxes may not be available on all graphing calculators.

The rectangular screen where a portion of the rectangular coordinate system is displayed is called a **window**. We call it a **standard window** for graphing when both the x- and y-axes show coordinates between -10 and 10. This information is often displayed in the window menu on a graphing calculator as

Xmin $= -10$
Xmax $= 10$
Xscl $= 1$ The scale on the x-axis is one unit per tick mark.
Ymin $= -10$
Ymax $= 10$
Yscl $= 1$ The scale on the y-axis is one unit per tick mark.

To use a graphing calculator to graph the equation $y = 2x + 3$, press the ⌷Y=⌷ key and enter the keystrokes ⌷2⌷ ⌷x⌷ ⌷+⌷ ⌷3⌷. The top row should now read $Y_1 = 2x + 3$. Next press the ⌷GRAPH⌷ key, and the display should look like this:

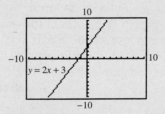

Use a standard window and graph the following linear equations. (Unless otherwise stated, use a standard window when graphing.)

1. $y = -3x + 7$ **2.** $y = -x + 5$ **3.** $y = 2.5x - 7.9$

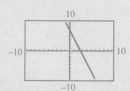   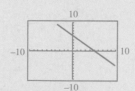   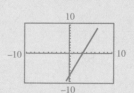

4. $y = -1.3x + 5.2$ **5.** $y = -\dfrac{3}{10}x + \dfrac{32}{5}$ **6.** $y = \dfrac{2}{9}x - \dfrac{22}{3}$

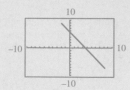   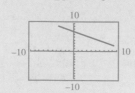   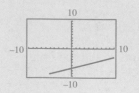

TEACHING TIP

Point out that some graphing calculators need to have coefficients which are fractions entered with parentheses. For instance, $y = \dfrac{1}{4}x - 2$ may need to be entered as $y = (1/4)x - 2$.

EXERCISE SET 6.2

A *For each equation, find three ordered pair solutions by completing the table. Then use the ordered pairs to graph the equation. See Examples 1 through 6.*

1. $x - y = 6$

x	y
	0
4	
	−1

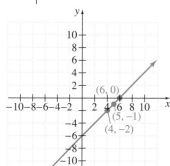

2. $x - y = 4$

x	y
0	
	2
−1	

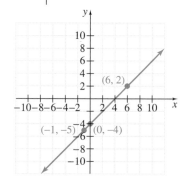

3. $y = -4x$

x	y
1	
0	
−1	

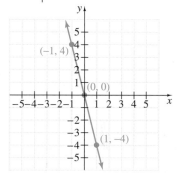

4. $y = -5x$

x	y
1	
0	
−1	

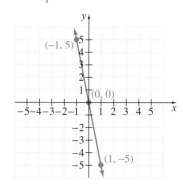

5. $y = \frac{1}{3}x$

x	y
0	
6	
−3	

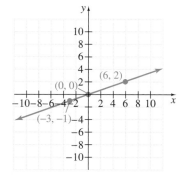

6. $y = \frac{1}{2}x$

x	y
0	
−4	
2	

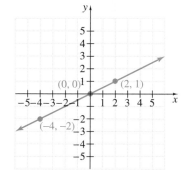

7. $y = -4x + 3$

x	y
0	
1	
2	

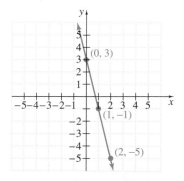

8. $y = -5x + 2$

x	y
0	
1	
2	

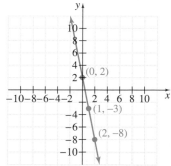

Name _____

Graph each linear equation. See Examples 1 through 6.

9. $x + y = 1$

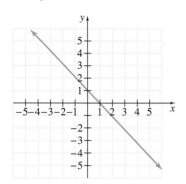

10. $x + y = 7$

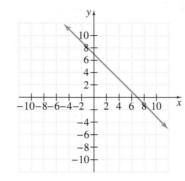

11. $x - y = -2$

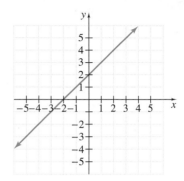

12. $-x + y = 6$

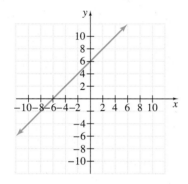

13. $x - 2y = 6$

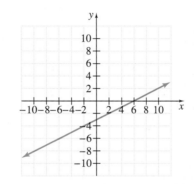

14. $-x + 5y = 5$

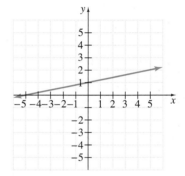

15. $y = 6x + 3$

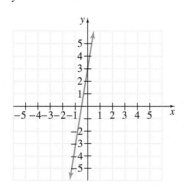

16. $y = -2x + 7$

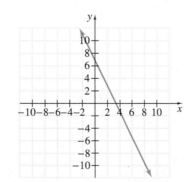

17. $x = -4$

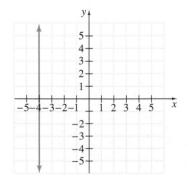

18. $y = 5$

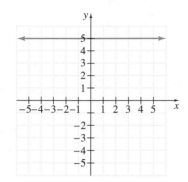

19. $y = 3$

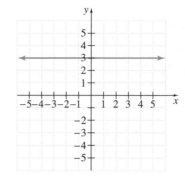

20. $x = -1$

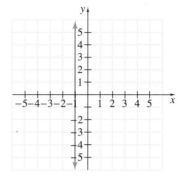

21. $y = x$

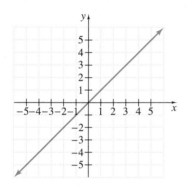

22. $y = -x$

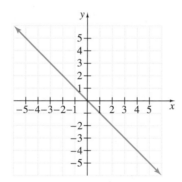

23. $y = 5x$

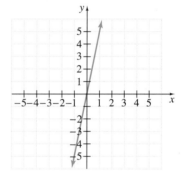

24. $y = 4x$

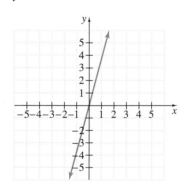

25. $x + 3y = 9$

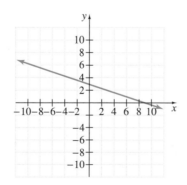

26. $2x + y = 2$

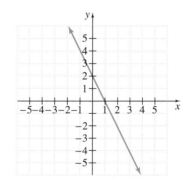

27. $y = \frac{1}{2}x - 1$

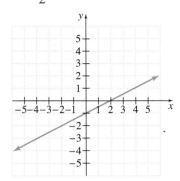

28. $y = \frac{1}{4}x + 3$

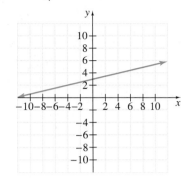

29. $3x - 2y = 12$

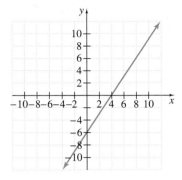

30. $2x - 7y = 14$

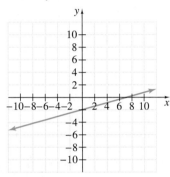

Graph each pair of linear equations on the same set of axes. Discuss how the graphs are similar and how they are different. See Example 5.

31. $y = 5x$
   $y = 5x + 4$

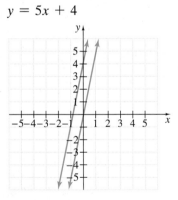

32. $y = 2x$
   $y = 2x + 5$

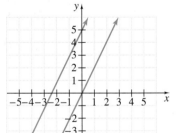

33. $y = -2x$
   $y = -2x - 3$

34. $y = x$
   $y = x - 7$

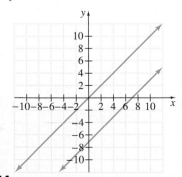

Name _____

REVIEW AND PREVIEW

35. The coordinates of three vertices of a rectangle are $(-2, 5)$, $(4, 5)$, and $(-2, -1)$. Find the coordinates of the fourth vertex. See Section 6.1.

36. The coordinates of two vertices of a square are $(-3, -1)$ and $(2, -1)$. Find the coordinates of two pairs of points possible for the third and fourth vertices. See Section 6.1.

36. $(-3, 4), (2, 4);$
$(-3, -6), (2, -6)$

Complete each table. See Section 6.1.

37. $x - y = -3$

x	y
0	
	0

38. $y - x = 5$

x	y
0	
	0

39. $y = 2x$

x	y
0	
	0

40. $x = -3y$

x	y
0	
	0

37. $(0, 3), (-3, 0)$

COMBINING CONCEPTS

41. Graph the nonlinear equation $y = x^2$ by completing the table shown. Plot the ordered pairs and connect them with a smooth curve.

x	y
0	0
1	1
−1	1
2	4
−2	4

42. Graph the nonlinear equation $y = |x|$ by completing the table shown. Plot the ordered pairs and connect them. This curve is "V" shaped.

x	y
0	0
1	1
−1	1
2	2
−2	2

38. $(0, 5), (-5, 0)$

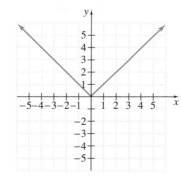

39. $(0, 0), (0, 0)$

40. $(0, 0), (0, 0)$

Name _____

43. The perimeter of the trapezoid is 22 centimeters. Write a linear equation in two variables for the perimeter. Find y if x is 3 cm.

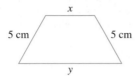

44. If (a, b) is an ordered pair solution of $x + y = 5$, is (b, a) also a solution? Explain why or why not.

45. One of the top five occupations in terms of growth in the next few years is expected to be registered nursing. The number of people y in thousands employed as registered nurses in the United States can be estimated by the linear equation $y = 43x + 2035$, where x is the number of years after 1997. (*Source:* Based on data from the Bureau of Labor Statistics)
   a. Graph the linear equation. The break in the vertical axis means that the numbers between 0 and 2000 have been skipped.

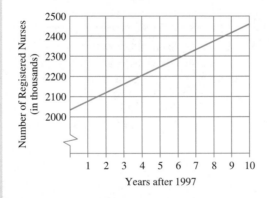

46. The average price in dollars y of an acre of farmland in Maryland can be approximated by the linear equation $y = 392x + 3707$, where x is the number of years after 1995. (*Source:* Based on data from the National Agricultural Statistics Service)
   a. Graph the linear equation.

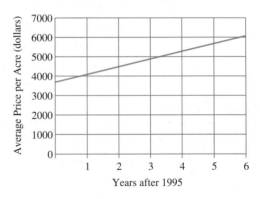

b. Does the point $(5, 5667)$ lie on the line? If so, what does this ordered pair mean?

b. Does the point $(8, 2379)$ lie on the line? If so, what does this ordered pair mean?

6.3 INTERCEPTS

A IDENTIFYING INTERCEPTS

The graph of $y = 4x - 8$ is shown below. Notice that this graph crosses the y-axis at the point $(0, -8)$. This point is called the **y-intercept point** and -8 is called the **y-intercept**. Likewise the graph crosses the x-axis at $(2, 0)$. This point is called the **x-intercept point** and 2 is the **x-intercept**.

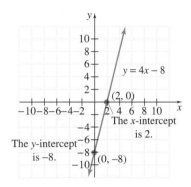

The intercept points are $(2, 0)$ and $(0, -8)$.

HELPFUL HINT

If a graph crosses the x-axis at $(-3, 0)$ and the y-axis at $(0, 7)$, then

Intercept points

$$\overbrace{(-3, 0) \qquad (0, 7)}$$
$$\quad\uparrow\qquad\qquad\uparrow$$
x-intercept y-intercept

Notice that if y is 0, the corresponding x-value is the x-intercept. Likewise, if x is 0, the corresponding y-value is the y-intercept.

Examples   Identify the x- and y-intercepts and the intercept points.

1.

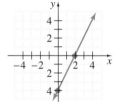

Solution: x-intercept: -3
y-intercept: 2
intercept points: $(-3, 0)$ and $(0, 2)$

Objectives

A Identify intercepts of a graph.
B Graph a linear equation by finding and plotting intercept points.
C Identify and graph vertical and horizontal lines.

SSM   CD-ROM   Video
6.3

TEACHING TIP

Start this lesson by showing students a graph of the line $y = 2x + 8$. Remind them that any two points on the line could be used to describe this line. Ask them which two they would use. Hopefully students will choose the intercepts. Ask what is special about these points.

Practice Problems 1–3

Identify the x- and y-intercepts and the intercept points.

1.

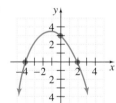

2.

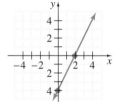

3.

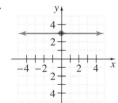

Answers

1. x-intercept: 2; y-intercept: -4; intercept points: $(2, 0)$ and $(0, -4)$,   **2.** x-intercepts: $-4, 2$; y-intercept: 3; intercept points: $(-4, 0)$, $(2, 0)$, and $(0, 3)$,   **3.** x-intercept: none; y-intercept: 3; intercept point: $(0, 3)$

Some questions you may want to ask your students:

▲ Can the same point ever be the x-intercept point and the y-intercept point of a graph?

▲ Excluding the line $y = 0$, what is the maximum number of x-intercepts a line will have?

▲ Will a line always have at least one y-intercept?

2.

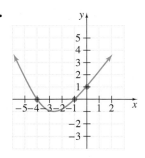

Solution: x-intercepts: $-4, -1$
y-intercept: 1
intercept points: $(-4, 0), (-1, 0), (0, 1)$

3.

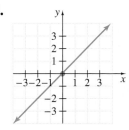

Solution: x-intercept: 0
y-intercept: 0
intercept point: $(0, 0)$

B FINDING AND PLOTTING INTERCEPT POINTS

Given an equation of a line, intercept points are usually easy to find since one coordinate is 0.

One way to find the y-intercept of a line, given its equation, is to let $x = 0$, since a point on the y-axis has an x-coordinate of 0. To find the x-intercept of a line, let $y = 0$, since a point on the x-axis has a y-coordinate of 0.

FINDING x- AND y-INTERCEPTS

To find the x-intercept, let $y = 0$ and solve for x.
To find the y-intercept, let $x = 0$ and solve for y.

Practice Problem 4

Graph $2x - y = 4$ by finding and plotting its intercept points.

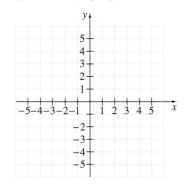

Example 4   Graph $x - 3y = 6$ by finding and plotting its intercept points.

Solution: We let $y = 0$ to find the x-intercept and $x = 0$ to find the y-intercept.

Let $y = 0$.	Let $x = 0$.
$x - 3y = 6$	$x - 3y = 6$
$x - 3(0) = 6$	$0 - 3y = 6$
$x - 0 = 6$	$-3y = 6$
$x = 6$	$y = -2$

The x-intercept is 6 and the y-intercept is -2. We find a third ordered pair solution to check our work. If we let $y = -1$, then $x = 3$. We plot the points $(6, 0), (0, -2)$,

Answer

4.

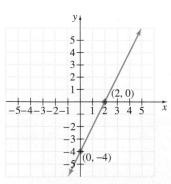

and $(3, -1)$. The graph of $x - 3y = 6$ is the line drawn through these points as shown.

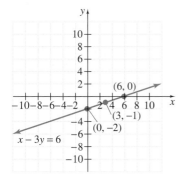

x	y
6	0
0	-2
3	-1

Example 5   Graph $x = -2y$ by finding and plotting its intercept points.

Solution: We let $y = 0$ to find the x-intercept and $x = 0$ to find the y-intercept.

Let $y = 0$.   Let $x = 0$.

$x = -2y$ $x = -2y$

$x = -2(0)$ $0 = -2y$

$x = 0$ $0 = y$

Both the x-intercept and y-intercept are 0. In other words, when $x = 0$, then $y = 0$, which gives the ordered pair $(0, 0)$. Also, when $y = 0$, then $x = 0$, which gives the same ordered pair $(0, 0)$. This happens when the graph passes through the origin. Since two points are needed to determine a line, we must find at least one more ordered pair that satisfies $x = -2y$. We let $y = -1$ to find a second ordered pair solution and let $y = 1$ as a checkpoint.

Let $y = -1$.   Let $y = 1$.

$x = -2(-1)$ $x = -2(1)$

$x = 2$ $x = -2$

The ordered pairs are $(0, 0)$, $(2, -1)$, and $(-2, 1)$. We plot these points to graph $x = -2y$.

x	y
0	0
2	-1
-2	1

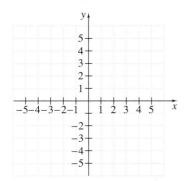

C GRAPHING VERTICAL AND HORIZONTAL LINES

The equation $x = 2$, for example, is a linear equation in two variables because it can be written in the form $x + 0y = 2$. The graph of this equation is a vertical line, as shown in the next example.

Practice Problem 5

Graph $y = 3x$ by finding and plotting its intercept points.

Answer

5.

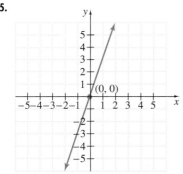

Practice Problem 6

Graph: $x = -3$

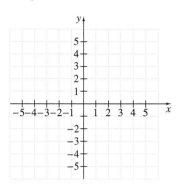

Practice Problem 7

Graph: $y = 4$

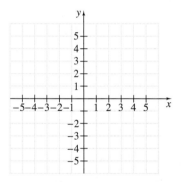

Answers

6.

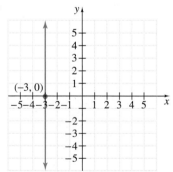

7.

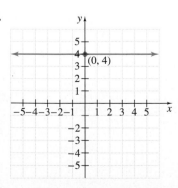

◘ Example 6   Graph: $x = 2$.

Solution: The equation $x = 2$ can be written as $x + 0y = 2$. For any y-value chosen, notice that x is 2. No other value for x satisfies $x + 0y = 2$. Any ordered pair whose x-coordinate is 2 is a solution of $x + 0y = 2$. We will use the ordered pair solutions $(2, 3)$, $(2, 0)$, and $(2, -3)$ to graph $x = 2$.

x	y
2	3
2	0
2	−3

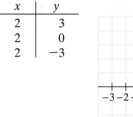

The graph is a vertical line with x-intercept 2. Note that this graph has no y-intercept because x is never 0.

In general, we have the following.

VERTICAL LINES

The graph of $x = c$, where c is a real number, is a vertical line with x-intercept c.

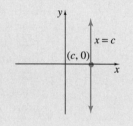

◘ Example 7   Graph: $y = -3$

Solution: The equation $y = -3$ can be written as $0x + y = -3$. For any x-value chosen, y is -3. If we chose 4, 1, and -2 as x-values, the ordered pair solutions are $(4, -3)$, $(1, -3)$, and $(-2, -3)$. We use these ordered pairs to graph $y = -3$. The graph is a horizontal line with y-intercept -3 and no x-intercept.

x	y
4	−3
1	−3
−2	−3

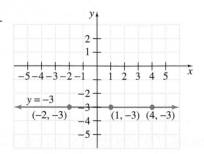

In general, we have the following.

HORIZONTAL LINES

The graph of $y = c$, where c is a real number, is a horizontal line with y-intercept c.

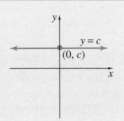

GRAPHING CALCULATOR EXPLORATIONS

You may have noticed that to use the $\boxed{Y=}$ key on a grapher to graph an equation, the equation must be solved for y. For example, to graph $2x + 3y = 7$, we solve this equation for y.

$$2x + 3y = 7$$

$$3y = -2x + 7 \qquad \text{Subtract } 2x \text{ from both sides.}$$

$$\frac{3y}{3} = -\frac{2x}{3} + \frac{7}{3} \qquad \text{Divide both sides by 3.}$$

$$y = -\frac{2}{3}x + \frac{7}{3} \qquad \text{Simplify.}$$

To graph $2x + 3y = 7$ or $y = -\frac{2}{3}x + \frac{7}{3}$, press the $\boxed{Y=}$ key and enter

$$Y_1 = -\frac{2}{3}x + \frac{7}{3}$$

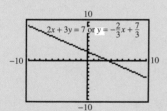

Graph each linear equation.

1. $x = 3.78y$

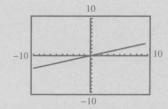

2. $-2.61y = x$

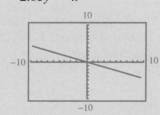

3. $-2.2x + 6.8y = 15.5$

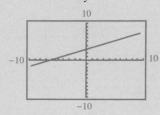

4. $5.9x - 0.8y = -10.4$

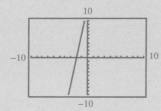

Focus On The Real World

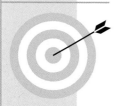

READING A MAP

How do you find a location on a map? Most maps we use today have a grid that is based on the rectangular coordinate system we use in algebra. After finding the coordinates of cities and other landmarks from the map index, the grid can help us find places on the map. To eliminate confusion, many maps use letters to label the grid along one edge and numbers along the other. However, the coordinates are still pairs of numbers and letters. For instance, the coordinates for Toledo on the map are A-2.

CRITICAL THINKING

1. Find the coordinates of the following cities: Hamilton, Columbus, Youngstown, and Cincinnati.
2. What cities correspond to the following coordinates: F-3, A-3, B-4, and D-2?
3. How are the map's coordinate system and the rectangular coordinate system we use in algebra the same? How are they different? What are the advantages of each?

Example 5   Find the slope of the line $x = 5$.

Solution: Recall that the graph of $x = 5$ is a vertical line with x-intercept 5. To find the slope, we find two ordered pair solutions of $x = 5$. Ordered pair solutions of $x = 5$ must have an x-value of 5. We will use $(5, 0)$ and $(5, 4)$. We let $(x_1, y_1) = (5, 0)$ and $(x_2, y_2) = (5, 4)$.

$$m = \frac{y_2 - y_1}{x_2 - x_1} = \frac{4 - 0}{5 - 5} = \frac{4}{0}$$

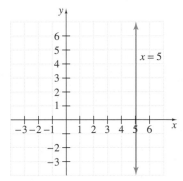

Since $\frac{4}{0}$ is undefined, we say the slope of the vertical line $x = 5$ is undefined. Since the x-values will have a difference of 0 for all vertical lines, we can say that all **vertical lines have undefined slope.**

HELPFUL HINT

Slope of 0 and undefined slope are not the same. Vertical lines have undefined slope or no slope, while horizontal lines have a slope of 0.

D SLOPES OF PARALLEL AND PERPENDICULAR LINES

Two lines in the same plane are **parallel** if they do not intersect. Slopes of lines can help us determine whether lines are parallel. Since parallel lines have the same steepness, it follows that they have the same slope.

For example, the graphs of

$$y = -2x + 4$$

and

$$y = -2x - 3$$

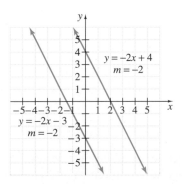

are shown. These lines have the same slope, -2. They also have different y-intercepts, so the lines are parallel. (If the y-intercepts are the same also, the lines are the same.)

Practice Problem 5

Find the slope of the line $x = -2$.

TEACHING TIP

Remind students that $\frac{4}{0}$ is undefined because division by zero is undefined.

Answer

5. undefined slope

> **PARALLEL LINES**
>
> Nonvertical parallel lines have the same slope and different *y*-intercepts.

    Two lines are **perpendicular** if they lie in the same plane and meet at a 90° (right) angle. How do the slopes of perpendicular lines compare? The product of the slopes of two perpendicular lines is -1.

    For example, the graphs of

$$y = 4x + 1$$

and

$$y = -\frac{1}{4}x - 3$$

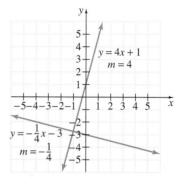

are shown. The slopes of the lines are 4 and $-\frac{1}{4}$. Their product is $4\left(-\frac{1}{4}\right) = -1$, so the lines are perpendicular.

> **PERPENDICULAR LINES**
>
> If the product of the slopes of two lines is -1, then the lines are perpendicular. (Two nonvertical lines are perpendicular if the slope of one is the negative reciprocal of the slope of the other.)

HELPFUL HINT

Here are examples of numbers that are negative reciprocals.

Number	Negative Reciprocal
$\frac{1}{3}$	$-\frac{3}{1}$ or -3
-5 or $-\frac{5}{1}$	$\frac{1}{5}$

HELPFUL HINT

Here are a few reminders about vertical and horizontal lines.

- ▲ Two distinct vertical lines are parallel.
- ▲ Two distinct horizontal lines are parallel.
- ▲ A horizontal line and a vertical line are always perpendicular.

TEACHING TIP

Some questions you might want to ask your students:

- ▲ Can parallel lines have the same *y*-intercept?
- ▲ Can perpendicular lines have the same *y*-intercept?
- ▲ Do perpendicular lines always have the same *y*-intercept?

Example 6   Determine whether each pair of lines is parallel, perpendicular, or neither.

a. $y = -\dfrac{1}{5}x + 1$ **b.** $x + y = 3$ **c.** $3x + y = 5$

$2x + 10y = 3$ $-x + y = 4$ $2x + 3y = 6$

Solution: **a.** The slope of the line $y = -\dfrac{1}{5}x + 1$ is $-\dfrac{1}{5}$. We find the slope of the second line by solving it for y.

$$2x + 10y = 3$$
$$10y = -2x + 3 \qquad \text{Subtract } 2x \text{ from both sides.}$$
$$y = \dfrac{-2}{10}x + \dfrac{3}{10} \qquad \text{Divide both sides by 10.}$$
$$y = -\dfrac{1}{5}x + \dfrac{3}{10} \qquad \text{Simplify.}$$

The slope of this line is $-\dfrac{1}{5}$ also. Since the lines have the same slope and different y-intercepts, they are parallel, as shown in the figure.

b. To find each slope, we solve each equation for y.

$$x + y = 3 \qquad\qquad -x + y = 4$$
$$y = -x + 3 \qquad\qquad y = x + 4$$

The slope is -1.   The slope is 1.

The slopes are not the same, so the lines are not parallel. Next we check the product of the slopes: $(-1)(1) = -1$. Since the product is -1, the lines are perpendicular, as shown in the figure.

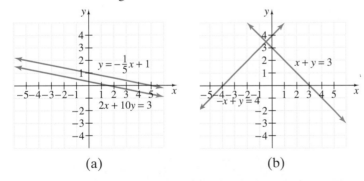

(a)   (b)

c. We solve each equation for y to find each slope. The slopes are -3 and $-\dfrac{2}{3}$. The slopes are not the same and their product is not -1. Thus, the lines are neither parallel nor perpendicular. ▬▬▬

TRY THE CONCEPT CHECK IN THE MARGIN.

Practice Problem 6

Determine whether each pair of lines is parallel, perpendicular, or neither.

a. $x + y = 5$
   $2x + y = 5$

b. $5y = 2x - 3$
   $5x + 2y = 1$

c. $y = 2x + 1$
   $4x - 2y = 8$

✓ CONCEPT CHECK

Write the equations of three parallel lines.

Answers

6. a. neither, **b.** perpendicular, **c.** parallel

✓ Concept Check

For example, $y = 2x + 3$, $y = 2x - 1$, $y = 2x$

GRAPHING CALCULATOR EXPLORATIONS

It is possible to use a grapher and sketch the graph of more than one equation on the same set of axes. This feature can be used to see parallel lines with the same slope. For example, graph the equations $y = \frac{2}{5}x$, $y = \frac{2}{5}x + 7$, and $y = \frac{2}{5}x - 4$ on the same set of axes. To do so, press the $\boxed{Y=}$ key and enter the equations on the first three lines.

$$Y_1 = \left(\frac{2}{5}\right)x$$

$$Y_2 = \left(\frac{2}{5}\right)x + 7$$

$$Y_3 = \left(\frac{2}{5}\right)x - 4$$

The displayed equations should look like:

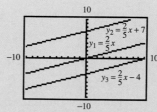

These lines are parallel as expected since they all have a slope of $\frac{2}{5}$. The graph of $y = \frac{2}{5}x + 7$ is the graph of $y = \frac{2}{5}x$ moved 7 units upward with a y-intercept of 7. Also, the graph of $y = \frac{2}{5}x - 4$ is the graph of $y = \frac{2}{5}x$ moved 4 units downward with a y-intercept of -4.

Graph the parallel lines on the same set of axes. Describe the similarities and differences in their graphs.

1. $y = 3.8x$, $y = 3.8x - 3$, $y = 3.8x + 9$

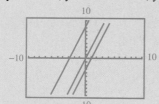

2. $y = -4.9x$, $y = -4.9x + 1$, $y = -4.9x + 8$

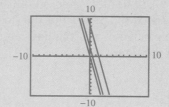

3. $y = \frac{1}{4}x$, $y = \frac{1}{4}x + 5$, $y = \frac{1}{4}x - 8$

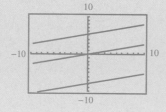

4. $y = -\frac{3}{4}x$, $y = -\frac{3}{4}x - 5$, $y = -\frac{3}{4}x + 6$

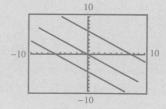

TEACHING TIP

Consider exploring slopes with a graphing calculator. Have students graph the following equations on a graphing calculator and sketch all the results of each set on the same axes.

First Set	Second Set	First Set	Second Set
$y = 10x$	$y = -\frac{1}{10}x$	$y = \frac{1}{2}x$	$y = -2x$
$y = 5x$	$y = -\frac{1}{5}x$	$y = \frac{1}{5}x$	$y = -5x$
$y = 2x$	$y = -\frac{1}{2}x$	$y = \frac{1}{10}x$	$y = -10x$
$y = x$	$y = -1x$		

Then ask how a line of slope 0 would be positioned.

Name _____ **Section** _____ **Date** _____

MENTAL MATH

Decide whether a line with the given slope is upward-sloping, downward-sloping, horizontal, or vertical.

1. $m = \dfrac{7}{6}$ **2.** $m = -3$ **3.** $m = 0$ **4.** m is undefined

EXERCISE SET 6.4

A *Use the points shown on each graph to find the slope of each line. See Examples 1 and 2.*

1.

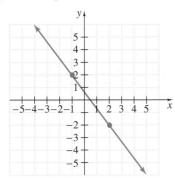

2.

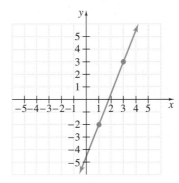

3.

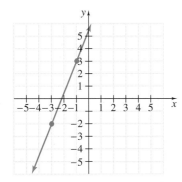

4.

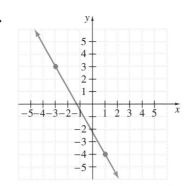

Find the slope of the line that passes through the given points. See Examples 1 and 2.

5. $(0, 0)$ and $(7, 8)$ **6.** $(-1, 5)$ and $(0, 0)$ ▣**7.** $(-1, 5)$ and $(6, -2)$

8. $(-1, 9)$ and $(-3, 4)$ **9.** $(1, 4)$ and $(5, 3)$ **10.** $(3, 1)$ and $(2, 6)$

11. $(-2, 8)$ and $(1, 6)$ **12.** $(4, -3)$ and $(2, 2)$ ▣**13.** $(5, 1)$ and $(-2, 1)$

14. $(5, 4)$ and $(5, 0)$

ANSWERS

1. $m = -\dfrac{4}{3}$

2. $m = \dfrac{5}{2}$

3. $m = \dfrac{5}{2}$

4. $m = -\dfrac{7}{4}$

5. $m = \dfrac{8}{7}$

6. $m = -5$

7. $m = -1$

8. $m = \dfrac{5}{2}$

9. $m = -\dfrac{1}{4}$

10. $m = -5$

11. $m = -\dfrac{2}{3}$

12. $m = -\dfrac{5}{2}$

13. $m = 0$

14. undefined slope

477

15. line 1

16. line 1

17. line 2

18. line 1

19. $m = 5$

20. $m = -2$

21. $m = -2$

22. $m = 5$

23. $m = \dfrac{2}{3}$

24. $m = -\dfrac{3}{4}$

25. $m = \dfrac{1}{2}$

26. $m = -\dfrac{1}{4}$

27. undefined slope

28. $m = 0$

29. undefined slope

30. $m = 0$

31. $m = 0$

32. undefined slope

For each graph, determine which line has the steeper slope.

15.

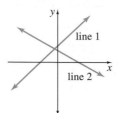

16.

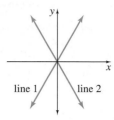

17.

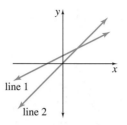

18.

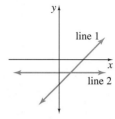

B *Find the slope of each line. See Example 3.*

19. $y = 5x - 2$

20. $y = -2x + 6$

21. $2x + y = 7$

22. $-5x + y = 10$

23. $2x - 3y = 10$

24. $-3x - 4y = 6$

25. $x = 2y$

26. $x = -4y$

C *Find the slope of each line. See Examples 4 and 5.*

27.

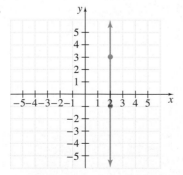

28.

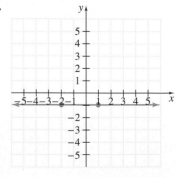

29. $x = 1$ **30.** $y = -2$ **31.** $y = -3$ **32.** $x = 5$

Name _____

33. neither

D *Determine whether each pair of lines is parallel, perpendicular, or neither. See Example 6.*

33. $x - 3y = -6$
$3x - y = 0$

34. $-5x + y = -6$
$x + 5y = 5$

35. $10 + 3x = 5y$
$5x + 3y = 1$

36. $y = 4x - 2$
$4x + y = 5$

37. $6x = 5y + 1$
$-12x + 10y = 1$

38. $-x + 2y = -2$
$2x = 4y + 3$

*Find the slope of the line that is (**a**) parallel and (**b**) perpendicular to the line through each pair of points. See Example 6.*

39. $(-3, -3)$ and $(0, 0)$

40. $(6, -2)$ and $(1, 4)$

41. $(-8, -4)$ and $(3, 5)$

42. $(6, -1)$ and $(-4, -10)$

REVIEW AND PREVIEW

Solve each equation for y. See Section 2.6.

43. $y - (-6) = 2(x - 4)$

44. $y - 7 = -9(x - 6)$

45. $y - 1 = -6(x - (-2))$

46. $y - (-3) = 4(x - (-5))$

---

33. neither

34. perpendicular

35. perpendicular

36. neither

37. parallel

38. parallel

39. a. 1

b. -1

40. a. $-\dfrac{6}{5}$

b. $\dfrac{5}{6}$

41. a. $\dfrac{9}{11}$

b. $-\dfrac{11}{9}$

42. a. $\dfrac{9}{10}$

b. $-\dfrac{10}{9}$

43. $y = 2x - 14$

44. $y = -9x + 61$

45. $y = -6x - 11$

46. $y = 4x + 17$

Name _____

Match each line with its slope.

A. $m = 0$ **B.** undefined slope   **C.** $m = 3$

D. $m = 1$ **E.** $m = -\dfrac{1}{2}$ **F.** $m = -\dfrac{3}{4}$

47.
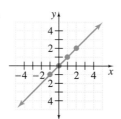

48.
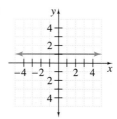

49.
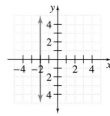

50.
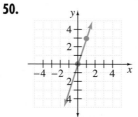

51.
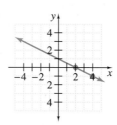

52.
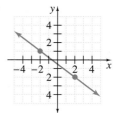

The pitch of a roof is its slope. Find the pitch of the roof shown.

53.

54. Find x so that the pitch of the roof shown is $\dfrac{1}{3}$.

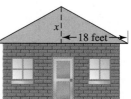

The grade of a road is its slope written as a percent. Find the grade of each road shown.

55.

16 feet
100 feet

56.
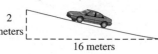
2 meters
16 meters

Name _____

The following line graph shows average miles per gallon for cars during the years shown. Use this graph to answer Exercises 57–62.

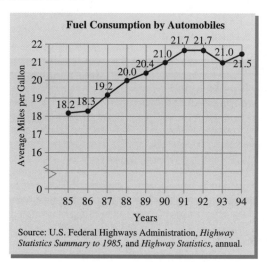

57. In 1988, cars averaged how many miles per gallon of fuel?

58. In 1991, cars averaged how many miles per gallon of fuel?

59. Find the increase of average miles per gallon for automobiles for the years 1985 to 1994.

60. Do you notice any trends from this graph?

61. What line segment has the greatest slope?

62. What line segment has the least slope?

57. 20 mpg

58. 21.7 mpg

59. 3.3 mpg

60. answers may vary

61. 86 to 87

62. 91 to 92

63. the line becomes steeper

64. the line becomes steeper

65. a. (0, 18,000) (10, 0); answers may vary

b. −1800

c. Between 1991 and 1995, the number of hepatitis cases is decreasing at a rate of 1800 per year.

66. a. (1980, 100) and (1996, 2342)

b. $\dfrac{1121}{8}$ or 140.125

c. From 1980 to 1996, the number of heart transplants increased at a rate of 1121 transplants every 8 years or 140.125 transplants per year

63. The graph of $y = -\frac{1}{3}x + 2$ has a slope of $-\frac{1}{3}$. The graph of $y = -2x + 2$ has a slope of -2. The graph of $y = -4x + 2$ has a slope of -4. Graph all three equations on a single coordinate system. As the absolute value of the slope becomes larger, how does the steepness of the line change?

64. The graph of $y = \frac{1}{2}x$ has a slope of $\frac{1}{2}$. The graph of $y = 3x$ has a slope of 3. The graph of $y = 5x$ has a slope of 5. Graph all three equations on a single coordinate system. As slope becomes larger, how does the steepness of the line change?

65. The number of hepatitis B cases y reported in the United States between 1991 and 1995 can be modeled by the linear equation $y = -1800x + 18,000$, where x represents the number of years after 1991. (*Source*: Based on data from the Center for Disease Control and Prevention)
   a. Find and interpret the intercept points.

   b. Find the slope of this line.

   c. What does the slope mean in this context?

66. There were 100 heart transplants performed in the United States in 1980. In 1996, the number of heart transplants performed in the United States rose to 2342. (*Source*: Bureau of Health Resources Development)
   a. Write two ordered pairs of the form (year, number of heart transplants).

   b. Find the slope of the line between the two points.

   c. What does the slope mean in this context?

6.5   EQUATIONS OF LINES

We know that when a linear equation is solved for y, the coefficient of x is the slope of the line. For example, the slope of the line whose equation is $y = 3x + 1$ is 3. In this equation, $y = 3x + 1$, what does 1 represent? To find out, let $x = 0$ and watch what happens.

$$y = 3x + 1$$
$$y = 3 \cdot 0 + 1 \quad \text{Let } x = 0.$$
$$y = 1$$

We now have the ordered pair $(0, 1)$, which means that 1 is the y-intercept.
   This is true in general. To see this, let $x = 0$ and solve for y in $y = mx + b$.

$$y = m \cdot 0 + b \quad \text{Let } x = 0.$$
$$y = b$$

We obtain the ordered pair $(0, b)$, which means that b is the y-intercept.
   The form $y = mx + b$ is appropriately called the **slope-intercept form** of a linear equation.

slope   y-intercept

SLOPE-INTERCEPT FORM

When a linear equation in two variables is written in slope-intercept form,

$$y = mx + b$$

then m is the slope of the line and b is the y-intercept of the line.

A   USING THE SLOPE-INTERCEPT FORM TO WRITE AN EQUATION

The slope-intercept form can be used to write the equation of a line when we know its slope and y-intercept.

Example 1   Find an equation of the line with y-intercept -3 and slope of $\frac{1}{4}$.

Solution: We are given the slope and the y-intercept. We let $m = \frac{1}{4}$ and $b = -3$ and write the equation in slope-intercept form, $y = mx + b$.

$$y = mx + b$$
$$y = \frac{1}{4}x + (-3) \quad \text{Let } m = \frac{1}{4} \text{ and } b = -3.$$
$$y = \frac{1}{4}x - 3 \quad \text{Simplify.}$$

Objectives

A Use the slope-intercept form to write an equation of a line.

B Use the slope-intercept form to graph a linear equation.

C Use the point-slope form to find an equation of a line given its slope and a point of the line.

D Use the point-slope form to find an equation of a line given two points of the line.

E Use the point-slope form to solve problems.

SSM   CD-ROM   Video 6.5

TEACHING TIP

If you are using graphing calculators, have students make a conjecture about b after exploring the following equations on a graphing calculator.
$$y = 3x + 0$$
$$y = 3x + 5$$
$$y = 3x + 2$$
$$y = 3x - 5$$

Practice Problem 1

Find an equation of the line with y-intercept -2 and slope of $\frac{3}{5}$.

Answer

1. $y = \frac{3}{5}x - 2$

Practice Problem 2

Graph the equation $y = \frac{2}{3}x - 4$.

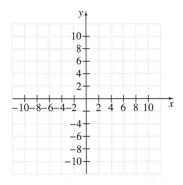

Practice Problem 3

Use the slope-intercept form to graph $3x + y = 2$.

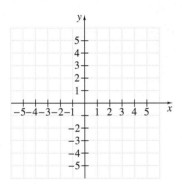

Answers

2.

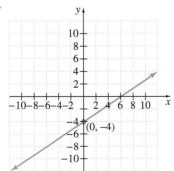

3.

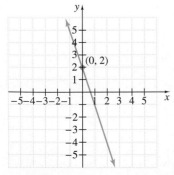

B USING THE SLOPE-INTERCEPT FORM TO GRAPH AN EQUATION

We also can use the slope-intercept form of the equation of a line to graph a linear equation.

Example 2   Use the slope-intercept form to graph the equation

$$y = \frac{3}{5}x - 2$$

Solution: Since the equation $y = \frac{3}{5}x - 2$ is written in slope-intercept form $y = mx + b$, the slope of its graph is $\frac{3}{5}$ and the y-intercept is -2. To graph this equation, we begin by plotting the intercept point $(0, -2)$. From this point, we can find another point of the graph by using the slope $\frac{3}{5}$ and recalling that slope is $\frac{\text{rise}}{\text{run}}$. We start at the intercept point and move 3 units up since the numerator of the slope is 3; then we move 5 units to the right since the denominator of the slope is 5. We stop at the point $(5, 1)$. The line through $(0, -2)$ and $(5, 1)$ is the graph of $y = \frac{3}{5}x - 2$.

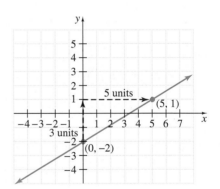

Example 3   Use the slope-intercept form to graph the equation $4x + y = 1$.

Solution: First we write the given equation in slope-intercept form.

$$4x + y = 1$$
$$y = -4x + 1$$

The graph of this equation will have slope -4 and y-intercept 1. To graph this line, we first plot the intercept point $(0, 1)$. To find another point of the graph, we use the slope -4, which can be written as $\frac{-4}{1}$ $\left(\frac{4}{-1} \text{ could also be used}\right)$. We start at the point $(0, 1)$ and move 4 units down (since the numerator of the slope is -4), then 1 unit to the right (since the denominator of the slope is 1). We arrive at the point $(1, -3)$. The line through $(0, 1)$ and $(1, -3)$ is the graph of $4x + y = 1$.

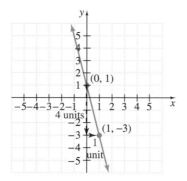

HELPFUL HINT

In Example 3, if we interpret the slope of -4 as $\dfrac{4}{-1}$, we arrive at $(-1, 5)$ for a second point. Notice that this point is also on the line.

C WRITING AN EQUATION GIVEN ITS SLOPE AND A POINT

Thus far, we have seen that we can write an equation of a line if we know its slope and y-intercept. We can also write an equation of a line if we know its slope and any point on the line. To see how we do this, let m represent slope and (x_1, y_1) represent the point on the line. Then if (x, y) is any other point of the line, we have that

$$\frac{y - y_1}{x - x_1} = m$$

or

$$y - y_1 = m(x - x_1) \quad \text{Multiply both sides by } (x - x_1).$$

This is the **point-slope form** of the equation of a line.

POINT-SLOPE FORM OF THE EQUATION OF A LINE

The point-slope form of the equation of a line is $y - y_1 = m(x - x_1)$, where m is the slope of the line and (x_1, y_1) is a point on the line.

Example 4   Find an equation of the line passing through $(-1, 5)$ with slope -2. Write the equation in the form $Ax + By = C$.

Solution: Since the slope and a point on the line are given, we use point-slope form $y - y_1 = m(x - x_1)$ to write the equation. Let $m = -2$ and $(-1, 5) = (x_1, y_1)$.

$$y - y_1 = m(x - x_1)$$
$$y - 5 = -2[x - (-1)] \quad \text{Let } m = -2 \text{ and } (x_1, y_1) = (-1, 5).$$
$$y - 5 = -2(x + 1) \quad \text{Simplify.}$$
$$y - 5 = -2x - 2 \quad \text{Use the distributive property.}$$
$$y = -2x + 3 \quad \text{Add 5 to both sides.}$$
$$2x + y = 3 \quad \text{Add } 2x \text{ to both sides.}$$

Practice Problem 4

Find an equation of the line that passes through $(2, -4)$ with slope -3. Write the equation in the form $Ax + By = C$.

Answer
4. $3x + y = 2$

D WRITING AN EQUATION GIVEN TWO POINTS

We can also find the equation of a line when we are given any two points of the line.

Practice Problem 5

Find an equation of the line through $(1, 3)$ and $(5, -2)$. Write the equation in the form $Ax + By = C$.

Example 5   Find an equation of the line through $(2, 5)$ and $(-3, 4)$. Write the equation in the form $Ax + By = C$.

Solution: First we find the slope of the line. Let (x_1, y_1) be $(2, 5)$ and (x_2, y_2) be $(-3, 4)$.

$$m = \frac{y_2 - y_1}{x_2 - x_1} = \frac{4 - 5}{-3 - 2} = \frac{-1}{-5} = \frac{1}{5}$$

Next we use the slope $\frac{1}{5}$ and either one of the given points to write the equation in point-slope form. We use $(2, 5)$. Let $x_1 = 2$, $y_1 = 5$, and $m = \frac{1}{5}$.

$y - y_1 = m(x - x_1)$	Use point-slope form.
$y - 5 = \frac{1}{5}(x - 2)$	Let $x_1 = 2$, $y_1 = 5$, and $m = \frac{1}{5}$.
$5(y - 5) = 5 \cdot \frac{1}{5}(x - 2)$	Multiply both sides by 5 to clear fractions.
$5y - 25 = x - 2$	Use the distributive property and simplify.
$-x + 5y - 25 = -2$	Subtract x from both sides.
$-x + 5y = 23$	Add 25 to both sides.

HELPFUL HINT

When you multiply both sides of the equation from Example 5, $-x + 5y = 23$ by -1, it becomes $x - 5y = -23$. Both $-x + 5y = 23$ and $x - 5y = -23$ are in the form $Ax + By = C$ and both are equations of the same line.

E USING THE POINT-SLOPE FORM TO SOLVE PROBLEMS

Problems occurring in many fields can be modeled by linear equations in two variables. The next example is from the field of marketing and shows how consumer demand of a product depends on the price of the product.

Practice Problem 6

The Pool Entertainment Company learned that by pricing a new pool toy at $10, local sales will reach 200 a week. Lowering the price to $9 will cause sales to rise to 250 a week.

a. Assume that the relationship between sales price and number of toys sold is linear, and write an equation describing this relationship. Write the equation in slope-intercept form. Use ordered pairs of the form (sales price, number sold).

b. Predict the weekly sales of the toy if the price is $7.50.

Example 6   The Whammo Company has learned that by pricing a newly released Frisbee at $6, sales will reach 2000 Frisbees per day. Raising the price to $8 will cause the sales to fall to 1500 Frisbees per day.

a. Assume that the relationship between sales price and number of Frisbees sold is linear and write an equation describing this relationship. Write the equation in slope-intercept form. Use ordered pairs of the form (sales price, number sold).

b. Predict the daily sales of Frisbees if the price is $7.50.

Answers

5. $5x + 4y = 17$,   **6. a.** $y = -50x + 700$,
b. 325

Solution: **a.** We use the given information and write two ordered pairs. Our ordered pairs are $(6, 2000)$ and $(8, 1500)$. To use the point-slope form to write an equation, we find the slope of the line that contains these points.

$$m = \frac{2000 - 1500}{6 - 8} = \frac{500}{-2} = -250$$

Next we use the slope and either one of the points to write the equation in point-slope form. We use $(6, 2000)$.

$y - y_1 = m(x - x_1)$ Use point-slope form.

$y - 2000 = -250(x - 6)$ Let $x_1 = 6$, $y_1 = 2000$, and $m = -250$.

$y - 2000 = -250x + 1500$ Use the distributive property.

$y = -250x + 3500$ Write in slope-intercept form.

b. To predict the sales if the price is $7.50, we find y when $x = 7.50$.

$y = -250x + 3500$

$y = -250(7.50) + 3500$ Let $x = 7.50$.

$y = -1875 + 3500$

$y = 1625$

If the price is $7.50, sales will reach 1625 Frisbees per day. ▬▬▬▬

Example 6 could also have been solved by using ordered pairs of the form (number sold, sales price).

Here is a summary of our discussion on linear equations thus far.

FORMS OF LINEAR EQUATIONS

$Ax + By = C$ **Standard form** of a linear equation. A and B are not both 0.

$y = mx + b$ **Slope-intercept form** of a linear equation. The slope is m and the y-intercept is b.

$y - y_1 = m(x - x_1)$ **Point-slope form** of a linear equation. The slope is m and (x_1, y_1) is a point on the line.

$y = c$ **Horizontal line** The slope is 0 and the y-intercept is c.

$x = c$ **Vertical line** The slope is undefined and the x-intercept is c.

PARALLEL AND PERPENDICULAR LINES

Nonvertical parallel lines have the same slope. The product of the slopes of two nonvertical perpendicular lines is -1.

TEACHING TIP

Have students write their strategy for various scenarios. How would you find the equation if you knew . . .
▲ the slope and y-intercept?
▲ two points?
▲ a point and the slope?
▲ a point and the y-intercept?
▲ the x-intercept and the y-intercept?
How much information must be given in order to determine a line?

GRAPHING CALCULATOR EXPLORATIONS

A graphing calculator is a very useful tool for discovering patterns. To discover the change in the graph of a linear equation caused by a change in slope, try the following. Use a standard window and graph a linear equation in the form $y = mx + b$. Recall that the graph of such an equation will have slope m and y-intercept b.

First graph $y = x + 3$. To do so, press the $\boxed{Y=}$ key and enter $Y_1 = x + 3$. Notice that this graph has slope 1 and that the y-intercept is 3. Next, on the same set of axes, graph $y = 2x + 3$ and $y = 3x + 3$ by pressing $\boxed{Y=}$ and entering $Y_2 = 2x + 3$ and $Y_3 = 3x + 3$.

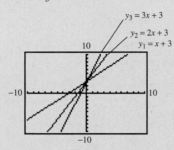

Notice the difference in the graph of each equation as the slope changes from 1 to 2 to 3. How would the graph of $y = 5x + 3$ appear? To see the change in the graph caused by a change in negative slope, try graphing $y = -x + 3$, $y = -2x + 3$, and $y = -3x + 3$ on the same set of axes.

Use a graphing calculator to graph the following equations. For each exercise, graph the first equation and use its graph to predict the appearance of the other equations. Then graph the other equations on the same set of axes and check your prediction.

1. $y = x$; $y = 6x$, $y = -6x$

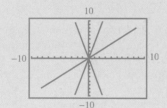

2. $y = -x$; $y = -5x$, $y = -10x$

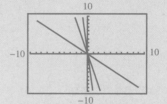

3. $y = \dfrac{1}{2}x + 2$; $y = \dfrac{3}{4}x + 2$, $y = x + 2$

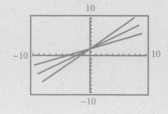

4. $y = x + 1$; $y = \dfrac{5}{4}x + 1$, $y = \dfrac{5}{2}x + 1$

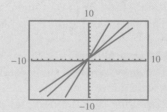

Name _____ **Section** _____ **Date** _____

MENTAL MATH

Identify the slope and the y-intercept of the graph of each equation.

1. $y = 2x - 1$ **2.** $y = -7x + 3$ **3.** $y = x + \dfrac{1}{3}$

4. $y = -x - \dfrac{2}{9}$ **5.** $y = \dfrac{5}{7}x - 4$ **6.** $y = -\dfrac{1}{4}x + \dfrac{3}{5}$

1. $m = 2; b = -1$

2. $m = -7; b = 3$

3. $m = 1; b = \dfrac{1}{3}$

4. $m = -1, b = -\dfrac{2}{9}$

5. $m = \dfrac{5}{7}; b = -4$

6. $m = -\dfrac{1}{4}; b = \dfrac{3}{5}$

EXERCISE SET 6.5

A *Write an equation of the line with each given slope, m, and y-intercept, b. See Example 1.*

1. $m = 5, b = 3$ **2.** $m = 2, b = \dfrac{3}{4}$ **3.** $m = \dfrac{2}{3}, b = 0$

4. $m = 0, b = -2$ **5.** $m = -\dfrac{1}{5}, b = \dfrac{1}{9}$ **6.** $m = -3, b = -3$

B *Use the slope-intercept form to graph each equation. See Examples 2 and 3.*

7. $y = 2x + 1$

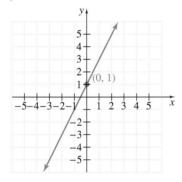

8. $y = -4x - 1$

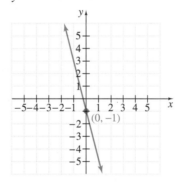

9. $y = \dfrac{2}{3}x + 5$

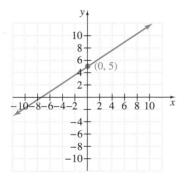

10. $y = \dfrac{1}{4}x - 3$

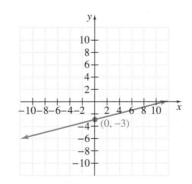

ANSWERS

1. $y = 5x + 3$

2. $y = 2x + \dfrac{3}{4}$

3. $y = \dfrac{2}{3}x$

4. $y = -2$

5. $y = -\dfrac{1}{5}x + \dfrac{1}{9}$

6. $y = -3x - 3$

17. $-6x + y = -10$

18. $-4x + y = -1$

19. $8x + y = -13$

20. $2x + y = -34$

21. $x - 2y = 17$

22. $2x - 3y = -43$

23. $x + 2y = -3$

24. $x + 5y = 4$

Name _____

 **11.** $y = -5x$

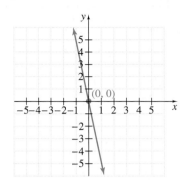

12. $y = 6x$

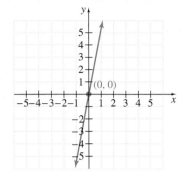

13. $4x + y = 6$

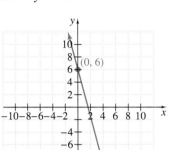

14. $-3x + y = 2$

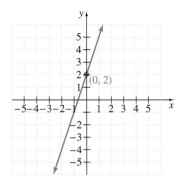

 **15.** $4x - 7y = -14$

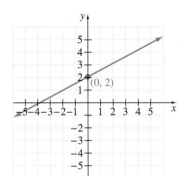

16. $3x - 4y = 4$

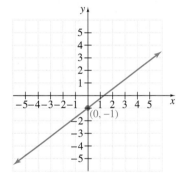

C *Find an equation of each line with the given slope that passes through the given point. Write the equation in the form $Ax + By = C$. See Example 4.*

17. $m = 6$; $(2, 2)$ **18.** $m = 4$; $(1, 3)$ **19.** $m = -8$; $(-1, -5)$

20. $m = -2$; $(-11, -12)$ **21.** $m = \dfrac{1}{2}$; $(5, -6)$ **22.** $m = \dfrac{2}{3}$; $(-8, 9)$

23. $m = -\dfrac{1}{2}$; $(-3, 0)$ **24.** $m = -\dfrac{1}{5}$; $(4, 0)$

D *Find an equation of the line passing through each pair of points. Write the equation in the form Ax + By = C. See Example 5.*

25. $(3, 2)$ and $(5, 6)$ **26.** $(6, 2)$ and $(8, 8)$ **27.** $(-1, 3)$ and $(-2, -5)$

28. $(-4, 0)$ and $(6, -1)$ **29.** $(2, 3)$ and $(-1, -1)$ **30.** $(0, 0)$ and $\left(\frac{1}{2}, \frac{1}{3}\right)$

31. $(10, 7)$ and $(7, 10)$ **32.** $(5, -6)$ and $(-6, 5)$

E *Solve. See Example 6.*

33. A rock is dropped from the top of a 400-foot cliff. After 1 second, the rock is traveling 32 feet per second. After 3 seconds, the rock is traveling 96 feet per second.

a. Assume that the relationship between time and speed is linear and write an equation describing this relationship. Use ordered pairs of the form (time, speed).

b. Use this equation to determine the speed of the rock 4 seconds after it was dropped.

34. Del Monte Fruit Company is studying the sales of a pineapple sauce to see if this product is to be continued. At the end of its first year, profits on this product amounted to $30,000. At the end of the fourth year, profits are $66,000.

a. Assume that the relationship between years on the market and profit is linear and write an equation describing this relationship. Use ordered pairs of the form (years on the market, profit).

b. Use this equation to predict the profit at the end of 7 years.

---

25. $2x - y = 4$

26. $3x - y = 16$

27. $8x - y = -11$

28. $x + 10y = -4$

29. $4x - 3y = -1$

30. $2x - 3y = 0$

31. $x + y = 17$

32. $x + y = -1$

33. a. $s = 32t$

     b. 128 ft/sec

34. a. $p = 12,000t + 18,000$

     b. $102,000

35. a. $(0, 27.6); (3, 24.8)$

b. $y = -\dfrac{2.8}{3}x + 27.6$

c. 23.87 heads/sq. ft

36. a. $(0, 24.4); (4, 26.77)$

b. $y = \dfrac{2.37}{4}x + 24.4$

c. 27.36 million barrels/day

37. -1

38. 11

39. 5

40. 19

41. no

42. no

492

35. In 1992, durum yield in North Dakota was 27.6 heads per square foot. In 1995, the yield had dropped to 24.8 heads per square foot. (*Source*: United States Department of Agriculture)
a. Write two ordered pairs of the form (years after 1992, durum yield) for this situation.

b. The relationship between years after 1992 and durum yield is linear over this period. Use the ordered pairs from part (a) to write an equation of the line relating year to durum yield.

c. Use the linear equation from part (b) to estimate the durum yield in 1996.

36. In 1992, crude oil production by OPEC countries was 24.40 million barrels per day. In 1996, OPEC crude oil production had risen to 26.77 million barrels per day. (*Source*: Energy Information Administration)
a. Write two ordered pairs of the form (years after 1992, crude oil production) for this situation.

b. The relationship between years after 1992 and crude oil production is linear over this period. Use the ordered pairs from part (a) to write an equation of the line relating year to crude oil production.

c. Use the linear equation from part (b) to estimate the crude oil production by OPEC countries in 1997.

REVIEW AND PREVIEW

Find the value of $x^2 - 3x + 1$ for each given value of x. See Section 2.1.

37. 2   **38.** 5   **39.** -1 **40.** -3

For each graph, determine whether any x-values correspond to two or more y-values. See Section 6.1.

41.

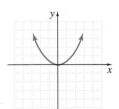

42.

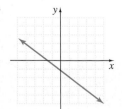

Name _____

43.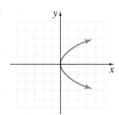

44.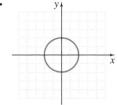

43. yes

44. yes

◆ COMBINING CONCEPTS

Match each linear equation with its graph.

45. $y = 2x + 1$ **46.** $y = -x + 1$ **47.** $y = -3x - 2$ **48.** $y = \dfrac{5}{3}x - 2$

45. B

A.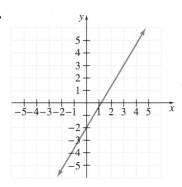

C.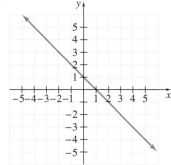

46. C

47. D

B.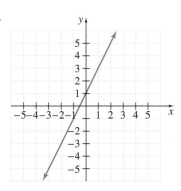

D.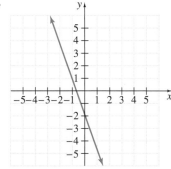

48. A

49. $3x - y = -5$

49. Write an equation of the line that contains the point $(-1, 2)$ and has the same slope as the line $y = 3x - 1$.

50. Write an equation of the line that contains the point $(4, 0)$ and has the same slope as the line $y = -2x + 3$.

50. $2x + y = 8$

Name _____

 Internet Excursions

Go to http://www.prenhall.com/martin-gay
The National Center for Education Statistics (NCES) is the federal agency responsible for collecting data on the state of education in the United States. The given World Wide Web address will provide you with access to the NCES's "Education at a Glance" Web site, or a related site, where you will have access to a wide variety of education statistics. Browse the links to find two different sets of paired data on education that are interesting to you. Then answer the questions below.

51. For one set of data that you have found, write down two ordered pairs. Describe what the ordered pairs represent. Then use the ordered pairs to find an equation for the line passing through these two points.

52. For your other set of data, write the data as a set of ordered pairs. Describe what the ordered pairs represent. Make a scatter diagram of the data. Do you think that a linear equation would represent the data well? Explain.

Name _____ **Section** _____ **Date** _____

ANSWERS

1. $m = 2$

INTEGRATED REVIEW—SUMMARY ON LINEAR EQUATIONS

Find the slope of each line.

1.

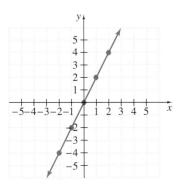

2.

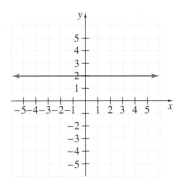

2. $m = 0$

3.

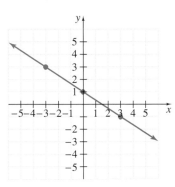

4.

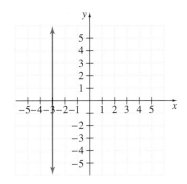

3. $m = -\dfrac{2}{3}$

4. slope is undefined

Graph each linear equation.

5. see graph

5. $y = -2x$

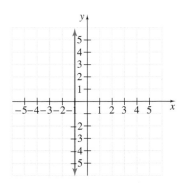

6. $x + y = 3$

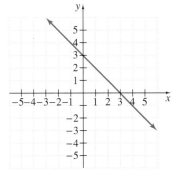

6. see graph

7. $x = -1$

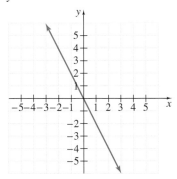

8. $y = 4$

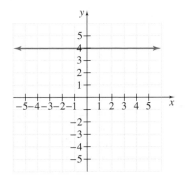

7. see graph

8. see graph

495

Name _____

9. $x - 2y = 6$

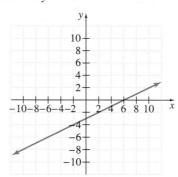

10. $y = 3x + 2$

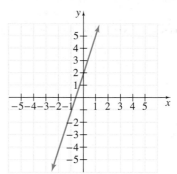

11. Write an equation of the line with slope $m = 2$ and y-intercept $b = -\dfrac{1}{3}$.

12. Find an equation of the line with slope $m = -4$ that passes through the point $(-1, 3)$. Write the equation in the form $Ax + By = C$.

13. Find an equation of the line that passes through the points $(2, 0)$ and $(-1, -3)$. Write the equation in the form $Ax + By = C$.

6.6 INTRODUCTION TO FUNCTIONS

A IDENTIFYING RELATIONS, DOMAINS, AND RANGES

In this chapter, we have studied paired data in the form of ordered pairs. For example, when we list an ordered pair such as $(3, 1)$, we are saying that when x is 3, then y is 1. In other words $x = 3$ and $y = 1$ are related to each other.

For this reason, we call a set of ordered pairs a **relation**. The set of all x-coordinates is called the **domain** of a relation, and the set of all y-coordinates is called the **range** of a relation.

Example 1   Find the domain and the range of the relation $\{(0, 2), (3, 3), (-1, 0), (3, -2)\}$.

Solution: The domain is the set of all x-values, or $\{-1, 0, 3\}$, and the range is the set of all y-values, or $\{-2, 0, 2, 3\}$.

B IDENTIFYING FUNCTIONS

Paired data occur often in real-life applications. Some special sets of paired data, or ordered pairs, are called **functions**.

FUNCTION

A function is a set of ordered pairs in which each x-coordinate has exactly one y-coordinate.

In other words, a function cannot have two ordered pairs with the same x-coordinate but different y-coordinates.

Example 2   Which of the following relations are also functions?

  a. $\{(-1, 1), (2, 3), (7, 3), (8, 6)\}$
  b. $\{(0, -2), (1, 5), (0, 3), (7, 7)\}$

Solution: **a.** Although the ordered pairs $(2, 3)$ and $(7, 3)$ have the same y-value, each x-value is assigned to only one y-value, so this set of ordered pairs is a function.
  b. The x-value 0 is paired with two y-values, -2 and 3, so this set of ordered pairs is not a function.

Relations and functions can be described by a graph of their ordered pairs.

Objectives

A Identify relations, domains, and ranges.
B Identify functions.
C Use the vertical line test.
D Use function notation.

SSM   CD-ROM   Video 6.6

Practice Problem 1

Find the domain and range of the relation $\{(-3, 5), (-3, 1), (4, 6), (7, 0)\}$

TEACHING TIP

Before beginning this lesson, discuss the nonmathematical meaning of the word *function* with your students using the following questions.
▲ What is the function of an oven?
▲ How does it work?
▲ If someone says, "That oven is functioning properly," what do they mean?
▲ Try to get them to verbalize that functions do something which is predictable.

Practice Problem 2

Which of the following relations are also functions?

a. $\{(2, 5), (-3, 7), (4, 5), (0, -1)\}$
b. $\{(1, 4), (6, 6), (1, -3), (7, 5)\}$

TEACHING TIP

Ask students to name some situations where paired data have exactly one y-value for each x-value (e.g., a person's weight versus time, the temperature versus time). Then have students name situations where paired data have more than one y-value for each value of x (e.g., age of employees at a company versus salary of employees at that company).

Answers

1. domain: $\{-3, 4, 7\}$; range: $\{0, 1, 5, 6\}$,
2. a. a function, **b.** not a function.

Practice Problem 3

Which graph is the graph of a function?

a.

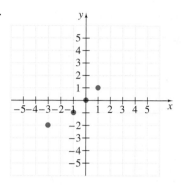

b.

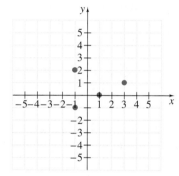

TEACHING TIP

Before discussing Objective C, ask students if it was easier in Examples 2 and 3 to determine if a relation was a function by looking at the list of points or by looking at the graph. Then have them explain their reasoning.

Example 3   Which graph is the graph of a function?

a.

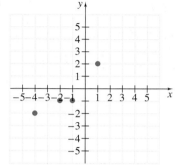

b.

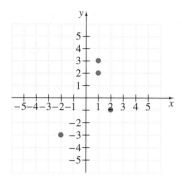

Solution: **a.** This is the graph of the relation $\{(-4, -2), (-2, -1), (-1, -1), (1, 2)\}$. Each x-coordinate has exactly one y-coordinate, so this is the graph of a function.

b. This is the graph of the relation $\{(-2, -3), (1, 2), (1, 3), (2, -1)\}$. The x-coordinate 1 is paired with two y-coordinates, 2 and 3, so this is not the graph of a function.

C USING THE VERTICAL LINE TEST

The graph in Example 3(b) was not the graph of a function because the x-coordinate 1 was paired with 2 y-coordinates, 2 and 3. Notice that when an x-coordinate is paired with more than one y-coordinate, a vertical line can be drawn that will intersect the graph at more than one point. We can use this fact to determine whether a relation is also a function. We call this the vertical line test.

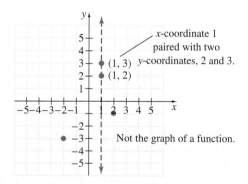

VERTICAL LINE TEST

If a vertical line can be drawn so that it intersects a graph more than once, the graph is not the graph of a function.

This vertical line test works for all types of graphs on the rectangular coordinate system.

Example 4 Use the vertical line test to determine whether each graph is the graph of a function.

a.
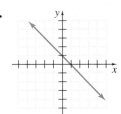

b.
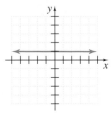

c.
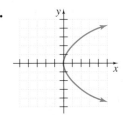

d.
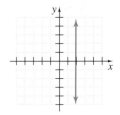

Solution: **a.** This graph is the graph of a function since no vertical line will intersect this graph more than once.
b. This graph is also the graph of a function; no vertical line will intersect it more than once.
c. This graph is not the graph of a function. Vertical lines can be drawn that intersect the graph in two points. An example of one is shown.

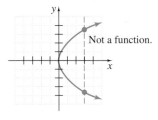

d. This graph is not the graph of a function. A vertical line can be drawn that intersects this line at every point.

Examples of functions can often be found in magazines, newspapers, books, and other printed material in the form of tables or graphs such as that in Example 5.

Example 5 The graph shows the sunrise time for Indianapolis, Indiana, for the year. Use this graph to answer the questions.

a. Approximate the time of sunrise on February 1.
b. Approximate the date(s) when the sun rises at 5 A.M.

Practice Problem 4

Determine whether each graph is the graph of a function.

a.
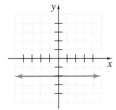

b.
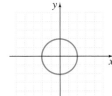

c.
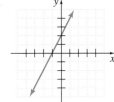

d.
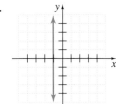

Practice Problem 5

Use the graph in Example 5 to answer the questions.

a. Approximate the time of sunrise on March 1.

b. Approximate the date(s) when the sun rises at 6 A.M.

Answers

4. a. a function, **b.** a function,
c. not a function, **d.** not a function,
5. a. 6:30 A.M., **b.** end of March and mid-October

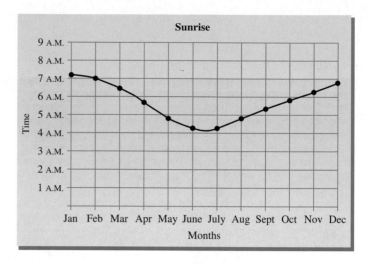

c. Is this the graph of a function?

Solution:

a. To approximate the time of sunrise on February 1, we find the mark on the horizontal axis that corresponds to February 1. From this mark, we move vertically upward until the graph is reached. From that point on the graph, we move horizontally to the left until the vertical axis is reached. The vertical axis there reads 7 A.M.

b. To approximate the date(s) when the sun rises at 5 A.M., we find 5 A.M. on the time axis and move horizontally to the right. Notice that we will reach the graph twice, corresponding to two dates for which the sun rises at 5 A.M.. We follow both points on the graph vertically downward until the horizontal axis is reached. The sun rises at 5 A.M. at approximately the end of the month of April and the middle of the month of August.

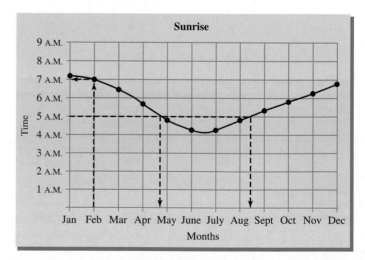

c. The graph is the graph of a function since it passes the vertical line test. In other words, for every day of the year in Indianapolis, there is exactly one sunrise time.

D USING FUNCTION NOTATION

The graph of the linear equation $y = 2x + 1$ passes the vertical line test, so we say that $y = 2x + 1$ is a function. In other words, $y = 2x + 1$ gives us a rule for writing ordered pairs where every x-coordinate is paired with at most one y-coordinate.

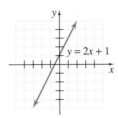

We often use letters such as f, g, and h to name functions. For example, the symbol $f(x)$ means *function of x* and is read "f of x." This notation is called **function notation**. The equation $y = 2x + 1$ can be written as $f(x) = 2x + 1$ using function notation, and these equations mean the same thing. In other words $y = f(x)$.

The notation $f(1)$ means to replace x with 1 and find the resulting y or function value. Since

$$f(x) = 2x + 1$$

then

$$f(1) = 2(1) + 1 = 3$$

This means that, when $x = 1$, y or $f(x) = 3$, and we have the ordered pair $(1, 3)$. Now let's find $f(2)$, $f(0)$, and $f(-1)$.

$f(x) = 2x + 1$	$f(x) = 2x + 1$	$f(x) = 2x + 1$
$f(2) = 2(2) + 1$	$f(0) = 2(0) + 1$	$f(-1) = 2(-1) + 1$
$= 4 + 1$	$= 0 + 1$	$= -2 + 1$
$= 5$	$= 1$	$= -1$

Ordered Pair: $(2, 5)$ $(0, 1)$ $(-1, -1)$

TEACHING TIP

Point out to your students that the variable for the function is given in parentheses. They can think of $f(x) =$ as defining a function with input (variable) x. When asked to evaluate $f(3)$, they need to evaluate the function for the specific case when the input equals 3.

TEACHING TIP

Point out that evaluating a function for a specific value is similar to evaluating an expression for a given value.

HELPFUL HINT

Note that $f(x)$ is a special symbol in mathematics used to denote a function. The symbol $f(x)$ is read "f of x." It does **not** mean $f \cdot x$ (f times x).

Practice Problem 6

Given $f(x) = x^2 + 1$, find the following and list the corresponding ordered pair.

a. $f(1)$

b. $f(-3)$

c. $f(0)$

Example 6   Given $f(x) = x^2 - 3$, find the following and list the corresponding ordered pair.

a. $f(2)$ b. $f(-2)$ c. $f(0)$

Solution:

a. $f(x) = x^2 - 3$
$f(2) = 2^2 - 3$
$ = 4 - 3$
$ = 1$

b. $f(x) = x^2 - 3$
$f(-2) = (-2)^2 - 3$
$ = 4 - 3$
$ = 1$

c. $f(x) = x^2 - 3$
$f(0) = 0^2 - 3$
$ = 0 - 3$
$ = -3$

Ordered
Pair $(2, 1)$ $(-2, 1)$ $(0, -3)$

TRY THE CONCEPT CHECK IN THE MARGIN.

✓ **CONCEPT CHECK**

Suppose that the value of a function f is -7 when the function is evaluated at 2. Write this situation in function notation.

Name _____ **Section** _____ **Date** _____

EXERCISE SET 6.6

A *Find the domain and the range of each relation. See Example 1.*

1. $\{(2, 4), (0, 0), (-7, 10), (10, -7)\}$ **2.** $\{(3, -6), (1, 4), (-2, -2)\}$

 3. $\{(0, -2), (1, -2), (5, -2)\}$ **4.** $\{(5, 0), (5, -3), (5, 4), (5, 3)\}$

B *Determine whether each relation is also a function. See Example 2.*

5. $\{(1, 1), (2, 2), (-3, -3), (0, 0)\}$ **6.** $\{(1, 2), (3, 2), (4, 2)\}$

7. $\{(-1, 0), (-1, 6), (-1, 8)\}$ **8.** $\{(11, 6), (-1, -2), (0, 0), (3, -2)\}$

C *Use the vertical line test to determine whether each graph is the graph of a function. See Examples 3 and 4.*

9.
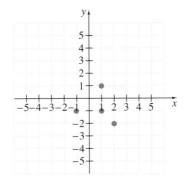

10.
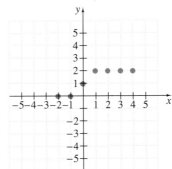

11.
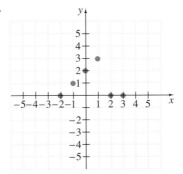

12.
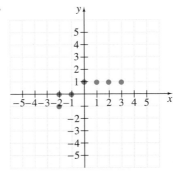

ANSWERS

1. $\{-7, 0, 2, 10\};$ $\{-7, 0, 4, 10\}$

2. $\{-2, 1, 3\};$ $\{-6, -2, 4\}$

3. $\{0, 1, 5\}; \{-2\}$

4. $\{5\}; \{-3, 0, 3, 4\}$

5. yes

6. yes

7. no

8. yes

9. no

10. yes

11. yes

12. no

503

13. yes

14. yes

15. no

16. no

17. 5:20 A.M.

18. Feb. 1st

19. answers may vary

20. mid-June; lowest point

504

📼 **13.**

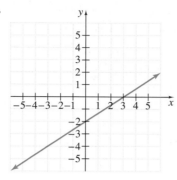

14.

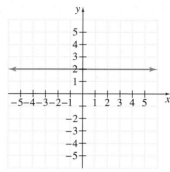

15.

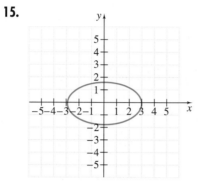

16.

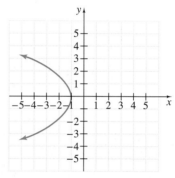

Use the graph in Example 5 to answer Exercises 17–20.

17. Approximate the time of sunrise on September 1 in Indianapolis.

18. Approximate the date(s) when the sun rises in Indianapolis at 7 A.M..

19. Describe the change in sunrise over the year for Indianapolis.

20. When, in Indianapolis, is the earliest sunrise? What point on the graph does this correspond to?

The graph shows the sunset times for Seward, Alaska. Use this graph to answer Exercises 21–26.

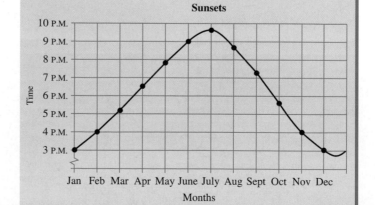

Name _____

21. Approximate the time of sunset on June 1.

22. Approximate the time of sunset on November 1.

23. Approximate the date(s) when the sunset is 3 P.M.

24. Approximate the date(s) when the sunset is 9 P.M.

25. Is this graph the graph of a function? Why or why not?

26. Do you think a graph of sunset times for any location will always be a function? Why or why not?

D *Find* $f(-2)$, $f(0)$, *and* $f(3)$ *for each function. See Example 6.*

27. $f(x) = 2x - 5$

28. $f(x) = 3 - 7x$

⊡ 29. $f(x) = x^2 + 2$

30. $f(x) = x^2 - 4$

31. $f(x) = 3x$

32. $f(x) = -3x$

33. $f(x) = |x|$

34. $f(x) = |2 - x|$

Find $h(-1)$, $h(0)$, *and* $h(4)$ *for each function. See Example 6.*

35. $h(x) = -5x$

36. $h(x) = -3x$

37. $h(x) = 2x^2 + 3$

38. $h(x) = 3x^2$

REVIEW AND PREVIEW

Solve each inequality. See Section 2.8.

39. $2x + 5 < 7$

40. $3x - 1 \geq 11$

41. $-x + 6 \leq 9$

42. $-2x + 3 > 3$

21. 9 P.M.

22. 4 P.M.

23. January 1st and December 1st

24. June 1st and end of July

25. yes; it passes the vertical line test

26. yes; every location has exactly 1 sunset time per day

27. $-9, -5, 1$

28. $17, 3, -18$

29. $6, 2, 11$

30. $0, -4, 5$

31. $-6, 0, 9$

32. $6, 0, -9$

33. $2, 0, 3$

34. $4, 2, 1$

35. $5, 0, -20$

36. $3, 0, -12$

37. $5, 3, 35$

38. $3, 0, 48$

39. $x < 1$

40. $x \geq 4$

41. $x \geq -3$

42. $x < 0$

505

43. $\dfrac{19}{2x}$ meters

44. $\dfrac{16}{x+2}$ in.

45. a. 166.38 cm

b. 148.25 cm

46. a. 190.4 mg

b. 380.8 mg

47. answers may vary

48. answers may vary

49. $f(x) = x + 7$

Name _____

Find the perimeter of each figure. See Section 5.4.

43.

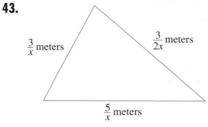

$\dfrac{3}{x}$ meters

$\dfrac{3}{2x}$ meters

$\dfrac{5}{x}$ meters

44.

$\dfrac{3}{x+2}$ inches

$\dfrac{5}{x+2}$ inches

$\dfrac{5}{x+2}$ inches

$\dfrac{3}{x+2}$ inches

◤ COMBINING CONCEPTS

45. Forensic scientists use the function

$$f(x) = 2.59x + 47.24$$

to estimate the height of a woman given the length x of her femur bone.

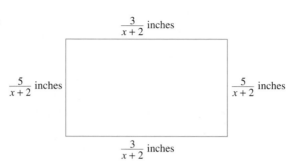

a. Estimate the height of a woman whose femur measures 46 centimeters.

b. Estimate the height of a woman whose femur measures 39 centimeters.

47. In your own words define (a) function; (b) domain; (c) range.

49. Since $y = x + 7$ is a function, rewrite the equation using function notation.

46. The dosage in milligrams of Ivermectin, a heartworm preventive for a dog who weighs x pounds, is given by the function

$$f(x) = \dfrac{136}{25}x$$

a. Find the proper dosage for a dog that weighs 35 pounds.

b. Find the proper dosage for a dog that weighs 70 pounds.

48. Explain the vertical line test and how it is used.

6.7 GRAPHING LINEAR INEQUALITIES IN TWO VARIABLES

Recall that a linear equation in two variables is an equation that can be written in the form $Ax + By = C$ where A, B, and C are real numbers and A and B are not both 0. A **linear inequality in two variables** is an inequality that can be written in one of the forms

$Ax + By < C$ $Ax + By \leq C$
$Ax + By > C$ $Ax + By \geq C$

where A, B, and C are real numbers and A and B are not both 0.

Objectives

A Determine whether an ordered pair is a solution of a linear inequality in two variables.

B Graph a linear inequality in two variables.

SSM   CD-ROM   Video
6.7

A DETERMINING SOLUTIONS OF LINEAR INEQUALITIES IN TWO VARIABLES

Just as for linear equations in x and y, an ordered pair is a **solution** of an inequality in x and y if replacing the variables with the coordinates of the ordered pair results in a true statement.

Example 1   Determine whether each ordered pair is a solution of the equation $2x - y < 6$.

  a. $(5, -1)$ **b.** $(2, 7)$

Solution: **a.** We replace x with 5 and y with -1 and see if a true statement results.

$$2x - y < 6$$
$$2(5) - (-1) < 6 \quad \text{Replace } x \text{ with 5 and } y \text{ with } -1.$$
$$10 + 1 < 6$$
$$11 < 6 \quad \text{False.}$$

The ordered pair $(5, -1)$ is not a solution since $11 < 6$ is a false statement.

b. We replace x with 2 and y with 7 and see if a true statement results.

$$2x - y < 6$$
$$2(2) - 7 < 6 \quad \text{Replace } x \text{ with 2 and } y \text{ with 7.}$$
$$4 - 7 < 6$$
$$-3 < 6 \quad \text{True.}$$

The ordered pair $(2, 7)$ is a solution since $-3 < 6$ is a true statement.

Practice Problem 1

Determine whether each ordered pair is a solution of $x - 4y > 8$.

a. $(-3, 2)$ b. $(9, 0)$

B GRAPHING LINEAR INEQUALITIES IN TWO VARIABLES

The linear equation $x - y = 1$ is graphed next. Recall that all points on the line correspond to ordered pairs that satisfy the equation $x - y = 1$. It can be shown that all the points above the line $x - y = 1$ have coordinates that satisfy the inequality $x - y < 1$. Similarly, all points below the line have coordinates that satisfy the inequality $x - y > 1$.

Answers

1. a. no,   **b.** yes

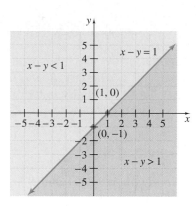

The region above the line and the region below the line are called **half-planes**. Every line divides the plane (similar to a sheet of paper extending indefinitely in all directions) into two half-planes; the line is called the **boundary**.

Recall that the inequality $x - y \leq 1$ means

$$x - y = 1 \qquad \text{or} \qquad x - y < 1$$

Thus, the graph of $x - y \leq 1$ is the half-plane $x - y < 1$ along with the boundary line $x - y = 1$.

To Graph a Linear Inequality in Two Variables

Step 1. Graph the boundary line found by replacing the inequality sign with an equal sign. If the inequality sign is $>$ or $<$, graph a dashed boundary line (indicating that the points on the line are not solutions of the inequality). If the inequality sign is $\geq$ or $\leq$, graph a solid boundary line (indicating that the points on the line are solutions of the inequality).

Step 2. Choose a point, not on the boundary line, as a test point. Substitute the coordinates of this test point into the original inequality.

Step 3. If a true statement is obtained in Step 2, shade the half-plane that contains the test point. If a false statement is obtained, shade the half-plane that does not contain the test point.

Practice Problem 2

Graph: $x - y > 3$

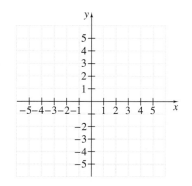

Answer

2.

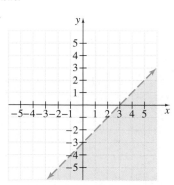

Example 2   Graph: $x + y < 7$

Solution: First we graph the boundary line by graphing the equation $x + y = 7$. We graph this boundary as a dashed line because the inequality sign is $<$, and thus the points on the line are not solutions of the inequality $x + y < 7$.

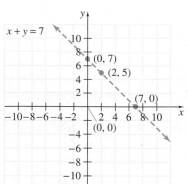

Next we choose a test point, being careful not to choose a point on the boundary line. We choose $(0, 0)$, and substitute the coordinates of $(0, 0)$ into $x + y < 7$.

$x + y < 7$ Original inequality

$0 + 0 < 7$ Replace x with 0 and y with 0.

$0 < 7$ True.

Since the result is a true statement, $(0, 0)$ is a solutuion of $x + y < 7$, and every point in the same half-plane as $(0, 0)$ is also a solution. To indicate this, we shade the entire half-plane containing $(0, 0)$, as shown.

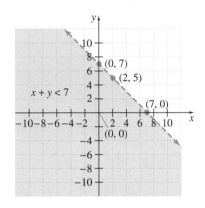

TEACHING TIP

Have students use the graph to write the inequality in slope-intercept form. Notice that if you look at any vertical slice, the y-values shaded are less than (or below) the line graphed so the inequality is $y < -x + 7$.

Example 3   Graph: $2x - y \geq 3$

Solution: We graph the boundary line by graphing $2x - y = 3$. We draw this line as a solid line because the inequality sign is $\geq$, and thus the points on the line are solutions of $2x - y \geq 3$. Once again, $(0, 0)$ is a convenient test point since it is not on the boundary line.

We substitute 0 for x and 0 for y into the original inequality.

$2x - y \geq 3$

$2(0) - 0 \geq 3$ Let $x = 0$ and $y = 0$.

$0 \geq 3$ False.

Since the statement is false, no point in the half-plane containing $(0, 0)$ is a solution. Therefore, we shade the half-plane that does not contain $(0, 0)$. Every point in the shaded half-plane and every point on the boundary line is a solution of $2x - y \geq 3$.

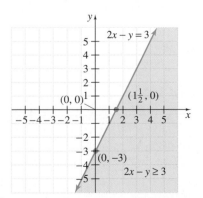

Practice Problem 3

Graph: $x - 4y \leq 4$

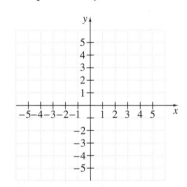

Answer

3.

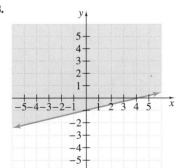

TEACHING TIP

For Examples 4 and 5, have your students use the graph to write the inequality in slope intercept form.

Practice Problem 4

Graph: $y < 3x$

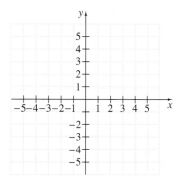

Practice Problem 5

Graph: $3x + 2y \geq 12$

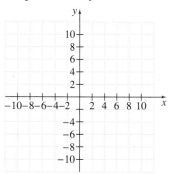

Answers

4.

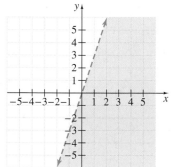

5.

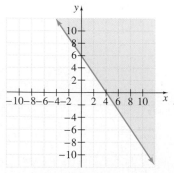

HELPFUL HINT

When graphing an inequality, make sure the test point is substituted into the **original inequality**. For Example 3, we substituted the test point $(0, 0)$ into the **original inequality** $2x - y \geq 3$, *not* $2x - y = 3$.

Example 4   Graph: $x > 2y$

Solution: We find the boundary line by graphing $x = 2y$. The boundary line is a dashed line since the inequality symbol is $>$. We cannot use $(0, 0)$ as a test point because it is a point on the boundary line. We choose instead $(0, 2)$.

$$x > 2y$$
$$0 > 2(2) \quad \text{Let } x = 0 \text{ and } y = 2.$$
$$0 > 4 \quad \text{False.}$$

Since the statement is false, we shade the half-plane that does not contain the test point $(0, 2)$, as shown.

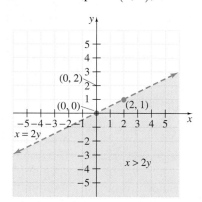

Example 5   Graph: $5x + 4y \leq 20$

Solution: We graph the solid boundary line $5x + 4y = 20$ and choose $(0, 0)$ as the test point.

$$5x + 4y \leq 20$$
$$5(0) + 4(0) \leq 20 \quad \text{Let } x = 0 \text{ and } y = 0.$$
$$0 \leq 20 \quad \text{True.}$$

We shade the half-plane that contains $(0, 0)$, as shown.

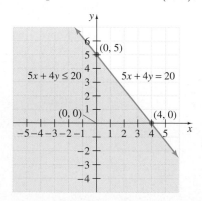

Example 6   Graph: $y > 3$

Solution: We graph the dashed boundary line $y = 3$ and choose $(0, 0)$ as the test point. (Recall that the graph of $y = 3$ is a horizontal line with y-intercept 3.)

$y > 3$

$0 > 3$ Let $y = 0$.

$0 > 3$ False.

We shade the half-plane that does not contain $(0, 0)$, as shown.

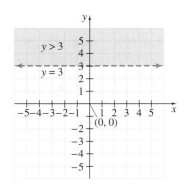

Practice Problem 6

Graph: $x < 2$

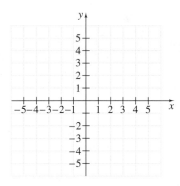

Answer

6.

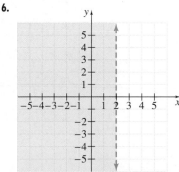

Focus On The Real World

MISLEADING GRAPHS

Graphs are very common in magazines and in newspapers such as *USA Today*. Graphs can be a convenient way to get an idea across because, as the old saying goes, "a picture is worth a thousand words." However, some graphs can be deceptive, which may or may not be intentional. It is important to know some of the ways that graphs can be misleading.

Beware of graphs like the one at the right. Notice that the graph shows a company's profit for various months. It appears that profit is growing quite rapidly. However, this impressive picture tells us little without knowing what units of profit are being graphed. Does the graph show profit in dollars or millions of dollars? An unethical company with profit increases of only a few pennies could use a graph like this one to make the profit increase seem much more substantial than it really is. A truthful graph describes the size of the units used along the vertical axis.

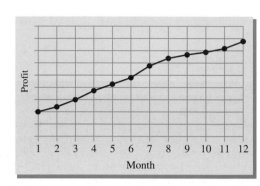

Another type of graph to watch for is one that misrepresents relationships. For example, the bar graph at the right shows the number of men and women employees in the accounting and shipping departments of a certain company. In the accounting department, the bar representing the number of women is shown twice as tall as the bar representing the number of men. However, the number of women (13) is not twice the number of men (10). This set of bars misrepresents the relationship between the number of men and women. Do you

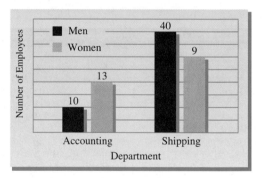

see how the relationship between the number of men and women in the shipping department is distorted by the heights of the bars used? A truthful graph will use bar heights that are proportional to the numbers they represent.

The impression a graph can give also depends on its vertical scale. The two graphs below represent exactly the same data. The only difference between the two graphs is the vertical scale—one shows enrollments from 246 to 260 students and the other shows enrollments between 0 and 300 students. If you were trying to convince readers that algebra enrollment at UPH had changed drastically over the period 1996–2000, which graph would you use? Which graph do you think gives the more honest representation?

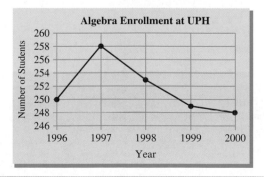

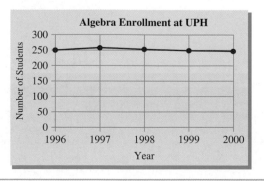

Name _____ **Section** _____ **Date** _____

MENTAL MATH

State whether the graph of each inequality includes its corresponding boundary line.

1. $y \geq x + 4$ **2.** $x - y > -7$ **3.** $y \geq x$ **4.** $x > 0$

Decide whether $(0, 0)$ *is a solution of each given inequality.*

5. $x + y > -5$ **6.** $2x + 3y < 10$ **7.** $x - y \leq -1$ **8.** $\dfrac{2}{3}x + \dfrac{5}{6}y > 4$

EXERCISE SET 6.7

A   *Determine whether the ordered pairs given are solutions of the linear inequality in two variables. See Example 1.*

1. $x - y > 3;\ (0, 3), (2, -1)$ **2.** $y - x < -2;\ (2, 1), (5, -1)$

3. $3x - 5y \leq -4;\ (2, 3), (-1, -1)$ **4.** $2x + y \geq 10;\ (0, 11), (5, 0)$

5. $x < -y;\ (0, 2), (-5, 1)$ **6.** $y > 3x;\ (0, 0), (1, 4)$

B   *Graph each inequality. See Examples 2 through 6.*

7. $x + y \leq 1$

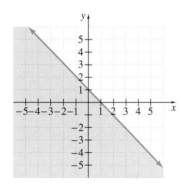

8. $x + y \geq -2$

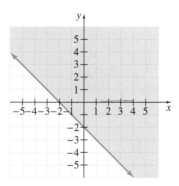

9. $2x - y > -4$

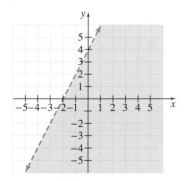

10. $x - 3y < 3$

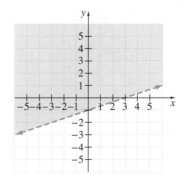

MENTAL MATH ANSWERS

1. yes

2. no

3. yes

4. no

5. yes

6. yes

7. no

8. no

ANSWERS

1. no; no

2. no; yes

3. yes; no

4. yes; yes

5. no; yes

6. no; yes

7. see graph

8. see graph

9. see graph

10. see graph

11. $y > 2x$

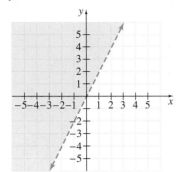

12. $y < 3x$

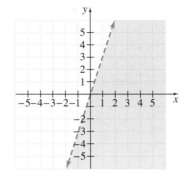

13. $x \le -3y$

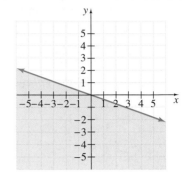

14. $x \ge -2y$

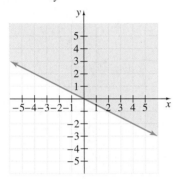

15. $y \ge x + 5$

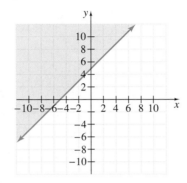

16. $y \le x + 1$

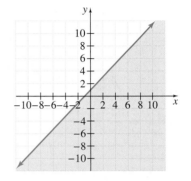

17. $y < 4$

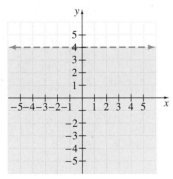

18. $y > 2$

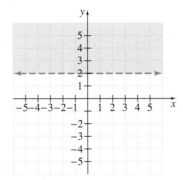

19. $x \ge -3$

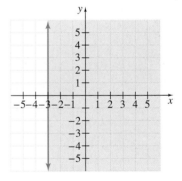

20. $x \le -1$

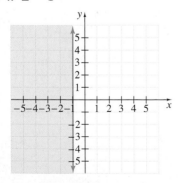

21. $5x + 2y \le 10$

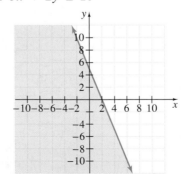

22. $4x + 3y \ge 12$

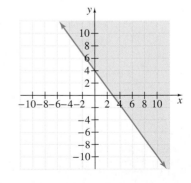

23. $x > y$

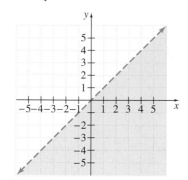

24. $x \le -y$

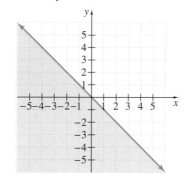

25. $x - y \le 6$

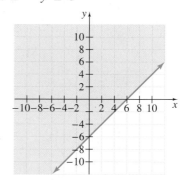

26. $x - y > 10$

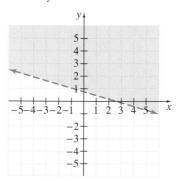

27. $x \ge 0$

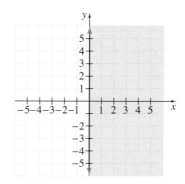

28. $y \le 0$

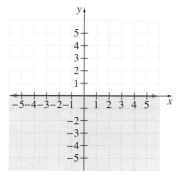

29. $2x + 7y > 5$

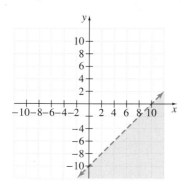

30. $3x + 5y \le -2$

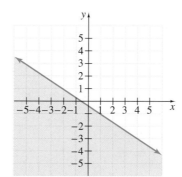

31. $(-2, 1)$

32. $(3, 0)$

33. $(-3, -1)$

34. $(-3, -3)$

35. a

36. c

37. b

38. d

39. answers may vary

516

REVIEW AND PREVIEW

Approximate the coordinates of each point of intersection. See Section 6.1.

31.

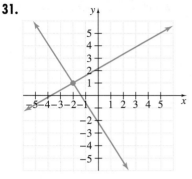

32.

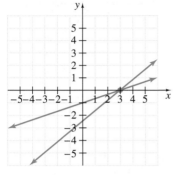

33.

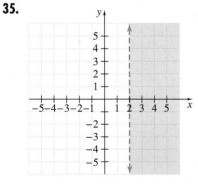

34.

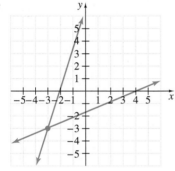

◆ **COMBINING CONCEPTS**

Match each inequality with its graph.

  a. $x > 2$ **b.** $y < 2$ **c.** $y \le 2x$ **d.** $y \le -3x$

35.

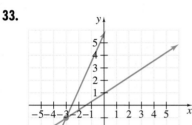

36.

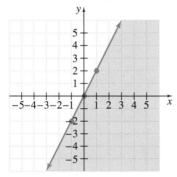

37.

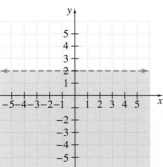

38.

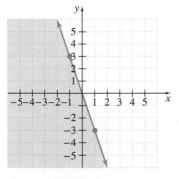

39. Explain why a point on the boundary line should not be chosen as the test point.

CHAPTER 6 ACTIVITY
FINANCIAL ANALYSIS

OPTIONAL MATERIALS:

▲ financial magazines
▲ annual reports

This activity may be completed by working in groups or individually.

Investment analysts must investigate a company's financial data, such as sales, net profit, debt, and assets, to evaluate whether investing in it is a wise choice. One way to analyze such data is to graph the data and identify trends in it visually over time. Another way is to find algebraically the rate at which the data are changing over time.

The table below gives the net profits in millions of dollars for the leading U.S. businesses in the food industry for the years 1995 and 1996. You have been asked to analyze the performances of these companies and, based on this information alone, make an investment recommendation.

Food Industry Net Profits (in millions of dollars)

Company	1995	1996
Archer Daniels Midland	795.9	695.9
Campbell Soup	698.0	802.0
ConAgra	471.6	180.3
General Mills	367.4	476.4
H.J. Heinz	591.0	659.3
IBP	257.9	198.7
RJR Nabisco Holdings	501.0	568.0
Sara Lee	776.0	889.0

(*Source:* Morningstar, Inc.)

1. Write the data for each company as two ordered pairs in the form (year, net profit). Assuming that the trends in net profit are linear, use your own graph paper and graph the line represented by the ordered pairs for each company. Describe the trend shown by each graph.
(1995, 795.9); (1996, 695.9)
(1995, 698.0); (1996, 802.0)
(1995, 471.6); (1996, 180.3)
(1995, 367.4); (1996, 476.4)
(1995, 591.0); (1996, 659.3)
(1995, 257.9); (1996, 198.7)
(1995, 501.0); (1996, 568.0)
(1995, 776.0); (1996, 889.0)

2. Find the slope of the line for each company.
Archer Daniels Midland: $m = -100$, Campbell Soup: $m = 104$, ConAgra: $m = -291.3$, General Mills: $m = 109$, H.J.Heinz: $m = 68.3$, IBP: $m = -59.2$, RJR Nabisco Holdings: $m = 67$, Sara Lee: $m = 113$

3. Which of the lines, if any, have positive slopes? What does that mean in this context? Which of the lines, if any, have negative slopes? What does that mean in this context? Campbell Soup, General Mills, H. J. Heinz, RJR Nabisco Holdings, and Sara Lee have positive slopes; profit is increasing. Archer Daniels Midland, ConAgra, and IBP have negative slopes; profit is decreasing.

4. Of these food industry companies, which one(s) would you recommend as an investment choice? Why? answers may vary

5. Do you think it is wise to make a decision after looking at only 2 years of net profits? What other factors do you think should be taken into consideration when making an investment choice? answers may vary

6. (Optional) Use financial magazines and/or company annual reports to find net profit, sales, or revenue information for two different years for two to four companies in the same industry. Analyze the sales and make an investment recommendation.

CHAPTER 6 HIGHLIGHTS

DEFINITIONS AND CONCEPTS	EXAMPLES

SECTION 6.1 THE RECTANGULAR COORDINATE SYSTEM

The **rectangular coordinate system** consists of a plane and a vertical and a horizontal number line intersecting at their 0 coordinate. The vertical number line is called the **y-axis** and the horizontal number line is called the **x-axis**. The point of intersection of the axes is called the **origin**.

To **plot** or **graph** an ordered pair means to find its corresponding point on a rectangular coordinate system.

To plot or graph an ordered pair such as $(3, -2)$, start at the origin. Move 3 units to the right and from there, 2 units down.

To plot or graph $(-3, 4)$; start at the origin. Move 3 units to the left and from there, 4 units up.

An ordered pair is a **solution** of an equation in two variables if replacing the variables with the coordinates of the ordered pair results in a true statement.

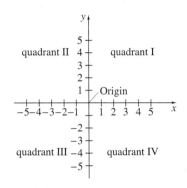

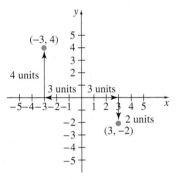

If one coordinate of an ordered pair solution is known, the other value can be determined by substitution.

Complete the ordered pair $(0, \ )$ for the equation $x - 6y = 12$.

$$x - 6y = 12$$
$$0 - 6y = 12 \qquad \text{Let } x = 0.$$
$$\frac{-6y}{-6} = \frac{12}{-6} \qquad \text{Divide by } -6.$$
$$y = -2$$

The ordered pair solution is $(0, -2)$.

SECTION 6.2 GRAPHING LINEAR EQUATIONS

A **linear equation in two variables** is an equation that can be written in the form $Ax + By = C$, where A and B are not both 0. The form $Ax + By = C$ is called **standard form**.

$$3x + 2y = -6 \qquad x = -5$$
$$y = 3 \qquad y = -x + 10$$

$x + y = 10$ is in standard form.

To graph a linear equation in two variables, find three ordered pair solutions. Plot the solution points and draw the line connecting the points.	Graph: $x - 2y = 5$ 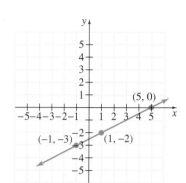

An **intercept point** of a graph is a point where the graph intersects an axis. If a graph intersects the x-axis at a, then a is the **x-intercept** and the corresponding intercept point is $(a, 0)$. If a graph intersects the y-axis at b, then b is the **y-intercept** and the corresponding intercept point is $(0, b)$.

The y-intercept is 3.
Intercept point: $(0, 3)$

The x-intercept is 5.
Intercept point: $(5, 0)$

To find the x-intercept, let $y = 0$ and solve for x.
To find the y-intercept, let $x = 0$ and solve for y.

Find the intercepts for $2x - 5y = -10$.

If $y = 0$, then
$$2x - 5 \cdot 0 = -10$$
$$2x = -10$$
$$\frac{2x}{2} = \frac{-10}{2}$$
$$x = -5$$
The x-intercept is -5.
Intercept point: $(-5, 0)$.

If $x = 0$, then
$$2 \cdot 0 - 5y = -10$$
$$-5y = -10$$
$$\frac{-5y}{-5} = \frac{-10}{-5}$$
$$y = 2$$
The y-intercept is 2.
Intercept point: $(0, 2)$.

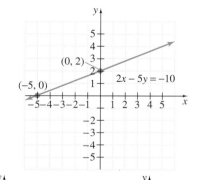

The graph of $x = c$ is a vertical line with x-intercept c.
The graph of $y = c$ is a horizontal line with y-intercept c.

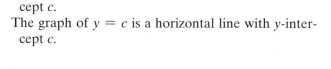

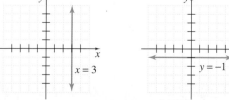

SECTION 6.4   SLOPE

The **slope m** of the line through points (x_1, y_1) and (x_2, y_2) is given by

$$m = \frac{y_2 - y_1}{x_2 - x_1} \quad \text{as long as } x_2 \neq x_1$$

A horizontal line has slope 0.
The slope of a vertical line is undefined.
Nonvertical parallel lines have the same slope.
Two nonvertical lines are perpendicular if the slope of one is the negative reciprocal of the slope of the other.

The slope of the line through points $(-1, 6)$ and $(-5, 8)$ is

$$m = \frac{y_2 - y_1}{x_2 - x_1} = \frac{8 - 6}{-5 - (-1)} = \frac{2}{-4} = -\frac{1}{2}$$

The slope of the line $y = -5$ is 0.
The line $x = 3$ has undefined slope.

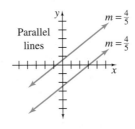

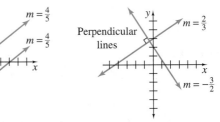

SECTION 6.5   EQUATION OF LINES

SLOPE-INTERCEPT FORM

$$y = mx + b$$

m is the slope of the line.
b is the y-intercept.

Find the slope and the y-intercept of the line $2x + 3y = 6$.
Solve for y: $2x + 3y = 6$

$$3y = -2x + 6 \qquad \text{Subtract } 2x.$$
$$y = -\frac{2}{3}x + 2 \qquad \text{Divide by 3.}$$

The slope of the line is $-\frac{2}{3}$ and the y-intercept is 2.

POINT-SLOPE FORM

$$y - y_1 = m(x - x_1)$$

m is the slope.
(x_1, y_1) is a point of the line.

Find an equation of the line with slope $\frac{3}{4}$ that contains the point $(-1, 5)$.

$$y - 5 = \frac{3}{4}(x - (-1))$$
$$4(y - 5) = 3(x + 1) \qquad \text{Multiply by 4.}$$
$$4y - 20 = 3x + 3 \qquad \text{Distribute.}$$
$$-3x + 4y = 23 \qquad \text{Subtract } 3x \text{ and add 20.}$$

SECTION 6.6   INTRODUCTION TO FUNCTIONS

A set of ordered pairs is a **relation**. The set of all x-coordinates is called the **domain** of the relation and the set of all y-coordinates is called the **range** of the relation.

The domain of the relation

$$\{(0, 5), (2, 5), (4, 5), (5, -2)\}$$

is $\{0, 2, 4, 5\}$. The range is $\{-2, 5\}$.

A **function** is a set of ordered pairs that assigns to each *x*-value exactly one *y*-value.

VERTICAL LINE TEST

If a vertical line can be drawn so that it intersects a graph more than once, the graph is not the graph of a function.

The symbol $f(x)$ means **function of x**. This notation is called **function** notation.

Which are graphs of functions?

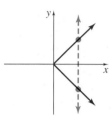 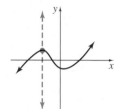

This graph is not the graph of a function.   This graph is the graph of a function.

If $f(x) = 3x - 7$, then

$$f(-1) = 3(-1) - 7$$
$$= -3 - 7$$
$$= -10$$

SECTION 6.7   GRAPHING LINEAR INEQUALITIES IN TWO VARIABLES

A **linear inequality in two variables** is an inequality that can be written in one of these forms:

$$Ax + By < C \qquad Ax + By \leq C$$
$$Ax + By > C \qquad Ax + By \geq C$$

TO GRAPH A LINEAR INEQUALITY

1. Graph the boundary line by graphing the related equation. Draw the line solid if the inequality symbol is $\leq$ or $\geq$. Draw the line dashed if the inequality symbol is $<$ or $>$.
2. Choose a test point not on the line. Substitute its coordinates into the original inequality.
3. If the resulting inequality is true, shade the half-plane that contains the test point. If the inequality is not true, shade the half-plane that does not contain the test point.

$$2x - 5y < 6 \qquad x \geq -5$$
$$y > -8x \qquad y \leq 2$$

Graph: $2x - y \leq 4$
1. Graph $2x - y = 4$. Draw a solid line because the inequality symbol is $\leq$.
2. Check the test point $(0, 0)$ in the original inequality, $2x - y \leq 4$.

$$2 \cdot 0 - 0 \leq 4 \qquad \text{Let } x = 0 \text{ and } y = 0.$$
$$0 \leq 4 \qquad \text{True.}$$

3. The inequality is true, so shade the half-plane containing $(0, 0)$ as shown.

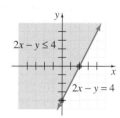

521

CHAPTER 6 REVIEW

(6.1) *Plot each pair on the same rectangular coordinate system.*

1. $(-7, 0)$

2. $\left(0, 4\frac{4}{5}\right)$

3. $(-2, -5)$

4. $(1, -3)$

5. $(0.7, 0.7)$

6. $(-6, 4)$

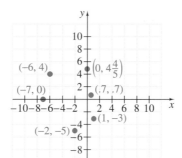

Complete each ordered pair so that it is a solution of the given equation.

7. $-2 + y = 6x; (7, \quad)$ $(7, 44)$

8. $y = 3x + 5; (\quad , -8)$ $\left(-\frac{13}{3}, -8\right)$

Complete the table of values for each given equation.

9. $9 = -3x + 4y$

x	y
	0
	3
9	

$(-3, 0)$
$(1, 3)$
$(9, 9)$

10. $y = 5$

x	y
7	
-7	
0	

$(7, 5)$
$(-7, 5)$
$(0, 5)$

11. $x = 2y$

x	y
	0
	5
	-5

$(0, 0)$
$(10, 5)$
$(-10, -5)$

12. The cost in dollars of producing x compact disc holders is given by $y = 5x + 2000$.
  a. Complete the table.

x	1	100	1000
y	2005	2500	7000

  b. Find the number of compact disc holders that can be produced for $6430.   886 compact discs

(6.2) *Graph each linear equation.*

13. $x - y = 1$

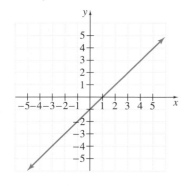

14. $x + y = 6$

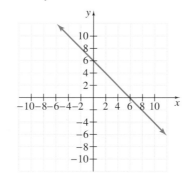

15. $x - 3y = 12$

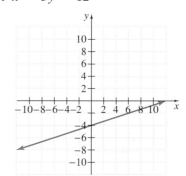

16. $5x - y = -8$

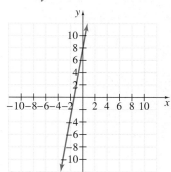

17. $x = 3y$

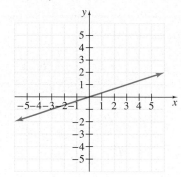

18. $y = -2x$

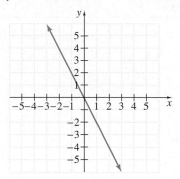

(6.3) *Identify the intercepts and intercept points in each graph.*

19.

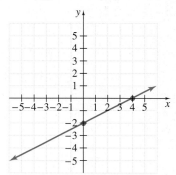

$x = 4; y = -2 ; (4, 0); (0, -2)$

20.

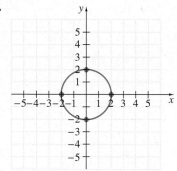

$x = -2; x = 2; y = 2; y = -2; (-2, 0); (2, 0);$
$(0, 2); (0, -2)$

Graph each linear equation.

21. $y = -3$

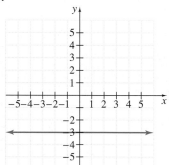

22. $x = 5$

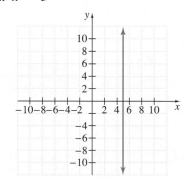

Find the intercept points for each equation.

23. $x - 3y = 12$ $(0, -4)\,(12, 0)$

24. $-4x + y = 8$ $(-2, 0)\,(0, 8)$

(6.4) *Find the slope of each line.*

25.

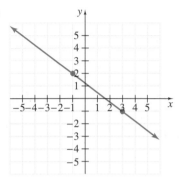

$m = -\dfrac{3}{4}$

26.

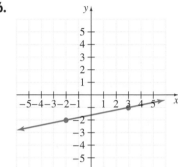

$m = \dfrac{1}{5}$

Match each line with its slope.

a.

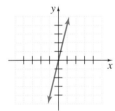

b.

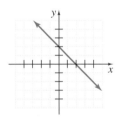

c.

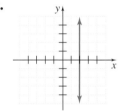

d.

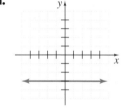

27. $m = 0$ d

28. $m = -1$ b

29. undefined slope   c

30. $m = 4$ a

Find the slope of the line that passes through each pair of points.

31. $(2, 5)$ and $(6, 8)$ $\dfrac{3}{4}$

32. $(4, 7)$ and $(1, 2)$ $\dfrac{5}{3}$

33. $(1, 3)$ and $(-2, -9)$ 4

34. $(-4, 1)$ and $(3, -6)$ -1

Find the slope of each line.

35. $y = 3x + 7$ 3

36. $x - 2y = 4$ $\dfrac{1}{2}$

37. $y = -2$ 0

38. $x = 0$ undefined slope

Determine whether each pair of lines are parallel, perpendicular, or neither.

39. $x - y = -6$ perpendicular
$\quad\;\; x + y = 3$

40. $3x + y = 7$ parallel
$\quad\;\; -3x - y = 10$

41. $y = 4x + \dfrac{1}{2}$
$\quad\;\; 4x + 2y = 1$ neither

(6.5) *Determine the slope and the y-intercept of the graph of each equation.*

42. $3x + y = 7$ $m = -3; b = 7$

43. $x - 6y = -1$ $m = \dfrac{1}{6}; b = \dfrac{1}{6}$

Write an equation of each line.

44. slope -5; y-intercept $\dfrac{1}{2}$ $y = -5x + \dfrac{1}{2}$

45. slope $\dfrac{2}{3}$; y-intercept 6 $y = \dfrac{2}{3}x + 6$

Match each equation with its graph.

46. $y = 2x + 1$ D   **47.** $y = -4x$ C   **48.** $y = 2x$ A   **49.** $y = 2x - 1$ B

A
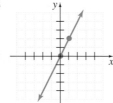

B
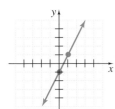

C
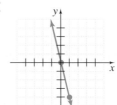

D
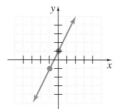

Write an equation of each line with the given slope that passes through the given point. Write the equation in the form $Ax + By = C.$

50. $m = 4; (2, 0)$ $-4x + y = -8$

51. $m = -3; (0, -5)$ $3x + y = -5$

52. $m = \dfrac{3}{5}$; $(1, 4)$ $-3x + 5y = 17$ **53.** $m = -\dfrac{1}{3}$; $(-3, 3)$ $x + 3y = 6$

Write an equation of the line passing through each pair of points. Write the equation in the form $Ax + By = C$.

54. $(1, 7)$ and $(2, -7)$ $14x + y = 21$ **55.** $(-2, 5)$ and $(-4, 6)$ $x + 2y = 8$

(6.6) *Determine whether each relation or graph is a function.*

56. $\{(7, 1), (7, 5), (2, 6)\}$ no   **57.** $\{(0, -1), (5, -1), (2, 2)\}$ yes

58.   yes   **59.**   yes

60.   no   **61.**   yes

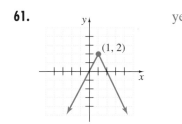

Find the indicated function value for the function.

62. Given $f(x) = -2x + 6$, find

   a. $f(0)$ 6   **b.** $f(-2)$ 10   **c.** $f\left(\dfrac{1}{2}\right)$ 5

(6.7) *Graph each inequality.*

63. $x + 6y < 6$

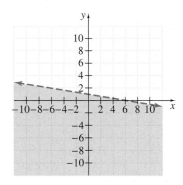

64. $x + y > -2$

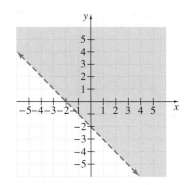

65. $y \geq -7$

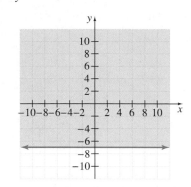

66. $y \leq -4$

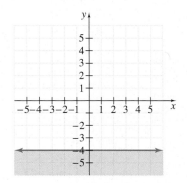

67. $-x \leq y$

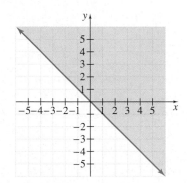

68. $x \geq -y$

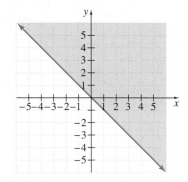

CHAPTER 6 TEST

Complete each ordered pair solution so that it is a solution of the given equation.

1. $12y - 7x = 5; (1, \ )$

2. $y = 17; (-4, \ )$

Find the slope of each line.

3.

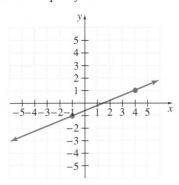

4.

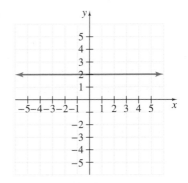

5. Passes through $(6, -5)$ and $(-1, 2)$

6. Passes through $(0, -8)$ and $(-1, -1)$

7. $-3x + y = 5$

8. $x = 6$

Graph.

9. $2x + y = 8$

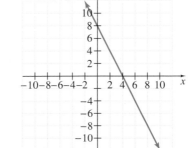

10. $-x + 4y = 5$

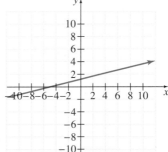

11. $x - y \geq -2$

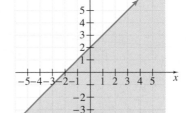

12. $y \geq -4x$

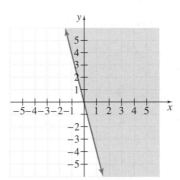

ANSWERS

1. $(1, 1)$

2. $(-4, 17)$

3. $m = \frac{2}{5}$

4. $m = 0$

5. $m = -1$

6. $m = -7$

7. $m = 3$

8. undefined slope

9. see graph

10. see graph

11. see graph

12. see graph

529

13. see graph

14. see graph

15. see graph

16. see graph

17. neither

18. $x + 4y = 10$

19. $7x + 6y = 0$

20. $8x + y = 11$

21. $x - 8y = -96$

22. yes

23. no

24. yes

25. yes

26. a. -8

b. -3.6

c. -4

27. a. 0

b. 0

c. 60

530

13. $5x - 7y = 10$

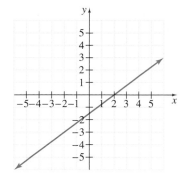

14. $2x - 3y > -6$

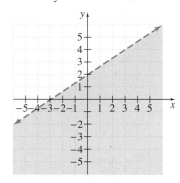

15. $6x + y > -1$

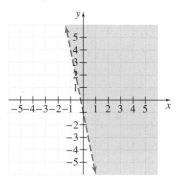

16. $y = -1$

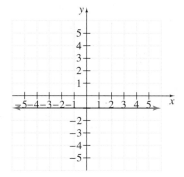

17. Determine whether the graphs of $y = 2x - 6$ and $-4x = 2y$ are parallel lines, perpendicular lines, or neither.

Find the equation of each line. Write the equation in the form $Ax + By = C$.

18. Slope $-\dfrac{1}{4}$, passes through $(2, 2)$

19. Passes through the origin and $(6, -7)$

20. Passes through $(2, -5)$ and $(1, 3)$

21. Slope $\dfrac{1}{8}$; y-intercept 12

Determine whether each relation is a function.

22. $\{(-1, 2), (-2, 4), (-3, 6), (-4, 8)\}$

23. $\{(-3, -3), (0, 5), (-3, 2), (0, 0)\}$

24. The graph shown in Exercise 3.

25. The graph shown in Exercise 4.

Find the indicated function values for each function.

26. $f(x) = 2x - 4$
  a. $f(-2)$

  b. $f(0.2)$

  c. $f(0)$

27. $f(x) = x^3 - x$
  a. $f(-1)$

  b. $f(0)$

  c. $f(4)$

CUMULATIVE REVIEW

Simplify each expression.

1. $6 \div 3 + 5^2$

2. $3[4(5 + 2) - 10]$

3. The bar graph shows Disney's top eight animated films and the amount of money they generated at theaters.

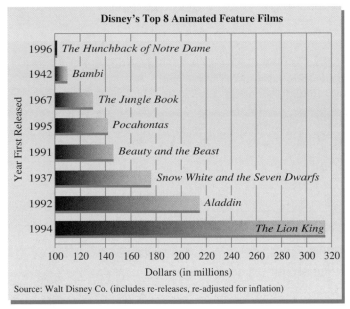

Disney's Top 8 Animated Feature Films

Year First Released

1996 | *The Hunchback of Notre Dame*
1942 | *Bambi*
1967 | *The Jungle Book*
1995 | *Pocahontas*
1991 | *Beauty and the Beast*
1937 | *Snow White and the Seven Dwarfs*
1992 | *Aladdin*
1994 | *The Lion King*

100 120 140 160 180 200 220 240 260 280 300 320
Dollars (in millions)

Source: Walt Disney Co. (includes re-releases, re-adjusted for inflation)

a. Find the film shown that generated the most income for Disney and approximate the income.

b. How much more money did the film *Aladdin* make than the film *Beauty and the Beast*?

Write each phrase as an algebraic expression and simplify if possible. Let x represent the unknown number.

4. Twice a number, added to 6.

5. The difference of a number and 4, divided by 7.

6. Five added to three times the sum of a number and 1.

7. Solve for x: $\frac{5}{2}x = 15$

8. Solve $2x < -4$. Graph the solutions.

-5 -4 -3 -2 -1 0 1 2 3 4 5

9. Find the degree of each polynomial and tell whether the polynomial is a monomial, binomial, trinomial, or none of these.

a. $-2t^2 + 3t + 6$

b. $15x - 10$

c. $7x + 3x^3 + 2x^2 - 1$

1. 27 (Sec. 1.2, Ex. 2)

2. 54 (Sec. 1.2, Ex. 4)

3. a. The Lion King; $315 million

b. $70 million (Sec. 1.8, Ex. 2)

4. $2x + 6$ (Sec. 2.1, Ex. 15)

5. $(x - 4) \div 7$ (Sec. 2.1, Ex. 16)

6. $8 + 3x$ (Sec. 2.1, Ex. 17)

7. $x = 6$ (Sec. 2.3, Ex. 1)

8. $x < -2$ (Sec. 2.8, Ex. 5)

9. a. 2; trinomial

b. 1; binomial

c. 3; none of these (Sec. 3.3, Ex. 3)

531

10. $\dfrac{-4x^2 + 6x + 2}{\text{(Sec. 3.4, Ex. 2)}}$

11. $\dfrac{9y^2 + 6y + 1}{\text{(Sec. 3.6, Ex. 4)}}$

12. $\dfrac{3a(-3a^4 + 6a - 1)}{\text{(Sec. 4.1, Ex. 3)}}$

13. $\dfrac{(x - 2)(x + 6)}{\text{(Sec. 4.2, Ex. 3)}}$

14. $\dfrac{(4x - 1)(2x - 5)}{\text{(Sec. 4.3, Ex. 2)}}$

15. $\dfrac{x = 4, x = 5}{\text{(Sec. 4.6, Ex. 5)}}$

16. $\dfrac{1 \text{ (Sec. 5.2, Ex. 7)}}{}$

17. $\dfrac{\frac{12ab^2}{27a^2b} \text{ (Sec. 5.3, Ex. 9)}}{}$

18. $\dfrac{\frac{2m + 1}{m + 1} \text{ (Sec. 5.4, Ex. 5)}}{}$

19. $\dfrac{x = -3, x = -2}{\text{(Sec. 5.5, Ex. 3)}}$

20. $\dfrac{\frac{x + 1}{x + 2y} \text{ (Sec. 5.7, Ex. 5)}}{}$

21. a. $\dfrac{(0, 12)}{}$

b. $\dfrac{(2, 6)}{}$

c. $\dfrac{(-1, 15)}{\text{(Sec. 6.1, Ex. 3)}}$

22. $\dfrac{}{\text{(Sec. 6.2, Ex. 1)}}$

23. $\dfrac{\frac{2}{3} \text{ (Sec. 6.4, Ex. 3)}}{}$

24. $\dfrac{2x + y = 3}{\text{(Sec. 6.5, Ex. 4)}}$

25. a. $\dfrac{1; (2, 1)}{}$

b. $\dfrac{1; (-2, 1)}{}$

c. $\dfrac{-3; (0, -3)}{\text{(Sec. 6.6, Ex. 6)}}$

532

Name _____

10. Add: $(-2x^2 + 5x - 1)$ and $(-2x^2 + x + 3)$

11. Multiply: $(3y + 1)^2$

12. Factor: $-9a^5 + 18a^2 - 3a$

13. Factor: $x^2 + 4x - 12$

14. Factor: $8x^2 - 22x + 5$

15. Solve: $x^2 - 9x = -20$

16. Divide: $\dfrac{2x^2 - 11x + 5}{5x - 25} \div \dfrac{4x - 2}{10}$

17. Write the rational expression as an equivalent rational expression with the given denominator.
$$\frac{4b}{9a} = \frac{}{27a^2b}$$

18. Add: $1 + \dfrac{m}{m + 1}$

19. Solve: $3 - \dfrac{6}{x} = x + 8$

20. Simplify: $\dfrac{\dfrac{x + 1}{y}}{\dfrac{x}{y} + 2}$

21. Complete each ordered pair solution so that it is a solution of the equation $3x + y = 12$.

a. $(0, \ \ )$

b. $(\ \ , 6)$

c. $(-1, \ \ )$

22. Graph: $2x + y = 5$

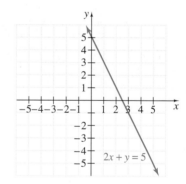

23. Find the slope of the line $-2x + 3y = 11$.

24. Find an equation of the line passing through $(-1, 5)$ with slope -2. Write the equation in the form $Ax + By = C$.

25. Given $f(x) = x^2 - 3$, find each function value and list the corresponding ordered pair.

a. $f(2)$

b. $f(-2)$

c. $f(0)$

Systems of Equations

In Chapter 6, we graphed equations containing two variables. As we have seen, equations like these are often needed to represent relationships between two different quantities. There are also many opportunities to compare and contrast two such equations, called a **system of equations**. This chapter presents **linear systems** and ways we solve these systems and apply them to real-life situations.

Today, television is a fact of life. In 1946, a year after World War II ended, the television industry came to life. Broadcasting was initially dominated by two radio companies, Columbia Broadcasting System (CBS) and National Broadcasting Company (NBC). Originally, the Federal Communications Commission (FCC) restricted television broadcasts in the U.S. to the very high frequency (VHF) channels 2–13. The use of channels 14–83, the ultra-high frequency (UHF) channels, was banned. Competition for the VHF channels was fierce, and together three major networks, ABC (American Broadcasting Company), CBS, and NBC, controlled 95% of viewership well into the 1970s. The FCC gradually began making UHF channels available for broadcasting and the number of television stations (particularly UHF stations) exploded. Exercise 45 on page 565 uses a system of equations to find the year in which the number of UHF stations caught up to and surpassed the number of VHF stations in the U.S.

CHAPTER 7 PRETEST

Solve each system by graphing.

1. $\begin{cases} x + y = 5 \\ x - y = 7 \end{cases}$

2. $\begin{cases} y = 4x \\ x = 1 \end{cases}$

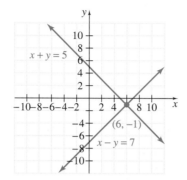

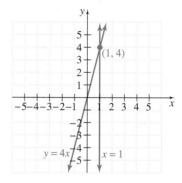

Solve each system using the substitution method.

3. $\begin{cases} x + y = 6 \\ x = 3y - 2 \end{cases}$

4. $\begin{cases} 5x + y = 13 \\ 4x - 5y = 22 \end{cases}$

5. $\begin{cases} 7y = x - 6 \\ 2x + 3y = -5 \end{cases}$

6. $\begin{cases} 4x = y + 6 \\ 8x - 2y = 12 \end{cases}$

7. $\begin{cases} x - 5 = 3y \\ 6y - 2x = 10 \end{cases}$

8. $\begin{cases} 4x + 6y = -14 \\ 6x + y = -1 \end{cases}$

9. $\begin{cases} \dfrac{1}{5}x - y = 3 \\ x - 5y = 15 \end{cases}$

10. $\begin{cases} y = 3x + 7 \\ y = 10x + 21 \end{cases}$

Solve each system by the addition method.

11. $\begin{cases} 2x + y = 11 \\ 3x - y = 29 \end{cases}$

12. $\begin{cases} 4x - 3y = 13 \\ 5x - 9y = 53 \end{cases}$

13. $\begin{cases} 6x + 8y = 92 \\ 5x - 3y = 9 \end{cases}$

14. $\begin{cases} 3x - 4y = 7 \\ -9x + 12y = 21 \end{cases}$

15. $\begin{cases} \dfrac{x}{2} + \dfrac{y}{3} = 2 \\ \dfrac{x}{6} - \dfrac{y}{4} = 5 \end{cases}$

16. $\begin{cases} 6x + 10y = -4 \\ -x + y = -1 \end{cases}$

17. $\begin{cases} 2x = 8 - 3y \\ 9y = 24 - 6x \end{cases}$

18. $\begin{cases} 11x = 5y + 30 \\ 3x + 4y = -24 \end{cases}$

Solve.

19. Two numbers have a sum of 97 and a difference of 65. Find the numbers.

20. Find the measures of two complementary angles if one angle is 6° less than twice the other.

Answers (side column)

1. see graph (7.1B)

2. see graph (7.1B)

3. $(4, 2)$; (7.2A)

4. $(3, -2)$; (7.2A)

5. $(-1, -1)$; (7.2A)

6. infinite number of solutions; (7.2A)

7. no solution; (7.2A)

8. $\left(\dfrac{1}{4}, -\dfrac{5}{2}\right)$; (7.2A)

9. infinite number of solutions; (7.2A)

10. $(-2, 1)$; (7.2A)

11. $(8, -5)$; (7.3A)

12. $(-2, -7)$; (7.3A)

13. $(6, 7)$; (7.3A)

14. no solution; (7.3A)

15. $(12, -12)$; (7.3A)

16. $\left(\dfrac{3}{8}, -\dfrac{5}{8}\right)$; (7.3A)

17. infinite number of solutions; (7.3A)

18. $(0, -6)$; (7.3A)

19. 81 and 16; (7.4A)

20. 32° and 58°; (7.4A)

7.1   SOLVING SYSTEMS OF LINEAR EQUATIONS BY GRAPHING

A **system of linear equations** consists of two or more linear equations. In this section, we focus on solving systems of linear equations containing two equations in two variables. Examples of such linear systems are

$$\begin{cases} 3x - 3y = 0 \\ x = 2y \end{cases} \quad \begin{cases} x - y = 0 \\ 2x + y = 10 \end{cases} \quad \begin{cases} y = 7x - 1 \\ y = 4 \end{cases}$$

A   DECIDING WHETHER AN ORDERED PAIR IS A SOLUTION

A **solution** of a system of two equations in two variables is an ordered pair of numbers that is a solution of both equations in the system.

Example 1   Determine whether $(12, 6)$ is a solution of the system

$$\begin{cases} 2x - 3y = 6 \\ x = 2y \end{cases}$$

Solution: To determine whether $(12, 6)$ is a solution of the system, we replace x with 12 and y with 6 in both equations.

$2x - 3y = 6$	First equation	$x = 2y$	Second equation
$2(12) - 3(6) \stackrel{?}{=} 6$	Let $x = 12$ and $y = 6$.	$12 \stackrel{?}{=} 2(6)$	Let $x = 12$ and $y = 6$.
$24 - 18 \stackrel{?}{=} 6$	Simplify.	$12 = 12$	True.
$6 = 6$	True.		

Since $(12, 6)$ is a solution of both equations, it is a solution of the system.   ▬▬▬

Example 2   Determine whether $(-1, 2)$ is a solution of the system

$$\begin{cases} x + 2y = 3 \\ 4x - y = 6 \end{cases}$$

Solution: We replace x with -1 and y with 2 in both equations.

$x + 2y = 3$	First equation	$4x - y = 6$	Second equation
$-1 + 2(2) \stackrel{?}{=} 3$	Let $x = -1$ and $y = 2$.	$4(-1) - 2 \stackrel{?}{=} 6$	Let $x = -1$ and $y = 2$.
$-1 + 4 \stackrel{?}{=} 3$	Simplify.	$-4 - 2 \stackrel{?}{=} 6$	Simplify.
$3 = 3$	True.	$-6 = 6$	False.

$(-1, 2)$ is not a solution of the second equation, $4x - y = 6$, so it is not a solution of the system.   ▬▬▬

B   SOLVING SYSTEMS OF EQUATIONS BY GRAPHING

Since a solution of a system of two equations in two variables is a solution common to both equations, it is also a point common to the graphs of both equations. Let's practice finding solutions of both equations in a system— that is, solutions of a system—by graphing and identifying points of intersection.

Objectives

A Decide whether an ordered pair is a solution of a system of linear equations.

B Solve a system of linear equations by graphing.

C Identify special systems: those with no solution and those with an infinite number of solutions.

SSM   CD-ROM   Video 7.1

Practice Problem 1

Determine whether $(3, 9)$ is a solution of the system

$$\begin{cases} 5x - 2y = -3 \\ y = 3x \end{cases}$$

TEACHING TIP

Please see page 539.

Practice Problem 2

Determine whether $(3, -2)$ is a solution of the system

$$\begin{cases} 2x - y = 8 \\ x + 3y = 4 \end{cases}$$

TEACHING TIP

If the desks in your classroom are arranged in columns and rows, consider beginning this lesson with a student demonstration. Tell students you will use the floor of the classroom as a coordinate plane with the back right corner as the origin and the rows and columns as unit measures of y and x, respectively. Then have students representing $y = x$ stand up, that is, the students whose row value equals their column value. While they remain standing, have students representing $y = 3$ stand up, that is, the students in row 3. Then ask if anyone is a member of both equations. What happens at that point?

Answers

1. $(3, 9)$ is a solution of the system,   **2.** $(3, -2)$ is not a solution of the system

Practice Problem 3

Solve the system of equations by graphing

$$\begin{cases} -3x + y = -10 \\ x - y = 6 \end{cases}$$

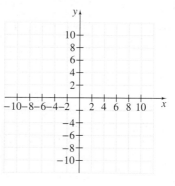

Example 3

Solve the system of equations by graphing.

$$\begin{cases} -x + 3y = 10 \\ x + y = 2 \end{cases}$$

Solution: On a single set of axes, graph each linear equation.

x	y
0	$\frac{10}{3}$
-4	2
2	4

x	y
0	2
2	0
1	1

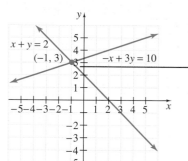

HELPFUL HINT

The point of intersection gives the solution of the system.

The two lines appear to intersect at the point $(-1, 3)$. To check, we replace x with -1 and y with 3 in both equations.

$-x + 3y = 10$	First equation	
$-(-1) + 3(3) \overset{?}{=} 10$	Let $x = -1$ and $y = 3$.	
$1 + 9 \overset{?}{=} 10$	Simplify.	
$10 = 10$	True.	

$x + y = 2$	Second equation	
$-1 + 3 \overset{?}{=} 2$	Let $x = -1$ and $y = 3$.	
$2 = 2$	True.	

$(-1, 3)$ checks, so it is the solution of the system.

HELPFUL HINT

Neatly drawn graphs can help when "guessing" the solution of a system of linear equations by graphing.

Practice Problem 4

Solve the system of equations by graphing.

$$\begin{cases} x + 3y = -1 \\ y = 1 \end{cases}$$

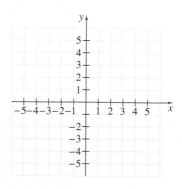

Example 4

Solve the system of equations by graphing.

$$\begin{cases} 2x + 3y = -2 \\ x = 2 \end{cases}$$

Solution: We graph each linear equation on a single set of axes.

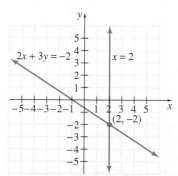

The two lines appear to intersect at the point $(2, -2)$. To determine whether $(2, -2)$ is the solution, we replace x with 2 and y with -2 in both equations.

Answers

3. please see page 538,   **4.** please see page 538

$$2x + 3y = -2 \qquad \text{First equation}$$
$$2(2) + 3(-2) \stackrel{?}{=} -2 \qquad \text{Let } x = 2 \text{ and } y = -2.$$
$$4 + (-6) \stackrel{?}{=} -2 \qquad \text{Simplify.}$$
$$-2 = -2 \qquad \text{True.}$$

$$x = 2 \qquad \text{Second equation}$$
$$2 = 2 \qquad \text{Let } x = 2.$$
$$2 = 2 \qquad \text{True.}$$

Since a true statement results in both equations, $(2, -2)$ is the solution of the system.

C   IDENTIFYING SPECIAL SYSTEMS OF LINEAR EQUATIONS

Not all systems of linear equations have a single solution. Some systems have no solution and some have an infinite number of solutions.

Example 5   Solve the system of equations by graphing.

$$\begin{cases} 2x + y = 7 \\ 2y = -4x \end{cases}$$

Solution: We graph the two equations in the system. The equations in slope-intercept form are $y = -2x + 7$ and $y = -2x$. Notice from the equations that the lines have the same slope, -2, and different y-intercepts. This means that the lines are parallel.

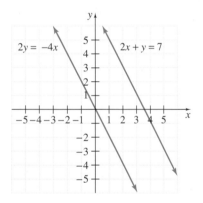

Since the lines are parallel, they do not intersect. This means that the system has *no solution*.

Example 6   Solve the system of equations by graphing.

$$\begin{cases} x - y = 3 \\ -x + y = -3 \end{cases}$$

Solution: We graph each equation.

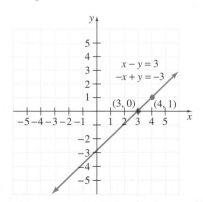

TEACHING TIP

Before discussing Example 5, do a student demonstration using the floor of the classroom as a coordinate plane with the back right corner as the origin. Have students representing $x = 2$ stand up and students representing $x = 4$ stand up. Then ask, "What is the solution of this system of equations?" Follow up by asking, "Is it possible for these two equations to intersect if we had a bigger room with more desks?" Why not?

Practice Problem 5

Solve the system of equations by graphing.

$$\begin{cases} 3x - y = 6 \\ 6x = 2y \end{cases}$$

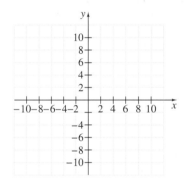

TEACHING TIP

Please see page 539.

Practice Problem 6

Solve the system of equations by graphing.

$$\begin{cases} 3x + 4y = 12 \\ 9x + 12y = 36 \end{cases}$$

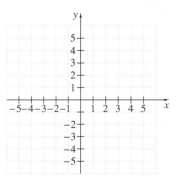

Answers

5. please see page 538,   **6.** please see page 538.

Answers

3. $(2, -4)$

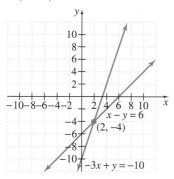

4. $(-4, 1)$

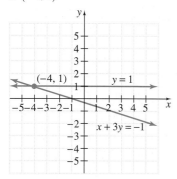

5. no solution

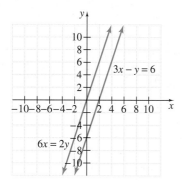

6. infinite number of solutions

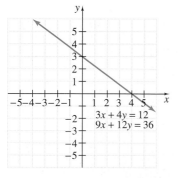

The graphs of the equations are the same line. To see this, notice that if both sides of the first equation in the system are multiplied by -1, the result is the second equation.

$$x - y = 3 \qquad \text{First equation}$$
$$-1(x - y) = -1(3) \qquad \text{Multiply both sides by } -1.$$
$$-x + y = -3 \qquad \text{Simplify. This is the second equation.}$$

This means that the system has an infinite number of solutions. Any ordered pair that is a solution of one equation is a solution of the other and is then a solution of the system.

Examples 5 and 6 are special cases of systems of linear equations. A system that has no solution is said to be an **inconsistent system**. If the graphs of the two equations of a system are identical—we call the equations **dependent equations**.

As we have seen, three different situations can occur when graphing the two lines associated with the equations in a linear system. These situations are shown in the figures.

One point of intersection: one solution

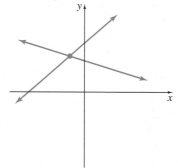

Parallel lines: no solution

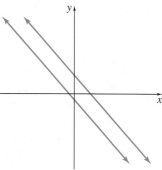

Same line: infinite number of solutions

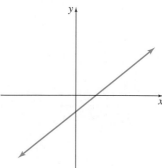

GRAPHING CALCULATOR EXPLORATIONS

A graphing calculator may be used to approximate solutions of systems of equations. For example, to approximate the solution of the system

$$\begin{cases} y = -3.14x - 1.35 \\ y = 4.88x + 5.25, \end{cases}$$

first graph each equation on the same set of axes. Then use the intersect feature of your calculator to approximate the point of intersection.

The approximate point of intersection is $(-0.82, 1.23)$.

Solve each system of equations. Approximate the solutions to two decimal places.

1. $\begin{cases} y = -2.68x + 1.21 \\ y = 5.22x - 1.68 \end{cases}$ $(0.37, 0.23)$

2. $\begin{cases} y = 4.25x + 3.89 \\ y = -1.88x + 3.21 \end{cases}$ $(-0.11, 3.42)$

3. $\begin{cases} 4.3x - 2.9y = 5.6 \\ 8.1x + 7.6y = -14.1 \end{cases}$ $(0.03, -1.89)$

4. $\begin{cases} -3.6x - 8.6y = 10 \\ -4.5x + 9.6y = -7.7 \end{cases}$ $(-0.41, -0.99)$

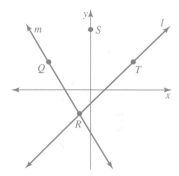

MENTAL MATH

Each rectangular coordinate system shows the graph of the equations in a system of equations. Use each graph to determine the number of solutions for each associated system. If the system has only one solution, give its coordinates.

1.

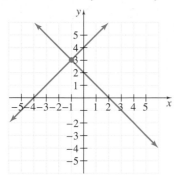

1 solution, $(-1, 3)$

2.

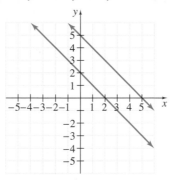

no solution

3.

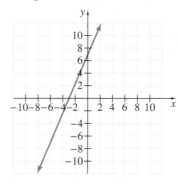

infinite number of solutions

4.

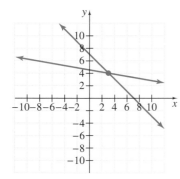

1 solution, $(3, 4)$

5.

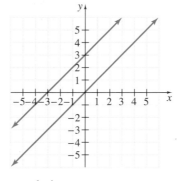

no solution

6.

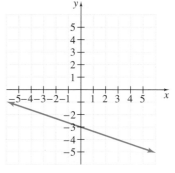

infinite number of solutions

7.

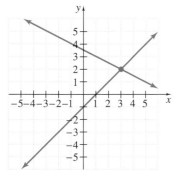

1 solution, $(3, 2)$

8.

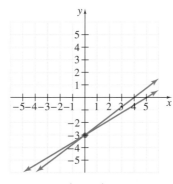

1 solution, $(0, -3)$

EXERCISE SET 7.1

A *Determine whether each ordered pair is a solution of the system of linear equations. See Examples 1 and 2.*

1. $\begin{cases} x + y = 8 \\ 3x + 2y = 21 \end{cases}$

  a. $(2, 4)$

  b. $(5, 3)$

2. $\begin{cases} 2x + y = 5 \\ x + 3y = 5 \end{cases}$

  a. $(5, 0)$

  b. $(2, 1)$

3. $\begin{cases} 3x - y = 5 \\ x + 2y = 11 \end{cases}$

  a. $(3, 4)$

  b. $(0, -5)$

ANSWERS

1. a. no

  b. yes

2. a. no

  b. yes

3. a. yes

  b. no

4. a. yes

b. no

5. a. yes

b. yes

6. a. no

b. no

4. $\begin{cases} 2x - 3y = 8 \\ x - 2y = 6 \end{cases}$

a. $(-2, -4)$

b. $(7, 2)$

5. $\begin{cases} 2y = 4x \\ 2x - y = 0 \end{cases}$

a. $(-3, -6)$

b. $(0, 0)$

6. $\begin{cases} 4x = 1 - y \\ x - 3y = -8 \end{cases}$

a. $(0, 1)$

b. $(1, -3)$

B C *Solve each system of linear equations by graphing. See Examples 3 through 6.*

7. $\begin{cases} x + y = 4 \\ x - y = 2 \end{cases}$

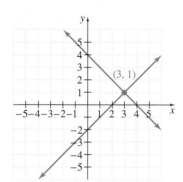

8. $\begin{cases} x + y = 3 \\ x - y = 5 \end{cases}$

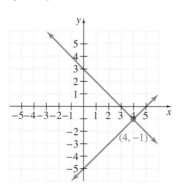

9. $\begin{cases} x + y = 6 \\ -x + y = -6 \end{cases}$

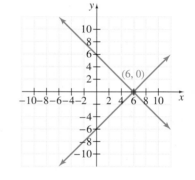

10. $\begin{cases} x + y = 1 \\ -x + y = -3 \end{cases}$

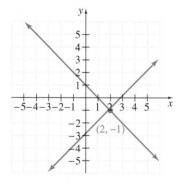

11. $\begin{cases} y = 2x \\ 3x - y = -2 \end{cases}$

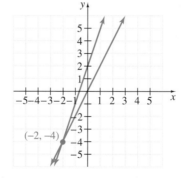

12. $\begin{cases} y = -3x \\ 2x - y = -5 \end{cases}$

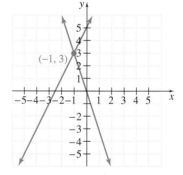

13. $\begin{cases} y = x + 1 \\ y = 2x - 1 \end{cases}$

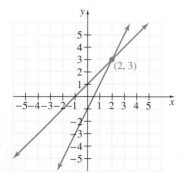

14. $\begin{cases} y = 3x - 4 \\ y = x + 2 \end{cases}$

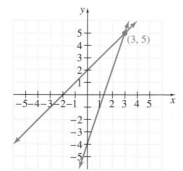

15. $\begin{cases} 2x + y = 0 \\ 3x + y = 1 \end{cases}$

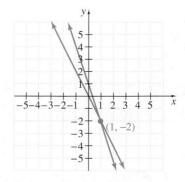

16. $\begin{cases} 2x + y = 1 \\ 3x + y = 0 \end{cases}$

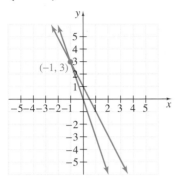

17. $\begin{cases} y = -x - 1 \\ y = 2x + 5 \end{cases}$

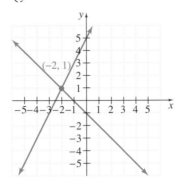

18. $\begin{cases} y = x - 1 \\ y = -3x - 5 \end{cases}$

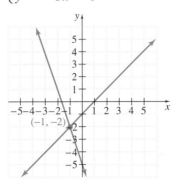

19. $\begin{cases} 2x - y = 6 \\ y = 2 \end{cases}$

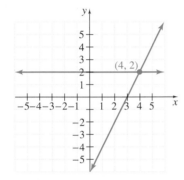

20. $\begin{cases} x + y = 5 \\ x = 4 \end{cases}$

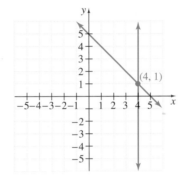

21. $\begin{cases} x + y = 5 \\ x + y = 6 \end{cases}$

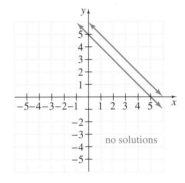

no solutions

22. $\begin{cases} x - y = 4 \\ x - y = 1 \end{cases}$

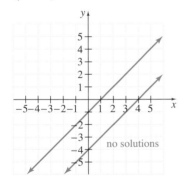

no solutions

23. $\begin{cases} 2x + y = 4 \\ x + y = 2 \end{cases}$

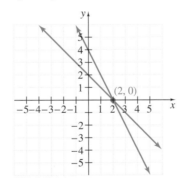

24. $\begin{cases} y + 2x = 3 \\ 4x = 2 - 2y \end{cases}$

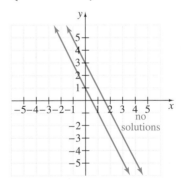

no solutions

25. $\begin{cases} x - 2y = 2 \\ 3x + 2y = -2 \end{cases}$

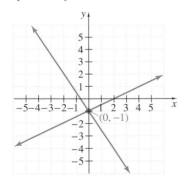

26. $\begin{cases} x + 3y = 7 \\ 2x - 3y = -4 \end{cases}$

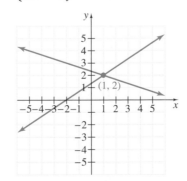

27. $\begin{cases} y - 3x = -2 \\ 6x - 2y = 4 \end{cases}$

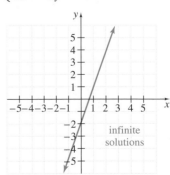

infinite solutions

28. $\begin{cases} x - 2y = -6 \\ -2x + 4y = 12 \end{cases}$

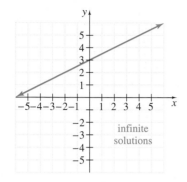

infinite solutions

29. $\begin{cases} x = 3 \\ y = -1 \end{cases}$

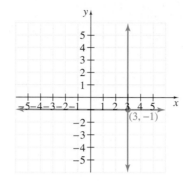

$(3, -1)$

30. $\begin{cases} x = -5 \\ y = 3 \end{cases}$

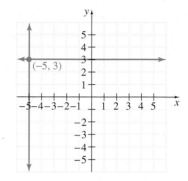

$(-5, 3)$

31. $\begin{cases} y = x - 2 \\ y = 2x + 3 \end{cases}$

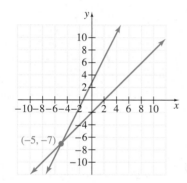

$(-5, -7)$

32. $\begin{cases} y = x + 5 \\ y = -2x - 4 \end{cases}$

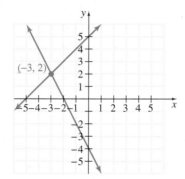

$(-3, 2)$

33. $\begin{cases} 2x - 3y = -2 \\ -3x + 5y = 5 \end{cases}$

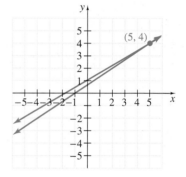

$(5, 4)$

34. $\begin{cases} 4x - y = 7 \\ 2x - 3y = -9 \end{cases}$

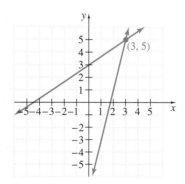

$(3, 5)$

35. Draw a graph of two linear equations whose associated system has the solution $(-1, 4)$.

possible answer:

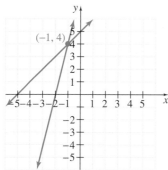

36. Draw a graph of two linear equations whose associated system has the solution $(3, -2)$.

possible answer:

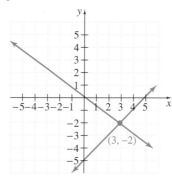

37. Draw a graph of two linear equations whose associated system has no solution.

possible answer:

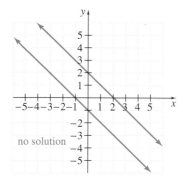

38. Draw a graph of two linear equations whose associated system has an infinite number of solutions.

possible answer:

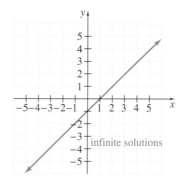

The double line graph below shows the number of pounds of fishery products from U.S. domestic catch and from imports. Use this graph to answer Exercises 39–41.

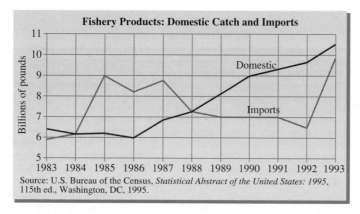

Source: U.S. Bureau of the Census, *Statistical Abstract of the United States: 1995*, 115th ed., Washington, DC, 1995.

39. In what year(s) was the number of pounds of imported fishery products equal to the number of pounds of domestic catch?

40. In what year(s) was the number of pounds of imported fishery products greater than or equal to the number of pounds of domestic catch?

39. 1984, 1988

40. 1984 through 1988

41. a. (4, 9)

b. yes

42. $x = 2$

43. $x = -1$

44. $y = -\dfrac{2}{5}$

45. $y = 3$

46. $a = 2$

47. $z = -7$

48. answers may vary

49. answers may vary

50. answers may vary

51. answers may vary

41. In the next column are tables of values for two linear equations. Using the tables,
  a. find a solution of the corresponding system.

x	y		x	y
1	3		1	6
2	5		2	7
3	7		3	8
4	9		4	9
5	11		5	10

  b. graph several ordered pairs from each table and sketch the two lines.

Does your graph confirm the solution from part (**a**)?

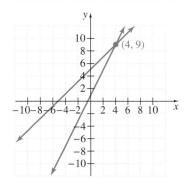

Review and Preview

Solve each equation. See Section 2.4.

42. $5(x - 3) + 3x = 1$

43. $-2x + 3(x + 6) = 17$

44. $4\left(\dfrac{y + 1}{2}\right) + 3y = 0$

45. $-y + 12\left(\dfrac{y - 1}{4}\right) = 3$

46. $8a - 2(3a - 1) = 6$

47. $3z - (4z - 2) = 9$

Combining Concepts

48. Construct a system of two linear equations that has (2, 5) as a solution.

49. Construct a system of two linear equations that has (0, 1) as a solution.

50. The ordered pair $(-2, 3)$ is a solution of all three linear equations:

$$x + y = 1$$
$$2x - y = -7$$
$$x + 3y = 7$$

If each equation has a distinct graph, describe the graph of all three equations on the same axes.

51. Explain how to use a graph to determine the number of solutions of a system.

7.2 Solving Systems of Linear Equations by Substitution

A Using the Substitution Method

You may have suspected by now that graphing alone is not an accurate way to solve a system of linear equations. For example, a solution of $\left(\frac{1}{2}, \frac{2}{9}\right)$ is unlikely to be read correctly from a graph. In this section, we discuss a second, more accurate method for solving systems of equations. This method is called the **substitution method** and is introduced in the next example.

Example 1   Solve the system:

$$\begin{cases} 2x + y = 10 & \text{First equation} \\ x = y + 2 & \text{Second equation} \end{cases}$$

Solution: The second equation in this system is $x = y + 2$. This tells us that x and $y + 2$ have the same value. This means that we may substitute $y + 2$ for x in the first equation.

$$2x + y = 10 \quad \text{First equation}$$

$$2(y + 2) + y = 10 \quad \text{Substitute } y + 2 \text{ for } x \text{ since } x = y + 2.$$

Notice that this equation now has one variable, y. Let's now solve this equation for y.

> **HELPFUL HINT**
>
> Don't forget the distributive property.

$$2(y + 2) + y = 10$$
$$2y + 4 + y = 10 \quad \text{Use the distributive property.}$$
$$3y + 4 = 10 \quad \text{Combine like terms.}$$
$$3y = 6 \quad \text{Subtract 4 from both sides.}$$
$$y = 2 \quad \text{Divide both sides by 3.}$$

Now we know that the y-value of the ordered pair solution of the system is 2. To find the corresponding x-value, we replace y with 2 in the equation $x = y + 2$ and solve for x.

$$x = y + 2$$
$$x = 2 + 2 \quad \text{Let } y = 2.$$
$$x = 4$$

The solution of the system is the ordered pair (4, 2). Since an ordered pair solution must satisfy both linear equations in the system, we could have chosen the equation $2x + y = 10$ to find the corresponding x-value. The resulting x-value is the same.

Check: We check to see that (4, 2) satisfies both equations of the original system.

First Equation	**Second Equation**
$2x + y = 10$	$x = y + 2$
$2(4) + 2 \stackrel{?}{=} 10$	$4 \stackrel{?}{=} 2 + 2 \quad$ Let $x = 4$ and $y = 2$.
$10 = 10 \quad$ True.	$4 = 4 \quad$ True.

Objective

A Use the substitution method to solve a system of linear equations.

SSM   CD-ROM   Video 7.2

Practice Problem 1

Use the substitution method to solve the system:

$$\begin{cases} 2x + 3y = 13 \\ x = y + 4 \end{cases}$$

TEACHING TIP

Consider beginning this lesson with the following activity: Graph the system of equations

$$\begin{cases} y = x \\ y = 4x - 7 \end{cases}$$

on graph chart paper or an overhead projector. Ask a student to put a sticky dot on the solution(s) to the system. Then ask students to estimate the solution as a point. Ask why it is difficult to give an exact answer. Lead students to notice this is because the intersection point does not occur on intersecting grid lines. Point out that in this section students will learn to find points of intersection without graphing.

Answer

1. (5, 1)

The solution of the system is (4, 2).

A graph of the two equations shows the two lines intersecting at the point (4, 2).

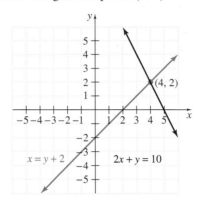

Practice Problem 2

Use the substitution method to solve the system:

$$\begin{cases} 4x - y = 2 \\ y = 5x \end{cases}$$

Example 2   Solve the system:

$$\begin{cases} 5x - y = -2 \\ y = 3x \end{cases}$$

Solution: The second equation is solved for y in terms of x. We substitute $3x$ for y in the first equation.

$$5x - y = -2 \quad \text{First equation}$$

$$5x - \left(3x\right) = -2$$

Now we solve for x.

$$5x - 3x = -2$$

$$2x = -2 \quad \text{Combine like terms.}$$

$$x = -1 \quad \text{Divide both sides by 2.}$$

The x-value of the ordered pair solution is -1. To find the corresponding y-value, we replace x with -1 in the equation $y = 3x$.

$$y = 3x$$

$$y = 3(-1) \quad \text{Let } x = -1.$$

$$y = -3$$

Check to see that the solution of the system is $(-1, -3)$.

TEACHING TIP

After discussing Example 2, ask students how the substitution helps us solve the first equation for the variable x. Help students see that with substitution, the equation in two variables is written as an equation in one variable.

To solve a system of equations by substitution, we first need an equation solved for one of its variables.

Practice Problem 3

Solve the system:

$$\begin{cases} 3x + y = 5 \\ 3x - 2y = -7 \end{cases}$$

Example 3   Solve the system:

$$\begin{cases} x + 2y = 7 \\ 2x + 2y = 13 \end{cases}$$

Solution: We choose one of the equations and solve for x or y. We will solve the first equation for x by subtracting $2y$ from both sides.

$$x + 2y = 7 \quad \text{First equation}$$

$$x = 7 - 2y \quad \text{Subtract } 2y \text{ from both sides.}$$

Answers

2. $(-2, -10)$, **3.** $\left(\frac{1}{3}, 4\right)$

Since $x = 7 - 2y$, we now substitute $7 - 2y$ for x in the second equation and solve for y.

$$2x + 2y = 13 \qquad \text{Second equation}$$

HELPFUL HINT

Don't forget to insert parentheses.

$$2(7 - 2y) + 2y = 13 \qquad \text{Let } x = 7 - 2y.$$

$$14 - 4y + 2y = 13 \qquad \text{Apply the distributive property.}$$

$$14 - 2y = 13 \qquad \text{Simplify.}$$

$$-2y = -1 \qquad \text{Subtract 14 from both sides.}$$

$$y = \frac{1}{2} \qquad \text{Divide both sides by } -2.$$

To find x, we let $y = \frac{1}{2}$ in the equation $x = 7 - 2y$.

$$x = 7 - 2y$$

$$x = 7 - 2\left(\frac{1}{2}\right) \qquad \text{Let } y = \frac{1}{2}.$$

$$x = 7 - 1$$

$$x = 6$$

Check the solution in both equations of the original system. The solution is $\left(6, \frac{1}{2}\right)$.

The following steps summarize how to solve a system of equations by the substitution method.

TO SOLVE A SYSTEM OF TWO LINEAR EQUATIONS BY THE SUBSTITUTION METHOD

Step 1. Solve one of the equations for one of its variables.

Step 2. Substitute the expression for the variable found in Step 1 into the other equation.

Step 3. Solve the equation from Step 2 to find the value of one variable.

Step 4. Substitute the value found in Step 3 in any equation containing both variables to find the value of the other variable.

Step 5. Check the proposed solution in the original system.

TRY THE CONCEPT CHECK IN THE MARGIN.

Example 4   Solve the system:
$$\begin{cases} 7x - 3y = -14 \\ -3x + y = 6 \end{cases}$$

Solution: To avoid introducing fractions, we will solve the second equation for y.

$$-3x + y = 6 \qquad \text{Second equation}$$

$$y = 3x + 6$$

Next, we substitute $3x + 6$ for y in the first equation.

✓ CONCEPT CHECK

As you solve the system
$$\begin{cases} 2x + y = -5 \\ x - y = 5 \end{cases}$$
you find that $y = -5$. Is this the solution of the system?

Practice Problem 4

Solve the system:
$$\begin{cases} 5x - 2y = 6 \\ -3x + y = -3 \end{cases}$$

Answers

4. $(0, -3)$

✓ **Concept Check:** No, the solution will be an ordered pair.

$$7x - 3y = -14 \quad \text{First equation}$$
$$7x - 3(3x + 6) = -14 \quad \text{Let } y = 3x + 6.$$
$$7x - 9x - 18 = -14 \quad \text{Use the distributive property.}$$
$$-2x - 18 = -14 \quad \text{Simplify.}$$
$$-2x = 4 \quad \text{Add 18 to both sides.}$$
$$\frac{-2x}{-2} = \frac{4}{-2} \quad \text{Divide both sides by } -2.$$
$$x = -2$$

To find the corresponding y-value, we substitute -2 for x in the equation $y = 3x + 6$. Then $y = 3(-2) + 6$ or $y = 0$. The solution of the system is $(-2, 0)$. Check this solution in both equations of the system. ▬▬▬

HELPFUL HINT

When solving a system of equations by the substitution method, begin by solving an equation for one of its variables. If possible, solve for a variable that has a coefficient of 1 or -1 to avoid working with time-consuming fractions.

Practice Problem 5

Solve the system:

$$\begin{cases} -x + 3y = 6 \\ y = \dfrac{1}{3}x + 2 \end{cases}$$

Example 5   Solve the system: $\begin{cases} \dfrac{1}{2}x - y = 3 \\ x = 6 + 2y \end{cases}$

Solution: The second equation is already solved for x in terms of y. Thus we substitute $6 + 2y$ for x in the first equation and solve for y.

$$\frac{1}{2}x - y = 3 \quad \text{First equation}$$

$$\frac{1}{2}(6 + 2y) - y = 3 \quad \text{Let } x = 6 + 2y.$$
$$3 + y - y = 3 \quad \text{Apply the distributive property.}$$
$$3 = 3 \quad \text{Simplify.}$$

Arriving at a true statement such as $3 = 3$ indicates that the two linear equations in the original system are equivalent. This means that their graphs are identical, as shown in the figure. There is an infinite number of solutions to the system, and any solution of one equation is also a solution of the other.

Answer

5. infinite number of solutions

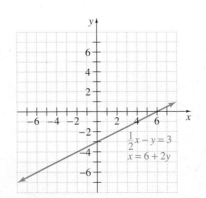

Example 6   Solve the system:

$$\begin{cases} 6x + 12y = 5 \\ -4x - 8y = 0 \end{cases}$$

Solution: We choose the second equation and solve for y.

$$-4x - 8y = 0 \quad \text{Second equation}$$

$$-8y = 4x \quad \text{Add } 4x \text{ to both sides.}$$

$$\frac{-8y}{-8} = \frac{4x}{-8} \quad \text{Divide both sides by } -8.$$

$$y = -\frac{1}{2}x \quad \text{Simplify.}$$

Now we replace y with $-\frac{1}{2}x$ in the first equation.

$$6x + 12y = 5 \quad \text{First equation}$$

$$6x + 12\left(-\frac{1}{2}x\right) = 5 \quad \text{Let } y = -\frac{1}{2}x.$$

$$6x + (-6x) = 5 \quad \text{Simplify.}$$

$$0 = 5 \quad \text{Combine like terms.}$$

The false statement $0 = 5$ indicates that this system has no solution. The graph of the linear equations in the system is a pair of parallel lines, as shown in the figure.

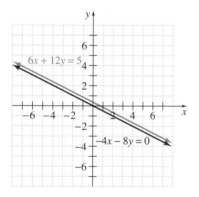

TRY THE CONCEPT CHECK IN THE MARGIN.

Practice Problem 6

Solve the system:

$$\begin{cases} 2x - 3y = 6 \\ -4x + 6y = -12 \end{cases}$$

TEACHING TIP

Ask students, "What types of linear systems have no solutions?" Have them name an equation whose graph is a line parallel to $y = 3x + 5$. Now use substitution to solve the resulting system. Because a false statement results, no solution of one equation will satisfy the other.

✓ **CONCEPT CHECK**

Describe how the graphs of the equations in a system appear if the system has

a. no solution

b. one solution

c. an infinite number of solutions

Answers

6. infinite number of solutions

✓ Concept Check

a. parallel lines
b. intersect at one point
c. identical graphs

Focus On Business and Career

BREAK-EVEN POINT

When a business sells a new product, it generally does not start making a profit right away. There are usually many expenses associated with producing a new product. These expenses might include an advertising blitz to introduce the product to the public. These start-up expenses might also include the cost of market research and product development or any brand-new equipment needed to manufacture the product. Start-up costs like these are generally called *fixed costs* because they don't depend on the number of items manufactured. Expenses that depend on the number of items manufactured, such as the cost of materials and shipping, are called *variable costs*. The total cost of manufacturing the new product is given by the cost equation: Total cost = Fixed costs + Variable costs.

For instance, suppose a greeting card company is launching a new line of greeting cards. The company spent $7000 doing product research and development for the new line and spent $15,000 on advertising the new line. The company did not need to buy any new equipment to manufacture the cards, but the paper and ink needed to make each card will cost $0.20 per card. The total cost y in dollars for manufacturing x cards is $y = 22{,}000 + 0.20x$.

Once a business sets a price for the new product, the company can find the product's expected *revenue*. Revenue is the amount of money the company takes in from the sales of its product. The revenue from selling a product is given by the revenue equation: Revenue = Price per item × Number of items sold.

For instance, suppose that the card company plans to sell its new cards for $1.50 each. The revenue y, in dollars, that the company can expect to receive from the sales of x cards is $y = 1.50x$.

If the total cost and revenue equations are graphed on the same coordinate system, the graphs should intersect. The point of intersection is where total cost equals revenue and is called the *break-even point*. The break-even point gives the number of items x that must be sold for the company to recover its expenses. If fewer than this number of items is sold, the company loses money. If more than this number of items is sold, the company makes a profit. In the case of the greeting card company, approximately 16,923 cards must be sold for the company to break even on this new card line. The total cost and revenue of producing and selling 16,923 cards is the same. It is approximately $25,385.

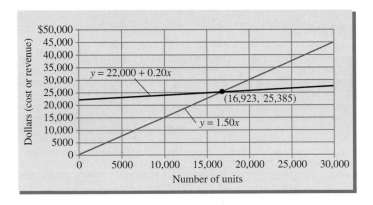

GROUP ACTIVITY

Suppose your group is starting a small business near your campus.
  a. Choose a business and decide what campus-related product or service you will provide.
  b. Research the fixed costs of starting up such a business.
  c. Research the variable costs of producing such a product or providing such a service.
  d. Decide how much you would charge per unit of your product or service.
  e. Find a system of equations for the total cost and revenue of your product or service.
  f. How many units of your product or service must be sold before your business will break even?

EXERCISE SET 7.2

A *Solve each system of equations by the substitution method. See Examples 1 and 2.*

1. $\begin{cases} x + y = 3 \\ x = 2y \end{cases}$

2. $\begin{cases} x + y = 20 \\ x = 3y \end{cases}$

📼 3. $\begin{cases} x + y = 6 \\ y = -3x \end{cases}$

4. $\begin{cases} x + y = 6 \\ y = -4x \end{cases}$

5. $\begin{cases} 3x + 2y = 16 \\ x = 3y - 2 \end{cases}$

6. $\begin{cases} 2x + 3y = 18 \\ x = 2y - 5 \end{cases}$

7. $\begin{cases} 3x - 4y = 10 \\ x = 2y \end{cases}$

8. $\begin{cases} 3x - 4y = 10 \\ y = 2x \end{cases}$

9. $\begin{cases} y = 3x + 1 \\ 4y - 8x = 12 \end{cases}$

10. $\begin{cases} y = 2x + 3 \\ 5y - 7x = 18 \end{cases}$

11. $\begin{cases} y = 2x + 9 \\ y = 7x + 10 \end{cases}$

12. $\begin{cases} y = 5x - 3 \\ y = 8x + 4 \end{cases}$

Solve each system of equations by the substitution method. See Examples 3 through 6.

13. $\begin{cases} x + 2y = 6 \\ 2x + 3y = 8 \end{cases}$

14. $\begin{cases} x + 3y = -5 \\ 2x + 2y = 6 \end{cases}$

15. $\begin{cases} 2x - 5y = 1 \\ 3x + y = -7 \end{cases}$

16. $\begin{cases} 4x + 2y = 5 \\ 2x + y = -4 \end{cases}$

17. $\begin{cases} 2y = x + 2 \\ 6x - 12y = 0 \end{cases}$

18. $\begin{cases} 3y = x + 6 \\ 4x + 12y = 0 \end{cases}$

19. $\begin{cases} 4x + y = 11 \\ 2x + 5y = 1 \end{cases}$

20. $\begin{cases} 3x + y = -14 \\ 4x + 3y = -22 \end{cases}$

21. $\begin{cases} 2x - 3y = -9 \\ 3x = y + 4 \end{cases}$

ANSWERS

1. $(2, 1)$

2. $(15, 5)$

3. $(-3, 9)$

4. $(-2, 8)$

5. $(4, 2)$

6. $(3, 4)$

7. $(10, 5)$

8. $(-2, -4)$

9. $(2, 7)$

10. $(1, 5)$

11. $\left(-\dfrac{1}{5}, \dfrac{43}{5}\right)$

12. $\left(-\dfrac{7}{3}, -\dfrac{44}{3}\right)$

13. $(-2, 4)$

14. $(7, -4)$

15. $(-2, -1)$

16. no solution

17. no solution

18. $(-3, 1)$

19. $(3, -1)$

20. $(-4, -2)$

21. $(3, 5)$

22. (1, 4)

23. $\left(\dfrac{2}{3}, -\dfrac{1}{3}\right)$

24. $\left(-\dfrac{9}{5}, \dfrac{3}{5}\right)$

25. (−1, −4)

26. (−5, −3)

27. (−6, 2)

28. (−3, 5)

29. (2, 1)

30. infinite solutions

31. no solution

32. (3, −2)

33. infinite solutions

34. infinite solutions

35. answers may vary

36. answers may vary

37. −6x − 4y = −12

38. −5x + 5y = 50

39. −12x + 3y = 9

40. −20a + 28b = 16

41. 5n

42. 16y

43. −15b

44. 0

554

22. $\begin{cases} 8x - 3y = -4 \\ 7x = y + 3 \end{cases}$ **23.** $\begin{cases} 6x - 3y = 5 \\ x + 2y = 0 \end{cases}$ **24.** $\begin{cases} 10x - 5y = -21 \\ x + 3y = 0 \end{cases}$

25. $\begin{cases} 3x - y = 1 \\ 2x - 3y = 10 \end{cases}$ **26.** $\begin{cases} 2x - y = -7 \\ 4x - 3y = -11 \end{cases}$ **27.** $\begin{cases} -x + 2y = 10 \\ -2x + 3y = 18 \end{cases}$

28. $\begin{cases} -x + 3y = 18 \\ -3x + 2y = 19 \end{cases}$ **29.** $\begin{cases} 5x + 10y = 20 \\ 2x + 6y = 10 \end{cases}$ **30.** $\begin{cases} 2x + 4y = 6 \\ 5x + 10y = 15 \end{cases}$

31. $\begin{cases} 3x + 6y = 9 \\ 4x + 8y = 16 \end{cases}$ **32.** $\begin{cases} 6x + 3y = 12 \\ 9x + 6y = 15 \end{cases}$

33. $\begin{cases} \dfrac{1}{3}x - y = 2 \\ x - 3y = 6 \end{cases}$ **34.** $\begin{cases} \dfrac{1}{4}x - 2y = 1 \\ x - 8y = 4 \end{cases}$

35. Explain how to identify a system with no solution when using the substitution method.

36. Occasionally, when using the substitution method, the equation 0 = 0 is obtained. Explain how this result indicates that the graphs of the equations in the system are identical.

REVIEW AND PREVIEW

Write equivalent equations by multiplying both sides of each given equation by the given nonzero number. See Section 2.3.

37. 3x + 2y = 6 by −2   **38.** −x + y = 10 by 5

39. −4x + y = 3 by 3   **40.** 5a − 7b = −4 by −4

Add the binomials. See Section 3.4.

41. $\begin{array}{r} 3n + 6m \\ 2n - 6m \\ \hline \end{array}$ **42.** $\begin{array}{r} -2x + 5y \\ 2x + 11y \\ \hline \end{array}$ **43.** $\begin{array}{r} -5a - 7b \\ 5a - 8b \\ \hline \end{array}$ **44.** $\begin{array}{r} 9q + p \\ -9q - p \\ \hline \end{array}$

Name _____

COMBINING CONCEPTS

Solve each system by the substitution method. First simplify each equation by combining like terms.

45. $\begin{cases} -5y + 6y = 3x + 2(x - 5) - 3x + 5 \\ 4(x + y) - x + y = -12 \end{cases}$

46. $\begin{cases} 5x + 2y - 4x - 2y = 2(2y + 6) - 7 \\ 3(2x - y) - 4x = 1 + 9 \end{cases}$

Use a graphing calculator to solve each system.

47. $\begin{cases} y = 5.1x + 14.56 \\ y = -2x - 3.9 \end{cases}$

48. $\begin{cases} y = 3.1x - 16.35 \\ y = -9.7x + 28.45 \end{cases}$

49. $\begin{cases} 3x + 2y = 14.05 \\ 5x + y = 18.5 \end{cases}$

50. $\begin{cases} x + y = -15.2 \\ -2x + 5y = -19.3 \end{cases}$

51. For the years 1960 through 1995, the annual percentage y of U.S. households that used fuel oil to heat their homes is given by the equation $y = -0.65x + 32.02$, where x is the number of years since 1960. For the same period, the annual percentage y of U.S. households that used electricity to heat their homes is given by the equation $y = 0.78x + 1.32$, where x is the number of years since 1960. (*Source:* Based on data from the U.S. Bureau of the Census)

a. Use the substitution method to solve this system of equations. (Round your final results to the nearest whole numbers.)

c. Sketch a graph of the system of equations. Write a sentence describing the use of fuel oil and electricity for heating homes between 1960 and 1995.

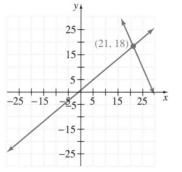

(21, 18)

b. Explain the meaning of your answer to part (a).

45. $(1, -3)$

46. $(5, 0)$

47. $(-2.6, 1.3)$

48. $(3.5, -5.5)$

49. $(3.28, 2.11)$

50. $(-8.1, -7.1)$

51. a. $(21, 18)$

b. answers may vary

c. see graph

555

Name _____

52. The number of music CDs (in millions) shipped to retailers in the United States from 1987 through 1996 is given by the equation $y = 79.3x + 57.86$, where x is the number of years since 1987. The number y of music cassettes (in millions) shipped to retailers in the United States from 1987 through 1996 is given by the equation $y = -22.5x + 467.0$, where x is the number of years since 1987. (*Source:* Based on data from the Recording Industry Association of America)

a. Use the substitution method to solve this system of equations. (Round your final results to the nearest whole numbers.)

b. Explain the meaning of your answer to part (a).

c. Sketch a graph of the system of equations. Write a sentence describing the trends in the popularity of these two types of music formats.

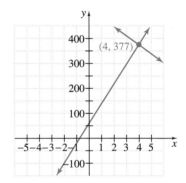

7.3 SOLVING SYSTEMS OF LINEAR EQUATIONS BY ADDITION

A USING THE ADDITION METHOD

We have seen that substitution is an accurate method for solving a system of linear equations. Another accurate method is the **addition** or **elimination method**. The addition method is based on the addition property of equality: Adding equal quantities to both sides of an equation does not change the solution of the equation. In symbols,

if $A = B$ and $C = D$, then $A + C = B + D$

Example 1   Solve the system: $\begin{cases} x + y = 7 \\ x - y = 5 \end{cases}$

Solution: Since the left side of each equation is equal to its right side, we are adding equal quantities when we add the left sides of the equations together and the right sides of the equations together. This adding eliminates the variable y and gives us an equation in one variable, x. We can then solve for x.

$$
\begin{array}{ll}
x + y = 7 & \text{First equation} \\
\underline{x - y = 5} & \text{Second equation} \\
2x \phantom{{}+ y} = 12 & \text{Add the equations to eliminate } y. \\
x = 6 & \text{Divide both sides by 2.}
\end{array}
$$

The x-value of the solution is 6. To find the corresponding y-value, we let $x = 6$ in either equation of the system. We will use the first equation.

$$
\begin{array}{ll}
x + y = 7 & \text{First equation} \\
6 + y = 7 & \text{Let } x = 6. \\
y = 1 & \text{Solve for } y.
\end{array}
$$

The solution is $(6, 1)$.

Check: Check the solution in both equations.

First Equation

$$
\begin{array}{ll}
x + y = 7 & \\
6 + 1 \stackrel{?}{=} 7 & \text{Let } x = 6 \text{ and } y = 1. \\
7 = 7 & \text{True.}
\end{array}
$$

Second Equation

$$
\begin{array}{ll}
x - y = 5 & \\
6 - 1 \stackrel{?}{=} 5 & \text{Let } x = 6 \text{ and } y = 1. \\
5 = 5 & \text{True.}
\end{array}
$$

Thus, the solution of the system is $(6, 1)$ and the graphs of the two equations intersect at the point $(6, 1)$, as shown.

Objective

A Use the addition method to solve a system of linear equations.

SSM   CD-ROM   Video
7.3

Practice Problem 1

Use the addition method to solve the system: $\begin{cases} x + y = 13 \\ x - y = 5 \end{cases}$

Answer

1. $(9, 4)$

Practice Problem 2

Solve the system: $\begin{cases} 2x - y = -6 \\ -x + 4y = 17 \end{cases}$

Example 2   Solve the system: $\begin{cases} -2x + y = 2 \\ -x + 3y = -4 \end{cases}$

Solution: If we simply add these two equations, the result is still an equation in two variables. However, our goal is to eliminate one of the variables so that we have an equation in the other variable. To do this, notice what happens if we multiply *both sides* of the first equation by -3. We are allowed to do this by the multiplication property of equality. Then the system

$$\begin{cases} -3(-2x + y) = -3(2) \\ -x + 3y = -4 \end{cases} \quad \text{simplifies to} \quad \begin{cases} 6x - 3y = -6 \\ -x + 3y = -4 \end{cases}$$

When we add the resulting equations, the y variable is eliminated.

$$
\begin{array}{rl}
6x - 3y = -6 & \\
\underline{-x + 3y = -4} & \\
5x \phantom{{}- 3y} = -10 & \text{Add.} \\
x = -2 & \text{Divide both sides by 5.}
\end{array}
$$

To find the corresponding y-value, we let $x = -2$ in any of the preceding equations containing both variables. We use the first equation of the original system.

$$
\begin{array}{rl}
-2x + y = 2 & \text{First equation} \\
-2(-2) + y = 2 & \text{Let } x = -2. \\
4 + y = 2 & \\
y = -2 &
\end{array}
$$

Check the ordered pair $(-2, -2)$ in both equations of the *original* system. The solution is $(-2, -2)$. ■

In Example 2, the decision to multiply the first equation by -3 was no accident. **To eliminate a variable** when adding two equations, **the coeffi-**

cient of the variable in one equation must be the opposite of its coefficient in the other equation.

HELPFUL HINT

Be sure to multiply *both sides* of an equation by a chosen number when solving by the addition method. A common mistake is to multiply only the side containing the variables.

Example 3   Solve the system: $\begin{cases} 2x - y = 7 \\ 8x - 4y = 1 \end{cases}$

Solution: When we multiply both sides of the first equation by -4, the resulting coefficient of x is -8. This is the opposite of 8, the coefficient of x in the second equation. Then the system

HELPFUL HINT

Don't forget to multiply both sides by -4.

$\begin{cases} -4(2x - y) = -4(7) \\ 8x - 4y = 1 \end{cases}$ simplifies to

$\begin{cases} -8x + 4y = -28 \\ \underline{8x - 4y = 1} \\ 0 = -27 \end{cases}$ Add the equations.

When we add the equations, both variables are eliminated and we have $0 = -27$, a false statement. This means that the system has no solution. The equations, if graphed, are parallel lines.   ▬▬▬

Example 4   Solve the system: $\begin{cases} 3x - 2y = 2 \\ -9x + 6y = -6 \end{cases}$

Solution: First we multiply both sides of the first equation by 3, then we add the resulting equations.

$\begin{cases} 3(3x - 2y) = 3 \cdot 2 \\ -9x + 6y = -6 \end{cases}$ simplifies to $\begin{cases} 9x - 6y = 6 \\ \underline{-9x + 6y = -6} \\ 0 = 0 \end{cases}$ Add the equations.

Both variables are eliminated and we have $0 = 0$, a true statement. This means that the system has an infinite number of solutions.   ▬▬▬

TEACHING TIP

Before discussing Example 3, have students predict what will happen when using the addition method to solve a system of two parallel lines. Have students check their prediction with the following system:

$\begin{cases} 3x - 4y = 12 \\ 3x - 4y = 16 \end{cases}$

Practice Problem 3

Solve the system: $\begin{cases} x - 3y = -2 \\ -3x + 9y = 5 \end{cases}$

TEACHING TIP

You may also want to have them predict what will happen when using the addition method to solve a system of two equivalent equations. Have students check their prediction with the following system:

$\begin{cases} 3x - 4y = 12 \\ -3x + 4y = -12 \end{cases}$

You may want students to warm up with the following exercise: Find the value for m which makes each statement true:

$m(3x) - (9x) = 0$

$-12x + m(2x) = 0$

$20x + m(5x) = 0$

$m(-6x) - 30x = 0$

Practice Problem 4

Solve the system: $\begin{cases} 2x + 5y = 1 \\ -4x - 10y = -2 \end{cases}$

Answers

3. no solution,   **4.** infinite number of solutions

✓ Concept Check

Suppose you are solving the system

$$\begin{cases} 3x + 8y = -5 \\ 2x - 4y = 3 \end{cases}$$

You decide to use the addition method by multiplying both sides of the second equation by 2. In which of the following was the multiplication performed correctly? Explain.

a. $4x - 8y = 3$
b. $4x - 8y = 6$

Practice Problem 5

Solve the system: $\begin{cases} 4x + 5y = 14 \\ 3x - 2y = -1 \end{cases}$

TEACHING TIP

You may want students to warm up with the following exercises: Find values for m and n which make each statement true.

$m(3x) + n(7x) = 0$

$m(-4x) + n(-5x) = 0$

$m(2x) - n(9x) = 0$

$m(-11x) - n(3x) = 0$

✓ Concept Check

Suppose you are solving the system

$$\begin{cases} -4x + 7y = 6 \\ x + 2y = 5 \end{cases}$$

by the addition method.

a. What step(s) should you take if you wish to eliminate x when adding the equations?

b. What step(s) should you take if you wish to eliminate y when adding the equations?

Try the Concept Check in the Margin.

Example 5   Solve the system: $\begin{cases} 3x + 4y = 13 \\ 5x - 9y = 6 \end{cases}$

Solution: We can eliminate the variable y by multiplying the first equation by 9 and the second equation by 4. Then we add the resulting equations.

$$\begin{cases} 9(3x + 4y) = 9(13) \\ 4(5x - 9y) = 4(6) \end{cases} \text{ simplifies to } \begin{cases} 27x + 36y = 117 \\ 20x - 36y = 24 \end{cases}$$

$$\begin{array}{ll} 47x = 141 & \text{Add the equations.} \\ x = 3 & \text{Solve for } x. \end{array}$$

To find the corresponding y-value, we let $x = 3$ in any equation in this example containing two variables. Doing so in any of these equations will give $y = 1$. Check to see that $(3, 1)$ satisfies each equation in the original system. The solution is $(3, 1)$.   ▬▬

If we had decided to eliminate x instead of y in Example 5, the first equation could have been multiplied by 5 and the second by -3. Try solving the original system this way to check that the solution is $(3, 1)$.

The following steps summarize how to solve a system of linear equations by the addition method.

> **To Solve a System of Two Linear Equations by the Addition Method**
>
> **Step 1.** Rewrite each equation in standard from $Ax + By = C$.
>
> **Step 2.** If necessary, multiply one or both equations by a nonzero number so that the coefficients of a chosen variable in the system are opposites.
>
> **Step 3.** Add the equations.
>
> **Step 4.** Find the value of one variable by solving the resulting equation from Step 3.
>
> **Step 5.** Find the value of the second variable by substituting the value found in Step 4 into either of the original equations.
>
> **Step 6.** Check the proposed solution in the orignal system.

Try the Concept Check in the Margin.

Answer

5. $(1, 2)$

✓ Concept Check: b

a. Multiply the second equation by 4.
b. Possible answer: Multiply the first equation by -2 and the second equation by 7.

Example 6   Solve the system: $\begin{cases} -x - \dfrac{y}{2} = \dfrac{5}{2} \\ -\dfrac{x}{2} + \dfrac{y}{4} = 0 \end{cases}$

Solution: We begin by clearing each equation of fractions. To do so, we multiply both sides of the first equation by the LCD 2 and both sides of the second equation by the LCD 4. Then the system

$$\begin{cases} 2\left(-x - \dfrac{y}{2}\right) = 2\left(\dfrac{5}{2}\right) \\ 4\left(-\dfrac{x}{2} + \dfrac{y}{4}\right) = 4(0) \end{cases} \quad \text{simplifies to} \quad \begin{cases} -2x - y = 5 \\ -2x + y = 0 \end{cases}$$

Now we add the resulting equations in the simplified system.

$$\begin{array}{rl} -2x - y &= 5 \\ -2x + y &= 0 \\ \hline -4x &= 5 \quad \text{Add.} \\ x &= -\dfrac{5}{4} \end{array}$$

To find y, we could replace x with $-\dfrac{5}{4}$ in one of the equations with two variables. Instead, let's go back to the simplified system and multiply by appropriate factors to eliminate the variable x and solve for y. To do this, we multiply the first equation in the simplified system by -1. Then the system

$$\begin{cases} -1(-2x - y) = -1(5) \\ -2x + y = 0 \end{cases} \quad \text{simplifies to} \quad \begin{array}{rl} 2x + y &= -5 \\ -2x + y &= 0 \\ \hline 2y &= -5 \quad \text{Add.} \\ y &= -\dfrac{5}{2} \end{array}$$

Check the ordered pair $\left(-\dfrac{5}{4}, -\dfrac{5}{2}\right)$ in both equations of the original system. The solution is $\left(-\dfrac{5}{4}, -\dfrac{5}{2}\right)$. ▬▬▬

Practice Problem 6

Solve the system: $\begin{cases} -\dfrac{x}{3} + y = \dfrac{4}{3} \\ \dfrac{x}{2} - \dfrac{5}{2}y = -\dfrac{1}{2} \end{cases}$

TEACHING TIP

Before attempting Example 6, remind students to use what they know to make this system as easy as possible to solve. Then ask how Example 6 could be written as a simpler system.

TEACHING TIP

Conclude this section by having students describe the three methods they have learned for solving systems of linear equations and list the pros and cons of each.

Answers

6. $\left(-\dfrac{17}{2}, -\dfrac{3}{2}\right)$

Focus On History

The oldest known arithmetic book is a Chinese textbook called *Nine Chapters on the Mathematical Art*. No one knows for sure who wrote this text or when it was first written. Experts believe that it was a collection of works written by many different people. It was probably written over the course of several centuries. Even though no one knows the original date of the *Nine Chapters*, we do know that it existed in 213 BC. In that year, all of the original copies of the *Nine Chapters*, along with many other books, were burned when the first emperor of the Qin Dynasty (221–206 BC) tried to erase all traces of previous rulers and dynasties.

The Qin Emperor was not quite successful in destroying all of the *Nine Chapters*. Pieces of the text were found and many Chinese mathematicians filled in the missing material. In 263 AD, the Chinese mathematician Liu Hui wrote a summary of the *Nine Chapters*, adding his own solutions to its problems. Liu Hui's version was studied in China for over a thousand years. At one point, the Chinese government even adopted *Nine Chapters* as the official study aid for university students to use when preparing for civil service exams.

The *Nine Chapters* is a guide to everyday math in ancient China. It contains a total of 246 problems covering widely encountered problems like field measurement, rice exchange, fair taxation, and construction. It includes the earliest known use of negative numbers and shows the first development of solving systems of linear equations. The following problem appears in Chapter 7, "Excess and Deficiency," of the *Nine Chapters*. (Note: A *wen* is a unit of currency.)

> A certain number of people are purchasing some chickens together. If each person contributes 9 wen, there is an excess of 11 wen. If each person contributes just 6 wen, there is a deficiency of 16 wen. Find the number of people and the total price of the chickens. (Adapted from *The History of Mathematics: An Introduction*, second edition, David M. Burton, 1991, Wm. C. Brown Publishers, p. 164)

CRITICAL THINKING

The information in the excess/deficiency problem from *Nine Chapters* can be translated into two equations in two variables. Let c represent the total price of the chickens, and let x represent the number of people pooling their money to buy the chickens. In this situation, an excess of 11 wen can be interpreted as 11 more than the price of the chickens. A deficiency of 16 wen can be interpreted as 16 less than the price of the chickens.

1. Use what you have learned so far in this book about translating sentences into equations to write two equations in two variables for the excess/deficiency problem.
2. Solve the problem from *Nine Chapters* by solving the system of equations you wrote in Question 1. How many people pooled their money? What was the price of the chickens?
3. Write a modern-day excess/deficiency problem of your own.

Name _____ Section _____ Date _____

EXERCISE SET 7.3

A *Solve each system of equations by the addition method. See Example 1.*

1. $\begin{cases} 3x + y = 5 \\ 6x - y = 4 \end{cases}$
2. $\begin{cases} 4x + y = 13 \\ 2x - y = 5 \end{cases}$

3. $\begin{cases} x - 2y = 8 \\ -x + 5y = -17 \end{cases}$
4. $\begin{cases} x - 2y = -11 \\ -x + 5y = 23 \end{cases}$

5. $\begin{cases} 3x + 2y = 11 \\ 5x - 2y = 29 \end{cases}$
6. $\begin{cases} 4x + 2y = 2 \\ 3x - 2y = 12 \end{cases}$

7. $\begin{cases} x + y = 6 \\ x - y = 6 \end{cases}$
8. $\begin{cases} x - y = 1 \\ -x + 2y = 0 \end{cases}$

Solve each system of equations by the addition method. See Examples 2 through 5.

9. $\begin{cases} 3x + y = -11 \\ 6x - 2y = -2 \end{cases}$
10. $\begin{cases} 4x + y = -13 \\ 6x - 3y = -15 \end{cases}$
11. $\begin{cases} x + 5y = 18 \\ 3x + 2y = -11 \end{cases}$

12. $\begin{cases} x + 4y = 14 \\ 5x + 3y = 2 \end{cases}$
13. $\begin{cases} 2x - 5y = 4 \\ 3x - 2y = 4 \end{cases}$
14. $\begin{cases} 6x - 5y = 7 \\ 4x - 6y = 7 \end{cases}$

15. $\begin{cases} 2x + 3y = 0 \\ 4x + 6y = 3 \end{cases}$
16. $\begin{cases} -x + 5y = -1 \\ 3x - 15y = 3 \end{cases}$
17. $\begin{cases} 3x + y = 4 \\ 9x + 3y = 6 \end{cases}$

18. $\begin{cases} 2x + y = 6 \\ 4x + 2y = 12 \end{cases}$
19. $\begin{cases} 3x - 2y = 7 \\ 5x + 4y = 8 \end{cases}$
20. $\begin{cases} 6x - 5y = 25 \\ 4x + 15y = 13 \end{cases}$

21. $\begin{cases} \frac{2}{3}x + 4y = -4 \\ 5x + 6y = 18 \end{cases}$
22. $\begin{cases} \frac{3}{2}x + 4y = 1 \\ 9x + 24y = 5 \end{cases}$
23. $\begin{cases} 4x - 6y = 8 \\ 6x - 9y = 12 \end{cases}$

1. $(1, 2)$
2. $(3, 1)$
3. $(2, -3)$
4. $(-3, 4)$
5. $(5, -2)$
6. $(2, -3)$
7. $(6, 0)$
8. $(2, 1)$
9. $(-2, -5)$
10. $(-3, -1)$
11. $(-7, 5)$
12. $(-2, 4)$
13. $\left(\frac{12}{11}, -\frac{4}{11}\right)$
14. $\left(\frac{7}{16}, -\frac{7}{8}\right)$
15. no solution
16. infinite number of solutions
17. no solution
18. infinite number of solutions
19. $\left(2, -\frac{1}{2}\right)$
20. $\left(4, -\frac{1}{5}\right)$
21. $(6, -2)$
22. no solution
23. infinite number of solutions

563

24. no solution

25. $(-2, 0)$

26. $(0, -4)$

27. answers may vary

28. answers may vary

29. $\left(\dfrac{3}{2}, 3\right)$

30. $(3, 12)$

31. $(1, 6)$

32. $(3, 12)$

33. infinite number of solutions

34. no solution

35. $2x + 6 = x - 3$

36. $x + (x + 1) + (x + 2) = 66$

37. $20 - 3x = 2$

38. $2(8 + x) = x - 20$

39. $4(x + 6) = 2x$

40. $\dfrac{1}{x} - \dfrac{2x}{7} = 2$

Name _____

24. $\begin{cases} 9x - 3y = 12 \\ 12x - 4y = 18 \end{cases}$

25. $\begin{cases} 8x = -11y - 16 \\ 2x + 3y = -4 \end{cases}$

26. $\begin{cases} 10x + 3y = -12 \\ 5x = -4y - 16 \end{cases}$

27. When solving a system of equations by the addition method, how do we know when the system has no solution?

28. To solve the system $\begin{cases} 2x - 3y = 5 \\ 5x + 2y = 6 \end{cases}$ explain why the addition method might be preferred rather than the substitution method.

Solve each system of equations by the addition method. See Example 6.

29. $\begin{cases} \dfrac{x}{3} + \dfrac{y}{6} = 1 \\ \dfrac{x}{2} - \dfrac{y}{4} = 0 \end{cases}$

30. $\begin{cases} \dfrac{x}{2} + \dfrac{y}{8} = 3 \\ x - \dfrac{y}{4} = 0 \end{cases}$

31. $\begin{cases} x - \dfrac{y}{3} = -1 \\ -\dfrac{x}{2} + \dfrac{y}{8} = \dfrac{1}{4} \end{cases}$

32. $\begin{cases} 2x - \dfrac{3y}{4} = -3 \\ x + \dfrac{y}{9} = \dfrac{13}{3} \end{cases}$

33. $\begin{cases} \dfrac{x}{3} - y = 2 \\ -\dfrac{x}{2} + \dfrac{3y}{2} = -3 \end{cases}$

34. $\begin{cases} \dfrac{x}{2} + \dfrac{y}{4} = 1 \\ -\dfrac{x}{4} - \dfrac{y}{8} = 1 \end{cases}$

REVIEW AND PREVIEW

Rewrite each sentence using mathematical symbols. Do not solve the equations. See Sections 2.3 and 2.5.

35. Twice a number, added to 6, is 3 less than the number.

36. The sum of three consecutive integers is 66.

37. Three times a number, subtracted from 20, is 2.

38. Twice the sum of 8 and a number is the difference of the number and 20.

39. The product of 4 and the sum of a number and 6 is twice the number.

40. If the quotient of twice a number and 7 is subtracted from the reciprocal of the number, the result is 2.

► COMBINING CONCEPTS

41. Use the system of linear equations below to answer the questions:

$$\begin{cases} x + y = 5 \\ 3x + 3y = b \end{cases}$$

a. Find the value of b so that the system has an infinite number of solutions.

b. Find a value of b so that there are no solutions to the system.

42. Use the system of linear equations below to answer the questions.

$$\begin{cases} x + y = 4 \\ 2x + by = 8 \end{cases}$$

a. Find the value of b so that the system has an infinite number of solutions.

b. Find a value of b so that the system has a single solution.

Solve each system by the addition method.

▦ 43. $\begin{cases} 2x + 3y = 14 \\ 3x - 4y = -69.1 \end{cases}$

▦ 44. $\begin{cases} 5x - 2y = -19.8 \\ -3x + 5y = -3.7 \end{cases}$

▦ 45. Commercial broadcast television stations can be divided into VHF stations (channels 2 through 13) and UHF stations (channels 14 through 83). The number y of VHF stations in the United States from 1988 through 1994 is given by the equation $21x - 7y = -3779$, where x is the number of years since 1988. The number y of UHF stations in the United States from 1988 through 1994 is given by the equation $-445x + 28y = 14{,}017$, where x is the number of years since 1988. (*Source*: Based on data from the Television Bureau of Advertising, Inc.)

a. Use the addition method to solve this system of equations. (Round your final results to the nearest whole numbers.)

b. Interpret your solution from part (a).

c. During which years were there more UHF commercial television stations than VHF stations?

46. In recent years, the number of daily newspapers printed as morning editions has been increasing and the number of daily newspapers printed as evening editions has been decreasing. The number y of daily morning newspapers in existence from 1987 through 1995 is given by the equation $-3363x + 180y = 90{,}748$, where x is the number of years since 1987. The number y of daily evening newspapers in existence from 1987 through 1995 is given by the equation $179x + 5y = 53{,}114$, where x is the number of years since 1987. (*Source*: Based on data from the *Editor and Publisher International Year Book*, Editor and Publisher Co., New York, NY, annual)

a. Suppose these trends continue in the future. Use the addition method to predict the year in which the number of morning newspapers will equal the number of evening newspapers. (Round to the nearest whole.)

b. How many of each type of newspaper will be in existence in that year?

b. about 3974

INTEGRATED REVIEW—SUMMARY ON SOLVING SYSTEMS OF EQUATIONS

Solve each system by either the addition method or the substitution method.

1. $\begin{cases} 2x - 3y = -11 \\ y = 4x - 3 \end{cases}$

2. $\begin{cases} 4x - 5y = 6 \\ y = 3x - 10 \end{cases}$

3. $\begin{cases} x + y = 3 \\ x - y = 7 \end{cases}$

4. $\begin{cases} x - y = 20 \\ x + y = -8 \end{cases}$

5. $\begin{cases} x + 2y = 1 \\ 3x + 4y = -1 \end{cases}$

6. $\begin{cases} x + 3y = 5 \\ 5x + 6y = -2 \end{cases}$

7. $\begin{cases} y = x + 3 \\ 3x - 2y = -6 \end{cases}$

8. $\begin{cases} y = -2x \\ 2x - 3y = -16 \end{cases}$

9. $\begin{cases} y = 2x - 3 \\ y = 5x - 18 \end{cases}$

10. $\begin{cases} y = 6x - 5 \\ y = 4x - 11 \end{cases}$

ANSWERS

1. $(2, 5)$

2. $(4, 2)$

3. $(5, -2)$

4. $(6, -14)$

5. $(-3, 2)$

6. $(-4, 3)$

7. $(0, 3)$

8. $(-2, 4)$

9. $(5, 7)$

10. $(-3, -23)$

11. $\left(\dfrac{1}{3}, 1\right)$

11. $\begin{cases} x + \dfrac{1}{6}y = \dfrac{1}{2} \\ 3x + 2y = 3 \end{cases}$

12. $\begin{cases} x + \dfrac{1}{3}y = \dfrac{5}{12} \\ 8x + 3y = 4 \end{cases}$

12. $\left(-\dfrac{1}{4}, 2\right)$

13. $\begin{cases} x - 5y = 1 \\ -2x + 10y = 3 \end{cases}$

14. $\begin{cases} -x + 2y = 3 \\ 3x - 6y = -9 \end{cases}$

13. no solution

15. For the system of equations:

$\begin{cases} 3x + 2y = -2 \\ y = -2x \end{cases}$

which method, substitution or addition, would you prefer to use to solve the system? Explain your reasoning.

16. For the system of equations

$\begin{cases} 3x - 2y = -3 \\ 6x + 2y = 12 \end{cases}$

which method, substitution or addition, would you prefer to use to solve the system? Explain your reasoning.

14. infinite number of solutions

15. answers may vary

16. answers may vary

7.4 SYSTEMS OF LINEAR EQUATIONS AND PROBLEM SOLVING

A USING A SYSTEM OF EQUATIONS FOR PROBLEM SOLVING

Many of the word problems solved earlier using one-variable equations can also be solved using two equations in two variables. We use the same problem-solving steps that have been used throughout this text. The only difference is that two variables are assigned to represent the two unknown quantities and that the problem is translated into two equations.

PROBLEM-SOLVING STEPS

1. UNDERSTAND the problem. During this step, become comfortable with the problem. Some ways of doing this are to

  Read and reread the problem.
  Choose two variables to represent the two unknowns.
  Construct a drawing.
  Propose a solution and check. Pay careful attention to how you check your proposed solution. This will help when writing an equation to model the problem.

2. TRANSLATE the problem into two equations.
3. SOLVE the system of equations.
4. INTERPRET the results: *Check* the proposed solution in the stated problem and *state* your conclusion.

Example 1   Finding Unknown Numbers

Find two numbers whose sum is 37 and whose difference is 21.

Solution:

1. UNDERSTAND. Read and reread the problem. Suppose that one number is 20. If their sum is 37, the other number is 17 because $20 + 17 = 37$. Is their difference 21? No; $20 - 17 = 3$. Our proposed solution is incorrect, but we now have a better understanding of the problem.

Since we are looking for two numbers, we let

$x =$ first number

$y =$ second number

2. TRANSLATE. Since we have assigned two variables to this problem, we translate our problem into two equations.

In words:	two numbers whose sum	is	37
	↓	↓	↓
Translate:	$x + y$	$=$	37

In words:	two numbers whose difference	is	21
	↓	↓	↓
Translate:	$x - y$	$=$	21

Objectives

A Use a system of equations to solve problems.

SSM   CD-ROM   Video 7.4

Practice Problem 1

Find two numbers whose sum is 50 and whose difference is 22.

TEACHING TIP

Consider beginning Example 1 by having students list some numbers whose sum is 37. Ask them, "How many such pairs exist?" Then have them list some numbers whose difference is 21. Again ask them, "How many such pairs exist?" Point out that finding numbers which satisfy both conditions could take a while using the list approach. Solving the problem with a system of equations can be more efficient.

Answer

1. The numbers are 36 and 14.

TEACHING TIP

After Example 1, point out to students
that the system involved 2 equations with
2 unknowns. One equation was the sum,
one equation was the difference, and the
two unknowns represented the two numbers that were to be found. After reading
Example 2, have students identify the
unknowns (price of an adult's ticket and
price of a child's ticket) and the two equations which can be written using these
unknowns.

Practice Problem 2

Admission prices at a local weekend fair were $5 for children and $7
for adults. The total money collected
was $3379, and 587 people attended
the fair. How many children and how
many adults attended the fair?

3. SOLVE. Now we solve the system

$$\begin{cases} x + y = 37 \\ x - y = 21 \end{cases}$$

Notice that the coefficients of the variable y are
opposites. Let's then solve by the addition method and
begin by adding the equations.

$$
\begin{array}{l}
x + y = 37 \\
\underline{x - y = 21} \\
2x \quad\;\; = 58
\end{array}
\qquad \text{Add the equations.}
$$

$$x = \frac{58}{2} = 29 \qquad \text{Divide both sides by 2.}$$

Now we let $x = 29$ in the first equation to find y.

$$
\begin{array}{ll}
x + y = 37 & \text{First equation} \\
29 + y = 37 & \\
\quad\;\; y = 37 - 29 = 8 &
\end{array}
$$

4. INTERPRET. The solution of the system is $(29, 8)$.

Check: Notice that the sum of 29 and 8 is $29 + 8 = 37$, the
required sum. Their difference is $29 - 8 = 21$, the
required difference.

State: The numbers are 29 and 8.   ▬▬▬

Example 2   Solving a Problem about Prices

The Barnum and Bailey Circus is in town. Admission for
4 adults and 2 children is $22, while admission for 2
adults and 3 children is $16.

a. What is the price of an adult's ticket?
b. What is the price of a child's ticket?
c. A special rate of $60 is charged for groups of
20 persons. Should a group of 4 adults and 16 children
use the group rate? Why or why not?

Solution: 1. UNDERSTAND. Read and reread the problem and
guess a solution. Let's suppose that the price of an
adult's ticket is $5 and the price of a child's ticket is
$4. To check our proposed solution, let's see if
admission for 4 adults and 2 children is $22. Admission
for 4 adults is 4($5) or $20 and admission for
2 children is 2($4) or $8. This gives a total admission
of $20 + $8 = $28, not the required $22. Again
though, we have accomplished the purpose of this
process: We have a better understanding of the
problem. To continue, we let

A = the price of an adult's ticket
C = the price of a child's ticket

2. TRANSLATE. We translate the problem into two
equations using both variables.

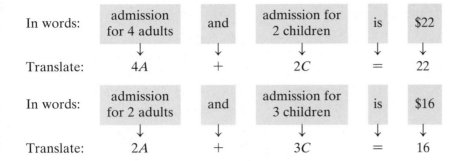

In words:	admission for 4 adults	and	admission for 2 children	is	$22
	↓	↓	↓	↓	↓
Translate:	4A	+	2C	=	22

In words:	admission for 2 adults	and	admission for 3 children	is	$16
	↓	↓	↓	↓	↓
Translate:	2A	+	3C	=	16

3. SOLVE. We solve the system

$$\begin{cases} 4A + 2C = 22 \\ 2A + 3C = 16 \end{cases}$$

Since both equations are written in standard form, we solve by the addition method. First we multiply the second equation by -2 to eliminate the variable A. Then the system

$$\begin{cases} 4A + 2C = 22 \\ -2(2A + 3C) = -2(16) \end{cases} \quad \text{simplifies to} \quad \begin{cases} 4A + 2C = 22 \\ \underline{-4A - 6C = -32} \\ \qquad -4C = -10 \end{cases}$$

Add the equations.

$$-4C = -10$$

$$C = \frac{5}{2} = 2.5 \text{ or } \$2.50, \text{ the children's ticket price.}$$

To find A, we replace C with 2.5 in the first equation.

$4A + 2C = 22$	First equation
$4A + 2(2.5) = 22$	Let $C = 2.5$.
$4A + 5 = 22$	
$4A = 17$	
$A = \dfrac{17}{4} = $	4.25 or \$4.25, the adult's ticket price.

4. INTERPRET.

Check: Notice that 4 adults and 2 children will pay $4(\$4.25) + 2(\$2.50) = \$17 + \$5 = \$22$, the required amount. Also, the price for 2 adults and 3 children is $2(\$4.25) + 3(\$2.50) = \$8.50 + \$7.50 = \$16$, the required amount.

State: Answer the three original questions.

a. Since $A = 4.25$, the price of an adult's ticket is \$4.25.
b. Since $C = 2.5$, the price of a child's ticket is \$2.50.
c. The regular admission price for 4 adults and 16 children is

$$4(\$4.25) + 16(\$2.50) = \$17.00 + \$40.00$$
$$= \$57.00$$

This is \$3 less than the special group rate of \$60, so they should *not* request the group rate.

Practice Problem 3

Two cars are 440 miles apart and traveling toward each other. They meet in 3 hours. If one car's speed is 10 miles per hour faster than the other car's speed, find the speed of each car.

	r ·	t =	d
Faster car			
Slower car			

TEACHING TIP

Before beginning Example 3, review the relationship between distance, rate, and time by asking students the following:

▲ Suppose a car traveled at 30 mph for 4 hours, what distance did it travel? (120 miles)
▲ Suppose a car traveled 400 miles in 8 hours, what was its average speed? (50 mph)
▲ Suppose a car traveled 240 miles at 60 miles per hour, how long did it take? (4 hours)

Describe the relationship between distance, rate, and time as an equation.

Answer

3. One car's speed is $68\frac{1}{3}$ mph and the other car's speed is $78\frac{1}{3}$ mph.

Example 3   Finding Rates

Albert and Louis live 15 miles away from each other. They decide to meet one day by walking toward one another. After 2 hours they meet. If Louis walks one mile per hour faster than Albert, find both walking speeds.

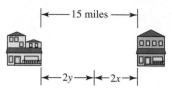

Solution

1. UNDERSTAND. Read and reread the problem. Let's propose a solution and use the formula $d = r \cdot t$ to check. Suppose that Louis' rate is 4 miles per hour. Since Louis' rate is 1 mile per hour faster, Albert's rate is 3 miles per hour. To check, see if they can walk a total of 15 miles in 2 hours. Louis' distance is rate · time = $4(2) = 8$ miles and Albert's distance is rate · time = $3(2) = 6$ miles. Their total distance is 8 miles + 6 miles = 14 miles, not the required 15 miles. Now that we have a better understanding of the problem, let's model it with a system of equations.

First, we let

x = Albert's rate in miles per hour

y = Louis' rate in miles per hour

Now we use the facts stated in the problem and the formula $d = rt$ to fill in the following chart.

	r ·	t =	d
Albert	x	2	$2x$
Louis	y	2	$2y$

2. TRANSLATE. We translate the problem into two equations using both variables.

In words:	Albert's distance	+	Louis' distance	=	15
Translate:	$2x$	+	$2y$	=	15

In words:	Louis' rate	is	1 mile per hour faster than Albert's
Translate:	y	=	$x + 1$

3. SOLVE. The system of equations we are solving is

$$\begin{cases} 2x + 2y = 15 \\ y = x + 1 \end{cases}$$

Let's use substitution to solve the system since the second equation is solved for y.

$$2x + 2y = 15 \quad \text{First equation}$$

$$2x + 2(x + 1) = 15 \quad \text{Replace } y \text{ with } x + 1.$$

$$2x + 2x + 2 = 15$$

$$4x = 13$$

$$x = \frac{13}{4} = 3.25$$

$$y = x + 1 = 3.25 + 1 = 4.25$$

4. INTERPRET. Albert's proposed rate is 3.25 miles per hour and Louis' proposed rate is 4.25 miles per hour.

Check: Use the formula $d = rt$ and find that in 2 hours, Albert's distance is $(3.25)(2)$ miles or 6.5 miles. In 2 hours, Louis' distance is $(4.25)(2)$ miles or 8.5 miles. The total distance walked is 6.5 miles $+$ 8.5 miles or 15 miles, the given distance.

State: Albert walks at a rate of 3.25 miles per hour and Louis walks at a rate of 4.25 miles per hour. ▬▬▬

Example 4   Finding Amounts of Solutions

Eric Daly, a chemistry teaching assistant, needs 10 liters of a 20% saline solution (salt water) for his 2 p.m. laboratory class. Unfortunately, the only mixtures on hand are a 5% saline solution and a 25% saline solution. How much of each solution should he mix to produce the 20% solution?

Solution: **1.** UNDERSTAND. Read and reread the problem. Suppose that we need 4 liters of the 5% solution. Then we need $10 - 4 = 6$ liters of the 25% solution. To see if this gives us 10 liters of a 20% saline solution, let's find the amount of pure salt in each solution.

	concentration rate	$\times$	amount of solution	$=$	amount of pure salt
5% solution:	0.05	$\times$	4 liters	$=$	0.2 liters
25% solution:	0.25	$\times$	6 liters	$=$	1.5 liters
20% solution:	0.20	$\times$	10 liters	$=$	2 liters

Since 0.2 liters $+$ 1.5 liters $=$ 1.7 liters, not 2 liters, our proposed solution is incorrect. But we have gained some insight into how to model and check this problem.

We let

x = number of liters of 5% solution

y = number of liters of 25% solution

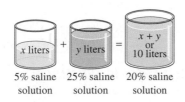

5% saline   25% saline   20% saline
solution   solution   solution

Practice Problem 4

Barb Hayes, a pharmacist, needs 50 liters of a 60% alcohol solution. She currently has available a 20% solution and a 70% solution. How many liters of each does she need to make the needed 50 liters of 60% alcohol solution?

Answer

4. 10 liters of the 20% alcohol solution and 40 liters of the 70% alcohol solution

Now we use a table to organize the given data.

	Concentration Rate	Liters of Solution	Liters of Pure Salt
First solution	5%	x	$0.05x$
Second solution	25%	y	$0.25y$
Mixture needed	20%	10	$(0.20)(10)$

2. TRANSLATE. We translate into two equations using both variables.

In words:
$$\boxed{\text{liters of 5\% solution}} + \boxed{\text{liters of 25\% solution}} = \boxed{10}$$

$$\downarrow \qquad\qquad\qquad \downarrow \qquad\qquad\quad \downarrow$$

Translate:
$$x \qquad + \qquad y \qquad = \quad 10$$

In words:
$$\boxed{\text{salt in 5\% solution}} + \boxed{\text{salt in 25\% solution}} = \boxed{\text{salt in mixture}}$$

$$\downarrow \qquad\qquad\qquad \downarrow \qquad\qquad\quad \downarrow$$

Translate:
$$0.05x \qquad + \qquad 0.25y \qquad = \quad (0.20)(10)$$

3. SOLVE. Here we solve the system

$$\begin{cases} x + y = 10 \\ 0.05x + 0.25y = 2 \end{cases}$$

To solve by the addition method, we first multiply the first equation by -25 and the second equation by 100. Then the system

$$\begin{cases} -25(x + y) = -25(10) \\ 100(0.05x + 0.25y) = 100(2) \end{cases} \quad \begin{matrix}\text{simplifies}\\ \text{to}\end{matrix} \quad \begin{cases} -25x - 25y = -250 \\ \underline{5x + 25y = 200} \\ -20x = -50 \quad \text{Add.}\\ x = 2.5 \end{cases}$$

To find y, we let $x = 2.5$ in the first equation of the original system.

$$x + y = 10$$
$$2.5 + y = 10 \quad \text{Let } x = 2.5.$$
$$y = 7.5$$

4. INTERPRET. Thus, we propose that Eric needs to mix 2.5 liters of 5% saline solution with 7.5 liters of 25% saline solution.

Check: Notice that $2.5 + 7.5 = 10$, the required number of liters. Also, the sum of the liters of salt in the two solutions equals the liters of salt in the required mixture:

$$0.05(2.5) + 0.25(7.5) = 0.20(10)$$
$$0.125 + 1.875 = 2$$

State: Eric needs 2.5 liters of the 5% saline solution and 7.5 liters of the 25% solution.   ▬▬

TRY THE CONCEPT CHECK IN THE MARGIN.

✓ CONCEPT CHECK

Suppose you mix an amount of 30% acid solution with an amount of 50% acid solution. Which of the following acid strengths would be possible for the resulting acid mixture?
a. 22%   b. 44%   c. 63%

✓ Concept Check: b

Name _____ Section _____ Date _____

MENTAL MATH

Without actually solving each problem, choose each correct solution by deciding which choice satisfies the given conditions.

1. The length of a rectangle is 3 feet longer than the width. The perimeter is 30 feet. Find the dimensions of the rectangle.
   a. length = 8 feet; width = 5 feet
   b. length = 8 feet; width = 7 feet
   c. length = 9 feet; width = 6 feet

2. An isosceles triangle, a triangle with two sides of equal length, has a perimeter of 20 inches. Each of the equal sides is one inch longer than the third side. Find the lengths of the three sides.
   a. 6 inches, 6 inches, and 7 inches
   b. 7 inches, 7 inches, and 6 inches
   c. 6 inches, 7 inches, and 8 inches

3. Two computer disks and three notebooks cost $17. However, five computer disks and four notebooks cost $32. Find the price of each.
   a. notebook = $4;
       computer disk = $3
   b. notebook = $3;
       computer disk = $4
   c. notebook = $5;
       computer disk = $2

4. Two music CDs and four music cassette tapes cost a total of $40. However, three music CDs and five cassette tapes cost $55. Find the price of each.
   a. CD = $12; cassette = $4
   b. CD = $15; cassette = $2
   c. CD = $10; cassette = $5

5. Kesha has a total of 100 coins, all of which are either dimes or quarters. The total value of the coins is $13.00. Find the number of each type of coin.
   a. 80 dimes; 20 quarters
   b. 20 dimes; 44 quarters
   c. 60 dimes; 40 quarters

6. Yolanda has 28 gallons of saline solution available in two large containers at her pharmacy. One container holds three times as much as the other container. Find the capacity of each container.
   a. 15 gallons; 5 gallons
   b. 20 gallons; 8 gallons
   c. 21 gallons; 7 gallons

EXERCISE SET 7.4

A *Write a system of equations describing each situation. Do not solve the system. See Example 1.*

1. Two numbers add up to 15 and have a difference of 7.

2. The total of two numbers is 16. The first number plus 2 more than 3 times the second equals 18.

3. Keiko has a total of $6500, which she has invested in two accounts. The larger account is $800 greater than the smaller account.

4. Dominique has four times as much money in his savings account as in his checking account. The total amount is $2300.

Solve. See Example 1.

5. Two numbers total 83 and have a difference of 17. Find the two numbers.

6. The sum of two numbers is 76 and their difference is 52. Find the two numbers.

7. A first number plus twice a second number is 8. Twice the first number plus the second totals 25. Find the numbers.

8. One number is 4 more than twice the second number. Their total is 25. Find the numbers.

9. The highest scorer during the WNBA 1997 regular season was Cynthia Cooper of the Houston Comets. Over the season, Cooper scored 174 more points than the second-highest scorer, Ruthie Bolton-Holifield of the Sacramento Monarchs. Together, Cooper and Bolton-Holifield scored 1068 points during the 1997 regular season. How many points did each player score over the course of the season? (*Source:* Women's National Basketball Association)

10. During the 1996–1997 regular NHL season, Teemu Selanne, of the Mighty Ducks of Anaheim, scored 13 fewer points than the Pittsburgh Penguins' Mario Lemieux. Together, they scored 231 points during the 1996–1997 regular season. How many points each did Selanne and Lemieux score? (*Source:* National Hockey League)

Solve. See Example 2.

11. Ann Marie Jones has been pricing Amtrak train fares for a group trip to New York. Three adults and four children must pay $159. Two adults and three children must pay $112. Find the price of an adult's ticket, and find the price of a child's ticket.

12. Last month, Jerry Papa purchased five cassettes and two compact discs at Wall-to-Wall Sound for $65. This month he bought three cassettes and four compact discs for $81. Find the price of each cassette, and find the price of each compact disc.

Name _____

13. Johnston and Betsy Waring have a jar containing 80 coins, all of which are either quarters or nickels. The total value of the coins is $14.60. How many of each type of coin do they have?

14. Art and Bette Meish purchased 40 stamps, a mixture of 32¢ and 19¢ stamps. Find the number of each type of stamp if they spent $12.15.

15. Fred and Staci Whittingham own 40 shares of Procter & Gamble stock and 25 shares of Microsoft stock. At the close of the markets on September 25, 1997, their stock portfolio was worth $6058.75. The closing price of Microsoft stock was $64.25 more per share than the closing price of Procter & Gamble stock on that day. What was the price of each stock on September 25, 1997? (*Source:* Based on data from Standard & Poor's ComStock)

16. Edie Hall has an investment in Kroger and General Motors stock. On September 25, 1997, Kroger stock closed at $28.3125 per share and General Motors stock closed at $65.125 per share. Edie's portfolio was worth $9513.75 at the end of the day. If Edie owns 60 more shares of General Motors stock than Kroger stock, how many of each type of stock does she own? (*Source:* Based on data from Standard & Poor's ComStock)

Solve. See Example 3.

17. Pratap Puri rowed 18 miles down the Delaware River in 2 hours, but the return trip took him $4\frac{1}{2}$ hours. Find the rate Pratap could row in still water, and find the rate of the current.

	d	$=$ r	$\cdot$ t
Downstream	18	$x + y$	2
Upstream	18	$x - y$	$4\frac{1}{2}$

18. The Jonathan Schultz family took a canoe 10 miles down the Allegheny River in 1 hour and 15 minutes. After lunch it took them 4 hours to return. Find the rate of the current.

	d	$=$ r	$\cdot$ t
Downstream	10	$x + y$	$1\frac{1}{4}$
Upstream	10	$x - y$	4

19. Dave and Sandy Hartranft are frequent flyers with Delta Airlines. They often fly from Philadelphia to Chicago, a distance of 780 miles. On one particular trip they fly into the wind, and the flight takes 2 hours. The return trip, with the wind behind them, only takes $1\frac{1}{2}$ hours. Find the speed of the wind and find the speed of the plane in still air.

20. With a strong wind behind it, a United Airlines jet flies 2400 miles from Los Angeles to Orlando in 4 hours and 45 minutes. The return trip takes 6 hours, as the plane flies into the wind. Find the speed of the plane in still air, and find the wind speed to the nearest tenth mile per hour.

21. 12% solution: $7\frac{1}{2}$ oz; 4% solution: $4\frac{1}{2}$ oz

22. 40% solution: $7\frac{1}{2}$ liters; 10% solution: $7\frac{1}{2}$ liters

23. $4.95 beans: 113 lbs; $2.65 beans: 87 lbs

24. macadamia: 4.1 lbs; standard mix: 35.9 lbs

25. 16

26. 9

27. $36x^2$

28. $121y^2$

29. $100y^6$

30. $64x^{10}$

31. 60°, 30°

32. 32°, 148°

Name _____

Solve. See Example 4.

21. Dorren Schmidt is a chemist with Gemco Pharmaceutical. She needs to prepare 12 ounces of a 9% hydrochloric acid solution. Find the amount of 4% and the amount of 12% solution she should mix to get this solution.

22. Elise Everly is preparing 15 liters of a 25% saline solution. Elise has two other saline solutions with strengths of 40% and 10%. Find the amount of 40% solution and the amount of 10% solution she should mix to get 15 liters of a 25% solution.

23. Wayne Osby blends coffee for Maxwell House. He needs to prepare 200 pounds of blended coffee beans selling for $3.95 per pound. He intends to do this by blending together a high-quality bean costing $4.95 per pound and a cheaper bean costing $2.65 per pound. To the nearest pound, find how much high-quality coffee bean and how much cheaper coffee bean he should blend.

24. Macadamia nuts cost an astounding $16.50 per pound, but research by Planter's Peanuts says that mixed nuts sell better if macadamias are included. The standard mix costs $9.25 per pound. Find how many pounds of macadamias and how many pounds of the standard mix should be combined to produce 40 pounds that will cost $10 per pound. Find the amounts to the nearest tenth of a pound.

REVIEW AND PREVIEW

Find the square of each expression. For example, the square of 7 is 7^2 or 49. The square of 5x is $(5x)^2$ or $25x^2$. See Section 3.1.

25. 4

26. 3

27. $6x$

28. $11y$

29. $10y^3$

30. $8x^5$

COMBINING CONCEPTS

31. Find the measures of two complementary angles if one angle is twice the other. (Recall that two angles are complementary if their sum is 90°.)

32. Find the measures of two supplementary angles if one angle is 20° more than four times the other. (Recall that two angles are supplementary if their sum is 180°.)

33. Find the measures of two complementary angles if one angle is 10° more than three times the other.

34. Find the measures of two supplementary angles if one angle is 18° more than twice the other.

35. Carrie and Raymond McCormick had a pottery stand at the annual Skippack Craft Fair. They sold some of their pottery at the original price of $9.50 each, but later decreased the price of each by $2. If they sold all 90 pieces and took in $721, find how many they sold at the original price and how many they sold at the reduced price.

36. Trinity Church held its annual spaghetti supper and fed a total of 387 people. They charged $6.80 for adults and half-price for children. If they took in $2444.60, find how many adults and how many children attended the supper.

37. Dale Maxfield has decided to fence off a garden plot behind his house, using his house as the "fence" along one side of the garden. The length (which runs parallel to the house) is 3 feet less than twice the width. Find the dimensions if 33 feet of fencing is used along the three sides requiring it.

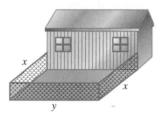

38. Judy McElroy plans to erect 152 feet of fencing around her rectangular horse pasture. A river bank serves as one side of the rectangle. If each width is 4 feet longer than half the length, find the dimensions.

Internet Excursions

Go to http://www.prenhall.com/martin-gay
Major League Soccer (MLS) had its inaugural season in the United States in 1996 with 10 teams. The given World Wide Web address will provide you with access to the Major League Soccer homepage, or a related site, for a listing of statistics for the current regular season.

39. Using actual data for the current (or most recent) season, write a problem similar to Exercises 9 and 10 involving statistics for the MLS leading scorers. When you have finished writing your problem, trade with another student in your class and solve each other's problem. Then check each other's work.

40. Using actual data for the current (or most recent) season, write a problem similar to Exercises 9 and 10 involving statistics for MLS game attendance. When you have finished writing your problem, trade with another student in your class and solve each other's problem. Then check each other's work.

33. 20°, 70°

34. 54°, 126°

35. number sold at $9.50: 23; number sold at $7.50: 67

36. adults: 332; children: 55

37. width: 9 ft; length: 15 ft

38. width: 40 ft; length: 72 ft

39. answers may vary

40. answers may vary

Focus On Business and Career

When you finish your present course of study, you will probably look for a job. When your job search has paid off and you receive a job offer, how will you decide whether to take the job? How do you decide between two or more job offers? These decisions are an important part of the job search and may not be easy to make. To evaluate the job offer, you should consider the nature of the work involved in the job, the type of company or organization that has offered the job, and the salary and benefits offered by the employer. You may also need to compare the compensation packages of two or more job offers. The following hints on assessing a job's compensation package were included in an article by Max Carry in the *Winter, 1990–91 Occupational Outlook Handbook Quarterly*.

Most companies will not talk about pay until they have decided to hire you. In order to know if their offer is reasonable, you need a rough estimate of what the job should pay. You may have to go to several sources for this information. Talk to friends who recently were hired in similar jobs. Ask your teachers and the staff in the college placement office about starting pay for graduates with your qualifications. Scan the help-wanted ads in newspapers. Check the library or your school's career center for salary surveys, such as the College Council Salary Survey and the Bureau of Labor Statistics wage surveys. If you are considering the salary and benefits for a job in another geographic area, make allowances for differences in the cost of living. Use the research to come up with a base salary range for yourself, the top being the best you can hope to get and the bottom being the least you will take. When negotiating, aim for the top of your estimated salary range, but be prepared to settle for less. Entry level salaries sometimes are not negotiable, particularly in government agencies. If you are not pleased with an offer, however, what harm could come from asking for more?

An employer cannot be specific about the amount of pay if it includes commissions and bonuses. The way the pay plan works, however, should be explained. The employer also should be able to tell you what most people in the job are earning. You should also learn the organization's policy regarding overtime. Depending on the job, you may or may not be exempt from laws requiring the employer to compensate you for overtime. Find out how many hours you will be expected to work each week and whether evening and weekend work is required or expected. Will you receive overtime pay or time off for working more than the specified number of hours in a week? Also take into account that the starting salary is just that, the start. Your salary should be reviewed on a regular basis—many organizations do it every 12 months. If the employer is pleased with your performance, how much can you expect to make after one year? Two years? Three years and so on?

Don't think of salary as the only compensation you will receive. Consider benefits. What on the surface looks like a great salary could be accompanied by little else. Do you know the value of the employer's contribution to your benefits? According to a 1989 Bureau of Labor Statistics study, for each dollar employers in private industry spent on straight-time wages and salaries, they contributed on average another 38 cents to employee benefits, including contributions required by law. Benefits can add a lot to your base pay. Your benefit package probably will consist of health insurance, life insurance, a pension plan, and paid vacations, holidays, and sick leave. It also may include items as diverse as profit sharing, moving expenses, parking space, a company car, and on-site day care. Do you know exactly what the benefit package includes and how much of the cost must be borne by you? Depending on your circumstances, you might want to increase or decrease particular benefits.

CRITICAL THINKING

1. Suppose you have been searching for a position as an electronics sales associate. You have received two job offers. The first job pays a monthly salary of $1500 per month plus a commission of 4% on all sales made. The second pays a monthly salary of $1800 per month plus a commission of 2% on all sales made. At what level of monthly sales would the jobs pay the same amount? Based only on the given information about the jobs, which job would you choose? Why? What other information would you want to have about the jobs before making a decision?

2. Suppose you have been searching for an entry-level bookkeeping position. You have received two job offers. The first company offers you a starting hourly wage of $6.50 per hour and says that each year entry-level workers receive a raise of $0.75 per hour. The second company offers you a starting hourly wage of $7.50 per hour and says that you can expect a $0.50 per hour raise each year. After how many years will the two jobs pay the same hourly wage? Based only on the given information about the jobs, which job would you choose? Why? What other information would you want to have about the jobs before making a decision?

MATERIALS:

▲ Ruler, graphing calculator (optional)

This activity may be completed by working in groups or individually.

From overhead photographs or satellite imagery of ships on the ocean, defense analysts can tell a lot about a ship's immediate course by looking at its wake. Assuming that two ships will maintain their present courses, it is possible to extend their paths, based on the wakes visible in the photograph, and find possible points of collision.

Investigate the courses and possibility of collision of the two ships shown in the figure. Assume that the ships will maintain their present courses.

1. Using each ship's wake as a guide, extend the paths of the ships on the figure. Estimate the coordinates of the point of intersection of the ships' courses from the grid. If the ships continue in these courses, they could possibly collide at the point of intersection of their paths.
    Answers should be close to (46, 90).

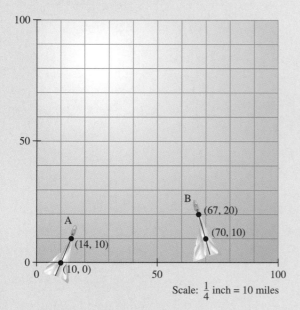

Scale: $\frac{1}{4}$ inch = 10 miles

2. Using the coordinates labeled on each ship's wake, find a linear equation that describes each path.
    Ship A: $y = \frac{5}{2}x - 25$; Ship B: $y = -\frac{10}{3}x + 243\frac{1}{3}$

3. (Optional) Use a graphing calculator to graph both equations in the same window. Use the Intersect or Trace feature to estimate the point of intersection of the two paths. Compare this estimate to your estimate in Question 1.
    (46, 90)

4. Solve the system of two linear equations using one of the methods in this chapter. The solution is the point of intersection of the two paths. Compare your answer to your estimates from Questions 1 and 3.
    (46, 90)

5. Plot the point of intersection you found in Question 4 on the figure. Use the figure's scale to find each ship's distance from this point of collision by measuring from the bow (tip) of each ship with a ruler. Suppose that the speed of ship A is r_1 and the speed of ship B is r_2. Given the present positions and courses of the two ships, find a relationship between their speeds that would ensure their collision.
    answers may vary

DEFINITIONS AND CONCEPTS	EXAMPLES

SECTION 7.1 SOLVING SYSTEMS OF LINEAR EQUATIONS BY GRAPHING

A **system of linear equations** consists of two or more linear equations.

A **solution** of a system of two equations in two variables is an ordered pair of numbers that is a solution of both equations in the system.

$$\begin{cases} 2x + y = 6 \\ x = -3y \end{cases} \quad \begin{cases} -3x + 5y = 10 \\ x - 4y = -2 \end{cases}$$

Determine whether $(-1, 3)$ is a solution of the system.

$$\begin{cases} 2x - y = -5 \\ x = 3y - 10 \end{cases}$$

Replace x with -1 and y with 3 in both equations.

$$2x - y = -5$$
$$2(-1) - 3 \stackrel{?}{=} -5$$
$$-5 = -5 \quad \text{True.}$$

$$x = 3y - 10$$
$$-1 \stackrel{?}{=} 3(3) - 10$$
$$-1 = -1 \quad \text{True.}$$

$(-1, 3)$ is a solution of the system.

Graphically, a solution of a system is a point common to the graphs of both equations.

Solve by graphing: $\begin{cases} 3x - 2y = -3 \\ x + y = 4 \end{cases}$

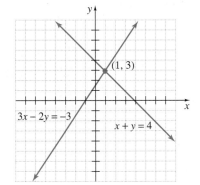

Three different situations can occur when graphing the two lines associated with the equations in a linear system.

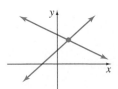

One point of inter-section; one solution

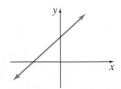

Same line; infinite number of solutions

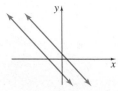

Parallel lines; no solution

SECTION 7.2   SOLVING SYSTEMS OF LINEAR EQUATIONS BY SUBSTITUTION

TO SOLVE A SYSTEM OF LINEAR EQUATIONS BY THE SUBSTITUTION METHOD

Step 1. Solve one equation for a variable.

Step 2. Substitute the expression for the variable into the other equation.

Step 3. Solve the equation from Step 2 to find the value of one variable.

Step 4. Substitute the value from Step 3 in either original equation to find the value of the other variable.

Step 5. Check the solution in both equations.

Solve by substitution.

$$\begin{cases} 3x + 2y = 1 \\ x = y - 3 \end{cases}$$

Substitute $y - 3$ for x in the first equation

$$3x + 2y = 1$$
$$3(y - 3) + 2y = 1$$
$$3y - 9 + 2y = 1$$
$$5y = 10$$
$$y = 2 \quad \text{Divide by 5.}$$

To find x, substitute 2 for y in $x = y - 3$ so that $x = 2 - 3$ or -1. The solution $(-1, 2)$ checks.

SECTION 7.3   SOLVING SYSTEMS OF LINEAR EQUATIONS BY ADDITION

TO SOLVE A SYSTEM OF LINEAR EQUATIONS BY THE ADDITION METHOD

Step 1. Rewrite each equation in standard form $Ax + By = C$.

Step 2. Multiply one or both equations by a nonzero number so that the coefficients of a variable are opposites.

Step 3. Add the equations.

Step 4. Find the value of one variable by solving the resulting equation.

Step 5. Substitute the value from Step 4 into either original equation to find the value of the other variable.

Step 6. Check the solution in both equations.

If solving a system of linear equations by substitution or addition yields a true statement such as $-2 = -2$, then the graphs of the equations in the system are identical and there is an infinite number of solutions of the system.

If solving a system of linear equations yields a false statement such as $0 = 3$, the graphs of the equations in the system are parallel lines and the system has no solution.

Solve by addition

$$\begin{cases} x - 2y = 8 \\ 3x + y = -4 \end{cases}$$

Multiply both sides of the first equation by -3.

$$\begin{cases} -3x + 6y = -24 \\ \underline{3x + y = -4} \end{cases}$$
$$7y = -28 \quad \text{Add.}$$
$$y = -4 \quad \text{Divide by 7.}$$

To find x, let $y = -4$ in an original equation.

$$x - 2(-4) = 8 \quad \text{First equation}$$
$$x + 8 = 8$$
$$x = 0$$

The solution $(0, -4)$ checks.

Solve: $\begin{cases} 2x - 6y = -2 \\ x = 3y - 1 \end{cases}$

Substitute $3y - 1$ for x in the first equation.

$$2(3y - 1) - 6y = -2$$
$$6y - 2 - 6y = -2$$
$$-2 = -2 \quad \text{True.}$$

The system has an infinite number of solutions.

Solve: $\begin{cases} 5x - 2y = 6 \\ -5x + 2y = -3 \end{cases}$

$$\underline{}$$
$$0 = 3 \quad \text{False.}$$

The system has no solution.

PROBLEM-SOLVING STEPS

1. UNDERSTAND. Read and reread the problem.

Two angles are supplementary if their sum is 180°. The larger of two supplementary angles is three times the smaller, decreased by twelve. Find the measure of each angle. Let

$$x = \text{measure of smaller angle}$$
$$y = \text{measure of larger angle}$$

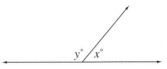

2. TRANSLATE.

In words:

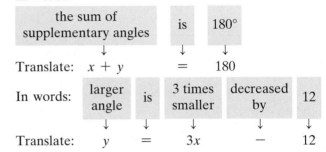

the sum of supplementary angles	is	180°
↓	↓	↓

Translate: $x + y$ = 180

In words:

larger angle	is	3 times smaller	decreased by	12
↓	↓	↓	↓	↓

Translate: y = $3x$ − 12

3. SOLVE.

Solve the system

$$\begin{cases} x + y = 180 \\ y = 3x - 12 \end{cases}$$

Use the substitution method and replace y with $3x - 12$ in the first equation.

$$x + y = 180$$
$$x + (3x - 12) = 180$$
$$4x = 192$$
$$x = 48$$

4. INTERPRET.

Since $y = 3x - 12$, then $y = 3 \cdot 48 - 12$ or 132.

The solution checks. The smaller angle measures 48° and the larger angle measures 132°.

CHAPTER 7 REVIEW

(7.1) *Determine whether each ordered pair is a solution of the system of linear equations.*

1. $\begin{cases} 2x - 3y = 12 \\ 3x + 4y = 1 \end{cases}$
    2. $\begin{cases} 4x + y = 0 \\ -8x - 5y = 9 \end{cases}$
    3. $\begin{cases} 5x - 6y = 18 \\ 2y - x = -4 \end{cases}$
    4. $\begin{cases} 2x + 3y = 1 \\ 3y - x = 4 \end{cases}$

  a. $(12, 4)$ no
                  a. $\left(\dfrac{3}{4}, -3\right)$ yes
          a. $(-6, -8)$ no
          a. $(2, 2)$ no

  b. $(3, -2)$ yes
                b. $(-2, 8)$ no
          b. $\left(3, \dfrac{5}{2}\right)$ no
          b. $(-1, 1)$ yes

Solve each system of equations by graphing.

5. $\begin{cases} x + y = 5 \\ x - y = 1 \end{cases}$
                                   6. $\begin{cases} x + y = 3 \\ x - y = -1 \end{cases}$

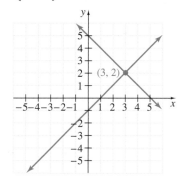
        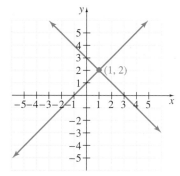

7. $\begin{cases} x = 5 \\ y = -1 \end{cases}$
                                     8. $\begin{cases} x = -3 \\ y = 2 \end{cases}$

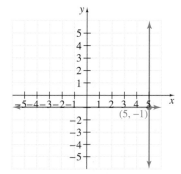
        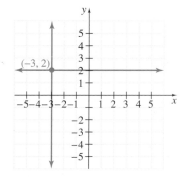

9. $\begin{cases} 2x + y = 5 \\ x = -3y \end{cases}$

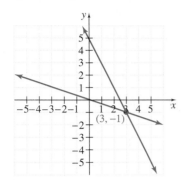

(3, −1)

10. $\begin{cases} 3x + y = -2 \\ y = -5x \end{cases}$

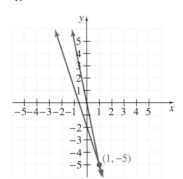

(1, −5)

11. $\begin{cases} y = 2x + 4 \\ y = -x - 5 \end{cases}$

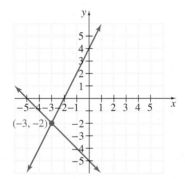

(−3, −2)

12. $\begin{cases} y = x - 5 \\ y = -2x + 2 \end{cases}$

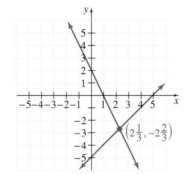

$\left(2\tfrac{1}{3}, -2\tfrac{2}{3}\right)$

13. $\begin{cases} y = 3x \\ -6x + 2y = 6 \end{cases}$

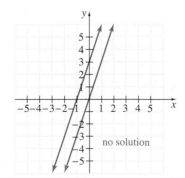

no solution

14. $\begin{cases} x - 2y = 2 \\ -2x + 4y = -4 \end{cases}$

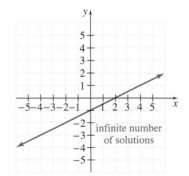

infinite number of solutions

(7.2) *Solve each system of equations by the substitution method.*

15. $\begin{cases} x = 2y \\ 2x - 3y = 2 \end{cases}$ $(4, 2)$

16. $\begin{cases} x = 5y \\ x - 4y = 1 \end{cases}$ $(5, 1)$

17. $\begin{cases} y = 2x + 6 \\ 3x - 2y = -11 \end{cases}$ $(-1, 4)$

18. $\begin{cases} y = 3x - 7 \\ 2x - 3y = 7 \end{cases}$ $(2, -1)$

19. $\begin{cases} x + 3y = -3 \\ 2x + y = 4 \end{cases}$ $(3, -2)$

20. $\begin{cases} 3x + y = 11 \\ x + 2y = 12 \end{cases}$ $(2, 5)$

21. $\begin{cases} 4y = 2x - 3 \\ x - 2y = 4 \end{cases}$ no solution

22. $\begin{cases} 2x = 3y - 18 \\ x + 4y = 2 \end{cases}$ $(-6, 2)$

23. $\begin{cases} x + y = 6 \\ y = -x - 4 \end{cases}$ no solution

24. $\begin{cases} -3x + y = 6 \\ y = 3x + 2 \end{cases}$ no solution

(7.3) *Solve each system of equations by the addition method.*

25. $\begin{cases} x + y = 14 \\ x - y = 18 \end{cases}$ $(16, -2)$

26. $\begin{cases} x + y = 9 \\ x - y = 13 \end{cases}$ $(11, -2)$

27. $\begin{cases} 2x + 3y = -6 \\ x - 3y = -12 \end{cases}$ $(-6, 2)$

28. $\begin{cases} 4x + y = 15 \\ -4x + 3y = -19 \end{cases}$ $(4, -1)$

29. $\begin{cases} 2x - 3y = -15 \\ x + 4y = 31 \end{cases}$ $(3, 7)$

30. $\begin{cases} x - 5y = -22 \\ 4x + 3y = 4 \end{cases}$ $(-2, 4)$

31. $\begin{cases} 2x - 6y = -1 \\ -x + 3y = \dfrac{1}{2} \end{cases}$ infinite number of solutions

32. $\begin{cases} -4x - 6y = 8 \\ 2x + 3y = -3 \end{cases}$ no solution

33. $\begin{cases} \dfrac{3}{4}x + \dfrac{2}{3}y = 2 \\ 3x + y = 18 \end{cases}$ $(8, -6)$

34. $\begin{cases} \dfrac{2}{5}x + \dfrac{3}{4}y = 1 \\ x + 3y = -2 \end{cases}$ $(10, -4)$

35. $\begin{cases} 5x + 2y = 9 \\ 3x + 4y = 25 \end{cases}$ $(-1, 7)$

36. $\begin{cases} 6x - 3y = -15 \\ 4x - 2y = -10 \end{cases}$ infinite number of solutions

(7.4) *Solve each problem by writing and solving a system of linear equations.*

37. The sum of two numbers is 16. Three times the larger number decreased by the smaller number is 72. Find the two numbers.   −6 and 22

38. The Forrest Theater can seat a total of 360 people. They take in $15,150 when every seat is sold. If orchestra section tickets cost $45 and balcony tickets cost $35, find the number of seats in the orchestra section and the number of seats in the balcony. orchestra: 255 seats; balcony: 105 seats

39. A riverboat can head 340 miles upriver in 19 hours, but the return trip takes only 14 hours. Find the current of the river and find the speed of the ship in still water to the nearest tenth of a mile. current of river: 3.2 mph; speed in still water: 21.1 mph

40. Sam and Cynthia Abney invested $9000 one year ago. Part of the money was invested at 6%, the rest at 10%. If the total interest earned in one year was $652.80, find how much was invested at each rate. amount invested at 6%: $6,180; amount invested at 10%: $2,820

	d =	r ·	t
Upriver	340	$x - y$	19
Downriver	340	$x + y$	14

41. Ancient Greeks thought that a picture had the most pleasing dimensions if the length was approximately 1.6 times as long as the width. This ratio is known as the Golden Ratio. If Sandreka Walker has 6 feet of framing material, find the dimensions of the largest frame she can make that satisfies the Golden Ratio. Find the dimensions to the nearest hundredth of a foot.   length: 1.85 ft; width: 1.15 ft

42. Find the amount of 6% acid solution and the amount of 14% acid solution Pat Mayfield should combine to prepare 50 cc (cubic centimeters) of a 12% solution.   6% solution: $12\frac{1}{2}$ cc; 14% solution: $37\frac{1}{2}$ cc

43. A deli charges $3.80 for a breakfast of three eggs and four strips of bacon. The charge is $2.75 for two eggs and three strips of bacon. Find the cost of each egg and the cost of each strip of bacon.   egg: 40¢; strip of bacon: 65¢

44. An exercise enthusiast alternates between jogging and walking. He traveled 15 miles during the past 3 hours. He jogs at a rate of 7.5 miles per hour and walks at a rate of 4 miles per hour. Find how much time, to the nearest hundredth of an hour, he actually spent jogging and how much time he spent walking.   jogging: 0.86 hour; walking: 2.14 hours

Name _____ **Section** _____ **Date** _____

CHAPTER 7 TEST

Solve each system by graphing.

1. $\begin{cases} x - y = 2 \\ 3x - y = -2 \end{cases}$

2. $\begin{cases} y = -3x \\ 3x + y = 6 \end{cases}$

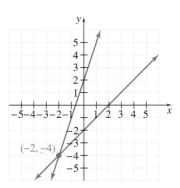

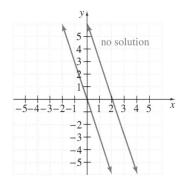

Solve each system by the substitution method.

3. $\begin{cases} 3x - 2y = -14 \\ y = x + 5 \end{cases}$

4. $\begin{cases} 3x + y = 7 \\ 4x + 3y = 1 \end{cases}$

5. $\begin{cases} x - y = 4 \\ x - 2y = 11 \end{cases}$

6. $\begin{cases} 8x - 4y = 12 \\ y = 2x - 3 \end{cases}$

Solve each system by the addition method.

7. $\begin{cases} x + y = 28 \\ x - y = 12 \end{cases}$

8. $\begin{cases} y - x = 6 \\ y + 2x = -6 \end{cases}$

9. $\begin{cases} 5x - 6y = 7 \\ 7x - 4y = 12 \end{cases}$

10. $\begin{cases} x - \dfrac{2}{3}y = 3 \\ -2x + 3y = 10 \end{cases}$

589

11. 78, 46

$1 bills: 20;
12. $5 bills: 42

30% solution: 7.5 liters; 70% solution:
13. 2.5 liters

14. 3 mph; 6 mph

15. 1992

16. 1993–1995

Name ______________________________

Solve each problem by writing and using a system of linear equations.

11. Two numbers have a sum of 124 and a difference of 32. Find the numbers.

12. Lisa has a bundle of money consisting of $1 bills and $5 bills. There are 62 bills in the bundle. The total value of the bundle is $230. Find the number of $1 bills and the number of $5 bills.

13. A 30% alcohol solution is to be mixed with a 70% alcohol solution. How much of each is needed to make 10 liters of 40% solution?

14. Two hikers start at opposite ends of the St. Tammany Trails and walk toward each other. The trail is 36 miles long and they meet in 4 hours. If one hiker is twice as fast as the other, find both hiking speeds.

The graph below shows the average hourly earnings of production workers in manufacturing industries for Illinois and Wisconsin. (Source: U.S. Bureau of Labor Statistics) Use this graph to answer Questions 15 and 16.

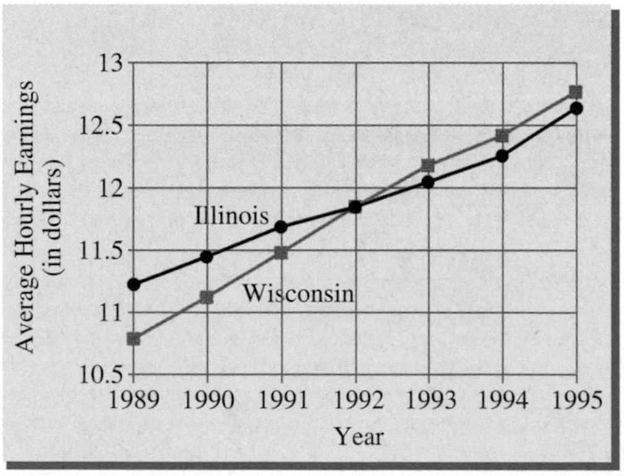

15. In what year were the average hourly earnings of production workers in manufacturing industries in Wisconsin equal to the average hourly earnings of production workers in Illinois?

16. In what year(s) were the average hourly earnings of production workers in Illinois less than the average hourly earnings of production workers in Wisconsin?

Name _____ **Section** _____ **Date** _____

Cumulative Review

1. Simplify each expression.

    a. $-14 - 8 + 10 - (-6)$

    b. $1.6 - (-10.3) + (-5.6)$

Find the reciprocal of each number.

2. 22

3. $\dfrac{3}{16}$

4. -10

5. $-\dfrac{9}{13}$

6. 1.7

7. **a.** The sum of two numbers is 8. If one number is 3, find the other number.

8. Solve:
   $$-2(x - 5) + 10 = -3(x + 2) + x$$

7. **b.** The sum of two numbers is 8. If one number is x, write an expression representing the other number.

9. Write a ratio for each phrase. Use fractional notation.

    a. The ratio of 2 parts salt to 5 parts water

10. Solve $-5x + 7 < 2(x - 3)$. Graph the solutions.

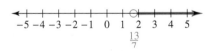

9. **b.** The ratio of 18 inches to 2 feet

Simplify each expression.

11. $\left(\dfrac{m}{n}\right)^7$

12. $\left(\dfrac{2x^4}{3y^5}\right)^4$

13. Subtract: $(2x^3 + 8x^2 - 6x) - (2x^3 - x^2 + 1)$

1. **a.** -6

    b. 6.3 (Sec. 1.4, Ex. 6)

2. $\dfrac{1}{22}$ (Sec. 1.6, Ex. 1)

3. $\dfrac{16}{3}$; (Sec. 1.6, Ex. 2)

4. $-\dfrac{1}{10}$ (Sec. 1.6, Ex. 3)

5. $-\dfrac{13}{9}$ (Sec. 1.6, Ex. 4)

6. $\dfrac{1}{1.7}$ (Sec. 1.6, Ex. 5)

7. **a.** 5

    b. $8 - x$ (Sec. 2.2, Ex. 8)

8. no solution (Sec. 2.4, Ex. 6)

9. **a.** $\dfrac{2}{5}$

    b. $\dfrac{3}{4}$ (Sec. 2.7, Ex. 4)

10. $x > \dfrac{13}{7}$ (Sec. 2.8, Ex. 7)

11. $\dfrac{m^7}{n^7}, n \neq 0$ (Sec. 3.1, Ex. 19)

12. $\dfrac{16x^{16}}{81y^{20}}, y \neq 0$ (Sec. 3.1, Ex. 20)

13. $9x^2 - 6x - 1$ (Sec. 3.4, Ex. 5)

Name _____

14. Divide $6x^2 + 10x - 5$ by $3x - 1$.

15. Solve: $x(2x - 7) = 4$

16. Find the lengths of the sides of a right triangle if the lengths can be expressed by three consecutive even integers.

17. Subtract: $\dfrac{2y}{2y - 7} - \dfrac{7}{2y - 7}$

18. Simplify: $\dfrac{\dfrac{x}{y} + \dfrac{3}{2x}}{\dfrac{x}{2} + y}$

19. Find the slope of the line $y = -1$.

20. Find an equation of the line through $(2, 5)$ and $(-3, 4)$. Write the equation in the form $Ax + By = C$.

21. Find the domain and the range of the relation $\{(0, 2), (3, 3), (-1, 0), (3, -2)\}$.

22. Determine whether $(12, 6)$ is a solution of the system $\begin{cases} 2x - 3y = 6 \\ x = 2y \end{cases}$

23. Solve the system $\begin{cases} x + 2y = 7 \\ 2x + 2y = 13 \end{cases}$

24. Solve the system $\begin{cases} -x - \dfrac{y}{2} = \dfrac{5}{2} \\ -\dfrac{x}{2} + \dfrac{y}{4} = 0 \end{cases}$

25. Find two numbers whose sum is 37 and whose difference is 21.

Roots and Radicals

Having spent the last chapter studying equations, we return now to algebraic expressions. We expand on our skills of operating on expressions—adding, subtracting, multiplying, dividing, and raising to powers—to include finding roots. Just as subtraction is defined by addition and division by multiplication, finding roots is defined by raising to powers. As we master finding roots, we will work with equations that contain roots and solve problems that can be modeled by such equations.

Nearly 85% of the gold recovered by humankind in all of recorded history is still in use. It's likely that the gold we use today in electronics or jewelry was once mined by the ancient Egyptians or retrieved from the New World by Christopher Columbus in the name of Spain. Gold's scarcity and durability made it a natural form of currency. By the 17th century, merchants who had accumulated gold were finding it bulky to transport and difficult to store safely. They began leaving their gold with goldsmiths in return for a receipt. Trading gold receipts having the same value as the gold itself became a popular way to conduct business and the idea of gold-backed paper money was born. In Exercise 65 on page 602, roots are used to find the dimensions of a cube representing all of the gold found by humankind since the beginning of recorded history.

1. −7; (8.1A)

2. $\frac{2}{5}$; (8.1A)

3. −4; (8.1B)

4. $2\sqrt{30}$; (8.2A)

5. $\frac{2\sqrt{6}}{y^3}$; (8.2C)

6. $2\sqrt[3]{14}$; (8.2D)

7. $-3\sqrt{15}$; (8.3A)

8. 0; (8.3B)

9. $\frac{7\sqrt{7}}{10}$; (8.3B)

10. $6\sqrt{3}$; (8.4A)

11. $2\sqrt{7} - \sqrt{10}$; (8.4A)

12. $y - 6\sqrt{y} + 9$; (8.4A)

13. $2x\sqrt{7}$; (8.4B)

14. $\frac{\sqrt{55}}{11}$; (8.4C)

15. $\frac{8\sqrt{2a}}{a}$; (8.4C)

16. $\frac{6 + 3\sqrt{x}}{4 - x}$; (8.4D)

17. $x = 49$; (8.5A)

18. $x = \frac{9}{4}$; (8.5B)

19. $4\sqrt{10}$ cm; (8.6A)

20. 2.52 in.; (8.6B)

Name _____ Section _____ Date _____

CHAPTER 8 PRETEST

Simplify the following. Indicate if the expression is not a real number. Assume that x represents a positive number.

1. $-\sqrt{49}$

2. $\sqrt{\dfrac{4}{25}}$

3. $\sqrt[3]{-64}$

4. $\sqrt{120}$

5. $\sqrt{\dfrac{24}{y^6}}$

6. $\sqrt[3]{112}$

Perform each indicated operation.

7. $\sqrt{15} + 2\sqrt{15} - 6\sqrt{15}$

8. $3\sqrt{12} - 2\sqrt{27}$

9. $\sqrt{\dfrac{7}{4}} + \sqrt{\dfrac{7}{25}}$

10. $\sqrt{6} \cdot \sqrt{18}$

11. $\sqrt{2}(\sqrt{14} - \sqrt{5})$

12. $(\sqrt{y} - 3)^2$

13. $\dfrac{\sqrt{56x^5}}{\sqrt{2x^3}}$

Rationalize each denominator.

14. $\sqrt{\dfrac{5}{11}}$

15. $\dfrac{16}{\sqrt{2a}}$

16. $\dfrac{3}{2 - \sqrt{x}}$

Solve each of the following radical equations.

17. $\sqrt{x} + 9 = 16$

18. $\sqrt{x + 4} = \sqrt{x} + 1$

19. Find the length of the unknown leg of the right triangle. Give an exact answer.

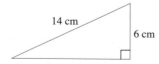

20. The formula $r = \sqrt{\dfrac{S}{4\pi}}$ can be used to find the radius of a sphere given its surface area S. Use this formula to approximate the radius of a sphere if its surface area is 80 square inches. Round to two decimal places.

8.1 Introduction to Radicals

A Finding Square Roots

In this section, we define finding the **root** of a number by its reverse operation, raising a number to a power. We begin with squares and square roots.

The square of 5 is $5^2 = 25$.
The square of -5 is $(-5)^2 = 25$.
The square of $\frac{1}{2}$ is $\left(\frac{1}{2}\right)^2 = \frac{1}{4}$.

The reverse operation of squaring a number is finding the **square root** of a number. For example,

A square root of 25 is 5, because $5^2 = 25$.
A square root of 25 is also -5, because $(-5)^2 = 25$.
A square root of $\frac{1}{4}$ is $\frac{1}{2}$, because $\left(\frac{1}{2}\right)^2 = \frac{1}{4}$.

In general, the number b is a square root of a number a if $b^2 = a$.

The symbol $\sqrt{}$ is used to denote the **positive** or **principal square root** of a number. For example,

$$\sqrt{25} = 5 \text{ only, since } 5^2 = 25 \text{ and } 5 \text{ is positive.}$$

The symbol $-\sqrt{}$ is used to denote the **negative square root**. For example,

$$-\sqrt{25} = -5$$

The symbol $\sqrt{}$ is called a **radical** or **radical sign**. The expression within or under a radical sign is called the **radicand**. An expression containing a radical is called a **radical expression**.

Square Root

If a is a positive number, then

$\sqrt{a}$ is the **positive square root** of a
$-\sqrt{a}$ is the **negative square root** of a

Also, $\sqrt{0} = 0$.

📼 Examples

Find each square root.

1. $\sqrt{36} = 6$, because $6^2 = 36$ and 6 is positive.
2. $\sqrt{64} = 8$, because $8^2 = 64$ and 8 is positive.
3. $-\sqrt{25} = -5$. The negative sign in front of the radical indicates the negative square root of 25.
4. $\sqrt{\frac{9}{100}} = \frac{3}{10}$ because $\left(\frac{3}{10}\right)^2 = \frac{9}{100}$ and $\frac{3}{10}$ is positive.
5. $\sqrt{0} = 0$ because $0^2 = 0$. ━━━

Is the square root of a negative number a real number? For example, is $\sqrt{-4}$ a real number? To answer this question, we ask ourselves, is there a real number whose square is -4? Since there is no real number whose square is -4, we say that $\sqrt{-4}$ is not a real number. In general,

Objectives

A Find square roots.
B Find cube roots.
C Find nth roots.
D Approximate square roots.
E Simplify radicals containing variables.

SSM   CD-ROM   Video
8.1

Practice Problems 1–5

Find each square root.

1. $\sqrt{100}$

2. $\sqrt{9}$

3. $-\sqrt{36}$

4. $\sqrt{\frac{25}{81}}$

5. $\sqrt{1}$

Answers

1. 10, **2.** 3, **3.** -6, **4.** $\frac{5}{9}$, **5.** 1

A square root of a negative number is not a real number.

B FINDING CUBE ROOTS

We can find roots other than square roots. For example, since $2^3 = 8$, we call 2 the **cube root** of 8. In symbols, we write

$$\sqrt[3]{8} = 2 \quad \text{The number 3 is called the } \textbf{index}.$$

Also,

$$\sqrt[3]{27} = 3 \qquad \text{Since } 3^3 = 27$$
$$\sqrt[3]{-64} = -4 \quad \text{Since } (-4)^3 = -64$$

Notice that unlike the square root of a negative number, the cube root of a negative number is a real number. This is so because while we cannot find a real number whose *square* is negative, we *can* find a real number whose *cube* is negative. In fact, the cube of a negative number is a negative number. Therefore, the cube root of a negative number is a negative number.

Examples   Find each cube root.

> **6.** $\sqrt[3]{1} = 1$ because $1^3 = 1$.
> **7.** $\sqrt[3]{-27} = -3$ because $(-3)^3 = -27$.
> **8.** $\sqrt[3]{\dfrac{1}{125}} = \dfrac{1}{5}$ because $\left(\dfrac{1}{5}\right)^3 = \dfrac{1}{125}$.

C FINDING nTH ROOTS

Just as we can raise a real number to powers other than 2 or 3, we can find roots other than square roots and cube roots. In fact, we can take the *n*th root of a number where *n* is any natural number. An **nth root** of a number *a* is a number whose *n*th power is *a*.

In symbols, the *n*th root of *a* is written as $\sqrt[n]{a}$. Recall that *n* is called the **index**. The index 2 is usually omitted for square roots.

> **HELPFUL HINT**
>
> If the index is even, such as $\sqrt{}$, $\sqrt[4]{}$, $\sqrt[6]{}$, and so on, the radicand must be nonnegative for the root to be a real number. For example,
>
> $$\sqrt[4]{81} = 3 \text{ but } \sqrt[4]{-81} \text{ is not a real number.}$$
> $$\sqrt[6]{64} = 2 \text{ but } \sqrt[6]{-64} \text{ is not a real number.}$$

TRY THE CONCEPT CHECK IN THE MARGIN.

Examples   Find each root.

> **9.** $\sqrt[4]{16} = 2$ because $2^4 = 16$ and 2 is positive.
> **10.** $\sqrt[5]{-32} = -2$ because $(-2)^5 = -32$.
> **11.** $-\sqrt[3]{8} = -2$ because $\sqrt[3]{8} = 2$.
> **12.** $\sqrt[4]{-81}$ is not a real number since the index 4 is even and the radicand -81 is negative. In other words, there is no real number that when raised to the 4th power gives -81.

Practice Problems 6–8

Find each cube root.

6. $\sqrt[3]{27}$

7. $\sqrt[3]{-8}$

8. $\sqrt[3]{\dfrac{1}{64}}$

✓ CONCEPT CHECK

Which of the following is a real number?

a. $\sqrt{-64}$

b. $\sqrt[4]{-64}$

c. $\sqrt[5]{-64}$

d. $\sqrt[6]{-64}$

Practice Problems 9–12

Find each root.

9. $\sqrt[4]{-16}$

10. $\sqrt[5]{-1}$

11. $\sqrt[4]{81}$

12. $\sqrt[6]{-64}$

Answers

6. 3, **7.** -2, **8.** $\dfrac{1}{4}$, **9.** not a real number,

10. -1, **11.** 3, **12.** not a real number

✓ Concept Check: c

D APPROXIMATING SQUARE ROOTS

Recall that numbers such as 1, 4, 9, 25, and $\frac{4}{25}$ are called **perfect squares**, since $1^2 = 1, 2^2 = 4, 3^2 = 9, 5^2 = 25$, and $\left(\frac{2}{5}\right)^2 = \frac{4}{25}$. Square roots of perfect square radicands simplify to rational numbers.

What happens when we try to simplify a root such as $\sqrt{3}$? Since 3 is not a perfect square, $\sqrt{3}$ is not a rational number. It cannot be written as a quotient of integers. It is called an **irrational number** and we can find a decimal **approximation** of it. To find decimal approximations, use a calculator or Appendix C. (For calculator help, see the next example or the box at the end of this section.)

Example 13   Use a calculator or Appendix C to approximate $\sqrt{3}$ to three decimal places.

Solution: We may use Appendix C or a calculator to approximate $\sqrt{3}$. To use a calculator, find the square root key $\boxed{\sqrt{}}$.

$$\sqrt{3} \approx 1.732050808$$

To three decimal places, $\sqrt{3} \approx 1.732.$ ▬▬▬

Practice Problem 13

Use a calculator or Appendix C to approximate $\sqrt{10}$ to three decimal places.

TEACHING TIP

Show students how to estimate a square root from the nearby perfect squares. For instance, $\sqrt{55}$ is between $\sqrt{49}$ and $\sqrt{64}$, so $\sqrt{55}$ is between 7 and 8.

E SIMPLIFYING RADICALS CONTAINING VARIABLES

Radicals can also contain variables. Since the square root of a negative number is not a real number, we want to make sure variables in the radicand do not have replacement values that would make the radicand negative. To avoid negative radicands, we assume for the rest of this chapter that **if a variable appears in the radicand of a radical expression, it represents positive numbers only.** Then

$$\sqrt{y^2} = y \quad \text{Because } (y)^2 = y^2$$
$$\sqrt{x^8} = x^4 \quad \text{Because } (x^4)^2 = x^8$$
$$\sqrt{9x^2} = 3x \quad \text{Because } (3x)^2 = 9x^2$$

Examples   Simplify each expression. Assume that all variables represent positive numbers.

14. $\sqrt{x^2} = x$ because $(x)^2 = x^2$.
15. $\sqrt{x^6} = x^3$ because $(x^3)^2 = x^6$.
16. $\sqrt[3]{27y^6} = 3y^2$ because $(3y^2)^3 = 27y^6$.
17. $\sqrt{16x^{16}} = 4x^8$ because $(4x^8)^2 = 16x^{16}$.   ▬▬▬

Practice Problems 14–17

Simplify each expression. Assume that all variables represent positive numbers.

14. $\sqrt{x^8}$

15. $\sqrt{x^{20}}$

16. $\sqrt{4x^6}$

17. $\sqrt[3]{8y^{12}}$

Answers

13. 3.162, **14.** x^4, **15.** x^{10}, **16.** $2x^3$, **17.** $2y^4$

CALCULATOR EXPLORATIONS

To simplify or approximate square roots using a calculator, locate the key marked $\boxed{\sqrt{}}$. To simplify $\sqrt{25}$ using a scientific calculator, press $\boxed{25}\ \boxed{\sqrt{}}$. The display should read $\boxed{5}$. To simplify $\sqrt{25}$ using a graphing calculator, press $\boxed{\sqrt{}}\ \boxed{25}$.

To approximate $\sqrt{30}$, press $\boxed{30}\ \boxed{\sqrt{}}$ (or $\boxed{\sqrt{}}\ \boxed{30}$). The display should read $\boxed{5.4772256}$. This is an approximation for $\sqrt{30}$. A three-decimal-place approximation is

$$\sqrt{30} \approx 5.477$$

Is this answer reasonable? Since 30 is between perfect squares 25 and 36, $\sqrt{30}$ is between $\sqrt{25} = 5$ and $\sqrt{36} = 6$. The calculator result is then reasonable since 5.4772256 is between 5 and 6.

Use a calculator to approximate each expression to three decimal places. Decide whether each result is reasonable.

1. $\sqrt{7}$ 2.646   **2.** $\sqrt{14}$ 3.742   **3.** $\sqrt{11}$ 3.317

4. $\sqrt{200}$ 14.142   **5.** $\sqrt{82}$ 9.055   **6.** $\sqrt{46}$ 6.782

Many scientific calculators have a key, such as $\boxed{\sqrt[x]{y}}$, that can be used to approximate roots other than square roots. To approximate these roots using a graphing calcuator, look under the $\boxed{\text{MATH}}$ menu or consult your manual.

Use a calculator to approximate each expression to three decimal places. Decide whether each result is reasonable.

7. $\sqrt[3]{40}$ 3.420   **8.** $\sqrt[3]{71}$ 4.141   **9.** $\sqrt[4]{20}$ 2.115

10. $\sqrt[4]{15}$ 1.968   **11.** $\sqrt[5]{18}$ 1.783   **12.** $\sqrt[6]{2}$ 1.122

EXERCISE SET 8.1

A *Find each square root. See Examples 1 through 5.*

1. $\sqrt{16}$

2. $\sqrt{9}$

3. $\sqrt{81}$

4. $\sqrt{49}$

5. $\sqrt{\dfrac{1}{25}}$

6. $\sqrt{\dfrac{1}{64}}$

7. $-\sqrt{100}$

8. $-\sqrt{36}$

9. $\sqrt{-4}$

10. $\sqrt{-25}$

11. $-\sqrt{121}$

12. $-\sqrt{49}$

13. $\sqrt{\dfrac{9}{25}}$

14. $\sqrt{\dfrac{4}{81}}$

B *Find each cube root. See Examples 6 through 8.*

15. $\sqrt[3]{125}$

16. $\sqrt[3]{64}$

17. $\sqrt[3]{-64}$

18. $\sqrt[3]{-27}$

19. $-\sqrt[3]{8}$

20. $-\sqrt[3]{27}$

21. $\sqrt[3]{\dfrac{1}{8}}$

22. $\sqrt[3]{\dfrac{1}{64}}$

23. $\sqrt[3]{-125}$

24. $\sqrt[3]{-27}$

25. Explain why the square root of a negative number is not a real number.

26. Explain why the cube root of a negative number is a real number.

ANSWERS

1. 4

2. 3

3. 9

4. 7

5. $\dfrac{1}{5}$

6. $\dfrac{1}{8}$

7. -10

8. -6

9. not a real number

10. not a real number

11. -11

12. -7

13. $\dfrac{3}{5}$

14. $\dfrac{2}{9}$

15. 5

16. 4

17. -4

18. -3

19. -2

20. -3

21. $\dfrac{1}{2}$

22. $\dfrac{1}{4}$

23. -5

24. -3

25. answers may vary

26. answers may vary

Name _____

C *Find each root. See Examples 9 through 12.*

27. $\sqrt[5]{32}$ **28.** $\sqrt[4]{81}$ **29.** $\sqrt[4]{-16}$ **30.** $\sqrt{-9}$

31. $-\sqrt[4]{625}$ **32.** $-\sqrt[5]{32}$ **33.** $\sqrt[6]{1}$ **34.** $\sqrt[5]{1}$

D *Approximate each square root to three decimal places. See Example 13.*

35. $\sqrt{37}$ **36.** $\sqrt{27}$ **37.** $\sqrt{136}$ **38.** $\sqrt{8}$

39. A standard baseball diamond is a square with 90-foot sides connecting the bases. The distance from home plate to second base is $90 \cdot \sqrt{2}$ feet. Approximate $\sqrt{2}$ to two decimal places and use your result to approximate the distance $90 \cdot \sqrt{2}$ feet.

40. The roof of the warehouse shown needs to be shingled. The total area of the roof is exactly $240 \cdot \sqrt{41}$ square feet. Approximate this area to the nearest whole number.

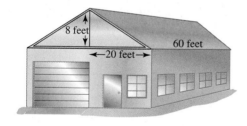

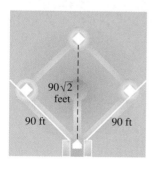

E *Find each root. Assume that all variables represent positive numbers. See Examples 14 through 17.*

41. $\sqrt{z^2}$ **42.** $\sqrt{y^{10}}$ **43.** $\sqrt{x^4}$ **44.** $\sqrt{x^6}$

45. $\sqrt{9x^8}$ **46.** $\sqrt{36x^{12}}$ **47.** $\sqrt{81x^2}$ **48.** $\sqrt{100z^4}$

Name _____

REVIEW AND PREVIEW

Write each integer as a product of two integers such that one of the factors is a perfect square. For example, in $18 = 9 \cdot 2$, 9 is a perfect square.

49. 50   **50.** 8   **51.** 32   **52.** 75

53. 28   **54.** 44   **55.** 27   **56.** 90

COMBINING CONCEPTS

57. Simplify $\sqrt{\sqrt{81}}$.

58. Simplify $\sqrt[3]{\sqrt[3]{1}}$.

59. Graph $y = \sqrt{x}$. (Complete the table below, plot the ordered pair solutions, and draw a smooth curve through the points. Remember that since the radicand cannot be negative, this particular graph begins at the point with coordinates $(0, 0)$.)

x	y	
0	0	
1	1	
3	1.7	(approximate)
4	2	
9	3	See App. C

60. Graph $y = \sqrt[3]{x}$. (Complete the table below, plot the ordered pair solutions, and draw a smooth curve through the points.)

x	y	
-8	-2	
-2	-1.3	(approximate)
-1	-1	
0	0	
1	1	
2	1.3	(approximate)
8	2	See App. C

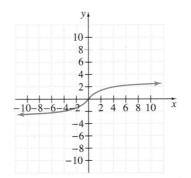

49. $25 \cdot 2$

50. $4 \cdot 2$

51. $16 \cdot 2$ or $4 \cdot 8$

52. $25 \cdot 3$

53. $4 \cdot 7$

54. $4 \cdot 11$

55. $9 \cdot 3$

56. $9 \cdot 10$

57. 3

58. 1

59. see graph

60. see graph

61. $(2, 0)$ _____

62. $(-3, 0)$ _____

63. $(-4, 0)$ _____

64. $(5, 0)$ _____

65. 58 ft _____

Use a graphing calculator and graph each function. Observe the graph from left to right and give the ordered pair that corresponds to the "beginning" of the graph. Then tell why the graph starts at that point.

61. $y = \sqrt{x - 2}$

62. $y = \sqrt{x + 3}$

63. $y = \sqrt{x + 4}$

64. $y = \sqrt{x - 5}$

65. If the amount of gold discovered by humankind could be assembled in one place, it would make a cube with a volume of 195,112 cubic feet. Each side of the cube would be $\sqrt[3]{195,112}$ feet long. How long would one side of the cube be? (*Source: Reader's Digest*, June 1996)

TEACHING TIP

Have students use their observations from Exercises 61–64 to make a conjecture about the graph $y = \sqrt{x - c}$ where c is a constant.

8.2 SIMPLIFYING RADICALS

A SIMPLIFYING RADICALS USING THE PRODUCT RULE

A square root is simplified when the radicand contains no perfect square factors (other than 1). For example, $\sqrt{20}$ is not simplified because $\sqrt{20} = \sqrt{4 \cdot 5}$ and 4 is a perfect square.

To begin simplifying square roots, we notice the following pattern.

$$\sqrt{9 \cdot 16} = \sqrt{144} = 12$$
$$\sqrt{9} \cdot \sqrt{16} = 3 \cdot 4 = 12$$

Since both expressions simplify to 12, we can write

$$\sqrt{9 \cdot 16} = \sqrt{9} \cdot \sqrt{16}$$

This suggests the following product rule for square roots.

PRODUCT RULE FOR SQUARE ROOTS

If $\sqrt{a}$ and $\sqrt{b}$ are real numbers, then

$$\sqrt{a \cdot b} = \sqrt{a} \cdot \sqrt{b}$$

In other words, the square root of a product is equal to the product of the square roots.

To simplify $\sqrt{20}$, for example, we factor 20 so that one of its factors is a perfect square factor.

$$\sqrt{20} = \sqrt{4 \cdot 5} \quad \text{Factor 20.}$$
$$= \sqrt{4} \cdot \sqrt{5} \quad \text{Use the product rule.}$$
$$= 2\sqrt{5} \quad \text{Write } \sqrt{4} \text{ as 2.}$$

The notation $2\sqrt{5}$ means $2 \cdot \sqrt{5}$. Since the radicand 5 has no perfect square factor other than 1, $2\sqrt{5}$ is in simplest form.

HELPFUL HINT

A radical expression in simplest form does *not mean* a decimal approximation. The simplest form of a radical expression is an exact form and may still contain a radical.

$$\underbrace{\sqrt{20} = 2\sqrt{5}}_{\text{exact}} \qquad \underbrace{\sqrt{20} \approx 4.47}_{\text{decimal approximation}}$$

Examples Simplify.

1. $\sqrt{54} = \sqrt{9 \cdot 6}$ Factor 54 so that one factor is a perfect square. 9 is a perfect square.
$= \sqrt{9} \cdot \sqrt{6}$ Use the product rule.
$= 3\sqrt{6}$ Write $\sqrt{9}$ as 3.

2. $\sqrt{12} = \sqrt{4 \cdot 3}$ Factor 12 so that one factor is a perfect square. 4 is a perfect square.
$= \sqrt{4} \cdot \sqrt{3}$ Use the product rule.
$= 2\sqrt{3}$ Write $\sqrt{4}$ as 2.

3. $\sqrt{200} = \sqrt{100 \cdot 2}$ Factor 200 so that one factor is a perfect square. 100 is a perfect square.
$= \sqrt{100} \cdot \sqrt{2}$ Use the product rule.
$= 10\sqrt{2}$ Write $\sqrt{100}$ as 10.

Objectives

A Use the product rule to simplify radicals.

B Use the quotient rule to simplify radicals.

C Use both rules to simplify radicals containing variables.

D Simplify roots other than square roots.

SSM   CD-ROM   Video 8.2

TEACHING TIP

Have students make a list of the squares of numbers from 1–25 so they become familiar with the perfect square factors.

Practice Problems 1–4

Simplify.

1. $\sqrt{40}$

2. $\sqrt{18}$

3. $\sqrt{700}$

4. $\sqrt{15}$

Answers

1. $2\sqrt{10}$, **2.** $3\sqrt{2}$, **3.** $10\sqrt{7}$, **4.** $\sqrt{15}$

4. $\sqrt{35}$ The radicand 35 contains no perfect square factors other than 1. Thus $\sqrt{35}$ is in simplest form.

In Example 3, 100 is the largest perfect square factor of 200. What happens if we don't use the largest perfect square factor? Although using the largest perfect square factor saves time, the result is the same no matter what perfect square factor is used. For example, it is also true that $200 = 4 \cdot 50$. Then

$$\sqrt{200} = \sqrt{4} \cdot \sqrt{50}$$
$$= 2 \cdot \sqrt{50}$$

Since $\sqrt{50}$ is not in simplest form, we continue.

$$\sqrt{200} = 2 \cdot \sqrt{50}$$
$$= 2 \cdot \sqrt{25} \cdot \sqrt{2}$$
$$= 2 \cdot 5 \cdot \sqrt{2}$$
$$= 10\sqrt{2}$$

B SIMPLIFYING RADICALS USING THE QUOTIENT RULE

Next, let's examine the square root of a quotient.

$$\sqrt{\frac{16}{4}} = \sqrt{4} = 2$$

Also,

$$\frac{\sqrt{16}}{\sqrt{4}} = \frac{4}{2} = 2$$

Since both expressions equal 2, we can write

$$\sqrt{\frac{16}{4}} = \frac{\sqrt{16}}{\sqrt{4}}$$

This suggests the following quotient rule.

> **QUOTIENT RULE FOR SQUARE ROOTS**
>
> If $\sqrt{a}$ and $\sqrt{b}$ are real numbers and $b \neq 0$, then
>
> $$\sqrt{\frac{a}{b}} = \frac{\sqrt{a}}{\sqrt{b}}$$

In other words, the square root of a quotient is equal to the quotient of the square roots.

Examples   Use the quotient rule to simplify.

5. $\sqrt{\frac{25}{36}} = \frac{\sqrt{25}}{\sqrt{36}} = \frac{5}{6}$

6. $\sqrt{\frac{3}{64}} = \frac{\sqrt{3}}{\sqrt{64}} = \frac{\sqrt{3}}{8}$

7. $\sqrt{\frac{40}{81}} = \frac{\sqrt{40}}{\sqrt{81}}$ Use the quotient rule.

$$= \frac{\sqrt{4} \cdot \sqrt{10}}{9}$$ Use the product rule and write $\sqrt{81}$ as 9.

$$= \frac{2\sqrt{10}}{9}$$ Write $\sqrt{4}$ as 2.

TEACHING TIP

Consider showing your students that the Product Rule gives the same result as the Quotient Rule with the following example:

$$\sqrt{\frac{16}{4}} = \sqrt{16 \cdot \frac{1}{4}}$$
$$= \sqrt{16} \cdot \sqrt{\frac{1}{4}}$$
$$= 4 \cdot \frac{1}{2}$$
$$= \frac{4}{2}$$
$$= 2$$

Practice Problems 5–7

Use the quotient rule to simplify.

5. $\sqrt{\frac{16}{81}}$

6. $\sqrt{\frac{2}{25}}$

7. $\sqrt{\frac{45}{49}}$

Answers

5. $\frac{4}{9}$,   **6.** $\frac{\sqrt{2}}{5}$,   **7.** $\frac{3\sqrt{5}}{7}$

C SIMPLIFYING RADICALS CONTAINING VARIABLES

Recall that $\sqrt{x^6} = x^3$ because $(x^3)^2 = x^6$. If an odd exponent occurs, we write the exponential expression so that one factor is the greatest even power contained in the expression. Then we use the product rule to simplify.

Examples   Simplify each radical. Assume that all variables represent positive numbers.

8. $\sqrt{x^5} = \sqrt{x^4 \cdot x} = \sqrt{x^4} \cdot \sqrt{x} = x^2\sqrt{x}$

9. $\sqrt{8y^2} = \sqrt{4 \cdot 2 \cdot y^2} = \sqrt{4y^2 \cdot 2} = \sqrt{4y^2} \cdot \sqrt{2} = 2y\sqrt{2}$

10. $\sqrt{\dfrac{45}{x^6}} = \dfrac{\sqrt{45}}{\sqrt{x^6}} = \dfrac{\sqrt{9 \cdot 5}}{x^3} = \dfrac{\sqrt{9} \cdot \sqrt{5}}{x^3} = \dfrac{3\sqrt{5}}{x^3}$

Practice Problems 8–10

Simplify each radical. Assume that all variables represent positive numbers.

8. $\sqrt{x^{11}}$

9. $\sqrt{18x^4}$

10. $\sqrt{\dfrac{27}{x^8}}$

D SIMPLIFYING ROOTS OTHER THAN SQUARE ROOTS

The product and quotient rules also apply to roots other than square roots. For example, to simplify cube roots, we look for perfect cube factors of the radicand. Recall that 8 is a perfect cube since $2^3 = 8$. Therefore, to simplify $\sqrt[3]{48}$, we factor 48 as $8 \cdot 6$.

$\sqrt[3]{48} = \sqrt[3]{8 \cdot 6}$ Factor 48.

$\qquad = \sqrt[3]{8} \cdot \sqrt[3]{6}$ Use the product rule.

$\qquad = 2\sqrt[3]{6}$ Write $\sqrt[3]{8}$ as 2.

$2\sqrt[3]{6}$ is in simplest form since the radicand 6 contains no perfect cube factors other than 1.

TEACHING TIP

Have students make a list of the cubes of numbers from 1–10 so they become familiar with the perfect cube factors.

Examples   Simplify each radical.

11. $\sqrt[3]{54} = \sqrt[3]{27 \cdot 2} = \sqrt[3]{27} \cdot \sqrt[3]{2} = 3\sqrt[3]{2}$

12. $\sqrt[3]{18}$. The number 18 contains no perfect cube factors, so $\sqrt[3]{18}$ cannot be simplified further.

13. $\sqrt[3]{\dfrac{7}{8}} = \dfrac{\sqrt[3]{7}}{\sqrt[3]{8}} = \dfrac{\sqrt[3]{7}}{2}$

14. $\sqrt[3]{\dfrac{40}{27}} = \dfrac{\sqrt[3]{40}}{\sqrt[3]{27}} = \dfrac{\sqrt[3]{8 \cdot 5}}{3} = \dfrac{\sqrt[3]{8} \cdot \sqrt[3]{5}}{3} = \dfrac{2\sqrt[3]{5}}{3}$

Practice Problems 11–14

Simplify each radical.

11. $\sqrt[3]{40}$

12. $\sqrt[3]{50}$

13. $\sqrt[3]{\dfrac{10}{27}}$

14. $\sqrt[3]{\dfrac{81}{8}}$

Answers

8. $x^5\sqrt{x}$,   **9.** $3x^2\sqrt{2}$,   **10.** $\dfrac{3\sqrt{3}}{x^4}$,   **11.** $2\sqrt[3]{5}$,

12. $\sqrt[3]{50}$,   **13.** $\dfrac{\sqrt[3]{10}}{3}$,   **14.** $\dfrac{3\sqrt[3]{3}}{2}$

Focus On Mathematical Connections

GRAPHING AND THE DISTANCE FORMULA

One application of radicals is finding the distance between two points in the coordinate plane. This can be very useful in graphing.

The distance d between two points with coordinates (x_1, y_1) and (x_2, y_2) is given by the **distance formula**
$d = \sqrt{(x_2 - x_1)^2 + (y_2 - y_1)^2}.$

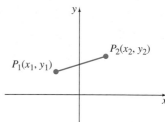

Suppose we want to find the distance between the two points $(-1, 9)$ and $(3, 5)$. We can use the distance formula with $(x_1, y_1) = (-1, 9)$ and $(x_2, y_2) = (3, 5)$. We have

$$d = \sqrt{(x_2 - x_1)^2 + (y_2 - y_1)^2}$$
$$= \sqrt{[3 - (-1)]^2 + (5 - 9)^2}$$
$$= \sqrt{(4)^2 + (-4)^2}$$
$$= \sqrt{16 + 16}$$
$$= \sqrt{32} = 4\sqrt{2}$$

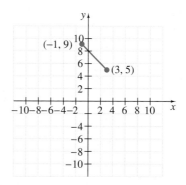

The distance between the two points is exactly $4\sqrt{2}$ units or approximately 5.66 units.

GROUP ACTIVITY

Brainstorm to come up with several disciplines or activities in which the distance formula might be useful. Make up an example that shows how the distance formula would be used in one of the activities on your list. Then present your example to the rest of the class.

Sample answers:
[aviation (using coordinate locations to find distance from airplane to airport, etc.); sports (finding distance that a golf ball or football has traveled over a grid); archaeology (using a grid to fix the exact locations of items in an archaeological dig and finding distance between objects to describe their relative locations); computer science (finding the distance between two pixels on a computer screen), etc.]

Name _____ **Section** _____ **Date** _____

MENTAL MATH

Simplify each radical. Assume that all variables represent positive numbers.

1. $\sqrt{4 \cdot 9}$ **2.** $\sqrt{9 \cdot 36}$ **3.** $\sqrt{x^2}$ **4.** $\sqrt{y^4}$

5. $\sqrt{0}$ **6.** $\sqrt{1}$ **7.** $\sqrt{25x^4}$ **8.** $\sqrt{49x^2}$

EXERCISE SET 8.2

A *Use the product rule to simplify each radical. See Examples 1 through 4.*

1. $\sqrt{20}$ **2.** $\sqrt{44}$ **3.** $\sqrt{18}$ **4.** $\sqrt{45}$

5. $\sqrt{50}$ **6.** $\sqrt{28}$ **7.** $\sqrt{33}$ **8.** $\sqrt{98}$

9. $\sqrt{60}$ **10.** $\sqrt{90}$ **11.** $\sqrt{180}$ **12.** $\sqrt{150}$

13. $\sqrt{52}$ **14.** $\sqrt{75}$

B *Use the quotient rule and the product rule to simplify each radical. See Examples 5 through 7.*

15. $\sqrt{\dfrac{8}{25}}$ **16.** $\sqrt{\dfrac{63}{16}}$ **17.** $\sqrt{\dfrac{27}{121}}$ **18.** $\sqrt{\dfrac{24}{169}}$

19. $\sqrt{\dfrac{9}{4}}$ **20.** $\sqrt{\dfrac{100}{49}}$ **21.** $\sqrt{\dfrac{125}{9}}$ **22.** $\sqrt{\dfrac{27}{100}}$

23. $\sqrt{\dfrac{11}{36}}$ **24.** $\sqrt{\dfrac{30}{49}}$ **25.** $-\sqrt{\dfrac{27}{144}}$ **26.** $-\sqrt{\dfrac{84}{121}}$

MENTAL MATH ANSWERS

1. 6
2. 18
3. x
4. y^2
5. 0
6. 1
7. $5x^2$
8. $7x$

ANSWERS

1. $2\sqrt{5}$
2. $2\sqrt{11}$
3. $3\sqrt{2}$
4. $3\sqrt{5}$
5. $5\sqrt{2}$
6. $2\sqrt{7}$
7. $\sqrt{33}$
8. $7\sqrt{2}$
9. $2\sqrt{15}$
10. $3\sqrt{10}$
11. $6\sqrt{5}$
12. $5\sqrt{6}$
13. $2\sqrt{13}$
14. $5\sqrt{3}$
15. $\dfrac{2\sqrt{2}}{5}$
16. $\dfrac{3\sqrt{7}}{4}$
17. $\dfrac{3\sqrt{3}}{11}$
18. $\dfrac{2\sqrt{6}}{13}$
19. $\dfrac{3}{2}$
20. $\dfrac{10}{7}$
21. $\dfrac{5\sqrt{5}}{3}$
22. $\dfrac{3\sqrt{3}}{10}$
23. $\dfrac{\sqrt{11}}{6}$
24. $\dfrac{\sqrt{30}}{7}$
25. $-\dfrac{\sqrt{3}}{4}$
26. $-\dfrac{2\sqrt{21}}{11}$

607

Name _____

C *Simplify each radical. Assume that all variables represent positive numbers. See Examples 8 through 10.*

27. $\sqrt{x^7}$ 28. $\sqrt{y^3}$ 🔊 29. $\sqrt{x^{13}}$ 30. $\sqrt{y^{17}}$

31. $\sqrt{75x^2}$ 32. $\sqrt{72y^2}$ 33. $\sqrt{96x^4}$ 34. $\sqrt{40y^{10}}$

🔊 35. $\sqrt{\dfrac{12}{y^2}}$ 36. $\sqrt{\dfrac{63}{x^4}}$ 37. $\sqrt{\dfrac{9x}{y^2}}$ 38. $\sqrt{\dfrac{6y^2}{x^4}}$

39. $\sqrt{\dfrac{88}{x^4}}$ 40. $\sqrt{\dfrac{x^{11}}{81}}$

D *Simplify each radical. See Examples 11 through 14.*

41. $\sqrt[3]{24}$ 42. $\sqrt[3]{81}$ 🔊 43. $\sqrt[3]{250}$ 44. $\sqrt[3]{40}$

🔊 45. $\sqrt[3]{\dfrac{5}{64}}$ 46. $\sqrt[3]{\dfrac{32}{125}}$ 47. $\sqrt[3]{\dfrac{7}{8}}$ 48. $\sqrt[3]{\dfrac{10}{27}}$

49. $\sqrt[3]{\dfrac{15}{64}}$ 50. $\sqrt[3]{\dfrac{4}{27}}$ 51. $\sqrt[3]{80}$ 52. $\sqrt[3]{108}$

Name _____

REVIEW AND PREVIEW

Perform each indicated operation. See Sections 3.3 and 3.4.

53. $6x + 8x$ **54.** $(6x)(8x)$ **55.** $(2x + 3)(x - 5)$

56. $(2x + 3) + (x - 5)$ **57.** $9y^2 - 9y^2$ **58.** $(9y^2)(-8y^2)$

COMBINING CONCEPTS

Simplify each radical. Assume that all variables represent positive numbers.

59. $\sqrt{x^6 y^3}$ **60.** $\sqrt{98x^5 y^4}$

61. $\sqrt{x^2 + 4x + 4}$ (Hint: Factor the trinomial first.)   ▄ **62.** $\sqrt[3]{-8x^6}$

63. If a cube is to have a volume of 80 cubic inches, then each side must be $\sqrt[3]{80}$ inches long. Simplify the radical representing the side length.

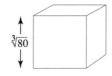

64. Jeannie Boswell is swimming across a 40-foot-wide river, trying to head straight across to the opposite shore. However, the current is strong enough to move her downstream 100 feet by the time she reaches land. (See the figure.) Because of the current, the actual distance she swam is $\sqrt{11,600}$ feet. Simplify this radical.

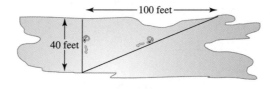

100 feet

40 feet

---

53. $14x$

54. $48x^2$

55. $2x^2 - 7x - 15$

56. $3x - 2$

57. 0

58. $-72y^4$

59. $x^3 y \sqrt{y}$

60. $7x^2 y^2 \sqrt{2x}$

61. $x + 2$

62. $-2x^2$

63. $2\sqrt[3]{10}$ in.

64. $20\sqrt{29}$ ft

609

65. $1700

66. $1493.70

67. answers may vary

68. answers may vary

The cost C in dollars per day to operate a small delivery service is given by $C = 100\sqrt[3]{n} + 700$, _where n is the number of deliveries per day._

65. Find the cost if the number of deliveries is 1000.

66. Approximate the cost if the number of deliveries is 500.

67. By using replacement values for a and b, show that $\sqrt{a^2 + b^2}$ does not equal $a + b$.

68. By using replacement values for a and b, show that $\sqrt{a + b}$ does not equal $\sqrt{a} + \sqrt{b}$.

TEACHING TIP

For Exercises 67 and 68, point out to students that if $a = 0$ and $b = 0$, the two expressions are equal but to be equivalent expressions (interchangeable expressions), they must be true for all values of a and b. If you can find 1 pair of values in which the expressions are not equal, the expressions are not equivalent.

8.3   ADDING AND SUBTRACTING RADICALS

Objectives

A Add or subtract like radicals.
B Simplify radical expressions, and then add or subtract any like radicals.

SSM   CD-ROM   Video
8.3

A ADDING AND SUBTRACTING RADICALS

Recall that to combine like terms, we use the distributive property.

$$5x + 3x = (5 + 3)x = 8x$$

The distributive property can also be applied to expressions containing the same radicals. For example,

$$5\sqrt{2} + 3\sqrt{2} = (5 + 3)\sqrt{2} = 8\sqrt{2}$$

Also,

$$9\sqrt{5} - 6\sqrt{5} = (9 - 6)\sqrt{5} = 3\sqrt{5}$$

Radical terms such as $5\sqrt{2}$ and $3\sqrt{2}$ are **like radicals**, as are $9\sqrt{5}$ and $6\sqrt{5}$. Like radicals have the same index and radicand.

TEACHING TIP

If students need practice identifying like radicals, consider having them find which of the following pairs are like radicals.
a. $-3\sqrt{2}, 2\sqrt{3}$, **b.** $-2\sqrt{3}, 3\sqrt{3}$,
c. $4\sqrt{2}, 4\sqrt{3}$, **d.** $8\sqrt{5}, 5\sqrt{8}$, **e.** $\sqrt{7}, 3\sqrt{7}$

Examples   Add or subtract as indicated.

1. $4\sqrt{5} + 3\sqrt{5} = (4 + 3)\sqrt{5} = 7\sqrt{5}$
2. $\sqrt{10} - 6\sqrt{10} = 1\sqrt{10} - 6\sqrt{10} = (1 - 6)\sqrt{10} = -5\sqrt{10}$
3. $2\sqrt{6} + 2\sqrt{5}$ cannot be simplified further since the radicands are not the same.
4. $\sqrt{15} + \sqrt{15} = 1\sqrt{15} + 1\sqrt{15} = (1 + 1)\sqrt{15} = 2\sqrt{15}$ ▬▬▬

Practice Problems 1–4

Add or subtract as indicated.

1. $6\sqrt{11} + 9\sqrt{11}$

2. $\sqrt{7} - 3\sqrt{7}$

3. $\sqrt{2} + \sqrt{2}$

4. $3\sqrt{3} - 3\sqrt{2}$

TRY THE CONCEPT CHECK IN THE MARGIN.

B SIMPLIFYING RADICALS BEFORE ADDING OR SUBTRACTING

At first glance, it appears that the expression $\sqrt{50} + \sqrt{8}$ cannot be simplified further because the radicands are different. However, the product rule can be used to simplify each radical, and then further simplification might be possible.

✓ CONCEPT CHECK

Which is true?

a. $2 + 3\sqrt{5} = 5\sqrt{5}$

b. $2\sqrt{3} + 2\sqrt{7} = 2\sqrt{10}$

c. $\sqrt{3} + \sqrt{5} = \sqrt{8}$

d. None of the above is true. In each case, the left-hand side cannot be simplified further.

Answers

✓ Concept Check: d

1. $15\sqrt{11}$, **2.** $-2\sqrt{7}$, **3.** $2\sqrt{2}$,
4. $3\sqrt{3} - 3\sqrt{2}$

Practice Problems 5–8

Add or subtract by first simplifying each radical.

5. $\sqrt{27} + \sqrt{75}$

6. $3\sqrt{20} - 7\sqrt{45}$

7. $\sqrt{36} - \sqrt{48} - 4\sqrt{3} - \sqrt{9}$

8. $\sqrt{9x^4} - \sqrt{36x^3} + \sqrt{x^3}$

Examples

Add or subtract by first simplifying each radical.

5. $\sqrt{50} + \sqrt{8} = \sqrt{25 \cdot 2} + \sqrt{4 \cdot 2}$ Factor radicands.

$\qquad\qquad = \sqrt{25} \cdot \sqrt{2} + \sqrt{4} \cdot \sqrt{2}$ Use the product rule.

$\qquad\qquad = 5\sqrt{2} + 2\sqrt{2}$ Simplify $\sqrt{25}$ and $\sqrt{4}$.

$\qquad\qquad = 7\sqrt{2}$ Add like radicals.

6. $7\sqrt{12} - \sqrt{75} = 7\sqrt{4 \cdot 3} - \sqrt{25 \cdot 3}$ Factor radicands.

$\qquad\qquad = 7\sqrt{4} \cdot \sqrt{3} - \sqrt{25} \cdot \sqrt{3}$ Use the product rule.

$\qquad\qquad = 7 \cdot 2\sqrt{3} - 5\sqrt{3}$ Simplify $\sqrt{4}$ and $\sqrt{25}$.

$\qquad\qquad = 14\sqrt{3} - 5\sqrt{3}$ Multiply.

$\qquad\qquad = 9\sqrt{3}$ Subtract like radicals.

7. $\sqrt{25} - \sqrt{27} - 2\sqrt{18} - \sqrt{16}$

$\qquad = 5 - \sqrt{9 \cdot 3} - 2\sqrt{9 \cdot 2} - 4$ Factor radicands and simplify $\sqrt{25}$ and $\sqrt{16}$.

$\qquad = 5 - \sqrt{9} \cdot \sqrt{3} - 2\sqrt{9} \cdot \sqrt{2} - 4$ Use the product rule.

$\qquad = 5 - 3\sqrt{3} - 2 \cdot 3\sqrt{2} - 4$ Simplify.

$\qquad = 1 - 3\sqrt{3} - 6\sqrt{2}$ Write $5 - 4$ as 1 and $2 \cdot 3$ as 6.

8. $2\sqrt{x^2} - \sqrt{25x} + \sqrt{x}$

$\qquad = 2x - \sqrt{25} \cdot \sqrt{x} + \sqrt{x}$ Write $\sqrt{x^2}$ as x and use the product rule.

$\qquad = 2x - 5\sqrt{x} + 1\sqrt{x}$ Simplify.

$\qquad = 2x - 4\sqrt{x}$ Add like radicals.

Answers

5. $8\sqrt{3}$,   **6.** $-15\sqrt{5}$,   **7.** $3 - 8\sqrt{3}$,
8. $3x^2 - 5x\sqrt{x}$

Name _____ **Section** _____ **Date** _____

MENTAL MATH

Simplify each expression by combining like radicals.

1. $3\sqrt{2} + 5\sqrt{2}$ **2.** $3\sqrt{5} + 7\sqrt{5}$ **3.** $5\sqrt{x} + 2\sqrt{x}$

4. $8\sqrt{x} + 3\sqrt{x}$ **5.** $5\sqrt{7} - 2\sqrt{7}$ **6.** $8\sqrt{6} - 5\sqrt{6}$

EXERCISE SET 8.3

A *Add or subtract as indicated. See Examples 1 through 4.*

1. $4\sqrt{3} - 8\sqrt{3}$ **2.** $\sqrt{5} - 9\sqrt{5}$

3. $3\sqrt{6} + 8\sqrt{6} - 2\sqrt{6} - 5$ **4.** $12\sqrt{2} - 3\sqrt{2} + 8\sqrt{2} + 10$

5. $6\sqrt{5} - 5\sqrt{5} + \sqrt{2}$ **6.** $4\sqrt{3} + \sqrt{5} - 3\sqrt{3}$

7. $2\sqrt{3} + 5\sqrt{3} - \sqrt{3}$ **8.** $8\sqrt{14} + 2\sqrt{14} + 4$

9. $2\sqrt{2} - 7\sqrt{2} - 6$ **10.** $5\sqrt{7} + 2 - 11\sqrt{7}$

11. $12\sqrt{5} - \sqrt{5} - 4\sqrt{5}$ **12.** $\sqrt{5} + \sqrt{15}$

13. $\sqrt{5} + \sqrt{5}$ **14.** $4 + 8\sqrt{2} - 9$

15. $6 - 2\sqrt{3} - \sqrt{3}$ **16.** $8 - \sqrt{2} - 5\sqrt{2}$

17. answers may vary

18. answers may vary

19. $5\sqrt{3}$

20. $8\sqrt{2}$

21. $9\sqrt{5}$

22. $14\sqrt{2}$

23. $4\sqrt{6} + \sqrt{5}$

24. $-\sqrt{2} + 4\sqrt{3}$

25. $x + \sqrt{x}$

26. $-5x + 2\sqrt{x}$

27. 0

28. $\sqrt{x}$

29. $x\sqrt{x}$

30. $(4 - x)\sqrt{x}$

31. $9\sqrt{3}$

32. $5\sqrt{5}$

33. $5\sqrt{2} + 12$

34. $3\sqrt{6} + 9$

35. $\dfrac{4\sqrt{5}}{9}$

614

⬧**17.** In your own words, describe like radicals.

⬧**18.** In the expression $\sqrt{5} + 2 - 3\sqrt{5}$, explain why 2 and -3 cannot be combined.

B *Add or subtract by first simplifying each radical and then combining any like radicals. Assume that all variables represent positive numbers. See Examples 5 through 8.*

📼 **19.** $\sqrt{12} + \sqrt{27}$

20. $\sqrt{50} + \sqrt{18}$

21. $\sqrt{45} + 3\sqrt{20}$

22. $5\sqrt{32} - \sqrt{72}$

23. $2\sqrt{54} - \sqrt{20} + \sqrt{45} - \sqrt{24}$

24. $2\sqrt{8} - \sqrt{128} + \sqrt{48} + \sqrt{18}$

25. $4x - 3\sqrt{x^2} + \sqrt{x}$

26. $x - 6\sqrt{x^2} + 2\sqrt{x}$

27. $\sqrt{25x} + \sqrt{36x} - 11\sqrt{x}$

28. $\sqrt{9x} - \sqrt{16x} + 2\sqrt{x}$

29. $3\sqrt{x^3} - x\sqrt{4x}$

📼 **30.** $\sqrt{16x} - \sqrt{x^3}$

31. $\sqrt{75} + \sqrt{48}$

32. $2\sqrt{80} - \sqrt{45}$

33. $\sqrt{8} + \sqrt{9} + \sqrt{18} + \sqrt{81}$

34. $\sqrt{6} + \sqrt{16} + \sqrt{24} + \sqrt{25}$

35. $\sqrt{\dfrac{5}{9}} + \sqrt{\dfrac{5}{81}}$

36. $\sqrt{\dfrac{3}{64}} + \sqrt{\dfrac{3}{16}}$ ⬛

37. $\sqrt{\dfrac{3}{4}} - \sqrt{\dfrac{3}{64}}$

38. $\sqrt{\dfrac{2}{25}} + \sqrt{\dfrac{2}{9}}$

39. $2\sqrt{45} - 2\sqrt{20}$

40. $5\sqrt{18} + 2\sqrt{32}$

41. $\sqrt{35} - \sqrt{140}$

42. $\sqrt{15} - \sqrt{135}$

43. $3\sqrt{9x} + 2\sqrt{x}$

44. $5\sqrt{x} + 4\sqrt{4x}$

45. $\sqrt{9x^2} + \sqrt{81x^2} - 11\sqrt{x}$

46. $\sqrt{100x^2} + 3\sqrt{x} - \sqrt{36x^2}$

47. $\sqrt{3x^3} + 3x\sqrt{x}$

48. $x\sqrt{4x} + \sqrt{9x^3}$

49. $\sqrt{32x^2} + \sqrt{32x^2} + \sqrt{4x^2}$

50. $\sqrt{18x^2} + \sqrt{24x^3} + \sqrt{2x^2}$

51. $\sqrt{40x} + \sqrt{40x^4} - 2\sqrt{10x} - \sqrt{5x^4}$

52. $\sqrt{72x^2} + \sqrt{54x} - x\sqrt{50} - 3\sqrt{2x}$

Review and Preview

Square each binomial. See Section 3.5.

53. $(x + 6)^2$ **54.** $(3x + 2)^2$ **55.** $(2x - 1)^2$ **56.** $(x - 5)^2$

Solve each system of linear equations. See Section 7.2.

57. $\begin{cases} x = 2y \\ x + 5y = 14 \end{cases}$

58. $\begin{cases} y = -5x \\ x + y = 16 \end{cases}$

36. $\dfrac{3\sqrt{3}}{8}$

37. $\dfrac{3\sqrt{3}}{8}$

38. $\dfrac{8\sqrt{2}}{15}$

39. $2\sqrt{5}$

40. $23\sqrt{2}$

41. $-\sqrt{35}$

42. $-2\sqrt{15}$

43. $11\sqrt{x}$

44. $13\sqrt{x}$

45. $12x - 11\sqrt{x}$

46. $4x + 3\sqrt{x}$

47. $x\sqrt{3x} + 3x\sqrt{x}$

48. $5x\sqrt{x}$

49. $8x\sqrt{2} + 2x$

50. $4x\sqrt{2} + 2x\sqrt{6x}$

51. $2x^2\sqrt{10} - x^2\sqrt{5}$

52. $x\sqrt{2} + 3\sqrt{6x} - 3\sqrt{2x}$

53. $x^2 + 12x + 36$

54. $9x^2 + 12x + 4$

55. $4x^2 - 4x + 1$

56. $x^2 - 10x + 25$

57. $(4, 2)$

58. $(-4, 20)$

Name _____

59. Find the perimeter of the rectangular picture frame.

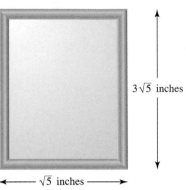

$3\sqrt{5}$ inches

$\sqrt{5}$ inches

60. Find the perimeter of the plot of land.

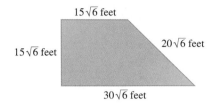

$15\sqrt{6}$ feet

$15\sqrt{6}$ feet

$20\sqrt{6}$ feet

$30\sqrt{6}$ feet

61. A water trough is to be made of wood. Each of the two triangular end pieces has an area of $\frac{3\sqrt{27}}{4}$ square feet. The two side panels are both rectangular. In simplest radical form, find the total area of the wood needed.

3 feet   8 feet

3 feet   3 feet

62. Eight wooden braces are to be attached along the diagonals of the vertical sides of a storage bin. Each of four of these diagonals has a length of $\sqrt{52}$ feet, while each of the other four has a length of $\sqrt{80}$ feet. In simplest radical form, find the total length of the wood needed for these braces.

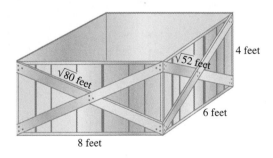

4 feet

$\sqrt{52}$ feet

$\sqrt{80}$ feet

6 feet

8 feet

8.4 MULTIPLYING AND DIVIDING RADICALS

A MULTIPLYING RADICALS

In Section 8.2, we used the product and quotient rules for radicals to help us simplify radicals. In this section, we use these rules to simplify products and quotients of radicals.

> **PRODUCT RULE FOR RADICALS**
>
> If $\sqrt{a}$ and $\sqrt{b}$ are real numbers, then
> $$\sqrt{a} \cdot \sqrt{b} = \sqrt{a \cdot b}$$

In other words, the product of the square roots of two numbers is the square root of the product of the two numbers. For example,

$$\sqrt{3} \cdot \sqrt{2} = \sqrt{3 \cdot 2} = \sqrt{6}$$

Examples   Multiply. Then simplify each product if possible.

1. $\sqrt{7} \cdot \sqrt{3} = \sqrt{7 \cdot 3}$
$= \sqrt{21}$

2. $\sqrt{3} \cdot \sqrt{15} = \sqrt{45}$ Use the product rule.
$= \sqrt{9 \cdot 5}$ Factor the radicand.
$= \sqrt{9} \cdot \sqrt{5}$ Use the product rule.
$= 3\sqrt{5}$ Simplify $\sqrt{9}$.

3. $\sqrt{2x^3} \cdot \sqrt{6x} = \sqrt{2x^3 \cdot 6x}$ Use the product rule.
$= \sqrt{12x^4}$ Multiply.
$= \sqrt{4x^4 \cdot 3}$ Write $12x^4$ so that one factor is a perfect square.
$= \sqrt{4x^4} \cdot \sqrt{3}$ Use the product rule.
$= 2x^2\sqrt{3}$ Simplify.

When multiplying radical expressions containing more than one term, we use the same techniques we use to multiply other algebraic expressions with more than one term.

Example 4   Multiply.

  a. $\sqrt{5}(\sqrt{5} - \sqrt{2})$
  b. $(\sqrt{x} + \sqrt{2})(\sqrt{3} - \sqrt{2})$

Solution: **a.** Using the distributive property, we have

$$\sqrt{5}(\sqrt{5} - \sqrt{2}) = \sqrt{5} \cdot \sqrt{5} - \sqrt{5} \cdot \sqrt{2}$$
$$= \sqrt{25} - \sqrt{10}$$
$$= 5 - \sqrt{10}$$

b. Using the FOIL method of multiplication, we have

$$(\sqrt{x} + \sqrt{2})(\sqrt{3} - \sqrt{2})$$
$$\overset{F}{=} \sqrt{x} \cdot \sqrt{3} \overset{O}{-} \sqrt{x} \cdot \sqrt{2} \overset{I}{+} \sqrt{2} \cdot \sqrt{3} \overset{L}{-} \sqrt{2} \cdot \sqrt{2}$$
$$= \sqrt{3x} - \sqrt{2x} + \sqrt{6} - \sqrt{4}$$ Use the product rule.
$$= \sqrt{3x} - \sqrt{2x} + \sqrt{6} - 2$$ Simplify.

From Example 4, we found that

$$\sqrt{5} \cdot \sqrt{5} = 5 \quad \text{and} \quad \sqrt{2} \cdot \sqrt{2} = 2$$

Objectives

A Multiply radicals.
B Divide radicals.
C Rationalize denominators.
D Rationalize using conjugates.

SSM   CD-ROM   Video
                          8.4

Practice Problems 1–3

Multiply. Then simplify each product if possible.

1. $\sqrt{5} \cdot \sqrt{2}$

2. $\sqrt{6} \cdot \sqrt{3}$

3. $\sqrt{10x} \cdot \sqrt{2x}$

Practice Problem 4

Multiply.

a. $\sqrt{7}(\sqrt{7} - \sqrt{3})$

b. $(\sqrt{x} + \sqrt{5})(\sqrt{x} - \sqrt{3})$

Answers

1. $\sqrt{10}$, **2.** $3\sqrt{2}$, **3.** $2x\sqrt{5}$, **4. a.** $7 - \sqrt{21}$,
b. $x - \sqrt{3x} + \sqrt{5x} - \sqrt{15}$

✓ CONCEPT CHECK

Identify the true statement(s).
a. $\sqrt{7} \cdot \sqrt{7} = 7$

b. $\sqrt{2} \cdot \sqrt{3} = 6$

c. $\sqrt{131} \cdot \sqrt{131} = 131$

d. $\sqrt{5x} \cdot \sqrt{5x} = 5x$ (Here x is a positive number.)

Practice Problem 5

Multiply.

a. $(\sqrt{3} + 6)(\sqrt{3} - 6)$
b. $(\sqrt{5x} + 4)^2$

TEACHING TIP

Use substitution to show students that these problems can be solved using the skills they have already developed. For instance, in Example 5a

Let $x = \sqrt{5}$ and $y = 7$
$(\sqrt{5} - 7)(\sqrt{5} + 7) = (x - y)(x + y)$
$= x^2 - y^2$
$= (\sqrt{5})^2 - 7^2$
$= 5 - 49$
$= -44$

Practice Problems 6–8

Divide. Then simplify the quotient if possible.

6. $\dfrac{\sqrt{15}}{\sqrt{3}}$

7. $\dfrac{\sqrt{90}}{\sqrt{2}}$

8. $\dfrac{\sqrt{75x^3}}{\sqrt{5x}}$

Answers

✓ **Concept Check:**   a, c, d

5. a. -33,   **b.** $5x + 8\sqrt{5x} + 16$,   **6.** $\sqrt{5}$,
7. $3\sqrt{5}$,   **8.** $x\sqrt{15}$

This is true in general.

> If a is a positive number,
> $$\sqrt{a} \cdot \sqrt{a} = a$$

TRY THE CONCEPT CHECK IN THE MARGIN.

Special products also can be used to multiply expressions containing radicals.

Example 5   Multiply.

   a. $(\sqrt{5} - 7)(\sqrt{5} + 7)$
   b. $(\sqrt{7x} + 2)^2$

Solution:

a. $(\sqrt{5} - 7)(\sqrt{5} + 7) = (\sqrt{5})^2 - 7^2$ Recall that $(a-b)(a+b) = a^2 - b^2$.
$$= 5 - 49$$
$$= -44$$

b. $(\sqrt{7x} + 2)^2$
$= (\sqrt{7x})^2 + 2(\sqrt{7x})(2) + (2)^2$ Recall that $(a+b)^2 = a^2 + 2ab + b^2$.
$= 7x + 4\sqrt{7x} + 4$ ▬▬▬

B   DIVIDING RADICALS

To simplify quotients of rational expressions, we use the quotient rule.

> **QUOTIENT RULE FOR RADICALS**
>
> If $\sqrt{a}$ and $\sqrt{b}$ are real numbers and $b \neq 0$, then
> $$\frac{\sqrt{a}}{\sqrt{b}} = \sqrt{\frac{a}{b}}$$

Examples   Divide. Then simplify the quotient if possible.

6. $\dfrac{\sqrt{14}}{\sqrt{2}} = \sqrt{\dfrac{14}{2}} = \sqrt{7}$

7. $\dfrac{\sqrt{100}}{\sqrt{5}} = \sqrt{\dfrac{100}{5}} = \sqrt{20} = \sqrt{4 \cdot 5} = \sqrt{4} \cdot \sqrt{5} = 2\sqrt{5}$

8. $\dfrac{\sqrt{12x^3}}{\sqrt{3x}} = \sqrt{\dfrac{12x^3}{3x}} = \sqrt{4x^2} = 2x$ ▬▬▬

C   RATIONALIZING DENOMINATORS

It is sometimes easier to work with radical expressions if the denominator does not contain a radical. To get rid of the radical in the denominator of a radical expression, we use the fact that we can multiply the numerator and the denominator of a fraction by the same nonzero number without changing the value of the expression. This is the same as multiplying the fraction by 1. For example, to get rid of the radical in the denominator of $\dfrac{\sqrt{5}}{\sqrt{2}}$, we multiply the numerator and the denominator by $\sqrt{2}$. Then

$$\frac{\sqrt{5}}{\sqrt{2}} = \frac{\sqrt{5} \cdot \sqrt{2}}{\sqrt{2} \cdot \sqrt{2}} = \frac{\sqrt{10}}{2}$$

This process is called **rationalizing** the denominator.

Example 9   Rationalize the denominator of $\frac{2}{\sqrt{7}}$.

Solution: To get rid of the radical in the denominator of $\frac{2}{\sqrt{7}}$, we multiply the numerator and the denominator by $\sqrt{7}$.

$$\frac{2}{\sqrt{7}} = \frac{2 \cdot \sqrt{7}}{\sqrt{7} \cdot \sqrt{7}} = \frac{2\sqrt{7}}{7}$$

Example 10   Rationalize the denominator of $\frac{\sqrt{5}}{\sqrt{12}}$.

Solution: We can multiply the numerator and denominator by $\sqrt{12}$, but see what happens if we simplify first.

$$\frac{\sqrt{5}}{\sqrt{12}} = \frac{\sqrt{5}}{\sqrt{4 \cdot 3}} = \frac{\sqrt{5}}{2\sqrt{3}}$$

To rationalize the denominator now, we multiply the numerator and the denominator by $\sqrt{3}$.

$$\frac{\sqrt{5}}{2\sqrt{3}} = \frac{\sqrt{5} \cdot \sqrt{3}}{2\sqrt{3} \cdot \sqrt{3}} = \frac{\sqrt{15}}{2 \cdot 3} = \frac{\sqrt{15}}{6}$$

Example 11   Rationalize the denominator of $\sqrt{\frac{1}{18x}}$.

Solution: First we simplify.

$$\sqrt{\frac{1}{18x}} = \frac{\sqrt{1}}{\sqrt{18x}} = \frac{1}{\sqrt{9} \cdot \sqrt{2x}} = \frac{1}{3\sqrt{2x}}$$

Now to rationalize the denominator, we multiply the numerator and denominator by $\sqrt{2x}$.

$$\frac{1}{3\sqrt{2x}} = \frac{1 \cdot \sqrt{2x}}{3\sqrt{2x} \cdot \sqrt{2x}} = \frac{\sqrt{2x}}{3 \cdot 2x} = \frac{\sqrt{2x}}{6x}$$

D   RATIONALIZING DENOMINATORS USING CONJUGATES

To rationalize a denominator that is a sum or a difference, such as the denominator in

$$\frac{2}{4 + \sqrt{3}}$$

we multiply the numerator and the denominator by $4 - \sqrt{3}$. The expressions $4 + \sqrt{3}$ and $4 - \sqrt{3}$ are called **conjugates** of each other. When a radical expression such as $4 + \sqrt{3}$ is multiplied by its conjugate $4 - \sqrt{3}$, the product simplifies to an expression that contains no radicals.

$$(a + b)(a - b) = a^2 - b^2$$
$$(4 + \sqrt{3})(4 - \sqrt{3}) = 4^2 - (\sqrt{3})^2 = 16 - 3 = 13$$

Then

$$\frac{2}{4 + \sqrt{3}} = \frac{2(4 - \sqrt{3})}{(4 + \sqrt{3})(4 - \sqrt{3})} = \frac{2(4 - \sqrt{3})}{13}$$

TEACHING TIP
Consider asking students, "Why doesn't the value of the expression change when we multiply the numerator and denominator by $\sqrt{2}$?"

Practice Problem 9

Rationalize the denominator of $\frac{5}{\sqrt{3}}$.

Practice Problem 10

Rationalize the denominator of $\frac{\sqrt{7}}{\sqrt{20}}$.

TEACHING TIP

Use Example 10 to show students that if the expression is not simplified first, it must be simplified at the end.

Practice Problem 11

Rationalize the denominator of $\sqrt{\frac{2}{45x}}$.

TEACHING TIP

Before beginning a discussion of conjugates, ask students to simplify the following expressions:

$(1 + \sqrt{3})(1 - \sqrt{2})$
$(1 + \sqrt{3})(1 + \sqrt{3})$
$(1 + \sqrt{3})(1 - \sqrt{3})$

Lead students to notice that when multiplying binomials including radicals, the product does not always contain a radical. Ask them to guess what $3 + \sqrt{5}$ could be multiplied by to obtain a product without radicals.

Answers

9. $\frac{5\sqrt{3}}{3}$,   10. $\frac{\sqrt{35}}{10}$,   11. $\frac{\sqrt{10x}}{15x}$

Practice Problem 12

Rationalize the denominator of
$\dfrac{3}{1 + \sqrt{7}}$.

Example 12   Rationalize the denominator of $\dfrac{2}{1 + \sqrt{3}}$.

    Solution: We multiply the numerator and the denominator of this fraction by the conjugate of $1 + \sqrt{3}$, that is, by $1 - \sqrt{3}$.

$$\frac{2}{1 + \sqrt{3}} = \frac{2(1 - \sqrt{3})}{(1 + \sqrt{3})(1 - \sqrt{3})}$$

$$= \frac{2(1 - \sqrt{3})}{1^2 - (\sqrt{3})^2}$$

HELPFUL HINT

Don't forget that
$(\sqrt{3})^2 = \sqrt{3} \cdot \sqrt{3} = 3$

$$= \frac{2(1 - \sqrt{3})}{1 - 3}$$

$$= \frac{2(1 - \sqrt{3})}{-2}$$

$$= -\frac{2(1 - \sqrt{3})}{2} \qquad \frac{a}{-b} = -\frac{a}{b}$$

$$= -1(1 - \sqrt{3}) \qquad \text{Simplify.}$$

$$= -1 + \sqrt{3} \qquad \text{Multiply.}$$

Practice Problem 13

Rationalize the denominator of
$\dfrac{\sqrt{2} + 5}{\sqrt{2} - 1}$.

Example 13   Rationalize the denominator of $\dfrac{\sqrt{5} + 4}{\sqrt{5} - 1}$.

    Solution:

$$\frac{\sqrt{5} + 4}{\sqrt{5} - 1} = \frac{(\sqrt{5} + 4)(\sqrt{5} + 1)}{(\sqrt{5} - 1)(\sqrt{5} + 1)} \quad \text{Multiply the numerator and denominator by } \sqrt{5} + 1, \text{ the conjugate of } \sqrt{5} - 1.$$

$$= \frac{5 + \sqrt{5} + 4\sqrt{5} + 4}{5 - 1} \quad \text{Multiply.}$$

$$= \frac{9 + 5\sqrt{5}}{4} \quad \text{Simplify.}$$

Practice Problem 14

Rationalize the denominator of
$\dfrac{7}{2 - \sqrt{x}}$.

Example 14   Rationalize the denominator of $\dfrac{3}{1 + \sqrt{x}}$.

    Solution:

$$\frac{3}{1 + \sqrt{x}} = \frac{3(1 - \sqrt{x})}{(1 + \sqrt{x})(1 - \sqrt{x})} \quad \text{Multiply the numerator and denominator by } 1 - \sqrt{x}, \text{ the conjugate of } 1 + \sqrt{x}.$$

$$= \frac{3(1 - \sqrt{x})}{1 - x}$$

Answers

12. $\dfrac{-1 + \sqrt{7}}{2}$,   **13.** $7 + 6\sqrt{2}$,

14. $\dfrac{7(2 + \sqrt{x})}{4 - x}$

Name _____ **Section** _____ **Date** _____

MENTAL MATH

Multiply. Assume that all variables represent positive numbers.

1. $\sqrt{2} \cdot \sqrt{3}$ 2. $\sqrt{5} \cdot \sqrt{7}$ 3. $\sqrt{1} \cdot \sqrt{6}$

4. $\sqrt{7} \cdot \sqrt{x}$ 5. $\sqrt{10} \cdot \sqrt{y}$ 6. $\sqrt{x} \cdot \sqrt{y}$

EXERCISE SET 8.4

A *Multiply and simplify. See Examples 1 through 5.*

1. $\sqrt{8} \cdot \sqrt{2}$ 2. $\sqrt{3} \cdot \sqrt{12}$ 3. $\sqrt{10} \cdot \sqrt{5}$

4. $\sqrt{2} \cdot \sqrt{14}$ 5. $\sqrt{6} \cdot \sqrt{6}$ 6. $\sqrt{10} \cdot \sqrt{10}$

7. $\sqrt{2x} \cdot \sqrt{2x}$ 8. $\sqrt{5y} \cdot \sqrt{5y}$ 9. $(2\sqrt{5})^2$

10. $(3\sqrt{10})^2$ 11. $(6\sqrt{x})^2$ 12. $(8\sqrt{y})^2$

13. $\sqrt{3y} \cdot \sqrt{6x}$ 14. $\sqrt{21y} \cdot \sqrt{3x}$ 15. $\sqrt{2xy^2} \cdot \sqrt{8xy}$

16. $\sqrt{18x^2y^2} \cdot \sqrt{2x^2y}$ 17. $\sqrt{2}(\sqrt{5} + 1)$ 18. $\sqrt{3}(\sqrt{2} - 1)$

19. $\sqrt{10}(\sqrt{2} + \sqrt{5})$ 20. $\sqrt{6}(\sqrt{3} + \sqrt{2})$ 21. $\sqrt{6}(\sqrt{5} + \sqrt{7})$

22. $\sqrt{10}(\sqrt{3} - \sqrt{7})$ 23. $(\sqrt{3} + 6)(\sqrt{3} - 6)$ 24. $(\sqrt{5} + 2)(\sqrt{5} - 2)$

MENTAL MATH ANSWERS

1. $\sqrt{6}$

2. $\sqrt{35}$

3. $\sqrt{6}$

4. $\sqrt{7x}$

5. $\sqrt{10y}$

6. $\sqrt{xy}$

ANSWERS

1. 4

2. 6

3. $5\sqrt{2}$

4. $2\sqrt{7}$

5. 6

6. 10

7. $2x$

8. $5y$

9. 20

10. 90

11. $36x$

12. $64y$

13. $3\sqrt{2xy}$

14. $3\sqrt{7xy}$

15. $4xy\sqrt{y}$

16. $6x^2y\sqrt{y}$

17. $\sqrt{10} + \sqrt{2}$

18. $\sqrt{6} - \sqrt{3}$

19. $2\sqrt{5} + 5\sqrt{2}$

20. $3\sqrt{2} + 2\sqrt{3}$

21. $\sqrt{30} + \sqrt{42}$

22. $\sqrt{30} - \sqrt{70}$

23. -33

24. 1

621

25. $\dfrac{\sqrt{6} - \sqrt{15}}{+ \sqrt{10} - 5}$

26. $\dfrac{\sqrt{14} - \sqrt{35}}{+ \sqrt{10} - 5}$

27. $16 - 11\sqrt{11}$

28. $13 - 3\sqrt{3}$

29. $x - 36$

30. $y - 25$

31. $x - 14\sqrt{x} + 49$

32. $x + 8\sqrt{x} + 16$

33. $6y + 2\sqrt{6y} + 1$

34. $3y - 4\sqrt{3y} + 4$

35. 4

36. 2

37. $\sqrt{7}$

38. $\sqrt{11}$

39. $3\sqrt{2}$

40. $2\sqrt{3}$

41. $5y^2$

42. $2x^3$

43. $5\sqrt{3}$

44. $2\sqrt{10}$

45. $2y\sqrt{6}$

46. $3x\sqrt{3}$

47. $2xy\sqrt{3y}$

48. $4xy\sqrt{2x}$

49. $\dfrac{\sqrt{15}}{5}$

50. $\dfrac{\sqrt{6}}{3}$

51. $\dfrac{7\sqrt{2}}{2}$

52. $\dfrac{8\sqrt{11}}{11}$

53. $\dfrac{\sqrt{6y}}{6y}$

54. $\dfrac{\sqrt{10z}}{10z}$

55. $\dfrac{\sqrt{10}}{6}$

56. $\dfrac{\sqrt{21}}{6}$

25. $(\sqrt{3} + \sqrt{5})(\sqrt{2} - \sqrt{5})$

26. $(\sqrt{7} + \sqrt{5})(\sqrt{2} - \sqrt{5})$

27. $(2\sqrt{11} + 1)(\sqrt{11} - 6)$

28. $(5\sqrt{3} + 2)(\sqrt{3} - 1)$

📼 **29.** $(\sqrt{x} + 6)(\sqrt{x} - 6)$ **30.** $(\sqrt{y} + 5)(\sqrt{y} - 5)$ **31.** $(\sqrt{x} - 7)^2$

32. $(\sqrt{x} + 4)^2$ **33.** $(\sqrt{6y} + 1)^2$ **34.** $(\sqrt{3y} - 2)^2$

B *Divide and simplify. See Examples 6 through 8.*

35. $\dfrac{\sqrt{32}}{\sqrt{2}}$ **36.** $\dfrac{\sqrt{40}}{\sqrt{10}}$ **37.** $\dfrac{\sqrt{21}}{\sqrt{3}}$ **38.** $\dfrac{\sqrt{55}}{\sqrt{5}}$

📼 **39.** $\dfrac{\sqrt{90}}{\sqrt{5}}$ **40.** $\dfrac{\sqrt{96}}{\sqrt{8}}$ 📼 **41.** $\dfrac{\sqrt{75y^5}}{\sqrt{3y}}$ **42.** $\dfrac{\sqrt{24x^7}}{\sqrt{6x}}$

43. $\dfrac{\sqrt{150}}{\sqrt{2}}$ **44.** $\dfrac{\sqrt{120}}{\sqrt{3}}$ **45.** $\dfrac{\sqrt{72y^5}}{\sqrt{3y^3}}$ **46.** $\dfrac{\sqrt{54x^3}}{\sqrt{2x}}$

47. $\dfrac{\sqrt{24x^3y^4}}{\sqrt{2xy}}$ **48.** $\dfrac{\sqrt{96x^5y^3}}{\sqrt{3x^2y}}$

C *Rationalize each denominator and simplify. See Examples 9 through 11.*

📼 **49.** $\dfrac{\sqrt{3}}{\sqrt{5}}$ **50.** $\dfrac{\sqrt{2}}{\sqrt{3}}$ **51.** $\dfrac{7}{\sqrt{2}}$ **52.** $\dfrac{8}{\sqrt{11}}$

53. $\dfrac{1}{\sqrt{6y}}$ **54.** $\dfrac{1}{\sqrt{10z}}$ **55.** $\sqrt{\dfrac{5}{18}}$ **56.** $\sqrt{\dfrac{7}{12}}$

57. $\sqrt{\dfrac{3}{x}}$ **58.** $\sqrt{\dfrac{5}{x}}$ **59.** $\sqrt{\dfrac{1}{8}}$ **60.** $\sqrt{\dfrac{1}{27}}$

61. $\sqrt{\dfrac{2}{15}}$ **62.** $\sqrt{\dfrac{11}{14}}$ **63.** $\sqrt{\dfrac{3}{20}}$ **64.** $\sqrt{\dfrac{3}{50}}$

65. $\dfrac{3x}{\sqrt{2x}}$ **66.** $\dfrac{5y}{\sqrt{3y}}$ **67.** $\dfrac{8y}{\sqrt{5}}$ **68.** $\dfrac{7x}{\sqrt{2}}$

69. $\sqrt{\dfrac{y}{12x}}$ **70.** $\sqrt{\dfrac{x}{20y}}$

D *Rationalize each denominator and simplify. See Examples 12 through 14.*

71. $\dfrac{3}{\sqrt{2}+1}$ **72.** $\dfrac{6}{\sqrt{5}+2}$ **73.** $\dfrac{4}{2-\sqrt{5}}$ **74.** $\dfrac{2}{\sqrt{10}-3}$

75. $\dfrac{\sqrt{5}+1}{\sqrt{6}-\sqrt{5}}$ **76.** $\dfrac{\sqrt{3}+1}{\sqrt{3}-\sqrt{2}}$ **77.** $\dfrac{\sqrt{3}+1}{\sqrt{2}-1}$ **78.** $\dfrac{\sqrt{2}-2}{2-\sqrt{3}}$

79. $\dfrac{5}{2+\sqrt{x}}$ **80.** $\dfrac{9}{3+\sqrt{x}}$ **81.** $\dfrac{3}{\sqrt{x}-4}$ **82.** $\dfrac{4}{\sqrt{x}-1}$

REVIEW AND PREVIEW

Solve each equation. See Sections 2.4 and 4.6.

83. $x+5=7^2$ **84.** $2y-1=3^2$

85. $4z^2+6z-12=(2z)^2$ **86.** $16x^2+x+9=(4x)^2$

87. $9x^2+5x+4=(3x+1)^2$ **88.** $x^2+3x+4=(x+2)^2$

---

57. $\dfrac{\sqrt{3x}}{x}$

58. $\dfrac{\sqrt{5x}}{x}$

59. $\dfrac{\sqrt{2}}{4}$

60. $\dfrac{\sqrt{3}}{9}$

61. $\dfrac{\sqrt{30}}{15}$

62. $\dfrac{\sqrt{154}}{14}$

63. $\dfrac{\sqrt{15}}{10}$

64. $\dfrac{\sqrt{6}}{10}$

65. $\dfrac{3\sqrt{2x}}{2}$

66. $\dfrac{5\sqrt{3y}}{3}$

67. $\dfrac{8y\sqrt{5}}{5}$

68. $\dfrac{7x\sqrt{2}}{2}$

69. $\dfrac{\sqrt{3xy}}{6x}$

70. $\dfrac{\sqrt{5xy}}{10y}$

71. $3\sqrt{2}-3$

72. $6\sqrt{5}-12$

73. $-8-4\sqrt{5}$

74. $2\sqrt{10}+6$

75. $5+\sqrt{30}+\sqrt{6}+\sqrt{5}$

76. $3+\sqrt{6}+\sqrt{3}+\sqrt{2}$

77. $\sqrt{6}+\sqrt{3}+\sqrt{2}+1$

78. $2\sqrt{2}+\sqrt{6}-2\sqrt{3}-4$

79. $\dfrac{10-5\sqrt{x}}{4-x}$

80. $\dfrac{27-9\sqrt{x}}{9-x}$

81. $\dfrac{3\sqrt{x}+12}{x-16}$

82. $\dfrac{4\sqrt{x}+4}{x-1}$

83. $x=44$

84. $y=5$

85. $z=2$

86. $x=-9$

87. $x=3$

88. $x=0$

Name _____

89. Find the area of a rectangular room whose length is $13\sqrt{2}$ meters and width is $5\sqrt{6}$ meters.

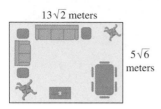

13√2 meters

5√6 meters

90. Find the volume of a microwave oven whose length is $\sqrt{3}$ feet, width is $\sqrt{2}$ feet, and height is $\sqrt{2}$ feet.

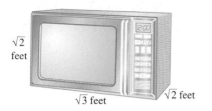

√2 feet

√3 feet   √2 feet

91. When rationalizing the denominator of $\dfrac{\sqrt{2}}{\sqrt{3}}$, explain why both the numerator and the denominator must be multiplied by $\sqrt{3}$.

92. If a circle has area A, then the formula for the

radius r of the circle is

$$r = \sqrt{\dfrac{A}{\pi}}$$

Simplify this expression by rationalizing the denominator.

It is often more convenient to work with a radical expression whose numerator is rationalized. Rationalize the numerator of each expression by multiplying the numerator and denominator by the conjugate of the numerator.

93. $\dfrac{\sqrt{3}+1}{\sqrt{2}-1}$

94. $\dfrac{\sqrt{2}-2}{2-\sqrt{3}}$

Name _____ **Section** _____ **Date** _____

Integrated Review—Simplifying Radicals

Simplify. Assume that all variables represent positive numbers.

1. $\sqrt{36}$

2. $\sqrt{48}$

3. $\sqrt{x^4}$

4. $\sqrt{y^7}$

5. $\sqrt{16x^2}$

6. $\sqrt{18x^{11}}$

7. $\sqrt[3]{8}$

8. $\sqrt[4]{81}$

9. $\sqrt[3]{-27}$

10. $\sqrt{-4}$

11. $\sqrt{\dfrac{11}{9}}$

12. $\sqrt[3]{\dfrac{7}{64}}$

Add or subtract as indicated.

13. $5\sqrt{7} + \sqrt{7}$

14. $\sqrt{50} - \sqrt{8}$

15. $2\sqrt{x} + \sqrt{25x} - \sqrt{36x} + 3x$

Multiply and simplify if possible.

16. $\sqrt{2} \cdot \sqrt{15}$

17. $\sqrt{3} \cdot \sqrt{3}$

18. $\sqrt{3}(\sqrt{11} + 1)$

19. $(\sqrt{x} - 5)(\sqrt{x} + 2)$

20. $(3 + \sqrt{2})^2$

1. 6

2. $4\sqrt{3}$

3. x^2

4. $y^3\sqrt{y}$

5. $4x$

6. $3x^5\sqrt{2x}$

7. 2

8. 3

9. -3

10. not a real number

11. $\dfrac{\sqrt{11}}{3}$

12. $\dfrac{\sqrt[3]{7}}{4}$

13. $6\sqrt{7}$

14. $3\sqrt{2}$

15. $\sqrt{x} + 3x$

16. $\sqrt{30}$

17. 3

18. $\sqrt{33} + \sqrt{3}$

19. $x - 3\sqrt{x} - 10$

20. $11 + 6\sqrt{2}$

21. 2 _____

Divide and simplify if possible.

21. $\dfrac{\sqrt{8}}{\sqrt{2}}$ **22.** $\dfrac{\sqrt{45}}{\sqrt{15}}$ **23.** $\dfrac{\sqrt{24x^5}}{\sqrt{2x}}$

22. $\sqrt{3}$ _____

Rationalize each denominator.

24. $\sqrt{\dfrac{1}{6}}$ **25.** $\dfrac{x}{\sqrt{20}}$ **26.** $\dfrac{4}{\sqrt{6}+1}$ **27.** $\dfrac{\sqrt{2}+1}{\sqrt{x}-5}$

23. $2x^2\sqrt{3}$ _____

24. $\dfrac{\sqrt{6}}{6}$ _____

25. $\dfrac{x\sqrt{5}}{10}$ _____

26. $\dfrac{4\sqrt{6}-4}{5}$ _____

27. $\dfrac{\sqrt{2x}+5\sqrt{2}+\sqrt{x}+5}{x-25}$ _____

8.5   Solving Equations Containing Radicals

A   Using the Squaring Property of Equality Once

In this section, we solve **radical equations** such as

$$\sqrt{x + 3} = 5 \quad \text{and} \quad \sqrt{2x + 1} = \sqrt{3x}$$

Radical equations contain variables in the radicand. To solve these equations, we rely on the following squaring property.

The Squaring Property of Equality

If $a = b$, then $a^2 = b^2$

Unfortunately, this squaring property does not guarantee that all solutions of the new equation are solutions of the original equation. For example, if we square both sides of the equation

$$x = 2$$

we have

$$x^2 = 4$$

This new equation has two solutions, 2 and -2, while the original equation $x = 2$ has only one solution. For this reason, we must **always check proposed solutions of radical equations in the original equation.**

Example 1   Solve: $\sqrt{x + 3} = 5$

Solution: To solve this radical equation, we use the squaring property of equality and square both sides of the equation.

$$\sqrt{x + 3} = 5$$
$$(\sqrt{x + 3})^2 = 5^2 \quad \text{Square both sides.}$$
$$x + 3 = 25 \quad \text{Simplify.}$$
$$x = 22 \quad \text{Subtract 3 from both sides.}$$

Check: We replace x with 22 in the original equation.

> **HELPFUL HINT**
>
> Don't forget to check the proposed solutions of radical equations in the original equation.

$$\sqrt{x + 3} = 5 \quad \text{Original equation}$$
$$\sqrt{22 + 3} \stackrel{?}{=} 5 \quad \text{Let } x = 22.$$
$$\sqrt{25} \stackrel{?}{=} 5$$
$$5 = 5 \quad \text{True.}$$

Since a true statement results, 22 is the solution.

Example 2   Solve: $\sqrt{x} + 6 = 4$

Solution: First we set the radical by itself on one side of the equation. Then we square both sides.

$$\sqrt{x} + 6 = 4$$
$$\sqrt{x} = -2 \quad \text{Subtract 6 from both sides to get the radical by itself.}$$

Recall that $\sqrt{x}$ is the principal or nonnegative square root of x so that $\sqrt{x}$ cannot equal -2 and thus this equa-

---

Objectives

A   Solve radical equations by using the squaring property of equality once.

B   Solve radical equations by using the squaring property of equality twice.

SSM   CD-ROM   Video 8.5

TEACHING TIP

Consider beginning this lesson by having students verify that $x = 1$ is a solution of $\sqrt{2x + 1} = \sqrt{3x}$.

Practice Problem 1

Solve: $\sqrt{x - 2} = 7$

TEACHING TIP

Before doing Example 2, ask students, "What is the first step in solving $x + 6 = 4$?" Help students see that the same approach applies to the equation $\sqrt{x} + 6 = 4$.

Practice Problem 2

Solve: $\sqrt{x} + 9 = 2$

Answers

1. $x = 51$,   **2.** no solution

tion has no solution. We arrive at the same conclusion if we continue by applying the squaring property.

$$\sqrt{x} = -2$$
$$(\sqrt{x})^2 = (-2)^2 \quad \text{Square both sides.}$$
$$x = 4 \quad \text{Simplify.}$$

Check: We replace x with 4 in the original equation.

$$\sqrt{x} + 6 = 4 \quad \text{Original equation}$$
$$\sqrt{4} + 6 \stackrel{?}{=} 4 \quad \text{Let } x = 4.$$
$$2 + 6 = 4 \quad \text{False.}$$

Since 4 *does not* satisfy the original equation, this equation has no solution. ▬▬▬

Example 2 makes it very clear that we *must* check proposed solutions in the original equation to determine if they are truly solutions. If a proposed solution does not work, we say that the value is an **extraneous solution**.

The following steps can be used to solve radical equations containing square roots.

TO SOLVE A RADICAL EQUATION CONTAINING SQUARE ROOTS

Step 1. Arrange terms so that one radical is by itself on one side of the equation. That is, isolate a radical.

Step 2. Square both sides of the equation.

Step 3. Simplify both sides of the equation.

Step 4. If the equation still contains a radical term, repeat Steps 1 through 3.

Step 5. Solve the equation.

Step 6. Check all solutions in the original equation for extraneous solutions.

Practice Problem 3

Solve: $\sqrt{6x - 1} = \sqrt{x}$

Example 3   Solve: $\sqrt{x} = \sqrt{5x - 2}$

Solution: Each of the radicals is already isolated, since each is by itself on one side of the equation. So we begin solving by squaring both sides.

$$\sqrt{x} = \sqrt{5x - 2} \quad \text{Original equation}$$
$$(\sqrt{x})^2 = (\sqrt{5x - 2})^2 \quad \text{Square both sides.}$$
$$x = 5x - 2 \quad \text{Simplify.}$$
$$-4x = -2 \quad \text{Subtract } 5x \text{ from both sides.}$$
$$x = \frac{-2}{-4} = \frac{1}{2} \quad \text{Divide both sides by } -4 \text{ and simplify.}$$

Check: We replace x with $\frac{1}{2}$ in the original equation.

$$\sqrt{x} = \sqrt{5x - 2} \quad \text{Original equation}$$
$$\sqrt{\frac{1}{2}} \stackrel{?}{=} \sqrt{5 \cdot \frac{1}{2} - 2} \quad \text{Let } x = \frac{1}{2}.$$
$$\sqrt{\frac{1}{2}} \stackrel{?}{=} \sqrt{\frac{5}{2} - 2} \quad \text{Multiply.}$$
$$\sqrt{\frac{1}{2}} \stackrel{?}{=} \sqrt{\frac{5}{2} - \frac{4}{2}} \quad \text{Write 2 as } \frac{4}{2}.$$
$$\sqrt{\frac{1}{2}} = \sqrt{\frac{1}{2}} \quad \text{True.}$$

This statement is true, so the solution is $\frac{1}{2}$. ▬▬▬

Example 4   Solve: $\sqrt{4y^2 + 5y - 15} = 2y$

Solution: The radical is already isolated, so we start by squaring both sides.

$$\sqrt{4y^2 + 5y - 15} = 2y$$
$$(\sqrt{4y^2 + 5y - 15})^2 = (2y)^2 \quad \text{Square both sides.}$$
$$4y^2 + 5y - 15 = 4y^2 \quad \text{Simplify.}$$
$$5y - 15 = 0 \quad \text{Subtract } 4y^2 \text{ from both sides.}$$
$$5y = 15 \quad \text{Add 15 to both sides.}$$
$$y = 3 \quad \text{Divide both sides by 5.}$$

Check: We replace y with 3 in the original equation.

$$\sqrt{4y^2 + 5y - 15} = 2y \quad \text{Original equation}$$
$$\sqrt{4 \cdot 3^2 + 5 \cdot 3 - 15} \stackrel{?}{=} 2 \cdot 3 \quad \text{Let } y = 3.$$
$$\sqrt{4 \cdot 9 + 15 - 15} \stackrel{?}{=} 6 \quad \text{Simplify.}$$
$$\sqrt{36} \stackrel{?}{=} 6$$
$$6 = 6 \quad \text{True.}$$

This statement is true, so the solution is 3. ▬▬▬

Example 5   Solve: $\sqrt{x + 3} - x = -3$

Solution: First we isolate the radical by adding x to both sides. Then we square both sides.

$$\sqrt{x + 3} - x = -3$$
$$\sqrt{x + 3} = x - 3 \quad \text{Add } x \text{ to both sides.}$$
$$(\sqrt{x + 3})^2 = (x - 3)^2 \quad \text{Square both sides.}$$
$$x + 3 = x^2 - 6x + 9$$

┌───┐
| **HELPFUL HINT**
|
| Don't forget that $(x - 3)^2 = (x - 3)(x - 3) = x^2 - 6x + 9$
└───┘

To solve the resulting quadratic equation, we write the equation in standard form by subtracting x and 3 from both sides.

$$3 = x^2 - 7x + 9 \quad \text{Subtract } x \text{ from both sides.}$$
$$0 = x^2 - 7x + 6 \quad \text{Subtract 3 from both sides.}$$
$$0 = (x - 6)(x - 1) \quad \text{Factor.}$$
$$0 = x - 6 \quad \text{or} \quad 0 = x - 1 \quad \text{Set each factor equal to zero.}$$
$$6 = x \qquad\qquad 1 = x \quad \text{Solve for } x.$$

Check: We replace x with 6 and then x with 1 in the original equation.

Let $x = 6$.

$$\sqrt{x + 3} - x = -3$$
$$\sqrt{6 + 3} - 6 \stackrel{?}{=} -3$$
$$\sqrt{9} - 6 \stackrel{?}{=} -3$$
$$3 - 6 \stackrel{?}{=} -3$$
$$-3 = -3 \quad \text{True.}$$

Let $x = 1$.

$$\sqrt{x + 3} - x = -3$$
$$\sqrt{1 + 3} - 1 \stackrel{?}{=} -3$$
$$\sqrt{4} - 1 \stackrel{?}{=} -3$$
$$2 - 1 \stackrel{?}{=} -3$$
$$1 = -3 \quad \text{False.}$$

Practice Problem 4

Solve: $\sqrt{9y^2 + 2y - 10} = 3y$

TEACHING TIP

Before solving Example 5, ask students, "Will squaring both sides of an equation always eliminate the radical sign?" Demonstrate that the radical is not eliminated when both sides of $\sqrt{x + 3} - x = -3$ are squared directly.

Practice Problem 5

Solve: $\sqrt{x + 1} - x = -5$

Answers

4. $y = 5$,   **5.** $x = 8$

Since replacing x with 1 resulted in a false statement, 1 is an extraneous solution. The only solution is 6.   ▬▬▬

B USING THE SQUARING PROPERTY OF EQUALITY TWICE

If a radical equation contains two radicals, we may need to use the squaring property twice.

Practice Problem 6

Solve: $\sqrt{x} + 3 = \sqrt{x + 15}$

Example 6   Solve: $\sqrt{x - 4} = \sqrt{x} - 2$

Solution:

$$\sqrt{x - 4} = \sqrt{x} - 2$$

$$(\sqrt{x - 4})^2 = (\sqrt{x} - 2)^2 \qquad \text{Square both sides.}$$

$$x - 4 = \underbrace{x - 4\sqrt{x} + 4}$$

$$-8 = -4\sqrt{x}$$

$$2 = \sqrt{x} \qquad \text{Divide both sides by } -4.$$

$$4 = x \qquad \text{Square both sides again.}$$

> **HELPFUL HINT**
>
> $$(\sqrt{x} - 2)^2 = (\sqrt{x} - 2)(\sqrt{x} - 2)$$
> $$= \sqrt{x} \cdot \sqrt{x} - 2\sqrt{x} - 2\sqrt{x} + 4$$
> $$= x - 4\sqrt{x} + 4$$

Check the proposed solution in the original equation. The solution is 4.   ▬▬▬

EXERCISE SET 8.5

A *Solve each equation. See Examples 1 through 3.*

1. $\sqrt{x} = 9$ **2.** $\sqrt{x} = 4$ **▣ 3.** $\sqrt{x + 5} = 2$

4. $\sqrt{x + 12} = 3$ **5.** $\sqrt{2x + 6} = 4$ **6.** $\sqrt{3x + 7} = 5$

7. $\sqrt{x} - 2 = 5$ **8.** $4\sqrt{x} - 7 = 5$ **▣ 9.** $3\sqrt{x} + 5 = 2$

10. $3\sqrt{x} + 5 = 8$ **▣ 11.** $\sqrt{x + 6} + 1 = 3$ **12.** $\sqrt{x + 5} + 2 = 5$

13. $\sqrt{2x + 1} + 3 = 5$ **14.** $\sqrt{3x - 1} + 4 = 1$ **15.** $\sqrt{x} + 3 = 7$

16. $\sqrt{x} + 5 = 10$ **17.** $\sqrt{x + 6} + 5 = 3$ **18.** $\sqrt{2x - 1} + 7 = 1$

19. $\sqrt{4x - 3} = \sqrt{x + 3}$ **20.** $\sqrt{5x - 4} = \sqrt{x + 8}$ **21.** $\sqrt{x} = \sqrt{3x - 8}$

22. $\sqrt{x} = \sqrt{4x - 3}$ **23.** $\sqrt{4x} = \sqrt{2x + 6}$ **24.** $\sqrt{5x + 6} = \sqrt{8x}$

Solve each equation. See Examples 4 and 5.

25. $\sqrt{9x^2 + 2x - 4} = 3x$ **26.** $\sqrt{4x^2 + 3x - 9} = 2x$

27. $\sqrt{16x^2 - 3x + 6} = 4x$ **28.** $\sqrt{9x^2 - 2x + 8} = 3x$

29. $\sqrt{16x^2 + 2x + 2} = 4x$ **30.** $\sqrt{4x^2 + 3x - 2} = 2x$

ANSWERS

1. $x = 81$

2. $x = 16$

3. $x = -1$

4. $x = -3$

5. $x = 5$

6. $x = 6$

7. $x = 49$

8. $x = 9$

9. no solution

10. $x = 1$

11. $x = -2$

12. $x = 4$

13. $x = \dfrac{3}{2}$

14. no solution

15. $x = 16$

16. $x = 25$

17. no solution

18. no solution

19. $x = 2$

20. $x = 3$

21. $x = 4$

22. $x = 1$

23. $x = 3$

24. $x = 2$

25. $x = 2$

26. $x = 3$

27. $x = 2$

28. $x = 4$

29. no solution

30. $x = \dfrac{2}{3}$

31. $x = 0, -3$

32. $x = 0, -2$

31. $\sqrt{2x^2 + 6x + 9} = 3$ **32.** $\sqrt{3x^2 + 6x + 4} = 2$

33. $x = -3$

34. $x = 4$

33. $\sqrt{x + 7} = x + 5$ **34.** $\sqrt{x + 5} = x - 1$ **35.** $\sqrt{x} = x - 6$

35. $x = 9$

36. no solution

36. $\sqrt{x} = x + 6$ **37.** $\sqrt{2x + 1} = x - 7$ **38.** $\sqrt{2x + 5} = x - 5$

37. $x = 12$

38. $x = 10$

39. $x = \sqrt{2x - 2} + 1$ **40.** $\sqrt{1 - 8x} + 2 = x$ ⌨ **41.** $\sqrt{1 - 8x} - x = 4$

39. $x = 3, 1$

40. no solution

41. $x = -1$

42. $\sqrt{3x + 7} - x = 3$ **43.** $\sqrt{2x + 5} - 1 = x$ **44.** $x = \sqrt{4x - 7} + 1$

42. $x = -1, -2$

43. $x = 2$

B Solve each equation. See Example 6.

44. $x = 2, 4$

45. $\sqrt{x - 7} = \sqrt{x} - 1$ **46.** $\sqrt{x} + 2 = \sqrt{x + 24}$ **47.** $\sqrt{x} + 3 = \sqrt{x + 15}$

45. $x = 16$

46. $x = 25$

48. $\sqrt{x - 8} = \sqrt{x} - 2$ **49.** $\sqrt{x + 8} = \sqrt{x} + 2$ **50.** $\sqrt{x} + 1 = \sqrt{x + 15}$

47. $x = 1$

48. $x = 9$

49. $x = 1$

REVIEW AND PREVIEW

50. $x = 49$

Translate each sentence into an equation and then solve. See Section 2.5.

51. $3x - 8 = 19; x = 9$

51. If 8 is subtracted from the product of 3 and x, the result is 19. Find x.

52. If 3 more than x is subtracted from twice x, the result is 11. Find x.

52. $2x - (x + 3) = 11;$ $x = 14$

53. $2(2x + x) = 24;$ length = 8 in.

53. The length of a rectangle is twice the width. The perimeter is 24 inches. Find the length.

54. The length of a rectangle is 2 inches longer than the width. The perimeter is 24 inches. Find the length.

54. $2x + 2(x + 2) = 24;$ length: 7 in.

◇ **COMBINING CONCEPTS**

55. The formula $b = \sqrt{\frac{V}{2}}$ can be used to determine the length b of a side of the base of a square-based pyramid with height 6 units and volume V cubic units.

  a. Find the length of the side of the base that produces a pyramid with each volume. (Round to the nearest tenth of a unit.)

V	20	200	2000
b			

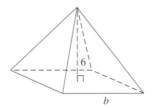

  b. Notice in the table that volume V has been increased by a factor of 10 each time. Does the corresponding length b of a side increase by a factor of 10 each time also?

56. The formula $r = \sqrt{\frac{V}{2\pi}}$ can be used to determine the radius r of a cylinder with height 2 units and volume V cubic units.

  a. Find the radius needed to manufacture a cylinder with each volume. (Round to the nearest tenth of a unit.)

V	10	100	1000
r			

2 units

  b. Notice in the table that volume V has been increased by a factor of 10 each time. Does the corresponding radius increase by a factor of 10 each time also?

57. Explain why proposed solutions of radical equations must be checked in the original equation.

58. 7.30

59. 2.43

60. 0.76

61. 0.48

Graphing calculators can be used to solve equations. To solve $\sqrt{x-2} = x - 5$, for example, graph $y_1 = \sqrt{x-2}$ and $y_2 = x - 5$ on the same set of axes. Use the Trace and Zoom features or an Intersect feature to find the point of intersection of the graphs. The x-value of the point is the solution of the equation. Use a graphing calculator to solve the equations below. Approximate solutions to the nearest hundredth.

58. $\sqrt{x-2} = x - 5$

59. $\sqrt{x+1} = 2x - 3$

60. $-\sqrt{x+4} = 5x - 6$

61. $-\sqrt{x+5} = -7x + 1$

8.6 RADICAL EQUATIONS AND PROBLEM SOLVING

A USING THE PYTHAGOREAN THEOREM

Applications of radicals can be found in geometry, finance, science, and other areas of technology. Our first application involves the Pythagorean theorem, giving a formula that relates the lengths of the three sides of a right triangle. We first studied the Pythagorean theorem in Chapter 4 and we review it here.

THE PYTHAGOREAN THEOREM

If a and b are lengths of the legs of a right triangle and c is the length of the hypotenuse, then $a^2 + b^2 = c^2$.

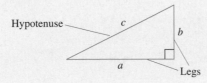

Example 1 Find the length of the hypotenuse of a right triangle whose legs are 6 inches and 8 inches long.

Solution: Because this is a right triangle, we use the Pythagorean theorem. We let $a = 6$ inches and $b = 8$ inches. Length c must be the length of the hypotenuse.

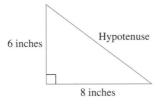

$a^2 + b^2 = c^2$ Use the Pythagorean theorem.
$6^2 + 8^2 = c^2$ Substitute the lengths of the legs.
$36 + 64 = c^2$ Simplify.
$100 = c^2$

Since c represents a length, we know that c is positive and is the principal square root of 100.

$100 = c^2$
$\sqrt{100} = c$ Use the definition of principal square root.
$10 = c$ Simplify.

The hypotenuse has a length of 10 inches.

Example 2 Find the length of the leg of the right triangle shown. Give the exact length and a two-decimal-place approximation.

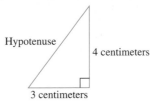

Solution: We let $a = 2$ meters and b be the unknown length of the other leg. The hypotenuse is $c = 5$ meters.

$a^2 + b^2 = c^2$ Use the Pythagorean theorem.
$2^2 + b^2 = 5^2$ Let $a = 2$ and $c = 5$.
$4 + b^2 = 25$
$b^2 = 21$
$b = \sqrt{21} \approx 4.58$ meters

Objectives

A Use the Pythagorean formula to solve problems.

B Solve problems using formulas containing radicals.

SSM   CD-ROM   Video 8.6

Practice Problem 1

Find the length of the hypotenuse of the right triangle shown.

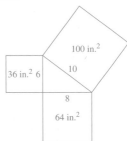

TEACHING TIP
After Example 1, demonstrate the Pythagorean Theorem by drawing squares on each side of the triangle, finding the area of each square, and noting that the sum of the area of the squares of each leg equals the area of the square of the hypotenuse.

Practice Problem 2

Find the length of the leg of the right triangle shown. Give the exact length and a two-decimal-place approximation.

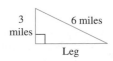

Answers

1. 5 cm, **2.** $3\sqrt{3}$ mi; 5.20 mi

The length of the leg is exactly $\sqrt{21}$ meters or approximately 4.58 meters. ▬▬▬

Practice Problem 3

Evan Saacks wants to determine the distance at certain points across a pond on his property. He is able to measure the distances shown on the following diagram. Find how wide the pond is to the nearest tenth of a foot.

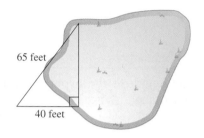

65 feet

40 feet

Example 3 ## Finding a Distance

A surveyor must determine the distance across a lake at points P and Q as shown in the figure. To do this, she finds a third point R perpendicular to line PQ. If the length of $\overline{PR}$ is 320 feet and the length of $\overline{QR}$ is 240 feet, what is the distance across the lake? Approximate this distance to the nearest whole foot.

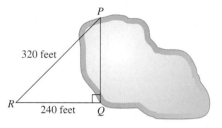

P

320 feet

R   240 feet   Q

Solution:

1. UNDERSTAND. Read and reread the problem. We will set up the problem using the Pythagorean theorem. By creating a line perpendicular to line PQ, the surveyor deliberately constructed a right triangle. The hypotenuse, $\overline{PR}$, has a length of 320 feet, so we let $c = 320$ in the Pythagorean theorem. The side $\overline{QR}$ is one of the legs, so we let $a = 240$ and $b =$ the unknown length.

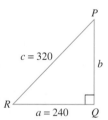

P

$c = 320$

b

R $a = 240$ Q

2. TRANSLATE.

$a^2 + b^2 = c^2$ Use the Pythagorean theorem.

$240^2 + b^2 = 320^2$ Let $a = 240$ and $c = 320$.

3. SOLVE.

$57{,}600 + b^2 = 102{,}400$

$b^2 = 44{,}800$ Subtract 57,600 from both sides.

$b = \sqrt{44{,}800}$ Use the definition of principal square root.

4. INTERPRET.

Check: See that $240^2 + (\sqrt{44{,}800})^2 = 320^2$.

State: The distance across the lake is *exactly* $\sqrt{44{,}800}$ feet. The surveyor can now use a calculator to find that $\sqrt{44{,}800}$ feet is *approximately* 211.6601 feet, so the distance across the lake is roughly 212 feet. ▬▬▬

TEACHING TIP

Consider using the following activity: Have students work in groups to create a word problem involving the Pythagorean Theorem. Encourage groups to be creative. Then ask groups to exchange their problems with another group to solve.

Answer

3. 51.2 ft

B USING FORMULAS CONTAINING RADICALS

The Pythagorean theorem is an extremely important result in mathematics and should be memorized. But there are other applications involving formulas containing radicals that are not quite as well known, such as the velocity formula used in the next example.

Example 4   A formula used to determine the velocity v, in feet per second, of an object (neglecting air resistance) after it has fallen a certain height is $v = \sqrt{2gh}$, where g is the acceleration due to gravity, and h is the height the object has fallen. On Earth, the acceleration g due to gravity is approximately 32 feet per second per second. Find the velocity of a watermelon after it has fallen 5 feet.

Solution: We are told that $g = 32$ feet per second per second. To find the velocity v when $h = 5$ feet, we use the velocity formula.

$$v = \sqrt{2gh} \qquad \text{Use the velocity formula.}$$
$$= \sqrt{2 \cdot 32 \cdot 5} \qquad \text{Substitute known values.}$$
$$= \sqrt{320}$$
$$= 8\sqrt{5} \qquad \text{Simplify the radicand.}$$

The velocity of the watermelon after it falls 5 feet is *exactly* $8\sqrt{5}$ feet per second, or *approximately* 17.9 feet per second.

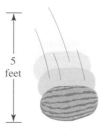

5 feet

Practice Problem 4

Use the formula from Example 4 and find the velocity of an object after it has fallen 20 feet.

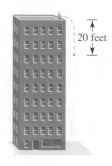

20 feet

TEACHING TIP

After discussing Example 4, ask students to think of the advantages and disadvantages of giving an exact answer such as $8\sqrt{5}$ versus an approximate answer such as 17.9.

Answer

4. $16\sqrt{5}$ ft per sec ≈ 35.8 ft per sec

Focus On The Real World

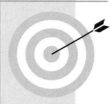

ESCAPE VELOCITY

Each planet in the solar system has a minimum speed that an object must reach to escape the planet's pull of gravity. This speed is called the **escape velocity**. For instance, Earth's escape velocity is 11.19 kilometers per second. A rocket launched from Earth's surface must be going at least 11.19 kilometers per second to leave the planet for outer space. If the rocket goes slower than 11.19 kilometers per second, it will either fall back to Earth or go into orbit around Earth.

The escape velocity v from a planet with mass m and radius r is given by the formula

$$v = \sqrt{\frac{2Gm}{r}}$$

where G is a constant called the *universal constant of gravitation*. The escape velocity of each planet in our solar system, along with its mass and radius, is given in the table.

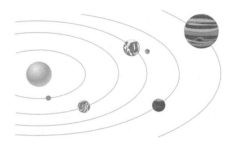

Planet	Mass (kilograms)	Radius (kilometers)	Escape Velocity (kilometers per second)
Mercury	3.302×10^{23}	2440	4.30
Venus	4.869×10^{24}	6052	10.36
Earth	5.974×10^{24}	6371	11.19
Moon	7.349×10^{22}	1738	2.38
Mars	6.419×10^{23}	3390	5.03
Jupiter	1.899×10^{27}	69,911	59.50
Saturn	5.685×10^{26}	58,232	35.50
Uranus	8.683×10^{25}	25,362	21.30
Neptune	1.024×10^{26}	24,624	23.50
Pluto	1.250×10^{22}	1137	1.10

Source: National Space Science Data Center

CRITICAL THINKING

1. List the planets in order of decreasing mass.
2. List the planets in order of decreasing escape velocity.
3. What do you notice about your lists from Questions 1 and 2? In general, what could you conclude about the relationship between mass and escape velocity?
4. Use the formula for escape velocity and the data in the table to estimate the value of the universal constant of gravitation G. Explain how you found your estimate.

EXERCISE SET 8.6

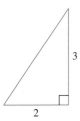 **A** *Use the Pythagorean theorem to find the length of the unknown side of each right triangle. Give an exact answer and a two-decimal-place approximation. See Examples 1 and 2.*

1.

3

2

2.

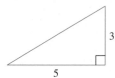

3

5

3.

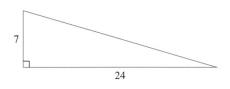

3

6

4.

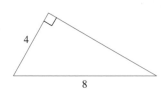

4

8

5.

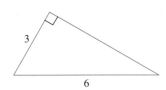

7

24

6.

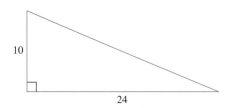

10

24

ANSWERS

1. $\sqrt{13}$; 3.61

2. $\sqrt{34}$; 5.83

3. $3\sqrt{3}$; 5.20

4. $4\sqrt{3}$; 6.93

5. 25

6. 26

639

640

Name

7.

8.

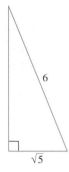

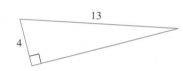 **9.**

10.

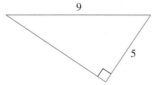

Find the length of the unknown side of each right triangle with sides a, b, and c, where c is the hypotenuse. See Examples 1 and 2. Give an exact answer and a two-decimal-place approximation.

11. $a = 4, b = 5$ **12.** $a = 2, b = 7$ **13.** $b = 2, c = 6$

14. $b = 1, c = 5$ **15.** $a = \sqrt{10}, c = 10$ **16.** $a = \sqrt{7}, c = \sqrt{35}$

Name _____

Solve each problem. See Example 3.

17. A wire is used to anchor a 20-foot-high pole. One end of the wire is attached to the top of the pole. The other end is fastened to a stake five feet away from the bottom of the pole. Find the length of the wire, to the nearest tenth of a foot.

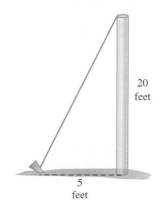

20 feet

5 feet

18. Jim Spivey needs to connect two underground pipelines, which are off-set by 3 feet, as pictured in the diagram. Neglecting the joints needed to join the pipes, find the length of the shortest possible connecting pipe rounded to the nearest hundredth of a foot.

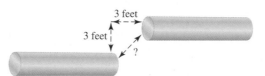

3 feet

3 feet

?

19. Robert Weisman needs to attach a diagonal brace to a rectangular frame in order to make it structurally sound. If the framework is 6 feet by 10 feet, find how long the brace needs to be to the nearest tenth of a foot.

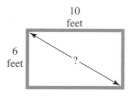

10 feet

6 feet

?

20. Elizabeth Kaster is flying a kite. She let out 80 feet of string and attached the string to a stake in the ground. The kite is now directly above her brother Mike, who is 32 feet away from Elizabeth. Find the height of the kite to the nearest foot.

80 feet

32 feet

B *Solve each problem. See Example 4.*

21. For a square-based pyramid, the formula $b = \sqrt{\frac{3V}{h}}$ describes the relationship between the length b of one side of the base, the volume V, and the height h. Find the volume if each side of the base is 6 feet long, and the pyramid is 2 feet high.

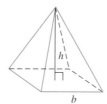

h

b

22. The formula $t = \frac{\sqrt{d}}{4}$ relates the distance d, in feet, that an object falls in t seconds, assuming that air resistance does not slow down the object. Find how long, to the nearest hundredth of a second, it takes an object to reach the ground from the top of the Sears Tower in Chicago, a distance of 1450 feet. (*Source:* World Almanac and Book of Facts, 1997)

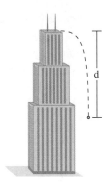

23. Police use the formula $s = \sqrt{30fd}$ to estimate the speed s of a car in miles per hour. In this formula, d represents the distance the car skidded in feet and f represents the coefficient of friction. The value of f depends on the type of road surface, and for wet concrete f is 0.35. Find how fast a car was moving if it skidded 280 feet on wet concrete, to the nearest mile per hour.

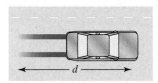

24. The coefficient of friction of a certain dry road is 0.95. Use the formula in Exercise 23 to find how far a car will skid on this dry road if it is traveling at a rate of 60 mph. Round the length to the nearest foot.

25. The formula $v = \sqrt{2.5r}$ can be used to estimate the maximum safe velocity, v, in miles per hour, at which a car can travel if it is driven along a curved road with a **radius of curvature**, r, in feet. To the nearest whole number, find the maximum safe speed if a cloverleaf exit on an expressway has a radius of curvature of 300 feet.

26. Use the formula from Exercise 25 to find the radius of curvature if the safe velocity is 30 mph.

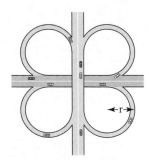

27. The maximum distance d in kilometers that you can see from a height of h meters is given by $d = 3.5\sqrt{h}$. Find how far you can see from the top of the Texas Commerce Tower in Houston, a height of 305.4 meters. Round to the nearest tenth of a kilometer. (*Source:* World Almanac and Book of Facts, 1997)

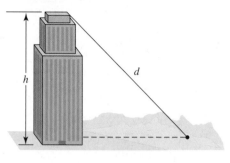

28. Use the formula from Exericse 27 to determine how high above the ground you need to be to see 40 kilometers. Round to the nearest tenth of a meter.

REVIEW AND PREVIEW

Find two numbers whose square is the given number. See Section 8.1.

29. 9

30. 25

31. 100

32. 49

33. 64

34. 121

 ## COMBINING CONCEPTS

For each triangle, find the length of x.

35.

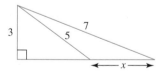

36.

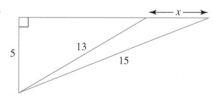

37. Mike and Sandra Hallahan leave the seashore at the same time. Mike drives northward at a rate of 30 miles per hour, while Sandra drives west at 60 mph. Find how far apart they are after 3 hours to the nearest mile.

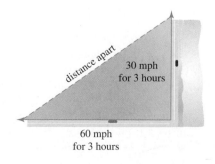

distance apart

30 mph
for 3 hours

60 mph
for 3 hours

27. 61.2 km

28. 130.6 m

29. 3

30. 5

31. 10

32. 7

33. 8

34. 11

35. $x = 2\sqrt{10} - 4$

36. $x = 10\sqrt{2} - 12$

37. 201 miles

Name _____

38. Railroad tracks are invariably made up of relatively short sections of rail connected by expansion joints. To see why this construction is necessary, consider a single rail 100 feet long (or 1200 inches). On an extremely hot day, suppose it expands 1 inch in the hot sun to a new length of 1201 inches. Theorectically, the track would bow upward as pictured.

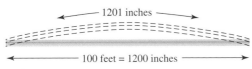

39. Based on the results of Exercise 38, explain why railroads use short sections of rail connected by expansion joints. Let us approximate the bulge in the railroad this way.

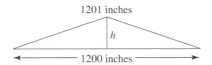

Calculate the height of h of the bulge to the nearest tenth of an inch.

Internet Excursions

Go to http://www.prenhall.com/martin-gay
The Central Intelligence Agency of the United States publishes an annual factbook as a basic reference for 267 nations around the world. It includes a variety of information for each nation, such as descriptions of and data on the nation's geography, people, government, transportation, economy, communications, and defense. The given World Wide Web address will provide you with access to the CIA Factbook, or a related site. You can look up data for any country. The section on geography in a country's listing gives information on the highest and lowest points in the country.

40. Choose any country in North America. Look up that country on the CIA Factbook Website. Using data in the geography section, complete the following information: Country: _____; Highest point: _____; Elevation: _____. Use the formula $d = 3.5\sqrt{h}$, where d is the maximum distance (in kilometers) that you can see from a height of h meters above the surface of the Earth, to find the distance that can be seen from the highest point in the country that you chose.

41. Choose any country in Asia. Look up that country on the CIA Factbook Website. Using data in the geography section, complete the following information: Country: _____; Highest point: _____; Elevation: _____. Use the formula from Exercise 40 to find the distance that can be seen from the highest point in the country that you chose.

CHAPTER 8 ACTIVITY

INVESTIGATING THE DIMENSIONS OF CYLINDERS

MATERIALS:

▲ calculator
▲ several empty cans of different sizes
▲ 2-cup (16-fluid-ounce) transparent measuring cup with metric markings (in milliliters)
▲ metric ruler
▲ water

This activity may be completed by working in groups or individually.

The radius r of a cylinder is related to its volume V (in cubic units) and its height h by the formula $r = \sqrt{\frac{V}{\pi h}}$. You will investigate the radii of several cylindrical cans by completing the following table.

Can	Volume (ml)	Height (cm)	Calculated Radius (cm)	Measured Radius (cm)
A				
B				
C				
D				

1. For each can, measure its volume by filling it with water and pouring the water into the measuring cup. Find the volume of the water in milliliters (ml). Record the volumes of the cans in the table. (Remember that 1 ml = 1 cm^3).
   answers may vary

2. For each can, use a ruler to measure its height in centimeters (cm). Record the heights in the table.   answers may vary

3. Use the formula $r = \sqrt{\frac{V}{\pi h}}$ to calculate an estimate of each can's radius. Record these calculated radii in the table.   answers may vary

4. Try to measure the radius of each can and record these measured radii in the table. (Remember that radius $= \frac{1}{2}$ diameter.)
   answers may vary

5. How close are the values of the calculated radius and the measured radius of each can? What factors could account for the differences?
   answers may vary

CHAPTER 8 HIGHLIGHTS

DEFINITIONS AND CONCEPTS	EXAMPLES

SECTION 8.1   INTRODUCTION TO RADICALS

The **positive or principal square root** of a positive number a is written as $\sqrt{a}$. The **negative square root** of a is written as $-\sqrt{a}$. $\sqrt{a} = b$ only if $b^2 = a$ and $b > 0$. A square root of a negative number is not a real number.	$\sqrt{25} = 5 \qquad\qquad \sqrt{100} = 10$ $-\sqrt{9} = -3 \qquad\qquad \sqrt{\dfrac{4}{49}} = \dfrac{2}{7}$ $\sqrt{-4}$ is not a real number.
The **cube root** of a real number a is written as $\sqrt[3]{a}$ and $\sqrt[3]{a} = b$ only if $b^3 = a$.	$\sqrt[3]{64} = 4 \qquad \sqrt[3]{-8} = -2$
The **nth root** of a number a is written as $\sqrt[n]{a}$ and $\sqrt[n]{a} = b$ only if $b^n = a$.	$\sqrt[4]{81} = 3$ $\sqrt[5]{-32} = -2$
The natural number n is called the **index**, the symbol $\sqrt{}$ is called a **radical,** and the expression within the radical is called the **radicand.** (Note: If the index is even, the radicand must be non-negative for the root to be a real number.)	$\underset{\text{index}}{\searrow} \sqrt[n]{a} \,{\underset{\text{radicand}}{\nwarrow}}$

SECTION 8.2   SIMPLIFYING RADICALS

PRODUCT RULE FOR RADICALS If $\sqrt{a}$ and $\sqrt{b}$ are real numbers, then $\qquad \sqrt{a \cdot b} = \sqrt{a} \cdot \sqrt{b}$ A square root is in **simplified form** if the radicand contains no perfect square factors other than 1. To simplify a square root, factor the radicand so that one of its factors is a perfect square factor.	$\begin{aligned} \sqrt{45} &= \sqrt{9 \cdot 5} \\ &= \sqrt{9} \cdot \sqrt{5} \\ &= 3\sqrt{5} \end{aligned}$
QUOTIENT RULE FOR RADICALS If $\sqrt{a}$ and $\sqrt{b}$ are real numbers and $b \neq 0$, then $\qquad \sqrt{\dfrac{a}{b}} = \dfrac{\sqrt{a}}{\sqrt{b}}$	$\sqrt{\dfrac{18}{x^6}} = \dfrac{\sqrt{9 \cdot 2}}{\sqrt{x^6}} = \dfrac{\sqrt{9} \cdot \sqrt{2}}{x^3} = \dfrac{3\sqrt{2}}{x^3}$

SECTION 8.3   ADDING AND SUBTRACTING RADICALS

Like radicals are radical expressions that have the same index and the same radicand.	$5\sqrt{2}, \ -7\sqrt{2}, \ \sqrt{2}$
To **combine like radicals** use the distributive property.	$2\sqrt{7} - 13\sqrt{7} = (2 - 13)\sqrt{7} = -11\sqrt{7}$ $\sqrt{8} + \sqrt{50} = 2\sqrt{2} + 5\sqrt{2} = 7\sqrt{2}$

The product and quotient rules for radicals may be used to simplify products and quotients of radicals.

Perform each indicated operation and simplify.
Multiply.

$$\sqrt{2} \cdot \sqrt{8} = \sqrt{16} = 4$$

$$(\sqrt{3x} + 1)(\sqrt{5} - \sqrt{3})$$
$$= \sqrt{15x} - \sqrt{9x} + \sqrt{5} - \sqrt{3}$$
$$= \sqrt{15x} - 3\sqrt{x} + \sqrt{5} - \sqrt{3}$$

Divide.

$$\frac{\sqrt{20}}{\sqrt{2}} = \sqrt{\frac{20}{2}} = \sqrt{10}$$

The process of eliminating the radical in the denominator of a radical expression is called **rationalizing the denominator**.

Rationalize the denominator.

$$\frac{5}{\sqrt{11}} = \frac{5 \cdot \sqrt{11}}{\sqrt{11} \cdot \sqrt{11}} = \frac{5\sqrt{11}}{11}$$

The **conjugate** of $a + b$ is $a - b$.

To rationalize a denominator that is a sum or difference of radicals, multiply the numerator and the denominator by the conjugate of the denominator.

The conjugate of $2 + \sqrt{3}$ is $2 - \sqrt{3}$.

Rationalize the denominator.

$$\frac{5}{6 - \sqrt{5}} = \frac{5(6 + \sqrt{5})}{(6 - \sqrt{5})(6 + \sqrt{5})}$$
$$= \frac{5(6 + \sqrt{5})}{36 - 5}$$
$$= \frac{5(6 + \sqrt{5})}{31}$$

TO SOLVE A RADICAL EQUATION CONTAINING SQUARE ROOTS

Step 1. Get one radical by itself on one side of the equation.

Step 2. Square both sides of the equation.

Step 3. Simplify both sides of the equation.

Step 4. If the equation still contains a radical term, repeat Steps 1 through 3.

Step 5. Solve the equation.

Step 6. Check solutions in the original equation.

Solve:

$$\sqrt{2x - 1} - x = -2$$
$$\sqrt{2x - 1} = x - 2$$
$$(\sqrt{2x - 1})^2 = (x - 2)^2 \quad \text{Square both sides.}$$
$$2x - 1 = x^2 - 4x + 4$$
$$0 = x^2 - 6x + 5$$
$$0 = (x - 1)(x - 5) \quad \text{Factor.}$$
$$x - 1 = 0 \quad \text{or} \quad x - 5 = 0$$
$$x = 1 \qquad x = 5 \quad \text{Solve.}$$

Check both proposed solutions in the original equation. Here, 5 checks but 1 does not. The only solution is 5.

PROBLEM-SOLVING STEPS

1. UNDERSTAND. Read and reread the problem.

A rain gutter is to be mounted on the eaves of a house 15 feet above the ground. A garden is adjacent to the house so that the closest a ladder can be placed to the house is 6 feet. How long a ladder is needed for installing the gutter?

Let x = the length of the ladder.

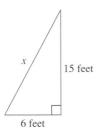

2. TRANSLATE.

Here, we use the Pythagorean theorem. The unknown length x is the hypotenuse.

In words:

$$(\text{leg})^2 + (\text{leg})^2 = (\text{hypotenuse})^2$$

3. SOLVE.

Translate:
$$6^2 + 15^2 = x^2$$
$$36 + 225 = x^2$$
$$261 = x^2$$
$$\sqrt{261} = x \quad \text{or} \quad x = 3\sqrt{29}$$

4. INTERPRET.

Check and state. The ladder needs to be $3\sqrt{29}$ feet or approximately 16.2 feet long.

Chapter 8 Review

(8.1) *Find each root.*

1. $\sqrt{81}$ 9

2. $-\sqrt{49}$ -7

3. $\sqrt[3]{27}$ 3

4. $\sqrt[4]{16}$ 2

5. $-\sqrt{\dfrac{9}{64}}$ $-\dfrac{3}{8}$

6. $\sqrt{\dfrac{36}{81}}$ $\dfrac{2}{3}$

7. $\sqrt[4]{16}$ 2

8. $\sqrt[3]{-8}$ -2

9. Which radical(s) is not a real number?   c

   a. $\sqrt{4}$

   b. $-\sqrt{4}$

   c. $\sqrt{-4}$

   d. $\sqrt[3]{-4}$

10. Which radical(s) is not a real number?   a, c

   a. $\sqrt{-5}$

   b. $\sqrt[3]{-5}$

   c. $\sqrt[4]{-5}$

   d. $\sqrt[5]{-5}$

Find each root. Assume that all variables represent positive numbers.

11. $\sqrt{x^{12}}$ x^6

12. $\sqrt{x^8}$ x^4

13. $\sqrt{9y^2}$ $3y$

14. $\sqrt{25x^4}$ $5x^2$

(8.2) *Simplify each expression using the product rule. Assume that all variables represent positive numbers.*

15. $\sqrt{40}$ $2\sqrt{10}$

16. $\sqrt{24}$ $2\sqrt{6}$

17. $\sqrt{54}$ $3\sqrt{6}$

18. $\sqrt{88}$ $2\sqrt{22}$

19. $\sqrt{x^5}$ $x^2\sqrt{x}$

20. $\sqrt{y^7}$ $y^3\sqrt{y}$

21. $\sqrt{20x^2}$ $2x\sqrt{5}$

22. $\sqrt{50y^4}$ $5y^2\sqrt{2}$

23. $\sqrt[3]{54}$ $3\sqrt[3]{2}$

24. $\sqrt[3]{88}$ $2\sqrt[3]{11}$

Name _____

Simplify each expression using the quotient rule. Assume that all variables represent positive numbers.

25. $\sqrt{\dfrac{18}{25}}$ $\dfrac{3\sqrt{2}}{5}$

26. $\sqrt{\dfrac{75}{64}}$ $\dfrac{5\sqrt{3}}{8}$

27. $-\sqrt{\dfrac{50}{9}}$ $\dfrac{-5\sqrt{2}}{3}$

28. $-\sqrt{\dfrac{12}{49}}$ $\dfrac{-2\sqrt{3}}{7}$

29. $\sqrt{\dfrac{11}{x^2}}$ $\dfrac{\sqrt{11}}{x}$

30. $\sqrt{\dfrac{7}{y^4}}$ $\dfrac{\sqrt{7}}{y^2}$

31. $\sqrt{\dfrac{y^5}{100}}$ $\dfrac{y^2\sqrt{y}}{10}$

32. $\sqrt{\dfrac{x^3}{81}}$ $\dfrac{x\sqrt{x}}{9}$

(8.3) *Add or subtract by combining like radicals.*

33. $5\sqrt{2} - 8\sqrt{2}$ $-3\sqrt{2}$

34. $\sqrt{3} - 6\sqrt{3}$ $-5\sqrt{3}$

35. $6\sqrt{5} + 3\sqrt{6} - 2\sqrt{5} + \sqrt{6}$ $4\sqrt{5} + 4\sqrt{6}$

36. $-\sqrt{7} + 8\sqrt{2} - \sqrt{7} - 6\sqrt{2}$ $-2\sqrt{7} + 2\sqrt{2}$

Add or subtract by simplifying each radical and then combining like terms. Assume that all variables represent positive numbers.

37. $\sqrt{28} + \sqrt{63} + \sqrt{56}$ $5\sqrt{7} + 2\sqrt{14}$

38. $\sqrt{75} + \sqrt{48} - \sqrt{16}$ $9\sqrt{3} - 4$

39. $\sqrt{\dfrac{5}{9}} - \sqrt{\dfrac{5}{36}}$ $\dfrac{\sqrt{5}}{6}$

40. $\sqrt{\dfrac{11}{25}} + \sqrt{\dfrac{11}{16}}$ $\dfrac{9\sqrt{11}}{20}$

41. $\sqrt{45x^2} + 3\sqrt{5x^2} - 7x\sqrt{5} + 10$ $10 - x\sqrt{5}$

42. $\sqrt{50x} - 9\sqrt{2x} + \sqrt{72x} - \sqrt{3x}$ $2\sqrt{2x} - \sqrt{3x}$

(8.4) *Multiply and simplify if possible.*

43. $\sqrt{3} \cdot \sqrt{6}$ $3\sqrt{2}$

44. $\sqrt{5} \cdot \sqrt{15}$ $5\sqrt{3}$

45. $\sqrt{2}(\sqrt{5} - \sqrt{7})$ $\sqrt{10} - \sqrt{14}$

46. $\sqrt{5}(\sqrt{11} + \sqrt{3})$ $\sqrt{55} + \sqrt{15}$

47. $(\sqrt{3} + 2)(\sqrt{6} - 5)$ $3\sqrt{2} - 5\sqrt{3} + 2\sqrt{6} - 10$

48. $(\sqrt{5} + 1)(\sqrt{5} - 3)$ $2 - 2\sqrt{5}$

49. $(\sqrt{x} - 2)^2$ $x - 4\sqrt{x} + 4$

50. $(\sqrt{y} + 4)^2$ $y + 8\sqrt{y} + 16$

650

Name _____

Divide and simplify if possible. Assume that all variables represent positive numbers.

51. $\dfrac{\sqrt{27}}{\sqrt{3}}$ $\quad 3$

52. $\dfrac{\sqrt{20}}{\sqrt{5}}$ $\quad 2$

53. $\dfrac{\sqrt{160}}{\sqrt{8}}$ $\quad 2\sqrt{5}$

54. $\dfrac{\sqrt{96}}{\sqrt{3}}$ $\quad 4\sqrt{2}$

55. $\dfrac{\sqrt{30x^6}}{\sqrt{2x^3}}$ $\quad x\sqrt{15x}$

56. $\dfrac{\sqrt{54x^5y^2}}{\sqrt{3xy^2}}$ $\quad 3x^2\sqrt{2}$

Rationalize each denominator and simplify.

57. $\dfrac{\sqrt{2}}{\sqrt{11}}$ $\quad \dfrac{\sqrt{22}}{11}$

58. $\dfrac{\sqrt{3}}{\sqrt{13}}$ $\quad \dfrac{\sqrt{39}}{13}$

59. $\sqrt{\dfrac{5}{6}}$ $\quad \dfrac{\sqrt{30}}{6}$

60. $\sqrt{\dfrac{7}{10}}$ $\quad \dfrac{\sqrt{70}}{10}$

61. $\dfrac{1}{\sqrt{5x}}$ $\quad \dfrac{\sqrt{5x}}{5x}$

62. $\dfrac{5}{\sqrt{3y}}$ $\quad \dfrac{5\sqrt{3y}}{3y}$

63. $\sqrt{\dfrac{3}{x}}$ $\quad \dfrac{\sqrt{3x}}{x}$

64. $\sqrt{\dfrac{6}{y}}$ $\quad \dfrac{\sqrt{6y}}{y}$

65. $\dfrac{3}{\sqrt{5}-2}$ $\quad 3\sqrt{5}+6$

66. $\dfrac{8}{\sqrt{10}-3}$ $\quad 8\sqrt{10}+24$

67. $\dfrac{\sqrt{2}+1}{\sqrt{3}-1}$ $\quad \dfrac{\sqrt{6}+\sqrt{2}+\sqrt{3}+1}{2}$

68. $\dfrac{\sqrt{3}-2}{\sqrt{5}+2}$ $\quad \sqrt{15}-2\sqrt{3}-2\sqrt{5}+4$

69. $\dfrac{10}{\sqrt{x}+5}$ $\quad \dfrac{10\sqrt{x}-50}{x-25}$

70. $\dfrac{8}{\sqrt{x}-1}$ $\quad \dfrac{8\sqrt{x}+8}{x-1}$

(8.5) *Solve each radical equation.*

71. $\sqrt{2x}=6$ $\quad x=18$

72. $\sqrt{x+3}=4$ $\quad x=13$

73. $\sqrt{x}+3=8$ $\quad x=25$

74. $\sqrt{x}+8=3$ $\quad$ no solution

75. $\sqrt{2x+1}=x-7$ $\quad x=12$

76. $\sqrt{3x+1}=x-1$ $\quad x=5$

77. $\sqrt{x}+3=\sqrt{x+15}$ $\quad x=1$

78. $\sqrt{x-5}=\sqrt{x}-1$ $\quad x=9$

Name _____

(8.6) *Use the Pythagorean theorem to find the length of each unknown side. Give an exact answer and a two-decimal-place approximation.*

79.

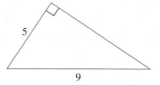

$2\sqrt{14}$; 7.48

80.

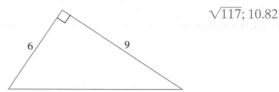

$\sqrt{117}$; 10.82

81. Romeo is standing 20 feet away from the wall below Juliet's balcony during a school play. Juliet is on the balcony, 12 feet above the ground. Find how far apart Romeo and Juliet are. $4\sqrt{34}$ ft

82. The diagonal of a rectangle is 10 inches long. If the width of the rectangle is 5 inches, find the length of the rectangle. $5\sqrt{3}$ in.

Use the formula $r = \sqrt{\dfrac{S}{4\pi}}$, *where* $r =$ *the radius of a sphere and* $S =$ *the surface area of the sphere, for Exercises 83 and 84.*

83. Find the radius of a sphere to the nearest tenth of an inch if the surface area is 72 square inches.
2.4 in.

84. Find the exact surface area of a sphere if its radius is 6 inches. (Do not approximate π.) 144π sq. in.

652

CHAPTER 8 TEST

Simplify each radical. Indicate if the radical is not a real number. Assume that x represents a positive number.

1. $\sqrt{16}$ **2.** $\sqrt[3]{125}$ **3.** $\sqrt[4]{81}$

4. $\sqrt{\dfrac{9}{16}}$ **5.** $\sqrt[4]{-81}$ **6.** $\sqrt{x^{10}}$

Simplify each radical. Assume that all variables represent positive numbers.

7. $\sqrt{54}$ **8.** $\sqrt{92}$ **9.** $\sqrt{y^7}$ **10.** $\sqrt{24x^8}$

11. $\sqrt[3]{27}$ **12.** $\sqrt[3]{16}$ **13.** $\sqrt{\dfrac{5}{16}}$ **14.** $\sqrt{\dfrac{y^3}{25}}$

Perform each indicated operation.

15. $\sqrt{13} + \sqrt{13} - 4\sqrt{13}$ **16.** $\sqrt{18} - \sqrt{75} + 7\sqrt{3} - \sqrt{8}$

17. $\sqrt{\dfrac{3}{4}} + \sqrt{\dfrac{3}{25}}$ **18.** $\sqrt{7} \cdot \sqrt{14}$ **19.** $\sqrt{2}(\sqrt{6} - \sqrt{5})$

20. $(\sqrt{x} + 2)(\sqrt{x} - 3)$ **21.** $\dfrac{\sqrt{50}}{\sqrt{10}}$ **22.** $\dfrac{\sqrt{40x^4}}{\sqrt{2x}}$

1. 4

2. 5

3. 3

4. $\dfrac{3}{4}$

5. not a real number

6. x^5

7. $3\sqrt{6}$

8. $2\sqrt{23}$

9. $y^3\sqrt{y}$

10. $2x^4\sqrt{6}$

11. 3

12. $2\sqrt[3]{2}$

13. $\dfrac{\sqrt{5}}{4}$

14. $\dfrac{y\sqrt{y}}{5}$

15. $-2\sqrt{13}$

16. $\sqrt{2} + 2\sqrt{3}$

17. $\dfrac{7\sqrt{3}}{10}$

18. $7\sqrt{2}$

19. $2\sqrt{3} - 10$

20. $x - \sqrt{x} - 6$

21. $\sqrt{5}$

22. $2x\sqrt{5x}$

23. $\dfrac{\sqrt{6}}{3}$

Rationalize each denominator.

23. $\sqrt{\dfrac{2}{3}}$ **24.** $\dfrac{8}{\sqrt{5y}}$ **25.** $\dfrac{8}{\sqrt{6}+2}$ **26.** $\dfrac{1}{3-\sqrt{x}}$

24. $\dfrac{8\sqrt{5y}}{5y}$

25. $4\sqrt{6}-8$

Solve each radical equation.

27. $\sqrt{x}+8=11$ **28.** $\sqrt{3x-6}=\sqrt{x+4}$ **29.** $\sqrt{2x-2}=x-5$

26. $\dfrac{3+\sqrt{x}}{9-x}$

30. Find the length of the unknown leg of the right triangle shown. Give an exact answer.

27. $x=9$

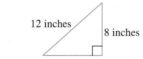

12 inches — 8 inches

31. The formula $r=\sqrt{\dfrac{A}{\pi}}$ can be used to find the radius r of a circle given its area A. Use this formula to approximate the radius of the given circle. Round to two decimal places.

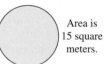

Area is 15 square meters.

28. $x=5$

29. $x=9$

30. $4\sqrt{5}$ in.

31. 2.19 m

CUMULATIVE REVIEW

Multiply.

1. $-5(-10)$

2. $-\dfrac{2}{3} \cdot \dfrac{4}{7}$

3. Solve: $4(2x - 3) + 7 = 3x + 5$.

4. The circle graph below shows how much money homeowners in the United States spend annually on maintaining their homes. Use this graph to answer questions below.

Yearly Home Maintenance in the U.S.

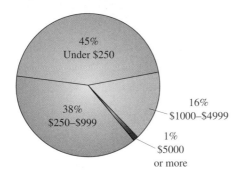

a. What percent of homeowners spend under $250 on yearly home maintenance?

b. What percent of homeowners spend less than $1000 per year on maintenance?

c. How many of the 22,000 homeowners in a town called Fairview might we expect to spend under $250 a year on home maintenance?

5. Write the following numbers in standard notation, without exponents.

a. 1.02×10^5

b. 7.358×10^{-3}

c. 8.4×10^7

d. 3.007×10^{-5}

6. Find the product: $(3x + 2)(2x - 5)$

7. Factor $xy + 2x + 3y + 6$ by grouping.

8. Factor: $3x^2 + 11x + 6$

1. 50 (Sec. 1.5; Ex. 3)

2. $-\dfrac{8}{21}$ (Sec. 1.5; Ex. 4)

3. $x = 2$; (Sec. 2.4; Ex. 1)

4. a. 45%

b. 83%

c. 9900 homeowners (Sec. 2.7; Ex. 3)

5. a. 102,000

b. 0.007358

c. 84,000,000

d. 0.00003007 (Sec. 3.2; Ex. 16)

6. $6x^2 - 11x - 10$ (Sec. 3.5; Ex. 7)

7. $(y + 2)(x + 3)$ (Sec. 4.1; Ex. 8)

8. $(3x + 2)(x + 3)$ (Sec. 4.3; Ex. 1)

Name _____

9. Are there any values for x for which each expression is undefined?

a. $\dfrac{x}{x - 3}$

b. $\dfrac{x^2 + 2}{x^2 - 3x + 2}$

c. $\dfrac{x^3 - 6x^2 - 10x}{3}$

10. Simplify: $\dfrac{x^2 + 4x + 4}{x^2 + 2x}$

11. Perform each indicated operation.

a. $\dfrac{a}{4} - \dfrac{2a}{8}$

b. $\dfrac{3}{10x^2} + \dfrac{7}{25x}$

12. Solve: $\dfrac{4x}{x^2 - 25} + \dfrac{2}{x - 5} = \dfrac{1}{x + 5}$

13. Graph $y = -3$.

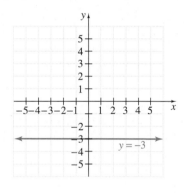

14. Find an equation of the line with y-intercept -3 and slope of $\dfrac{1}{4}$.

15. Solve the system $\begin{cases} 3x + 4y = 13 \\ 5x - 9y = 6 \end{cases}$

16. Albert and Louis live 15 miles away from each other. They decide to meet one day by walking toward one another. After 2 hours they meet. If Louis walks one mile per hour faster than Albert, find both walking speeds.

Name _____

Find each cube root.

17. $\sqrt[3]{1}$ **18.** $\sqrt[3]{-27}$ **19.** $\sqrt[3]{\dfrac{1}{125}}$

Simplify.

20. $\sqrt{54}$ **21.** $\sqrt{200}$

Add or subtract by first simplifying each radical.

22. $7\sqrt{12} - \sqrt{75}$ **23.** $2\sqrt{x^2} - \sqrt{25x} + \sqrt{x}$

24. Rationalize the denominator of $\dfrac{2}{\sqrt{7}}$.   **25.** Solve: $\sqrt{x} = \sqrt{5x - 2}$

17. 1 (Sec. 8.1; Ex. 6)

18. -3 (Sec. 8.1; Ex. 7)

19. $\dfrac{1}{5}$ (Sec. 8.1; Ex. 8)

20. $3\sqrt{6}$ (Sec. 8.2; Ex. 1)

21. $10\sqrt{2}$ (Sec. 8.2; Ex. 3)

22. $9\sqrt{3}$ (Sec. 8.3; Ex. 6)

23. $2x - 4\sqrt{x}$ (Sec. 8.3; Ex. 8)

24. $\dfrac{2\sqrt{7}}{7}$ (Sec. 8.4; Ex. 9)

25. $x = \dfrac{1}{2}$ (Sec. 8.5; Ex. 3)

Quadratic Equations

An important part of the study of algebra is learning to use methods for solving equations. In Chapter 2, we presented techniques for solving linear equations in one variable. In Chapter 4, we solved quadratic equations in one variable by factoring the quadratic expressions. We now present other methods for solving quadratic equations in one variable.

It's a bird—it's a plane—no, . . . it's a Goodyear blimp! These widely recognized blimps are one of the best-known corporate symbols in the United States. Since 1925, the Goodyear Tire and Rubber Company has maintained a fleet of helium-filled lighter-than-air ships to carry out advertising and public relations functions. Today, Goodyear's fleet includes four blimps. The largest blimp, the Spirit of Akron, based at the Goodyear Headquarters in Akron, Ohio, is 205.5 feet long and has a top speed of 65 mph in the air. The Spirit of Akron, along with its sister ships the Eagle, Stars & Stripes, and Eagle Azteca, fly over 200,000 miles each year in their roles as aerial ambassadors for Goodyear. In Exercise 60 on page 680, we will use a method called the **quadratic formula** to predict when the net income of the Goodyear Tire and Rubber Company will reach a certain level.

ANSWERS

1. $a = 0, 6$; (9.1A)

2. $x = -\dfrac{1}{2}, 6$; (9.1A)

3. $b = \pm 12$; (9.1B)

4. $y = \dfrac{7 \pm 2\sqrt{6}}{2}$; (9.1B)

5. $x = 6, 8$; (9.2A)

6. $x = -\dfrac{1}{3}, 2$; (9.2B)

7. $x = -3, 9$; (9.3A)

8. $m = \dfrac{7 \pm \sqrt{145}}{8}$; (9.3A)

9. $x = \dfrac{-1 \pm \sqrt{73}}{4}$; (9.3A)

10. $x = \dfrac{-3 \pm 3\sqrt{2}}{5}$; (9.1B)

11. $x = -\dfrac{3}{4}, -\dfrac{3}{2}$; (9.1A)

12. $m = 3 \pm \sqrt{6}$; (9.2A)

13. $x = \dfrac{-4 \pm 3\sqrt{2}}{2}$; (9.3A)

14. $y = -7 \pm \sqrt{5}$; (9.1B)

15. see graph; (9.4A)

16. see graph; (9.4B)

17. see graph; (9.4B)

18. see graph; (9.4B)

Name _____ **Section** _____ **Date** _____

CHAPTER 9 PRETEST

Solve by factoring.

1. $a^2 - 6a = 0$ **2.** $2x^2 - 11x = 6$

Solve using the square root property.

3. $b^2 = 144$ **4.** $(2y - 7)^2 = 24$

Solve by completing the square.

5. $x^2 - 14x + 48 = 0$ **6.** $3x^2 - 5x = 2$

Solve using the quadratic formula.

7. $x^2 - 6x - 27 = 0$ **8.** $m^2 - \dfrac{7}{4}m - \dfrac{3}{2} = 0$

Solve by the most appropriate method.

9. $(2x + 3)(x - 1) = 6$ **10.** $(5x + 3)^2 = 18$ **11.** $8x^2 + 18x + 9 = 0$

12. $m^2 - 6m = -3$ **13.** $\dfrac{1}{4}x^2 + x - \dfrac{1}{8} = 0$ **14.** $(y + 7)^2 - 5 = 0$

Graph the quadratic equations. Label the vertex and the intercept points with their coordinates.

15. $y = -3x^2$ **16.** $y = x^2 + 3$

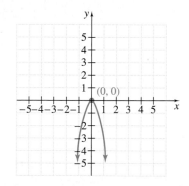
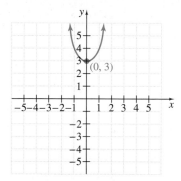

17. $y = x^2 + 4x$ **18.** $y = x^2 + 2x - 3$

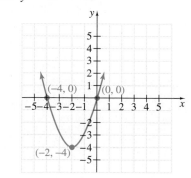
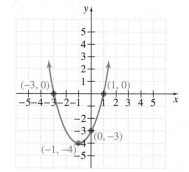

9.1 SOLVING QUADRATIC EQUATIONS BY THE SQUARE ROOT PROPERTY

Recall that a quadratic equation is an equation that can be written in the form

$$ax^2 + bx + c = 0$$

where a, b, and c are real numbers and $a \neq 0$.

A SOLVING QUADRATIC EQUATIONS BY FACTORING

To solve quadratic equations by factoring, we use the **zero factor property**: If the product of two numbers is zero, then at least one of the two numbers is zero. Examples 1 and 2 review the process of solving quadratic equations by factoring.

Example 1   Solve: $x^2 - 4 = 0$

Solution:
$$x^2 - 4 = 0$$
$$(x + 2)(x - 2) = 0 \quad \text{Factor.}$$
$$x + 2 = 0 \quad \text{or} \quad x - 2 = 0 \quad \text{Use the zero factor property.}$$
$$x = -2 \qquad\qquad x = 2 \quad \text{Solve each equation.}$$

The solutions are -2 and 2.

Example 2   Solve: $3y^2 + 13y = 10$

Solution: Recall that to use the zero factor property, one side of the equation must be 0 and the other side must be factored.

$$3y^2 + 13y = 10$$
$$3y^2 + 13y - 10 = 0 \quad \text{Subtract 10 from both sides.}$$
$$(3y - 2)(y + 5) = 0 \quad \text{Factor.}$$
$$3y - 2 = 0 \quad \text{or} \quad y + 5 = 0 \quad \text{Use the zero factor property.}$$
$$3y = 2 \qquad\qquad y = -5 \quad \text{Solve each equation.}$$
$$y = \frac{2}{3}$$

The solutions are $\frac{2}{3}$ and -5.

B USING THE SQUARE ROOT PROPERTY

Consider solving Example 1, $x^2 - 4 = 0$, another way. First, add 4 to both sides of the equation.

$$x^2 - 4 = 0$$
$$x^2 = 4 \quad \text{Add 4 to both sides.}$$

Now we see that the value for x must be a number whose square is 4. Therefore $x = \sqrt{4} = 2$ or $x = -\sqrt{4} = -2$. This reasoning is an example of the square root property.

Objectives

A Review factoring to solve quadratic equations.

B Use the square root property to solve quadratic equations.

SSM   CD-ROM   Video
                       9.1

Practice Problem 1

Solve: $x^2 - 25 = 0$

Practice Problem 2

Solve: $2x^2 - 3x = 9$

TEACHING TIP

Consider pointing out the similarities and differences in structure of $x - 4 = 0$, $\sqrt{x} - 4 = 0$, and $x^2 - 4 = 0$.

Answers

1. 5 and -5,   **2.** $-\frac{3}{2}$ and 3

TEACHING TIP

Before doing Example 3, you may want students to warm up by evaluating $(-5)^2$, 5^2, $(-4)^2$, and 4^2. Then ask what number(s) squared give 100. Be sure students give all possible answers.

Practice Problem 3

Use the square root property to solve $x^2 - 16 = 0$.

SQUARE ROOT PROPERTY

If $x^2 = a$ for $a \geq 0$, then
$$x = \sqrt{a} \quad \text{or} \quad x = -\sqrt{a}$$

Example 3   Use the square root property to solve $x^2 - 9 = 0$.

Solution: First we solve for x^2 by adding 9 to both sides.

$$x^2 - 9 = 0$$
$$x^2 = 9 \quad \text{Add 9 to both sides.}$$

Next we use the square root property.

$$x = \sqrt{9} \quad \text{or} \quad x = -\sqrt{9}$$
$$x = 3 \qquad\qquad x = -3$$

Check:

$x^2 - 9 = 0$	Original equation	$x^2 - 9 = 0$	Original equation
$3^2 - 9 \stackrel{?}{=} 0$	Let $x = 3$.	$(-3)^2 - 9 \stackrel{?}{=} 0$	Let $x = -3$.
$0 = 0$	True.	$0 = 0$	True.

The solutions are 3 and -3.

Practice Problem 4

Use the square root property to solve $3x^2 = 11$.

Example 4   Use the square root property to solve $2x^2 = 7$.

Solution: First we solve for x^2 by dividing both sides by 2. Then we use the square root property.

$$2x^2 = 7$$
$$x^2 = \frac{7}{2} \qquad\qquad \text{Divide both sides by 2.}$$
$$x = \sqrt{\frac{7}{2}} \quad \text{or} \quad x = -\sqrt{\frac{7}{2}} \qquad \text{Use the square root property.}$$
$$x = \frac{\sqrt{7} \cdot \sqrt{2}}{\sqrt{2} \cdot \sqrt{2}} \qquad x = -\frac{\sqrt{7} \cdot \sqrt{2}}{\sqrt{2} \cdot \sqrt{2}} \qquad \text{Rationalize the denominator.}$$
$$x = \frac{\sqrt{14}}{2} \qquad\quad x = -\frac{\sqrt{14}}{2} \qquad \text{Simplify.}$$

Remember to check both solutions in the original equation. The solutions are $\dfrac{\sqrt{14}}{2}$ and $-\dfrac{\sqrt{14}}{2}$.

Practice Problem 5

Use the square root property to solve $(x - 4)^2 = 49$.

Example 5   Use the square root property to solve $(x - 3)^2 = 16$.

Solution: Instead of x^2, here we have $(x - 3)^2$. But the square root property can still be used.

$$(x - 3)^2 = 16$$
$$x - 3 = \sqrt{16} \quad \text{or} \quad x - 3 = -\sqrt{16} \qquad \text{Use the square root property.}$$
$$x - 3 = 4 \qquad\qquad x - 3 = -4 \qquad \text{Write } \sqrt{16} \text{ as 4 and } -\sqrt{16} \text{ as } -4.$$
$$x = 7 \qquad\qquad\quad x = -1 \qquad \text{Solve.}$$

Answers

3. 4 and -4,   **4.** $\dfrac{\sqrt{33}}{3}$ and $\dfrac{-\sqrt{33}}{3}$,

5. 11 and -3

Check:

$(x - 3)^2 = 16$ Original equation $(x - 3)^2 = 16$ Original equation

$(7 - 3)^2 \overset{?}{=} 16$ Let $x = 7$. $(-1 - 3)^2 \overset{?}{=} 16$ Let $x = -1$.

$4^2 \overset{?}{=} 16$ Simplify. $(-4)^2 \overset{?}{=} 16$ Simplify.

$16 = 16$ True. $16 = 16$ True.

Both 7 and -1 are solutions.

Example 6

Use the square root property to solve $(x + 1)^2 = 8$.

Solution: $(x + 1)^2 = 8$

$x + 1 = \sqrt{8}$ or $x + 1 = -\sqrt{8}$ Use the square root property.

$x + 1 = 2\sqrt{2}$ $x + 1 = -2\sqrt{2}$ Simplify the radical.

$x = -1 + 2\sqrt{2}$ $x = -1 - 2\sqrt{2}$ Solve for x.

Check both solutions in the original equation. The solutions are $-1 + 2\sqrt{2}$ and $-1 - 2\sqrt{2}$ This can be written compactly as $-1 \pm 2\sqrt{2}$. The notation $\pm$ is read as "plus or minus."

HELPFUL HINT

read "plus or minus"

The notation $-1 \pm \sqrt{5}$, for example, is just a shorthand notation for both $-1 + \sqrt{5}$ and $-1 - \sqrt{5}$.

Example 7

Use the square root property to solve $(x - 1)^2 = -2$.

Solution: This equation has no real solution because the square root of -2 is not a real number.

Example 8

Use the square root property to solve $(5x - 2)^2 = 10$.

Solution: $(5x - 2)^2 = 10$

$5x - 2 = \sqrt{10}$ or $5x - 2 = -\sqrt{10}$ Use the square root property.

$5x = 2 + \sqrt{10}$ $5x = 2 - \sqrt{10}$ Add 2 to both sides.

$x = \dfrac{2 + \sqrt{10}}{5}$ $x = \dfrac{2 - \sqrt{10}}{5}$ Divide both sides by 5.

Check both solutions in the original equation. The solutions are $\dfrac{2 + \sqrt{10}}{5}$ and $\dfrac{2 - \sqrt{10}}{5}$, which can be written as $\dfrac{2 \pm \sqrt{10}}{5}$.

Practice Problem 6

Use the square root property to solve $(x - 5)^2 = 18$.

TEACHING TIP

For Example 7, encourage students to look at the big picture by asking, "What does this equation state?" Lead students to notice that a square is set equal to a negative number. Ask whether that situation could ever be possible.

Practice Problem 7

Use the square root property to solve $(x + 3)^2 = -5$.

Practice Problem 8

Use the square root property to solve $(4x + 1)^2 = 15$.

Answers

6. $5 \pm 3\sqrt{2}$,   **7.** no real solution,

8. $\dfrac{-1 \pm \sqrt{15}}{4}$

Focus On the Real World

USES OF PARABOLAS

We learned in Chapter 6 that the graph of an equation in two variables of the form $y = mx + b$ is a straight line. Later in this chapter, we will find that the graph of a quadratic equation in two variables of the form $y = ax^2 + bx + c$ is a shape called a **parabola**. The figure to the right shows the general shape of a parabola.

The shape of a parabola shows up in many situations in the world around us, both natural and human-made.

NATURAL SITUATIONS

▲ **Hurricanes**   The paths of many hurricanes are roughly shaped like a parabola. In the northern hemisphere, hurricanes generally begin moving to the northwest. Then, as they move further from the equator, they swing around to head in a northeastern direction.

▲ **Projectiles**   The force of the earth's gravity acts on a projectile launched into the air. The resulting path of the projectile, anything from a bullet to a football, is generally shaped like a parabola.

▲ **Orbits**   There are several different possible shapes of orbits of satellites, planets, moons, and comets in outer space. One of the possible types of orbits is in the shape of a parabola. A parabolic orbit is probably most often seen with comets.

HUMAN-MADE SITUATIONS

▲ **Telescopes**   Because a parabola has nice reflecting properties, the shape of a parabola is used in many kinds of telescopes. The largest non-steerable radio telescope is Arecibo Observatory in Puerto Rico. This telescope consists of a huge parabolic dish built into a valley. The dish is about 1000 feet across.

▲ **Training Astronauts**   Astronauts must be able to work in zero-gravity conditions on missions in space. However, it's nearly impossible to escape the force of gravity on earth. To help astronauts train to work in weightlessness, a specially modified jet can be flown in a parabolic path. At the top of the parabola, weightlessness can be simulated for up to 30 seconds at a time.

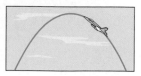

▲ **Architecture**   The reinforced concrete arches used in many modern buildings are based on the shape of a parabola.

▲ **Music**   The design of the modern flute incorporates a parabolic head joint.

Name _____ Section _____ Date _____

EXERCISE SET 9.1

A *Solve each equation by factoring. See Examples 1 and 2.*

1. $k^2 - 9 = 0$ **2.** $k^2 - 49 = 0$ **3.** $m^2 + 2m = 15$ **4.** $m^2 + 6m = 7$

5. $2x^2 - 32 = 0$ **6.** $3p^4 - 9p^3 = 0$ **7.** $4a^2 - 36 = 0$

8. $7a^2 - 175 = 0$ **9.** $x^2 + 7x = -10$ **10.** $x^2 + 10x = -24$

B *Use the square root property to solve each quadratic equation. See Examples 3 and 4.*

⊞ 11. $x^2 = 64$ **12.** $x^2 = 121$ **13.** $x^2 = 21$ **14.** $x^2 = 22$

15. $x^2 = \dfrac{1}{25}$ **16.** $x^2 = \dfrac{1}{16}$ **⊞ 17.** $x^2 = -4$ **18.** $x^2 = -25$

19. $3x^2 = 13$ **20.** $5x^2 = 2$ **21.** $7x^2 = 4$ **22.** $2x^2 = 9$

23. $x^2 - 2 = 0$ **24.** $x^2 - 15 = 0$

25. Explain why the equation $x^2 = -9$ has no real solution.   **26.** Explain why the equation $x^2 = 9$ has two solutions.

Use the square root property to solve each quadratic equation. See Examples 5 through 8.

27. $(x - 5)^2 = 49$ **28.** $(x + 2)^2 = 25$ **⊞ 29.** $(x + 2)^2 = 7$

30. $(x - 7)^2 = 2$ **31.** $\left(m - \dfrac{1}{2}\right)^2 = \dfrac{1}{4}$ **32.** $\left(m + \dfrac{1}{3}\right)^2 = \dfrac{1}{9}$

33. $(p + 2)^2 = 10$ **34.** $(p - 7)^2 = 13$ **35.** $(3y + 2)^2 = 100$

36. $(4y - 3)^2 = 81$ **37.** $(z - 4)^2 = -9$ **38.** $(z + 7)^2 = -20$

39. $(2x - 11)^2 = 50$ **40.** $(3x - 17)^2 = 28$ **⊞ 41.** $(3x - 7)^2 = 32$

42. $(5x - 11)^2 = 54$

665

43. $(x + 3)^2$

44. $(y + 5)^2$

45. $(x - 2)^2$

46. $(x - 10)^2$

47. $x = 2, -6$

48. $y = 5 \pm \sqrt{11}$

49. $r = 6$ in.

50. $x = \dfrac{27\sqrt{2}}{2}$ in.

51. 5 sec

52. $x = \pm 1.33$

53. $x = -1.02, 3.76$

54. 1997

55. 1999

666

Name _____

REVIEW AND PREVIEW

Factor each perfect square trinomial. See Section 4.5.

43. $x^2 + 6x + 9$ **44.** $y^2 + 10y + 25$ **45.** $x^2 - 4x + 4$ **46.** $x^2 - 20x + 100$

◤ COMBINING CONCEPTS

Solve each quadratic equation by first factoring the perfect square trinomial on the left side. Then apply the square root property.

47. $x^2 + 4x + 4 = 16$

48. $y^2 - 10y + 25 = 11$

49. The area of a circle is found by the equation $A = \pi r^2$. If the area A of a certain circle is 36π square inches, find its radius r.

36π square inches

50. A 27-inch-square TV is advertised in the local paper. If 27 inches is the measure of the diagonal of the picture tube, use the Pythagorean theorem to find the measure of the side of the picture tube.

27
x
x

51. Neglecting air resistance, the distance d in feet that an object falls in t seconds is given by the equation $d = 16t^2$. If a sandblaster drops his goggles from a bridge 400 feet from the water below, find how long it takes for the goggles to hit the water.

400 feet

Solve each quadratic equation by using the square root property. Use a calculator and round each solution to the nearest hundredth.

▤ **52.** $x^2 = 1.78$

▤ **53.** $(x - 1.37)^2 = 5.71$

54. The number of hogs and pigs y (in thousands) on farms in Iowa from 1992 through 1994 is given by the equation $y = 100x^2 + 14,900$. In this equation, $x = 0$ represents the year 1993. Assume that this trend continues and find the year in which there were 16,500 thousand hogs and pigs on Iowa farms. (*Hint:* Replace y with 16,500 in the equation and solve for x.) (*Source:* Based on data from the U.S. Department of Agriculture)

55. The soybean yield y (in bushels per acre) in Missouri from 1992 through 1994 is given by the equation $y = 5x^2 + 33$. In this equation, $x = 0$ represents the year 1993. Assume that this trend continues and predict the year in which the Missouri soybean yield will be 213 bushels per acre. (*Hint:* Replace y with 213 in the equation and solve for x.) (*Source:* Based on data from the U.S. Department of Agriculture)

9.2 SOLVING QUADRATIC EQUATIONS BY COMPLETING THE SQUARE

Objectives

A Solve quadratic equations of the form $x^2 + bx + c = 0$ by completing the square.

B Solve quadratic equations of the form $ax^2 + bx + c = 0$ by completing the square.

SSM   CD-ROM   Video
9.2

A COMPLETING THE SQUARE TO SOLVE $x^2 + bx + c = 0$

In the last section, we used the square root property to solve equations such as

$$(x + 1)^2 = 8 \quad \text{and} \quad (5x - 2)^2 = 3$$

Notice that one side of each equation is a quantity squared and that the other side is a constant. To solve

$$x^2 + 2x = 4$$

notice that if we add 1 to both sides of the equation, the left side is a perfect square trinomial that can be factored.

$$x^2 + 2x + 1 = 4 + 1 \quad \text{Add 1 to both sides.}$$
$$(x + 1)^2 = 5 \quad \text{Factor.}$$

Now we can solve this equation as we did in the previous section by using the square root property.

$$x + 1 = \sqrt{5} \quad \text{or} \quad x + 1 = -\sqrt{5} \quad \text{Use the square root property.}$$
$$x = -1 + \sqrt{5} \qquad\qquad x = -1 - \sqrt{5} \quad \text{Solve.}$$

The solutions are $-1 \pm \sqrt{5}$.

Adding a number to $x^2 + 2x$ to form a perfect square trinomial is called **completing the square** on $x^2 + 2x$.

In general, we have the following:

COMPLETING THE SQUARE

To complete the square on $x^2 + bx$, add $\left(\dfrac{b}{2}\right)^2$. To find $\left(\dfrac{b}{2}\right)^2$, **find half the coefficient of x, then square the result.**

Example 1   Solve $x^2 + 6x + 3 = 0$ by completing the square.

Solution: First we get the variable terms alone by subtracting 3 from both sides of the equation.

$$x^2 + 6x + 3 = 0$$
$$x^2 + 6x = -3 \quad \text{Subtract 3 from both sides.}$$

Next we find half the coefficient of the x-term, then square it. Add this result to *both sides* of the equation. This will make the left side a perfect square trinomial. The coefficient of x is 6, and half of 6 is 3. So we add 3^2 or 9 to both sides.

$$x^2 + 6x + 9 = -3 + 9 \quad \text{Complete the square.}$$
$$(x + 3)^2 = 6 \quad \text{Factor the trinomial } x^2 + 6x + 9.$$
$$x + 3 = \sqrt{6} \quad \text{or} \quad x + 3 = -\sqrt{6} \quad \text{Use the square root property.}$$
$$x = -3 + \sqrt{6} \qquad\qquad x = -3 - \sqrt{6} \quad \text{Subtract 3 from both sides.}$$

Check by substituting $-3 + \sqrt{6}$ and $-3 - \sqrt{6}$ in the original equation. The solutions are $-3 \pm \sqrt{6}$. ■

Practice Problem 1

Solve $x^2 + 8x + 1 = 0$ by completing the square.

TEACHING TIP

Point out to students that in Examples 1 and 2 the coefficient of the x^2-term is 1.

Answer

1. $-4 \pm \sqrt{15}$

Practice Problem 2

Solve $x^2 - 14x = -32$ by completing the square.

Example 2   Solve $x^2 - 10x = -14$ by completing the square.

Solution: The variable terms are already alone on one side of the equation. The coefficient of x is -10. Half of -10 is -5, and $(-5)^2 = 25$. So we add 25 to both sides.

$$x^2 - 10x = -14$$
$$x^2 - 10x + 25 = -14 + 25$$

$$(x - 5)^2 = 11 \quad \text{Factor the trinomial and simplify } -14 + 25.$$
$$x - 5 = \sqrt{11} \quad \text{or} \quad x - 5 = -\sqrt{11} \quad \text{Use the square root property.}$$
$$x = 5 + \sqrt{11} \qquad\qquad x = 5 - \sqrt{11} \quad \text{Add 5 to both sides.}$$

The solutions are $5 \pm \sqrt{11}$.

TEACHING TIP

If students are having difficulty, consider having pairs of students work together to complete the squares in several equations like those below until they are comfortable with the process. To save time in class, groups need not solve the equations.
$x^2 - 3x = 5$
$y^2 + 8y = 6$
$3x^2 + 6x = 4$

B COMPLETING THE SQUARE TO SOLVE $ax^2 + bx + c = 0$

The method of completing the square can be used to solve *any* quadratic equation whether the coefficient of the squared variable is 1 or not. When the coefficient of the squared variable is not 1, we first divide both sides of the equation by the coefficient of the squared variable so that the coefficient is 1. Then we complete the square.

Practice Problem 3

Solve $4x^2 - 16x - 9 = 0$ by completing the square.

Example 3   Solve $4x^2 - 8x - 5 = 0$ by completing the square.

Solution: $4x^2 - 8x - 5 = 0$

$$x^2 - 2x - \frac{5}{4} = 0 \quad \text{Divide both sides by 4.}$$

$$x^2 - 2x = \frac{5}{4} \quad \text{Get the variable terms alone on one side of the equation.}$$

The coefficient of x is -2. Half of -2 is -1, and $(-1)^2 = 1$. So we add 1 to both sides.

$$x^2 - 2x + 1 = \frac{5}{4} + 1$$

$$(x - 1)^2 = \frac{9}{4} \quad \text{Factor } x^2 - 2x + 1 \text{ and simplify } \frac{5}{4} + 1.$$

$$x - 1 = \sqrt{\frac{9}{4}} \quad \text{or} \quad x - 1 = -\sqrt{\frac{9}{4}} \quad \text{Use the square root property.}$$

$$x = 1 + \frac{3}{2} \qquad\qquad x = 1 - \frac{3}{2} \quad \text{Add 1 to both sides and simplify the radical.}$$

$$x = \frac{5}{2} \qquad\qquad x = -\frac{1}{2} \quad \text{Simplify.}$$

Both $\frac{5}{2}$ and $-\frac{1}{2}$ are solutions.

Answers

2. $7 \pm \sqrt{17}$,   **3.** $\frac{9}{2}$ and $-\frac{1}{2}$

The following steps may be used to solve a quadratic equation in x by completing the square.

To Solve a Quadratic Equation in x by Completing the Square

Step 1. If the coefficient of x^2 is 1, go to Step 2. If not, divide both sides of the equation by the coefficient of x^2.

Step 2. Get all terms with variables on one side of the equation and constants on the other side.

Step 3. Find half the coefficient of x and then square the result. Add this number to both sides of the equation.

Step 4. Factor the resulting perfect square trinomial.

Step 5. Use the square root property to solve the equation.

Example 4   Solve $2x^2 + 6x = -7$ by completing the square.

Solution: The coefficient of x^2 is not 1. We divide both sides by 2, the coefficient of x^2.

$$2x^2 + 6x = -7$$

$$x^2 + 3x = -\frac{7}{2} \quad \text{Divide both sides by 2.}$$

$$x^2 + 3x + \frac{9}{4} = -\frac{7}{2} + \frac{9}{4} \quad \text{Add } \left(\frac{3}{2}\right)^2 \text{ or } \frac{9}{4} \text{ to both sides.}$$

$$\left(x + \frac{3}{2}\right)^2 = -\frac{5}{4} \quad \begin{array}{l}\text{Factor the left side and simplify the}\\\text{right.}\end{array}$$

There is no real solution to this equation since the square root of a negative number is not a real number. ▬

Example 5   Solve $2x^2 = 10x + 1$ by completing the square.

Solution: First we divide both sides of the equation by 2, the coefficient of x^2.

$$2x^2 = 10x + 1$$

$$x^2 = 5x + \frac{1}{2} \quad \text{Divide both sides by 2.}$$

Next we get the variable terms alone by subtracting $5x$ from both sides.

$$x^2 - 5x = \frac{1}{2}$$

$$x^2 - 5x + \frac{25}{4} = \frac{1}{2} + \frac{25}{4} \quad \text{Add } \left(-\frac{5}{2}\right)^2 \text{ or } \frac{25}{4} \text{ to both sides.}$$

$$\left(x - \frac{5}{2}\right)^2 = \frac{27}{4} \quad \text{Factor the left side and simplify the right side.}$$

$$x - \frac{5}{2} = \sqrt{\frac{27}{4}} \quad \text{or} \quad x - \frac{5}{2} = -\sqrt{\frac{27}{4}} \quad \text{Use the square root property.}$$

$$x - \frac{5}{2} = \frac{3\sqrt{3}}{2} \qquad\qquad x - \frac{5}{2} = -\frac{3\sqrt{3}}{2} \quad \text{Simplify.}$$

$$x = \frac{5}{2} + \frac{3\sqrt{3}}{2} \qquad\qquad x = \frac{5}{2} - \frac{3\sqrt{3}}{2}$$

The solutions are $\dfrac{5 \pm 3\sqrt{3}}{2}$.

Practice Problem 4

Solve $2x^2 + 10x = -13$ by completing the square.

Practice Problem 5

Solve $2x^2 = -3x + 2$ by completing the square.

Answers

4. no real solution,   **5.** $\frac{1}{2}$ and -2

Focus On Business and Career

MODELING THE SIZE OF THE FEDEX VEHICLE FLEET

How would you like to drive 2.5 million miles in a day? The couriers who work for FedEx drive over 2.5 million miles *each* day to deliver packages in the United States alone. Founded in 1973, Federal Express, or FedEx as it is now known, was the first company to offer overnight package delivery in the United States. Today, it delivers to 212 countries and is the world's largest express transportation company. To deliver the 2.9 million packages it handles daily worldwide, FedEx uses a fleet of nearly 600 airplanes and an ever-growing fleet of motor vehicles for local delivery.

The number of motor vehicles y in the constantly expanding FedEx fleet from 1992 through 1995 can be modeled by the equation $y = 1825x^2 - 3545x + 30,255$, where $x = 0$ represents 1992. (*Source*: Based on data from Federal Express)

GROUP ACTIVITY

- ▲ Use the given quadratic equation to complete the table of values.
- ▲ Graph the equation for the years 1992 through 1995 on the given set of axes.
- ▲ Use the graph to estimate when the fleet of FedEx motor vehicles was the smallest during this period. Estimate the smallest number of vehicles in the FedEx fleet during this period. What feature of the parabola you graphed gives this information?
- ▲ Use the model to predict the year in which the FedEx vehicle fleet consists of 45,275 vehicles.

Year	x	y
1992	0	
1993		
1994		
1995		

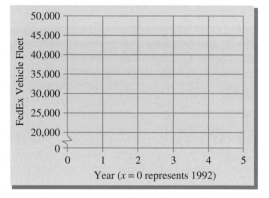

Internet Excursions

Go to http://www.prenhall.com/martin-gay
Most publicly held corporations publish an annual report summarizing annual earnings and financial positions, as well as various operating data, for their stockholders. Many corporations make their annual reports available on their web sites. Federal Express is one such corporation. The given World Wide Web address will provide you with access to the FedEx Web site, or a related site, where you will find selected consolidated financial data from its current annual report for the most recent 10 years.

1. Locate data for the actual size of the FedEx vehicle fleet from 1992 through 1995. Plot the data on the above graph you made of the model. How well do you think the given model fits the actual data? Explain.
   answers may vary

2. In the annual report data, look up the year for which you predicted 45,275 vehicles. How many vehicles were actually in the FedEx vehicle fleet that year? How close is this number to 45,275 vehicles? Do you think the model for 1992–1995 will make accurate predictions in the future? Explain.
   answers may vary

Name _____ Section _____ Date _____

MENTAL MATH

Determine the number to add to make each expression a perfect square trinomial.

1. $p^2 + 8p$ **2.** $p^2 + 6p$ **3.** $x^2 + 20x$

4. $x^2 + 18x$ **5.** $y^2 + 14y$ **6.** $y^2 + 2y$

EXERCISE SET 9.2

A Solve each quadratic equation by completing the square. See Examples 1 and 2.

⊟ 1. $x^2 + 8x = -12$ **2.** $x^2 - 10x = -24$ **3.** $x^2 + 2x - 5 = 0$

4. $z^2 + 6z - 9 = 0$ **5.** $x^2 - 6x = 0$ **6.** $y^2 + 4y = 0$

7. $z^2 + 5z = 7$ **8.** $x^2 - 7x = 5$ **⊟ 9.** $x^2 - 2x - 1 = 0$

10. $x^2 - 4x + 2 = 0$ **11.** $y^2 + 5y + 4 = 0$ **12.** $y^2 - 5y + 6 = 0$

13. $x^2 + 6x - 25 = 0$ **14.** $x^2 - 6x + 7 = 0$

15. $x(x + 3) = 18$ **16.** $x(x - 3) = 18$

B Solve each quadratic equation by completing the square. See Examples 3 through 5.

17. $4x^2 - 24x = 13$ **18.** $2x^2 + 8x = 10$ **19.** $5x^2 + 10x + 6 = 0$

20. $3x^2 - 12x + 14 = 0$ **21.** $2x^2 = 6x + 5$ **22.** $4x^2 = -20x + 3$

23. $3x^2 - 6x = 24$ **24.** $2x^2 + 18x = -40$ **⊟ 25.** $2y^2 + 8y + 5 = 0$

26. $3z^2 + 6z + 4 = 0$ **27.** $2y^2 - 3y + 1 = 0$ **28.** $2y^2 - y - 1 = 0$

671

29. $-\dfrac{1}{2}$

30. $\dfrac{7}{5}$

31. -1

32. $\dfrac{1}{5}$

33. $3 + 2\sqrt{5}$

34. $5 - 10\sqrt{3}$

35. $\dfrac{1 - 3\sqrt{2}}{2}$

36. $\dfrac{3 - 2\sqrt{7}}{4}$

37. $k = 8 \text{ or } k = -8$

38. $k = 10 \text{ or } k = -10$

39. 1999

40. 1998

41. $x = -6, -2$

42. $x = 4, 6$

43. $x \approx -0.68, 3.68$

44. $x \approx -0.65, 7.65$

Name _____

REVIEW AND PREVIEW

Simplify each expression. See Section 8.2.

29. $\dfrac{3}{4} - \sqrt{\dfrac{25}{16}}$ **30.** $\dfrac{3}{5} + \sqrt{\dfrac{16}{25}}$ **31.** $\dfrac{1}{2} - \sqrt{\dfrac{9}{4}}$ **32.** $\dfrac{9}{10} - \sqrt{\dfrac{49}{100}}$

Simplify each expression. See Section 8.4.

33. $\dfrac{6 + 4\sqrt{5}}{2}$ **34.** $\dfrac{10 - 20\sqrt{3}}{2}$ **35.** $\dfrac{3 - 9\sqrt{2}}{6}$ **36.** $\dfrac{12 - 8\sqrt{7}}{16}$

COMBINING CONCEPTS

37. Find a value of k that will make $x^2 + kx + 16$ a perfect square trinomial.

38. Find a value of k that will make $x^2 + kx + 25$ a perfect square trinomial.

39. The number of Home Depot stores y operating in North America from 1994 through 1996 is given by the equation $y = 3x^2 + 80x + 340$. In this equation, x is the number of years after 1994. Assume that this trend continues and predict the year after 1994 in which the number of Home Depot stores will be 815. (*Source:* Based on data from The Home Depot, Inc.)

40. The average price of tin y (in cents per pound) from 1993 through 1995 is given by the equation $y = 16x^2 + 3x + 350$. In this equation, x is the number of years after 1993. Assume that this trend continues and find the year after 1993 in which the price of tin will be 765 cents per pound. (*Source:* Based on data from the U.S. Bureau of Mines)

Recall that a graphing calculator may be used to solve an equation. For example, to solve $x^2 + 8x = -12$ (Exercise 1), graph

$$y_1 = x^2 + 8x \quad \textit{(left side of equation) and}$$
$$y_2 = -12 \qquad \textit{(right side of equation)}$$

The x-coordinate of the point of intersection of the graphs is the solution. Use a graphing calculator and solve each equation. Round solutions to the nearest hundredth.

41. Exercise 1

42. Exercise 2

43. Exercise 21

44. Exercise 8

9.3 SOLVING QUADRATIC EQUATIONS BY THE QUADRATIC FORMULA

A USING THE QUADRATIC FORMULA

We can use the technique of completing the square to develop a formula to find solutions of any quadratic equation. We develop and use the **quadratic formula** in this section.

Recall that a quadratic equation in **standard form** is

$$ax^2 + bx + c = 0, \quad a \neq 0$$

To develop the quadratic formula, let's complete the square for this quadratic equation in standard form.

First we divide both sides of the equation by the coefficient of x^2 and then get the variable terms alone on one side of the equation.

$$x^2 + \frac{b}{a}x + \frac{c}{a} = 0 \qquad \text{Divide by } a; \text{ recall that } a \text{ cannot be 0.}$$

$$x^2 + \frac{b}{a}x = -\frac{c}{a} \qquad \text{Get the variable terms alone on one side of the equation.}$$

The coefficient of x is $\frac{b}{a}$. Half of $\frac{b}{a}$ is $\frac{b}{2a}$ and $\left(\frac{b}{2a}\right)^2 = \frac{b^2}{4a^2}$. So we add $\frac{b^2}{4a^2}$ to both sides of the equation.

$$x^2 + \frac{b}{a}x + \frac{b^2}{4a^2} = -\frac{c}{a} + \frac{b^2}{4a^2} \qquad \text{Add } \frac{b^2}{4a^2} \text{ to both sides.}$$

$$\left(x + \frac{b}{2a}\right)^2 = -\frac{c}{a} + \frac{b^2}{4a^2} \qquad \text{Factor the left side.}$$

$$\left(x + \frac{b}{2a}\right)^2 = -\frac{4ac}{4a^2} + \frac{b^2}{4a^2} \qquad \text{Multiply } -\frac{c}{a} \text{ by } \frac{4a}{4a} \text{ so that both terms on the the right side have a common denominator.}$$

$$\left(x + \frac{b}{2a}\right)^2 = \frac{b^2 - 4ac}{4a^2} \qquad \text{Simplify the right side.}$$

Now we use the square root property.

$$x + \frac{b}{2a} = \sqrt{\frac{b^2 - 4ac}{4a^2}} \quad \text{or} \quad x + \frac{b}{2a} = -\sqrt{\frac{b^2 - 4ac}{4a^2}} \qquad \text{Use the square root property.}$$

$$x + \frac{b}{2a} = \frac{\sqrt{b^2 - 4ac}}{2a} \qquad\qquad x + \frac{b}{2a} = -\frac{\sqrt{b^2 - 4ac}}{2a} \qquad \text{Simplify the radical.}$$

$$x = -\frac{b}{2a} + \frac{\sqrt{b^2 - 4ac}}{2a} \qquad\qquad x = -\frac{b}{2a} - \frac{\sqrt{b^2 - 4ac}}{2a}$$

$$\text{Subtract } \frac{b}{2a} \text{ from both sides.}$$

$$x = \frac{-b + \sqrt{b^2 - 4ac}}{2a} \qquad\qquad x = \frac{-b - \sqrt{b^2 - 4ac}}{2a} \qquad \text{Simplify.}$$

The solutions are $\dfrac{-b \pm \sqrt{b^2 - 4ac}}{2a}$. This final equation is called the **quadratic formula** and gives the solutions of any quadratic equation.

QUADRATIC FORMULA

If a, b, and c are real numbers and $a \neq 0$, a quadratic equation written in the form $ax^2 + bx + c = 0$ has solutions

$$x = \frac{-b \pm \sqrt{b^2 - 4ac}}{2a}$$

Objective

A Use the quadratic formula to solve quadratic equations.

SSM   CD-ROM   Video
9.3

HELPFUL HINT

Don't forget that to correctly identify a, b, and c in the quadratic formula, you should write the equation in standard form.

Quadratic Equations in Standard Form

$$5x^2 - 6x + 2 = 0 \qquad a = 5, b = -6, c = 2$$
$$4y^2 - 9 = 0 \qquad a = 4, b = 0, c = -9$$
$$x^2 + x = 0 \qquad a = 1, b = 1, c = 0$$
$$\sqrt{2}x^2 + \sqrt{5}x + \sqrt{3} = 0 \qquad a = \sqrt{2}, b = \sqrt{5}, c = \sqrt{3}$$

Practice Problem 1

Solve $2x^2 - x - 5 = 0$ using the quadratic formula.

TEACHING TIP

Help students develop their confidence applying the quadratic formula by having them solve the following equation by factoring, completing the square, and by the quadratic formula.
$x^2 + 7x + 12 = 0$
$(x = -3, x = -4)$

Practice Problem 2

Solve $3x^2 + 8x = 3$ using the quadratic formula.

Example 1   Solve $3x^2 + x - 3 = 0$ using the quadratic formula.

Solution: This equation is in standard form with $a = 3$, $b = 1$, and $c = -3$. By the quadratic formula, we have

$$x = \frac{-b \pm \sqrt{b^2 - 4ac}}{2a}$$

$$x = \frac{-1 \pm \sqrt{1^2 - 4 \cdot 3 \cdot (-3)}}{2 \cdot 3} \qquad \text{Let } a = 3, b = 1, \text{ and } c = -3.$$

$$= \frac{-1 \pm \sqrt{1 + 36}}{6} \qquad \text{Simplify.}$$

$$= \frac{-1 \pm \sqrt{37}}{6}$$

Check both solutions in the original equation. The solutions are $\dfrac{-1 + \sqrt{37}}{6}$ and $\dfrac{-1 - \sqrt{37}}{6}$.   ▬▬▬

Example 2   Solve $2x^2 - 9x = 5$ using the quadratic formula.

Solution: First we write the equation in standard form by subtracting 5 from both sides.

$$2x^2 - 9x = 5$$
$$2x^2 - 9x - 5 = 0$$

Next we note that $a = 2$, $b = -9$, and $c = -5$. We substitute these values into the quadratic formula.

HELPFUL HINT

Notice that the fraction bar is under the entire numerator of $-b \pm \sqrt{b^2 - 4ac}$.

$$x = \frac{-b \pm \sqrt{b^2 - 4ac}}{2a}$$

$$x = \frac{-(-9) \pm \sqrt{(-9)^2 - 4 \cdot 2 \cdot (-5)}}{2 \cdot 2} \qquad \begin{array}{l}\text{Substitute} \\ \text{in the} \\ \text{formula.}\end{array}$$

$$= \frac{9 \pm \sqrt{81 + 40}}{4} \qquad \text{Simplify.}$$

$$= \frac{9 \pm \sqrt{121}}{4} = \frac{9 \pm 11}{4}$$

Then,

$$x = \frac{9 - 11}{4} = -\frac{1}{2} \qquad \text{or} \qquad x = \frac{9 + 11}{4} = 5$$

Check $-\dfrac{1}{2}$ and 5 in the original equation. Both $-\dfrac{1}{2}$ and 5 are solutions.   ▬▬▬

Answers

1. $\dfrac{1 + \sqrt{41}}{4}$ and $\dfrac{1 - \sqrt{41}}{4}$,   **2.** $\dfrac{1}{3}$ and -3

The following steps may be useful when solving a quadratic equation by the quadratic formula.

✓ **CONCEPT CHECK**

TO SOLVE A QUADRATIC EQUATION BY THE QUADRATIC FORMULA

Step 1. Write the quadratic equation in standard form: $ax^2 + bx + c = 0$.

Step 2. If necessary, clear the equation of fractions to simplify calculations.

Step 3. Identify a, b, and c.

Step 4. Replace a, b, and c in the quadratic formula with the identified values, and simplify.

For the quadratic equation $2x^2 - 5 = 7x$, if $a = 2$ and $c = -5$ in the quadratic formula, the value of b is

a. $\dfrac{7}{2}$

b. 7

c. -5

d. -7

TRY THE CONCEPT CHECK IN THE MARGIN.

Example 3   Solve $7x^2 = 1$ using the quadratic formula.

Solution: First we write the equation in standard form by subtracting 1 from both sides.

$$7x^2 = 1$$
$$7x^2 - 1 = 0$$

Next we replace a, b, and c with the identified values: $a = 7, b = 0, c = -1$.

$$x = \frac{0 \pm \sqrt{0^2 - 4 \cdot 7 \cdot (-1)}}{2 \cdot 7} \quad \text{Substitute in the formula.}$$

$$= \frac{\pm\sqrt{28}}{14} \quad\quad \text{Simplify.}$$

$$= \frac{\pm 2\sqrt{7}}{14}$$

$$= \pm\frac{\sqrt{7}}{7}$$

The solutions are $\dfrac{\sqrt{7}}{7}$ and $-\dfrac{\sqrt{7}}{7}$.

Practice Problem 3

Solve $5x^2 = 2$ using the quadratic formula.

Notice that the equation in Example 3, $7x^2 = 1$, could have been easily solved by dividing both sides by 7 and then using the square root property. We solved the equation by the quadratic formula to show that this formula can be used to solve any quadratic equation.

Example 4   Solve $x^2 = -x - 1$ using the quadratic formula.

Solution: First we write the equation in standard form.

$$x^2 + x + 1 = 0$$

Next we replace a, b, and c in the quadratic formula with $a = 1, b = 1$, and $c = 1$.

$$x = \frac{-1 \pm \sqrt{1^2 - 4 \cdot 1 \cdot 1}}{2 \cdot 1} \quad \text{Substitute in the formula.}$$

$$= \frac{-1 \pm \sqrt{-3}}{2} \quad\quad \text{Simplify.}$$

There is no real number solution because $\sqrt{-3}$ is not a real number.

Practice Problem 4

Solve $x^2 = -2x - 3$ by using the quadratic formula.

Answers

✓ **Concept Check:** d

3. $\dfrac{\sqrt{10}}{5}$ and $-\dfrac{\sqrt{10}}{5}$,   **4.** no real solution

Practice Problem 5

Solve $\frac{1}{3}x^2 - x = 1$ by using the quadratic formula.

TEACHING TIP

For Example 5, point out that the fractions were cleared to get coefficients which are easier to work with. The original coefficients, $a = \frac{1}{2}$, $b = -1$, and $c = -2$, would give the same results.

Example 5   Solve $\frac{1}{2}x^2 - x = 2$ by using the quadratic formula.

Solution: We write the equation in standard form and then clear the equation of fractions by multiplying both sides by the LCD, 2.

$$\frac{1}{2}x^2 - x = 2$$

$$\frac{1}{2}x^2 - x - 2 = 0 \quad \text{Write in standard form.}$$

$$x^2 - 2x - 4 = 0 \quad \text{Multiply both sides by 2.}$$

Here, $a = 1$, $b = -2$, and $c = -4$, so we substitute these values into the quadratic formula.

$$x = \frac{-(-2) \pm \sqrt{(-2)^2 - 4 \cdot 1 \cdot (-4)}}{2 \cdot 1}$$

$$= \frac{2 \pm \sqrt{20}}{2} = \frac{2 \pm 2\sqrt{5}}{2} \quad \text{Simplify.}$$

$$= \frac{2(1 \pm \sqrt{5})}{2} = 1 \pm \sqrt{5} \quad \text{Factor and simplify.}$$

The solutions are $1 - \sqrt{5}$ and $1 + \sqrt{5}$. ▬

HELPFUL HINT

When simplifying expressions such as

$$\frac{3 \pm 6\sqrt{2}}{6}$$

first factor out a common factor from the terms of the numerator and then simplify.

$$\frac{3 \pm 6\sqrt{2}}{6} = \frac{3(1 \pm 2\sqrt{2})}{2 \cdot 3} = \frac{1 \pm 2\sqrt{2}}{2}$$

Answer

5. $\frac{3 \pm \sqrt{21}}{2}$

Name _____ Section _____ Date _____

MENTAL MATH

Identify the value of a, b, and c in each quadratic equation.

1. $2x^2 + 5x + 3 = 0$
2. $5x^2 - 7x + 1 = 0$
3. $10x^2 - 13x - 2 = 0$
4. $x^2 + 3x - 7 = 0$
5. $x^2 - 6 = 0$
6. $9x^2 - 4 = 0$

EXERCISE SET 9.3

A *Use the quadratic formula to solve each quadratic equation. See Examples 1 through 4.*

1. $x^2 - 3x + 2 = 0$
2. $x^2 - 5x - 6 = 0$
3. $3k^2 + 7k + 1 = 0$

4. $7k^2 + 3k - 1 = 0$
5. $49x^2 - 4 = 0$
6. $25x^2 - 15 = 0$

7. $5z^2 - 4z + 3 = 0$
8. $3x^2 + 2x + 1 = 0$
9. $y^2 = 7y + 30$

10. $y^2 = 5y + 36$
11. $2x^2 = 10$
12. $5x^2 = 15$

13. $m^2 - 12 = m$
14. $m^2 - 14 = 5m$
15. $3 - x^2 = 4x$

16. $10 - x^2 = 2x$
17. $6x^2 + 9x = 2$
18. $3x^2 - 9x = 8$

19. $7p^2 + 2 = 8p$
20. $11p^2 + 2 = 10p$
21. $a^2 - 6a + 2 = 0$

22. $a^2 - 10a + 19 = 0$
23. $2x^2 - 6x + 3 = 0$
24. $5x^2 - 8x + 2 = 0$

25. $x = -1, \dfrac{1}{3}$

26. $y = \dfrac{4}{5}, -1$

27. $y = \dfrac{3 \pm \sqrt{13}}{4}$

28. no real solution

29. $y = \dfrac{1}{5}, -\dfrac{3}{4}$

30. $z = 1, \dfrac{3}{2}$

31. no real solution

32. no real solution

33. no real solution

34. no real solution

35. $m = 1 \pm \sqrt{2}$

36. $m = 3 \pm \sqrt{7}$

37. $p = -\dfrac{3}{4}, -\dfrac{1}{2}$

38. $p = \dfrac{1}{2}, \dfrac{3}{4}$

39. $x = \dfrac{7 \pm \sqrt{129}}{20}$

40. $x = \dfrac{5 \pm \sqrt{137}}{8}$

41. no real solution

42. $x = \dfrac{3 \pm \sqrt{13}}{2}$

43. $z = \dfrac{1 \pm \sqrt{2}}{5}$

44. $z = \dfrac{-2 \pm \sqrt{3}}{3}$

45. see graph

46. see graph

678

---

25. $3x^2 = 1 - 2x$ **26.** $5y^2 = 4 - y$ **27.** $4y^2 = 6y + 1$

28. $6z^2 + 3z + 2 = 0$ **29.** $20y^2 = 3 - 11y$ **30.** $2z^2 = z + 3$

31. $x^2 + x + 2 = 0$ **32.** $k^2 + 2k + 5 = 0$

Use the quadratic formula to solve each quadratic equation. See Example 5.

33. $3p^2 - \dfrac{2}{3}p + 1 = 0$ **34.** $\dfrac{5}{2}p^2 - p + \dfrac{1}{2} = 0$ **35.** $\dfrac{m^2}{2} = m + \dfrac{1}{2}$

36. $\dfrac{m^2}{2} = 3m - 1$ **37.** $4p^2 + \dfrac{3}{2} = -5p$ **38.** $4p^2 + \dfrac{3}{2} = 5p$

39. $5x^2 = \dfrac{7}{2}x + 1$ **40.** $2x^2 = \dfrac{5}{2}x + \dfrac{7}{2}$ **41.** $28x^2 + 5x + \dfrac{11}{4} = 0$

42. $\dfrac{2}{3}x^2 - 2x - \dfrac{2}{3} = 0$ **43.** $5z^2 - 2z = \dfrac{1}{5}$ **44.** $9z^2 + 12z = -1$

REVIEW AND PREVIEW

Graph the following linear equations in two variables. See Section 6.2.

45. $y = -3$ **46.** $x = 4$

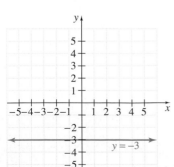

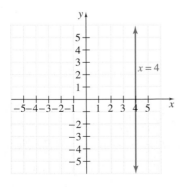

Name _____

47. $y = 3x - 2$

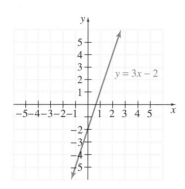

48. $y = 2x + 3$

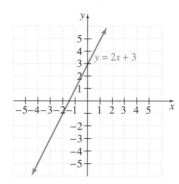

Find the length of the unknown side of each triangle.

49.

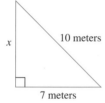

10 meters

x

7 meters

50.

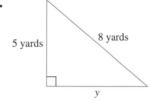

5 yards

8 yards

y

 COMBINING CONCEPTS

Solve each equation using the quadratic formula.

51. $x^2 + 3\sqrt{2}x - 5 = 0$

52. $y^2 - 2\sqrt{5}y - 1 = 0$

53. Explain how the quadratic formula is developed and why it is useful.

Use the quadratic formula and a calculator to solve each equation. Round solutions to the nearest tenth.

54. $x^2 + x = 15$

55. $y^2 - y = 11$

56. $1.2x^2 - 5.2x - 3.9 = 0$

57. $7.3z^2 + 5.4z - 1.1 = 0$

47. see graph

48. see graph

49. $x = \sqrt{51}$ meters

50. $y = \sqrt{39}$ yd

51. $x = \dfrac{-3\sqrt{2} \pm \sqrt{38}}{2}$

52. $y = \sqrt{5} \pm \sqrt{6}$

53. answers may vary

54. $x = -4.4, 3.4$

55. $y = 3.9, -2.9$

56. $x = -0.7, 5.0$

57. $z = -0.9, 0.2$

A rocket is launched from the top of an 80-foot cliff with an initial velocity of 120 feet per second. The height, h, of the rocket after t seconds is given by the equation $h = -16t^2 + 120t + 80$.

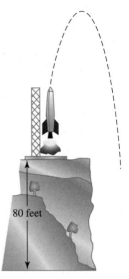

80 feet

58. How long after the rocket is launched will it be 30 feet from the ground? Round to the nearest tenth of a second.

59. How long after the rocket is launched will it strike the ground? Round to the nearest tenth of a second. (*Hint*: The rocket will strike the ground when its height $h = 0$.)

60. The net income y (in millions of dollars) of Goodyear Tire and Rubber Company from 1994 through 1996 is given by the equation $y = 10x^2 + 34x + 567$, where $x = 0$ represents 1994. Assume that this trend continues and predict the year in which Goodyear's net income will be $1295 million. (*Source*: Based on data from the Goodyear Tire and Rubber Company)

61. The sales y (in billions of dollars) of Wal-Mart Stores from 1995 through 1997 is given by the equation $y = -0.5x^2 + 12.5x + 82$, where $x = 0$ represents 1994. Assume that this trend continues and predict the year in which Wal-Mart sales will first be $150 billion. (*Source*: Based on data from Wal-Mart Stores, Inc.)

INTEGRATED REVIEW—SUMMARY ON QUADRATIC EQUATIONS

An important skill in mathematics is learning when to use one technique in favor of another. We now practice this by deciding which method to use when solving quadratic equations. Although both the quadratic formula and completing the square can be used to solve any quadratic equation, the quadratic formula is usually less tedious and thus preferred. The following steps may be used to solve a quadratic equation.

TO SOLVE A QUADRATIC EQUATION

Step 1. If the equation is in the form $ax^2 = c$ or $(ax + b)^2 = c$, use the square root property and solve. If not, go to Step 2.

Step 2. Write the equation in standard form: $ax^2 + bx + c = 0$.

Step 3. Try to solve the equation by the factoring method. If not possible, go to Step 4.

Step 4. Solve the equation by the quadratic formula.

Choose and use a method to solve each equation.

1. $5x^2 - 11x + 2 = 0$ **2.** $5x^2 + 13x - 6 = 0$ **3.** $x^2 - 1 = 2x$

4. $x^2 + 7 = 6x$ **5.** $a^2 = 20$ **6.** $a^2 = 72$

7. $x^2 - x + 4 = 0$ **8.** $x^2 - 2x + 7 = 0$ **9.** $3x^2 - 12x + 12 = 0$

10. $5x^2 - 30x + 45 = 0$ **11.** $9 - 6p + p^2 = 0$ **12.** $49 - 28p + 4p^2 = 0$

13. $4y^2 - 16 = 0$ **14.** $3y^2 - 27 = 0$ **15.** $x^4 - 3x^3 + 2x^2 = 0$

16. $x^3 + 7x^2 + 12x = 0$ **▭17.** $(2z + 5)^2 = 25$ **18.** $(3z - 4)^2 = 16$

ANSWERS

1. $x = \dfrac{1}{5}, 2$

2. $x = -3, \dfrac{2}{5}$

3. $x = 1 \pm \sqrt{2}$

4. $x = 3 \pm \sqrt{2}$

5. $a = \pm 2\sqrt{5}$

6. $a = \pm 6\sqrt{2}$

7. no real solution

8. no real solution

9. $x = 2$

10. $x = 3$

11. $p = 3$

12. $p = \dfrac{7}{2}$

13. $y = \pm 2$

14. $y = \pm 3$

15. $x = 0, 1, 2$

16. $x = 0, -3, -4$

17. $z = -5, 0$

18. $z = 0, \dfrac{8}{3}$

19. $x = \dfrac{3 \pm \sqrt{7}}{5}$

20. $x = \dfrac{3 \pm \sqrt{5}}{2}$

21. $x = \dfrac{3}{2}, -1$

22. $m = \dfrac{2}{5}, -2$

23. $x = \dfrac{5 \pm \sqrt{105}}{20}$

24. $x = \dfrac{-1 \pm \sqrt{3}}{4}$

25. $x = 5, \dfrac{7}{4}$

26. $x = \dfrac{7}{9}, 1$

27. $x = \dfrac{7 \pm 3\sqrt{2}}{5}$

28. $x = \dfrac{5 \pm 5\sqrt{3}}{4}$

29. $z = \dfrac{7 \pm \sqrt{193}}{6}$

30. $z = \dfrac{-7 \pm \sqrt{193}}{12}$

31. $x = 11, -10$

32. $x = -8, 7$

33. $x = -\dfrac{2}{3}, 4$

34. $x = 2, -\dfrac{4}{5}$

35. $x = 0.5, 0.1$

36. $x = 0.3, \ -0.2$

37. $x = \dfrac{11 \pm \sqrt{41}}{20}$

38. $x = \dfrac{11 \pm \sqrt{41}}{40}$

39. $z = \dfrac{4 \pm \sqrt{10}}{2}$

40. $z = \dfrac{5 \pm \sqrt{185}}{4}$

41. answers may vary

Name _____

19. $30x = 25x^2 + 2$ **20.** $12x = 4x^2 + 4$ **21.** $\dfrac{2}{3}m^2 - \dfrac{1}{3}m - 1 = 0$

22. $\dfrac{5}{8}m^2 + m - \dfrac{1}{2} = 0$ **23.** $x^2 - \dfrac{1}{2}x - \dfrac{1}{5} = 0$ **24.** $x^2 + \dfrac{1}{2}x - \dfrac{1}{8} = 0$

25. $4x^2 - 27x + 35 = 0$ **26.** $9x^2 - 16x + 7 = 0$ **27.** $(7 - 5x)^2 = 18$

28. $(5 - 4x)^2 = 75$ **29.** $3z^2 - 7z = 12$ **30.** $6z^2 + 7z = 6$

31. $x = x^2 - 110$ **32.** $x = 56 - x^2$ **33.** $\dfrac{3}{4}x^2 - \dfrac{5}{2}x - 2 = 0$

34. $x^2 - \dfrac{6}{5}x - \dfrac{8}{5} = 0$ **35.** $x^2 - 0.6x + 0.05 = 0$ **36.** $x^2 - 0.1x - 0.06 = 0$

37. $10x^2 - 11x + 2 = 0$ **38.** $20x^2 - 11x + 1 = 0$

39. $\dfrac{1}{2}z^2 - 2z + \dfrac{3}{4} = 0$ **40.** $\dfrac{1}{5}z^2 - \dfrac{1}{2}z - 2 = 0$

41. Explain how you will decide what method to use when solving quadratic equations.

CHAPTER 9 ACTIVITY
MODELING A PHYSICAL SITUATION

Figure 1

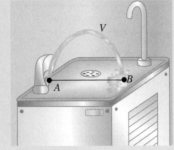

Figure 2

This activity may be completed by working in groups or individually.

Model the physical situation of the parabolic path of water from a water fountain. For simplicity, use the given Figure 2 to investigate the following questions.

Data Table

	x	y
Point A	0	0
Point B		
Point V		

1. Collect data for the x-intercepts of the parabolic path. Let points A and B in Figure 2 be on the x-axis and let the coordinates of point A be $(0, 0)$. Use a ruler to measure the distance between points A and B **on Figure 2** to the nearest even one-tenth centimeter, and use this information to determine the coordinates of point B. Record this data in the data table. (*Hint:* If the distance from A to B measures 8 one-tenth centimeters, then the coordinates of point B are $(8, 0)$.) answers should be close to $(20, 0)$

2. Next, collect data for the vertex V of the parabolic path. What is the relationship between the x-coordinate of the vertex and the x-intercepts found in Question 1? What is the line of symmetry? To locate point V in Figure 2, find the midpoint of the line segment joining points A and B and mark point V on the path of water directly above the midpoint. To approximate the y-coordinate of the vertex, use a ruler to measure its distance from the x-axis to the nearest one-tenth

centimeter. Record this data in the data table. answers may vary; $x = 10$; vertex should be close to $(10, 13)$

3. Plot the points from the data table on a rectangular coordinate system. Sketch the parabola through your points A, B, and V.

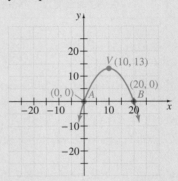

4. Which of the following models best fits the data you collected? Explain your reasoning.   d

   a. $y = 16x + 18$

   b. $y = -13x^2 + 20x$

   c. $y = 0.13x^2 - 2.6x$

   d. $y = -0.13x^2 + 2.6x$

5. (Optional) Enter your data into a graphing calculator and use the quadratic curve-fitting feature to find a model for your data. How does the model compare with your selection from Question 4? answers may vary

CHAPTER 9 HIGHLIGHTS

DEFINITIONS AND CONCEPTS	EXAMPLES

SECTION 9.1 SOLVING QUADRATIC EQUATIONS BY THE SQUARE ROOT PROPERTY

SQUARE ROOT PROPERTY

If $x^2 = a$ for $a \geq 0$, then $x = \sqrt{a}$ or $x = -\sqrt{a}$.

Solve the equation.
$$(x - 1)^2 = 15$$
$$x - 1 = \sqrt{15} \quad \text{or} \quad x - 1 = -\sqrt{15}$$
$$x = 1 + \sqrt{15} \qquad x = 1 - \sqrt{15}$$

SECTION 9.2 SOLVING QUADRATIC EQUATIONS BY COMPLETING THE SQUARE

TO SOLVE A QUADRATIC EQUATION BY COMPLETING THE SQUARE

Step 1. If the coefficient of x^2 is not 1, divide both sides of the equation by the coefficient.

Step 2. Get all terms with variables alone on one side.

Step 3. Complete the square by adding the square of half of the coefficient of x to both sides.

Step 4. Factor the perfect square trinomial.

Step 5. Use the square root property to solve.

Solve $2x^2 + 12x - 10 = 0$ by completing the square.

$$\frac{2x^2}{2} + \frac{12x}{2} - \frac{10}{2} = \frac{0}{2} \quad \text{Divide by 2.}$$
$$x^2 + 6x - 5 = 0 \quad \text{Simplify.}$$
$$x^2 + 6x = 5 \quad \text{Add 5.}$$

The coefficient of x is 6. Half of 6 is 3 and $3^2 = 9$. Add 9 to both sides.

$$x^2 + 6x + 9 = 5 + 9$$
$$(x + 3)^2 = 14 \quad \text{Factor.}$$
$$x + 3 = \sqrt{14} \qquad \text{or} \quad x + 3 = -\sqrt{14}$$
$$x = -3 + \sqrt{14} \qquad x = -3 - \sqrt{14}$$

SECTION 9.3 SOLVING QUADRATIC EQUATIONS BY THE QUADRATIC FORMULA

QUADRATIC FORMULA

If a, b, and c are real numbers and $a \neq 0$, the quadratic equation $ax^2 + bx + c = 0$ has solutions

$$x = \frac{-b \pm \sqrt{b^2 - 4ac}}{2a}$$

Identify a, b, and c in the quadratic equation

$$4x^2 - 6x = 5$$

First, subtract 5 from both sides

$$4x^2 - 6x - 5 = 0$$

$a = 4$, $b = -6$, and $c = -5$.

TO SOLVE A QUADRATIC EQUATION BY THE QUADRATIC FORMULA

Step 1. Write the equation in standard form: $ax^2 + bx + c = 0$.

Step 2. If necessary, clear the equation of fractions.

Step 3. Identify a, b, and c.

Step 4. Replace a, b, and c in the quadratic formula with the identified values, and simplify.

Solve $3x^2 - 2x - 2 = 0$.

In this equation, $a = 3$, $b = -2$, and $c = -2$.

$$x = \frac{-(-2) \pm \sqrt{(-2)^2 - 4(3)(-2)}}{2 \cdot 3}$$
$$= \frac{2 \pm \sqrt{4 - (-24)}}{6}$$
$$= \frac{2 \pm \sqrt{28}}{6} = \frac{2 \pm \sqrt{4 \cdot 7}}{6} = \frac{2 \pm 2\sqrt{7}}{6}$$
$$= \frac{2(1 \pm \sqrt{7})}{2 \cdot 3} = \frac{1 \pm \sqrt{7}}{3}$$

The graph of a quadratic equation $y = ax^2 + bx + c, a \neq 0$, is called a **parabola**. The lowest point on a parabola opening upward or the highest point on a parabola opening downward is called the **vertex**. The vertical line through the vertex is the **line of symmetry**.

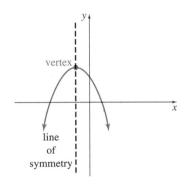

VERTEX FORMULA

The vertex of the parabola $y = ax^2 + bx + c$ has x-coordinate $\dfrac{-b}{2a}$. To find the corresponding y-coordinate, substitute the x-coordinate into the original equation and solve for y.

Graph: $y = 2x^2 - 6x + 4$

The x-coordinate of the vertex is

$$x = \frac{-b}{2a} = \frac{-(-6)}{2(2)} = \frac{6}{4} = \frac{3}{2}$$

The y-coordinate is

$$y = 2\left(\frac{3}{2}\right)^2 - 6\left(\frac{3}{2}\right) + 4 = 2\left(\frac{9}{4}\right) - 9 + 4$$
$$= -\frac{1}{2}$$

The vertex is $\left(\dfrac{3}{2}, -\dfrac{1}{2}\right)$.

The y-intercept is

$$y = 2 \cdot 0^2 - 6 \cdot 0 + 4 = 4$$

The x-intercepts are

$$0 = 2x^2 - 6x + 4$$
$$0 = 2(x - 2)(x - 1)$$
$$x = 2 \quad \text{or} \quad x = 1$$

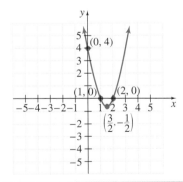

CHAPTER 9 REVIEW

(9.1) *Solve each quadradic equation by factoring.*

1. $(x - 4)(5x + 3) = 0$ $x = 4, -\frac{3}{5}$

2. $(x + 7)(3x + 4) = 0$ $x = -7, -\frac{4}{3}$

3. $3m^2 - 5m = 2$ $m = -\frac{1}{3}, 2$

4. $7m^2 + 2m = 5$ $m = \frac{5}{7}, -1$

5. $6x^3 - 54x = 0$ $x = 0, 3, -3$

6. $2x^2 - 8 = 0$ $x = -2, 2$

Use the square root property to solve each quadratic equation.

7. $x^2 = 36$ $x = 6, -6$

8. $x^2 = 81$ $x = 9, -9$

9. $k^2 = 50$ $k = \pm 5\sqrt{2}$

10. $k^2 = 45$ $k = \pm 3\sqrt{5}$

11. $(x - 11)^2 = 49$ $x = 4, 18$

12. $(x + 3)^2 = 100$ $x = 7, -13$

13. $(4p + 2)^2 = 100$ $p = 2, -3$

14. $(3p + 6)^2 = 81$ $p = 1, -5$

15. The formula $A = P(1 + r)^2$ is used to find the amount of money A in an account after P dollars have been invested in the account paying r annual interest rate for 2 years. Find the interest rate r if $1000 invested grows to $1690 in 2 years.   30%

16. Use the formula in Exercise 15 and find the interest rate r if $1000 invested grows to $1210 in 2 years.   10%

(9.2) *Solve each quadratic equation by completing the square.*

17. $x^2 + 4x = 1$ $x = -2 \pm \sqrt{5}$

18. $x^2 - 8x = 3$ $x = 4 \pm \sqrt{19}$

19. $x^2 - 6x + 7 = 0$ $x = 3 \pm \sqrt{2}$

20. $x^2 + 6x + 7 = 0$ $x = -3 \pm \sqrt{2}$

21. $2y^2 + y - 1 = 0$ $y = \frac{1}{2}, -1$

22. $y^2 + 3y - 1 = 0$ $y = \frac{-3 \pm \sqrt{13}}{2}$

(9.3) *Use the quadratic formula to solve each quadratic equation.*

23. $x^2 - 10x + 7 = 0$ $x = 5 \pm 3\sqrt{2}$

24. $x^2 + 4x - 7 = 0$ $x = -2 \pm \sqrt{11}$

25. $2x^2 + x - 1 = 0$ $x = \frac{1}{2}, -1$

26. $x^2 + 3x - 1 = 0$ $x = \frac{-3 \pm \sqrt{13}}{2}$

27. $9x^2 + 30x + 25 = 0$ $x = -\frac{5}{3}$

28. $16x^2 - 72x + 81 = 0$ $x = \frac{9}{4}$

29. $15x^2 + 2 = 11x$ $x = \frac{2}{5}, \frac{1}{3}$

30. $15x^2 + 2 = 13x$ $x = \frac{2}{3}, \frac{1}{5}$

31. $2x^2 + x + 5 = 0$ no real solution

32. $7x^2 - 3x + 1 = 0$ no real solution

33. If a line segment AB is divided by a point C into two segments, AC and CB, such that the proportion $\frac{AB}{AC} = \frac{AC}{CB}$ is true, this ratio $\frac{AB}{AC}\left(\text{or } \frac{AC}{CB}\right)$ is called the golden ratio. If AC is 1 unit, find the length of AB. (*Hint:* Let x be the unknown length as shown and substitute into the given proportion.)

$\frac{1 + \sqrt{5}}{2}$

(9.4) *Graph each quadratic equation by finding and plotting intercept points.*

34. $y = 3x^2$

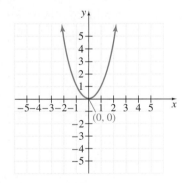

35. $y = -\frac{1}{2}x^2$

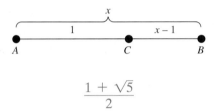

Graph each quadratic equation. Label the vertex and the intercept points with their coordinates.

36. $y = x^2 - 25$

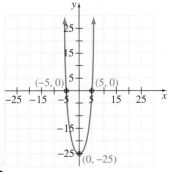

37. $y = x^2 - 36$

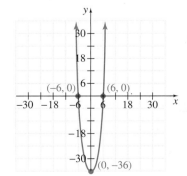

38. $y = x^2 + 3$

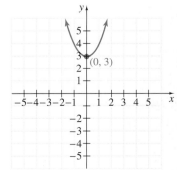

FROM 2284 FAIR OAKS ROAD
DECATUR GA 30033

SHIP TO
CHRISTINA JOINER, ADULT LIT
LANIER TECHNICAL INSTITUTE
2990 LANDRUM EDUC DR,
P O BOX 58

OAKWOOD GA 30566

BLK RT BND PRINTED MTR
U.S. POSTAGE PAID
WEST NYACK, NY
PERMIT NO. 76

PPL#:14543
LBL#: 1232

CARRIER

STD B BMC (ATLANTA)

L706

INFORMATION ONLY - NOT FOR DELIVERY PURPOSES

SENT WITH THE COMPLIMENTS OF YOUR PH REPRESENTATIVE~C

HORTON, CAROLE 2284 FAIR OAKS ROAD

TELEPHONE (404) 327-9743 DECATUR GA 30033

FPK AREA

013862467-4

QTY : 1

ORD#:FC19942084

PO #:0181202333

A/I/E INTRODCTRY ALGEBR

MARTIN-GAY

3835519298

39. $y = x^2 + 8$

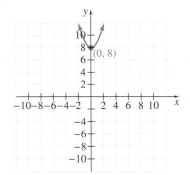

40. $y = -4x^2 + 8$

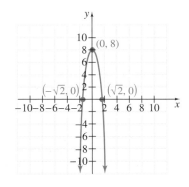

41. $y = -3x^2 + 9$

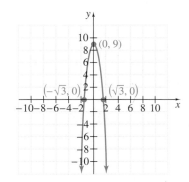

42. $y = x^2 + 3x - 10$

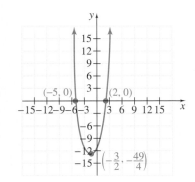

43. $y = x^2 + 3x - 4$

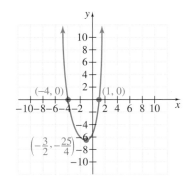

44. $y = -x^2 - 5x - 6$

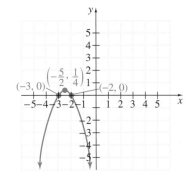

45. $y = -x^2 + 4x + 8$

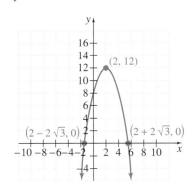

46. $y = 2x^2 - 11x - 6$

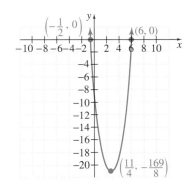

47. $y = 3x^2 - x - 2$

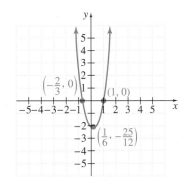

Name _____

Quadratic equations in the form $y = ax^2 + bx + c$ are graphed below. Determine the number of real solutions for the related equation $0 = ax^2 + bx + c$ from each graph. List the solutions.

48. 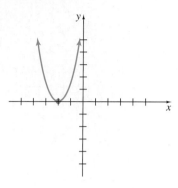 $x = -2$

49. 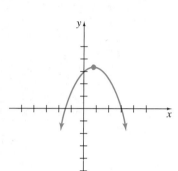 $x = -\dfrac{3}{2}, 3$

50. 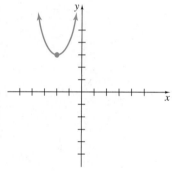   no real solution

51. 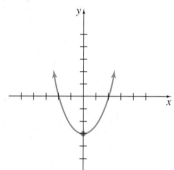 $x = -2, 2$

Match each quadratic equation with its graph.

52. $y = 2x^2$ A

53. $y = -x^2$ D

54. $y = x^2 + 4x + 4$ B

55. $y = x^2 + 5x + 4$ C

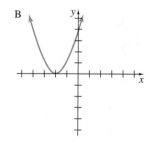 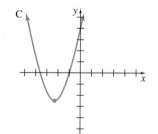 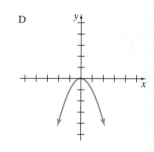

A   B   C   D

Name _____ **Section** _____ **Date** _____

CHAPTER 9 TEST

Solve by factoring.

1. $2x^2 - 11x = 21$

2. $x^4 + x^3 - 2x^2 = 0$

Solve using the square root property.

3. $5k^2 = 80$

4. $(3m - 5)^2 = 8$

Solve by completing the square.

5. $x^2 - 26x + 160 = 0$

6. $5x^2 + 9x = 2$

Solve using the quadratic formula.

7. $x^2 - 3x - 10 = 0$

8. $p^2 - \dfrac{5}{3}p - \dfrac{1}{3} = 0$

Solve by the most appropriate method.

9. $(3x - 5)(x + 2) = -6$ **10.** $(3x - 1)^2 = 16$ **11.** $3x^2 - 7x - 2 = 0$

12. $x^2 - 4x - 5 = 0$ **13.** $3x^2 - 7x + 2 = 0$ **14.** $2x^2 - 6x + 1 = 0$

15. The height of a triangle is 4 times the length of the base. The area of the triangle is 18 square feet. Find the height and base of the triangle.

4x

x

ANSWERS

1. $x = -\dfrac{3}{2}, 7$

2. $x = -2, 0, 1$

3. $k = \pm 4$

4. $m = \dfrac{5 \pm 2\sqrt{2}}{3}$

5. $x = 10, 16$

6. $x = -2, \dfrac{1}{5}$

7. $x = -2, 5$

8. $p = \dfrac{5 \pm \sqrt{37}}{6}$

9. $x = 1, -\dfrac{4}{3}$

10. $x = -1, \dfrac{5}{3}$

11. $x = \dfrac{7 \pm \sqrt{73}}{6}$

12. $x = -1, 5$

13. $x = 2, \dfrac{1}{3}$

14. $x = \dfrac{3 \pm \sqrt{7}}{2}$

15. base $= 3$ ft; height $= 12$ ft

Name _____

Graph each quadratic equation. Label the vertex and the intercept points with their coordinates.

16. $y = -5x^2$

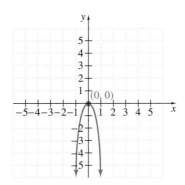

17. $y = x^2 - 4$

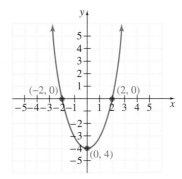

18. $y = x^2 - 7x + 10$

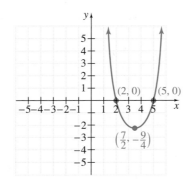

19. $y = 2x^2 + 4x - 1$

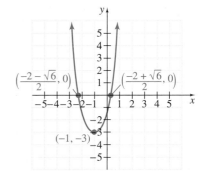

20. The number of diagonals d that a polygon with n sides has is given by the formula

$$d = \frac{n^2 - 3n}{2}$$

Find the number of sides of a polygon if it has 9 diagonals.

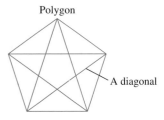

Polygon

A diagonal

Name _____ **Section** _____ **Date** _____

Cumulative Review

1. Solve: $y + 0.6 = -1.0$

2. Solve: $8(2 - t) = -5t$

3. A local cellular phone company charges Elaine Chapoton $50 per month and $0.36 per minute of phone use in her usage category. If Elaine was charged $99.68 for a month's cellular phone use, determine the number of whole minutes of phone use.

Simplify the following expressions.

4. 3^0

5. $(5x^3y^2)^0$

6. -4^0

7. Multiply: $(3y + 2)^2$

8. Divide $x^2 + 7x + 12$ by $x + 3$ using long division

9. Factor: $r^2 - r - 42$

10. Factor: $10x^2 - 13xy - 3y^2$

11. Factor $8x^2 - 14x + 5$ by grouping.

12. Factor the difference of squares.
   a. $4x^3 - 49x$

   b. $162x^4 - 2$

13. Solve: $(5x - 1)(2x^2 + 15x + 18) = 0$

14. Simplify: $\dfrac{x^2 + 8x + 7}{x^2 - 4x - 5}$

15. The quotient of a number and 6 minus $\dfrac{5}{3}$ is the quotient of the number and 2. Find the number.

16. Complete the table for the equation $y = 3x$.

	x	y
a.	-1	
b.		0
c.		-9

Name _____

17. Determine whether each pair of lines is parallel, perpendicular, or neither.

  a. $y = -\dfrac{1}{5}x + 1$

  $2x + 10y = 3$

  b. $x + y = 3$

  $-x + y = 4$

  c. $3x + y = 5$

  $2x + 3y = 6$

18. Which of the following relations are also functions?

  a. $\{(-1, 1), (2, 3), (7, 3), (8, 6)\}$

  b. $\{(0, -2), (1, 5), (0, 3), (7, 7)\}$

19. Solve the system

  $\begin{cases} 2x + y = 10 \\ x = y + 2 \end{cases}$

20. Solve the system

  $\begin{cases} 2x - y = 7 \\ 8x - 4y = 1 \end{cases}$

Find each square root.

21. $\sqrt{36}$

22. $\sqrt{\dfrac{9}{100}}$

23. Rationalize the denominator of $\dfrac{2}{1 + \sqrt{3}}$.

24. Use the square root property to solve $(x - 3)^2 = 16$.

25. Solve $\dfrac{1}{2}x^2 - x = 2$ by using the quadratic formula.

APPENDIX A

Review of Angles, Lines, and Special Triangles

The word **geometry** is formed from the Greek words, **geo**, meaning earth, and **metron**, meaning measure. Geometry literally means to measure the earth.

This appendix contains a review of some basic geometric ideas. It will be assumed that fundamental ideas of geometry such as point, line, ray, and angle are known. In this appendix, the notation $\angle 1$ is read "angle 1" and the notation $m\angle 1$ is read "the measure of angle 1."

We first review types of angles.

ANGLES

An angle whose measure is greater than 0° but less than 90° is called an **acute angle**.

A **right angle** is an angle whose measure is 90°. A right angle can be indicated by a square drawn at the vertex of the angle, as shown below.

An angle whose measure is greater than 90° but less than 180° is called an **obtuse angle**.

An angle whose measure is 180° is called a **straight angle**.

Two angles are said to be **complementary** if the sum of their measures is 90°. Each angle is called the **complement** of the other.

Two angles are said to be **supplementary** if the sum of their measures is 180°. Each angle is called the **supplement** of the other.

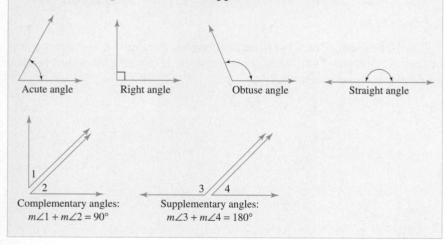

Example 1 If an angle measures 28°, find its complement.

Solution: Two angles are complementary if the sum of their measures is 90°. The complement of a 28° angle is an angle whose measure is $90° - 28° = 62°$. To check, notice that $28° + 62° = 90°$. ▬▬▬▬

Plane is an undefined term that we will describe. A plane can be thought of as a flat surface with infinite length and width, but no thickness. A plane is two dimensional. The arrows in the following diagram indicate that a plane extends indefinitely and has no boundaries.

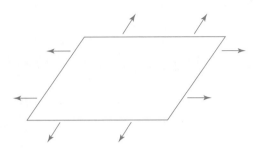

Figures that lie on a plane are called **plane figures**. (See the description of common plane figures in Appendix B.) Lines that lie in the same plane are called **coplanar**.

LINES

Two lines are **parallel** if they lie in the same plane but never meet. **Intersecting lines** meet or cross in one point.

Two lines that form right angles when they intersect are said to be **perpendicular**.

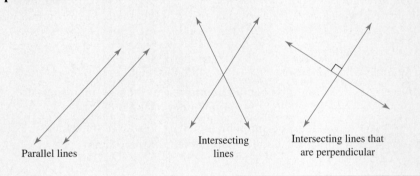

Parallel lines　　　　Intersecting lines　　　Intersecting lines that are perpendicular

---

Two intersecting lines form **vertical angles**. Angles 1 and 3 are vertical angles. Also angles 2 and 4 are vertical angles. It can be shown that **vertical angles have equal measures**.

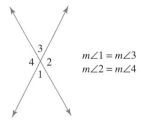

$$m\angle 1 = m\angle 3$$
$$m\angle 2 = m\angle 4$$

Adjacent angles have the same vertex and share a side. Angles 1 and 2 are adjacent angles. Other pairs of adjacent angles are angles 2 and 4, angles 3 and 4, and angles 3 and 1.

A **transversal** is a line that intersects two or more lines in the same plane. Line l is a transversal that intersects lines m and n. The eight angles formed are numbered and certain pairs of these angles are given special names.

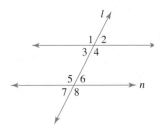

Corresponding angles: ∠1 and ∠5, ∠3 and ∠7, ∠2 and ∠6, and ∠4, and ∠8.
Exterior angles: ∠1, ∠2, ∠7, and ∠8.
Interior angles: ∠3, ∠4, ∠5, and ∠6.
Alternate interior angles: ∠3 and ∠6, ∠4 and ∠5.

These angles and parallel lines are related in the following manner.

PARALLEL LINES CUT BY A TRANSVERSAL

1. If two parallel lines are cut by a transversal, then
   a. corresponding angles are equal and
   b. alternate interior angles are equal.
2. If corresponding angles formed by two lines and a transversal are equal, then the lines are parallel.
3. If alternate interior angles formed by two lines and a transversal are equal, then the lines are parallel.

Example 2   Given that lines *m* and *n* are parallel and that the measure of angle 1 is 100°, find the measures of angles 2, 3, and 4.

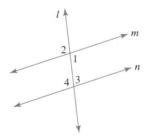

Solution:

$m\angle 2 = 100°$ since angles 1 and 2 are vertical angles.

$m\angle 4 = 100°$ since angles 1 and 4 are alternate interior angles.

$m\angle 3 = 180° - 100° = 80°$ since angles 4 and 3 are supplementary angles.

A **polygon** is the union of three or more coplanar line segments that intersect each other only at each end point, with each end point shared by exactly two segments.

A **triangle** is a polygon with three sides. The sum of the measures of the three angles of a triangle is 180°. In the following figure, $m\angle 1 + m\angle 2 + m\angle 3 = 180°$.

Example 3   Find the measure of the third angle of the triangle shown.

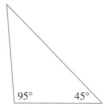

Solution: The sum of the measures of the angles of a triangle is 180°. Since one angle measures 45° and the other angle measures 95°, the third angle measures $180° - 45° - 95° = 40°$. ▬▬▬

Two triangles are **congruent** if they have the same size and the same shape. In congruent triangles, the measures of corresponding angles are equal and the lengths of corresponding sides are equal. The following triangles are congruent.

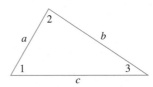 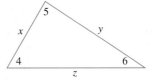

Corresponding angles are equal: $m\angle 1 = m\angle 4$, $m\angle 2 = m\angle 5$, and $m\angle 3 = m\angle 6$. Also, lengths of corresponding sides are equal: $a = x$, $b = y$, and $c = z$.

Any one of the following may be used to determine whether two triangles are congruent.

CONGRUENT TRIANGLES

1. If the measures of two angles of a triangle equal the measures of two angles of another triangle and the lengths of the sides between each pair of angles are equal, the triangles are congruent.

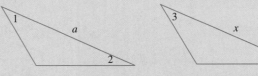

$$m\angle 1 = m\angle 3$$
$$m\angle 2 = m\angle 4$$
and
$$a = x$$

2. If the lengths of the three sides of a triangle equal the lengths of corresponding sides of another triangle, the triangles are congruent.

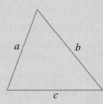

$$a = x$$
$$b = y$$
and
$$c = z$$

3. If the lengths of two sides of a triangle equal the lengths of corresponding sides of another triangle, and the measures of the angles between each pair of sides are equal, the triangles are congruent.

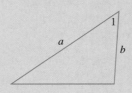 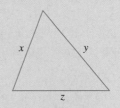

$$a = x$$
$$b = y$$
and
$$m\angle 1 = m\angle 2$$

Two triangles are **similar** if they have the same shape but not necessarily the same size. In similar triangles, the measures of corresponding angles are equal and corresponding sides are in proportion. The following triangles are similar. (All similar triangles drawn in this appendix will be oriented the same.)

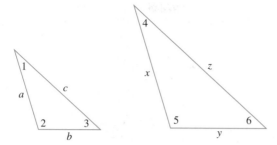

Corresponding angles are equal: $m\angle 1 = m\angle 4$, $m\angle 2 = m\angle 5$, and

$m\angle 3 = m\angle 6$. Also, corresponding sides are proportional: $\dfrac{a}{x} = \dfrac{b}{y} = \dfrac{c}{z}$.

Any one of the following may be used to determine whether two triangles are similar.

SIMILAR TRIANGLES

1. If the measures of two angles of a triangle equal the measures of two angles of another triangle, the triangles are similar.

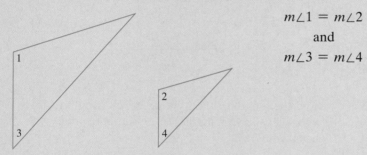

$$m\angle 1 = m\angle 2$$
$$\text{and}$$
$$m\angle 3 = m\angle 4$$

2. If three sides of one triangle are proportional to three sides of another triangle, the triangles are similar.

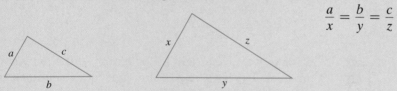

$$\frac{a}{x} = \frac{b}{y} = \frac{c}{z}$$

3. If two sides of a triangle are proportional to two sides of another triangle and the measures of the included angles are equal, the triangles are similar.

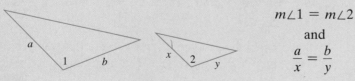

$$m\angle 1 = m\angle 2$$
$$\text{and}$$
$$\frac{a}{x} = \frac{b}{y}$$

Example 4   Given that the following triangles are similar, find the missing length x.

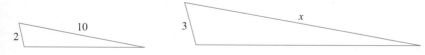

Solution: Since the triangles are similar, corresponding sides are in proportion. Thus, $\frac{2}{3} = \frac{10}{x}$. To solve this equation for x, we cross multiply.

$$\frac{2}{3} = \frac{10}{x}$$
$$2x = 30$$
$$x = 15$$

The missing length is 15 units.   ▬▬▬

A **right triangle** contains a right angle. The side opposite the right angle is called the **hypotenuse**, and the other two sides are called the **legs**. The **Pythagorean theorem** gives a formula that relates the lengths of the three sides of a right triangle.

THE PYTHAGOREAN THEOREM

If a and b are the lengths of the legs of a right triangle, and c is the length of the hypotenuse, then $a^2 + b^2 = c^2$.

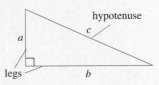

Example 5   Find the length of the hypotenuse of a right triangle whose legs have lengths of 3 centimeters and 4 centimeters.

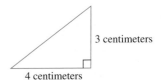

Solution: Because we have a right triangle, we use the Pythagorean theorem. The legs are 3 centimeters and 4 centimeters, so let $a = 3$ and $b = 4$ in the formula.

$$a^2 + b^2 = c^2$$
$$3^2 + 4^2 = c^2$$
$$9 + 16 = c^2$$
$$25 = c^2$$

Since c represents a length, we assume that c is positive. Thus, if c^2 is 25, c must be 5. The hypotenuse has a length of 5 centimeters.   ▬▬▬

Name _____ **Section** _____ **Date** _____

APPENDIX A EXERCISE SET

Find the complement of each angle. See Example 1.

1. 19°　　　　　　**2.** 65°　　　　　　**3.** 70.8°

4. $45\frac{2}{3}°$　　　　**5.** $11\frac{1}{4}°$　　　　**6.** 19.6°

Find the supplement of each angle.

7. 150°　　　　　**8.** 90°　　　　　**9.** 30.2°

10. 81.9°　　　　**11.** $79\frac{1}{2}°$　　　**12.** $165\frac{8}{9}°$

13. If lines *m* and *n* are parallel, find the measures of angles 1 through 7. See Example 2.

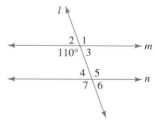

$m\angle 1 = m\angle 5 = m\angle 7 = 110°$,
$m\angle 2 = m\angle 3 = m\angle 4 = m\angle 6 = 70°$

14. If lines *m* and *n* are parallel, find the measures of angles 1 through 5. See Example 2.

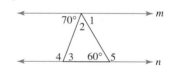

$m\angle 1 = 60°$, $m\angle 2 = 50°$, $m\angle 3 = 70°$,
$m\angle 4 = 110°$, $m\angle 5 = 120°$

In each of the following, the measures of two angles of a triangle are given. Find the measure of the third angle. See Example 3.

15. 11°, 79°　　　　**16.** 8°, 102°　　　　**17.** 25°, 65°

18. 44°, 19°　　　　**19.** 30°, 60°　　　　**20.** 67°, 23°

ANSWERS

1. 71°

2. 25°

3. 19.2°

4. $44\frac{1}{3}°$

5. $78\frac{3}{4}°$

6. 70.4°

7. 30°

8. 90°

9. 149.8°

10. 98.1°

11. $100\frac{1}{2}°$

12. $14\frac{1}{9}°$

13. answer below problem

14. answer below problem

15. 90°

16. 70°

17. 90°

18. 117°

19. 90°

20. 90°

21. 45°, 90°

22. 30°, 90°

23. 73°, 90°

24. 60°, 90°

25. $50\frac{1}{4}°$, 90°

26. 17.4°, 90°

27. $x = 6$

28. $x = 6$

29. $x = 4.5$

30. $x = 36$

31. 10

32. 13

33. 12

34. 16

Name _____

In each of the following, the measure of one angle of a right triangle is given. Find the measures of the other two angles.

21. 45°

22. 60°

23. 17°

24. 30°

25. $39\frac{3}{4}°$

26. 72.6°

Given that each of the following pairs of triangles is similar, find the missing length x. See Example 4.

27.

28.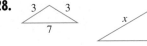

29.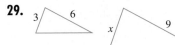

30.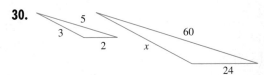

Use the Pythagorean theorem to find the missing lengths in the right triangles. See Example 5.

31.

32.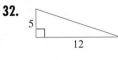

33.

34.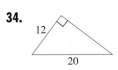

Appendix B

Review of Geometric Figures

PLANE FIGURES HAVE LENGTH AND WIDTH BUT NO THICKNESS OR DEPTH.		
Name	Description	Figure
POLYGON	Union of three or more coplanar line segments that intersect with each other only at each end point, with each end point shared by two segments.	
TRIANGLE	Polygon with three sides (sum of measures of three angles is 180°).	
SCALENE TRIANGLE	Triangle with no sides of equal length.	
ISOSCELES TRIANGLE	Triangle with two sides of equal length.	
EQUILATERAL TRIANGLE	Triangle with all sides of equal length.	
RIGHT TRIANGLE	Triangle that contains a right angle.	leg, hypotenuse, leg
QUADRILATERAL	Polygon with four sides (sum of measures of four angles is 360°).	
TRAPEZOID	Quadrilateral with exactly one pair of opposite sides parallel.	base, leg, parallel sides, leg, base
ISOSCELES TRAPEZOID	Trapezoid with legs of equal length.	
PARALLELOGRAM	Quadrilateral with both pairs of opposite sides parallel.	
RHOMBUS	Parallelogram with all sides of equal length.	

Name	Description	Figure
RECTANGLE	Parallelogram with four right angles.	
SQUARE	Rectangle with all sides of equal length.	
CIRCLE	All points in a plane the same distance from a fixed point called the **center**.	

SOLID FIGURES HAVE LENGTH, WIDTH, AND HEIGHT OR DEPTH.		
Name	Description	Figure
RECTANGULAR SOLID	A solid with six sides, all of which are rectangles.	
CUBE	A rectangular solid whose six sides are squares.	
SPHERE	All points the same distance from a fixed point, called the **center**.	
RIGHT CIRCULAR CYLINDER	A cylinder consisting of two circular bases that are perpendicular to its altitude.	
RIGHT CIRCULAR CONE	A cone with a circular base that is perpendicular to its altitude.	

Appendix C

Table of Squares and Square Roots

n	n^2	$\sqrt{n}$	n	n^2	$\sqrt{n}$
1	1	1.000	51	2,601	7.141
2	4	1.414	52	2,704	7.211
3	9	1.732	53	2,809	7.280
4	16	2.000	54	2,916	7.348
5	25	2.236	55	3,025	7.416
6	36	2.449	56	3,136	7.483
7	49	2.646	57	3,249	7.550
8	64	2.828	58	3,364	7.616
9	81	3.000	59	3,481	7.681
10	100	3.162	60	3,600	7.746
11	121	3.317	61	3,721	7.810
12	144	3.464	62	3,844	7.874
13	169	3.606	63	3,969	7.937
14	196	3.742	64	4,096	8.000
15	225	3.873	65	4,225	8.062
16	256	4.000	66	4,356	8.124
17	289	4.123	67	4,489	8.185
18	324	4.243	68	4,624	8.246
19	361	4.359	69	4,761	8.307
20	400	4.472	70	4,900	8.367
21	441	4.583	71	5,041	8.426
22	484	4.690	72	5,184	8.485
23	529	4.796	73	5,329	8.544
24	576	4.899	74	5,476	8.602
25	625	5.000	75	5,625	8.660
26	676	5.099	76	5,776	8.718
27	729	5.196	77	5,929	8.775
28	784	5.292	78	6,084	8.832
29	841	5.385	79	6,241	8.888
30	900	5.477	80	6,400	8.944
31	961	5.568	81	6,561	9.000
32	1,024	5.657	82	6,724	9.055
33	1,089	5.745	83	6,889	9.110
34	1,156	5.831	84	7,056	9.165
35	1,225	5.916	85	7,225	9.220
36	1,296	6.000	86	7,396	9.274
37	1,369	6.083	87	7,569	9.327
38	1,444	6.164	88	7,744	9.381
39	1,521	6.245	89	7,921	9.434
40	1,600	6.325	90	8,100	9.487
41	1,681	6.403	91	8,281	9.539
42	1,764	6.481	92	8,464	9.592
43	1,849	6.557	93	8,649	9.644
44	1,936	6.633	94	8,836	9.695
45	2,025	6.708	95	9,025	9.747
46	2,116	6.782	96	9,216	9.798
47	2,209	6.856	97	9,409	9.849
48	2,304	6.928	98	9,604	9.899
49	2,401	7.000	99	9,801	9.950
50	2,500	7.071	100	10,000	10.000

APPENDIX D

Mean, Median, and Mode

It is sometimes desirable to be able to describe a set of data, or a set of numbers, by a single "middle" number. Three such **measures of central tendency** are the mean, the median, and the mode.

The most common measure of central tendency is the mean (sometimes called the arithmetic mean or the average). The **mean** of a set of data items, denoted by $\overline{x}$, is the sum of the items divided by the number of items.

Example 1   Seven students in a psychology class conducted an experiment on mazes. Each student was given a pencil and asked to successfully complete the same maze. The timed results are below.

Student	Ann	Thanh	Carlos	Jesse	Melinda	Ramzi	Dayni
Time (seconds)	13.2	11.8	10.7	16.2	15.9	13.8	18.5

  a. Who completed the maze in the shortest time? Who completed the maze in the longest time?
  b. Find the mean.
  c. How many students took longer than the mean time? How many students took shorter than the mean time?

Solution: **a.** Carlos completed the maze in 10.7 seconds, the shortest time. Dayni completed the maze in 18.5 seconds, the longest time.
  b. To find the mean, $\overline{x}$, find the sum of the data items and divide by 7, the number of items.

$$\overline{x} = \frac{13.2 + 11.8 + 10.7 + 16.2 + 15.9 + 13.8 + 18.5}{7} = \frac{100.1}{7} = 14.3$$

  c. Three students, Jesse, Melinda, and Dayni, had times longer than the mean time. Four students, Ann, Thanh, Carlos, and Ramzi, had times shorter than the mean time. ▬▬▬

Two other measures of central tendency are the median and the mode.

The **median** of an ordered set of numbers is the middle number. If the number of items is even, the median is the mean of the two middle numbers. The **mode** of a set of numbers is the number that occurs most often. It is possible for a data set to have no mode or more than one mode.

Example 2   Find the median and the mode of the following set of numbers. These numbers were high temperatures for fourteen consecutive days in a city in Montana.

76, 80, 85, 86, 89, 87, 82, 77, 76, 79, 82, 89, 89, 92

Solution: First, write the numbers in order.

76, 76, 77, 79, 80, 82, 82, 85, 86, 87, 89, 89, 89, 92

two middle numbers   mode

Since there are an even number of items, the median is the mean of the two middle numbers.

$$\text{median} = \frac{82 + 85}{2} = 83.5$$

The mode is 89, since 89 occurs most often. ▬▬▬

Name _____ **Section** _____ **Date** _____

APPENDIX D EXERCISE SET

For each of the following data sets, find the mean, the median, and the mode. If necessary, round the mean to one decimal place.

1. 21, 28, 16, 42, 38

2. 42, 35, 36, 40, 50

3. 7.6, 8.2, 8.2, 9.6, 5.7, 9.1

4. 4.9, 7.1, 6.8, 6.8, 5.3, 4.9

5. 0.2, 0.3, 0.5, 0.6, 0.6, 0.9, 0.2, 0.7, 1.1

6. 0.6, 0.6, 0.8, 0.4, 0.5, 0.3, 0.7, 0.8, 0.1

7. 231, 543, 601, 293, 588, 109, 334, 268

8. 451, 356, 478, 776, 892, 500, 467, 780

The ten tallest buildings in the United States are listed below. Use this table for Exercises 9–12.

Building	Height (Feet)
Sears Tower, Chicago, IL	1454
One World Trade Center (1972), New York, NY	1368
One World Trade Center (1973), New York, NY	1362
Empire State, New York, NY	1250
Amoco, Chicago, IL	1136
John Hancock Center, Chicago, IL	1127
First Interstate World Center, Los Angeles, CA	1107
Chrysler, New York, NY	1046
NationsBank Tower, Atlanta, GA	1023
Texas Commerce Tower, Houston, TX	1002

9. Find the mean height for the five tallest buildings.

10. Find the median height for the five tallest buildings.

11. Find the median height for the ten tallest buildings.

12. Find the mean height for the ten tallest buildings.

ANSWERS

1. mean: 29, median: 28, no mode

2. mean: 40.6, median: 40, no mode

3. mean: 8.1, median: 8.2, mode: 8.2

4. mean: 6.0, median: 6.05, mode: 6.8 and 4.9

5. mean: 0.6, median: 0.6, mode: 0.2 and 0.6

6. mean: 0.5, median: 0.6, mode: 0.6 and 0.8

7. mean: 370.9, median: 313.5, no mode

8. mean: 587.5, median: 489, no mode

9. 1314 feet

10. 1362 feet

11. 1131.5 feet

12. 1187.5 feet

A-15

13. 6.8

14. 6.95

15. 6.9

During an experiment, the following times (in seconds) were recorded: 7.8, 6.9, 7.5, 4.7, 6.9, 7.0.

13. Find the mean. Round to the nearest tenth.

14. Find the median.

15. Find the mode.

16. 84.67

17. 85.5

In a mathematics class, the following test scores were recorded for a student: 86, 95, 91, 74, 77, 85.

16. Find the mean. Round to the nearest hundredth.

17. Find the median.

18. Find the mode.

18. no mode

The following pulse rates were recorded for a group of fifteen students: 78, 80, 66, 68, 71, 64, 82, 71, 70, 65, 70, 75, 77, 86, 72.

19. Find the mean.

20. Find the median.

21. Find the mode.

19. 73

20. 71

22. How many rates were higher than the mean?

23. How many rates were lower than the mean?

21. 70 and 71

22. 6

24. Have each student in your algebra class take his/her pulse rate. Record the data and find the mean, the median, and the mode.

23. 9

24. answers may vary

Find the missing numbers in each set of numbers. (These numbers are not necessarily in numerical order.)

25. ___, ___, 16, 18, ___
The mode is 21.
The median is 20.

26. ___, ___, ___, ___, 40
The mode is 35.
The median is 37.
The mean is 38.

25. 21, 21, 24

26. 35, 35, 37, 43

Answers to Selected Exercises

Chapter R Prealgebra Review

Chapter R Pretest

1. 1, 2, 3, 4, 6, 12; R.1A **2.** $2 \cdot 3 \cdot 5 \cdot 5$; R.1B **3.** 280; R.1C **4.** $\dfrac{35}{40}$; R.2A **5.** $\dfrac{3}{5}$; R.2B **6.** $\dfrac{12}{25}$; R.2B **7.** $\dfrac{1}{12}$; R.2C

8. $\dfrac{13}{12}$; R.2D **9.** $\dfrac{30}{49}$; R.2C **10.** $\dfrac{1}{9}$; R.2D **11.** 78.53; R.3B **12.** 5.33; R.3B **13.** 2.432; R.3B **14.** 34.9; R.3B

15. $\dfrac{716}{100}$; R.3A **16.** 0.1875; R.3D **17.** $0.8\overline{3}$; R.3D **18.** 78.6; R.3C **19.** 78.62; R.3C **20.** 0.806; R.3E **21.** 30%; R.3E

Section R.1

1. 1, 3, 9 **3.** 1, 2, 3, 4, 6, 8, 12, 24 **5.** 1, 2, 3, 6, 7, 14, 21, 42 **7.** 1, 2, 4, 5, 8, 10, 16, 20, 40, 80 **9.** prime **11.** composite
13. prime **15.** composite **17.** $2 \cdot 3 \cdot 3$ **19.** $2 \cdot 2 \cdot 5$ **21.** $2 \cdot 2 \cdot 2 \cdot 7$ **23.** $2 \cdot 2 \cdot 3 \cdot 5 \cdot 5$ **25.** $3 \cdot 3 \cdot 3 \cdot 3$
27. $2 \cdot 2 \cdot 3 \cdot 7 \cdot 7$ **29.** 42 **31.** 12 **33.** 60 **35.** 35 **37.** 12 **39.** 60 **41.** 350 **43.** 72 **45.** 60 **47.** 30 **49.** 360
51. 24 **53.** 2520 **55.** every 35 days **57.** every 140 days

Section R.2

1. $\dfrac{21}{30}$ **3.** $\dfrac{4}{18}$ **5.** $\dfrac{16}{20}$ **7.** $\dfrac{1}{2}$ **9.** $\dfrac{2}{3}$ **11.** $\dfrac{3}{7}$ **13.** 1 **15.** 5 **17.** $\dfrac{3}{5}$ **19.** $\dfrac{4}{5}$ **21.** $\dfrac{11}{8}$ **23.** $\dfrac{30}{61}$ **25.** $\dfrac{3}{8}$ **27.** $\dfrac{1}{2}$

29. $\dfrac{6}{7}$ **31.** 15 **33.** $\dfrac{1}{6}$ **35.** $\dfrac{3}{80}$ **37.** 1 **39.** $\dfrac{3}{5}$ **41.** $\dfrac{9}{35}$ **43.** $\dfrac{1}{3}$ **45.** $\dfrac{23}{21}$ **47.** $\dfrac{65}{21}$ **49.** $\dfrac{5}{7}$ **51.** $\dfrac{5}{66}$ **53.** $\dfrac{7}{5}$

55. $\dfrac{17}{18}$ **57.** $\dfrac{1}{5}$ **59.** $\dfrac{3}{8}$ **61.** $\dfrac{3}{4}$ years **63. a.** $\dfrac{1}{5}$ **b.** $\dfrac{13}{50}$ **c.** $\dfrac{1}{4}$ **d.** $\dfrac{11}{20}$ **65.** answers may vary

Section R.3

1. $\dfrac{6}{10}$ **3.** $\dfrac{186}{100}$ **5.** $\dfrac{114}{1000}$ **7.** $\dfrac{1231}{10}$ **9.** 6.83 **11.** 27.0578 **13.** 6.5 **15.** 15.22 **17.** 56.431 **19.** 598.23 **21.** 0.12
23. 67.5 **25.** 43.274 **27.** 84.97593 **29.** 0.094 **31.** 70 **33.** 5.8 **35.** 840 **37.** 0.6 **39.** 0.23 **41.** 0.594
43. 98,207.2 **45.** 12.35 **47.** 0.75 **49.** $0.\overline{3} \approx 0.333$ **51.** 0.4375 **53.** $0.\overline{54} \approx 0.55$ **55.** 0.28 **57.** 0.031 **59.** 1.35
61. 0.9655 **63.** 0.61 **65.** 68% **67.** 87.6% **69.** 100% **71.** 50% **73.** 6.98 years **75.** 64%

Chapter R Review

1. $2 \cdot 3 \cdot 7$ **2.** $2 \cdot 2 \cdot 2 \cdot 2 \cdot 2 \cdot 5 \cdot 5$ **3.** 60 **4.** 42 **5.** 60 **6.** 70 **7.** $\dfrac{15}{24}$ **8.** $\dfrac{40}{60}$ **9.** $\dfrac{2}{5}$ **10.** $\dfrac{3}{20}$ **11.** 2 **12.** 1

13. $\dfrac{8}{77}$ **14.** $\dfrac{11}{20}$ **15.** $\dfrac{1}{20}$ **16.** $\dfrac{11}{18}$ **17.** $\dfrac{11}{20}$ sq mile **18.** $\dfrac{5}{16}$ sq meter **19.** $\dfrac{181}{100}$ **20.** $\dfrac{35}{1000}$ **21.** 95.118 **22.** 36.785
23. 13.38 **24.** 691.573 **25.** 91.2 **26.** 46.816 **27.** 28.6 **28.** 230 **29.** 0.77 **30.** 25.6 **31.** 0.5 **32.** 0.375
33. $0.\overline{36} \approx 0.364$ **34.** $0.8\overline{3} \approx 0.833$ **35.** 0.29 **36.** 0.014 **37.** 39% **38.** 120% **39.** 0.708 **40.** b

Chapter R Test

1. $2 \cdot 2 \cdot 2 \cdot 3 \cdot 3$ **2.** 180 **3.** $\dfrac{25}{60}$ **4.** $\dfrac{3}{4}$ **5.** $\dfrac{12}{25}$ **6.** $\dfrac{13}{10}$ **7.** $\dfrac{53}{40}$ **8.** $\dfrac{18}{49}$ **9.** $\dfrac{1}{20}$ **10.** $\dfrac{29}{36}$ **11.** 45.11 **12.** 65.88

13. 12.688 **14.** 320 **15.** 23.73 **16.** 0.875 **17.** $0.1\overline{6} \approx 0.167$ **18.** 0.581 **19.** 7% **20.** 75% **21.** $\dfrac{3}{4}$ **22.** $\dfrac{1}{200}$

23. $\dfrac{49}{200}$ **24.** $\dfrac{199}{200}$

Chapter 1 Real Numbers and Introduction to Algebra

Chapter 1 Pretest

1. >; 1.1A **2.** <; 1.1A **3.** >; 1.1A **4.** 5; 1.1D **5.** 1.2; 1.1D **6.** 0; 1.1D **7.** 6; 1.2B **8.** $2x - 10$; 1.2D

9. 64; 1.2A **10.** −9; 1.5A **11.** $\dfrac{3}{5}$; 1.3C **12.** 8; 1.6A **13.** 53; 1.2A **14.** 3; 1.3A **15.** −27; 1.4A **16.** 56; 1.5A

17. -70; 1.6B   **18.** 4; 1.6B   **19.** -40; 1.5B   **20.** not a solution; 1.4C   **21.** solution; 1.6D   **22.** 36°F; 1.4D
23. $5 + 2y$; 1.7A   **24.** $12 + 8t$; 1.7B   **25.** additive inverse property; 1.7C

SECTION 1.1

1. <   **3.** >   **5.** =   **7.** <   **9.** $32 < 212$ **11.** true   **13.** false   **15.** false   **17.** true   **19.** $30 \le 45$ **21.** $20 \le 25$
23. $6 > 0$ **25.** $-12 < -10$ **27.** $8 < 12$ **29.** $5 \ge 4$ **31.** $15 \ne -2$ **33.** 535; -8 **35.** $-398{,}000$ **37.** 350; -126
39.

$-4 \ -3 \ -2 \ -1 \ 0 \ 1 \ 2 \ 3 \ 4 \ 5$

41.

$-2 \quad -\frac{1}{4} \ \frac{1}{2} \qquad 4$

$-4 \ -3 \ -2 \ -1 \ 0 \ 1 \ 2 \ 3 \ 4$

43.

$-2.5 \ -\frac{3}{2} \qquad \frac{7}{4} \ 3.25$

$-4 \ -3 \ -2 \ -1 \ 0 \ 1 \ 2 \ 3 \ 4$

45. whole, integers, rational, real   **47.** integers, rational, real   **49.** natural, whole, integers, rational, real   **51.** rational, real
53. false   **55.** true   **57.** true   **59.** true   **61.** >   **63.** =   **65.** <   **67.** <   **69.** false   **71.** true   **73.** false   **75.** true
77. 90   **79.** $70 \le 90$ **81.** $-0.04 > -26.7$ **83.** sun   **85.** sun   **87.** answers may vary

CALCULATOR EXPLORATIONS

1. 125   **3.** 59,049   **5.** 30   **7.** 9857   **9.** 2376

SECTION 1.2

1. 243   **3.** 27   **5.** 1   **7.** 5   **9.** $\frac{1}{125}$ **11.** $\frac{16}{81}$ **13.** 49   **15.** 1.44   **17.** 5^2 sq meters   **19.** 17   **21.** 20   **23.** 10

25. 21   **27.** 45   **29.** 0   **31.** $\frac{2}{7}$ **33.** 30   **35.** 2   **37.** $\frac{7}{18}$ **39.** $\frac{27}{10}$ **41.** $\frac{7}{5}$ **43.** no   **45.** 9   **47.** 1   **49.** 1   **51.** 11

53. 8   **55.** 45   **57.** 15   **59.** 3   **61.** 6   **63.** 16; 64; 144; 256   **65.** solution   **67.** not a solution   **69.** not a solution

71. solution   **73.** not a solution   **75.** $x + 15$ **77.** $x - 5$ **79.** $3x + 22$ **81.** $1 + 2 = 9 \div 3$ **83.** $3 \ne 4 \div 2$

85. $5 + x = 20$ **87.** $13 - 3x = 13$ **89.** $\frac{12}{x} = \frac{1}{2}$ **91.** $(20 - 4) \cdot 4 \div 2$ **93.** answers may vary   **95.** answers may vary

SECTION 1.3

1. 9   **3.** -14 **5.** 1   **7.** -12 **9.** -5 **11.** -12 **13.** -4 **15.** 7   **17.** -2 **19.** 0   **21.** -19 **23.** 31   **25.** -47

27. -2.1 **29.** -8 **31.** 38   **33.** -13.1 **35.** $\frac{2}{8} = \frac{1}{4}$ **37.** $-\frac{3}{16}$ **39.** $-\frac{13}{10}$ **41.** -8 **43.** -59 **45.** -9 **47.** 5

49. 11   **51.** -18 **53.** 19   **55.** -7 **57.** answers may vary   **59.** $-6°$ **61.** -638 feet   **63.** $-4\frac{1}{8}$ points   **65.** -10

67. $-\$33.2$ million   **69.** -6 **71.** 2   **73.** 0   **75.** -6 **77.** answers may vary   **79.** -2 **81.** 0   **83.** $-\frac{2}{3}$ **85.** Tuesday

87. 7°   **89.** 1°   **91.** negative   **93.** positive

SECTION 1.4

1. -10 **3.** -5 **5.** 19   **7.** $\frac{1}{6}$ **9.** 2   **11.** -11 **13.** 11   **15.** 5   **17.** 37   **19.** -6.4 **21.** -71 **23.** 0   **25.** 4.1

27. $\frac{2}{11}$ **29.** $-\frac{22}{24} = -\frac{11}{12}$ **31.** 8.92   **33.** sometimes positive and sometimes negative   **35.** 13   **37.** -5 **39.** -1 **41.** -23

43. -26 **45.** -24 **47.** 3   **49.** -45 **51.** -4 **53.** 13   **55.** 6   **57.** 9   **59.** -9 **61.** -7 **63.** $\frac{7}{5}$ **65.** 21   **67.** $\frac{1}{4}$

69. not a solution   **71.** not a solution   **73.** solution   **75.** 100°   **77.** lost 23 yards   **79.** 384 B.C.   **81.** -308 feet
83. 22,965 feet   **85.** 130°   **87.** 30°   **89.** $7°, 4°, -9°, 3°, -6°, 6°$ **91.** Wednesday   **93.** true   **95.** true   **97.** negative, -2.6466

SECTION 1.5

1. -24 **3.** -2 **5.** 50   **7.** -12 **9.** 42   **11.** -18 **13.** $\frac{3}{10}$ **15.** $\frac{24}{36} = \frac{2}{3}$ **17.** -7 **19.** 0.14   **21.** -800 **23.** -28

25. 25   **27.** $-\frac{8}{27}$ **29.** -121 **31.** $-\frac{100}{400} = -\frac{1}{4}$ **33.** 0.84   **35.** -30 **37.** 90   **39.** 16   **41.** -36 **43.** -125

45. -16 **47.** 18   **49.** -30 **51.** -24 **53.** $\frac{9}{16}$ **55.** true   **57.** false   **59.** -21 **61.** 41   **63.** -134 **65.** solution

67. solution   **69.** not a solution   **71.** solution   **73.** solution   **75.** positive   **77.** can't determine   **79.** negative
81. Saturday, October 2   **83.** no; answers may vary

CALCULATOR EXPLORATIONS

1. 38   **3.** -441 **5.** $163\frac{1}{3}$ **7.** 54,499   **9.** 15,625

Section 1.6

1. $\dfrac{1}{9}$ **3.** $\dfrac{3}{2}$ **5.** $-\dfrac{1}{14}$ **7.** $-\dfrac{11}{3}$ **9.** $\dfrac{1}{0.2}$ **11.** -6.3 **13.** $1, -1$ **15.** -9 **17.** 4 **19.** -4 **21.** 0 **23.** -5

25. undefined **27.** 3 **29.** -15 **31.** $-\dfrac{18}{7}$ **33.** $\dfrac{20}{27}$ **35.** -1 **37.** $-\dfrac{20}{24} = -\dfrac{5}{6}$ **39.** -40 **41.** 160 **43.** $-\dfrac{9}{2}$

45. -4 **47.** 16 **49.** -3 **51.** $-\dfrac{16}{7}$ **53.** 2 **55.** $\dfrac{6}{5}$ **57.** -5 **59.** $\dfrac{3}{2}$ **61.** 3 **63.** -1 **65.** $\dfrac{8}{9}$ **67.** solution

69. not a solution **71.** not a solution **73.** $\dfrac{0}{5} - 7 = -7$ **75.** $-8(-5) + (-1) = 39$ **77.** $\dfrac{-8}{-20} = \dfrac{2}{5}$ **79.** negative

81. negative **83.** answers may vary

Integrated Review

1. -35 **2.** 30 **3.** 5 **4.** -5 **5.** 10 **6.** -18 **7.** -2 **8.** -2 **9.** $\dfrac{3}{8}$ **10.** $-\dfrac{11}{42}$ **11.** -60 **12.** 1.9 **13.** -42

14. -7 **15.** 2 **16.** -39 **17.** 64 **18.** -81 **19.** -27 **20.** 16 **21.** 48 **22.** -30 **23.** -26 **24.** 6 **25.** 4

26. -3 **27.** 2 **28.** 16 **29.** 0 **30.** $-\dfrac{32}{15}$

Section 1.7

1. $16 + x$ **3.** $y \cdot (-4)$ **5.** yx **7.** $13 + 2x$ **9.** $x \cdot (yz)$ **11.** $(2 + a) + b$ **13.** $4a \cdot (b)$ **15.** $a + (b + c)$ **17.** $17 + b$
19. $24y$ **21.** y **23.** $26 + a$ **25.** $-72x$ **27.** s **29.** answers may vary **31.** $4x + 4y$ **33.** $9x - 54$ **35.** $6x + 10$
37. $28x - 21$ **39.** $18 + 3x$ **41.** $-2y + 2z$ **43.** $-21y - 35$ **45.** $5x + 20m + 10$ **47.** $-4 + 8m - 4n$ **49.** $-5x - 2$
51. $-r + 3 + 7p$ **53.** $3x + 4$ **55.** $-x + 3y$ **57.** $6r + 8$ **59.** $-36x - 70$ **61.** $-16x - 25$ **63.** $4(1 + y)$ **65.** $11(x + y)$
67. $-1(5 + x)$ **69.** $30(a + b)$ **71.** -16 **73.** 8 **75.** -1.2 **77.** 2 **79.** $\dfrac{3}{2}$ **81.** $-\dfrac{6}{5}$ **83.** $\dfrac{6}{23}$ **85.** $-\dfrac{1}{2}$

87. commutative property of multiplication **89.** associative property of addition **91.** distributive property
93. associative property of multiplication **95.** identity property of addition **97.** distributive property

99. commutative and associative properties of multiplication **101.** $-8; \dfrac{1}{8}$ **103.** $-x; \dfrac{1}{x}$ **105.** $2x; -2x$

107. a. commutative property of addition **b.** commutative property of addition **c.** associative property of addition
109. Answers may vary.

Section 1.8

1. approx. 7.8 million **3.** 2002 **5.** PGA/LPGA tours **7.** Major League Baseball, NBA **9.** approx. 15 million
11. Cleveland Indians **13.** tied **15.** approx. 3 years **17.** approx. 142 million **19.** *Snow White and the Seven Dwarfs*
21. answers may vary **23.** 1994 **25.** 1986, 1987 or 1989, 1990 **27.** approx. 52% **29.** 1994 **31.** approx. 59 **33.** approx. 26
35. 1992 **37.** 1992, approx. $450 **39.** 30° north, 90° west **41.** answers may vary

Chapter 1 Review

1. $<$ **2.** $>$ **3.** $>$ **4.** $>$ **5.** $<$ **6.** $>$ **7.** $=$ **8.** $=$ **9.** $>$ **10.** $<$ **11.** $4 \geq -3$ **12.** $6 \neq 5$ **13.** $0.03 < 0.3$

14. $50 > 40$ **15. a.** $1, 3$ **b.** $0, 1, 3$ **c.** $-6, 0, 1, 3$ **d.** $-6, 0, 1, 1\dfrac{1}{2}, 3, 9.62$ **e.** π **f.** all numbers in set

16. a. $2, 5$ **b.** $2, 5$ **c.** $-3, 2, 5$ **d.** $-3, -1.6, 2, 5, \dfrac{11}{2}, 15.1$ **e.** $\sqrt{5}, 2\pi$ **f.** all numbers in set **17.** Friday

18. Wednesday **19.** c **20.** b **21.** 37 **22.** 41 **23.** $\dfrac{18}{7}$ **24.** 80 **25.** $20 - 12 = 2 \cdot 4$ **26.** $\dfrac{9}{2} > -5$ **27.** 18

28. 108 **29.** 5 **30.** 24 **31.** 63° **32.** solution **33.** not a solution **34.** 9 **35.** $-\dfrac{2}{3}$ **36.** -2 **37.** 7 **38.** -11

39. -17 **40.** $-\dfrac{3}{16}$ **41.** -5 **42.** -13.9 **43.** 3.9 **44.** -14 **45.** -11.5 **46.** 5 **47.** -11 **48.** -19 **49.** 4

50. a **51.** d **52.** $51 **53.** $-\dfrac{1}{6}$ **54.** $\dfrac{5}{3}$ **55.** -48 **56.** 28 **57.** 3 **58.** -14 **59.** -36 **60.** 0 **61.** undefined

62. $-\dfrac{1}{2}$ **63.** commutative property of addition **64.** multiplicative identity property **65.** distributive property

66. additive inverse property **67.** associative property of addition **68.** commutative property of multiplication
69. distributive property **70.** associative property of multiplication **71.** multiplicative inverse property
72. additive identity property **73.** commutative property of addition **74.** $1,800 million **75.** $400 million **76.** 1994
77. revenue is increasing

Chapter 1 Test

1. $|-7| > 5$ **2.** $(9 + 5) \geq 4$ **3.** -5 **4.** -11 **5.** -14 **6.** -39 **7.** 12 **8.** -2 **9.** undefined **10.** -8 **11.** $-\dfrac{1}{3}$

12. $4\dfrac{5}{8}$ **13.** $\dfrac{51}{40}$ **14.** -32 **15.** -48 **16.** 3 **17.** 0 **18.** $>$ **19.** $>$ **20.** $>$ **21.** $=$ **22. a.** $\{1, 7\}$ **b.** $\{0, 1, 7\}$

c. $\{-5, -1, 0, 1, 7\}$ **d.** $\{-5, -1, \frac{1}{4}, 0, 1, 7, 11.6\}$ **e.** $\{\sqrt{7}, 3\pi\}$ **f.** $\{-5, -1, \frac{1}{4}, 0, 1, 7, 11.6, \sqrt{7}, 3\pi\}$ **23.** 40   **24.** 12   **25.** 22
26. -1 **27.** associative property of addition   **28.** commutative property of multiplication   **29.** distributive property
30. multiplicative inverse   **31.** 9   **32.** -3 **33.** second down   **34.** yes   **35.** 17°   **36.** loss of $420   **37.** $8 billion
38. $3 billion   **39.** $5.5 billion   **40.** 1994

Chapter 2   Equations, Inequalities, and Problem Solving

Chapter 2 Pretest

1. $9c - 13$; 2.1B   **2.** $-17y + 16$; 2.1C   **3.** $x = 15$; 2.2A   **4.** $b = 2$; 2.2B   **5.** $m = -12$; 2.3A   **6.** $y = -3$; 2.3B   **7.** $x = -3$; 2.4A
8. $x = 22.5$; 2.4B   **9.** no solution; 2.4C   **10.** 4; 2.5A   **11.** 20, 22; 2.5A   **12.** $A = 45$; 2.6A   **13.** 9 feet; 2.6A

14. $y = 8 - 2x$; 2.6B   **15.** 19.8; 2.7A   **16.** $\frac{1}{5}$; 2.7C   **17.** $x = \frac{24}{7}$; 2.7D

18. $x \le 6$; 2.8B   **19.** $y < -4$; 2.8C   **20.** $x \ge 3$; 2.8D

Mental Math

1. -7 **3.** 1   **5.** 17   **7.** like   **9.** unlike   **11.** like

Section 2.1

1. $15y$ **3.** $13w$ **5.** $-7b - 9$ **7.** $-m - 6$ **9.** -8 **11.** $7.2x - 5.2$ **13.** $k - 6$ **15.** $-15x + 18$ **17.** $4x - 3$ **19.** $5x^2$
21. -11 **23.** $1.3x + 3.5$ **25.** $5y + 20$ **27.** $-2x - 4$ **29.** $-10x + 15y - 30$ **31.** $-3x + 2y - 1$ **33.** $7d - 11$ **35.** 16
37. $x + 5$ **39.** $x + 2$ **41.** $2k + 10$ **43.** $-3x + 5$ **45.** $2x + 14$ **47.** $0.9m + 1$ **49.** answers may vary   **51.** $10x - 3$
53. $-4x - 9$ **55.** $-4m - 3$ **57.** $2x - 4$ **59.** $\frac{3}{4}x + 12$ **61.** $12x - 2$ **63.** $8x + 48$ **65.** $x - 10$ **67.** 2   **69.** -23
71. -25 **73.** balanced   **75.** balanced   **77.** $(18x - 2)$ feet   **79.** $(15x + 23)$ in.

Mental Math

1. $x = 2$ **3.** $n = 12$ **5.** $b = 17$

Section 2.2

1. $x = 3$ **3.** $x = -2$ **5.** $x = -14$ **7.** $r = 0.5$ **9.** $f = \frac{5}{12}$ **11.** $b = -0.7$ **13.** $x = 3$ **15.** answers may vary   **17.** $x = -3$

19. $y = -10$ **21.** $x = 11$ **23.** $x = -1$ **25.** $y = -9$ **27.** $x = 13$ **29.** $x = -17.9$ **31.** $x = -\frac{1}{2}$ **33.** $x = 11$ **35.** $w = -30$

37. $x = -7$ **39.** $n = 2$ **41.** $x = -12$ **43.** $n = 21$ **45.** $x = 25$ **47.** $20 - p$ **49.** $(10 - x)$ ft   **51.** $(180 - x)°$

53. $(n + 30{,}898)$ votes   **55.** $\frac{8}{5}$ **57.** $\frac{1}{2}$ **59.** -9 **61.** x **63.** y **65.** x **67.** $x = -145.478$ **69.** $(173 - 3x)°$

Mental Math

1. $a = 9$ **3.** $b = 2$ **5.** $x = -5$

Section 2.3

1. $x = -4$ **3.** $x = 0$ **5.** $x = 12$ **7.** $x = -12$ **9.** $d = 3$ **11.** $a = 2$ **13.** $k = 0$ **15.** $x = 6.3$ **17.** $x = 6$ **19.** $x = -5.5$
21. $p = \frac{14}{3}$ **23.** $x = -9$ **25.** $x = 10$ **27.** $x = -20$ **29.** $a = 0$ **31.** $x = -5$ **33.** $k = 0$ **35.** $x = -\frac{3}{2}$ **37.** $x = -21$
39. $x = \frac{11}{2}$ **41.** $z = 1$ **43.** $x = -\frac{1}{4}$ **45.** $x = -30$ **47.** $z = \frac{9}{10}$ **49.** $2x + 2$ **51.** $2x + 2$ **53.** $7x - 12$ **55.** $12z + 44$
57. 1   **59.** $x = 2$ **61.** answers may vary   **63.** answers may vary

Calculator Explorations

1. solution   **3.** not a solution   **5.** solution

Section 2.4

1. $y = -6$ **3.** $x = -3$ **5.** $x = 1$ **7.** $n = \frac{9}{2}$ **9.** $x = \frac{3}{2}$ **11.** $x = 0$ **13.** $a = 1$ **15.** $x = 4$ **17.** $y = -4$ **19.** $t = \frac{19}{6}$

21. $x = 2$ **23.** $x = -5$ **25.** $x = 10$ **27.** $z = 18$ **29.** $x = 3$ **31.** $x = 13$ **33.** $x = 50$ **35.** $y = 0.2$ **37.** $x = 1$ **39.** $x = \frac{7}{3}$
41. answers may vary   **43.** all real numbers   **45.** no solution   **47.** no solution   **49.** no solution   **51.** answers may vary
53. $(6x - 8)$ meters   **55.** $-8 - x$ **57.** $-3 + 2x$ **59.** $9(x + 20)$ **61.** $x = 15.3$ **63.** $x = -0.2$ **65.** $x = 4$ cm, $2x = 8$ cm

1. $x = 6$ **2.** $y = -17$ **3.** $y = 12$ **4.** $x = -26$ **5.** $x = -3$ **6.** $y = -1$ **7.** $x = 13.5$ **8.** $z = 12.5$ **9.** $r = 8$ **10.** $y = -64$
11. $x = 2$ **12.** $y = -3$ **13.** $x = 5$ **14.** $y = -1$ **15.** $a = -2$ **16.** $b = -2$ **17.** $x = -\dfrac{5}{6}$ **18.** $y = \dfrac{1}{6}$ **19.** $n = 1$ **20.** $m = 6$
21. $c = 4$ **22.** $t = 1$ **23.** $z = \dfrac{9}{5}$ **24.** $w = -\dfrac{6}{5}$ **25.** all real numbers   **26.** all real numbers   **27.** $t = 0$ **28.** $m = -1.6$

Section 2.5

1. 1   **3.** -25 **5.** $-\dfrac{3}{4}$ **7.** -16 **9.** governor of Nebraska = \$65,000; governor of New York = \$130,000
11. 1st piece: 5 in.; 2nd piece: 10 in.; 3rd piece: 25 in.   **13.** 172 miles   **15.** 1st angle: 37.5°; 2nd angle: 37.5°; 3rd angle: 105°
17. Little: 46,895 votes; Brown: 63,845 votes   **19.** smaller angle: 45°; larger angle: 135°   **21.** shorter piece: 5 ft; longer piece: 12 ft
23. height: 34 in.; diameter: 49 in.   **25.** Midway   **27.** 145   **29.** $\dfrac{1}{2}(x - 1) = 37$ **31.** $\dfrac{3(x + 2)}{5} = 0$ **33.** 34   **35.** 225π
37. $-11, -9$ **39.** Bulls: 90; Jazz 88   **41.** 10, 11, 12   **43.** Mali Republic: 223; Côte d'Ivoire: 225; Niger: 227   **45.** There are none.
47. answers may vary

Section 2.6

1. $h = 3$ **3.** $h = 3$ **5.** $h = 20$ **7.** $c = 12$ **9.** $r = 2.5$ **11.** $T = 3$ **13.** $h = 15$ **15.** 131 ft   **17.** 2000 mph   **19.** $-10°C$
21. 96 piranhas   **23.** 2.25 hours   **25.** 6.25 hrs   **27.** 2 bags   **29.** 800 ft³   **31.** one 16-in. pizza   **33.** $-109.3°F$
35. 500 sec or $8\dfrac{1}{3}$ min   **37.** 33,493,333,333 cubic miles   **39.** 10.8   **41.** 44.3 sec   **43.** $h = \dfrac{f}{5g}$ **45.** $W = \dfrac{V}{LH}$ **47.** $y = 7 - 3x$
49. $R = \dfrac{A - p}{PT}$ **51.** $A = \dfrac{3V}{h}$ **53.** $a = P - b - c$ **55.** $h = \dfrac{S - 2\pi r^2}{2\pi r}$ **57.** 0.32   **59.** 2.00 or 2   **61.** 17%
63. 720%   **65.** $V = G(N - R)$ **67.** multiplies the volume by 8   **69.** $-40°$

Section 2.7

1. 11.2   **3.** 55%   **5.** 180   **7.** 4.6   **9.** 50   **11.** 30%   **13.** \$39 decrease; \$117 sale price   **15.** 647.5 ft   **17.** 55.40%
19. 54 people   **21.** No, many people use several medications.   **23.** 31 men   **25.** 48%; 30%; 22%; 100%   **27.** $\dfrac{2}{15}$ **29.** $\dfrac{5}{6}$ **31.** $\dfrac{5}{12}$
33. $\dfrac{1}{10}$ **35.** $\dfrac{7}{20}$ **37.** $\dfrac{19}{18}$ **39.** answers may vary   **41.** $x = 4$ **43.** $x = \dfrac{50}{9}$ **45.** $x = \dfrac{21}{4}$ **47.** $x = 7$ **49.** $x = -3$
51. $x = \dfrac{14}{9}$ **53.** $x = 5$ **55.** 123 lb   **57.** 165 calories   **59.** 3833 women   **61.** 9 gal   **63.** 110 oz for \$5.79   **65.** 8 oz for \$0.90
67. $>$ **69.** $=$ **71.** $>$ **73.** 9.6%   **75.** 26.9%; yes   **77.** 17.1%

Mental Math

1. $x > 2$ **3.** $x \geq 8$ **5.** -5 **7.** 4.1

Section 2.8

1. -1 **3.** $\dfrac{1}{2}$ **5.** 4   **7.** -2

9. $x \geq -5$ **11.** $y < 9$ **13.** $x > -3$ **15.** $x \leq 1$

-5 9 -3 1

17. $x < -3$ **19.** $x \geq -2$ **21.** $x < 0$ **23.** $y \geq -\dfrac{8}{3}$

-3 -2 0 $-\dfrac{8}{3}$

25. $y > 3$ **27.** when multiplying or dividing by a negative number   **29.** $x > -3$ **31.** $x \geq -\dfrac{2}{3}$

3

33. $x \leq -2$ **35.** $x \leq -8$ **37.** $x > 4$ **39.** $x > 3$ **41.** $x > -13$ **43.** $x \geq 0$ **45.** $x > -10$ **47.** 35 cm   **49.** 193
51. final exam ≥ 78.5 **53.** 8   **55.** 1   **57.** $\dfrac{16}{49}$ **59.** 37 million   **61.** 1992, 1993, 1994, or 1995   **63.** answers may vary

Chapter 2 Review

1. $6x$ **2.** $-11.8z$ **3.** $4x - 2$ **4.** $2y + 3$ **5.** $3n - 18$ **6.** $4w - 6$ **7.** $-6x + 7$ **8.** $-0.4y + 2.3$ **9.** $3x - 7$
10. $5x + 5.6$ **11.** $x = 4$ **12.** $y = -3$ **13.** $x = 6$ **14.** $x = -6$ **15.** $x = 0$ **16.** $x = -9$ **17.** $n = -23$ **18.** $x = 28$
19. b   **20.** a   **21.** c   **22.** $x = -12$ **23.** $x = 4$ **24.** $x = 0$ **25.** $y = -7$ **26.** $x = 0.75$ **27.** $x = -3$ **28.** $x = -6$
29. $x = -1$ **30.** $x = 3$ **31.** $y = 0$ **32.** $x = -\dfrac{1}{5}$ **33.** $3x + 3$ **34.** $x = -4$ **35.** $x = 2$ **36.** $x = -3$ **37.** no solution

38. no solution   **39.** $z = \dfrac{3}{4}$ **40.** $n = -\dfrac{8}{9}$ **41.** $n = 20$ **42.** $a = -\dfrac{6}{23}$ **43.** $c = \dfrac{23}{7}$ **44.** $a = -\dfrac{2}{5}$ **45.** $x = 102$ **46.** $y = 0.25$
47. 1052 ft   **48.** short piece: 4 ft; long piece: 8 ft   **49.** 1st area code: 307; 2nd area code: 955   **50.** $-39, -38, -37$ **51.** 3   **52.** -4
53. $w = 9$ **54.** $h = 4$ **55.** $m = \dfrac{y - b}{x}$ **56.** $s = \dfrac{r + 5}{vt}$ **57.** $x = \dfrac{2y - 7}{5}$ **58.** $y = \dfrac{2 + 3x}{6}$ **59.** $\pi = \dfrac{C}{D}$ **60.** $\pi = \dfrac{C}{2r}$
61. 15 meters   **62.** 40°C   **63.** 1 hour and 20 min   **64.** 20%   **65.** 70%   **66.** 110   **67.** 1280   **68.** 6300   **69.** 6%
70. eat from the Minibar   **71.** 120 travelers   **72.** no; answers may vary   **73.** $\dfrac{1}{5}$ **74.** $\dfrac{2}{3}$ **75.** $x = 6$ **76.** $x = 500$
77. $x = 312.5$ **78.** $c = 50$ **79.** $x = 9$ **80.** no solution   **81.** $y = 3$ **82.** no solution   **83.** 10 oz for $1.29
84. 15 oz for $1.63   **85.** 675 parts   **86.** $33.75   **87.** 157 letters
88. $x \le -2$ **89.** $x > 0$ **90.** $x \le 1$ **91.** $x > -5$ **92.** $x \le 10$ **93.** $x < -4$
94. $x < -4$ **95.** $x \le 4$ **96.** $y > 9$ **97.** $y \ge -15$
98. $x < \dfrac{7}{4}$ **99.** $x \le \dfrac{19}{3}$ **100.** at least $2,500   **101.** score must be less than 83

Chapter 2 Test

1. $y - 10$ **2.** $5.9x + 1.2$ **3.** $-2x + 10$ **4.** $-15y + 1$ **5.** $x = -5$ **6.** $n = 8$ **7.** $y = \dfrac{7}{10}$ **8.** $z = 0$ **9.** $x = 27$
10. $y = -\dfrac{19}{6}$ **11.** $x = 3$ **12.** $y = -6$ **13.** $y = \dfrac{3}{11}$ **14.** $x = 0.25$ **15.** $a = \dfrac{25}{7}$ **16.** 21   **17.** 7 gal   **18.** 6 oz for $1.19
19. 18 bulbs   **20.** $x = 6$ **21.** $h = \dfrac{V}{\pi r^2}$ **22.** $y = \dfrac{3x - 10}{4}$
23. $x < -2$ **24.** $x < 4$ **25.** $x \le -8$ **26.** $x \ge 11$ **27.** $x > \dfrac{2}{5}$ **28.** 80.8%
29. 7.33584 billion   **30.** 40%   **31.** New York: 758; Indiana: 238

Cumulative Review

1. True; Sec. 1.1, Ex. 3   **2.** True; Sec. 1.1, Ex. 4   **3.** False; Sec. 1.1, Ex. 5   **4.** True; Sec. 1.1, Ex. 6   **5. a.** $<$ **b.** $=$ **c.** $>$ **d.** $<$
e. $>$; Sec. 1.1, Ex. 12   **6.** $\dfrac{8}{3}$; Sec. 1.2, Ex. 5   **7.** -19; Sec. 1.3, Ex. 7   **8.** 8; Sec. 1.3, Ex. 8   **9.** -0.3; Sec. 1.3, Ex. 9   **10. a.** -12
b. -3; Sec. 1.4, Ex. 7   **11. a.** 0   **b.** -24 **c.** 45   **d.** -8 **e.** 54; Sec. 1.5, Ex. 7   **12. a.** -6 **b.** 7
c. -5; Sec. 1.6, Ex. 6   **13.** $15 - 10z$; Sec. 1.7, Ex. 8   **14.** $12x + 38$; Sec. 1.7, Ex. 12   **15. a.** $70   **b.** 280 miles; Sec. 1.8, Ex. 3
16. a. unlike   **b.** like   **c.** like   **d.** like; Sec. 2.1, Ex. 2   **17.** $-2x - 1$; Sec. 2.1, Ex. 14   **18.** 17; Sec. 2.2, Ex. 1
19. -10; Sec. 2.3, Ex. 7   **20.** 0; Sec. 2.4, Ex. 4   **21.** 54 Republicans, 46 Democrats; Sec. 2.5, Ex. 3   **22.** 79.2 years; Sec. 2.6, Ex. 1
23. 87.5%; Sec. 2.7, Ex. 1   **24.** 63; Sec. 2.7, Ex. 5   **25.** ←———○———→; Sec. 2.8, Ex. 2   **26.** $x \ge 1$; Sec. 2.8, Ex. 8

Chapter 3   Exponents and Polynomials

Chapter 3 Pretest

1. $\dfrac{9}{16}$; 3.1A   **2.** $8y^{13}$; 3.1B   **3.** $\dfrac{b^{11}}{a^3}$; 3.1D   **4.** 3; 3.1E   **5.** -216; 3.2A   **6.** $\dfrac{m^{16}}{n^{18}}$; 3.2B   **7.** $4x^2 + 10x - 5$; 3.3D
8. 8.14×10^{-7}; 3.2C   **9.** 5; 3.3B   **10.** 1; 3.3C   **11.** $10x^2 + 1$; 3.4A   **12.** $9y^2 - 5y - 3$; 3.4B   **13.** $-a^2 - 16b^2$; 3.4D
14. $-\dfrac{3}{8}n^9$; 3.5A   **15.** $-6t^7 - 8t^5 + 16t^2$; 3.5B   **16.** $10y^2 + 7y - 6$; 3.5C   **17.** $49a^2 - 70a + 25$; 3.6B   **18.** $16b^2 - 81$; 3.6C
19. $4p^3 - 2p^2 + 5p$; 3.7A   **20.** $5x + 2$; 3.7B

Mental Math

1. base: 3; exponent: 2   **3.** base: -3; exponent: 6   **5.** base: 4; exponent: 2   **7.** base: 5; exponent: 1; base: 3; exponent: 4
9. base: 5; exponent: 1; base: x; exponent: 2

Section 3.1

1. 49   **3.** -5 **5.** -16 **7.** 16   **9.** $\dfrac{1}{27}$ **11.** 112   **13.** answers may vary   **15.** 4   **17.** 135   **19.** 150   **21.** $\dfrac{32}{5}$
23. x^7 **25.** $(-3)^{12}$ **27.** $15y^5$ **29.** $-24z^{20}$ **31.** $20x^5$ sq ft   **33.** x^{36} **35.** p^7q^7 **37.** $8a^{15}$ **39.** $\dfrac{m^9}{n^9}$ **41.** $x^{10}y^{15}$
43. $\dfrac{4x^2z^2}{y^{10}}$ **45.** $64z^{10}$ sq decimeters   **47.** $27y^{12}$ cubic ft   **49.** x^2 **51.** 4   **53.** p^6q^5 **55.** $\dfrac{y^3}{2}$ **57.** 1   **59.** -2 **61.** 2
63. answers may vary   **65.** -25 **67.** $\dfrac{1}{64}$ **69.** z^8 **71.** $81x^2y^2$ **73.** 1   **75.** 40   **77.** b^6 **79.** a^9 **81.** $-16x^7$

83. $64a^3$ **85.** $36x^2y^2z^6$ **87.** $\dfrac{y^{15}}{8x^{12}}$ **89.** $3x$ **91.** $2x^2y$ **93.** -2 **95.** 5 **97.** -7 **99.** 343 cubic meters   **101.** volume

103. x^{9a} **105.** a^{5b} **107.** x^{5a} **109.** $\$1,045.85$

CALCULATOR EXPLORATIONS

1. 5.31 EE 03   **3.** 6.6 EE -09 **5.** 1.5×10^{13} **7.** 8.15×10^{19}

MENTAL MATH

1. $\dfrac{5}{x^2}$ **3.** y^6 **5.** $4y^3$

SECTION 3.2

1. $\dfrac{1}{64}$ **3.** $\dfrac{7}{x^3}$ **5.** -64 **7.** $\dfrac{5}{6}$ **9.** p^3 **11.** $\dfrac{q^4}{p^5}$ **13.** $\dfrac{1}{x^3}$ **15.** z^3 **17.** $\dfrac{4}{3}$ **19.** $\dfrac{1}{9}$ **21.** $-p^4$ **23.** -2 **25.** x^4 **27.** p^4

29. m^{11} **31.** r^6 **33.** $\dfrac{1}{x^{15}y^9}$ **35.** $\dfrac{1}{x^4}$ **37.** $\dfrac{1}{a^2}$ **39.** $4k^3$ **41.** $3m$ **43.** $-\dfrac{4a^5}{b}$ **45.** $-\dfrac{6x}{7y^2}$ **47.** $\dfrac{a^{30}}{b^{12}}$ **49.** $\dfrac{1}{x^{10}y^6}$ **51.** $\dfrac{z^2}{4}$

53. $\dfrac{1}{32x^5}$ **55.** $\dfrac{49a^4}{b^6}$ **57.** $a^{24}b^8$ **59.** x^9y^{19} **61.** $-\dfrac{y^8}{8x^2}$ **63.** $\dfrac{27}{z^3x^6}$ cubic in.   **65.** 7.8×10^4 **67.** 1.67×10^{-6}

69. 6.35×10^{-3} **71.** 1.16×10^6 **73.** 2.0×10^7 **75.** 9.3×10^7 **77.** 1.2×10^8 **79.** 0.0000000008673   **81.** 0.033   **83.** 20,320

85. 6,250,000,000,000,000,000   **87.** 9,460,000,000,000   **89.** 0.000036   **91.** 0.0000000000000000028   **93.** 0.0000005   **95.** 200,000

97. 1.512×10^{10} cubic ft   **99.** $-2x + 7$ **101.** $2y - 10$ **103.** $-x - 4$ **105.** -394.5 **107.** 1.3 sec   **109.** a^m **111.** $27y^{6z}$

113. answers may vary   **115.** answers may vary

SECTION 3.3

1. $1; -3x; 5$ **3.** $-5; 3.2; 1; -5$ **5.** 1; binomial   **7.** 3; none of these   **9.** 4; binomial   **11.** 1; binomial   **13.** answers may vary

15. answers may vary   **17. a.** 6   **b.** 5   **19. a.** -2 **b.** 4   **21. a.** -15 **b.** -16 **23.** 184 ft   **25.** 595.84 ft

27. 11.7 million   **29.** $23x^2$ **31.** $12x^2 - y$ **33.** $7s$ **35.** $-1.1y^2 + 4.8$ **37.** $5x + 3 + 4x + 3 + 2x + 6 + 3x + 7x; 21x + 12$

39. $4x^2 + 7x + x^2 + 5x; 5x^2 + 12x$ **41.** 2, 1, 1, 0; 2   **43.** 4, 0, 4, 3; 4   **45.** $9ab - 11a$ **47.** $4x^2 - 7xy + 3y^2$ **49.** $-3xy^2 + 4$

51. $14y^3 - 19 - 16a^2b^2$ **53.** $10x + 19$ **55.** $-x + 5$ **57.** answers may vary   **59.** $11.1x^2 - 7.97x + 10.76$

SECTION 3.4

1. $12x + 12$ **3.** $-3x^2 + 10$ **5.** $-3x^2 + 4$ **7.** $-y^2 - 3y - 1$ **9.** $8t^2 - 4$ **11.** $15a^3 + a^2 + 16$ **13.** $-x + 14$

15. $-2x + 9$ **17.** $2x^2 + 7x - 16$ **19.** $y^2 - 7$ **21.** $2x^2 + 11x$ **23.** $-2z^2 - 16z + 6$ **25.** $2u^5 - 10u^2 + 11u - 9$ **27.** $5x - 9$

29. $6y + 13$ **31.** $-2x^2 + 8x - 1$ **33.** $7x^2 + 14x + 18$ **35.** $3x - 3$ **37.** $7x^2 - 4x + 2$ **39.** $7x^2 - 2x + 2$

41. $4y^2 + 12y + 19$ **43.** $-2a - b + 1$ **45.** $3x^2 + 5$ **47.** $6x^2 - 2xy + 19y^2$ **49.** $8r^2s + 16rs - 8 + 7r^2s^2$ **51.** $6x^2$

53. $-12x^8$ **55.** $200x^3y^2$ **57.** $(x^2 + 7x + 4)$ ft   **59.** $(2x^2 - 2x + 2)$ cm   **61.** $-6.6x^2 - 1.8x - 1.8$

63. $-464.5x^2 + 2388.5x + 40,759$

MENTAL MATH

1. x^8 **3.** y^5 **5.** x^{14}

SECTION 3.5

1. $24x^3$ **3.** $-12.4x^{12}$ **5.** x^4 **7.** $-\dfrac{2}{15}y^3$ **9.** $-24x^8$ **11.** $6x^2 + 15x$ **13.** $7x^3 + 14x^2 - 7x$ **15.** $-2a^2 - 8a$

17. $6x^3 - 9x^2 - 12x$ **19.** $3a^3 + 6a$ **21.** $-6a^4 + 4a^3 - 6a^2$ **23.** $6x^5y - 3x^4y^3 + 24x^2y^4$ **25.** $x^2 + 3x$ **27.** $x^2 + 7x + 12$

29. $a^2 + 5a - 14$ **31.** $x^2 + \dfrac{1}{3}x - \dfrac{2}{9}$ **33.** $12x^2 + 25x + 7$ **35.** $12x^2 - 29x + 15$ **37.** $1 - 7a + 12a^2$ **39.** $4y^2 - 16y + 16$

41. $x^3 - 5x^2 + 13x - 14$ **43.** $x^4 + 5x^3 - 3x^2 - 11x + 20$ **45.** $10a^3 - 27a^2 + 26a - 12$ **47.** $49x^2y^2 - 14xy^2 + y^2$ **49.** $x^2 + 5x + 6$

51. $12x^2 - 64x - 11$ **53.** $2x^3 + 10x^2 + 11x - 3$ **55.** $x^4 - 2x^3 - 51x^2 + 4x + 63$ **57.** $25x^2$ **59.** $9y^6$ **61.** $\$7000$ **63.** $\$500$

65. answers may vary   **67.** $(4x^2 - 25)$ sq yds   **69.** $(6x^2 - 4x)$ sq in.   **71.** $6x + 12$; answers may vary   **73. a.** $a^2 - b^2$

b. $4x^2 - 9y^2$ **c.** $16x^2 - 49$; answers may vary

SECTION 3.6

1. $x^2 + 7x + 12$ **3.** $x^2 + 5x - 50$ **5.** $5x^2 + 4x - 12$ **7.** $4y^2 - 25y + 6$ **9.** $6x^2 + 13x - 5$ **11.** $6y^3 + 4y^2 + 42y + 28$

13. $x^2 + \dfrac{1}{3}x - \dfrac{2}{9}$ **15.** $8 - 26a + 15a^2$ **17.** $2x^2 + 9xy - 5y^2$ **19.** $x^2 + 4x + 4$ **21.** $4x^2 - 4x + 1$ **23.** $9a^2 - 30a + 25$

25. $x^4 + 10x^2 + 25$ **27.** $y^2 - \dfrac{4}{7}y + \dfrac{4}{49}$ **29.** $4a^2 - 12a + 9$ **31.** $25x^2 + 90x + 81$ **33.** $9x^2 - 42xy + 49y^2$

35. $16m^2 + 40mn + 25n^2$ **37.** answers may vary   **39.** $a^2 - 49$ **41.** $x^2 - 36$ **43.** $9x^2 - 1$ **45.** $x^4 - 25$ **47.** $4y^4 - 1$

49. $16 - 49x^2$ **51.** $9x^2 - \dfrac{1}{4}$ **53.** $81x^2 - y^2$ **55.** $4m^2 - 25n^2$ **57.** $a^2 + 9a + 20$ **59.** $a^2 - 14a + 49$ **61.** $12a^2 - a - 1$

63. $x^2 - 4$ **65.** $9a^2 + 6a + 1$ **67.** $x^2 + 3xy - y^2$ **69.** $a^2 - \dfrac{1}{4}y^2$ **71.** $6b^2 - b - 35$ **73.** $x^4 - 100$ **75.** $16x^2 - 25$

77. $25x^2 - 60xy + 36y^2$ **79.** $4r^2 - 9s^2$ **81.** $\dfrac{5b^5}{7}$ **83.** $-\dfrac{2a^{10}}{b^5}$ **85.** $\dfrac{2y^8}{3}$ **87.** $(4x^2 + 4x + 1)$ sq ft   **89.** $(24x^2 - 32x + 8)$ sq meters

A23

INTEGRATED REVIEW

1. $35x^5$ **2.** $32y^9$ **3.** $2x^2 - 9x - 5$ **4.** $3x^2 + 13x - 10$ **5.** $3x - 4$ **6.** $4x + 3$ **7.** $16y^2 - 9$ **8.** $49x^2 - 1$
9. $2x^2 - 2x - 6$ **10.** $6x^2 + 13x - 11$ **11.** $x^2 + 8x + 16$ **12.** $y^2 - 18x + 81$ **13.** $x^3 + 2x^2 - 16x + 3$
14. $x^3 - 2x^2 - 5x - 2$

MENTAL MATH

1. a^2 **3.** a^2 **5.** k^3

SECTION 3.7

1. $4x^2 + x + \dfrac{9}{5}$ **3.** $12x^3 + 3x$ **5.** $5p^2 + 6p$ **7.** $-\dfrac{3}{2x} + 3$ **9.** $-3x^2 + x - \dfrac{4}{x^3}$ **11.** $1 + \dfrac{3}{2x} - \dfrac{7}{4x^4}$ **13.** $ab - b^2$

15. $x + 4xy - \dfrac{y}{2}$ **17.** $x + 1$ **19.** $2x + 3$ **21.** $2x + 1 + \dfrac{7}{x - 4}$ **23.** $4x + 9$ **25.** $3a^2 - 3a + 1 + \dfrac{2}{3a + 2}$

27. $2b^2 + b + 2 - \dfrac{12}{b + 4}$ **29.** $4x + 3 - \dfrac{2}{2x + 1}$ **31.** $2x^2 + 6x - 5 - \dfrac{2}{x - 2}$ **33.** $x^2 + 3x + 9$ **35.** $-3x + 6 - \dfrac{11}{x + 2}$

37. $2b - 1 - \dfrac{6}{2b - 1}$ **39.** 3 **41.** -4 **43.** $3x$ **45.** $9x$ **47.** $x^3 - x^2 + x$ **49.** $(3x^3 + x - 4)$ ft   **51.** $(2x + 5)$ meters
53. Answers may vary.

CHAPTER 3 REVIEW

1. base: 3; exponent: 2   **2.** base: -5; exponent: 4   **3.** base: 5; exponent: 4   **4.** base: x; exponent: 6   **5.** 512   **6.** 36   **7.** -36
8. -65 **9.** 1   **10.** 1   **11.** y^9 **12.** x^{14} **13.** $-6x^{11}$ **14.** $-20y^7$ **15.** x^8 **16.** y^{15} **17.** $81y^{24}$ **18.** $8x^9$ **19.** x^5
20. z^7 **21.** a^4b^3 **22.** x^3y^5 **23.** $\dfrac{4}{x^3y^4}$ **24.** $\dfrac{x^6y^6}{4}$ **25.** $40a^{19}$ **26.** $36x^3$ **27.** 3   **28.** 9   **29.** b   **30.** c   **31.** $\dfrac{1}{49}$
32. $-\dfrac{1}{49}$ **33.** $\dfrac{2}{x^4}$ **34.** $\dfrac{1}{16x^4}$ **35.** 125   **36.** $\dfrac{9}{4}$ **37.** $\dfrac{17}{16}$ **38.** $\dfrac{1}{42}$ **39.** x^8 **40.** z^8 **41.** r **42.** y^3 **43.** c^4 **44.** $\dfrac{x^3}{y^3}$
45. $\dfrac{1}{x^6y^{13}}$ **46.** $\dfrac{a^{10}}{b^{10}}$ **47.** 2.7×10^{-4} **48.** 8.868×10^{-1} **49.** 8.08×10^7 **50.** -8.68×10^5 **51.** 3.188×10^7 **52.** 4.0×10^3
53. 867,000   **54.** 0.00386   **55.** 0.00086   **56.** 893,600   **57.** 100,000,000,000,000,000,000   **58.** 0.0000000000000000000000003
59. 0.016   **60.** 400,000,000,000   **61.** 5   **62.** 2   **63.** 5   **64.** 6   **65.** 22; 78; 154.02; 400   **66.** $2a^2$ **67.** $-4y$ **68.** $15a^2 + 4a$
69. $22x^2 + 3x + 6$ **70.** $-6a^2b - 3b^2 - q^2$ **71.** cannot be combined   **72.** $8x^2 + 3x + 6$
73. $2x^5 + 3x^4 + 4x^3 + 9x^2 + 7x + 6$ **74.** $-7y^2 - 1$ **75.** $-6m^7 - 3x^4 + 7m^6 - 4m^2$ **76.** $-x^2 - 6xy - 2y^2$
77. $-5x^2 + 5x + 1$ **78.** $-2x^2 - x + 20$ **79.** $6x + 30$ **80.** $9x - 63$ **81.** $8a + 28$ **82.** $54a - 27$ **83.** $-7x^3 - 35x$
84. $-32y^3 + 48y$ **85.** $-2x^3 + 18x^2 - 2x$ **86.** $-3a^3b - 3a^2b - 3ab^2$ **87.** $-6a^4 + 8a^2 - 2a$ **88.** $42b^4 - 28b^2 + 14b$
89. $2x^2 - 12x - 14$ **90.** $6x^2 - 11x - 10$ **91.** $4a^2 + 27a - 7$ **92.** $42a^2 + 11a - 3$ **93.** $x^4 + 7x^3 + 4x^2 + 23x - 35$
94. $x^6 + 2x^5 + x^2 + 3x + 2$ **95.** $x^4 + 4x^3 + 4x^2 - 16$ **96.** $x^6 + 8x^4 + 16x^2 - 16$ **97.** $x^3 + 21x^2 + 147x + 343$
98. $8x^3 - 60x^2 + 150x - 125$ **99.** $x^2 + 14x + 49$ **100.** $x^2 - 10x + 25$ **101.** $9x^2 - 42x + 49$ **102.** $16x^2 + 16x + 4$
103. $25x^2 - 90x + 81$ **104.** $25x^2 - 1$ **105.** $49x^2 - 16$ **106.** $a^2 - 4b^2$ **107.** $4x^2 - 36$ **108.** $16a^4 - 4b^2$ **109.** $\dfrac{1}{7} + \dfrac{3}{x} + \dfrac{7}{x^2}$
110. $-a^2 + 3b - 4$ **111.** $a + 1 + \dfrac{6}{a - 2}$ **112.** $4x + \dfrac{7}{x + 5}$ **113.** $a^2 + 3a + 8 + \dfrac{22}{a - 2}$ **114.** $3b^2 - 4b - \dfrac{1}{3b - 2}$
115. $2x^3 - x^2 + 2 - \dfrac{1}{2x - 1}$ **116.** $-x^2 - 16x - 117 - \dfrac{684}{x - 6}$

CHAPTER 3 TEST

1. 32   **2.** 81   **3.** -81 **4.** $\dfrac{1}{64}$ **5.** $-15x^{11}$ **6.** y^5 **7.** $\dfrac{1}{r^5}$ **8.** $\dfrac{y^{14}}{x^2}$ **9.** $\dfrac{1}{6xy^8}$ **10.** 5.63×10^5 **11.** 8.63×10^{-5}
12. 0.0015   **13.** 62,300   **14.** 0.036   **15.** 5   **16.** $-2x^2 + 12x + 11$ **17.** $16x^3 + 7x^2 - 3x - 13$ **18.** $-3x^3 + 5x^2 + 4x + 5$
19. $x^3 + 8x^2 + 3x - 5$ **20.** $3x^3 + 22x^2 + 41x + 14$ **21.** $2x^5 - 5x^4 + 12x^3 - 8x^2 + 4x + 7$ **22.** $3x^2 + 16x - 35$ **23.** $9x^2 - 49$
24. $16x^2 - 16x + 4$ **25.** $64x^2 + 48x + 9$ **26.** $x^4 - 81b^2$ **27.** 1001 ft; 985 ft; 857 ft; 601 ft   **28.** $\dfrac{x}{2y} + \dfrac{1}{4} - \dfrac{7}{8}y$ **29.** $x + 2$
30. $9x^2 - 6x + 4 - \dfrac{16}{3x + 2}$

CUMULATIVE REVIEW

1. a. $11, 112$ **b.** $0, 11, 112$ **c.** $-3, -2, 0, 11, 112$ **d.** $-3, -2, 0, \dfrac{1}{4}, 11, 112$ **e.** $\sqrt{2}$

f. $-2, 0, \dfrac{1}{4}, 112, -3, 11, \sqrt{2}$; Sec. 1.1, Ex. 10   **2. a.** 9   **b.** 125   **c.** 16   **d.** 7   **e.** $\dfrac{9}{49}$; Sec. 1.2, Ex. 1   **3.** $\dfrac{1}{4}$; Sec. 1.2, Ex. 3

4. a. $x + 3$ **b.** $3x$ **c.** $2x$ **d.** $10 - x$ **e.** $5x + 7$; Sec 1.2, Ex. 8   **5.** 6.7; Sec 1.3, Ex. 11   **6. a.** $\dfrac{1}{2}$ **b.** 9; Sec 1.4, Ex. 8
7. 3; Sec. 1.6, Ex. 7   **8.** -70; Sec. 1.6, Ex. 10   **9.** $5x + 10$; Sec. 2.1, Ex. 8   **10.** $-2y - 0.6z + 2$; Sec. 2.1, Ex. 9
11. $-x - y + 2z - 6$; Sec. 2.1, Ex. 10   **12.** $a = 19$; Sec. 2.2, Ex. 6   **13.** $y = 140$; Sec. 2.3, Ex. 4   **14.** $x = 4$; Sec. 2.4, Ex. 5
15. 10; Sec. 2.5, Ex. 1   **16.** 40 ft; Sec. 2.6, Ex. 2

17. 800; Sec. 2.7, Ex. 2   **18.** $x \le 4$; Sec. 2.8, Ex. 6   **19. a.** x^{11} **b.** $\dfrac{1}{16}$ **c.** $81y^{10}$; Sec. 3.1, Ex. 29

20. $\dfrac{b^3}{27a^6}$; Sec. 3.2, Ex. 10   **21.** $\dfrac{1}{25y^6}$; Sec. 3.2, Ex. 14   **22.** $10x^3$; Sec. 3.3, Ex. 8   **23.** $5x^2 - 3x - 3$; Sec. 3.3, Ex. 9

24. $7x^3 + 14x^2 + 35x$; Sec. 3.5, Ex. 4   **25.** $3x^3 - 4 + \dfrac{1}{x}$; Sec. 3.7, Ex. 2

Chapter 4 Factoring Polynomials

Chapter 4 Pretest

1. $2x^2 y(x - 3y)$; 4.1B   **2.** $(x - 4)(y + 6)$; 4.1C   **3.** $(a + 6)(a + 2)$; 4.2A   **4.** prime; 4.2A
5. $3x(x - 1)(x - 5)$; 4.2B   **6.** $(2x - 3)(x + 4)$; 4.3A   **7.** $7(2x + 5)(x + 2)$; 4.3B   **8.** $(3b - 2)(8b - 3)$; 4.4A
9. $(5y + 1)(3y + 7)$; 4.4A   **10.** $(x + 12)^2$; 4.5B   **11.** $(2x - 3y)^2$; 4.5B   **12.** $(a - 7b)(a + 7b)$; 4.5C
13. $(1 - 8t)(1 + 8t)$; 4.5C   **14.** prime; 4.5C   **15.** 18; 4.5A   **16.** $x = 12, x = -5$; 4.6A   **17.** $y = 0, y = 13$; 4.6A
18. $m = 0, m = 4, m = -3$; 4.6B   **19.** 8 in. × 15 in.; 4.7A   **20.** −16 or 15; 4.7A

Mental Math

1. 2   **3.** 3   **5.** 7

Section 4.1

1. y^2 **3.** xy^2 **5.** 4   **7.** $4y^3$ **9.** $5x^2$ **11.** $3x^3$ **13.** $9x^2 y$ **15.** $3(a + 2)$ **17.** $15(2x - 1)$ **19.** $x^2(x + 5)$
21. $2y(3y^3 - 1)$ **23.** $2x(16y - 9x)$ **25.** $4(x - 2y + 1)$ **27.** $3x(2x^2 - 3x + 4)$ **29.** $a^2 b^2(a^5 b^4 - a + b^3 - 1)$ **31.** $5xy(x^2 - 3x + 2)$
33. $4(2x^5 + 4x^4 - 5x^3 + 3)$ **35.** $\dfrac{1}{3}x(x^3 + 2x^2 - 4x + 1)$ **37.** $(x + 2)(y + 3)$ **39.** $(x + 2)(8 - y)$ **41.** answers may vary
43. $(x^2 + 5)(x + 2)$ **45.** $(x + 3)(5 + y)$ **47.** $(2x^2 + 5)(3x - 2)$ **49.** $(y - 4)(2 + x)$ **51.** $(2x + 1)(x^2 + 4)$ **53.** $(x - 2y)(4x - 3)$
55. answers may vary   **57.** $x^2 + 7x + 10$ **59.** $b^2 - 3b - 4$ **61.** 2, 6   **63.** −1, −8   **65.** −2, 5   **67.** −8, 3
69. $2(3x^2 - 1)(2y - 7)$ **71.** $12x^3 - 2x$; $2x(6x^2 - 1)$ **73.** $(n^3 - 6)$ units   **75. a.** 6 million   **b.** 12 million   **c.** $3(x^2 - 7x + 14)$

Mental Math

1. +5   **3.** −3   **5.** +2

Section 4.2

1. $(x + 6)(x + 1)$ **3.** $(x - 9)(x - 1)$ **5.** $(x - 6)(x + 3)$ **7.** $(x + 10)(x - 7)$ **9.** prime   **11.** $(x + 5y)(x + 3y)$
13. $(a^2 - 5)(a^2 + 3)$ **15.** $x^2 + 5x - 24$ **17.** answers may vary   **19.** $2(z + 8)(z + 2)$ **21.** $2x(x - 5)(x - 4)$
23. $(x - 4y)(x + y)$ **25.** $(x + 12)(x + 3)$ **27.** $(x - 2)(x + 1)$ **29.** $(r - 12)(r - 4)$ **31.** $(x + 2y)(x - y)$
33. $3(x + 5)(x - 2)$ **35.** $3(x - 18)(x - 2)$ **37.** $(x - 24)(x + 6)$ **39.** prime   **41.** $(x - 5)(x - 3)$ **43.** $6x(x + 4)(x + 5)$
45. $4y(x^2 + x - 3)$ **47.** $(x - 7)(x + 3)$ **49.** $(x + 5y)(x + 2y)$ **51.** $2(t + 8)(t + 4)$ **53.** $x(x - 6)(x + 4)$
55. $2t^3(t - 4)(t - 3)$ **57.** $5xy(x - 8y)(x + 3y)$ **59.** $2x^2 + 11x + 5$ **61.** $15y^2 - 17y + 4$ **63.** $9a^2 + 23a - 12$
65. $2x^2 + 28x + 66$; $2(x + 3)(x + 11)$ **67.** $(x + 1)(y - 5)(y + 3)$ **69.** 3; 4   **71.** 8; 16   **73.** $(x^n + 2)(x^n + 3)$

Section 4.3

1. $x + 4$ **3.** $10x - 1$ **5.** $4x - 3$ **7.** $(2x + 3)(x + 5)$ **9.** $(y - 1)(8y - 9)$ **11.** $(2x + 1)(x - 5)$ **13.** $(4r - 1)(5r + 8)$
15. $(5x + 1)(2x + 3)$ **17.** $(3x - 2)(x + 1)$ **19.** $(3x - 5y)(2x - y)$ **21.** $(3x - 5)(5x + 3)$ **23.** $(x - 4)(x - 5)$ **25.** $(2x + 11)(x - 9)$
27. $(7t + 1)(t - 4)$ **29.** $(3a + b)(a + 3b)$ **31.** $(7x + 1)(7x - 2)$ **33.** $(6x - 7)(3x + 2)$ **35.** $x(3x + 2)(4x + 1)$
37. $3(7x + 5)(x - 3)$ **39.** $(3x + 4)(4x - 3)$ **41.** $2y^2(3x - 10)(x + 3)$ **43.** $(2x - 7)(2x + 3)$ **45.** $3(x^2 - 14x + 21)$
47. $(4x + 9)(2x - 3)$ **49.** $x(4x + 3)(x - 3)$ **51.** $(4x - 9)(6x - 1)$ **53.** $b(8a - 3)(5a + 3)$ **55.** $(3x^2 + 2)(5x^2 + 3)$
57. $2y(3y + 5)(y - 3)$ **59.** $5x(2x - y)(x + 3y)$ **61.** $60,000 and above   **63.** answers may vary   **65.** $(y - 1)^2(4x^2 + 10x + 25)$
67. $-3xy^2(4x - 5)(x + 1)$ **69.** 5; 13   **71.** 4; 5

Section 4.4

1. $(x + 3)(x + 2)$ **3.** $(x - 4)(x + 7)$ **5.** $(y + 8)(y - 2)$ **7.** $(3x + 4)(x + 4)$ **9.** $(8x - 5)(x - 3)$ **11.** $(5x^2 - 3)(x^2 + 5)$
13. a. 9, 2   **b.** $9x + 2x$ **c.** $(2x + 3)(3x + 1)$ **15. a.** −20, −3   **b.** $-20x - 3x$ **c.** $(5x - 1)(3x - 4)$ **17.** $(3y + 2)(7y + 1)$
19. $(7x - 11)(x + 1)$ **21.** $(5x - 2)(2x - 1)$ **23.** $(2x - 5)(x - 1)$ **25.** $(2x + 3)(2x + 3)$ or $(2x + 3)^2$ **27.** $(2x + 3)(2x - 7)$
29. $(5x - 4)(2x - 3)$ **31.** $x(2x + 3)(x + 5)$ **33.** $2(8y - 9)(y - 1)$ **35.** $(2x - 3)(3x - 2)$ **37.** $3(3a + 2)(6a - 5)$
39. $a(4a + 1)(5a + 8)$ **41.** $3x(4x + 3)(x - 3)$ **43.** $x^2 - 4$ **45.** $y^2 + 8y + 16$ **47.** $81z^2 - 25$ **49.** $16x^2 - 24x + 9$
51. $(x^n + 2)(x^n + 3)$ **53.** $(3x^n - 5)(x^n + 7)$

Calculator Exploration

16, 14, 16; 16, 14, 16; 2.89, 0.89, 2.89; 171.61, 169.61, 171.61; 1, −1, 1

Mental Math

1. 1^2 **3.** 9^2 **5.** 3^2 **7.** $(3x)^2$ **9.** $(5a)^2$ **11.** $(6p^2)^2$

Section 4.5

1. yes **3.** no **5.** yes **7.** no **9.** no **11.** yes **13.** 8 **15.** $(x+11)^2$ **17.** $(x-8)^2$ **19.** $(4a-3)^2$ **21.** $(x^2+2)^2$
23. $2(n-7)^2$ **25.** $(4y+5)^2$ **27.** $(xy-5)^2$ **29.** $m(m+9)^2$ **31.** prime **33.** $(3x-4y)^2$ **35.** $(x+7y)^2$
37. answers may vary **39.** $(x-2)(x+2)$ **41.** $(9-p)(9+p)$ **43.** $(2r-1)(2r+1)$ **45.** $(3x-4)(3x+4)$ **47.** prime
49. $(-6+x)(6+x)$ **51.** $(m^2+1)(m+1)(m-1)$ **53.** $(x-13y)(x+13y)$ **55.** $2(3r-2)(3r+2)$ **57.** $x(3y-2)(3y+2)$
59. $25y^2(y-2)(y+2)$ **61.** $xy(x-2y)(x+2y)$ **63.** $9(5a-3b)(5a+3b)$ **65.** $3(2x-3)(2x+3)$ **67.** $(7a-4)(7a+4)$
69. $(13a-7b)(13a+7b)$ **71.** $(4-ab)(4+ab)$ **73.** $\left(y-\dfrac{1}{4}\right)\left(y+\dfrac{1}{4}\right)$ **75.** $\left(10-\dfrac{2}{9}n\right)\left(10+\dfrac{2}{9}n\right)$ **77.** $(5-y)$

79. $y=-5$ **81.** $x=3$ **83.** $a=-\dfrac{1}{2}$ **85.** 5 ft **87.** $(y-6-z)(y-6+z)$ **89.** $(m-3)(m+3)(n+8)$

91. $(x+1-6y)(x+1+6y)$ **93.** $(x^n+9)(x^n-9)$ **95.** perfect square trinomial **97. a.** 513 ft **b.** 273 ft
c. 6 seconds **d.** $(23-4t)(23+4t)$

Integrated Review

1. $(x-3)(x+4)$ **2.** $(x-2)(x-8)$ **3.** $(x+2)(x-3)$ **4.** $(x+1)^2$ **5.** $(x-3)^2$ **6.** $(x+2)(x-1)$
7. $(x+3)(x-2)$ **8.** $(x+3)(x+4)$ **9.** $(x-5)(x-2)$ **10.** $(x-6)(x+5)$ **11.** $2(x-7)(x+7)$ **12.** $3(x-5)(x+5)$
13. $(x+3)(x+5)$ **14.** $(y-7)(3+x)$ **15.** $(x+8)(x-2)$ **16.** $(x-7)(x+4)$ **17.** $4x(x+7)(x-2)$
18. $6x(x-5)(x+4)$ **19.** $2(3x+4)(2x+3)$ **20.** $(2a-b)(4a+5b)$ **21.** $(2a-b)(2a+b)$ **22.** $(x-5y)(x+5y)$
23. $(4-3x)(7+2x)$ **24.** $(5-2x)(4+x)$ **25.** prime **26.** prime **27.** $(3y+5)(2y-3)$ **28.** $(4x-5)(x+1)$
29. $9x(2x^2-7x+1)$ **30.** $4a(3a^2-6a+1)$ **31.** $(4a-7)^2$ **32.** $(5p-7)^2$ **33.** $(7-x)(2+x)$ **34.** $(3+x)(1-x)$
35. $3x^2y(x+6)(x-4)$ **36.** $2xy(x+5y)(x-y)$ **37.** $3xy(4x^2+81)$ **38.** $2xy^2(3x^2+4)$ **39.** $2xy(1-6x)(1+6x)$
40. $2x(x-3)(x+3)$ **41.** $(x-2)(x+2)(x+6)$ **42.** $(x-2)(x-6)(x+6)$ **43.** $2a^2(3a+5)$ **44.** $2n(2n-3)$
45. $(x^2+4)(3x-1)$ **46.** $(x-2)(x^2+3)$ **47.** $6(x+2y)(x+y)$ **48.** $2(x+4y)(6x-y)$ **49.** $(5+x)(x+y)$
50. $(x-y)(7+y)$ **51.** $(7t-1)(2t-1)$ **52.** prime **53.** $(3x+5)(x-1)$ **54.** $(7x-2)(x+3)$ **55.** $(1-10a)(1+2a)$
56. $(1+5a)(1-12a)$ **57.** $(x-3)(x+3)(x-1)(x+1)$ **58.** $(x-3)(x+3)(x-2)(x+2)$ **59.** $(x-15)(x-8)$
60. $(y+16)(y+6)$ **61.** answers may vary **62.** yes; $9(x^2+9y^2)$

Mental Math

1. $a=3, a=7$ **3.** $x=-8, x=-6$ **5.** $x=-1, x=3$

Section 4.6

1. $x=2, x=-1$ **3.** $x=6, x=7$ **5.** $x=-9, x=-17$ **7.** $x=0, x=-6$ **9.** $x=0, x=8$ **11.** $x=-\dfrac{3}{2}, x=\dfrac{5}{4}$

13. $x=\dfrac{7}{2}, x=-\dfrac{2}{7}$ **15.** $x=\dfrac{1}{2}, x=-\dfrac{1}{3}$ **17.** $x=-0.2, x=-1.5$ **19.** $(x-6)(x+1)=0$ **21.** $x=9, x=4$ **23.** $x=-4, x=2$

25. $x=0, x=7$ **27.** $x=0, x=-20$ **29.** $x=4, x=-4$ **31.** $x=8, x=-4$ **33.** $x=\dfrac{7}{3}, x=-2$ **35.** $x=\dfrac{8}{3}, x=-9$

37. $x=0, x=\dfrac{1}{2}, x=-\dfrac{1}{2}$ **39.** $x=\dfrac{17}{2}$ **41.** $x=\dfrac{3}{4}$ **43.** $y=\dfrac{1}{2}, y=-\dfrac{1}{2}$ **45.** $x=-\dfrac{3}{2}, x=-\dfrac{1}{2}, x=3$ **47.** $x=-5, x=3$

49. $x=0, x=16$ **51.** $x=-5, x=6$ **53.** $y=-\dfrac{4}{3}, y=5$ **55.** $y=-4, y=3$ **57.** $x=0, x=8, x=4$ **59.** $x^2-12x+35=0$

61. $\dfrac{47}{45}$ **63.** $\dfrac{17}{60}$ **65.** $\dfrac{7}{10}$ **67.** didn't write equation in standard form **69. a.** 300; 304; 276; 216; 124; 0; -156 **b.** 5 sec

c. 304 ft **71.** $x=0, x=\dfrac{1}{2}$ **73.** $x=0, x=-15$

Section 4.7

1. width $=x$; length $=x+4$ **3.** x and $x+2$ if x is an odd integer **5.** base $=x$; height $=4x+1$ **7.** 11 units
9. 15 cm, 13 cm, 70 cm, 22 cm **11.** base $=16$ mi; height $=6$ mi **13.** 5 sec **15.** length $=5$ cm; width $=6$ cm **17.** 54 diagonals
19. 10 sides **21.** -12 or 11 **23.** slow boat: 8 mph; fast boat: 15 mph **25.** 13 and 7 **27.** 5 in. **29.** 12 mm; 16 mm; 20 mm
31. 10 km **33.** 36 ft **35.** 6.25 seconds **37.** 175 acres **39.** 6.25 million **41.** 1966 **43.** answers may vary
45. width of pool: 29 meters; length of pool: 35 meters **47.** 60129: Esmond, IL; 60130: Forest Park, IL

Chapter 4 Review

1. $2x-5$ **2.** $2x^4+1-5x^3$ **3.** $5(m+6)$ **4.** $4x(5x^2+3x+6)$ **5.** $(2x+3)(3x-5)$ **6.** $(x+1)(5x-1)$ **7.** $(x-1)(3x+2)$
8. $(2x-1)(3x+5)$ **9.** $(a+3b)(3a+b)$ **10.** $(x+4)(x+2)$ **11.** $(x-8)(x-3)$ **12.** prime **13.** $(x-6)(x+1)$
14. $(x+4)(x-2)$ **15.** $(x+6y)(x-2y)$ **16.** $(x+5y)(x+3y)$ **17.** $2(3-x)(12+x)$ **18.** $4(8+3x-x^2)$ **19.** $5y(y-6)(y-4)$
20. $-48, 2$ **21.** factor out the GCF, 3 **22.** $(2x+1)(x+6)$ **23.** $(2x+3)(2x-1)$ **24.** $(3x+4y)(2x-y)$ **25.** prime
26. $(2x+3)(x-13)$ **27.** $(6x+5y)(3x-4y)$ **28.** $5y(2y-3)(y+4)$ **29.** $5x^2-9x-2; (5x+1)(x-2)$
30. $16x^2-28x+6; 2(4x-1)(2x-3)$ **31.** yes **32.** no **33.** no **34.** yes **35.** yes; $(x-3)(x+3)$ **36.** no **37.** yes
38. no **39.** $(x-9)(x+9)$ **40.** $(x+6)^2$ **41.** $(2x-3)(2x+3)$ **42.** $(3t-5s)(3t+5s)$ **43.** prime **44.** $(n-9)^2$
45. $3(r+6)^2$ **46.** $(3y-7)^2$ **47.** $5m^6(m-1)(m+1)$ **48.** $(2x-7y)^2$ **49.** $3y(x+y)^2$ **50.** $(2x-1)(2x+1)(4x^2+1)$

51. $x=-6, x=2$ **52.** $x=0, x=-1, x=\dfrac{2}{7}$ **53.** $x=-\dfrac{1}{5}, x=-3$ **54.** $x=-7, x=-1$ **55.** $x=-4, x=6$ **56.** $x=-5$

A26

57. $x = 2, x = 8$ **58.** $x = \dfrac{1}{3}$ **59.** $x = -\dfrac{2}{7}, x = \dfrac{3}{8}$ **60.** $m = 0, m = 6$ **61.** $r = 5, r = -5$ **62.** $x^2 - 9x + 20 = 0$ **63.** c   **64.** d
65. 9 units   **66.** 8 units, 13 units, 16 units, 10 units   **67.** width: 20 in.; length: 25 in.   **68.** 36 yd   **69.** 19 and 20
70. a. 17.5 sec and 10 sec; answers may vary   **b.** 27.5 sec   **71.** 32 cm

CHAPTER 4 TEST

1. $3x(3x - 1)$ **2.** $(x + 7)(x + 4)$ **3.** $(7 - m)(7 + m)$ **4.** $(y + 11)^2$ **5.** $(x - 2)(x + 2)(x^2 + 4)$ **6.** $(4 - y)(a + 3)$
7. prime   **8.** $(y - 12)(y + 4)$ **9.** $(3a - 7)(a + b)$ **10.** $(3x - 2)(x - 1)$ **11.** $5(6 - x)(6 + x)$ **12.** $3x(x - 5)(x - 2)$
13. $(6t + 5)(t - 1)$ **14.** $(y - 2)(y + 2)(x - 7)$ **15.** $x(1 - x)(1 + x)(1 + x^2)$ **16.** $(x + 12y)(x + 2y)$ **17.** $x = 3$ and $x = -9$
18. $x = -6$ and $x = -4$ **19.** $x = -7, x = 2$ **20.** $x = 0, x = \dfrac{3}{2}, x = -\dfrac{4}{3}$ **21.** $t = 0, t = 3, t = -3$ **22.** $x = 0, x = -4$
23. $t = -3, t = 5$ **24.** $x = -\dfrac{2}{3}, x = 1$ **25.** width: 6 units; length: 9 units   **26.** 17 ft   **27.** 8 and 9   **28.** 7 sec
29. hypotenuse: 25 cm; legs: 15 cm, 20 cm

CUMULATIVE REVIEW

1. a. $9 \le 11$ **b.** $8 > 1$ **c.** $3 \ne 4$; Sec. 1.1, Ex. 7   **2.** solution; Sec. 1.2, Ex. 7   **3.** -12; Sec. 1.4, Ex. 5
4. a. -6 **b.** -24; Sec. 1.5, Ex. 8   **5.** $5x + 7$; Sec. 2.1, Ex. 4   **6.** $-4a - 1$; Sec. 2.1, Ex. 5   **7.** $7.3x - 6$; Sec. 2.1, Ex. 7
8. $x = -11$; Sec. 2.3, Ex. 3   **9.** every real number; Sec. 2.4, Ex. 7   **10.** $l = \dfrac{V}{wh}$; Sec. 2.6, Ex.3   **11.** $x = \dfrac{31}{2}$; Sec. 2.7, Ex. 6
12. 5^{18}; Sec. 3.1, Ex. 14   **13.** y^{16}; Sec. 3.1, Ex. 15   **14.** x^6; Sec. 3.2, Ex. 9   **15.** $\dfrac{y^{18}}{z^{36}}$; Sec. 3.2, Ex. 11   **16.** $\dfrac{1}{x^{19}}$; Sec. 3.2, Ex. 13
17. $4x$; Sec. 3.3, Ex. 6   **18.** $13x^2 - 2$; Sec. 3.3, Ex. 7   **19.** $4x^2 - 4xy + y^2$; Sec. 3.5, Ex. 8   **20.** $t^2 + 4t + 4$; Sec. 3.6, Ex. 5
21. $x^4 - 14x^2y + 49y^2$; Sec. 3.6, Ex. 8   **22.** $2xy - 4 + \dfrac{1}{2y}$; Sec. 3.7, Ex. 3   **23.** $(x + 3)(5 + y)$; Sec. 4.1, Ex. 7
24. $(x^2 + 2)(x^2 + 3)$; Sec. 4.2, Ex. 7   **25.** $2(x - 2)(3x + 5)$; Sec. 4.4, Ex. 2

Chapter 5 RATIONAL EXPRESSIONS

CHAPTER 5 PRETEST

1. $x = -1, x = 10$; 5.1B   **2.** $\dfrac{4}{x + 2}$ 5.1C   **3.** $10(x + 2)(x + 3)$; 5.3B   **4.** 3; 5.2A   **5.** $\dfrac{5(x + 5)}{x^3(x - 5)}$; 5.2B   **6.** $\dfrac{1}{b - 11}$; 5.3A
7. $\dfrac{7 - 4x}{x - 1}$; 5.4A   **8.** $\dfrac{9}{x - 5}$; 5.4A   **9.** $\dfrac{a^2 + 12}{(a + 4)(a - 4)(a - 3)}$; 5.4A   **10.** $b = -7$; 5.5A   **11.** no solution; 5.5A
12. $y = -1$; 5.5A   **13.** $b = \dfrac{2A}{h}$; 5.5B   **14.** $\dfrac{15n^6}{m^3}$; 5.7A   **15.** $4a - 1$; 5.7A, B   **16.** $x = 5$; 5.6D   **17.** 2 or 5; 5.6A
18. $3\dfrac{1}{13}$ hr; 5.6B   **19.** 250 mph; 5.6C

MENTAL MATH

1. $x = 0$ **3.** $x = 0, x = 1$

SECTION 5.1

1. $\dfrac{7}{4}$ **3.** $-\dfrac{8}{3}$ **5.** $-\dfrac{11}{2}$ **7. a.** \$37.5 million   **b.** \$85.7 million   **c.** \$48.2 million   **9.** $x = 0$ **11.** $x = -2$ **13.** $x = 4$
15. $x = -2$ **17.** none   **19.** answers may vary   **21.** $\dfrac{1}{4(x + 2)}$ **23.** $\dfrac{1}{x + 2}$ **25.** can't simplify   **27.** 1   **29.** -1 **31.** -5
33. $\dfrac{1}{x - 9}$ **35.** $5x + 1$ **37.** $\dfrac{1}{x - 2}$ **39.** $x + 2$ **41.** $\dfrac{x + 5}{x - 5}$ **43.** $\dfrac{x + 2}{x + 4}$ **45.** $\dfrac{x + 2}{2}$ **47.** $\dfrac{11x}{6}$ **49.** -1 **51.** $\dfrac{x + 1}{x - 1}$
53. $\dfrac{m - 3}{m + 3}$ **55.** $\dfrac{3}{11}$ **57.** $\dfrac{50}{99}$ **59.** $\dfrac{4}{3}$ **61.** $\dfrac{117}{40}$ **63.** $x + y$ **65.** $\dfrac{5 - y}{2}$ **67.** answers may vary   **69.** 400 mg

71. no; $B \approx 24$

MENTAL MATH

1. $\dfrac{2x}{3y}$ **3.** $\dfrac{5y^2}{7x^2}$ **5.** $\dfrac{9}{5}$

SECTION 5.2

1. $\dfrac{21}{4y}$ **3.** x^4 **5.** $-\dfrac{b^2}{6}$ **7.** $\dfrac{x^2}{10}$ **9.** $\dfrac{1}{3}$ **11.** 1   **13.** $\dfrac{x + 5}{x}$ **15.** $\dfrac{2}{9(x - 5)}$ sq ft   **17.** x^4 **19.** $\dfrac{12}{y^6}$ **21.** $x(x + 4)$
23. $\dfrac{3(x + 1)}{x^3(x - 1)}$ **25.** $m^2 - n^2$ **27.** $-\dfrac{x + 2}{x - 3}$ **29.** $\dfrac{x + 2}{x - 3}$ **31.** $\dfrac{5}{6}$ **33.** $\dfrac{3x}{8}$ **35.** $\dfrac{3}{2}$ **37.** $\dfrac{3x + 4y}{2(x + 2y)}$ **39.** $\dfrac{2(x + 2)}{x - 2}$

41. $\dfrac{(a+5)(a+3)}{(a+2)(a+1)}$ **43.** 1440   **45.** 73   **47.** 3424.8 mph   **49.** $10.63   **51.** 1   **53.** $-\dfrac{10}{9}$ **55.** $-\dfrac{1}{5}$ **57.** $\dfrac{x}{2}$

59. $\dfrac{5a(2a+b)(3a-2b)}{b^2(a-b)(a+2b)}$ **61.** answers may vary   **63.** 50 8-oz cups

MENTAL MATH

1. 1   **3.** $\dfrac{7x}{9}$ **5.** $\dfrac{1}{9}$ **7.** $\dfrac{17y}{5}$

SECTION 5.3

1. $\dfrac{a+9}{13}$ **3.** $\dfrac{3m}{n}$ **5.** 4   **7.** $\dfrac{y+10}{3+y}$ **9.** 3   **11.** $\dfrac{1}{a+5}$ **13.** $\dfrac{1}{x-6}$ **15.** $\dfrac{20}{x-2}$ m   **17.** answers may vary   **19.** $4x^3$

21. $8x(x+2)$ **23.** $(x+3)(x-2)$ **25.** $3(x+6)$ **27.** $6(x+1)^2$ **29.** $x-8$ or $8-x$ **31.** $(x-1)(x+4)(x+3)$

33. answers may vary   **35.** $\dfrac{6x}{4x^2}$ **37.** $\dfrac{24b^2}{12ab^2}$ **39.** $\dfrac{18}{2(x+3)}$ **41.** $\dfrac{9ab+2b}{5b(a+2)}$ **43.** $\dfrac{x^2+x}{x(x+4)(x+2)(x+1)}$ **45.** $\dfrac{18y-2}{30x^2-60}$

47. $\dfrac{29}{21}$ **49.** $-\dfrac{5}{12}$ **51.** $\dfrac{7}{30}$ **53.** 3 packages hot dogs and 2 packages buns   **55.** answers may vary

SECTION 5.4

1. $\dfrac{5}{x}$ **3.** $\dfrac{75a+6b^2}{5b}$ **5.** $\dfrac{6x+5}{2x^2}$ **7.** $\dfrac{11}{x+1}$ **9.** $\dfrac{17x+30}{2(x-2)(x+2)}$ **11.** $\dfrac{35x-6}{4x(x-2)}$ **13.** $-\dfrac{2}{x-3}$ **15.** $-\dfrac{1}{x^2-1}$

17. $\dfrac{5+2x}{x}$ **19.** $\dfrac{6x-7}{x-2}$ **21.** $-\dfrac{y+4}{y+3}$ **23.** answers may vary   **25.** 2   **27.** $3x^3-4$ **29.** $\dfrac{x+2}{(x+3)^2}$ **31.** $\dfrac{9b-4}{5b(b-1)}$

33. $\dfrac{2+m}{m}$ **35.** $\dfrac{10}{1-2x}$ **37.** $\dfrac{15x-1}{(x+1)^2(x-1)}$ **39.** $\dfrac{x^2-3x-2}{(x-1)^2(x+1)}$ **41.** $\dfrac{a+2}{2(a+3)}$ **43.** $\dfrac{x-10}{2(x-2)}$ **45.** $\dfrac{-3-2y}{(y-2)(y-1)}$

47. $\dfrac{-5+23}{(x-2)(x-3)}$ **49.** $\dfrac{2x^2-2x-46}{(x+1)(x-6)(x-5)}$ **51.** $x=\dfrac{2}{3}$ **53.** $x=-\dfrac{1}{2}, x=1$ **55.** $x=-\dfrac{15}{2}$ **57.** $\dfrac{2x-16}{(x-4)(x+4)}$ in.

59. $\dfrac{11DA-DA^2-12D}{24(A+12)}$ **61.** answers may vary   **63.** $\dfrac{-3x^2+7x+2}{(x+5)(x+1)(x-1)}$ **65.** $\dfrac{73-7x}{(x-4)(x+1)(x+5)}$

MENTAL MATH

1. $x=10$ **3.** $z=36$

SECTION 5.5

1. $x=30$ **3.** $x=0$ **5.** $x=-2$ **7.** $x=-5, x=2$ **9.** $a=5$ **11.** $x=3$ **13.** $y=1$ **15.** $x=-3$ **17.** no solution

19. $a=1$ **21.** no solution   **23.** $x=3, x=-4$ **25.** $x=6, x=-4$ **27.** $y=5$ **29.** $x=0$ **31.** $t=8, t=-2$ **33.** $x=-2$

35. no solution   **37.** $r=3$ **39.** $x=-11, x=1$ **41.** $r=\dfrac{d}{t}$ **43.** $Q=\dfrac{V}{T}$ **45.** $t=\dfrac{A-Bi}{i}$ **47.** $G=\dfrac{V}{N-R}$ **49.** $r=\dfrac{C}{2\pi}$

51. $\dfrac{1}{x}$ **53.** $\dfrac{1}{x}+\dfrac{1}{2}$ **55.** $\dfrac{1}{3}$ **57.** $100°, 80°$ **59.** $22.5°, 67.5°$ **61.** $a=5$

INTEGRATED REVIEW

1. expression; $\dfrac{3+2x}{3x}$ **2.** expression; $\dfrac{18+5a}{6a}$ **3.** equation; $x=3$ **4.** equation; $a=18$ **5.** expression; $\dfrac{x-1}{x(x+1)}$

6. expression; $\dfrac{3(x+1)}{x(x-3)}$ **7.** equation; no solution   **8.** equation; $x=1$ **9.** expression; 10   **10.** expression; $\dfrac{z}{3(9z-5)}$

11. expression; $\dfrac{5x+7}{x-3}$ **12.** expression; $\dfrac{7p+5}{2p+7}$ **13.** equation; $x=23$ **14.** equation; $x=3$ **15.** expression; $\dfrac{25a}{9(a-2)}$

16. expression; $\dfrac{9}{4(x-1)}$ **17.** answers may vary   **18.** answers may vary

SECTION 5.6

1. 2   **3.** -3 **5.** 5   **7.** 2   **9.** $2\dfrac{2}{9}$ hr   **11.** $1\dfrac{1}{2}$ min   **13.** $108.00   **15.** 3 hr   **17.** 20 hr   **19.** 6 mph

21. 1st portion speed: 10 mph; cooldown speed: 8 mph   **23.** 30 mph   **25.** 8 mph   **27.** 63 mph   **29.** $x=6$ **31.** $x=5$

33. $y=21.25$ **35.** $x=4.4$ ft; $y=5.6$ ft   **37.** $\dfrac{1}{2}$ **39.** $\dfrac{3}{7}$ **41.** Zanardi's speed: 227.7 mph; Gugelmin's speed: 233.3 mph

SECTION 5.7

1. $\dfrac{2}{3}$ **3.** $\dfrac{2}{3}$ **5.** $-\dfrac{4x}{15}$ **7.** $\dfrac{4}{3}$ **9.** $\dfrac{27}{16}$ **11.** $\dfrac{m-n}{m+n}$ **13.** $\dfrac{2x(x-5)}{7x^2+10}$ **15.** $\dfrac{1}{y-1}$ **17.** $\dfrac{1}{6}$ **19.** $\dfrac{x+y}{x-y}$ **21.** $\dfrac{3}{7}$

23. $\dfrac{a}{x+b}$ **25.** $\dfrac{x+8}{2-x}$ or $-\dfrac{x+8}{x-2}$ **27.** $\dfrac{s^2+r^2}{s^2-r^2}$ **29.** answers may vary   **31.** *Thriller*

33. tie between *Born in the U.S.A.* and *Eagles Greatest Hits*   **35.** $\dfrac{13}{24}$ **37.** $\dfrac{R_1R_2}{R_2+R_1}$ **39.** $\dfrac{2x}{2-x}$ **41.** $\dfrac{1}{y^2-1}$

CHAPTER 5 REVIEW

1. $x = 2, x = -2$ **2.** $x = \dfrac{5}{2}, x = -\dfrac{3}{2}$ **3.** $\dfrac{4}{3}$ **4.** $\dfrac{11}{12}$ **5.** $\dfrac{2}{x}$ **6.** $\dfrac{3}{x}$ **7.** $\dfrac{1}{x-5}$ **8.** $\dfrac{1}{x+1}$ **9.** $\dfrac{x(x-2)}{x+1}$ **10.** $\dfrac{5(x-5)}{x-3}$

11. $\dfrac{x-3}{x-5}$ **12.** $\dfrac{x}{x+4}$ **13.** $\dfrac{x+a}{x-c}$ **14.** $\dfrac{x+5}{x-3}$ **15.** $\dfrac{3x^2}{y}$ **16.** $-\dfrac{9x^2}{8}$ **17.** $\dfrac{x-3}{x+2}$ **18.** $-\dfrac{2x(2x+5)}{(x-6)^2}$ **19.** $\dfrac{x+3}{x-4}$

20. $\dfrac{4x}{3y}$ **21.** $(x-6)(x-3)$ **22.** $\dfrac{2}{3}$ **23.** $\dfrac{1}{x-3}$ **24.** $\dfrac{x}{x+6}$ **25.** $\dfrac{1}{2}$ **26.** $\dfrac{3(x+2)}{3x+y}$ **27.** $\dfrac{1}{x+2}$ **28.** $\dfrac{1}{x-3}$

29. $\dfrac{2x-10}{3x^2}$ **30.** $\dfrac{2x+1}{2x^2}$ **31.** $14x$ **32.** $(x-8)(x+8)(x+3)$ **33.** $\dfrac{10x^2y}{14x^3y}$ **34.** $\dfrac{36y^2x}{16y^3x}$ **35.** $\dfrac{x^2-3x-10}{(x+2)(x-5)(x+9)}$

36. $\dfrac{3x^2+4x-15}{(x+2)^2(x+3)}$ **37.** $\dfrac{4y-30x^2}{5x^2y}$ **38.** $\dfrac{-2x+10}{(x-3)(x-1)}$ **39.** $\dfrac{14x+58}{(x+3)(x+7)}$ **40.** $\dfrac{-2x-2}{x+3}$

41. $\dfrac{5x+5}{(x+4)(x-2)(x-1)}$ **42.** $\dfrac{x-4}{3x}$ **43.** $-\dfrac{x}{x-1}$ **44.** $\dfrac{x^2+2x-3}{(x+2)^2}$ **45.** $\dfrac{x^2+2x+4}{4x}; \dfrac{x+2}{32}$

46. $\dfrac{29x}{12(x-1)}; \dfrac{3xy}{5(x-1)}$ **47.** $x=1$ **48.** $n=30$ **49.** $y=2$ **50.** $x=3, x=-4$ **51.** $a=-\dfrac{5}{2}$ **52.** no solution   **53.** $x=1$

54. $x=5$ **55.** $x=\dfrac{9}{7}$ **56.** $x=-6, x=1$ **57.** $b=\dfrac{4A}{5x^2}$ **58.** $y=\dfrac{560-8x}{7}$ **59.** 3 **60.** 2

61. fast car speed: 30 mph; slow car speed: 20 mph   **62.** 20 mph   **63.** $17\dfrac{1}{2}$ hr   **64.** $8\dfrac{4}{7}$ days   **65.** $x=15$ **66.** $x=6$ **67.** $x=15$

68. $x=60$ **69.** $-\dfrac{7}{18y}$ **70.** $\dfrac{2x}{x-3}$ **71.** $\dfrac{6}{7}$ **72.** $\dfrac{2x^2+1}{x+2}$ **73.** $\dfrac{3y-1}{2y-1}$ **74.** $-\dfrac{7+2x}{2x}$

CHAPTER 5 TEST

1. $x=-1, x=-3$ **2. a.** \$115   **b.** \$103   **3.** $\dfrac{3}{5}$ **4.** $\dfrac{1}{x-10}$ **5.** $\dfrac{1}{x+6}$ **6.** -1 **7.** $\dfrac{2m(m+2)}{m-2}$

8. $-\dfrac{1}{x+y}$ **9.** $\dfrac{(x-6)(x-7)}{(x+2)(x+7)}$ **10.** 15 **11.** $\dfrac{y-2}{4}$ **12.** $-\dfrac{1}{2x+5}$ **13.** $\dfrac{3a-4}{(a-3)(a+2)}$ **14.** $\dfrac{3}{x-1}$

15. $\dfrac{2(x+3)(x+5)}{x(x^2+4x+1)}$ **16.** $\dfrac{x^2+2x+35}{(x+9)(x+2)(x-5)}$ **17.** $\dfrac{4y^2+13y-15}{(y+5)(y+1)(y+4)}$ **18.** $y=\dfrac{30}{11}$ **19.** $y=-6$

20. no solution   **21.** no solution   **22.** $\dfrac{xz}{2y}$ **23.** $b-a$ **24.** $\dfrac{5y^2-1}{y+2}$ **25.** $x=12$

26. $x=1$ and $x=5$ **27.** 30 mph   **28.** $6\dfrac{2}{3}$ hr

CUMULATIVE REVIEW

1. a. $\dfrac{15}{x}=4$ **b.** $12-3=x$ **c.** $4x+17=21$; Sec. 1.2, Ex. 9   **2. a.** -12 **b.** -9; Sec. 1.3, Ex. 13

3. commutative property of multiplication; Sec. 1.7, Ex. 15   **4.** associative property of addition; Sec. 1.7, Ex. 16   **5.** $x=-4$; Sec. 2.2, Ex. 7

6. shorter piece, 2 feet; longer piece, 8 feet; Sec. 2.5, Ex. 2   **7.** $\dfrac{y-b}{m}=x$; Sec. 2.6, Ex. 4   **8.** $x \le -10$; Sec. 2.8, Ex. 3

9. x^3; Sec. 3.1, Ex. 21   **10.** $4^4=256$; Sec. 3.1, Ex. 22   **11.** -27; Sec. 3.1, Ex. 23   **12.** $2x^4y$; Sec. 3.1, Ex. 24   **13.** $\dfrac{2}{x^3}$; Sec. 3.2, Ex. 2

14. $\dfrac{1}{16}$; Sec. 3.2, Ex. 4   **15.** $10x^4+30x$; Sec. 3.5, Ex. 5   **16.** $-15x^4-18x^3+3x^2$; Sec. 3.5, Ex. 6

17. $4x^2-4x+6+\dfrac{-11}{2x+3}$; Sec. 3.7, Ex. 6   **18.** $(x+3)(x+4)$; Sec. 4.2, Ex. 1   **19.** $(5x+2y)^2$; Sec. 4.5, Ex. 5

20. $x=11, x=-2$; Sec. 4.6, Ex. 4   **21.** $\dfrac{2}{5}$; Sec. 5.2, Ex. 2   **22.** $3x-5$; Sec. 5.3, Ex. 3   **23.** $\dfrac{3}{x-2}$; Sec. 5.4, Ex. 2   **24.** $t=5$; Sec. 5.5, Ex. 2

25. $2\dfrac{1}{10}$ hr; Sec. 5.6, Ex. 2   **26.** $\dfrac{3}{z}$; Sec. 5.7, Ex. 3

Chapter 6 GRAPHING EQUATIONS AND INEQUALITIES

CHAPTER 6 PRETEST

1. 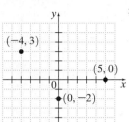 ; 6.1B **2.** $(-2, -6)$; 6.1C **3.** 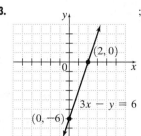 ; 6.2A **4.** 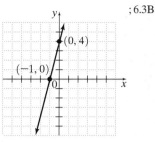 ; 6.3B

5. 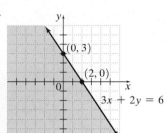 ; 6.7B **6.** $-\dfrac{3}{10}$; 6.4A **7.** $\dfrac{4}{5}$; 6.4B **8.** undefined slope; 6.4C

9. $x + 3y = -15$; 6.5C **10.** $x + 8y = 0$; 6.5D **11.** $2x - 7y = -98$; 6.5A

12. domain: $\{-3, 0, 2, 7\}$; range: $\{-1, 6, 8\}$; 6.6A **13.** function; 6.6B

14. not a function; 6.6B **15. a.** 11; 6.6D **b.** 8; 6.6D **c.** -22; 6.6D

SECTION 6.1

1. 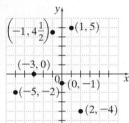 ; $(1, 5)$ is in quadrant I, $\left(-1, 4\dfrac{1}{2}\right)$ is in quadrant II, $(-5, -2)$ is in quadrant III, $(2, -4)$ is in

quadrant IV, $(-3, 0)$ lies on the x-axis, $(0, 1)$ lies on the y-axis

3. $a = b$ **5.** $A: (0, 0)$ **7.** $C: (3, 2)$ **9.** $E: (-2, -2)$ **11.** $G: (2, -1)$

13. $B: (0, -3)$ **15.** $D: (1, 3)$ **17.** $F: (-3, -1)$

19. a. $(1991, 80), (1992, 79), (1993, 77), (1994, 74),$
$(1995, 77), (1996, 85), (1997, 86)$

b.

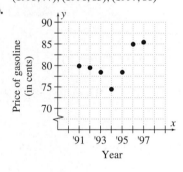

21. a. $(1994, 578), (1995, 613), (1996, 654),$
$(1997, 675), (1998, 717)$

b.

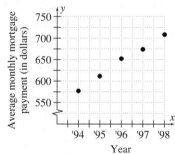

c. The scatter diagram shows a trend of increasing mortgage payments.

23. a. $(0.50, 10), (0.75, 12), (1.00, 15), (1.25, 16), (1.50, 18), (1.50, 19), (1.75, 19), (2.00, 20)$

b.

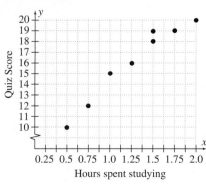

c. Minh might conclude that more time spent studying results in a better quiz score.

25. $(-4, -2); (4, 0)$ **27.** $(0, 9); (3, 0)$ **29.** $(11, -7)$; any x

31.

x	y
0	2
6	0
3	1

33.

x	y
0	−12
5	−2
3	−6

35.

x	y
0	$\frac{5}{7}$
$\frac{5}{2}$	0
−1	1

37.

x	y
3	0
3	−0.5
3	$\frac{1}{4}$

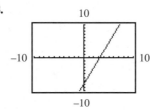

39.

x	y
0	0
−5	1
10	−2

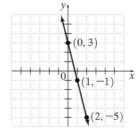

41. a. 13,000; 21,000; 29,000   **b.** 45 desks   **43.** $y = 5 - x$ **45.** $y = \dfrac{5 - 2x}{4}$ **47.** $y = -2x$ **49.** 26 units

51. $500 million; $1,500 million; $1,000 million; $1,500 million   **53.** answers may vary   **55. a.** 29.219; 45.599; 54.699   **b.** 1977

GRAPHING CALCULATOR EXPLORATIONS

1.

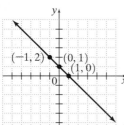

3.

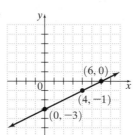

5.

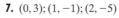

SECTION 6.2

1. $(6, 0); (4, -2); (5, -1)$

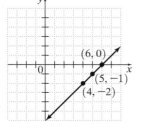

3. $(1, -4); (0, 0); (-1, 4)$

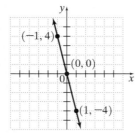

5. $(0, 0); (6, 2); (-3, -1)$

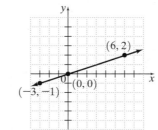

7. $(0, 3); (1, -1); (2, -5)$

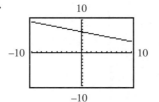

9.

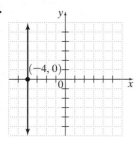

11.

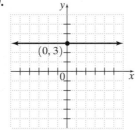

13.

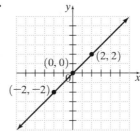

15.

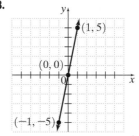

17.

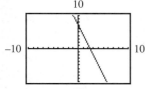

19.

21.

23.

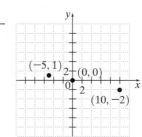

25.

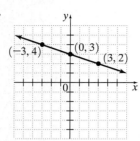

27.

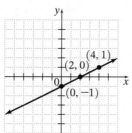

29.

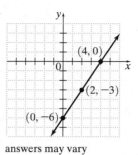

answers may vary

31.

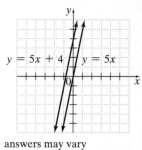

answers may vary

33.

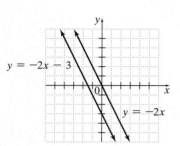

answers may vary

35. $(4, -1)$ **37.** $(0, 3); (-3, 0)$ **39.** $(0, 0)$

41. $0, 1, 1, 4, 4;$

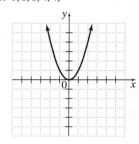

43. $x + y = 12; 9$ cm   **45. a.**

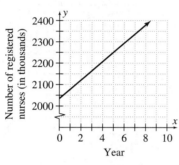

b. Yes; 8 years after 1997 (or in 2005) there will be 2379 thousand registered nurses.

GRAPHING CALCULATOR EXPLORATIONS

1.

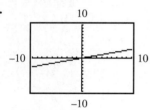

3.

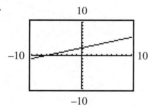

SECTION 6.3

1. $x = -1; y = 1; (-1, 0); (0, 1)$ **3.** $x = -2; x = 1; x = 3; y = 1; (-2, 0); (1, 0); (3, 0); (0, 1)$

5. infinite   **7.** 0

9.

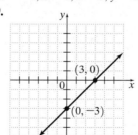

11.

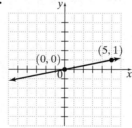

13.

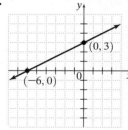

15.

17.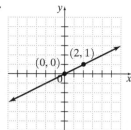

19.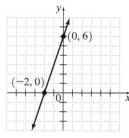

21.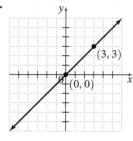

23.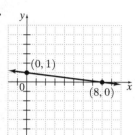

25.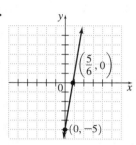

27.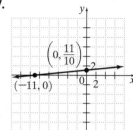

29.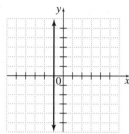

31.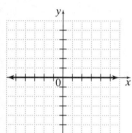

33.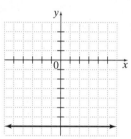

35.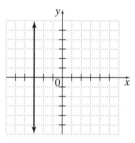

37. $\dfrac{3}{2}$ **39.** 6   **41.** $-\dfrac{6}{5}$ **43.** C   **45.** A   **47.** answers may vary

49. (a) $(0, 200)$; no chairs and 200 desks are manufactured   **(b)** $(400, 0)$; 400 chairs and no desks are manufactured   **(c)** 300 chairs

51. 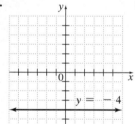 ; $y = -4$ **53. (a)** 23.6   **(b)** $(0, 23.6)$

(c) In 1995, U.S. farm expenses for livestock feed were $23.6 billion.

Calculator Explorations

1.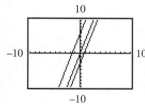

3.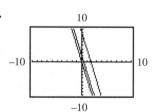

Mental Math

1. upward   **3.** horizontal

Section 6.4

1. $-\dfrac{4}{3}$ **3.** $\dfrac{5}{2}$ **5.** $\dfrac{8}{7}$ **7.** -1 **9.** $-\dfrac{1}{4}$ **11.** $-\dfrac{2}{3}$ **13.** 0   **15.** line 1   **17.** line 2   **19.** 5   **21.** -2 **23.** $\dfrac{2}{3}$ **25.** $\dfrac{1}{2}$

27. undefined slope   **29.** undefined slope   **31.** 0   **33.** neither   **35.** perpendicular   **37.** parallel   **39. a.** 1   **b.** -1

41. a. $\dfrac{9}{11}$ **b.** $-\dfrac{11}{9}$ **43.** $y = 2x - 14$ **45.** $y = -6x - 11$ **47.** D   **49.** B   **51.** E   **53.** $\dfrac{3}{5}$

55. 16%   **57.** 20 mpg   **59.** 3.3 mpg   **61.** 86 to 87   **63.** The line becomes steeper.

65. a. $(0, 18,000)$ and $(10, 0)$ **b.** -1800 **c.** The number cases is decreasing by an average of 1800 per year.

GRAPHING CALCULATOR EXPLORATIONS

1.

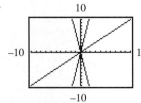

3.

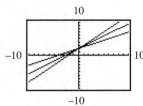

MENTAL MATH

1. $m = 2; (0, -1)$ **3.** $m = 1; \left(0, \dfrac{1}{3}\right)$ **5.** $m = \dfrac{5}{7}; (0, -4)$

SECTION 6.5

1. $y = 5x + 3$ **3.** $y = \dfrac{2}{3}x$ **5.** $y = -\dfrac{1}{5}x + \dfrac{1}{9}$

7.

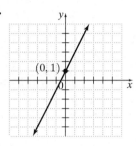

9.

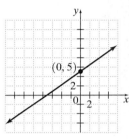

11.

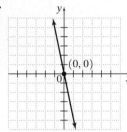

13.

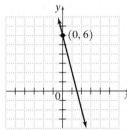

15.

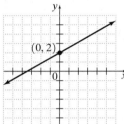

17. $-6x + y = -10$ **19.** $8x + y = -13$ **21.** $x - 2y = 17$ **23.** $x + 2y = -3$ **25.** $2x - y = 4$ **27.** $8x - y = -11$

29. $4x - 3y = -1$ **31.** $x + y = 17$ **33. a.** $s = 32t$ **b.** 128 ft/sec   **35. a.** $(0, 27.6)$ and $(3, 24.8)$ **b.** $y = -\dfrac{2.8}{3}x + 27.6$

c. 23.87 heads/sq ft   **37.** -1 **39.** 5   **41.** no   **43.** yes   **45.** B   **47.** D   **49.** $3x - y = -5$ **51.** answers may vary

INTEGRATED REVIEW

1. 2   **2.** 0   **3.** $-\dfrac{2}{3}$ **4.** undefined

5.

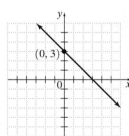

6.

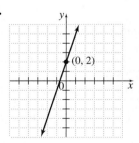

7.

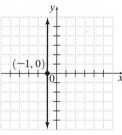

8.

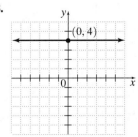

9.

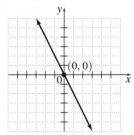

10.

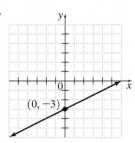

11. $y = 2x - \dfrac{1}{3}$ **12.** $4x + y = -1$ **13.** $-x + y = -2$

A34

Section 6.6

1. domain: $\{-7, 0, 2, 10\}$; range: $\{-7, 0, 4, 10\}$ **3.** domain: $\{0, 1, 5\}$; range: $\{-2\}$ **5.** yes   **7.** no   **9.** no   **11.** yes   **13.** yes
15. no   **17.** 5:20 A.M.   **19.** answers may vary   **21.** 9:00 P.M.   **23.** January 1st and December 1st
25. Yes; it passes the vertical line test.   **27.** $-9; -5; 1$ **29.** $6; 2; 11$ **31.** $-6; 0; 9$ **33.** $2; 0; 3$ **35.** $5; 0; -20$ **37.** $5; 3; 35$
39. $x < 1$ **41.** $x \geq -3$ **43.** $\dfrac{19}{2x}$ m   **45. a.** 166.38 cm   **b.** 148.25 cm   **47. a.** answers may vary   **b.** answers may vary
c. answers may vary   **49.** $f(x) = x + 7$

Mental Math

1. yes   **3.** yes   **5.** yes   **7.** no

Section 6.7

1. no; no
3. yes; no
5. no; yes

7.
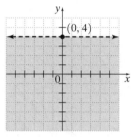

9.
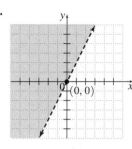

11.
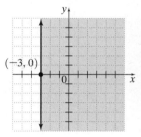

13.
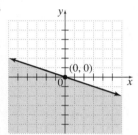

15.
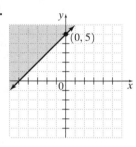

17.
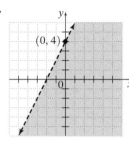

19.
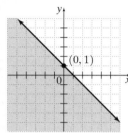

21.
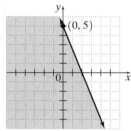

23.
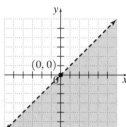

25.
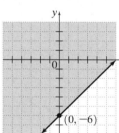

27.
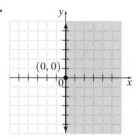

29.
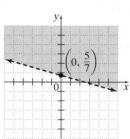

31. $(-2, 1)$ **33.** $(-3, -1)$ **35.** a   **37.** b   **39.** answers may vary

Chapter 6 Review

1–6.
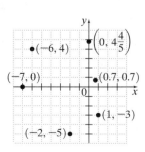

7. $(7, 44)$ **8.** $\left(-\dfrac{13}{3}, -8\right)$

9.

x	y
-3	0
1	3
9	9

10.

x	y
7	5
-7	5
0	5

11.

x	y
0	0
10	5
-10	-5

12. a. 2005; 2500; 7000   **b.** 886 compact disks

13.
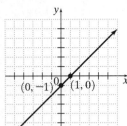

14.
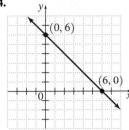

15.
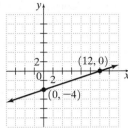

16.
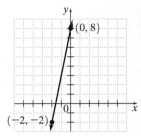

17.
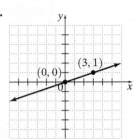

18.
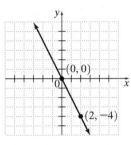

19. $x = 4; y = -2; (4, 0); (0, -2)$
20. $x = -2; x = 2; y = -2; y = 2; (-2, 0); (2, 0), (0, -2); (0, 2)$

21.
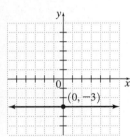

22.
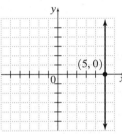

23. $(0, -4); (12, 0)$ **24.** $(0, 8); (-2, 0)$

25. $-\dfrac{3}{4}$ **26.** $\dfrac{1}{5}$ **27.** d   **28.** b   **29.** c   **30.** a   **31.** $\dfrac{3}{4}$ **32.** $\dfrac{5}{3}$ **33.** 4   **34.** -1 **35.** 3

36. $\dfrac{1}{2}$ **37.** 0   **38.** undefined   **39.** perpendicular   **40.** parallel   **41.** neither   **42.** $-3; (0, 7)$

43. $\dfrac{1}{6}; \left(0, \dfrac{1}{6}\right)$ **44.** $y = -5x + \dfrac{1}{2}$ **45.** $y = \dfrac{2}{3}x + 6$

46. D   **47.** C   **48.** A   **49.** B   **50.** $-4x + y = -8$ **51.** $3x + y = -5$ **52.** $3x - 5y = -17$ **53.** $x + 3y = 6$

54. $14x + y = 21$ **55.** $x + 2y = 8$ **56.** no   **57.** yes   **58.** yes   **59.** yes

60. no   **61.** yes   **62. a.** 6   **b.** 10   **c.** 5

63.
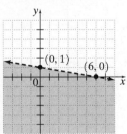

64.
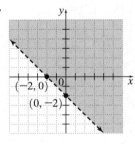

65.
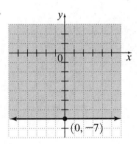

66.
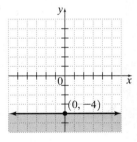

67.
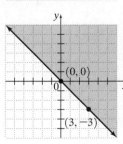

68.
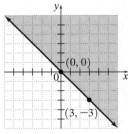

Chapter 6 Test

1. $(1, 1)$ **2.** $(-4, 17)$ **3.** $\dfrac{2}{5}$ **4.** 0   **5.** -1 **6.** -7 **7.** 3   **8.** undefined

ANSWERS

A36

9.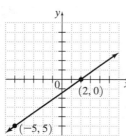

10.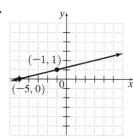

11.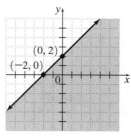

12.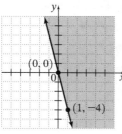

13.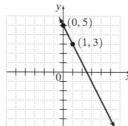

14.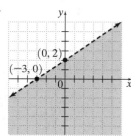

15.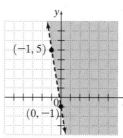

16.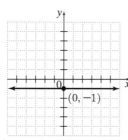

17. neither **18.** $x + 4y = 10$ **19.** $7x + 6y = 0$ **20.** $8x + y = 11$ **21.** $x - 8y = -96$ **22.** yes **23.** no **24.** yes **25.** yes
26. a. -8 **b.** -3.6 **c.** -4 **27. a.** 0 **b.** 0 **c.** 60

CUMULATIVE REVIEW

1. 27; Sec. 1.2, Ex. 2 **2.** 54; Sec. 1.2, Ex. 4 **3. a.** The Lion King; $315 million **b.** $70 million; Sec. 1.8, Ex. 2
4. $2x + 6$; Sec. 2.1, Ex. 15 **5.** $(x - 4) \div 7$; Sec. 2.1, Ex. 16 **6.** $8 + 3x$; Sec. 2.1, Ex. 17 **7.** $x = 6$; Sec. 2.3, Ex. 1
8. $x < -2$; Sec. 2.8, Ex. 5

9. a. 2; trinomial; Sec. 3.3, Ex. 3 **b.** 1; binomial; Sec. 3.3, Ex. 3 **c.** 3; none of these; Sec. 3.3, Ex. 3 **10.** $-4x^2 + 6x + 2$; Sec. 3.4, Ex. 2
11. $9y^2 + 6y + 1$; Sec. 3.6, Ex. 4 **12.** $3a(-3a^4 + 6a - 1)$; Sec. 4.1, Ex. 3 **13.** $(x - 2)(x + 6)$; Sec. 4.2, Ex. 3
14. $(4x - 1)(2x - 5)$; Sec. 4.3, Ex. 2 **15.** $x = 4, x = 5$; Sec. 4.6, Ex. 5 **16.** 1; Sec. 5.2, Ex. 7 **17.** $\dfrac{12ab^2}{27a^2b}$; Sec. 5.3, Ex. 9
18. $\dfrac{2m + 1}{m + 1}$; Sec. 5.4, Ex. 5 **19.** $x = -3, x = -2$; Sec. 5.5, Ex. 3 **20.** $\dfrac{x + 1}{x + 2y}$; Sec. 5.7, Ex. 5
21. a. $(0, 12)$; Sec. 6.1, Ex. 3 **b.** $(2, 6)$; Sec. 6.1, Ex. 3 **c.** $(-1, 15)$; Sec. 6.1, Ex. 3
22. 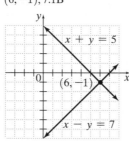 ; Sec. 6.2, Ex. 1 **23.** $\dfrac{2}{3}$; Sec. 6.4, Ex. 3 **24.** $2x + y = 3$; Sec. 6.5, Ex. 4
25. a. 1; $(2, 1)$; Sec. 6.6, Ex. 6 **b.** 1; $(-2, 1)$; Sec. 6.6, Ex. 6 **c.** -3; $(0, -3)$; Sec. 6.6, Ex. 6

Chapter 7 SYSTEMS OF EQUATIONS

CHAPTER 7 PRETEST

1. $(6, -1)$; 7.1B **2.** $(1, 4)$; 7.1B

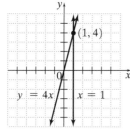

3. $(4, 2)$; 7.2A **4.** $(3, -2)$; 7.2A
5. $(-1, -1)$; 7.2A
6. infinite number of solutions; 7.2A
7. no solution; 7.2A **8.** $\left(\dfrac{1}{4}, -\dfrac{5}{2}\right)$; 7.2A
9. infinite number of solutions; 7.2A
10. $(-2, 1)$; 7.2A **11.** $(8, -5)$; 7.3A
12. $(-2, -7)$; 7.3A **13.** $(6, 7)$; 7.3A

14. no solution; 7.3A **15.** $(12, -12)$; 7.3A **16.** $\left(\dfrac{3}{8}, -\dfrac{5}{8}\right)$; 7.3A **17.** infinite number of solutions; 7.3A

18. $(0, -6)$; 7.3A **19.** 81 and 16; 7.4A **20.** $32°$ and $58°$; 7.4A

Calculator Explorations

1. $(0.37, 0.23)$ **3.** $(0.03, -1.89)$

Mental Math

1. 1 solution, $(-1, 3)$ **3.** infinite number of solutions **5.** no solution **7.** 1 solution, $(3, 2)$

Section 7.1

1. a. no **b.** yes **3. a.** yes **b.** no **5. a.** yes **b.** yes
7. $(3, 1)$

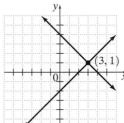

9. $(6, 0)$

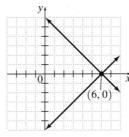

11. $(-2, -4)$

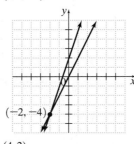

13. $(2, 3)$

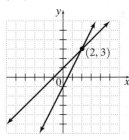

15. $(1, -2)$

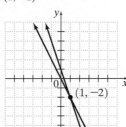

17. $(-2, 1)$

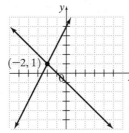

19. $(4, 2)$

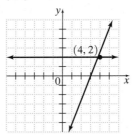

21. no solution

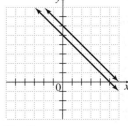

23. $(2, 0)$

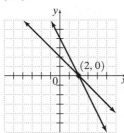

25. $(0, -1)$

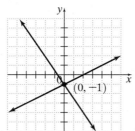

27. infinite number of solutions

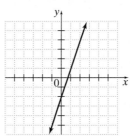

29. $(3, -1)$

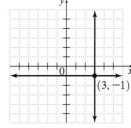

31. $(-5, -7)$

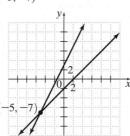

33. $(5, 4)$

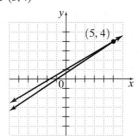

35. answers may vary

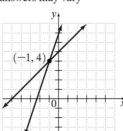

37. answers may vary

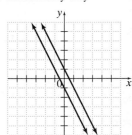

39. 1984, 1988 **41. a.** $(4, 9)$ **b.** yes

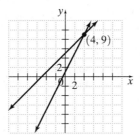

43. $x = -1$ **45.** $y = 3$ **47.** $z = -7$
49. answers may vary **51.** answers may vary

A38

Section 7.2

1. $(2,1)$ **3.** $(-3,9)$ **5.** $(4,2)$ **7.** $(10,5)$ **9.** $(2,7)$ **11.** $\left(-\dfrac{1}{5}, \dfrac{43}{5}\right)$ **13.** $(-2,4)$ **15.** $(-2,-1)$ **17.** no solution

19. $(3,-1)$ **21.** $(3,5)$ **23.** $\left(\dfrac{2}{3}, -\dfrac{1}{3}\right)$ **25.** $(-1,-4)$ **27.** $(-6,2)$ **29.** $(2,1)$ **31.** no solution

33. infinite number of solutions **35.** answers may vary **37.** $-6x - 4y = -12$ **39.** $-12x + 3y = 9$ **41.** $5n$ **43.** $-15b$

45. $(1,-3)$ **47.** $(-2.6, 1.3)$ **49.** $(3.28, 2.11)$ **51. a.** $(21,18)$ **b.** answers may vary **c.**

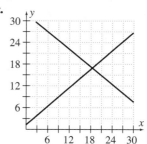

Section 7.3

1. $(1,2)$ **3.** $(2,-3)$ **5.** $(5,-2)$ **7.** $(6,0)$ **9.** $(-2,-5)$ **11.** $(-7,5)$ **13.** $\left(\dfrac{12}{11}, -\dfrac{4}{11}\right)$ **15.** no solution

17. no solution **19.** $\left(2, -\dfrac{1}{2}\right)$ **21.** $(6,-2)$ **23.** infinite number of solutions **25.** $(-2,0)$ **27.** answers may vary

29. $\left(\dfrac{3}{2}, 3\right)$ **31.** $(1,6)$ **33.** infinite number of solutions **35.** $2x + 6 = x - 3$ **37.** $20 - 3x = 2$ **39.** $4(x + 6) = 2x$

41. a. $b = 15$ **b.** any real number except 15 **43.** $(-8.9, 10.6)$ **45. a.** $(3, 549)$ **b.** answers may vary **c.** 1992 to 1994
46. a. 2173 **b.** about 3974

Integrated Review

1. $(2,5)$ **2.** $(4,2)$ **3.** $(5,-2)$ **4.** $(6,-14)$ **5.** $(-3,2)$ **6.** $(-4,3)$ **7.** $(0,3)$ **8.** $(-2,4)$ **9.** $(5,7)$ **11.** $\left(\dfrac{1}{3}, 1\right)$

12. $\left(-\dfrac{1}{4}, 2\right)$ **13.** no solution **14.** infinite number of solutions **15.** answers may vary **16.** answers may vary

Mental Math

1. c **3.** b **5.** a

Section 7.4

1. $\begin{cases} x + y = 15 \\ x - y = 7 \end{cases}$ **3.** $\begin{cases} x + y = 6500 \\ x = y + 800 \end{cases}$ **5.** 33 and 50 **7.** 14 and -3 **9.** Cooper: 621 points; Bolton-Holifield: 447 points
11. child's ticket = $18; adult's ticket: 29 **13.** quarters: 53; nickels: 27 **15.** Procter & Gamble: $68.50; Microsoft: $132.75

17. still water: 6.5 mph; current: 2.5 mph **19.** still air = 455 mph; wind: 65 mph **21.** 12% solution: $7\dfrac{1}{2}$ oz; 4% solution: $4\dfrac{1}{2}$ oz

23. $4.95 beans: 113 lbs; $2.65 beans: 87 lbs **25.** 16 **27.** $36x^2$ **29.** $100y^6$ **31.** $60°, 30°$ **33.** $20°, 70°$
35. number sold at $9.50: 23; number sold at $7.50: 67 **37.** width: 9 ft; length: 15 ft **39.** answers may vary

Chapter 7 Review

1. a. no **b.** yes **2. a.** yes **b.** no **3. a.** no **b.** no **4. a.** no **b.** yes
5. $(3,2)$ **6.** $(1,2)$ **7.** $(5,-1)$ **8.** $(-3,2)$

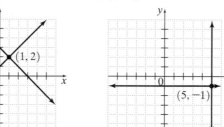

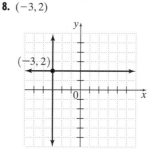

9. $(3, -1)$ **10.** $(1, -5)$ **11.** $(-3, -2)$ **12.** $\left(12\frac{1}{3}, -2\frac{2}{3}\right)$

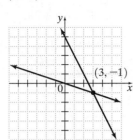 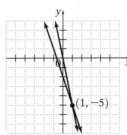 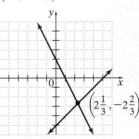

13. no solution   **14.** infinite number of solutions

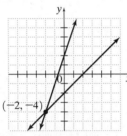 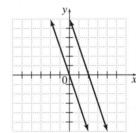

15. $(4, 2)$ **16.** $(5, 1)$ **17.** $(-1, 4)$ **18.** $(2, -1)$
19. $(3, -2)$ **20.** $(2, 5)$ **21.** no solution
22. $(-6, 2)$ **23.** no solution   **24.** no solution
25. $(16, -2)$ **26.** $(11, -2)$ **27.** $(-6, 2)$ **28.** $(4, -1)$
29. $(3, 7)$ **30.** $(-2, 4)$ **31.** infinite number of solutions
32. no solution   **33.** $(8, -6)$ **34.** $(10, -4)$
35. $(-1, 7)$ **36.** infinite number of solutions
37. -6 and 22

38. orchestra: 255 seats; balcony: 105 seats   **39.** current of river: 3.2 mph; speed in still water: 21.1 mph
40. amount invested at 6%: $6,180; amount invested at 10%: $2,820   **41.** length: 1.85 ft; width: 1.15 ft
42. 6% solution: $12\frac{1}{2}$ cc; 14% solution: $37\frac{1}{2}$ cc   **43.** egg: 40¢; strip of bacon: 65¢   **44.** jogging: 0.86 hour; walking: 2.14 hours

CHAPTER 7 TEST

1. $(-2, -4)$ **2.** no solution   **3.** $(-4, 1)$ **4.** $(4, -5)$ **5.** $(-3, -7)$
6. infinite number of solutions   **7.** $(20, 8)$ **8.** $(-4, 2)$
9. $\left(2, \frac{1}{2}\right)$ **10.** $\left(9\frac{2}{5}, 9\frac{3}{5}\right)$ **11.** $78, 46$

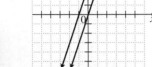

12. $1 bills: 20; $5 bills: 42
13. 30% solution: 7.5 liters; 70% solution: 2.5 liters
14. 3 mph; 6 mph   **15.** 1992   **16.** 1993–1995

CUMULATIVE REVIEW

1. a. -6; Sec. 1.4, Ex. 6   **b.** 6.3 **2.** $\frac{1}{22}$; Sec. 1.6, Ex. 1   **3.** $\frac{16}{3}$; Sec. 1.6, Ex. 2   **4.** $-\frac{1}{10}$; Sec. 1.6, Ex. 3   **5.** $-\frac{13}{9}$; Sec. 1.6, Ex. 4

6. $\frac{1}{1.7}$; Sec. 1.6, Ex. 5   **7. a.** 5 **b.** $8 - x$; Sec. 2.2, Ex. 8   **8.** no solution; Sec. 2.4, Ex. 6   **9. a.** $\frac{2}{5}$ **b.** $\frac{3}{4}$; Sec. 2.7, Ex. 4

10. $x > \frac{13}{7}$; Sec. 2.8, Ex. 7   **11.** $\frac{m^7}{n^7}$, $n \neq 0$; Sec. 3.1, Ex. 19   **12.** $\frac{16x^{16}}{81y^{20}}$, $y \neq 0$; Sec. 3.1, Ex. 20   **13.** $9x^2 - 6x - 1$; Sec. 3.4, Ex. 5

$$\xleftarrow{\qquad\overset{\displaystyle\circ}{\underset{\displaystyle\frac{13}{7}}{\quad}}\qquad\longrightarrow}$$

14. $2x + 4 + \frac{-1}{3x - 1}$; Sec. 3.7, Ex. 5   **15.** $x = -\frac{1}{2}$, $x = 4$; Sec. 4.6, Ex. 6   **16.** $6, 8, 10$; Sec. 4.7, Ex. 4   **17.** 1; Sec. 5.3, Ex. 2

18. $\frac{2x^2 + 3y}{x^2y + 2xy^2}$; Sec. 5.7, Ex. 6   **19.** $m = 0$; Sec. 6.4, Ex. 4   **20.** $-x + 5y = 23$; Sec. 6.5, Ex. 5

21. domain: $\{-1, 0, 3\}$; range: $\{-2, 0, 2, 3\}$; Sec. 6.6, Ex. 1   **22.** It is a solution.; Sec. 7.1, Ex. 1   **23.** $\left(6, \frac{1}{2}\right)$; Sec. 7.2, Ex. 3

24. $\left(-\frac{5}{4}, -\frac{5}{2}\right)$; Sec. 7.3, Ex. 6   **25.** 29 and 8; Sec. 7.4, Ex. 1

Chapter 8 ROOTS AND RADICALS

CHAPTER 8 PRETEST

1. -7; 8.1A **2.** $\dfrac{2}{5}$; 8.1A **3.** -4; 8.1B **4.** $2\sqrt{30}$; 8.2A **5.** $\dfrac{2\sqrt{6}}{y^3}$; 8.2C **6.** $2\sqrt[3]{14}$; 8.2D **7.** $-3\sqrt{15}$; 8.3A **8.** 0; 8.3B

9. $\dfrac{7\sqrt{7}}{10}$; 8.3B **10.** $6\sqrt{3}$; 8.4A **11.** $2\sqrt{7} - \sqrt{10}$; 8.4A **12.** $y - 6\sqrt{y} + 9$; 8.4A **13.** $2x\sqrt{7}$; 8.4B **14.** $\dfrac{\sqrt{55}}{11}$; 8.4C

15. $\dfrac{8\sqrt{2a}}{a}$; 8.4C **16.** $\dfrac{6 + 3\sqrt{x}}{4 - x}$; 8.4D **17.** $x = 49$; 8.5A **18.** $x = \dfrac{9}{4}$; 8.5B **19.** $4\sqrt{10}$ cm; 8.6A **20.** 2.52 in.; 8.6B

CALCULATOR EXPLORATIONS

1. 2.646 **3.** 3.317 **5.** 9.055 **7.** 3.420 **9.** 2.115 **11.** 1.783

SECTION 8.1

1. 4 **3.** 9 **5.** $\dfrac{1}{5}$ **7.** -10 **9.** not a real number **11.** -11 **13.** $\dfrac{3}{5}$ **15.** 5 **17.** -4 **19.** -2 **21.** $\dfrac{1}{2}$ **23.** -5
25. answers may vary **27.** 2 **29.** not a real number **31.** -5 **33.** 1 **35.** 6.083 **37.** 11.662 **39.** $\sqrt{2} \approx 1.41$; 126.90 ft
41. z **43.** x^2 **45.** $3x^4$ **47.** $9x$ **49.** $25 \cdot 2$ **51.** $16 \cdot 2$ or $4 \cdot 8$ **53.** $4 \cdot 7$ **55.** $9 \cdot 3$ **57.** 3
59. 1; 1.7; 2; 3 **61.** $(2, 0)$ **63.** $(-4, 0)$ **65.** 58 ft

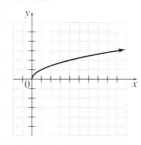

MENTAL MATH

1. 6 **3.** x **5.** 0 **7.** $5x^2$

SECTION 8.2

1. $2\sqrt{5}$ **3.** $3\sqrt{2}$ **5.** $5\sqrt{2}$ **7.** $\sqrt{33}$ **9.** $2\sqrt{15}$ **11.** $6\sqrt{5}$ **13.** $2\sqrt{13}$ **15.** $\dfrac{2\sqrt{2}}{5}$ **17.** $\dfrac{3\sqrt{3}}{11}$ **19.** $\dfrac{3}{2}$ **21.** $\dfrac{5\sqrt{5}}{3}$
23. $\dfrac{\sqrt{11}}{6}$ **25.** $-\dfrac{\sqrt{3}}{4}$ **27.** $x^3\sqrt{x}$ **29.** $x^6\sqrt{x}$ **31.** $5x\sqrt{3}$ **33.** $4x^2\sqrt{6}$ **35.** $\dfrac{2\sqrt{3}}{y}$ **37.** $\dfrac{3\sqrt{x}}{y}$ **39.** $\dfrac{2\sqrt{22}}{x^2}$ **41.** $2\sqrt[3]{3}$
43. $5\sqrt[3]{2}$ **45.** $\dfrac{\sqrt[3]{5}}{4}$ **47.** $\dfrac{\sqrt[3]{7}}{2}$ **49.** $\dfrac{\sqrt[3]{15}}{4}$ **51.** $2\sqrt[3]{10}$ **53.** $14x$ **55.** $2x^2 - 7x - 15$ **57.** 0 **59.** $x^3y\sqrt{y}$ **61.** $x + 2$
63. $2\sqrt[3]{80}$ in. **65.** \$1700 **67.** answers may vary

MENTAL MATH

1. $8\sqrt{2}$ **3.** $7\sqrt{x}$ **5.** $3\sqrt{7}$

SECTION 8.3

1. $-4\sqrt{3}$ **3.** $9\sqrt{6} - 5$ **5.** $\sqrt{5} + \sqrt{2}$ **7.** $6\sqrt{3}$ **9.** $-5\sqrt{2} - 6$ **11.** $7\sqrt{5}$ **13.** $2\sqrt{5}$ **15.** $6 - 3\sqrt{3}$
17. answers may vary **19.** $5\sqrt{3}$ **21.** $9\sqrt{5}$ **23.** $4\sqrt{6} + \sqrt{5}$ **25.** $x + \sqrt{x}$ **27.** 0 **29.** $x\sqrt{x}$ **31.** $9\sqrt{3}$ **33.** $5\sqrt{2} + 12$
35. $\dfrac{4\sqrt{5}}{9}$ **37.** $\dfrac{3\sqrt{3}}{8}$ **39.** $2\sqrt{5}$ **41.** $-\sqrt{35}$ **43.** $11\sqrt{x}$ **45.** $12x - 11\sqrt{x}$ **47.** $x\sqrt{3x} + 3x\sqrt{x}$ **49.** $8x\sqrt{2} + 2x$
51. $2x^2\sqrt{10} - x^2\sqrt{5}$ **53.** $x^2 + 12x + 36$ **55.** $4x^2 - 4x + 1$ **57.** $(4, 2)$ **59.** $8\sqrt{5}$ in. **61.** $\left(48 + \dfrac{9\sqrt{3}}{2}\right)$ sq ft

MENTAL MATH

1. $\sqrt{6}$ **3.** $\sqrt{6}$ **5.** $\sqrt{10y}$

SECTION 8.4

1. 4 **3.** $5\sqrt{2}$ **5.** 6 **7.** $2x$ **9.** 20 **11.** $36x$ **13.** $3\sqrt{2xy}$ **15.** $4xy\sqrt{y}$ **17.** $\sqrt{10} + \sqrt{2}$ **19.** $2\sqrt{5} + 5\sqrt{2}$
21. $\sqrt{30} + \sqrt{42}$ **23.** -33 **25.** $\sqrt{6} - \sqrt{15} + \sqrt{10} - 5$ **27.** $16 - 11\sqrt{11}$ **29.** $x - 36$ **31.** $x - 14\sqrt{x} + 49$
33. $6y + 2\sqrt{6y} + 1$ **35.** 4 **37.** $\sqrt{7}$ **39.** $3\sqrt{2}$ **41.** $5y^2$ **43.** $5\sqrt{3}$ **45.** $2y\sqrt{6}$ **47.** $2xy\sqrt{3y}$ **49.** $\dfrac{\sqrt{15}}{5}$ **51.** $\dfrac{7\sqrt{2}}{2}$
53. $\dfrac{\sqrt{6y}}{6y}$ **55.** $\dfrac{\sqrt{10}}{6}$ **57.** $\dfrac{\sqrt{3x}}{x}$ **59.** $\dfrac{\sqrt{2}}{4}$ **61.** $\dfrac{\sqrt{30}}{15}$ **63.** $\dfrac{\sqrt{15}}{10}$ **65.** $\dfrac{3\sqrt{2x}}{2}$ **67.** $\dfrac{8y\sqrt{5}}{5}$ **69.** $\dfrac{\sqrt{3xy}}{6x}$ **71.** $3\sqrt{2} - 3$

73. $-8 - 4\sqrt{5}$ **75.** $5 + \sqrt{30} + \sqrt{6} + \sqrt{5}$ **77.** $\sqrt{6} + \sqrt{3} + \sqrt{2} + 1$ **79.** $\dfrac{10 - 5\sqrt{x}}{4 - x}$ **81.** $\dfrac{3\sqrt{x} + 12}{x - 16}$ **83.** $x = 44$

85. $z = 2$ **87.** $x = 3$ **89.** $130\sqrt{3}$ sq meters   **91.** answers may vary   **93.** $\dfrac{2}{\sqrt{6} - \sqrt{2} - \sqrt{3} + 1}$

INTEGRATED REVIEW

1. 6   **2.** $4\sqrt{3}$ **3.** x^2 **4.** $y^3\sqrt{y}$ **5.** $4x$ **6.** $3x^5\sqrt{2x}$ **7.** 2   **8.** 3   **9.** -3 **10.** not a real number   **11.** $\dfrac{\sqrt{11}}{3}$

12. $\dfrac{\sqrt[3]{7}}{4}$ **13.** $6\sqrt{7}$ **14.** $3\sqrt{2}$ **15.** $\sqrt{x} + 3x$ **16.** $\sqrt{30}$ **17.** 3   **18.** $\sqrt{33} + \sqrt{3}$ **19.** $x - 3\sqrt{x} - 10$ **20.** $11 + 6\sqrt{2}$

21. 2   **22.** $\sqrt{3}$ **23.** $2x^2\sqrt{3}$ **24.** $\dfrac{\sqrt{6}}{6}$ **25.** $\dfrac{x\sqrt{5}}{10}$ **26.** $\dfrac{4\sqrt{6} - 4}{5}$ **27.** $\dfrac{\sqrt{2x} + 5\sqrt{2} + \sqrt{x} + 5}{x - 25}$

SECTION 8.5

1. $x = 81$ **3.** $x = -1$ **5.** $x = 5$ **7.** $x = 49$ **9.** no solution   **11.** $x = -2$ **13.** $x = \dfrac{3}{2}$ **15.** $x = 16$ **17.** no solution

19. $x = 2$ **21.** $x = 4$ **23.** $x = 3$ **25.** $x = 2$ **27.** $x = 2$ **29.** no solution   **31.** $x = 0, -3$ **33.** $x = -3$ **35.** $x = 9$

37. $x = 12$ **39.** $x = 3, 1$ **41.** $x = -1$ **43.** $x = 2$ **45.** $x = 16$ **47.** $x = 1$ **49.** $x = 1$ **51.** $3x - 8 = 19; x = 9$

53. $2(2x + x) = 24$; length $= 8$ in.   **55. a.** 3.2; 10; 31.6   **b.** no   **57.** answers may vary   **59.** 2.43   **61.** 0.48

SECTION 8.6

1. $\sqrt{13}$; 3.61   **3.** $3\sqrt{3}$; 5.20   **5.** 25   **7.** $\sqrt{22}$; 4.69   **9.** $3\sqrt{17}$; 12.37   **11.** $\sqrt{41}$; 6.40   **13.** $4\sqrt{2}$; 5.66   **15.** $3\sqrt{10}$; 9.49

17. 20.6 ft   **19.** 11.7 ft   **21.** 24 cubic ft   **23.** 54 mph   **25.** 27 mph   **27.** 61.2 km   **29.** 3   **31.** 10

33. 8   **35.** $x = 2\sqrt{10} - 4$ **37.** 201 miles   **39.** answers may vary   **41.** answers may vary

CHAPTER 8 REVIEW

1. 9   **2.** -7 **3.** 3   **4.** 2   **5.** $-\dfrac{3}{8}$ **6.** $\dfrac{2}{3}$ **7.** 2   **8.** -2 **9.** c   **10.** a, c   **11.** x^6 **12.** x^4 **13.** $3y$ **14.** $5x^2$

15. $2\sqrt{10}$ **16.** $2\sqrt{6}$ **17.** $3\sqrt{6}$ **18.** $2\sqrt{22}$ **19.** $x^2\sqrt{x}$ **20.** $y^3\sqrt{y}$ **21.** $2x\sqrt{5}$ **22.** $5y^2\sqrt{2}$ **23.** $3\sqrt[3]{2}$ **24.** $2\sqrt[3]{11}$

25. $\dfrac{3\sqrt{2}}{5}$ **26.** $\dfrac{5\sqrt{3}}{8}$ **27.** $\dfrac{-5\sqrt{2}}{3}$ **28.** $\dfrac{-2\sqrt{3}}{7}$ **29.** $\dfrac{\sqrt{11}}{x}$ **30.** $\dfrac{\sqrt{7}}{y^2}$ **31.** $\dfrac{y^2\sqrt{y}}{10}$ **32.** $\dfrac{x\sqrt{x}}{9}$ **33.** $-3\sqrt{2}$ **34.** $-5\sqrt{3}$

35. $4\sqrt{5} + 4\sqrt{6}$ **36.** $-2\sqrt{7} + 2\sqrt{2}$ **37.** $5\sqrt{7} + 2\sqrt{14}$ **38.** $9\sqrt{3} - 4$ **39.** $\dfrac{\sqrt{5}}{6}$ **40.** $\dfrac{9\sqrt{11}}{20}$ **41.** $10 - x\sqrt{5}$

42. $2\sqrt{2x} - \sqrt{3x}$ **43.** $3\sqrt{2}$ **44.** $5\sqrt{3}$ **45.** $\sqrt{10} - \sqrt{14}$ **46.** $\sqrt{55} + \sqrt{15}$ **47.** $3\sqrt{2} - 5\sqrt{3} + 2\sqrt{6} - 10$ **48.** $2 - 2\sqrt{5}$

49. $x - 4\sqrt{x} + 4$ **50.** $y + 8\sqrt{y} + 16$ **51.** 3   **52.** 2   **53.** $2\sqrt{5}$ **54.** $4\sqrt{2}$ **55.** $x\sqrt{15x}$ **56.** $3x^2\sqrt{2}$ **57.** $\dfrac{\sqrt{22}}{11}$

58. $\dfrac{\sqrt{39}}{13}$ **59.** $\dfrac{\sqrt{30}}{6}$ **60.** $\dfrac{\sqrt{70}}{10}$ **61.** $\dfrac{\sqrt{5x}}{5x}$ **62.** $\dfrac{5\sqrt{3y}}{3y}$ **63.** $\dfrac{\sqrt{3x}}{x}$ **64.** $\dfrac{\sqrt{6y}}{y}$ **65.** $3\sqrt{5} + 6$ **66.** $8\sqrt{10} + 24$

67. $\dfrac{\sqrt{6} + \sqrt{2} + \sqrt{3} + 1}{2}$ **68.** $\sqrt{15} - 2\sqrt{3} - 2\sqrt{5} + 4$ **69.** $\dfrac{10\sqrt{x} - 50}{x - 25}$ **70.** $\dfrac{8\sqrt{x} + 8}{x - 1}$ **71.** $x = 18$ **72.** $x = 13$

73. $x = 25$ **74.** no solution   **75.** $x = 12$ **76.** $x = 5$ **77.** $x = 1$ **78.** $x = 9$ **79.** $2\sqrt{14}$; 7.48   **80.** $\sqrt{117}$; 10.82

81. $4\sqrt{34}$ ft   **82.** $5\sqrt{3}$ in.   **83.** 2.4 in.   **84.** 144π sq in.

CHAPTER 8 TEST

1. 4   **2.** 5   **3.** 3   **4.** $\dfrac{3}{4}$ **5.** not a real number   **6.** x^5 **7.** $3\sqrt{6}$ **8.** $2\sqrt{23}$ **9.** $y^3\sqrt{y}$ **10.** $2x^4\sqrt{6}$ **11.** 3

12. $2\sqrt[3]{2}$ **13.** $\dfrac{\sqrt{5}}{4}$ **14.** $\dfrac{y}{5}\sqrt{y}$ **15.** $-2\sqrt{13}$ **16.** $\sqrt{2} + 2\sqrt{3}$ **17.** $\dfrac{7\sqrt{3}}{10}$ **18.** $7\sqrt{2}$ **19.** $2\sqrt{3} - \sqrt{10}$ **20.** $x - \sqrt{x} - 6$

21. $\sqrt{5}$ **22.** $2x\sqrt{5x}$ **23.** $\dfrac{\sqrt{6}}{3}$ **24.** $\dfrac{8\sqrt{5y}}{5y}$ **25.** $4\sqrt{6} - 8$ **26.** $\dfrac{3 + \sqrt{x}}{9 - x}$ **27.** $x = 9$ **28.** $x = 5$ **29.** $x = 9$

30. $4\sqrt{5}$ in.   **31.** 2.19 m

CUMULATIVE REVIEW

1. 50; Sec. 1.5, Ex. 3   **2.** $-\dfrac{8}{21}$; Sec. 1.5, Ex. 4   **3.** $x = 2$; Sec. 2.4, Ex. 1   **4. a.** 45%   **b.** 83%

c. 9900 homeowners; Sec. 2.7, Ex. 3   **5. a.** 102,000   **b.** 0.007358   **c.** 84,000,000   **d.** 0.00003007; Sec. 3.2, Ex. 16

6. $6x^2 - 11x - 10$; Sec. 3.5, Ex. 7   **7.** $(y + 2)(x + 3)$; Sec. 4.1, Ex. 7   **8.** $(3x + 2)(x + 3)$; Sec. 4.3, Ex. 1   **9. a.** $x = 3$ **b.** $x = 2, x = 1$

c. none; Sec. 5.1, Ex. 2   **10.** $\dfrac{x - 2}{x}$; Sec. 5.1, Ex. 5   **11. a.** 0   **b.** $\dfrac{15 + 14x}{50x^2}$; Sec. 5.4, Ex. 1   **12.** $x = -3$; Sec. 5.5, Ex. 4

13.

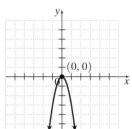

; Sec. 6.3, Ex. 10   **14.** $y = \frac{1}{4}x - 3$; Sec. 6.5, Ex. 1   **15.** $(3, 1)$; Sec. 7.3, Ex. 5

16. Albert: 3.25 mph; Louis: 4.25 mph; Sec. 7.4, Ex. 3   **17.** 1; Sec. 8.1, Ex. 6

18. -3; Sec. 8.1, Ex. 7   **19.** $\frac{1}{5}$; Sec. 8.1, Ex. 8   **20.** $3\sqrt{6}$; Sec. 8.2, Ex. 1   **21.** $10\sqrt{2}$; Sec. 8.2, Ex. 3

22. $9\sqrt{3}$; Sec. 8.3, Ex. 6   **23.** $2x - 4\sqrt{x}$; Sec. 8.3, Ex. 8   **24.** $\frac{2\sqrt{7}}{7}$; Sec. 8.4, Ex. 9

25. $x = \frac{1}{2}$; Sec. 8.5, Ex. 3

Chapter 9 QUADRATIC EQUATIONS

CHAPTER 9 PRETEST

1. $a = 0, 6$; 9.1A   **2.** $x = -\frac{1}{2}, 6$; 9.1A   **3.** $b = \pm 12$; 9.1B   **4.** $y = \frac{7 \pm 2\sqrt{6}}{2}$; 9.1B   **5.** $x = 6, 8$; 9.2A   **6.** $x = -\frac{1}{3}, 2$; 9.2B

7. $x = -3, 9$; 9.3A   **8.** $m = \frac{7 \pm \sqrt{145}}{8}$; 9.3A   **9.** $x = \frac{-1 \pm \sqrt{73}}{4}$; 9.3A   **10.** $x = \frac{-3 \pm 3\sqrt{2}}{5}$; 9.1B   **11.** $x = -\frac{3}{4}, -\frac{3}{2}$; 9.1A

12. $m = 3 \pm \sqrt{6}$; 9.2A   **13.** $x = \frac{-4 \pm 3\sqrt{2}}{2}$; 9.3A   **14.** $y = -70 \pm \sqrt{5}$; 9.1B

15.

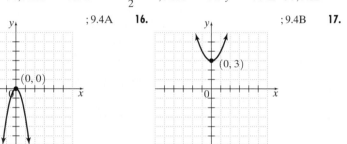

; 9.4A   **16.** ; 9.4B   **17.** ; 9.4B

18.

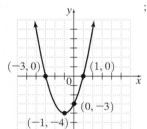

; 9.4B

SECTION 9.1

1. $k = \pm 3$ **3.** $m = -5, 3$ **5.** $x = \pm 4$ **7.** $a = \pm 3$ **9.** $x = -5, -2$ **11.** $x = \pm 8$ **13.** $x = \pm\sqrt{21}$ **15.** $x = \pm\frac{1}{5}$

17. no real solution   **19.** $x = \pm\frac{\sqrt{39}}{3}$ **21.** $x = \pm\frac{2\sqrt{7}}{7}$ **23.** $x = \pm\sqrt{2}$ **25.** answers may vary   **27.** $x = 12, -2$

29. $x = -2 \pm \sqrt{7}$ **31.** $m = 1, 0$ **33.** $p = -2 \pm \sqrt{10}$ **35.** $y = \frac{8}{3}, -4$ **37.** no real solution   **39.** $x = \frac{11 \pm 5\sqrt{2}}{2}$

41. $x = \frac{7 \pm 4\sqrt{2}}{3}$ **43.** $(x + 3)^2$ **45.** $(x - 2)^2$ **47.** $x = 2, -6$ **49.** $r = 6$ in.   **51.** 5 sec   **53.** $x = -1.02, 3.76$ **55.** 1999

MENTAL MATH

1. 16   **3.** 100   **5.** 49

SECTION 9.2

1. $x = -6, -2$ **3.** $x = -1 \pm \sqrt{6}$ **5.** $x = 0, 6$ **7.** $z = \frac{-5 \pm \sqrt{53}}{2}$ **9.** $x = 1 \pm \sqrt{2}$ **11.** $y = -1, -4$

13. $x = -3 \pm \sqrt{34}$ **15.** $x = -6, 3$ **17.** $x = \frac{13}{2}, -\frac{1}{2}$ **19.** no real solution   **21.** $x = \frac{3 \pm \sqrt{19}}{2}$ **23.** $x = -2, 4$

25. $y = -2 \pm \frac{\sqrt{6}}{2}$ **27.** $y = \frac{1}{2}, 1$ **29.** $-\frac{1}{2}$ **31.** -1 **33.** $3 + 2\sqrt{5}$ **35.** $\frac{1 - 3\sqrt{2}}{2}$ **37.** $k = 8$ or $k = -8$ **39.** 1999

41. $x = -6, -2$ **43.** $x \approx -0.68, 3.68$

1. $a = 2, b = 5, c = 3$ **3.** $a = 10, b = -13, c = -2$ **5.** $a = 1, b = 0, c = -6$

SECTION 9.3

1. $x = 2, 1$ **3.** $k = \dfrac{-7 \pm \sqrt{37}}{6}$ **5.** $x = \pm\dfrac{2}{7}$ **7.** no real solution   **9.** $y = 10, -3$ **11.** $x = \pm\sqrt{5}$ **13.** $m = -3, 4$

15. $x = -2 \pm \sqrt{7}$ **17.** $x = \dfrac{-9 \pm \sqrt{129}}{12}$ **19.** $p = \dfrac{4 \pm \sqrt{2}}{7}$ **21.** $a = 3 \pm \sqrt{7}$ **23.** $x = \dfrac{3 \pm \sqrt{3}}{2}$ **25.** $x = -1, \dfrac{1}{3}$

27. $y = \dfrac{3 \pm \sqrt{13}}{4}$ **29.** $y = \dfrac{1}{5}, -\dfrac{3}{4}$ **31.** no real solution   **33.** no real solution   **35.** $m = 1 \pm \sqrt{2}$ **37.** $p = -\dfrac{3}{4}, -\dfrac{1}{2}$

39. $x = \dfrac{7 \pm \sqrt{129}}{20}$ **41.** no real solution   **43.** $z = \dfrac{1 \pm \sqrt{2}}{5}$

45.
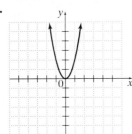

47.
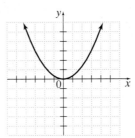

49. $x = \sqrt{51}$ meters   **51.** $x = \dfrac{-3\sqrt{2} \pm \sqrt{38}}{2}$

53. answers may vary   **55.** $y = 3.9, -2.9$

57. $z \approx -0.9, 0.2$ **59.** 8.1 sec   **61.** 2002

INTEGRATED REVIEW

1. $x = \dfrac{1}{5}, 2$ **2.** $x = -3, \dfrac{2}{5}$ **3.** $x = 1 \pm \sqrt{2}$ **4.** $x = 3 \pm \sqrt{2}$ **5.** $a = \pm 2\sqrt{5}$ **6.** $\pm 6\sqrt{2}$ **7.** no real solution

8. no real solution   **9.** $x = 2$ **10.** $x = 3$ **11.** $p = 3$ **12.** $p = \dfrac{7}{2}$ **13.** $y = \pm 2$ **14.** $y = \pm 3$ **15.** $x = 0, 1, 2$

16. $x = 0, -3, -4$ **17.** $z = -5, 0$ **18.** $z = 0, \dfrac{8}{3}$ **19.** $x = \dfrac{3 \pm \sqrt{7}}{5}$ **20.** $x = \dfrac{3 \pm \sqrt{5}}{2}$ **21.** $x = \dfrac{3}{2}, -1$ **22.** $m = \dfrac{2}{5}, -2$

23. $x = \dfrac{5 \pm \sqrt{105}}{20}$ **24.** $x = \dfrac{-1 \pm \sqrt{3}}{4}$ **25.** $x = 5, \dfrac{7}{4}$ **26.** $x = \dfrac{7}{9}, 1$ **27.** $x = \dfrac{7 \pm 3\sqrt{2}}{5}$ **28.** $x = \dfrac{5 \pm 5\sqrt{3}}{4}$

29. $z = \dfrac{7 \pm \sqrt{193}}{6}$ **30.** $z = \dfrac{-7 \pm \sqrt{193}}{12}$ **31.** $x = 11, -10$ **32.** $x = -8, 7$ **33.** $x = -\dfrac{2}{3}, 4$ **34.** $x = 2, -\dfrac{4}{5}$ **35.** $x = 0.1, 0.5$

36. $x = 0.3, -0.2$ **37.** $x = \dfrac{11 \pm \sqrt{41}}{20}$ **38.** $x = \dfrac{11 \pm \sqrt{41}}{40}$ **39.** $z = \dfrac{4 \pm \sqrt{10}}{2}$ **40.** $z = \dfrac{5 \pm \sqrt{185}}{4}$ **41.** answers may vary

CALCULATOR EXPLORATIONS

1. $x = -0.41, 7.41$ **3.** $x = 0.91, 2.38$ **5.** $x = -0.39, 0.84$

SECTION 9.4

1.
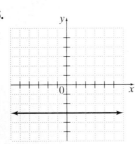

3.
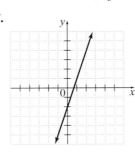

5.
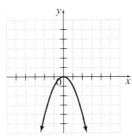

7.
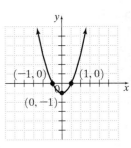

9.
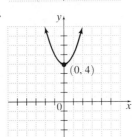

11.
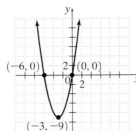

13.
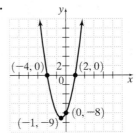

15.
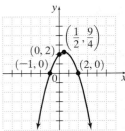

17.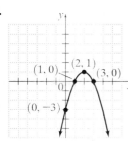

19.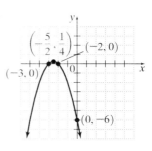

21. $\dfrac{5}{14}$ **23.** $\dfrac{x}{2}$ **25.** $\dfrac{2x^2}{x-1}$ **27.** $-4b$

29. a. 256 ft   **b.** $t = 4$ sec   **c.** $t = 8$ sec

31. E   **33.** C   **35.** B

Chapter 9 Review

1. $x = 4, -\dfrac{3}{5}$ **2.** $x = -7, -\dfrac{4}{3}$ **3.** $m = -\dfrac{1}{3}, 2$ **4.** $m = \dfrac{5}{7}, -1$ **5.** $x = 0, 3, -3$ **6.** $x = -2, 2$ **7.** $x = 6, -6$ **8.** $x = 9, -9$

9. $k = \pm 5\sqrt{2}$ **10.** $k = \pm 3\sqrt{5}$ **11.** $x = 4, 18$ **12.** $x = 7, -13$ **13.** $p = 2, -3$ **14.** $p = 1, -5$ **15.** 30%   **16.** 10%

17. $x = -2 \pm \sqrt{5}$ **18.** $x = 4 \pm \sqrt{19}$ **19.** $x = 3 \pm \sqrt{2}$ **20.** $x = -3 \pm \sqrt{2}$ **21.** $y = \dfrac{1}{2}, -1$ **22.** $y = \dfrac{-3 \pm \sqrt{13}}{2}$

23. $x = 5 \pm 3\sqrt{2}$ **24.** $x = -2 \pm \sqrt{11}$ **25.** $x = \dfrac{1}{2}, -1$ **26.** $x = \dfrac{-3 \pm \sqrt{13}}{2}$ **27.** $x = -\dfrac{5}{3}$ **28.** $x = \dfrac{9}{4}$ **29.** $x = \dfrac{2}{5}, \dfrac{1}{3}$

30. $x = \dfrac{2}{3}, \dfrac{1}{5}$ **31.** no real solution   **32.** no real solution   **33.** $\dfrac{1 \pm \sqrt{5}}{2}$

34.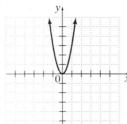

35.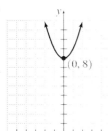

36.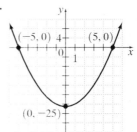

37.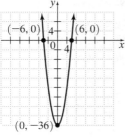

38.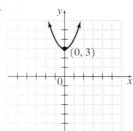

39.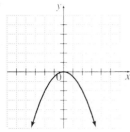

40.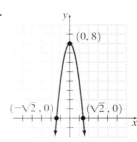

41.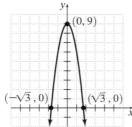

42.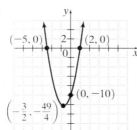

43.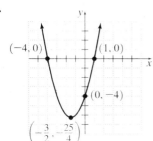

44.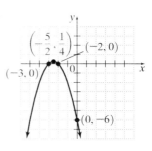

45.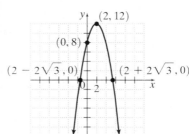

46.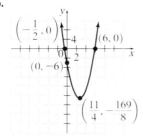

47.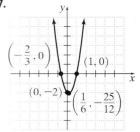

48. $x = -2$ **49.** $x = -\dfrac{3}{2}, 3$ **50.** no real solution **51.** $x = -2, 2$ **52.** A **53.** D **54.** B **55.** C

Chapter 9 Test

1. $x = -\dfrac{3}{2}, 7$ **2.** $x = -2, 0, 1$ **3.** $k = \pm 4$ **4.** $m = \dfrac{5 \pm 2\sqrt{2}}{3}$ **5.** $x = 10, 16$ **6.** $x = -2, \dfrac{1}{5}$ **7.** $x = -2, 5$ **8.** $p = \dfrac{5 \pm \sqrt{37}}{6}$

9. $x = 1, -\dfrac{4}{3}$ **10.** $x = -1, \dfrac{5}{3}$ **11.** $x = \dfrac{7 \pm \sqrt{73}}{6}$ **12.** $x = -1, 5$ **13.** $x = 2, \dfrac{1}{3}$ **14.** $x = \dfrac{3 \pm \sqrt{7}}{2}$

15. base = 3 ft; height = 12 ft

16.
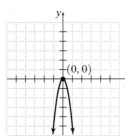

17.
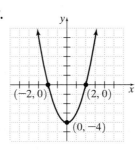

18.
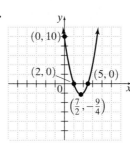

19.
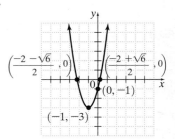

20. 6 sides

Cumulative Review

1. $y = -1.6$; Sec. 2.2, Ex. 2 **2.** $t = \dfrac{16}{3}$; Sec. 2.4, Ex. 2 **3.** 138 min; Sec. 2.5, Ex. 4 **4.** 1; Sec. 3.1, Ex. 25 **5.** 1; Sec. 3.1, Ex. 26

6. -1; Sec. 3.1, Ex. 28 **7.** $9y^2 + 12y + 4$; Sec. 3.6, Ex. 15 **8.** $x + 4$; Sec. 3.7, Ex. 4 **9.** $(r + 6)(r - 7)$; Sec. 4.2, Ex. 4

10. $(2x - 3y)(5x + y)$; Sec. 4.3, Ex. 4 **11.** $(2x - 1)(4x - 5)$; Sec. 4.4, Ex. 1 **12. a.** $x(2x + 7)(2x - 7)$; Sec. 4.5, Ex. 16

b. $2(9x^2 + 1)(3x + 1)(3x - 1)$; Sec. 4.5, Ex. 17 **13.** $x = \dfrac{1}{5}, -\dfrac{3}{2}, -6$; Sec. 4.6, Ex. 8 **14.** $\dfrac{x + 7}{x - 5}$; Sec. 5.1, Ex. 4 **15.** -5; Sec. 5.6, Ex. 1

16.

x	y
-1	-3
0	0
-3	-9

; Sec. 6.1, Ex. 4 **17. a.** parallel **b.** perpendicular **c.** neither; Sec. 6.4, Ex. 6

18. a. function **b.** not a function; Sec. 6.6, Ex. 2 **19.** $(4, 2)$; Sec. 7.2, Ex. 1

20. no solution; Sec. 7.3, Ex. 3

21. 6; Sec. 8.1, Ex. 1 **22.** $\dfrac{3}{10}$; Sec. 8.1, Ex. 4 **23.** $-1 + \sqrt{3}$; Sec. 8.4, Ex. 12 **24.** $x = 7, -1$; Sec. 9.1, Ex. 5

25. $x = 1 \pm \sqrt{5}$; Sec. 9.3, Ex. 5

Solutions to Selected Exercises

Chapter R

Section R.1

1. $9 = 1 \cdot 9, 9 = 3 \cdot 3$
The factors of 9 are 1, 3, and 9.

5. $42 = 1 \cdot 42, 42 = 21 \cdot 2, 42 = 3 \cdot 14, 42 = 6 \cdot 7$
The factors of 42 are 1, 2, 3, 6, 7, 14, and 21, 42.

9. 13 is a prime number. Its factors are 1 and 13 only.

13. 37 is a prime number. Its factors are 1 and 37 only.

17. $18 = 2 \cdot 3 \cdot 3$ **21.** $56 = 2 \cdot 2 \cdot 2 \cdot 7$

25. $81 = 3 \cdot 3 \cdot 3 \cdot 3$

29. $6 = 2 \cdot 3$
$14 = 2 \cdot 7$
$LCM = 2 \cdot 3 \cdot 7 = 42$

33. $20 = 2 \cdot 2 \cdot 5$
$30 = 2 \cdot 3 \cdot 5$
$LCM = 2 \cdot 2 \cdot 3 \cdot 5 = 60$

37. $6 = 2 \cdot 3$
$12 = 2 \cdot 2 \cdot 3$
$LCM = 2 \cdot 2 \cdot 3 = 12$

41. $50 = 2 \cdot 5 \cdot 5$
$70 = 2 \cdot 5 \cdot 7$
$LCM = 2 \cdot 5 \cdot 5 \cdot 7 = 350$

45. $5 = 5$
$10 = 2 \cdot 5$
$12 = 2 \cdot 2 \cdot 3$
$LCM = 2 \cdot 2 \cdot 3 \cdot 5 = 60$

49. $8 = 2 \cdot 2 \cdot 2$
$18 = 2 \cdot 3 \cdot 3$
$30 = 2 \cdot 3 \cdot 5$
$LCM = 2 \cdot 2 \cdot 2 \cdot 3 \cdot 3 \cdot 5 = 360$

53. $315 = 3 \cdot 3 \cdot 5 \cdot 7$
$504 = 2 \cdot 2 \cdot 2 \cdot 3 \cdot 3 \cdot 7$
$LCM = 2 \cdot 2 \cdot 2 \cdot 3 \cdot 3 \cdot 5 \cdot 7 = 2520$

Section R.2

1. $\dfrac{7}{10} = \dfrac{7 \cdot 3}{10 \cdot 3} = \dfrac{21}{30}$ **5.** $\dfrac{4}{5} = \dfrac{4 \cdot 4}{5 \cdot 4} = \dfrac{16}{20}$

9. $\dfrac{10}{15} = \dfrac{2 \cdot 5}{3 \cdot 5} = \dfrac{2}{3}$ **13.** $\dfrac{20}{20} = \dfrac{2 \cdot 2 \cdot 5}{2 \cdot 2 \cdot 5} = 1$

17. $\dfrac{18}{30} = \dfrac{2 \cdot 3 \cdot 3}{2 \cdot 3 \cdot 5} = \dfrac{3}{5}$

21. $\dfrac{66}{48} = \dfrac{2 \cdot 3 \cdot 11}{2 \cdot 2 \cdot 2 \cdot 2 \cdot 3} = \dfrac{11}{2 \cdot 2 \cdot 2} = \dfrac{11}{8}$

25. $\dfrac{1}{2} \cdot \dfrac{3}{4} = \dfrac{1 \cdot 3}{2 \cdot 4} = \dfrac{3}{8}$

29. $\dfrac{1}{2} \div \dfrac{7}{12} = \dfrac{1}{2} \cdot \dfrac{12}{7} = \dfrac{1 \cdot 2 \cdot 2 \cdot 3}{2 \cdot 7} = \dfrac{6}{7}$

33. $\dfrac{7}{10} \cdot \dfrac{5}{21} = \dfrac{7 \cdot 5}{2 \cdot 5 \cdot 3 \cdot 7} = \dfrac{1}{6}$

37. $\dfrac{4}{5} + \dfrac{1}{5} = \dfrac{4 + 1}{5} = \dfrac{5}{5} = 1$

41. $\dfrac{23}{105} + \dfrac{4}{105} = \dfrac{23 + 4}{105} = \dfrac{27}{105} = \dfrac{3 \cdot 9}{3 \cdot 35} = \dfrac{9}{35}$

45. $\dfrac{2}{3} + \dfrac{3}{7} = \dfrac{2 \cdot 7}{3 \cdot 7} + \dfrac{3 \cdot 3}{7 \cdot 3} = \dfrac{14}{21} + \dfrac{9}{21} = \dfrac{14 + 9}{21} = \dfrac{23}{21}$

49. $\dfrac{10}{21} + \dfrac{5}{21} = \dfrac{10 + 5}{21} = \dfrac{15}{21} = \dfrac{3 \cdot 5}{3 \cdot 7} = \dfrac{5}{7}$

53. $\dfrac{12}{5} - 1 = \dfrac{12}{5} - \dfrac{1 \cdot 5}{1 \cdot 5} = \dfrac{12}{5} - \dfrac{5}{5} = \dfrac{12 - 5}{5} = \dfrac{7}{5}$

57. $1 - \dfrac{3}{10} - \dfrac{5}{10} = \dfrac{1 \cdot 10}{1 \cdot 10} - \dfrac{3}{10} - \dfrac{5}{10}$
$= \dfrac{10}{10} - \dfrac{3}{10} - \dfrac{5}{10}$
$= \dfrac{10 - 3 - 5}{10}$
$= \dfrac{2}{10}$
$= \dfrac{1}{5}$

The unknown part is $\dfrac{1}{5}$.

61. $9\dfrac{1}{4} - 8\dfrac{1}{2} = \dfrac{37}{4} - \dfrac{17}{2}$
$= \dfrac{37}{4} - \dfrac{17 \cdot 2}{2 \cdot 2}$
$= \dfrac{37}{4} - \dfrac{34}{4}$
$= \dfrac{37 - 34}{4}$
$= \dfrac{3}{4} \text{ years}$

65. answers may vary

Section R.3

1. $0.6 = \dfrac{6}{10}$ **5.** $0.144 = \dfrac{114}{1000}$

9. $\begin{array}{r} 5.7 \\ + 1.13 \\ \hline 6.83 \end{array}$

13.
$$
\begin{array}{r}
8.8 \\
-\ 2.3 \\
\hline
6.5
\end{array}
$$

17.
$$
\begin{array}{r}
45.02 \\
3.006 \\
+\ 8.405 \\
\hline
56.431
\end{array}
$$

21.
$$
\begin{array}{r}
0.2 \\
\times\ 0.6 \\
\hline
0.12
\end{array}
$$

25.
$$
\begin{array}{r}
5.62 \\
\times\ 7.7 \\
\hline
3934 \\
3934 \\
\hline
43.274
\end{array}
$$

29.
$$
\begin{array}{r}
0.094 \\
5)\overline{0.470} \\
-\ 0 \\
\hline
47 \\
-\ 45 \\
\hline
20 \\
-\ 20
\end{array}
$$

33.
$$
\begin{array}{r}
5.8 \\
82)\overline{475.6} \\
-\ 410 \\
\hline
656 \\
-\ 656
\end{array}
$$

37. 0.6   **41.** 0.594

45. 12.35

49.
$$
\begin{array}{r}
0.333 \\
3)\overline{1.00} \\
-\ 9 \\
\hline
10 \\
-\ 9 \\
\hline
10 \\
-\ 9 \\
\hline
1
\end{array}
$$

This pattern will continue so that
$$\frac{1}{3} = 0.3333\ldots$$
$$\frac{1}{3} = 0.\overline{3} \approx 0.33$$

53.
$$
\begin{array}{r}
0.5454 \\
11)\overline{6.0000} \\
-\ 55 \\
\hline
50 \\
-\ 44 \\
\hline
60 \\
-\ 55 \\
\hline
50 \\
-\ 44 \\
\hline
6
\end{array}
$$

This pattern will continue so that
$$\frac{6}{11} = 0.5454\ldots$$
$$\frac{6}{11} = 0.\overline{54} \approx 0.55$$

57. $3.1\% = 0.031$ **61.** $96.55\% = 0.9655$

65. $0.68 = 68\%$ **69.** $1 = 100\%$

73.
$$
\begin{array}{r}
82.65 \\
-\ 75.67 \\
\hline
6.98
\end{array}
$$
Females are expected to live 6.98 years longer than males.

CHAPTER R TEST

1. $72 = 2 \cdot 2 \cdot 2 \cdot 3 \cdot 3$

5. $\dfrac{48}{100} = \dfrac{2 \cdot 2 \cdot 2 \cdot 2 \cdot 3}{2 \cdot 2 \cdot 5 \cdot 5} = \dfrac{2 \cdot 2 \cdot 3}{5 \cdot 5} = \dfrac{12}{25}$

9. $\dfrac{9}{10} \div 18 = \dfrac{9}{10} \cdot \dfrac{1}{18} = \dfrac{3 \cdot 3 \cdot 1}{2 \cdot 5 \cdot 2 \cdot 3 \cdot 3} = \dfrac{1}{2 \cdot 5 \cdot 2} = \dfrac{1}{20}$

13.
$$
\begin{array}{r}
7.93 \\
\times\ 1.6 \\
\hline
4758 \\
793 \\
\hline
12.688
\end{array}
$$

17.
$$
\begin{array}{r}
0.1666 \\
6)\overline{1.0000} \\
-\ 6 \\
\hline
40 \\
-\ 36 \\
\hline
40 \\
-\ 36 \\
\hline
40 \\
-\ 36 \\
\hline
4
\end{array}
$$

This pattern will continue so that
$$\frac{1}{6} = 0.1666\ldots$$
$$\frac{1}{6} = 0.1\overline{6} \approx 0.167$$

21. From the graph, we can see that $\dfrac{3}{4}$ of the fresh water is icecaps and glaciers.

Chapter 1

SECTION 1.1

1. $4 < 10$ **5.** $6.26 = 6.26$

9. $32 < 212$

13. False, since 10 is to the left of 11 on the number line.

17. True, since 7 is to the right of 0 on the number line.

21. $20 \le 25$ **25.** $-12 < -10$

29. $5 \ge 4$

33. The integer 535 represents 535 feet.
The integer -8 represents 8 feet below sea level.

37. The integer 350 represents a deposit of $350.
The integer -126 represents a withdrawal of $126.

41.

45. The number 0 belongs to the sets of: whole numbers, integers, rational numbers, and real numbers.

49. The number 6 belongs to the sets of: natural numbers, whole numbers, integers, rational numbers, and real numbers.

53. False; rational numbers can be either nonintegers, such as $\frac{1}{2}$, or integers, such as 2.

57. True; 0 corresponds to a point on the number line. Therefore, 0 is a real number.

61. $|-5| > -4$ since $|-5| = 5$ and $5 > -4$.

65. $|-2| < |-3|$ since $|-2| = 2$ and $|-3| = 3$ and $2 < 3$.

69. False, since $\frac{1}{2}$ is to the right of $\frac{1}{3}$ on the number line.

73. False, since -9.6 is to the left of -9.1 on the number line.

77. Bill's highest quiz score is 90.

81. $-0.04 > -26.7$

85. The sun; since on the number line -26.7 is to the left of all other numbers listed, and therefore, -26.7 is smaller than all other numbers listed.

Section 1.2

1. $3^5 = 3 \cdot 3 \cdot 3 \cdot 3 \cdot 3 = 243$

5. $1^5 = 1 \cdot 1 \cdot 1 \cdot 1 \cdot 1 = 1$

9. $\left(\frac{1}{5}\right)^3 = \left(\frac{1}{5}\right)\left(\frac{1}{5}\right)\left(\frac{1}{5}\right) = \frac{1 \cdot 1 \cdot 1}{5 \cdot 5 \cdot 5} = \frac{1}{125}$

13. $7^2 = 7 \cdot 7 = 49$

17. $(5 \cdot 5)$ square meters $= 5^2$ square meters

21. $4 \cdot 8 - 6 \cdot 2 = 32 - 12 = 20$

25. $2 + (5 - 2) + 4^2 = 2 + 3 + 4^2$
$= 2 + 3 + 16 = 5 + 16$
$= 21$

29. $\frac{1}{4} \cdot \frac{2}{3} - \frac{1}{6} = \frac{2}{12} - \frac{1}{6} = \frac{1}{6} - \frac{1}{6} = 0$

33. $2[5 + 2(8 - 3)] = 2[5 + 2(5)]$
$= 2[5 + 10]$
$= 2[15]$
$= 30$

37. $\dfrac{|6 - 2| + 3}{8 + 2 \cdot 5} = \dfrac{4 + 3}{8 + 10} = \dfrac{7}{18}$

41. $\dfrac{6 + |8 - 2| + 3^2}{18 - 3} = \dfrac{6 + |6| + 3^2}{15}$
$= \dfrac{6 + 6 + 9}{15}$
$= \dfrac{21}{15}$
$= \dfrac{3 \cdot 7}{3 \cdot 5}$
$= \dfrac{7}{5}$

45. Replace y with 3.
$3y = 3(3) = 9$

49. Replace x with 1.
$3x - 2 = 3(1) - 2 = 3 - 2 = 1$

53. Replace x with 1, y with 3, and z with 5.
$xy + z = (1)(3) + 5 = 3 + 5 = 8$

57. Replace z with 3.
$5z = 5(3) = 15$

61. Replace x with 2 and y with 6.
$\dfrac{y}{x} + \dfrac{y}{x} = \dfrac{6}{2} + \dfrac{6}{2} = 3 + 3 = 6$

65. Replace x with 5 and see if a true statement results.
$3x - 6 = 9$
$3(5) - 6 \stackrel{?}{=} 9$
$15 - 6 \stackrel{?}{=} 9$
$9 = 9$
Since we arrived at a true statement, 5 is a solution of the equation $3x - 6 = 9$.

69. Replace x with 8 and see if true statement results.
$2x - 5 = 5$
$2(8) - 5 \stackrel{?}{=} 5$
$16 - 5 \stackrel{?}{=} 5$
$11 \neq 5$
Because $2(8) - 5 = 5$ is not a true statement, 8 is not a solution of $2x - 5 = 5$.

73. Replace x with 0 and see if a true statement results.
$x = 5x + 15$
$0 \stackrel{?}{=} 5(0) + 15$
$0 \stackrel{?}{=} 0 + 15$
$0 \neq 15$
Because $0 = 5(0) + 15$ is not a true statement, 0 is not a solution of $x = 5x + 15$.

77. $x - 5$

81. $1 + 2 = 9 \div 3$

85. $5 + x = 20$

89. $\dfrac{12}{x} = \dfrac{1}{2}$

93. answers may vary

Section 1.3

1. $6 + 3 = 9$

5. $8 + (-7) = 1$

9. $-2 + (-3) = -5$

13. $-7 + 3 = -4$

17. $5 + (-7) = -2$

21. $27 + (-46) = -19$

25. $-33 + (-14) = -47$

29. $|-8| + (-16) = 8 + (-16) = -8$

33. $-9.6 + (-3.5) = -13.1$

37. $-\dfrac{7}{16} + \dfrac{1}{4} = -\dfrac{7}{16} + \dfrac{4\cdot 1}{4\cdot 4}$

$\qquad\qquad = -\dfrac{7}{16} + \dfrac{4}{16}$

$\qquad\qquad = -\dfrac{3}{16}$

41. $-15 + 9 + (-2) = -6 + (-2) = -8$

45. $-23 + 16 + (-2) = -7 + (-2) = -9$

49. $6 + (-4) + 9 = 2 + 9 = 11$

53. $|9 + (-12)| + |-16| = |-3| + |-16| = 3 + 16 = 19$

57. answers may vary

61. $-1296 + 658 = -638$

You are at an elevation of -638 feet (638 feet below sea level).

65. His total overall score is the sum of the scores for all four rounds of play.

$0 + (-4) + (-2) + (-4) = -4 + (-2) + (-4)$
$\qquad\qquad\qquad\qquad\quad = -6 + (-4)$
$\qquad\qquad\qquad\qquad\quad = -10$

His total overall score was -10 (10 below par).

69. The opposite of 6 is -6.

73. The opposite of 0 is 0.

77. answers may vary

81. $-|0| = -0 = 0$

85. The highest temperature is represented by the bar that is farthest above 0 degrees. From the graph, we can see that the highest temperature occurred on Tuesday.

89. To find the average of the five temperatures, we first find the sum and then divide by 5.

$\dfrac{-4 + 3 + 7 + (-2) + 1}{5} = \dfrac{-1 + 7 + (-2) + 1}{5}$

$\qquad\qquad\qquad\qquad = \dfrac{6 + (-2) + 1}{5}$

$\qquad\qquad\qquad\qquad = \dfrac{4 + 1}{5}$

$\qquad\qquad\qquad\qquad = \dfrac{5}{5}$

$\qquad\qquad\qquad\qquad = 1$

The average daily low temperature for Sunday through Thursday was $1°$.

93. $a + a$ is a positive number.

Section 1.4

1. $-6 - 4 = -6 + (-4) = -10$

5. $16 - (-3) = 16 + (3) = 19$

9. $-16 - (-18) = -16 + (18) = 2$

13. $7 - (-4) = 7 + (4) = 11$

17. $16 - (-21) = 16 + (21) = 37$

21. $-44 - 27 = -44 + (-27) = -71$

25. $-2.6 - (-6.7) = -2.6 + (6.7) = 4.1$

29. $-\dfrac{1}{6} - \dfrac{3}{4} = -\dfrac{1}{6} + \left(-\dfrac{3}{4}\right)$

$\qquad\qquad = -\dfrac{4\cdot 1}{4\cdot 6} + \left(-\dfrac{6\cdot 3}{6\cdot 4}\right)$

$\qquad\qquad = -\dfrac{4}{24} + \left(-\dfrac{18}{24}\right)$

$\qquad\qquad = -\dfrac{22}{24}$

$\qquad\qquad = -\dfrac{2\cdot 11}{2\cdot 12}$

$\qquad\qquad = -\dfrac{11}{12}$

33. Sometimes positive and sometimes negative. If a and b are positive numbers and $a \geq b$, then $a - b \geq 0$. If a and b are positive numbers and $a \leq b$, then $a - b \leq 0$.

37. $-6 - (-1) = -6 + (1) = -5$

41. $-8 - 15 = -8 + (-15) = -23$

45. $5 - 9 + (-4) - 8 - 8$
$\quad = 5 + (-9) + (-4) + (-8) + (-8)$
$\quad = -4 + (-4) + (-8) + (-8)$
$\quad = -8 + (-8) + (-8)$
$\quad = -16 + (-8)$
$\quad = -24$

49. $3^3 - 8\cdot 9 = 27 - 8\cdot 9$
$\qquad\qquad = 27 - 72$
$\qquad\qquad = 27 + (-72)$
$\qquad\qquad = -45$

53. $(3 - 6) + 4^2 = (3 + (-6)) + 4^2$
$\qquad\qquad\quad = (-3) + 4^2$
$\qquad\qquad\quad = -3 + 16$
$\qquad\qquad\quad = 13$

57. $|-3| + 2^2 + [-4 - (-6)] = 3 + 2^2 + [-4 + 6]$
$\qquad\qquad\qquad\qquad\quad = 3 + 2^2 + [2]$
$\qquad\qquad\qquad\qquad\quad = 3 + 4 + 2$
$\qquad\qquad\qquad\qquad\quad = 7 + 2$
$\qquad\qquad\qquad\qquad\quad = 9$

61. Replace x with -5, y with 4, and t with 10.
$|x| + 2t - 8y = |-5| + 2(10) - 8(4)$
$\qquad\qquad\quad = 5 + 20 - 32$
$\qquad\qquad\quad = 5 + 20 + (-32)$
$\qquad\qquad\quad = 25 + (-32)$
$\qquad\qquad\quad = -7$

65. Replace x with -5 and y with 4.
$y^2 - x = 4^2 - (-5)$
$\qquad\quad = 16 - (-5)$
$\qquad\quad = 16 + 5$
$\qquad\quad = 21$

69. Replace x with -4 and see if a true statement results.
$x - 9 = 5$
$-4 - 9 \stackrel{?}{=} 5$
$-4 + (-9) \stackrel{?}{=} 5$
$-13 \neq 5$
-4 is not a solution of $x - 9 = 5$.

73. Replace x with 2 and see if a true statement results.
$$-x - 13 = -15$$
$$-2 - 13 \stackrel{?}{=} -15$$
$$-2 + (-13) \stackrel{?}{=} -15$$
$$-15 = -15$$
2 is a solution of $-x - 13 = -15$.

77. The total gain or loss of yardage is the sum of the gains and losses. Gains are represented by positive numbers and losses are represented by negative numbers.
$$2 + (-5) + (-20) = -3 + (-20) = -23$$
The 49ers lost a total of 23 yards.

81. The overall vertical change is the sum of the gains and losses in altitude. Gains are represented by positive numbers and losses are represented by negative numbers.
$$-250 + 120 + (-178) = -130 + (-178) = -308$$
The overall vertical change is -308 feet.

85. These angles are supplementary, so their sum is $180°$.
$$y = 180° - 50° = 130°$$

89. The daily change in temperature is found by subtracting yesterday's temperature from today's temperature. A temperature increase is represented by a positive number. A temperature decrease is represented by a negative number.

On Monday, the temperature increased from $-4°$ to $3°$.
$$3 - (-4) = 3 + 4 = 7$$
On Monday, the temperature increased $7°$.

On Tuesday, the temperature increased from $3°$ to $7°$.
$$7 - 3 = 7 + (-3) = 4$$
On Tuesday, the temperature increased $4°$.

On Wednesday, the temperature decreased from $7°$ to $-2°$.
$$-2 - 7 = -2 + (-7) = -9$$
On Wednesday, the temperature decreased $9°$.

On Thursday, the temperature increased from $-2°$ to $1°$.
$$1 - (-2) = 1 + 2 = 3$$
On Thursday, the temperature increased $3°$.

On Friday, the temperature decreased from $1°$ to $-5°$.
$$-5 - 1 = -5 + (-1) = -6$$
On Friday, the temperature decreased $6°$.

Finally, on Saturday, the temperature increased from $-5°$ to $1°$.
$$1 - (-5) = 1 + 5 = 6$$
On Saturday, the temperature increased $6°$.

Day	Daily Increase or Decrease
Monday	$7°$
Tuesday	$4°$
Wednesday	$-9°$
Thursday	$3°$
Friday	$-6°$
Saturday	$6°$

93. True

97. $4.362 + (-7.0086)$
Negative since $7.0086 > 4.362$
$$4.362 - 7.0086 = -2.6466$$

Section 1.5

1. $-6(4) = -24$

5. $-5(-10) = 50$

9. $-6(-7) = 42$

13. $-\dfrac{1}{2}\left(-\dfrac{3}{5}\right) = -\left(-\dfrac{1 \cdot 3}{2 \cdot 5}\right) = -\left(-\dfrac{3}{10}\right) = \dfrac{3}{10}$

17. $5(-1.4) = -7$

21. $-10(80) = -800$

25. $(-5)(-5) = 25$

29. $-11(11) = -121$

33. $-2.1(-0.4) = 0.84$

37. $(2)(-1)(-3)(5)(3) = (-2)(-3)(5)(3)$
$$= (6)(5)(3)$$
$$= (30)(3)$$
$$= 90$$

41. $(-6)(3)(-2)(-1) = (-18)(-2)(-1)$
$$= (36)(-1)$$
$$= -36$$

45. $-4^2 = -(4)(4) = -16$

49. $6(3 - 8) = 6(3 + (-8)) = 6(-5) = -30$

53. $\left(-\dfrac{3}{4}\right)^2 = \left(-\dfrac{3}{4}\right)\left(-\dfrac{3}{4}\right)$
$$= -\left(-\dfrac{3 \cdot 3}{4 \cdot 4}\right)$$
$$= -\left(-\dfrac{9}{16}\right)$$
$$= \dfrac{9}{16}$$

57. False

61. Replace x with -5 and y with -3 and simplify.
$$2x^2 - y^2 = 2(-5)^2 - (-3)^2$$
$$= 2(25) - (9)$$
$$= 50 - 9$$
$$= 50 + (-9)$$
$$= 41$$

65. Replace x with 7 and see if a true statement results.
$$-5x = -35$$
$$-5(7) \stackrel{?}{=} -35$$
$$-35 = -35$$
Since $-35 = -35$ is a true statement, 7 is a solution of the equation.

69. Replace x with -1 and see if a true statement results.
$$9x + 1 = 14$$
$$9(-1) + 1 \stackrel{?}{=} 14$$
$$-9 + 1 \stackrel{?}{=} 14$$
$$-8 \neq 14$$
Since $-8 = 14$ is not a true statement, -1 is not a solution of the equation.

73. Replace x with -2 and see if a true statement results.
$$17 - 4x = x + 27$$
$$17 - 4(-2) \stackrel{?}{=} -2 + 27$$
$$17 - (-8) \stackrel{?}{=} -2 + 27$$
$$17 + 8 \stackrel{?}{=} 25$$
$$25 = 25$$
Since $25 = 25$ is a true statement, -2 is a solution of the equation.

77. Can't determine.

81. Since the share price decreases in value by $1.50 for each day that goes by, the stock decreases by $1 \cdot \$1.50 = \$1.50$ on the first day, $2 \cdot \$1.50 = \$3.00$ after 2 days, $3 \cdot \$1.50 = \$4.50$ after 3 days, and so on. We can express this trend in a more general way by letting x represent the number of days. Then $x \cdot \$1.50$ represents the decrease in value of the stock after x days. To find the final value of the stock we need to subtract the loss, $x \cdot \$1.50$, from the initial price, 38. Thus, the expression $38 - 1.50x$ can be used to find the price of the stock after x days. If we replace x with 4 we obtain

$$38 - 1.50x = 38 - 1.50(4)$$
$$= 38 - 6$$
$$= 38 + (-6)$$
$$= 32$$

Thus, 4 days after September 20, the value of the stock wil be $32. To find when the stock price will reach $20 per share, we continue evaluating the expression $38 - 1.50x$ with increasing values of x until we find the value of x that makes $38 - 1.50x = 20$ a true statement. If we do this, we find that when $x = 12$,

$$38 - 1.50x = 20 \text{ is a true statement}$$
$$38 - 1.50x = 20$$
$$38 - 1.50(12) \stackrel{?}{=} 20$$
$$38 - 18 \stackrel{?}{=} 20$$
$$38 + (-18) \stackrel{?}{=} 20$$
$$20 = 20$$

Therefore, 12 days after September 20, on Saturday, October 2, the stock will be worth $20 per share.

Section 1.6

1. The reciprocal of 9 is $\frac{1}{9}$ since $9 \cdot \frac{1}{9} = 1$.

5. The reciprocal of -14 is $-\frac{1}{14}$ since $(-14)\left(-\frac{1}{14}\right) = 1.$

9. The reciprocal of 0.2 is $\frac{1}{0.2}$ since $0.2 \cdot \frac{1}{0.2} = 1$.

13. $1, -1$

17. $\dfrac{-16}{-4} = -16 \cdot -\dfrac{1}{4} = 4$

21. $\dfrac{0}{-4} = 0 \cdot -\dfrac{1}{4} = 0$

25. $\dfrac{5}{0}$ is undefined

29. $\dfrac{30}{-2} = 30 \cdot -\dfrac{1}{2} = -15$

33. $-\dfrac{5}{9} \div \left(-\dfrac{3}{4}\right) = -\dfrac{5}{9} \cdot \left(-\dfrac{4}{3}\right) = \dfrac{20}{27}$

37. $-\dfrac{5}{8} \div \dfrac{3}{4} = -\dfrac{5}{8} \cdot \dfrac{4}{3} = -\dfrac{20}{24} = -\dfrac{5}{6}$

41. $-3.2 \div -0.02 = -3.2 \cdot -\dfrac{1}{0.02} = 160$

45. $\dfrac{12}{9 - 12} = \dfrac{12}{-3} = -4$

49. $\dfrac{8 + (-4)^2}{4 - 12} = \dfrac{8 + 16}{4 - 12} = \dfrac{24}{-8} = -3$

53. $\dfrac{-3 - 5^2}{2(-7)} = \dfrac{-3 - 25}{-14} = \dfrac{-28}{-14} = 2$

57. $\dfrac{-3 - 2(-9)}{-15 - 3(-4)} = \dfrac{-3 + 18}{-15 + 12} = \dfrac{15}{-3} = -5$

61. Replace x with -5 and y with -3 and simplify.
$$\dfrac{2x - 5}{y - 2} = \dfrac{2(-5) - 5}{-3 - 2} = \dfrac{-10 - 5}{-5} = \dfrac{-15}{-5} = 3$$

65. Replace x with -5 and y with -3 and simplify.
$$\dfrac{x + y}{3y} = \dfrac{-5 + (-3)}{3(-3)} = \dfrac{-8}{-9} = \dfrac{8}{9}$$

69. Replace x with 15.
$$\dfrac{x}{5} + 2 = -1$$
$$\dfrac{15}{5} + 2 - 1$$
$$3 + 2 - 1$$
$$5 \neq -1$$

Since $5 = -1$ is a false statement, 15 is not a solution of the equation.

73. $\dfrac{0}{5} - 7 = 0 - 7 = -7$

77. $\dfrac{-8}{-20} = \dfrac{2}{5}$

81. The quotient of two numbers with different signs is a negative number. If a is a positive number and b is a negative number, then $b + b$ is a negative number and $a + a$ is a positive number. Therefore $\dfrac{b + b}{a + a}$ is a negative number.

Section 1.7

1. $x + 16 = 16 + x$ **5.** $xy = yx$

9. $(xy) \cdot z = x \cdot (yz)$ **13.** $4 \cdot (ab) = 4a \cdot (b)$

17. $8 + (9 + b) = (8 + 9) + b = 17 + b$

21. $\dfrac{1}{5}(5y) = \left(\dfrac{1}{5} \cdot 5\right) \cdot y = 1 \cdot y = y$

25. $-9(8x) = (-9 \cdot 8) \cdot x = -72 \cdot x = -72x$

29. answers may vary

33. $9(x - 6) = 9(x) - 9(6) = 9x - 54$

37. $7(4x - 3) = 7(4x) - 7(3) = 28x - 21$

41. $-2(y - z) = -2(y) - (-2)(z) = -2y + 2z$

45. $5(x + 4m + 2) = 5(x) + 5(4m) + 5(2)$
$$= 5x + 20m + 10$$

49. $-(5x + 2) = -1(5x + 2)$
$$= (-1)(5x) + (-1)(2)$$
$$= -5x - 2$$

53. $\frac{1}{2}(6x + 8) = \frac{1}{2}(6x) + \frac{1}{2}(8)$

$$= \left(\frac{1}{2}\cdot 6\right)x + \left(\frac{1}{2}\cdot 8\right)$$
$$= 3x + 4$$

57. $3(2r + 5) - 7 = 3(2r) + 3(5) - 7$
$$= 6r + 15 - 7$$
$$= 6r + 8$$

61. $-4(4x + 5) - 5 = -4(4x) + (-4)(5) - 5$
$$= -16x - 20 - 5$$
$$= -16x - 25$$

65. $11x + 11y = 11(x + y)$

69. $30a + 30b = 30(a + b)$

73. The additive inverse of -8 is 8 since $(-8) + 8 = 0$.

77. The additive inverse of $-|-2|$ is 2 since $-|-2| + 2 = -2 + 2 = 0$

81. The multiplicative inverse of $-\frac{5}{6}$ is $-\frac{6}{5}$ since $-\frac{5}{6}\cdot\left(-\frac{6}{5}\right) = 1.$

85. The multiplicative inverse of -2 is $-\frac{1}{2}$ since $-2\cdot\left(-\frac{1}{2}\right) = 1.$

89. The associative property of addition

93. The associative property of multiplication

97. The distributive property

101.

Expression	Opposite	Reciprocal
8	−8	$\frac{1}{8}$

105.

Expression	Opposite	Reciprocal
2x	−2x	$\frac{1}{2x}$

109. answers may vary

Section 1.8

1. The number of teenagers expected to use the internet in 1999 is about 7.8 million.

5. Look for the shortest bar, which is the bar representing the PGA/LPGA tours. The PGA/LPGA tours spent the least amount of money on advertising.

9. The NBA spent about $15 million on advertising.

13. Each team has gone an equal number of years without being in a playoff.

17. *Pocahontas* generated approximately 142 million dollars, or $142,000,000.

21. answers may vary

25. 1986 and 1987 or 1989 and 1990

29. The greatest increase in the percent of arson fires started by juveniles occurred in 1994. Notice that the line graph is steepest between the years 1993 and 1994.

33. 5 minutes before lighting a cigarette, the pulse rate was approximately 59 beats per minute. 10 minutes after lighting a cigarette, the pulse rate had increased to about 85 beats per minute.
$85 - 59 = 26$
The pulse rate increased by 26 beats per minute between 5 minutes before and 10 minutes after lighting a cigarette.

37. The cost of newsprint was lowest in 1992, when it cost approximately $450 per metric ton.

41. answers may vary

Chapter 1 Test

1. $|-7| > 5$

5. $6\cdot 3 - 8\cdot 4 = 18 - 32 = 18 + (-32) = -14$

9. $\frac{-8}{0}$ is undefined

13. $-\frac{3}{5} + \frac{15}{8} = -\frac{8\cdot 3}{8\cdot 5} + \frac{5\cdot 15}{5\cdot 8} = -\frac{24}{40} + \frac{75}{40} = \frac{51}{40}$

17. $\frac{(-2)(0)(-3)}{-6} = \frac{0(-3)}{-6} = \frac{0}{-6} = 0$

21. $|-2| = -1 - (-3)$

25. Replace x with 6 and y with -2, then simplify.
$2 + 3x - y = 2 + 3(6) - (-2)$
$$= 2 + 18 + 2$$
$$= 20 + 2$$
$$= 22$$

29. The distributive property

33. Losses are represented by negative numbers. The greatest loss occurred on second down when the Saints lost 10 yards.

37. Intel's revenue in 1993 was $8 billion.

Chapter 2

Section 2.1

1. $7y + 8y = (7 + 8)y = 15y$

5. $3b - 5 - 10b - 4 = 3b - 10b + (-5 - 4)$
$$= (3 - 10)b + (-5 - 4)$$
$$= -7b - 9$$

9. $5g - 3 - 5 - 5g = 5g - 5g + (-3 - 5)$
$$= (5 - 5)g + (-3 - 5)$$
$$= 0\cdot g + (-3 - 5)$$
$$= -8$$

13. $2k - k - 6 = (2 - 1)k - 6 = k - 6$

17. $6x - 5x + x - 3 + 2x = 6x - 5x + x + 2x - 3$
$$= (6 - 5 + 1 + 2)x - 3$$
$$= 4x - 3$$

21. $3.4m - 4 - 3.4m - 7 = 3.4m - 3.4m + (-4 - 7)$
$$= (3.4 - 3.4)m + (-4 - 7)$$
$$= 0\cdot m - 11$$
$$= -11$$

25. $5(y + 4) = 5(y) + 5(4) = 5y + 20$

29. $-5(2x - 3y + 6)$
$$= -5(2x) + (-5)(-3y) + (-5)(6)$$
$$= -10x + 15y - 30$$

33. $7(d - 3) + 10 = 7d - 21 + 10 = 7d - 11$

37. $3(2x - 5) - 5(x - 4) = 6x - 15 - 5x + 20$
$$= 6x - 5x - 15 + 20$$
$$= x + 5$$

41. $5k - (3k - 10) = 5k - 3k + 10$
$$= 2k + 10$$

45. $5(x + 2) - (3x - 4) = 5x + 10 - 3x + 4$
$$= 5x - 3x + 10 + 4$$
$$= 2x + 14$$

49. answers may vary

53. $(3x - 8) - (7x + 1) = 3x - 8 - 7x - 1$
$$= 3x - 7x - 8 - 1$$
$$= -4x - 9$$

57. $2x - 4$

61. $(5x - 2) + 7x = 5x + 7x - 2 = 12x - 2$

65. $2x - (x + 10) = 2x - x - 10 = x - 10$

69. $a - b^2 = 2 - (-5)^2 = 2 - 25 = 2 + (-25) = -23$

73. To determine if the scale is balanced, find the number of cubes on each side of the scale and see if they are equal.

Left side	Right side
1 cone = 1 cube 1 cylinder = 2 cubes	3 cubes
Total = 3 cubes	Total = 3 cubes

The scale is balanced.

77. $5x + (4x - 1) + 5x + (4x - 1)$
$$= 5x + 4x - 1 + 5x + 4x - 1$$
$$= 5x + 4x + 5x + 4x + (-1 - 1)$$
$$= (5 + 4 + 5 + 4)x + (-1 - 1)$$
$$= 18x - 2$$
The perimeter of the rectangle is $(18x - 2)$ feet.

SECTION 2.2

1. $\quad x + 7 = 10$
$$x + 7 - 7 = 10 - 7$$
$$x = 3$$
Check: $\quad x + 7 = 10$
$$3 + 7 \overset{?}{=} 10$$
$$10 = 10$$
The solution is 3.

5. $\quad 3 + x = -11$
$$3 + x - 3 = -11 - 3$$
$$x = -14$$
Check: $\quad 3 + x = -11$
$$3 + (-14) \overset{?}{=} -11$$
$$11 = -11$$
The solution is -14.

9. $\quad \dfrac{1}{3} + f = \dfrac{3}{4}$
$$\dfrac{1}{3} + f - \dfrac{1}{3} = \dfrac{3}{4} - \dfrac{1}{3}$$
$$f = \dfrac{9}{12} - \dfrac{4}{12}$$
$$f = \dfrac{5}{12}$$
Check: $\quad \dfrac{1}{3} + f = \dfrac{3}{4}$
$$\dfrac{1}{3} + \dfrac{5}{12} \overset{?}{=} \dfrac{3}{4}$$
$$\dfrac{4}{12} + \dfrac{5}{12} \overset{?}{=} \dfrac{3}{4}$$
$$\dfrac{9}{12} \overset{?}{=} \dfrac{3}{4}$$
$$\dfrac{3}{4} = \dfrac{3}{4}$$
The solution is $\dfrac{5}{12}$.

13. $\quad 7x - 3 = 6x$
$$7x - 3 - 6x = 6x - 6x$$
$$x - 3 = 0$$
$$x - 3 + 3 = 0 + 3$$
$$x = 3$$
Check: $\quad 7x - 3 = 6x$
$$7(3) - 3 \overset{?}{=} 6(3)$$
$$21 - 3 \overset{?}{=} 18$$
$$18 = 18$$
The solution is 3.

17. $\quad 7x + 2 = 8x - 3$
$$9x = 8x - 3$$
$$9x - 8x = 8x - 3 - 8x$$
$$x = -3$$
Check: $\quad 7x + 2 = 8x - 3$
$$7(-3) + 2(-3) \overset{?}{=} 8(-3) - 3$$
$$-21 - 6 \overset{?}{=} -24 - 3$$
$$-27 = -27$$
The solution is -3.

21. $\quad 3x - 6 = 2x + 5$
$$3x - 6 - 2x = 2x + 5 - 2x$$
$$x - 6 = 5$$
$$x - 6 + 6 = 5 + 6$$
$$x = 11$$
Check: $\quad 3x - 6 = 2x + 5$
$$3(11) - 6 \overset{?}{=} 2(11) + 5$$
$$33 - 6 \overset{?}{=} 22 + 5$$
$$27 = 27$$
The solution is 11.

25. $\quad 8y + 2 - 6y = 3 + y - 10$
$$2y + 2 = y - 7$$
$$2y + 2 - y = y - 7 - y$$
$$y + 2 = -7$$
$$y + 2 - 2 = -7 - 2$$
$$y = -9$$
Check: $\quad 8y + 2 - 6y = 3 + y - 10$
$$8(-9) + 2 - 6(-9) \overset{?}{=} 3 + (-9) - 10$$
$$-72 + 2 + 54 \overset{?}{=} 3 - 9 - 10$$
$$-16 = -16$$
The solution is -9.

29. $-6.5 - 4x - 1.6 - 3x = -6x + 9.8$

$$-7x - 8.1 = -6x + 9.8$$
$$-7x - 8.1 + 6x = -6x + 9.8 + 6x$$
$$-x - 8.1 = 9.8$$
$$-x - 8.1 + 8.1 = 9.8 + 8.1$$
$$-x = 17.9$$

If $-x = 17.9$, then $x = -17.9$.

Check:
$$-6.5 - 4x - 1.6 - 3x = -6x + 9.8$$
$$-6.5 - 4(-17.9) - 1.6 - 3(-17.9)$$
$$\overset{?}{=} -6(-17.9) + 9.8$$
$$-6.5 + 71.6 - 1.6 + 53.7 \overset{?}{=} 107.4 + 9.8$$
$$117.2 = 117.2$$

The solution is -17.9.

33. $\qquad 2(x - 4) = x + 3$

$$2(x) + 2(-4) = x + 3$$
$$2x - 8 = x + 3$$
$$2x - 8 - x = x + 3 - x$$
$$x - 8 = 3$$
$$x - 8 + 8 = 3 + 8$$
$$x = 11$$

Check: $\qquad 2(x - 4) = x + 3$
$$2(11 - 4) \overset{?}{=} 11 + 3$$
$$2(7) \overset{?}{=} 14$$
$$14 = 14$$

The solution is 11.

37. $\qquad 10 - (2x - 4) = 7 - 3x$

$$10 - 1(2x - 4) = 7 - 3x$$
$$10 - 1(2x) - 1(-4) = 7 - 3x$$
$$10 - 2x + 4 = 7 - 3x$$
$$14 - 2x = 7 - 3x$$
$$14 - 2x + 3x = 7 - 3x + 3x$$
$$x + 14 = 7$$
$$x + 14 - 14 = 7 - 14$$
$$x = -7$$

Check: $\qquad 10 - (2x - 4) = 7 - 3x$
$$10 - (2(-7) - 4) \overset{?}{=} 7 - 3(-7)$$
$$10 - (-14 - 4) \overset{?}{=} 7 + 21$$
$$10 - (-18) \overset{?}{=} 28$$
$$10 + 18 \overset{?}{=} 28$$
$$28 = 28$$

The solution is -7.

41. $\qquad -3(x - 4) = -4x$

$$-3(x) - 3(-4) = -4x$$
$$-3x + 12 = -4x$$
$$-3x + 12 + 4x = -4x + 4x$$
$$x + 12 = 0$$
$$x + 12 - 12 = 0 - 12$$
$$x = -12$$

Check: $\qquad -3(x - 4) = -4x$
$$-3(-12 - 4) \overset{?}{=} -4(-12)$$
$$-3(-16) \overset{?}{=} 48$$
$$48 = 48$$

The solution is -12.

45. $\qquad -2(x + 6) + 3(2x - 5) = 3(x - 4) + 10$

$$-2(x) - 2(6) + 3(2x) + 3(-5)$$
$$= 3(x) + 3(-4) + 10$$
$$-2x - 12 + 6x - 15 = 3x - 12 + 10$$
$$4x - 27 = 3x - 2$$
$$4x - 27 - 3x = 3x - 2 - 3x$$
$$x - 27 = -2$$
$$x - 27 + 27 = -2 + 27$$
$$x = 25$$

Check:
$$-2(x + 6) + 3(2x - 5) = 3(x - 4) + 10$$
$$-2(25 + 6) + 3(2(25) - 5) \overset{?}{=} 3(25 - 4) + 10$$
$$-2(31) + 3(50 - 5) \overset{?}{=} 3(21) + 10$$
$$-62 + 3(45) \overset{?}{=} 63 + 10$$
$$-62 + 135 \overset{?}{=} 73$$
$$73 = 73$$

The solution is 25.

49. Since the sum of the lengths of the two pieces of board is 10 feet and one piece is x feet long, we can find the length of the other piece by subtracting x from 10. Therefore, the length of the other piece is $(10 - x)$ feet.

53. If Joseph Brennan received n votes and Susan Collins received 30,898 more votes than Brennan, we find the number of votes that Collins received by adding 30,898 to n. Thus, Susan Collins received $(n + 30,898)$ votes.

57. The reciprocal or multiplicative inverse of 2 is $\dfrac{1}{2}$ because $2 \cdot \dfrac{1}{2} = 1$.

61. $\dfrac{3x}{3} = \left(\dfrac{3}{3}\right)x = (1)x = x$

65. $\dfrac{3}{5}\left(\dfrac{5}{x}\right) = \left(\dfrac{3}{5} \cdot \dfrac{5}{3}\right)x = (1)x = x$

69. Since the sum of the angles in a triangle is $180°$, one angle measures $x°$, and a second angle measures $(2x + 7)°$, we find the measure of the third angle by subtracting x and $(2x + 7)$ from 180.

$$180 - [x + (2x + 7)] = 180 - [x + 2x + 7]$$
$$= 180 - [3x + 7]$$
$$= 180 - 1[3x + 7]$$
$$= 180 - 1(3x) - 1(7)$$
$$= 180 - 3x - 7$$
$$= 173 - 3x$$

The third angle measures $(173 - 3x)°$.

Section 2.3

1. $\qquad -5x = 20$

$$\dfrac{-5x}{-5} = \dfrac{20}{-5}$$
$$x = -4$$

Check: $\quad -5x = 20$
$$-5(-4) \overset{?}{=} 20$$
$$20 = 20$$

The solution is -4.

5.
$$-x = -12$$
$$\frac{-x}{-1} = \frac{-12}{-1}$$
$$x = 12$$
Check: $-x = -12$
$-(12) \stackrel{?}{=} -12$
$-12 = -12$
The solution is 12.

9. $\frac{1}{6}d = \frac{1}{2}$
$$6\left(\frac{1}{6}d\right) = 6\left(\frac{1}{2}\right)$$
$$d = 3$$
Check: $\frac{1}{6}d = \frac{1}{2}$
$\frac{1}{6}(3) \stackrel{?}{=} \frac{1}{2}$
$\frac{1}{2} = \frac{1}{2}$
The solution is 3.

13.
$$\frac{k}{-7} = 0$$
$$-7\left(\frac{k}{-7}\right) = -7(0)$$
$$k = 0$$
Check: $\frac{k}{-7} = 0$
$\frac{0}{-7} \stackrel{?}{=} 0$
$0 = 0$
The solution is 0.

17. $42 = 7x$
$$\frac{42}{7} = \frac{7x}{7}$$
$$6 = x$$
Check: $42 = 7x$
$42 \stackrel{?}{=} 7(6)$
$42 = 42$
The solution is 6.

21. $-\frac{3}{7}p = -2$
$$-\frac{7}{3}\left(-\frac{3}{7}p\right) = -\frac{7}{3}(-2)$$
$$p = \frac{14}{3}$$
Check: $-\frac{3}{7}p = -2$
$-\frac{3}{7}\left(\frac{14}{3}\right) \stackrel{?}{=} -2$
$-2 = -2$
The solution is $\frac{14}{3}$.

25.
$$2x - 4 = 16$$
$$2x - 4 + 4 = 16 + 4$$
$$2x = 20$$
$$\frac{2x}{2} = \frac{20}{2}$$
$$x = 10$$
Check: $2x - 4 = 16$
$2(10) - 4 \stackrel{?}{=} 16$
$20 - 4 \stackrel{?}{=} 16$
$16 = 16$
The solution is 10.

29.
$$6a + 3 = 3$$
$$6a + 3 - 3 = 3 - 3$$
$$6a = 0$$
$$\frac{6a}{6} = \frac{0}{6}$$
$$a = 0$$
Check: $6a + 3 = 3$
$6(0) + 3 \stackrel{?}{=} 3$
$0 + 3 \stackrel{?}{=} 3$
$3 = 3$
The solution is 0.

33.
$$5 - 0.3k = 5$$
$$5 - 0.3k - 5 = 5 - 5$$
$$-0.3k = 0$$
$$\frac{-0.3k}{-0.3} = \frac{0}{-0.3}$$
$$k = 0$$
Check: $5 - 0.3k = 5$
$5 - 0.3(0) \stackrel{?}{=} 5$
$5 - 0 \stackrel{?}{=} 5$
$5 = 5$
The solution is 0.

37.
$$\frac{x}{3} + 2 = -5$$
$$\frac{x}{3} + 2 - 2 = -5 - 2$$
$$\frac{x}{3} = -7$$
$$3\left(\frac{x}{3}\right) = 3(-7)$$
$$x = -21$$
Check: $\frac{x}{3} + 2 = -5$
$\frac{-21}{3} + 2 \stackrel{?}{=} -5$
$-7 + 2 \stackrel{?}{=} -5$
$-5 = -5$
The solution is -21.

SOLUTIONS

41. $6z - 8 - z + 3 = 0$
$$5z - 5 = 0$$
$$5z - 5 + 5 = 0 + 5$$
$$5z = 5$$
$$\frac{5z}{5} = \frac{5}{5}$$
$$z = 1$$

Check: $6z - 8 - z + 3 = 0$
$$6(1) - 8 - 1 + 3 = 0$$
$$6 - 8 - 1 + 3 \stackrel{?}{=} 0$$
$$0 = 0$$

The solution is 1.

45. $1 = 0.4x - 0.6x - 5$
$$1 = -0.2x - 5$$
$$1 + 5 = -0.2x - 5 + 5$$
$$6 = -0.2x$$
$$\frac{6}{-0.2} = \frac{-0.2x}{-0.2}$$
$$-30 = x$$

Check: $1 = 0.4x - 0.6x - 5$
$$1 \stackrel{?}{=} 0.4(-30) - 0.6(-30) - 5$$
$$1 \stackrel{?}{=} -12 + 18 - 5$$
$$1 = 1$$

The solution is -30.

49. If $x = $ the first odd integer, then
$x + 2 = $ the next odd integer. Their sum is
$x + (x + 2) = x + x + 2 = 2x + 2$

53. $5x + 2(x - 6) = 5x + 2(x) + 2(-6)$
$$= 5x + 2x - 12$$
$$= 7x - 12$$

57. $-(x - 1) + x = -1(x - 1) + x$
$$= -(x) - 1(-1) + x$$
$$= -x + 1 + x$$
$$= 1$$

61. Answers may vary. If we solve the equation for x, we
obtain the following.
$$3x + 6 = 2x + 10 + x - 4$$
$$3x + 6 = 3x + 6$$
$$3x + 6 - 6 = 3x + 6 - 6$$
$$3x = 3x$$
$$3x - 3x = 3x - 3x$$
$$0 = 0$$

Section 2.4

1. $-4y + 10 = -2(3y + 1)$
$$-4y + 10 = -6y - 2$$
$$-4y + 10 + 4y = -6y - 2 + 4y$$
$$10 = -2y - 2$$
$$10 + 2 = -2y - 2 + 2$$
$$12 = -2y$$
$$\frac{12}{-2} = \frac{-2y}{-2}$$
$$-6 = y$$

5. $-2(3x - 4) = 2x$
$$-6x + 8 = 2x$$
$$-6x + 8 + 6x = 2x + 6x$$
$$8 = 8x$$
$$\frac{8}{8} = \frac{8x}{8}$$
$$1 = x$$

9. $5(2x - 1) - 2(3x) = 1$
$$10x - 5 - 6x = 1$$
$$4x - 5 = 1$$
$$4x - 5 + 5 = 1 + 5$$
$$4x = 6$$
$$\frac{4x}{4} = \frac{6}{4}$$
$$x = \frac{3}{2}$$

13. $8 - 2(a - 1) = 7 + a$
$$8 - 2a + 2 = 7 + a$$
$$10 - 2a = 7 + a$$
$$10 - 2a + 2a = 7 + a + 2a$$
$$10 = 7 + 3a$$
$$10 - 7 = 7 + 3a - 7$$
$$3 = 3a$$
$$\frac{3}{3} = \frac{3a}{3}$$
$$1 = a$$

17. $-2y - 10 = 5y + 18$
$$-2y - 10 + 2y = 5y + 18 + 2y$$
$$-10 = 7y + 18$$
$$-10 - 18 = 7y + 18 - 18$$
$$-28 = 7y$$
$$\frac{-28}{7} = \frac{7y}{7}$$
$$-4 = y$$

21. $\frac{3}{4}x - \frac{1}{2} = 1$
$$4\left(\frac{3}{4}x - \frac{1}{2}\right) = 4(1)$$
$$4\left(\frac{3}{4}x\right) + 4\left(-\frac{1}{2}\right) = 4$$
$$3x - 2 = 4$$
$$3x - 2 + 2 = 4 + 2$$
$$3x = 6$$
$$\frac{3x}{3} = \frac{6}{3}$$
$$x = 2$$

25. $\frac{x}{2} - 1 = \frac{x}{5} + 2$
$$10\left(\frac{x}{2} - 1\right) = 10\left(\frac{x}{5} + 2\right)$$
$$10\left(\frac{x}{2}\right) + 10(-1) = 10\left(\frac{x}{5}\right) + 10(2)$$
$$5x - 10 = 2x + 20$$
$$5x - 10 - 2x = 2x + 20 - 2x$$
$$3x - 10 = 20$$
$$3x - 10 + 10 = 20 + 10$$
$$3x = 30$$
$$\frac{3x}{3} = \frac{30}{3}$$
$$x = 10$$

29.
$$0.06 - 0.01(x + 1) = -0.02(2 - x)$$
$$100[0.06 - 0.01(x + 1)] = 100[-0.02(2 - x)]$$
$$6 - 1(x + 1) = -2(2 - x)$$
$$6 - x - 1 = -4 + 2x$$
$$5 - x = -4 + 2x$$
$$5 - x + x = -4 + 2x + x$$
$$5 = -4 + 3x$$
$$5 + 4 = -4 + 3x + 4$$
$$9 = 3x$$
$$\frac{9}{3} = \frac{3x}{3}$$
$$3 = x$$

33.
$$0.05x + 0.15(70) = 0.25(142)$$
$$100[0.50x + 0.15(70)] = 100[0.25(142)]$$
$$50x + 15(70) = 25(142)$$
$$50x + 1050 = 3550$$
$$50x + 1050 - 1050 = 3550 - 1050$$
$$50x = 2500$$
$$\frac{50x}{50} = \frac{2500}{50}$$
$$x = 50$$

37.
$$\frac{2(x + 1)}{4} = 3x - 2$$
$$4\left[\frac{2(x + 1)}{4}\right] = 4(3x - 2)$$
$$2(x + 1) = 4(3x - 2)$$
$$2(x) + 2(1) = 4(3x) + 4(-2)$$
$$2x + 2 = 12x - 8$$
$$2x + 2 - 2x = 12x - 8 - 2x$$
$$2 = 10x - 8$$
$$2 + 8 = 10x - 8 + 8$$
$$10 = 10x$$
$$\frac{10}{10} = \frac{10x}{10}$$
$$1 = x$$

41. answers may vary

45.
$$\frac{x}{4} + 1 = \frac{x}{4}$$
$$4\left(\frac{x}{4} + 1\right) = 4\left(\frac{x}{4}\right)$$
$$x + 4 = x$$
$$x + 4 - x = x - x$$
$$4 = 0$$
There is no solution to the equation $\frac{x}{4} + 1 = \frac{x}{4}$.

49. $2(x + 3) - 5 = 5x - 3(1 + x)$
$$2x + 6 - 5 = 5x - 3 - 3x$$
$$2x + 1 = 2x - 3$$
$$2x + 1 - 1 = 2x - 3 - 1$$
$$2x = 2x - 4$$
$$2x - 2x = 2x - 4 - 2x$$
$$0 = -4$$
There is no solution to the equation
$2(x + 3) - 5 = 5x - 3(1 + x)$.

53. The perimeter of the lot is the sum of the lengths of the sides.
$$x + (2x - 3) + (3x - 5) = x + 2x - 3 + 3x - 5$$
$$= 6x - 8$$
The perimeter of the lot is $(6x - 8)$ meters.

57. $-3 + 2x$

A58

61.
$$1000(7x - 10) = 50(412 + 100x)$$
$$7000x - 10{,}000 = 20{,}600 + 5000x$$
$$7000x - 10{,}000 + 10{,}000$$
$$= 20{,}600 + 5000x + 10{,}000$$
$$7000x = 30{,}600 + 5000x$$
$$7000x - 5000x = 30{,}600 + 5000x - 5000x$$
$$2000x = 30{,}600$$
$$\frac{2000x}{2000} = \frac{30{,}600}{2000}$$
$$x = 15.3$$

65. Since we know the perimeter of the pentagon is 28 cm,
$$x + x + x + 2x + 2x = 28$$
$$(1 + 1 + 1 + 2 + 2)x = 28$$
$$7x = 28$$
$$\frac{7x}{7} = \frac{28}{7}$$
$$x = 4$$
If $x = 4$ cm, then $2x = 2(4) = 8$ cm.

Section 2.5

1. Let x represent the number.
$$2x + \frac{1}{5} = 3x - \frac{4}{5}$$
$$5\left(2x + \frac{1}{5}\right) = 5\left(3x - \frac{4}{5}\right)$$
$$10x + 1 = 15x - 4$$
$$10x + 1 - 10x = 15x - 4 - 10x$$
$$1 = 5x - 4$$
$$1 + 4 = 5x - 4 + 4$$
$$5 = 5x$$
$$\frac{5}{5} = \frac{5x}{5}$$
$$1 = x$$
The number is 1.

5. Let x represent the number.
$$2x \cdot 3 = 5x - \frac{3}{4}$$
$$6x = 5x - \frac{3}{4}$$
$$6x - 5x = 5x - \frac{3}{4} - 5x$$
$$x = -\frac{3}{4}$$
The number is $-\frac{3}{4}$.

9. Let $x =$ salary of the governor of Nebraska, then $2x =$ salary of the governor of New York.
$$x + 2x = 195{,}000$$
$$3x = 195{,}000$$
$$\frac{3x}{3} = \frac{195{,}000}{3}$$
$$x = 65{,}000$$
The salary of the governor of Nebraska is $65,000. The salary of the governor of New York is
$2 \cdot \$65{,}000 = \$130{,}000$.

13. The cost a renting the car is equal to the daily rental charge plus $0.29 per mile. Let x = number of miles.

$$2 \cdot 24.95 + 0.29x = 100$$
$$49.90 + 0.29x = 100$$
$$49.90 + 0.29x - 49.90 = 100 - 49.90$$
$$0.29x = 50.10$$
$$\frac{0.29x}{0.29} = \frac{50.10}{0.29}$$
$$x = 172$$

You can drive 172 whole miles on a budget of $100.

17. Let x = number of votes for Marc Little, then $x + 16,950$ = number of votes for Corrine Brown.

$$x + (x + 16,950) = 110,740$$
$$x + x + 16,950 = 110,740$$
$$2x + 16,950 = 110,740$$
$$2x + 16,950 - 16,950 = 110,740 - 16,950$$
$$2x = 93,790$$
$$\frac{2x}{2} = \frac{93,790}{2}$$
$$x = 46,895$$

Marc Little received 46,895 votes. Corrine Brown received $46,895 + 16,950 = 63,845$ votes.

21. If x = length of the shorter piece, then $2x + 2$ = length of the longer piece.

$$x + (2x + 2) = 17$$
$$x + 2x + 2 = 17$$
$$3x + 2 = 17$$
$$3x + 2 - 2 = 17 - 2$$
$$3x = 15$$
$$\frac{3x}{3} = \frac{15}{3}$$
$$x = 5$$

The shorter piece is 5 feet long. The longer piece is 12 feet long.

25. The tallest bar represents the most popular name which is Midway.

29. $\frac{1}{2}(x - 1) = 37$

33. $2W + 2L = 2(7) + 2(10) = 14 + 20 = 34$

37. Let x represent the first odd integer, then $x + 2$ represents the next consecutive odd integer.

$$2(x + 2) = 3x + 15$$
$$2x + 4 = 3x + 15$$
$$2x + 4 - 2x = 3x + 15 - 2x$$
$$4 = x + 15$$
$$4 - 15 = x + 15 - 15$$
$$-11 = x$$

The integers are -11 and -9.

41. Let x = the first integer, then $x + 1$ = the next consecutive integer, and $x + 2$ = third consecutive integer.

$$x + (x + 1) + (x + 2) = 2x + 13$$
$$x + x + 1 + x + 2 = 2x + 13$$
$$3x + 3 = 2x + 13$$
$$3x + 3 - 2x = 2x + 13 - 2x$$
$$x + 3 = 13$$
$$x + 3 - 3 = 13 - 3$$
$$x = 10$$

The integers are 10, 11, and 12.

45. Let x = the first odd integer, then $x + 2$ = the next consecutive odd integer.

$$7x - 54 = 5(x + 2)$$
$$7x - 54 = 5x + 10$$
$$7x - 54 + 54 = 5x + 10 + 54$$
$$7x = 5x + 64$$
$$7x - 5x = 5x + 64 - 5x$$
$$2x = 64$$
$$\frac{2x}{2} = \frac{64}{2}$$
$$x = 32$$

Since the solution is an even number, we conclude that there are no odd integers which satisfy these conditions.

SECTION 2.6

1. $A = bh$
$$45 = 15 \cdot h$$
$$\frac{45}{15} = \frac{15 \cdot h}{15}$$
$$3 = h$$

5. $A = \frac{1}{2}(B + b)h$
$$180 = \frac{1}{2}(11 + 7)h$$
$$180 = \frac{1}{2}(18)h$$
$$180 = 9h$$
$$\frac{180}{9} = \frac{9h}{9}$$
$$20 = h$$

9. $C = 2\pi r$
$$15.7 = 2\pi r$$
$$\frac{15.7}{2\pi} = \frac{2\pi r}{2\pi}$$
$$\frac{15.7}{6.28} = r$$
$$2.5 = r$$

13. $V = \frac{1}{3}\pi r^2 h$
$$565.2 = \frac{1}{3}\pi(6^2)h$$
$$565.2 = \frac{1}{3}\pi(36)h$$
$$565.2 = 12\pi h$$
$$\frac{565.2}{12\pi} = \frac{12\pi h}{12\pi}$$
$$\frac{565.2}{37.68} = h$$
$$15 = h$$

17. $d = rt$
$$3000 = r \cdot 1.5$$
$$\frac{3000}{1.5} = \frac{r \cdot 1.5}{1.5}$$
$$2000 = r$$

The plane travels at a rate of 2000 miles per hour.

21. $V = lwh$
$V = (8)(3)(6)$
$V = 144$
Since the tank has a volume of 144 cubic feet, and each piranha requires 1.5 cubic feet, the tank can hold

$\dfrac{144}{1.5} = 96$ piranhas.

25. $d = rt$
$25{,}000 = 4000 \cdot t$
$\dfrac{25{,}000}{4000} = \dfrac{4000 \cdot t}{4000}$
$6.25 = t$
It will take 6.25 hours.

29. $V = lwh$
$V = (10)(8)(10)$
$V = 800$
The minimum volume of the box must be 800 cubic feet.

33. Use $F = \left(\dfrac{9}{5}\right)C + 32$ with $C = -78.5$.

$F = \left(\dfrac{9}{5}\right)C + 32$

$F = \left(\dfrac{9}{5}\right)(-78.5) + 32 = -141.3 + 32 = -109.3$

$-78.5°C$ is the same as $-109.3°F$.

37. Use the formula for the volume of a sphere, $V = \dfrac{4}{3}\pi r^3$, with $r = 2000$.

$V = \dfrac{4}{3}\pi(2000)^3$

$= \dfrac{4}{3}\pi(8{,}000{,}000{,}000)$

$= 33{,}493{,}333{,}333$

The volume of the fireball was 33,493,333,333 cubic miles.

41. $\dfrac{20 \text{ miles}}{1 \text{ hour}} = \dfrac{20 \text{ miles}}{1 \text{ hour}} \cdot \dfrac{5280 \text{ feet}}{1 \text{ mile}} \cdot \dfrac{1 \text{ hour}}{3600 \text{ sec}}$

$= \dfrac{88}{3}$ feet per second

Use $d = rt$ with $d = 1300$ feet and $r = \dfrac{88}{3}$ feet per second.

$1300 = \dfrac{88}{3}t$

$\dfrac{3}{88}(1300) = \left(\dfrac{3}{88}\right)\left(\dfrac{88}{3}t\right)$

$44.3 = t$

It took about 44.3 seconds.

45. $V = LWH$
$\dfrac{V}{LH} = \dfrac{LWH}{LH}$
$\dfrac{V}{LH} = W$

49. $A = p + PRT$
$A - p = p + PRT - p$
$A - p = PRT$
$\dfrac{A - p}{PT} = \dfrac{PRT}{PT}$
$\dfrac{A - p}{PT} = R$

53. $P = a + b + c$
$P - b = a + b + c - b$
$P - b = a + c$
$P - b - c = a + c - c$
$P - b - c = a$

57. $32\% = 0.32$

61. $0.17 = 17\%$

65. $N = R + \dfrac{V}{G}$

$G(N) = G\left(R + \dfrac{V}{G}\right)$

$GN = GR + V$
$GN - GR = GR + V - GR$
$GN - GR = V$
$G(N - R) = V$

69. Use the formula $C = \left(\dfrac{5}{9}\right)(F - 32)$ to find when $C = F$.

$C = \left(\dfrac{5}{9}\right)(F - 32)$

$F = \left(\dfrac{5}{9}\right)(F - 32)$

$9(F) = 9\left[\left(\dfrac{5}{9}\right)(F - 32)\right]$

$9F = 5(F - 32)$
$9F = 5F - 160$
$9F - 5F = 5F - 160 - 5F$
$4F = -160$
$\dfrac{4F}{4} = \dfrac{-160}{4}$
$F = -40$

$-40°F$ is the same as $-40°C$.

Section 2.7

1. Let $x =$ the unknown number.
$x = 0.16 \cdot 70$
$x = 11.2$
The number 11.2 is 16% of 70.

5. Let $x =$ the unknown number.
$45 = 0.25 \cdot x$
$180 = x$
The number 45 is 25% of 180.

9. Let $x =$ the unknown number.
$40 = 0.80 \cdot x$
$50 = x$
The number 40 is 80% of 50.

13. To find the decrease in price, we find 25% of $156.
25% of $156 = 0.25(156) = 39$
The coat is selling for $39 off the original price. The sale price $156 - 39 = \$117$.

SOLUTIONS

A60

17. 55.40% of those surveyed have used over-the-counter drugs to combat the common cold.

21. No, because many people have used over-the-counter drugs for more than one of the categories listed.

25. The percent of total operating income from each division is the ratio of the operating income of the division to the total.

Creative Content: $\dfrac{1596}{3333} = 0.48 = 48\%$

Theme Parks: $\dfrac{990}{3333} = 0.30 = 30\%$

Broadcasting: $\dfrac{747}{3333} = 22 = 22\%$

Total: $48\% + 30\% + 22\% = 100\%$

29. $\dfrac{10}{12} = \dfrac{5}{6}$

33. 2 dollars $= 2 \cdot 20$ nickels $= 40$ nickels

The ratio of 4 nickels to 2 dollars is $\dfrac{4}{40} = \dfrac{1}{10}$.

37. 3 hours $= 3 \cdot 60$ minutes $= 180$ minutes

The ratio of 190 minutes to 3 hours is $\dfrac{190}{180} = \dfrac{19}{18}$.

41.
$$\dfrac{2}{3} = \dfrac{x}{6}$$
$$2 \cdot 6 = 3 \cdot x$$
$$\dfrac{12}{3} = \dfrac{3x}{3}$$
$$4 = x$$

45.
$$\dfrac{4x}{6} = \dfrac{7}{2}$$
$$4x \cdot 2 = 6 \cdot 7$$
$$\dfrac{8x}{8} = \dfrac{42}{8}$$
$$x = \dfrac{21}{4}$$

49.
$$\dfrac{x+1}{2x+3} = \dfrac{2}{3}$$
$$3(x+1) = 2(2x+3)$$
$$3x + 3 = 4x + 6$$
$$3x = 4x + 3$$
$$-x = 3$$
$$\dfrac{-x}{-1} = \dfrac{3}{-1}$$
$$x = -3$$

53.
$$\dfrac{3}{x+1} = \dfrac{5}{2x}$$
$$3 \cdot 2x = 5(x+1)$$
$$6x = 5x + 5$$
$$x = 5$$

57. Let x = number of calories in 42.6 grams.
$$\dfrac{110}{28.4} = \dfrac{x}{42.6}$$
$$110(42.6) = 28.4x$$
$$4686 = 28.4x$$
$$\dfrac{4686}{28.4} = \dfrac{28.4x}{28.4}$$
$$165 = x$$
There are 165 calories in 42.6 grams of the cereal.

61. Let x = number of gallons of water needed for 36 teaspoons of weed killer.
$$\dfrac{8}{2} = \dfrac{36}{x}$$
$$8x = 2(36)$$
$$8x = 72$$
$$\dfrac{8x}{8} = \dfrac{72}{8}$$
$$x = 9$$
9 gallons of water are needed.

65. Compare unit prices.

6 ounces: $\dfrac{\$0.69}{6} = \$0.115$

8 ounces: $\dfrac{\$0.90}{8} \approx \$0.113$

16 ounces: $\dfrac{\$1.89}{16} \approx \$0.118$

The 8-ounce size is the best buy.

69. Since $|-5| = 5$ and $-(-5) = 5$, $|-5| = -(-5)$.

73. $\dfrac{230}{2400} \approx 0.096 = 9.6\%$

About 9.6% of the daily value of sodium is contained in one serving.

77. Find the ratio of the calories from protein to the total calories. Each serving contains 12 g of protein.
$$12g = 12 \cdot 4 = 48 \text{ calories}$$
$$\dfrac{48}{280} \approx 0.171 = 17.1\%$$
About 17.1% of the calories in each serving come from protein.

Section 2.8

1. $x \le -1$

5. $y < 4$

9.
$$x - 2 \ge -7$$
$$x - 2 + 2 \ge -7 + 2$$
$$x \ge -5$$

13.
$$3x - 5 > 2x - 8$$
$$3x - 5 + 5 > 2x - 8 + 5$$
$$3x > 2x - 3$$
$$3x - 2x > 2x - 3 - 2x$$
$$x > -3$$

17. $2x < -6$

$$\frac{2x}{2} < \frac{-6}{2}$$

$$x < -3$$

21. $-x > 0$

$$\frac{-x}{-1} < \frac{0}{-1}$$

$$x < 0$$

25. $-0.6y < -1.8$

$$\frac{-0.6y}{-0.6} > \frac{-1.8}{-0.6}$$

$$y > 3$$

29.
$$3x - 7 < 6x + 2$$
$$3x - 7 - 6x < 6x + 2 - 6x$$
$$-3x - 7 < 2$$
$$-3x - 7 + 7 < 2 + 7$$
$$-3x < 9$$
$$\frac{-3x}{-3} > \frac{9}{-3}$$
$$x > -3$$

33.
$$-6x + 2 \geq 2(5 - x)$$
$$-6x + 2 \geq 10 - 2x$$
$$-6x + 2 + 2x \geq 10 - 2x + 2x$$
$$-4x + 2 \geq 10$$
$$-4x + 2 - 2 \geq 10 - 2$$
$$-4x \geq 8$$
$$\frac{-4x}{-4} \leq \frac{8}{-4}$$
$$x \leq -2$$

37. $3(x + 2) - 6 > -2(x - 3) + 14$
$$3x + 6 - 6 > -2x + 6 + 14$$
$$3x > -2x + 20$$
$$3x + 2x > -2x + 20 + 2x$$
$$5x > 20$$
$$\frac{5x}{5} > \frac{20}{5}$$
$$x > 4$$

41. $6\left[\frac{1}{2}(x - 5)\right] < 6\left[\frac{1}{3}(2x - 1)\right]$
$$3(x - 5) < 2(2x - 1)$$
$$3x - 15 < 4x - 2$$
$$3x - 15 - 4x < 4x - 2 - 4x$$
$$-x - 15 < -2$$
$$-x - 15 + 15 < -2 + 15$$
$$-x < 13$$
$$\frac{-x}{-1} > \frac{13}{-1}$$
$$x > -13$$

45. Let x = the unknown number.
$$2x + 6 > -14$$
$$2x + 6 - 6 > -14 - 6$$
$$2x > -20$$
$$\frac{2x}{2} > \frac{-20}{2}$$
$$x > -10$$
The statement is true for all numbers greater than –10.

49. Let x = score for the third game.
$$\frac{146 + 201 + x}{3} \geq 180$$
$$\frac{347 + x}{3} \geq 180$$
$$\left(\frac{347 + x}{3}\right) \geq 3(180)$$
$$347 + x \geq 540$$
$$347 + x - 347 \geq 540 - 347$$
$$x \geq 193$$
He must score at least 193 in his third game.

53. $(2)^3 = 2 \cdot 2 \cdot 2 = 8$

57. $\left(\frac{4}{7}\right)^2 = \frac{4}{7} \cdot \frac{4}{7} = \frac{16}{49}$

61. From the graph we can see that the greatest increase in the number of members occurred between the years 1992 and 1995. The membership increased by 3 million people each year.

Chapter 2 Test

1. $2y - 6 - y - 4 = 2y - y + (-6 - 4)$
$$= (2 - 1)y - 10$$
$$= y - 10$$

5. $-\frac{4}{5}x = 4$
$$-\frac{5}{4}\left(-\frac{4}{5}x\right) = -\frac{5}{4}(4)$$
$$x = -5$$

9. $\frac{2(x + 6)}{3} = x - 5$
$$3\left(\frac{2(x + 6)}{3}\right) = 3(x - 5)$$
$$2(x + 6) = 3(x - 5)$$
$$2(x) + 2(6) = 3(x) + 3(-5)$$
$$2x + 12 = 3x - 15$$
$$2x + 12 - 2x = 3x - 15 - 2x$$
$$12 = x - 15$$
$$12 + 15 = x - 15 + 15$$
$$27 = x$$

13.
$$\frac{1}{3}(y + 3) = 4y$$
$$3\left[\frac{1}{3}(y + 3)\right] = 3(4y)$$
$$y + 3 = 12y$$
$$y + 3 - y = 12y - y$$
$$3 = 11y$$
$$\frac{3}{11} = \frac{11y}{11}$$
$$\frac{3}{11} = 1y$$
$$\frac{3}{11} = y$$

17. The area of the rectangular deck is given by the formula $A = bh$. If $b = 20$ and $h = 35$, then
$$A = bh$$
$$A = (20)(35)$$
$$A = 700$$
Thus, the area of the deck is 700 square feet. Since 1 gallon covers 200 square feet, we can form the following proportion, where $x =$ number of gallons needed to cover 700 square feet.
$$\frac{1}{200} = \frac{x}{700}$$
$$1 \cdot 700 = 200 \cdot x$$
$$\frac{700}{200} = \frac{200x}{200}$$
$$3.5 = x$$
Since we are painting two coats we will need twice as much or $2 \cdot 3.5 = 7$ gallons of water seal.

21.
$$V = \pi r^2 h$$
$$\frac{1}{\pi r^2} \cdot V = \frac{1}{\pi r^2}(\pi r^2)h$$
$$\frac{V}{\pi r^2} = h$$
$$h = \frac{V}{\pi r^2}$$

25.
$$-0.3x \geq 2.4$$
$$-\frac{1}{0.3}(-0.3x) \leq -\frac{1}{0.3}(2.4)$$
$$x \leq -8$$

29. Since 5.1% of the income came from corporations, we find 5.1% of \$143.84 billion.
5.1% of \$143.84 billion $= 0.051(143.84)$ billion
$$= 7.33584 \text{ billion}$$
Thus, \$7.33584 billion came from corporations.

Chapter 3

Section 3.1

1. $7^2 = 7 \cdot 7 = 49$

5. $-2^4 = -2 \cdot 2 \cdot 2 \cdot 2 = -16$

9. $\left(\frac{1}{3}\right)^3 = \left(\frac{1}{3}\right)\left(\frac{1}{3}\right)\left(\frac{1}{3}\right) = \frac{1}{27}$

13. answers may vary

17. $5x^3 = 5(3)^3 = 5 \cdot 3 \cdot 3 \cdot 3 = 135$

21. $\frac{2z^4}{5} = \frac{2(-2)^4}{5} = \frac{2(-2)(-2)(-2)(-2)}{5} = \frac{32}{5}$

25. $(-3)^3 \cdot (-3)^9 = (-3)^{3+9} = (-3)^{12}$

29. $(4z^{10})(-6z^7)(z^3) = 4(-6)z^{10+7+3} = -24z^{20}$

33. $(x^9)^4 = x^{9 \cdot 4} = x^{36}$

37. $(2a^5)^3 = 2^3 a^{5 \cdot 3} = 8a^{15}$

41. $(x^2 y^3)^5 = x^{2 \cdot 5} y^{3 \cdot 5} = x^{10} y^{15}$

45. $(8z^5)^2 = 8^2 z^{5 \cdot 2} = 64z^{10}$
The area is $64z^{10}$ sq. decimeters.

49. $\frac{x^3}{x} = \frac{x^3}{x^1} = x^{3-1} = x^2$

53. $\frac{p^7 q^{20}}{pq^{15}} = p^{7-1} q^{20-15} = p^6 q^5$

57. $(2x)^0 = 1$

61. $5^0 + y^0 = 1 + 1 = 2$

65. $-5^2 = -5 \cdot 5 = -25$

69. $\frac{z^{12}}{z^4} = z^{12-4} = z^8$

73. $(6b)^0 = 1$

77. $b^4 b^2 = b^{4+2} = b^6$

81. $(2x^3)(-8x^4) = 2(-8)x^{3+4} = -16x^7$

85. $(-6xyz^3)^2 = (-6)^2 x^2 y^2 z^{3 \cdot 2} = 36x^2 y^2 z^6$

89. $\frac{3x^5}{x^4} = 3x^{5-4} = 3x$

93. $5 - 7 = 5 + (-7) = -2$

97. $-11 - (-4) = -11 + 4 = -7$

101. We use the volume formula.

105. $(a^b)^5 = a^{b \cdot 5} = a^{5b}$

109. $A = P\left(1 + \frac{r}{12}\right)^6$
$$A = 1000\left(1 + \frac{0.09}{12}\right)^6$$
$$A = 1000(1.0075)^6$$
$$A = 1045.85$$
You need \$1045.85 to pay off the loan.

Section 3.2

1. $4^{-3} = \frac{1}{4^3} = \frac{1}{64}$

5. $\left(-\frac{1}{4}\right)^{-3} = \frac{(-1)^{-3}}{(4)^{-3}} = \frac{4^3}{(-1)^3} = \frac{64}{-1} = -64$

9. $\frac{1}{p^{-3}} = p^3$

13. $\frac{x^{-2}}{x} = x^{-2-1} = x^{-3} = \frac{1}{x^3}$

17. $2^0 + 3^{-1} = 1 + \frac{1}{3} = \frac{3}{3} + \frac{1}{3} = \frac{4}{3}$

21. $\frac{-1}{p^{-4}} = -1(p^4) = -p^4$

25. $\frac{x^2 x^5}{x^3} = x^{2+5-3} = x^4$

29. $\dfrac{(m^5)^4 m}{m^{10}} = m^{5(4)+1-10} m^{20+1-10} = m^{11}$

33. $(x^5 y^3)^{-3} = x^{5(-3)} y^{3(-3)} = x^{-15} y^{-9} = \dfrac{1}{x^{15} y^9}$

37. $\dfrac{(a^5)^2}{(a^3)^4} = \dfrac{a^{10}}{a^{12}} = a^{10-12} = a^{-2} = \dfrac{1}{a^2}$

41. $\dfrac{-6m^4}{-2m^3} = \dfrac{-6}{-2} \cdot m^{4-3} = 3m$

45. $\dfrac{6x^2 y^3}{-7xy^5} = -\dfrac{6}{7} x^{2-1} y^{3-5} = -\dfrac{6}{7} x^1 y^{-2} = -\dfrac{6x}{7y^2}$

49. $\left(\dfrac{x^{-2} y^4}{x^3 y^7}\right)^2 = \dfrac{x^{-2(2)} y^{4(2)}}{x^{3(2)} y^{7(2)}}$

$= \dfrac{x^{-4} y^8}{x^6 y^{14}}$

$= x^{-4-6} y^{8-14}$

$= x^{-10} y^{-6}$

$= \dfrac{1}{x^{10} y^6}$

53. $\dfrac{2^{-3} x^{-4}}{2^2 x} = 2^{-3-2} x^{-4-1}$

$= 2^{-5} x^{-5}$

$= \dfrac{1}{2^5 x^5}$

$= \dfrac{1}{32 x^5}$

57. $\left(\dfrac{a^{-5} b}{ab^3}\right)^{-4} = \dfrac{a^{-5(-4)} b^{-4}}{a^{-4} b^{3(-4)}}$

$= \dfrac{a^{20} b^{-4}}{a^{-4} b^{-12}}$

$= a^{20-(-4)} b^{-4-(-12)}$

$= a^{24} b^8$

61. $\dfrac{(-2xy^{-3})^{-3}}{(xy^{-1})^{-1}} = \dfrac{(-2)^{-3} x^{-3} y^9}{x^{-1} y^1}$

$= (-2)^{-3} x^{-3-(-1)} y^{9-1}$

$= -\dfrac{y^8}{8x^2}$

65. $78{,}000 = 7.8 \times 10^4$

69. $0.00635 = 6.35 \times 10^{-3}$

73. $20{,}000{,}000 = 2.0 \times 10^7$

77. $120{,}000{,}000 = 1.2 \times 10^8$

81. $3.3 \times 10^{-2} = 0.033$

85. $6.25 \times 10^{18} = 6{,}250{,}000{,}000{,}000{,}000{,}000$

89. $(1.2 \times 10^{-3})(3 \times 10^{-2}) = 1.2 \cdot 3 \cdot 10^{-3} \cdot 10^{-2}$

$= 3.6 \times 10^{-5}$

$= 0.000036$

93. $\dfrac{8 \times 10^{-1}}{16 \times 10^5} = \dfrac{8}{16} \times 10^{-1-5}$

$= 0.5 \times 10^{-6}$

$= 0.0000005$

97. $3600 \times 4.2 \times 10^6 = 3.6 \times 10^3 \times 4.2 \times 10^6$

$= 3.6 \times 4.2 \times 10^3 \times 10^6$

$= 15.12 \times 10^9$

$= 1.512 \times 10^{10}$ cubic feet

101. $y - 10 + y = (1 + 1)y - 10 = 2y - 10$

105. $(2.63 \times 10^{12})(-1.5 \times 10^{-10})$

$= 2.63 \cdot (-1.5) \cdot 10^{12} \cdot 10^{-10}$

$= -3.945 \times 10^2$

$= -394.5$

109. $a^{-4m} \cdot a^{5m} = a^{-4m+5m} = a^m$

113. answers may vary

Section 3.3

1. $x^2 - 3x + 5$

Term	Coefficient
x^2	1
$-3x$	-3
5	5

5. $x + 2$
The degree is 1 since x is x^1. It is a binomial because it has two terms.

9. $12x^4 - x^2 - 12x^2 = 12x^4 - 13x^2$
The degree is 4, the greatest degree of any of its terms. It is a binomial because the simplified form has two terms.

13. answers may vary

17. a. $x + 6 = 0 + 6 = 6$

b. $x + 6 = -1 + 6 = 5$

21. a. $x^3 - 15 = 0^3 - 15 = -15$

b. $x^3 - 15 = (-1)^3 - 15 = -1 - 15 = -16$

25. $-16t^2 + 200t = -16(7.6)^2 + 200(7.6)$

$= -924.16 + 1520$

$= 595.84$ feet

29. $14x^2 + 9x^2 = (14 + 9)x^2 = 23x^2$

33. $8s - 5s + 4s = (8 - 5 + 4)s = 7s$

37. $5x + 3 + 4x + 3 + 2x + 6 + 3x + 7x$

$= (5x + 4x + 2x + 3x + 7x) + (3 + 3 + 6)$

$= 21x + 12$

41. $9ab - 6a + 5b - 3$

Terms	Degree	Degree of Polynomial
$9ab$	1 + 1 or 2	2 (highest degree)
$-6a$	1	
$5b$	1	
-3	0	

45. $3ab - 4a + 6ab - 7a = (3 + 6)ab + (-4 - 7)a$

$= 9ab - 11a$

49. $5x^2 y + 6xy^2 - 5yx^2 + 4 - 9y^2 x$

$= (5 - 5)x^2 y + (6 - 9)xy^2 + 4$

$= -3xy^2 + 4$

53. $4 + 5(2x + 3) = 4 + 10x + 15 = 10x + 19$

57. answers may vary

Section 3.4

1. $(3x + 7) + (9x + 5) = 3x + 7 + 9x + 5$

$= (3x + 9x) + (7 + 5)$

$= 12x + 12$

5. $(-5x^2 + 3) + (2x^2 + 1) = -5x^2 + 3 + 2x^2 + 1$
$$= (-5x^2 + 2x^2) + (3 + 1)$$
$$= -3x^2 + 4$$

9. $\begin{array}{r} 3t^2 + 4 \\ + 5t^2 - 8 \\ \hline 8t^2 - 4 \end{array}$

13. $(2x + 5) - (3x - 9) = (2x + 5) + (-3x + 9)$
$$= 2x + 5 + (-3x) + 9$$
$$= (2x - 3x) + (5 + 9)$$
$$= -x + 14$$

17. $(2x^2 + 3x - 9) - (-4x + 7)$
$$= (2x^2 + 3x - 9) + (4x - 7)$$
$$= 2x^2 + 3x - 9 + 4x - 7$$
$$= 2x^2 + (3x + 4x) + (-9 - 7)$$
$$= 2x^2 + 7x - 16$$

21. $(5x + 8) - (-2x^2 - 6x + 8)$
$$= (5x + 8) + (2x^2 + 6x - 8)$$
$$= 5x + 8 + 2x^2 + 6x - 8$$
$$= 2x^2 + (5x + 6x) + (8 - 8)$$
$$= 2x^2 + 11x$$

25. $\begin{array}{r} 5u^5 - 4u^2 + 3u - 7 \\ - (3u^5 + 6u^2 - 8u + 2) \\ \hline \end{array}$
$\begin{array}{r} 5u^5 - 4u^2 + 3u - 7 \\ + (-3u^5 - 6u^2 + 8u - 2) \\ \hline 2u^5 - 10u^2 + 11u - 9 \end{array}$

29. $(7y + 7) - (y - 6) = 7y + 7 - y + 6$
$$= 6y + 13$$

33. $(3x^2 + 5x - 8) + (5x^2 + 9x + 12) - (x^2 - 14)$
$$= 3x^2 + 5x - 8 + 5x^2 + 9x + 12 - x^2 + 14$$
$$= 7x^2 + 14x + 18$$

37. $(4x^2 - 6x + 1) + (3x^2 + 2x + 1)$
$$= 4x^2 - 6x + 1 + 3x^2 + 2x + 1$$
$$= 7x^2 - 4x + 2$$

41. $[(8y^2 + 7) + (6y + 9)] - (4y^2 - 6y - 3)$
$$= 8y^2 + 7 + 6y + 9 - 4y^2 + 6y + 3$$
$$= 4y^2 + 12y + 19$$

45. $(4x^2 + y^2 + 3) - (x^2 + y^2 - 2)$
$$= 4x^2 + y^2 + 3 - x^2 - y^2 + 2$$
$$= 3x^2 + 5$$

49. $(11r^2s + 16rs - 3 - 2r^2s^2) - (3sr^2 + 5 - 9r^2s^2)$
$$= 11r^2s + 16rs - 3 - 2r^2s^2 - 3sr^2 - 5 + 9r^2s^2$$
$$= 8r^2s + 16rs + 7r^2s^2 - 8$$

53. $(12x^3)(-x^5) = (12x^3)(-1x^5)$
$$= (12)(-1)(x^3)(x^5)$$
$$= -12x^8$$

57. $(-x^2 + 3x) + (2x^2 + 5) + (4x - 1)$
$$= -x^2 + 3x + 2x^2 + 5 + 4x - 1$$
$$= (x^2 + 7x + 4) \text{ feet}$$

61. $[(1.2x^2 - 3x + 9.1) - (7.8x^2 - 3.1 + 8)]$
$$+ (1.2x - 6)$$
$$= 1.2x^2 - 3x + 9.1 - 7.8x^2 + 3.1 - 8 + 1.2x - 6$$
$$= -6.6x^2 - 1.8x - 1.8$$

Section 3.5

1. $8x^2 \cdot 3x = (8 \cdot 3)(x^2 \cdot x) = 24x^3$

5. $(-x^3)(-x) = (-1)(-1)(x^3 \cdot x) = x^4$

9. $(2x)(-3x^2)(4x^5) = (2)(-3)(4)(x \cdot x^2 \cdot x^5) = -24x^8$

13. $7x(x^2 + 2x - 1) = 7x(x^2) + 7x(2x) + 7x(-1)$
$$= 7x^3 + 14x^2 - 7x$$

17. $3x(2x^2 - 3x + 4) = 3x(2x^2) + 3x(-3x) + 3x(4)$
$$= 6x^3 - 9x^2 + 12x$$

21. $-2a^2(3a^2 - 2a + 3)$
$$= -2a^2(3a^2) - 2a^2(-2a) - 2a^2(3)$$
$$= -6a^4 + 4a^3 - 6a^2$$

25. $x^2 + 3x = x(x + 3)$

29. $(a + 7)(a - 2) = a(a) + a(-2) + 7(a) + 7(-2)$
$$= a^2 - 2a + 7a - 14$$
$$= a^2 + 5a - 14$$

33. $(3x^2 + 1)(4x^2 + 7)$
$$= 3x^2(4x^2) + 3x^2(7) + 1(4x^2) = 1(7)$$
$$= 12x^4 + 21x^2 + 4x^2 + 7$$
$$= 12x^4 + 25x^2 + 7$$

37. $(1 - 3a)(1 - 4a)$
$$= 1(1) + 1(-4a) - 3a(1) - 3a(-4a)$$
$$= 1 - 4a - 3a + 12a^2$$
$$= 1 - 7a + 12a^2$$

41. $(x - 2)(x^2 - 3x + 7)$
$$= x(x^2) + x(-3x) + x(7) - 2(x^2)$$
$$\quad - 2(-3x) - 2(7)$$
$$= x^3 - 3x^2 + 7x - 2x^2 + 6x - 14$$
$$= x^3 - 5x^2 + 13x - 14$$

45. $(2a - 3)(5a^2 - 6a + 4)$
$$= 2a(5a^2) + 2a(-6a) + 2a(4)$$
$$\quad - 3(5a^2) - 3(-6a) - 3(4)$$
$$= 10a^3 - 12a^2 + 8a - 15a^2 + 18a - 12$$
$$= 10a^3 - 27a^2 + 26a - 12$$

49. $x^2 + 2x + 3x + 2(3) = x^2 + 5x + 6$

53. $\begin{array}{r} 2x^2 + 4x - 1 \\ x + 3 \\ \hline 6x^2 + 12x - 3 \\ 2x^3 + 4x^2 - x \\ \hline 2x^3 + 10x^2 + 11x - 3 \end{array}$

57. $(5x)^2 = 5^2x^2 = 25x^2$

61. At $t = 0$, value $= \$7000$

65. answers may vary

69. $\dfrac{1}{2}(3x - 2)(4x) = 2x(3x - 2)$
$$= 2x(3x) + 2x(-2)$$
$$= 6x^2 - 4x$$
$(6x^2 - 4x)$ square inches

73. a. $(a + b)(a - b) = a^2 - ab + ab - b^2$
$$= a^2 - b^2$$

b. $(2x + 3y)(2x - 3y)$
$$= (2x)^2 - 6xy + 6xy - (3y)^2$$
$$= 4x^2 - 9y^2$$

c. $(4x + 7)(4x - 7)$
$$= (4x)^2 - 28x + 28x - 7^2$$
$$= 16x^2 - 49$$
$(x + y)(x - y) = x^2 - y^2$

Section 3.6

1. $(x + 3)(x + 4) = x^2 + 4x + 3x + 12$
$$= x^2 + 7x + 12$$

5. $(5x - 6)(x + 2) = 5x^2 + 10x - 6x - 12$
$$= 5x^2 + 4x - 12$$

9. $(2x + 5)(3x - 1) = 6x^2 - 2x + 15x - 5$
$$= 6x^2 + 13x - 5$$

13. $\left(x - \dfrac{1}{3}\right)\left(x + \dfrac{2}{3}\right) = x^2 + \dfrac{2}{3}x - \dfrac{1}{3}x - \dfrac{2}{9}$
$$= x^2 + \dfrac{1}{3}x - \dfrac{2}{9}$$

17. $(x + 5y)(2x - y) = 2x^2 - xy + 10xy - 5y^2$
$$= 2x^2 + 9xy - 5y^2$$

21. $(2x - 1)^2 = (2x)^2 - 2(2x)(1) + (1)^2$
$$= 4x^2 - 4x + 1$$

25. $(x^2 + 5) = (x^2)^2 + 2(x^2)(5) + 5^2$
$$= x^4 + 10x^2 + 25$$

29. $(2a - 3)^2 = (2a)^2 - 2(2a)(3) + 3^2$
$$= 4a^2 - 12a + 9$$

33. $(3x - 7y)^2 = (3x)^2 - 2(3x)(7y) + (7y)^2$
$$= 9x^2 - 42xy + 49y^2$$

37. answers may vary

41. $(x + 6)(x - 6) = x^2 - 6^2 = x^2 - 36$

45. $(x^2 + 5)(x^2 - 5) = (x^2)^2 - 5^2 = x^4 - 25$

49. $(4 - 7x)(4 + 7x) = 4^2 - (7x)^2 = 16 - 49x^2$

53. $(9x + y)(9x - y) = (9x)^2 - y^2 = 81x^2 - y^2$

57. $(a + 5)(a + 4) = a^2 + 4a + 5a + 20$
$$= a^2 + 9a + 20$$

61. $(4a + 1)(3a - 1) = 12a^2 - 4a + 3a - 1$
$$= 12a^2 - a - 1$$

65. $(3a + 1)^2 = (3a)^2 + 2(3a)(1) + 1^2 = 9a^2 + 6a + 1$

69. $\left(a - \dfrac{1}{2}y\right)\left(1 + \dfrac{1}{2}y\right) = a^2 - \left(\dfrac{1}{2}y\right)^2 = a^2 - \dfrac{1}{4}y^2$

73. $(x^2 + 10)(x^2 - 10) = (x^2)^2 - (10)^2$
$$= x^4 - 100$$

77. $(5x - 6y)^2 = (5x)^2 - 2(5x)(6y) + (6y)^2$
$$= 25x^2 - 60xy + 36y^2$$

81. $\dfrac{50b^{10}}{70b^5} = \dfrac{10 \cdot 5 \cdot b^5 \cdot b^5}{10 \cdot 7 \cdot b^5} = \dfrac{5b^5}{7}$

85. $\dfrac{2x^4y^{12}}{3x^4y^4} = \dfrac{2 \cdot x^4 \cdot y^4 \cdot y^8}{3 \cdot x^4 \cdot y^4} = \dfrac{2y^8}{3}$

89. $(5x - 3)^2 - (x + 1)^2$
$$= [(5x)^2 - 2(5x)(3) + 3^2] - [x^2 + 2(x)(1) + 1^2]$$
$$= (25x^2 - 30x + 9) - (x^2 + 2x + 1)$$
$$= 25x^2 - 30x + 9 - x^2 - 2x - 1$$
$$= (24x^2 - 32x + 8) \text{ square meters}$$

Section 3.7

1. $\dfrac{20x^2 + 5x + 9}{5} = \dfrac{20x^2}{5} + \dfrac{5x}{5} + \dfrac{9}{5}$
$$= 4x^2 + x + \dfrac{9}{5}$$

5. $\dfrac{15p^3 + 18p^2}{3p} = \dfrac{15p^3}{3p} + \dfrac{18p^2}{3p} = 5p^2 + 6p$

9. $\dfrac{-9x^5 + 3x^4 - 12}{3x^3} = \dfrac{-9x^5}{3x^3} + \dfrac{3x^4}{3x^3} - \dfrac{12}{3x^3}$
$$= -3x^2 + x - \dfrac{4}{x^3}$$

13. $\dfrac{a^2b^2 - ab^3}{ab} = \dfrac{a^2b^2}{ab} - \dfrac{ab^3}{ab} = ab - b^2$

17.
$$
\begin{array}{r}
x + 1 \\
x + 3\overline{)x^2 + 4x + 3} \\
\underline{x^2 + 3x} \\
x + 3 \\
\underline{x + 3} \\
0
\end{array}
$$

$$\dfrac{x^2 + 4x + 3}{x + 3} = x + 1$$

21.
$$
\begin{array}{r}
2x + 1 \\
x - 4\overline{)2x^2 - 7x + 3} \\
\underline{2x^2 - 8x} \\
x + 3 \\
\underline{x - 4} \\
7
\end{array}
$$

$$\dfrac{2x^2 - 7x + 3}{x - 4} = 2x + 1 + \dfrac{7}{x - 4}$$

25.
$$
\begin{array}{r}
3a^2 - 3a + 1 \\
3a + 2\overline{)9a^3 - 3a^2 - 3a + 4} \\
\underline{9a^3 + 6a^2} \\
-9a^2 - 3a \\
\underline{-9a^2 - 6a} \\
3a + 4 \\
\underline{3a + 2} \\
2
\end{array}
$$

$$\dfrac{9a^3 - 3a^2 - 3a + 4}{3a + 2} = 3a^2 - 3a + 1 + \dfrac{2}{3a + 2}$$

29.
$$
\begin{array}{r}
4x + 3 \\
2x + 1\overline{)8x^2 + 10x + 1} \\
\underline{8x^2 + 4x} \\
6x + 1 \\
\underline{6x + 3} \\
-2
\end{array}
$$

$$\dfrac{8x^2 + 10x + 1}{2x + 1} = 4x + 3 - \dfrac{2}{2x + 1}$$

33.
$$
\begin{array}{r}
x^2 + 3x + 9 \\
x - 3\overline{)x^3 + 0x^2 + 0x - 27} \\
\underline{x^3 - 3x^2} \\
3x^2 + 0x \\
\underline{3x^2 - 9x} \\
9x - 27 \\
\underline{9x - 27} \\
0
\end{array}
$$

$$\dfrac{x^3 - 27}{x - 3} = x^2 + 3x + 9$$

37.
$$
\begin{array}{r}
2b - 1 \\
2b - 1 \overline{)\, 4b^2 - 4b - 5} \\
\underline{4b^2 - 2b} \\
-2b - 5 \\
\underline{-2b + 1} \\
-6
\end{array}
$$

$$\frac{-4b + 4b^2 - 5}{2b - 1} = 2b - 1 - \frac{6}{2b - 1}$$

41. $20 = -5 \cdot (-4)$ **45.** $36x^2 = 4x \cdot 9x$

49. $\dfrac{12x^3 + 4x - 16}{4} = \dfrac{12x^3}{4} + \dfrac{4x}{4} - \dfrac{16}{4}$
$= 3x^3 + x - 4$
Each side is $(3x^3 + x - 4)$ feet.

53. answers may vary

Chapter 3 Test

1. $2^5 = 2 \cdot 2 \cdot 2 \cdot 2 \cdot 2 = 32$

5. $(3x^2)(-5x^9) = (3)(-5)(x^2 \cdot x^9) = -15x^{11}$

9. $\dfrac{6^2 x^{-4} y^{-1}}{6^3 x^{-3} y^7} = 6^{2-3} x^{-4-(-3)} y^{-1-7}$
$= 6^{-1} x^{-1} y^{-8}$
$= \dfrac{1}{6xy^8}$

13. $6.23 \times 10^4 = 62{,}300$

17. $(8x^3 + 7x^2 + 4x - 7) + (8x^3 - 7x - 6)$
$= 8x^3 + 7x^2 + 4x - 7 + 8x^3 - 7x - 6$
$= 16x^3 + 7x^2 - 3x - 13$

21.
$$
\begin{array}{r}
x^3 - x^2 + x + 1 \\
2x^2 - 3x + 7 \\
\hline
7x^3 - 7x^2 + 7x + 7 \\
-3x^4 + 3x^3 - 3x^2 - 3x \\
2x^5 - 2x^4 + 2x^3 + 2x^2 \\
\hline
2x^5 - 5x^4 + 12x^3 - 8x^2 + 4x + 7
\end{array}
$$

25. $(8x + 3)^2 = (8x)^2 + 2(8x)(3) + 3^2$
$= 64x^2 + 48x + 9$

29.
$$
\begin{array}{r}
x + 2 \\
x + 5 \overline{)\, x^2 + 7x + 10} \\
\underline{x^2 + 5x} \\
2x + 10 \\
\underline{2x + 10} \\
0
\end{array}
$$

$(x^2 + 7x + 10) \div (x + 5) = x + 2$

Chapter 4

Section 4.1

1. y^2

5. $8x = 2 \cdot 2 \cdot 2 \cdot x$ **9.** $-10x^2 = -2 \cdot 5 \cdot x^2$
$4 = 2 \cdot 2$ $15x^3 = 3 \cdot 5 \cdot x^3$
GCF $= 2 \cdot 2 = 4$ GCF $= 5 \cdot x^2 = 5x^2$

13. $-18x^2 y = -2 \cdot 3 \cdot 3 \cdot x^2 \cdot y$
$9x^3 y^3 = 3 \cdot 3 \cdot x^3 \cdot y^3$
$36x^3 y = 2 \cdot 2 \cdot 3 \cdot 3 \cdot x^3 \cdot y$
GCF $= 3 \cdot 3 \cdot x^2 \cdot y = 9x^2 y$

17. $30x - 15 = 15(2x - 1)$

21. $6y^4 - 2y = 2y(3y^3 - 1)$

25. $4x - 8y + 4 = 4(x - 2y + 1)$

29. $a^7 b^6 - a^3 b^2 + a^2 b^5 - a^2 b^2$
$= a^2 b^2 (a^5 b^4 - a + b^3 - 1)$

33. $8x^5 + 16x^4 - 20x^3 + 12$
$= 4(2x^5 + 4x^4 - 5x^3 + 3)$

37. $y(x + 2) + 3(x + 2) = (x + 2)(y + 3)$

41. answers may vary

45. $5x + 15 + xy + 3y = 5(x + 3) + y(x + 3)$
$= (x + 3)(5 + y)$

49. $2y - 8 + xy - 4x = 2(y - 4) + x(y - 4)$
$= (y - 4)(2 + x)$

53. $4x^2 - 8xy - 3x + 6y = 4x(x - 2y) - 3(x - 2y)$
$= (x - 2y)(4x - 3)$

57. $(x + 2)(x + 5) = x^2 + 2x + 5x + 10$
$= x^2 + 7x + 10$

61. The two numbers are 2 and 6.
$2 \cdot 6 = 12; \ 2 + 6 = 8$

65. The two numbers are -2 and 5.
$-2 \cdot 5 = -10; \ -2 + 5 = 3$

69. $12x^2 y - 42x^2 - 4y + 14$
$= 2(6x^2 y - 21x^2 - 2y + 7)$
$= 2(3x^2(2y - 7) - 1(2y - 7))$
$= 2(3x^2 - 1)(2y - 7)$

73. Let $l =$ length of the rectangle.
$A = l \cdot w$
$$4n^4 - 24n = 4n \cdot l$$
$$4n(n^3 - 6) = 4n \cdot l$$
$$\frac{4n(n^3 - 6)}{4n} = \frac{4n \cdot l}{4n}$$
$$n^3 - 6 = l$$
The length is $(n^3 - 6)$ units.

Section 4.2

1. $x^2 + 7x + 6 = (x + 6)(x + 1)$

5. $x^2 - 3x - 18 = (x - 6)(x + 3)$

9. $x^2 + 5x + 2$ is a prime polynomial.

13. $a^4 - 2a^2 - 15 = (a^2 - 5)(a^2 + 3)$

17. answers may vary

21. $2x^3 - 18x^2 + 40x = 2x(x^2 - 9x + 20)$
$= 2x(x - 5)(x - 4)$

25. $x^2 + 15x + 36 = (x + 12)(x + 3)$

29. $r^2 - 16r + 48 = (r - 12)(r - 4)$

33. $3x^2 + 9x - 30 = 3(x^2 + 3x - 10)$
$= 3(x + 5)(x - 2)$

37. $x^2 - 18x - 144 = (x - 24)(x + 6)$

41. $x^2 - 8x + 15 = (x - 5)(x - 3)$

45. $4x^2 y + 4xy - 12y = 4y(x^2 + x - 3)$

49. $x^2 + 7xy + 10y^2 = (x + 5y)(x + 2y)$

53. $x^3 - 2x^2 - 24x = x(x^2 - 2x - 24)$
$= x(x - 6)(x + 4)$

57. $5x^3y - 25x^2y^2 - 120xy^3 = 5xy(x^2 - 5xy - 24y^2)$
$$= 5xy(x - 8y)(x + 3y)$$

61. $(5y - 4)(3y - 1) = 15y^2 - 12y - 5y + 4$
$$= 15y^2 - 17y + 4$$

65. $P = 2l + 2w$
$l = x^2 + 10x$ and $w = 4x + 33$, so
$P = 2(x^2 + 10x) + 2(4x + 33)$
$\qquad = 2x^2 + 20x + 8x + 66$
$\qquad = 2x^2 + 28x + 66$
$\qquad = 2(x^2 + 14x + 33)$
$\qquad = 2(x + 11)(x + 3)$
The perimeter of the rectangle is given by the polynomial $2x^2 + 28x + 66$ which factors as $2(x + 11)(x + 3)$.

69. $y^2 - 4y + c$ if factorable when c is 3 or 4.

73. $x^{2n} + 5x^n + 6 = (x^n + 2)(x^n + 3)$

Section 4.3

1. $5x^2 + 22x + 8 = (5x + 2)(x + 4)$

5. $20x^2 - 7x - 6 = (5x + 2)(4x - 3)$

9. $8y^2 - 17y + 9 = (y - 1)(8y - 9)$

13. $20r^2 + 27r - 8 = (4r - 1)(5r + 8)$

17. $3x^2 + x - 2 = (3x - 2)(x + 1)$

21. $15x^2 - 16x - 15 = (3x - 5)(5x + 3)$

25. $2x^2 - 7x - 99 = (2x + 11)(x - 9)$

29. $3a^2 + 10ab + 3b^2 = (3a + b)(a + 3b)$

33. $18x^2 - 9x - 14 = (6x - 7)(3x + 2)$

37. $21x^2 - 48x - 45 = 3(7x^2 - 16x - 15)$
$$= 3(7x + 5)(x - 3)$$

41. $6x^2y^2 - 2xy^2 - 60y^2 = 2y^2(3x^2 - x - 30)$
$$= 2y^2(3x - 10)(x + 3)$$

45. $3x^2 - 42x + 63 = 3(x^2 - 14x + 21)$

49. $4x^3 - 9x^2 - 9x = x(4x^2 - 9x - 9)$
$$= x(4x + 3)(x - 3)$$

53. $40a^2b + 9ab - 9b = b(40a^2 + 9a - 9)$
$$= b(8a - 3)(5a + 3)$$

57. $6y^3 - 8y^2 - 30y = 2y(3y^2 - 4y - 15)$
$$= 2y(3y + 5)(y - 3)$$

61. The greatest percentage of households having a personal computer corresponds to households having an income of $60,000 and above.

65. $4x^2(y - 1)^2 + 10x(y - 1)^2 + 25(y - 1)^2$
$= (y - 1)^2(4x^2 + 10x + 25)$

69. $2z^2 + bz - 7$ is factorable when b is 5 or 13.

Section 4.4

1. $x^2 + 3x + 2x + 6 = x(x + 3) + 2(x + 3)$
$$= (x + 3)(x + 2)$$

5. $y^2 + 8y - 2y - 16 = y(y + 8) - 2(y + 8)$
$$= (y + 8)(y - 2)$$

9. $8x^2 - 5x - 24x + 15 = x(8x - 5) - 3(8x - 5)$
$$= (8x - 5)(x - 3)$$

13. a. The numbers are 9 and 2.
$$9 \cdot 2 = 18$$
$$9 + 2 = 11$$

b. $9x + 2x = 11x$

c. $6x^2 + 11x + 3 = 6x^2 + 9x + 2x + 3$
$\qquad = 3x(2x + 3) + 1(2x + 3)$
$\qquad = (2x + 3)(3x + 1)$

17. $21y^2 + 17y + 2 = 21y^2 + 3y + 14y + 2$
$\qquad = 3y(7y + 1) + 2(7y + 1)$
$\qquad = (3y + 2)(7y + 1)$

21. $10x^2 - 9x + 2 = 10x^2 - 5x - 4x + 2$
$\qquad = 5x(2x - 1) - 2(2x - 1)$
$\qquad = (2x - 1)(5x - 2)$

25. $4x^2 + 12x + 9 = 4x^2 + 6x + 6x + 9$
$\qquad = 2x(2x + 3) + 3(2x + 3)$
$\qquad = (2x + 3)(2x + 3)$
$\qquad = (2x + 3)^2$

29. $10x^2 - 23x + 12 = 10x^2 - 15x - 8x + 12$
$\qquad = 5x(2x - 3) - 4(2x - 3)$
$\qquad = (2x - 3)(5x - 4)$

33. $16y^2 - 34y + 18 = 2(8y^2 - 17y + 9)$
$\qquad = 2(8y^2 - 8y - 9y + 9)$
$\qquad = 2[8y(y - 1) - 9(y - 1)]$
$\qquad = 2(y - 1)(8y - 9)$

37. $54a^2 - 9a - 30 = 3(18a^2 - 3a - 10)$
$\qquad = 3(18a^2 - 15a + 12a - 10)$
$\qquad = 3[3a(6a - 5) + 2(6a - 5)]$
$\qquad = 3(6a - 5)(3a + 2)$

41. $12x^3 - 27x^2 - 27x = 3x(4x^2 - 9x - 9)$
$\qquad = 3x(4x^2 - 12x + 3x - 9)$
$\qquad = 3x[4x(x - 3) + 3(x - 3)]$
$\qquad = 3x(x - 3)(4x + 3)$

45. $(y + 4)(y + 4) = y^2 + 4y + 4y + 16$
$$= y^2 + 8y + 16$$

49. $(4x - 3)^2 = 16x^2 + 2(4x)(-3) + 9$
$$= 16x^2 - 24x + 9$$

53. $3x^{2n} + 16x^n - 35 = 3x^{2n} - 5x^n + 21x^n - 35$
$\qquad = x^n(3x^n - 5) + 7(3x^n - 5)$
$\qquad = (3x^n - 5)(x^n + 7)$

Section 4.5

1. Yes; two terms, x^2 and 64, are squares $(64 = 8^2)$, and the third term of the trinomial, $16x$, is twice the product of x and 8 $(2 \cdot x \cdot 8 = 16x)$.

5. Yes; two terms, m^2 and 1, are squares $(1 = 1^2)$, and the third term of the trinomial, $-2m$, is the opposite of twice the product of m and $1(-(2 \cdot m \cdot 1) = -2m)$.

9. No; if we first factor out the GCF, 4, we find that only one of the terms, x^2, is a square.

13. $x^2 + 8x + 16$ is a perfect square trinomial because x^2 and 16 are squares $(16 = 4^2)$, and $8x$ is twice the product of x and 4 $(2 \cdot x \cdot 4 = 8x)$.

17. $x^2 - 16x + 64 = x^2 - 2 \cdot x \cdot 8 + 8^2$
$$= (x - 8)^2$$

21. $x^4 + 4x^2 + 4 = (x^2)^2 + 2 \cdot x^2 \cdot 2 + 2^2$
$$= (x^2 + 2)^2$$

25. $16y^2 + 40y + 25 = (4y)^2 + 2 \cdot 4y \cdot 5 + 5^2$
$$= (4y + 5)^2$$

29. $m^3 + 18m^2 + 81m = m(m^2 + 18m + 81)$
$$= m(m^2 + 2 \cdot m \cdot 9 + 9^2)$$
$$= m(m + 9)^2$$

33. $9x^2 - 24xy + 16y^2 = (3x)^2 - 2 \cdot 3x \cdot 4y + (4y)^2$
$$= (3x - 4y)^2$$

37. answers may vary

41. $81 - p^2 = 9^2 - p^2 = (9 + p)(9 - p)$

45. $9x^2 - 16 = (3x)^2 - 4^2 = (3x + 4)(3x - 4)$

49. $-36 + x^2 = -(6^2) + x^2 = (-6 + x)(6 + x)$

53. $x^2 - 169y^2 = x^2 - (13y)^2$
$$= (x + 13y)(x - 13y)$$

57. $9xy^2 - 4x = x(9y^2 - 4)$
$$= x((3y)^2 - 2^2)$$
$$= x(3y + 2)(3y - 2)$$

61. $x^3y - 4xy^3 = xy(x^2 - 4y^2)$
$$= xy(x^2 - (2y)^2)$$
$$= xy(x + 2y)(x - 2y)$$

65. $12x^2 - 27 = 3(4x^2 - 9)$
$$= 3((2x)^2 - 3^2)$$
$$= 3(2x + 3)(2x - 3)$$

69. $169a^2 - 49b^2 = (13a)^2 - (7b)^2$
$$= (13a + 7b)(13a - 7b)$$

73. $y^2 - \dfrac{1}{16} = y^2 - \left(\dfrac{1}{4}\right)^2 = \left(y + \dfrac{1}{4}\right)\left(y - \dfrac{1}{4}\right)$

77. $5 - y$, since
$(5 - y)(5 + y) = 25 - 5y + 5y - y^2$
$$= 25 - y^2$$
$$= 5^2 - y^2$$

81. $3x - 9 = 0$
$3x - 9 + 9 = 0 + 9$
$3x = 9$
$\dfrac{3x}{3} = \dfrac{9}{3}$
$x = 3$

85. The sail is shaped like a triangle. The area of a triangle is given by $A = \dfrac{1}{2}bh$. Use $b = 10$ feet and $h = x$ feet.

Then,
$A = \dfrac{1}{2}bh$
$25 = \dfrac{1}{2} \cdot 10 \cdot x$
$25 = 5x$
$\dfrac{25}{5} = \dfrac{5x}{5}$
$5 = x$

The height, x, is 5 feet.

89. $m^2(n + 8) - 9(n + 8) = (n + 8)(m^2 - 9)$
$$= (n + 8)(m^2 - 3^2)$$
$$= (n + 8)(m + 3)(m - 3)$$

93. $x^{2n} - 81 = (x^n)^2 - 9^2 = (x^n + 9)(x^n - 9)$

97. a. Let $t = 1$.
$529 - 16t^2 = 529 - 16(1^2)$
$$= 529 - 16(1)$$
$$= 529 - 16 = 513$$
After 1 second the height of the bolt is 513 feet.

b. Let $t = 4$.
$529 - 16t^2 = 529 - 16(4^2)$
$$= 529 - 16(16)$$
$$= 529 - 256 = 273$$
After 4 seconds the height of the bolt is 273 feet.

c. When the object hits the ground, its height is zero feet. Thus, to find the time, t, when the object's height is zero feet above the ground, we set the expression $529 - 16t^2$ equal to 0 and solve for t.
$529 - 16t^2 = 0$
$529 - 16t^2 + 16t^2 = 0 + 16t^2$
$529 = 16t^2$
$\dfrac{529}{16} = \dfrac{16t^2}{16}$
$33.0625 = t^2$
$\sqrt{33.06} = \sqrt{t^2}$
$5.75 = t$

Thus, the object will hit the ground after approximately 6 seconds.

Section 4.6

1. $(x - 2)(x + 1) = 0$
$x - 2 = 0$ or $x + 1 = 0$
$x = 2$ $x = -1$
The solutions are 2 and –1.

5. $(x + 9)(x + 17) = 0$
$x + 9 = 0$ or $x + 17 = 0$
$x = -9$ $x = -17$
The solutions are -9 and -17.

9. $3x(x - 8) = 0$
$3x = 0$ or $x - 8 = 0$
$x = 0$ $x = 8$
The solutions are 0 and 8.

13. $(2x - 7)(7x + 2) = 0$
$2x - 7 = 0$ or $7x + 2 = 0$
$2x = 7$ $7x = -2$
$x = \dfrac{7}{2}$ $x = -\dfrac{2}{7}$
The solutions are $\dfrac{7}{2}$ and $-\dfrac{2}{7}$.

17. $(x + 0.2)(x + 1.5) = 0$
$x + 0.2 = 0$ or $x + 1.5 = 0$
$x = -0.2$ $x = -1.5$
The solutions are -0.2 and -1.5.

21. $x^2 - 13x + 36 = 0$
$(x - 9)(x - 4) = 0$
$x - 9 = 0$ or $x - 4 = 0$
$x = 9$ $x = 4$
The solutions are 9 and 4.

25. $x^2 - 7x = 0$
$x(x - 7) = 0$
$x = 0$ or $x - 7 = 0$
                    $x = 7$
The solutions are 0 and 7.

29.
$$x^2 = 16$$
$$x^2 - 16 = 0$$
$$x^2 - 4^2 = 0$$
$$(x + 4)(x - 4) = 0$$
$$x + 4 = 0 \quad \text{or} \quad x - 4 = 0$$
$$x = -4 \qquad\qquad x = 4$$
The solutions are -4 and 4.

33.
$$x(3x - 1) = 14$$
$$3x^2 - x = 14$$
$$3x^2 - x - 14 = 0$$
$$(3x - 7)(x + 2) = 0$$
$$3x - 7 = 0 \quad \text{or} \quad x + 2 = 0$$
$$3x = 7 \qquad\qquad x = -2$$
$$x = \frac{7}{3}$$
The solutions are $\frac{7}{3}$ and -2.

37.
$$4x^3 - x = 0$$
$$x(4x^2 - 1) = 0$$
$$x(2x + 1)(2x - 1) = 0$$
$$x = 0 \quad \text{or} \quad 2x + 1 = 0 \quad \text{or} \quad 2x - 1 = 0$$
$$2x = -1 \qquad\qquad 2x = 1$$
$$x = -\frac{1}{2} \qquad\qquad x = \frac{1}{2}$$
The solutions are 0, $-\frac{1}{2}$, and $\frac{1}{2}$.

41.
$$(4x - 3)(16x^2 - 24x + 9) = 0$$
$$(4x - 3)(4x - 3)^2 = 0$$
$$(4x - 3)^3 = 0$$
$$4x - 3 = 0$$
$$4x = 3$$
$$x = \frac{3}{4}$$
The solution is $\frac{3}{4}$.

45.
$$(2x + 3)(2x^2 - 5x - 3) = 0$$
$$(2x + 3)(2x + 1)(x - 3) = 0$$
$$2x + 3 = 0 \quad \text{or} \quad 2x + 1 = 0 \quad \text{or} \quad x - 3 = 0$$
$$2x = -3 \qquad\qquad 2x = -1 \qquad\qquad x = 3$$
$$x = -\frac{3}{2} \qquad\qquad x = -\frac{1}{2}$$
The solutions are $-\frac{3}{2}$, $-\frac{1}{2}$, and 3.

49.
$$x^2 - 16x = 0$$
$$x(x - 16) = 0$$
$$x = 0 \quad \text{or} \quad x - 16 = 0$$
$$x = 16$$
The solutions are 0 and 16.

53.
$$6y^2 - 22y - 40 = 0$$
$$2(3y^2 - 11y - 20) = 0$$
$$2(3y + 4)(y - 5) = 0$$
$$3y + 4 = 0 \quad \text{or} \quad y - 5 = 0$$
$$3y = -4 \qquad\qquad y = 5$$
$$y = -\frac{4}{3}$$
The solutions are $-\frac{4}{3}$ and 5.

57.
$$x^3 - 12x^2 + 32x = 0$$
$$x(x^2 - 12x + 32) = 0$$
$$x(x - 8)(x - 4) = 0$$
$$x = 0 \quad \text{or} \quad x - 8 = 0 \quad \text{or} \quad x - 4 = 0$$
$$x = 8 \qquad\qquad x = 4$$
The solutions are 0, 8, and 4.

61.
$$\frac{3}{5} + \frac{4}{5} = \frac{3 \cdot 9}{5 \cdot 9} + \frac{4 \cdot 5}{9 \cdot 5}$$
$$= \frac{27}{45} + \frac{20}{45}$$
$$= \frac{27 + 20}{45}$$
$$= \frac{47}{45}$$

65.
$$\frac{4}{5} \cdot \frac{7}{8} = \frac{4 \cdot 7}{5 \cdot 8} = \frac{4 \cdot 7}{5 \cdot 2 \cdot 4} = \frac{7}{10}$$

69. a. When $x = 0$
$$y = -16x^2 + 20x + 300$$
$$y = -16(0^2) + 20(0) + 300$$
$$= -16(0) + 20(0) + 300$$
$$= 0 + 0 + 300$$
$$= 300$$
When $x = 1$:
$$y = -16x^2 + 20x + 300$$
$$y = -16(1)^2 + 20(1) + 300$$
$$= -16(1) + 20(1) + 300$$
$$= -16 + 20 + 300$$
$$= 304$$
When $x = 2$:
$$y = -16x^2 + 20x + 300$$
$$y = -16(2^2) + 20(2) + 300$$
$$= -16(4) + 20(2) + 300$$
$$= -64 + 40 + 300$$
$$= 276$$
When $x = 3$:
$$y = -16x^2 + 20x + 300$$
$$y = -16(3^2) + 20(3) + 300$$
$$= -16(9) + 20(3) + 300$$
$$= -144 + 60 + 300$$
$$= 216$$
When $x = 4$:
$$y = -16x^2 + 20x + 300$$
$$y = -16(4^2) + 20(4) + 300$$
$$= -16(16) + 20(4) + 300$$
$$= -256 + 80 + 300$$
$$= 124$$
When $x = 5$:
$$y = -16x^2 + 20x + 300$$
$$y = -16(5^2) + 20(5) + 300$$
$$= -16(25) + 20(5) + 300$$
$$= -400 + 100 + 300$$
$$= 0$$
When $x = 6$:
$$y = -16x^2 + 20x + 300$$
$$y = -16(6^2) + 20(6) + 300$$
$$= -16(36) + 20(6) + 300$$
$$= -576 + 120 + 300$$
$$= -156$$

b. The compass strikes the ground after 5 seconds, when the height, y, is zero feet.

c. The maximum height was approximately 304 feet.

73.
$$(2x - 3)(x + 8) = (x - 6)(x + 4)$$
$$2x^2 - 3x + 16x - 24 = x^2 - 6x + 4x - 24$$
$$2x^2 + 13x - 24 = x^2 - 2x - 24$$
$$x^2 + 15x = 0$$
$$x(x + 15) = 0$$
$$x = 0 \quad \text{or} \quad x + 15 = 0$$
$$x = -15$$
The solutions are 0 and –15.

Section 4.7

1. Let x = the width, then $x + 4$ = the length.

5. Let x = the base, then $4x + 1$ = the height.

9. The perimeter is the sum of the lengths of the sides.
$$120 = (x + 5) + (x^2 - 3x) + (3x - 8) + (x + 3)$$
$$120 = x + 5 + x^2 - 3x + 3x - 8 + x + 3$$
$$120 = x^2 + 2x$$
$$0 = x^2 + 2x - 120$$
$$x^2 + 2x - 120 = 0$$
$$(x + 12)(x - 10) = 0$$
$$x + 12 = 0 \quad \text{or} \quad x - 10 = 0$$
$$x = -12 \qquad\qquad x = 10$$
Since the dimensions cannot be negative, the lengths of the sides are: $10 + 5 = 15$ cm, $10^2 - 3(10) = 70$ cm, $3(10) - 8 = 22$ cm, and $10 + 3 = 13$ cm.

13. Find t when $h = 0$.
$$h = -16t^2 + 64t + 80$$
$$0 = -16t^2 + 64t + 80$$
$$0 = -16(t^2 - 4t - 5)$$
$$0 = -16(t - 5)(t + 1)$$
$$t - 5 = 0 \quad \text{or} \quad t + 1 = 0$$
$$t = 5 \qquad\qquad t = -1$$
Since the time t cannot be negative, the object hits the ground after 5 seconds.

17. Let $n = 12$.
$$D = \frac{1}{2}n(n - 3)$$
$$D = \frac{1}{2} \cdot 12(12 - 3) = 6(9) = 54$$
A polygon with 12 sides has 54 diagonals.

21. Let x = the unknown number.
$$x + x^2 = 132$$
$$x^2 + x - 132 = 0$$
$$(x + 12)(x - 11) = 0$$
$$x + 12 = 0 \quad \text{or} \quad x - 11 = 0$$
$$x = -12 \qquad\qquad x = 11$$
There are two numbers. They are -12 and 11.

25. Let x = the first number, then $20 - x$ = the other number.
$$x^2 + (20 - x)^2 = 218$$
$$x^2 + 400 - 40x + x^2 = 218$$
$$2x^2 - 40x + 400 = 218$$
$$2x^2 - 40x + 182 = 0$$
$$2(x^2 - 20x + 91) = 0$$
$$2(x - 13)(x - 7) = 0$$
$$x - 13 = 0 \quad \text{or} \quad x - 7 = 0$$
$$x = 13 \qquad\qquad x = 7$$
The numbers are 13 and 7.

29. Let x = the length of the shorter leg. Then $x + 4$ = the length of the longer leg and $x + 8$ = the length of the hypotenuse.
By the Pythagorean theorem
$$x^2 + (x + 4)^2 = (x + 8)^2$$
$$x^2 + x^2 + 8x + 16 = x^2 + 16x + 64$$
$$2x^2 + 8x + 16 = x^2 + 16x + 64$$
$$x^2 - 8x - 48 = 0$$
$$(x - 12)(x + 4) = 0$$
$$x - 12 = 0 \quad \text{or} \quad x + 4 = 0$$
$$x = 12 \qquad\qquad x = -4$$
Since the length cannot be negative, the sides of the triangle are 12 mm, $12 + 4 = 16$ mm, and $12 + 8 = 20$ mm.

33. Let x = the length of the shorter leg, then $x + 12$ = the length of the longer leg and $2x - 12$ = the length of the hypotenuse.
By the Pythagorean theorem
$$x^2 + (x + 12)^2 = (2x - 12)^2$$
$$x^2 + x^2 + 24x + 144 = 4x^2 - 48x + 144$$
$$2x^2 + 24x + 144 = 4x^2 - 48x + 144$$
$$0 = 2x^2 - 72x$$
$$0 = 2x(x - 36)$$
$$2x = 0 \quad \text{or} \quad x - 36 = 0$$
$$x = 0 \qquad\qquad x = 36$$
Since the length cannot be zero feet, the solution is 36. The shorter leg is 36 feet long.

37. The size of the average farm in 1940 was approximately 175 acres.

41. The lines intersect at approximately 1966.

45. Let x = the width of the pool, then
$x + 6$ = the length of the pool,
$x + 8$ = the combined width of the pool and the walk, and $x + 14$ = the combined length of the pool and the walk.
The area of the pool is $(x + 6)(x) = x^2 + 6x$.
The total area is $(x + 14)(x + 8) = x^2 + 22x + 112$
$$x^2 + 22x + 112 = (x^2 + 6x) + 576$$
$$16x = 464$$
$$x = 29$$
The pool is 29 meters wide and $29 + 6 = 35$ meters long.

Chapter 4 Test

1. $9x^2 - 3x = 3x(3x - 1)$

5. $x^4 - 16 = (x^2)^2 - 4^2$
$$= (x^2 + 4)(x^2 - 4)$$
$$= (x^2 + 4)(x^2 - 2^2)$$
$$= (x^2 + 4)(x + 2)(x - 2)$$

9. $3a^2 + 3ab - 7a - 7b = 3a(a + b) - 7(a + b)$
$$= (a + b)(3a - 7)$$

13. $6t^2 - t - 5 = (6t + 5)(t - 1)$

17. $(x - 3)(x + 9) = 0$
$$x - 3 = 0 \quad \text{or} \quad x + 9 = 0$$
$$x = 3 \qquad\qquad x = -9$$
The solutions are 3 and -9.

21.
$$5t^3 - 45t = 0$$
$$5t(t^2 - 9) = 0$$
$$5t(t + 3)(t - 3) = 0$$
$$5t = 0 \quad \text{or} \quad t + 3 = 0 \quad \text{or} \quad t - 3 = 0$$
$$t = 0 \qquad\qquad t = -3 \qquad\qquad t = 3$$
The solutions are 0, -3, and 3.

25. $x + 2 =$ the length of the rectangle and
$x - 1 =$ the width of the rectangle.
$$A = lw$$
$$54 = (x + 2)(x - 1)$$
$$54 = x^2 + x - 2$$
$$0 = x^2 + x - 56$$
$$0 = (x + 8)(x - 7)$$
$$x + 8 = 0 \qquad \text{or} \qquad x - 7 = 0$$
$$x = -8 \qquad\qquad\qquad x = 7$$
Since the dimensions cannot be negative, the length of the rectangle is $7 + 2 = 9$ units, and the width is $7 - 1 = 6$ units.

29. Let $x =$ the length of the shorter leg. Then $x + 10 =$ the length of the hypotenuse, and $x + 5 =$ the length of the longer leg. By the Pythagorean theorem
$$x^2 + (x + 5)^2 = (x + 10)^2$$
$$x^2 + x^2 + 10x + 25 = x^2 + 20x + 100$$
$$x^2 - 10x - 75 = 0$$
$$(x - 15)(x + 5) = 0$$
$$x - 15 = 0 \qquad \text{or} \qquad x + 5 = 0$$
$$x = 15 \qquad\qquad\qquad x = -5$$
Since the lengths cannot be negative, the length of the shorter leg is 15 cm, the longer leg is $15 + 5 = 20$ cm, and the hypotenuse is $5 + 10 = 25$ cm.

Chapter 5

SECTION 5.1

1. $\dfrac{x + 5}{x + 2} = \dfrac{2 + 5}{2 + 2} = \dfrac{7}{4}$

5. $\dfrac{x^2 + 8x + 2}{x^2 - x - 6} = \dfrac{2^2 + 8(2) + 2}{2^2 - 2 - 6}$
$$= \dfrac{4 + 16 + 2}{4 - 8}$$
$$= \dfrac{22}{-4}$$
$$= \dfrac{11 \cdot 2}{-2 \cdot 2}$$
$$= -\dfrac{11}{2}$$

9. $2x = 0$
$$x = 0$$
The expression is undefined when $x = 0$.

13. $2x - 8 = 0$
$$2x = 8$$
$$x = 4$$
The expression is undefined when $x = 4$.

17. The denominator is never zero so there are no values for which $\dfrac{x^2 - 5x - 2}{4}$ is undefined.

21. $\dfrac{2}{8x + 16} = \dfrac{2}{8(x + 2)} = \dfrac{2}{2 \cdot 4(x + 2)} = \dfrac{1}{4(x + 2)}$

25. $\dfrac{2x - 10}{3x - 30} = \dfrac{2(x - 5)}{3(x - 10)}$; does not simplify

29. $\dfrac{x - 7}{7 - x} = \dfrac{x - 7}{-1(x - 7)} = \dfrac{1}{-1} = -1$

33. $\dfrac{x + 5}{x^2 - 4x - 45} = \dfrac{x + 5}{(x - 9)(x + 5)} = \dfrac{1}{x - 9}$

37. $\dfrac{x + 7}{x^2 + 5x - 14} = \dfrac{x + 7}{(x - 2)(x + 7)} = \dfrac{1}{x - 2}$

41. $\dfrac{x^2 + 7x + 10}{x^2 - 3x - 10} = \dfrac{(x + 5)(x + 2)}{(x - 5)(x + 2)} = \dfrac{x + 5}{x - 5}$

45. $\dfrac{2x^2 - 8}{4x - 8} = \dfrac{2(x^2 - 4)}{4(x - 2)}$
$$= \dfrac{2(x + 2)(x - 2)}{2 \cdot 2(x - 2)}$$
$$= \dfrac{x + 2}{2}$$

49. $\dfrac{2 - x}{x - 2} = \dfrac{-1(x - 2)}{x - 2} = -1$

53. $\dfrac{m^2 - 6m + 9}{m^2 - 9} = \dfrac{(m - 3)(m - 3)}{(m + 3)(m - 3)} = \dfrac{m - 3}{m + 3}$

57. $\dfrac{5}{6} \cdot \dfrac{10}{11} \cdot \dfrac{2}{3} = \dfrac{5 \cdot 10 \cdot 2}{6 \cdot 11 \cdot 3}$
$$= \dfrac{5 \cdot 2 \cdot 5 \cdot 2}{3 \cdot 2 \cdot 11 \cdot 3}$$
$$= \dfrac{5 \cdot 5 \cdot 2}{3 \cdot 11 \cdot 3}$$
$$= \dfrac{50}{99}$$

61. $\dfrac{13}{20} \div \dfrac{2}{9} = \dfrac{13}{20} \cdot \dfrac{9}{2} = \dfrac{13 \cdot 9}{20 \cdot 2} = \dfrac{117}{40}$

65. $\dfrac{5x + 15 - xy - 3y}{2x + 6} = \dfrac{5(x + 3) - y(x + 3)}{2(x + 3)}$
$$= \dfrac{(x + 3)(5 - y)}{2(x + 3)}$$
$$= \dfrac{5 - y}{2}$$

69. $C = \dfrac{DA}{A + 12} = \dfrac{(1000)(8)}{8 + 12} = \dfrac{8000}{20} = \dfrac{400 \cdot 20}{20} = 400$
The child should receive 400 mg.

SECTION 5.2

1. $\dfrac{3x}{y^2} \cdot \dfrac{7y}{4x} = \dfrac{3x \cdot 7y}{y^2 \cdot 4x} = \dfrac{3 \cdot 7 \cdot x \cdot y}{4 \cdot x \cdot y \cdot y} = \dfrac{3 \cdot 7}{4 \cdot y} = \dfrac{21}{4y}$

5. $-\dfrac{5a^2b}{30a^2b^2} \cdot b^3 = -\dfrac{5a^2b \cdot b^3}{30a^2b^2}$
$$= -\dfrac{5 \cdot a^2 \cdot b \cdot b \cdot b^2}{5 \cdot 6 \cdot a^2 \cdot b^2}$$
$$= -\dfrac{b \cdot b}{6}$$
$$= -\dfrac{b^2}{6}$$

9. $\dfrac{6x+6}{5} \cdot \dfrac{10}{36x+36} = \dfrac{(6x+6) \cdot 10}{5 \cdot (36x+36)}$

$\qquad = \dfrac{6(x+1) \cdot 2 \cdot 5}{5 \cdot 36(x+1)}$

$\qquad = \dfrac{6 \cdot 5 \cdot 2 \cdot (x+1)}{6 \cdot 5 \cdot 2 \cdot 3 \cdot (x+1)}$

$\qquad = \dfrac{1}{3}$

13. $\dfrac{x^2-25}{x^2-3x-10} \cdot \dfrac{x+2}{x} = \dfrac{(x^2-25) \cdot (x+2)}{(x^2-3x-10) \cdot x}$

$\qquad = \dfrac{(x-5)(x+5) \cdot (x+2)}{(x-5)(x+2) \cdot x}$

$\qquad = \dfrac{x+5}{x}$

17. $\dfrac{5x^7}{2x^5} \div \dfrac{10x}{4x^3} = \dfrac{5x^7}{2x^5} \cdot \dfrac{4x^3}{10x}$

$\qquad = \dfrac{5 \cdot x^2 \cdot x^5 \cdot 2 \cdot 2x \cdot x^2}{2x^5 \cdot 2 \cdot 5 \cdot x}$

$\qquad = x^4$

21. $\dfrac{(x-6)(x+4)}{4x} \div \dfrac{2x-12}{8x^2}$

$\qquad = \dfrac{(x-6)(x+4)}{4x} \cdot \dfrac{8x^2}{2x-12}$

$\qquad = \dfrac{(x-6)(x+4) \cdot 2 \cdot 4 \cdot x \cdot x}{4x \cdot 2(x-6)}$

$\qquad = x(x+4)$

25. $\dfrac{m^2-n^2}{m+n} \div \dfrac{m}{m^2+nm} = \dfrac{m^2-n^2}{m+n} \cdot \dfrac{m^2+nm}{m}$

$\qquad = \dfrac{(m-n)(m+n) \cdot m(m+n)}{(m+n) \cdot m}$

$\qquad = (m-n)(m+n)$

$\qquad = m^2-n^2$

29. $\dfrac{x^2+7x+10}{x-1} \div \dfrac{x^2+2x-15}{x-1}$

$\qquad = \dfrac{x^2+7x+10}{x-1} \cdot \dfrac{x-1}{x^2+2x-15}$

$\qquad = \dfrac{(x+5)(x+2) \cdot (x-1)}{(x-1) \cdot (x+5)(x-3)}$

$\qquad = \dfrac{x+2}{x-3}$

33. $\dfrac{x^2+5x}{8} \cdot \dfrac{9}{3x+15} = \dfrac{x(x+5) \cdot 3 \cdot 3}{8 \cdot 3(x+5)} = \dfrac{3x}{8}$

37. $\dfrac{3x+4y}{x^2+4xy+4y^2} \cdot \dfrac{x+2y}{2} = \dfrac{(3x+4y) \cdot (x+2y)}{(x+2y)(x+2y) \cdot 2}$

$\qquad = \dfrac{3x+4y}{2(x+2y)}$

41. $\dfrac{a^2+7a+12}{a^2+5a+6} \cdot \dfrac{a^2+8a+15}{a^2+5a+4}$

$\qquad = \dfrac{(a+3)(a+4) \cdot (a+5)(a+3)}{(a+3)(a+2) \cdot (a+4)(a+1)}$

$\qquad = \dfrac{(a+5)(a+3)}{(a+2)(a+1)}$

45. $\dfrac{50 \text{ miles}}{1 \text{ hour}} \cdot \dfrac{1 \text{ hour}}{3600 \text{ seconds}} \cdot \dfrac{5280 \text{ feet}}{1 \text{ mile}} = \dfrac{50 \cdot 5280}{3600} \text{ feet/sec}$

$\qquad \approx 73 \text{ feet/sec}$

49. $\dfrac{1000 \text{ yen}}{1} \cdot \dfrac{1 \text{ US dollar}}{94.06 \text{ yen}} = \dfrac{1000}{94.06} \text{ U.S. dollars} = \$10.63$

53. $\dfrac{9}{9} - \dfrac{19}{9} = -\dfrac{10}{9}$

57. $\left(\dfrac{x^2-y^2}{x^2+y^2} \div \dfrac{x^2-y^2}{3x} \right) \cdot \dfrac{x^2+y^2}{6}$

$\qquad = \dfrac{x^2-y^2}{x^2+y^2} \cdot \dfrac{3x}{x^2-y^2} \cdot \dfrac{x^2+y^2}{6}$

$\qquad = \dfrac{(x^2-y^2) \cdot 3x \cdot (x^2+y^2)}{(x^2+y^2) \cdot (x^2-y^2) \cdot 2 \cdot 3}$

$\qquad = \dfrac{x}{2}$

61. answers may vary

Section 5.3

1. $\dfrac{a}{13} + \dfrac{9}{13} = \dfrac{a+9}{13}$

5. $\dfrac{4m}{m-6} - \dfrac{24}{m-6} = \dfrac{4m-24}{m-6} = \dfrac{4(m-6)}{m-6} = 4$

9. $\dfrac{5x+4}{x-1} - \dfrac{2x+7}{x-1} = \dfrac{5x+4-(2x+7)}{x-1}$

$\qquad = \dfrac{5x+4-2x-7}{x-1}$

$\qquad = \dfrac{3x-3}{x-1}$

$\qquad = \dfrac{3(x-1)}{x-1}$

$\qquad = 3$

13. $\dfrac{2x+3}{x^2-x-30} - \dfrac{x-2}{x^2-x-30} = \dfrac{2x+3-(x-2)}{x^2-x-30}$

$\qquad = \dfrac{2x+3-x+2}{x^2-x-30}$

$\qquad = \dfrac{x+5}{x^2-x-30}$

$\qquad = \dfrac{x+5}{(x-6)(x+5)}$

$\qquad = \dfrac{1}{x-6}$

17. answers may vary

21. $\quad 8x = 2^3 \cdot x$

$\quad 2x+4 = 2(x+2)$

$\quad \text{LCD} = 2^3 \cdot x \cdot (x+2) = 8x(x+2)$

25. $\quad x+6 = x+6$

$\quad 3x+18 = 3 \cdot (x+6)$

$\quad \text{LCD} = 3(x+6)$

29. $\quad x-8 = x-8$

$\quad 8-x = -(x-8)$

$\quad \text{LCD} = x-8 \text{ or } 8-x$

33. answers may vary

37. $\dfrac{6}{3a} = \dfrac{6(4b^2)}{3a(4b^2)} = \dfrac{24b^2}{12ab^2}$

41. $\dfrac{9a+2}{5a+10} = \dfrac{9a+2}{5(a+2)} = \dfrac{(9a+2)(b)}{5(a+2)(b)} = \dfrac{9ab+2b}{5b(a+2)}$

45. $\dfrac{9y-1}{15x^2-30} = \dfrac{(9y-1)(2)}{(15x^2-30)2} = \dfrac{18y-2}{30x^2-60}$

49. Since $6 = 2 \cdot 3$ and $4 = 2^2$, LCD $= 2^2 \cdot 3 = 12$.

$$\frac{2}{6} - \frac{3}{4} = \frac{2(2)}{6(2)} - \frac{3(3)}{4(3)} = \frac{4}{12} - \frac{9}{12} = \frac{4 - 9}{12} = -\frac{5}{12}$$

53. Since $8 = 2^3$ and $12 = 2^2 \cdot 3$, the least common multiple of 8 and 12 is $2^3 \cdot 3 = 24$. Since $8 \cdot 3 = 24$ and $12 \cdot 2 = 24$, buy three packages of hot dogs and two packages of buns.

Section 5.4

1. LCD $= 2 \cdot 3 \cdot x = 6x$

$$\frac{4}{2x} + \frac{9}{3x} = \frac{4(3)}{2x(3)} + \frac{9(2)}{3x(2)}$$
$$= \frac{12}{6x} + \frac{18}{6x}$$
$$= \frac{30}{6x}$$
$$= \frac{5(6)}{6x}$$
$$= \frac{5}{x}$$

5. LCD $= 2x^2$

$$\frac{3}{x} + \frac{5}{2x^2} = \frac{3(2x)}{x(2x)} + \frac{5}{2x^2} = \frac{6x}{2x^2} + \frac{5}{2x^2} = \frac{6x + 5}{2x^2}$$

9. $2x - 4 = 2(x - 2)$
$x^2 - 4 = (x - 2)(x + 2)$
$\quad$ LCD $= 2(x - 2)(x + 2)$

$$\frac{15}{2x - 4} + \frac{x}{x^2 - 4}$$
$$= \frac{15}{2(x - 2)} + \frac{x}{(x - 2)(x + 2)}$$
$$= \frac{15(x + 2)}{2(x - 2)(x + 2)} + \frac{x(2)}{(x - 2)(x + 2)(2)}$$
$$= \frac{15x + 30}{2(x - 2)(x + 2)} + \frac{2x}{2(x - 2)(x + 2)}$$
$$= \frac{15x + 30 + 2x}{2(x - 2)(x + 2)}$$
$$= \frac{17x + 30}{2(x - 2)(x + 2)}$$

13.
$$\frac{6}{x - 3} + \frac{8}{3 - x} = \frac{6}{x - 3} + \frac{8}{-(x - 3)}$$
$$= \frac{6}{x - 3} + \frac{-8}{x - 3}$$
$$= \frac{6 + (-8)}{x - 3}$$
$$= -\frac{2}{x - 3}$$

17. $\dfrac{5}{x} + 2 = \dfrac{5}{x} + \dfrac{2}{1} = \dfrac{5}{x} + \dfrac{2(x)}{1(x)} = \dfrac{5 + 2x}{x}$

21.
$$\frac{y + 2}{y + 3} - 2 = \frac{y + 2}{y + 3} - \frac{2}{1}$$
$$= \frac{y + 2}{y + 3} - \frac{2(y + 3)}{y + 3}$$
$$= \frac{y + 2}{y + 3} - \frac{2y + 6}{y + 3}$$
$$= \frac{y + 2 - (2y + 6)}{y + 3}$$
$$= \frac{y + 2 - 2y - 6}{y + 3}$$
$$= \frac{-y - 4}{y + 3}$$
$$= \frac{-(y + 4)}{y + 3}$$
$$= -\frac{y + 4}{y + 3}$$

25.
$$\frac{5x}{x + 2} - \frac{3x - 4}{x + 2} = \frac{5x - (3x - 4)}{x + 2}$$
$$= \frac{5x - 3x + 4}{x + 2}$$
$$= \frac{2x + 4}{x + 2}$$
$$= \frac{2(x + 2)}{x + 2}$$
$$= 2$$

29.
$$\frac{1}{x + 3} - \frac{1}{(x + 3)^2} = \frac{1(x + 3)}{(x + 3)(x + 3)} - \frac{1}{(x + 3)^2}$$
$$= \frac{x + 3}{(x + 3)^2} - \frac{1}{(x + 3)^2}$$
$$= \frac{x + 3 - 1}{(x + 3)^2}$$
$$= \frac{x + 2}{(x + 3)^2}$$

33. $\dfrac{2}{m} + 1 = \dfrac{2}{m} + \dfrac{1}{1} = \dfrac{2}{m} + \dfrac{1(m)}{1(m)} = \dfrac{2 + m}{m}$

37.
$$\frac{7}{(x + 1)(x - 1)} + \frac{8}{(x + 1)^2}$$
$$= \frac{7(x + 1)}{(x + 1)(x - 1)(x + 1)} + \frac{8(x - 1)}{(x + 1)^2(x - 1)}$$
$$= \frac{7x + 7}{(x + 1)^2(x - 1)} + \frac{8x - 8}{(x + 1)^2(x - 1)}$$
$$= \frac{7x + 7 + 8x - 8}{(x + 1)^2(x - 1)}$$
$$= \frac{15x - 1}{(x + 1)^2(x - 1)}$$

41.
$$\frac{3a}{2a + 6} - \frac{a - 1}{a + 3} = \frac{3a}{2(a + 3)} - \frac{a - 1}{a + 3}$$
$$= \frac{3a}{2(a + 3)} - \frac{(a - 1)(2)}{(a + 3)(2)}$$
$$= \frac{3a}{2(a + 3)} - \frac{2a - 2}{2(a + 3)}$$
$$= \frac{3a - (2a - 2)}{2(a + 3)}$$
$$= \frac{3a - 2a + 2}{2(a + 3)}$$
$$= \frac{a + 2}{2(a + 3)}$$

45. $\dfrac{-7}{y^2 - 3y + 2} - \dfrac{2}{y - 1} = \dfrac{-7}{(y-1)(y-2)} - \dfrac{2}{y-1}$

$$= \dfrac{-7}{(y-1)(y-2)} - \dfrac{2(y-2)}{(y-1)(y-2)}$$

$$= \dfrac{-7 - (2y - 4)}{(y-1)(y-2)}$$

$$= \dfrac{-7 - 2y + 4}{(y-1)(y-2)}$$

$$= \dfrac{-3 - 2y}{(y-2)(y-1)}$$

49. $\dfrac{x + 8}{x^2 - 5x - 6} + \dfrac{x + 1}{x^2 - 4x - 5} = \dfrac{x + 8}{(x-6)(x+1)} + \dfrac{x + 1}{(x-5)(x+1)}$

$$= \dfrac{(x+8)(x-5)}{(x-6)(x+1)(x-5)} + \dfrac{(x+1)(x-6)}{(x-5)(x+1)(x-6)}$$

$$= \dfrac{x^2 + 3x - 40 + x^2 - 5x - 6}{(x-6)(x+1)(x-5)}$$

$$= \dfrac{2x^2 - 2x - 46}{(x-6)(x+1)(x-5)}$$

53. $2x^2 - x - 1 = 0$

$(2x + 1)(x - 1) = 0$

$2x + 1 = 0 \qquad \text{or} \qquad x - 1 = 0$

$2x = -1 \qquad\qquad\qquad x = 1$

$x = -\dfrac{1}{2}$

The solutions are $x = -\dfrac{1}{2}$ and $x = 1$.

57. $\dfrac{3}{x + 4} - \dfrac{1}{x - 4} = \dfrac{3(x - 4)}{(x + 4)(x - 4)} - \dfrac{1(x + 4)}{(x - 4)(x + 4)}$

$$= \dfrac{3x - 12}{(x + 4)(x - 4)} - \dfrac{x + 4}{(x + 4)(x - 4)}$$

$$= \dfrac{3x - 12 - (x + 4)}{(x + 4)(x - 4)}$$

$$= \dfrac{3x - 12 - x - 4}{(x + 4)(x - 4)}$$

$$= \dfrac{2x - 16}{(x + 4)(x - 4)}$$

The length of the other board is $\dfrac{2x - 16}{(x + 4)(x - 4)}$ inches.

61. answers may vary

65. $\dfrac{10}{x^2 - 3x - 4} - \dfrac{8}{x^2 + 6x + 5} - \dfrac{9}{x^2 + x - 20} = \dfrac{10}{(x - 4)(x + 1)} - \dfrac{8}{(x + 5)(x + 1)} - \dfrac{9}{(x + 5)(x - 4)}$

$$= \dfrac{10(x + 5)}{(x - 4)(x + 1)(x + 5)} - \dfrac{8(x - 4)}{(x + 5)(x + 1)(x - 4)} - \dfrac{9(x + 1)}{(x + 5)(x - 4)(x + 1)}$$

$$= \dfrac{10(x + 5) - 8(x - 4) - 9(x + 1)}{(x - 4)(x + 1)(x + 5)}$$

$$= \dfrac{10x + 50 - 8x + 32 - 9x - 9}{(x - 4)(x + 1)(x + 5)}$$

$$= \dfrac{73 - 7x}{(x - 4)(x + 1)(x + 5)}$$

1.
$$\frac{x}{5} + 3 = 9$$

$$5\left(\frac{x}{5} + 3\right) = 5(9)$$

$$5\left(\frac{x}{5}\right) + 5(3) = 5(9)$$

$$x + 15 = 45$$

$$x = 30$$

Check:

$$\frac{x}{5} + 3 = 9$$

$$\frac{30}{5} + 3 \stackrel{?}{=} 9$$

$$6 + 3 \stackrel{?}{=} 9$$

$$9 = 9 \quad \text{True}$$

The solution is 30.

5.
$$2 - \frac{8}{x} = 6$$

$$x\left(2 - \frac{8}{x}\right) = x(6)$$

$$x(2) - x\left(\frac{8}{x}\right) = x(6)$$

$$2x - 8 = 6x$$

$$-8 = 4x$$

$$-2 = x$$

Check:

$$2 - \frac{8}{x} = 6$$

$$2 - \frac{8}{-2} \stackrel{?}{=} 6$$

$$2 - (-4) \stackrel{?}{=} 6$$

$$2 + 4 \stackrel{?}{=} 6$$

$$6 = 6 \quad \text{True}$$

The solution is −2.

9.
$$\frac{a}{5} = \frac{a - 3}{2}$$

$$10\left(\frac{a}{5}\right) = 10\left(\frac{a - 3}{2}\right)$$

$$2a = 5(a - 3)$$

$$2a = 5a - 15$$

$$-3a = -15$$

$$a = 5$$

Check:

$$\frac{a}{5} = \frac{a - 3}{2}$$

$$\frac{5}{5} \stackrel{?}{=} \frac{5 - 3}{2}$$

$$\frac{5}{5} \stackrel{?}{=} \frac{2}{2}$$

$$1 = 1 \quad \text{True}$$

The solution is 5.

13.
$$\frac{2}{y} + \frac{1}{2} = \frac{5}{2y}$$

$$2y\left(\frac{2}{y} + \frac{1}{2}\right) = 2y\left(\frac{5}{2y}\right)$$

$$2y\left(\frac{2}{y}\right) + 2y\left(\frac{1}{2}\right) = 2y\left(\frac{5}{2y}\right)$$

$$4 + y = 5$$

$$y = 1$$

Check:

$$\frac{2}{y} + \frac{1}{2} = \frac{5}{2y}$$

$$\frac{2}{1} + \frac{1}{2} \stackrel{?}{=} \frac{5}{2(1)}$$

$$\frac{4}{2} + \frac{1}{2} \stackrel{?}{=} \frac{5}{2}$$

$$\frac{5}{2} = \frac{5}{2} \quad \text{True}$$

The solution is 1.

17.
$$2 + \frac{3}{a - 3} = \frac{a}{a - 3}$$

$$(a - 3)\left(2 + \frac{3}{a - 3}\right) = (a - 3)\left(\frac{a}{a - 3}\right)$$

$$(a - 3)(2) + (a - 3)\left(\frac{3}{a - 3}\right) = (a - 3)\left(\frac{a}{a - 3}\right)$$

$$2a - 6 + 3 = a$$

$$2a - 3 = a$$

$$-3 = -a$$

$$3 = a$$

In the original equation, 3 makes a denominator 0. This equation has no solution.

21.
$$\frac{y}{y + 4} + \frac{4}{y + 4} = 3$$

$$(y + 4)\left(\frac{y}{y + 4} + \frac{4}{y + 4}\right) = (y + 4)(3)$$

$$(y + 4)\left(\frac{y}{y + 4}\right) + (y + 4)\left(\frac{4}{y + 4}\right) = (y + 4)(3)$$

$$y + 4 = 3y + 12$$

$$4 = 2y + 12$$

$$-8 = 2y$$

$$-\frac{8}{2} = y$$

$$-4 = y$$

In the original equation, −4 makes a denominator zero. This equation has no solution.

25.

$$\frac{2x}{x+2} - 2 = \frac{x-8}{x-2}$$

$$(x+2)(x-2)\left(\frac{2x}{x+2} - 2\right) = (x+2)(x-2)\left(\frac{x-8}{x-2}\right)$$

$$(x+2)(x-2)\left(\frac{2x}{x+2}\right) - (x+2)(x-2)(2) = (x+2)(x-2)\left(\frac{x-8}{x-2}\right)$$

$$2x(x-2) - 2(x^2 - 4) = (x+2)(x-8)$$
$$2x^2 - 4x - 2x^2 + 8 = x^2 - 6x - 16$$
$$-4x + 8 = x^2 - 6x - 16$$
$$0 = x^2 - 2x - 24$$
$$0 = (x-6)(x+4)$$

$$x - 6 = 0 \qquad \text{or} \qquad x + 4 = 0$$
$$x = 6 \qquad\qquad\qquad x = -4$$

Check:

$x = 6:$

$$\frac{2x}{x+2} - 2 = \frac{x-8}{x-2}$$

$$\frac{2(6)}{6+2} - 2 \overset{?}{=} \frac{6-8}{6-2}$$

$$\frac{12}{8} - 2 \overset{?}{=} -\frac{2}{4}$$

$$\frac{3}{2} - \frac{4}{2} \overset{?}{=} -\frac{1}{2}$$

$$\frac{3-4}{2} \overset{?}{=} -\frac{1}{2}$$

$$-\frac{1}{2} = -\frac{1}{2} \quad \text{True}$$

$x = -4:$

$$\frac{2x}{x+2} - 2 = \frac{x-8}{x-2}$$

$$\frac{2(-4)}{-4+2} - 2 \overset{?}{=} \frac{-4-8}{-4-2}$$

$$\frac{-8}{-2} - 2 \overset{?}{=} \frac{-12}{-6}$$

$$4 - 2 \overset{?}{=} 2$$

$$2 = 2 \qquad \text{True}$$

The solutions are 6 and -4.

29.

$$\frac{2}{x-2} + 1 = \frac{x}{x+2}$$

$$(x-2)(x+2)\left(\frac{2}{x-2} + 1\right) = (x-2)(x+2)\left(\frac{x}{x+2}\right)$$

$$(x-2)(x+2)\left(\frac{2}{x-2}\right) + (x-2)(x+2) = (x-2)(x+2)\left(\frac{x}{x+2}\right)$$

$$2(x+2) + (x-2)(x+2) = x(x-2)$$
$$2x + 4 + x^2 - 4 = x^2 - 2x$$
$$2x + x^2 = x^2 - 2x$$
$$2x = -2x$$
$$4x = 0$$
$$x = 0$$

Check:

$$\frac{2}{x-2} + 1 = \frac{x}{x+2}$$

$$\frac{2}{0-2} + 1 \overset{?}{=} \frac{0}{0+2}$$

$$\frac{2}{-2} + 1 \overset{?}{=} 0$$

$$-1 + 1 \overset{?}{=} 0$$

$$0 = 0 \qquad \text{True}$$

The solution is 0.

33.

$$\frac{x+1}{3} - \frac{x-1}{6} = \frac{1}{6}$$

$$6\left(\frac{x+1}{3} - \frac{x-1}{6}\right) = 6\left(\frac{1}{6}\right)$$

$$6\left(\frac{x+1}{3}\right) - 6\left(\frac{x-1}{6}\right) = 6\left(\frac{1}{6}\right)$$

$$2(x+1) - (x-1) = 1$$

$$2x + 2 - x + 1 = 1$$

$$x + 3 = 1$$

$$x = -2$$

Check:

$$\frac{x+1}{3} - \frac{x-1}{6} = \frac{1}{6}$$

$$\frac{-2+1}{3} - \frac{-2-1}{6} \stackrel{?}{=} \frac{1}{6}$$

$$-\frac{1}{3} - \frac{-3}{6} \stackrel{?}{=} \frac{1}{6}$$

$$-\frac{2}{6} - \frac{-3}{6} \stackrel{?}{=} \frac{1}{6}$$

$$\frac{-2 - (-3)}{6} \stackrel{?}{=} \frac{1}{6}$$

$$\frac{-2 + 3}{6} \stackrel{?}{=} \frac{1}{6}$$

$$\frac{1}{6} = \frac{1}{6} \quad \text{True}$$

The solution is -2.

37.

$$\frac{4r-4}{r^2 + 5r - 14} + \frac{2}{r+7} = \frac{1}{r-2}$$

$$\frac{4r-4}{(r+7)(r-2)} + \frac{2}{r+7} = \frac{1}{r-2}$$

$$(r+7)(r-2)\left(\frac{4r-4}{(r+7)(r-2)} + \frac{2}{r+7}\right) = (r+7)(r-2)\left(\frac{1}{r-2}\right)$$

$$(r+7)(r-2)\left(\frac{4r-4}{(r+7)(r-2)}\right) + (r+7)(r-2)\left(\frac{2}{r+7}\right) = (r+7)(r-2)\left(\frac{1}{r-2}\right)$$

$$4r - 4 + 2(r-2) = (r+7)(1)$$

$$4r - 4 + 2r - 4 = r + 7$$

$$6r - 8 = r + 7$$

$$5r = 15$$

$$r = 3$$

Check:

$$\frac{4r-4}{r^2 + 5r - 14} + \frac{2}{r+7} = \frac{1}{r-2}$$

$$\frac{4(3)-4}{3^2 + 5(3) - 14} + \frac{2}{3+7} \stackrel{?}{=} \frac{1}{3-2}$$

$$\frac{12-4}{9 + 15 - 14} + \frac{2}{10} \stackrel{?}{=} \frac{1}{1}$$

$$\frac{8}{10} + \frac{2}{10} \stackrel{?}{=} 1$$

$$\frac{8+2}{10} \stackrel{?}{=} 1$$

$$\frac{10}{10} \stackrel{?}{=} 1$$

$$1 = 1 \qquad \text{True}$$

The solution is 3.

41.

$$\frac{d}{r} = t$$

$$r\left(\frac{d}{r}\right) = r(t)$$

$$d = rt$$

$$\frac{d}{t} = \frac{rt}{t}$$

$$\frac{d}{t} = r$$

45.

$$i = \frac{A}{t + B}$$

$$(t + B)(i) = (t + B)\left(\frac{A}{t + B}\right)$$

$$ti + Bi = A$$

$$ti = A - Bi$$

$$\frac{ti}{i} = \frac{A - Bi}{i}$$

$$t = \frac{A - Bi}{i}$$

49.

$$\frac{C}{\pi r} = 2$$

$$\pi r\left(\frac{C}{\pi r}\right) = \pi r(2)$$

$$C = 2\pi r$$

$$\frac{C}{2\pi} = \frac{2\pi r}{2\pi}$$

$$\frac{C}{2\pi} = r$$

53. The reciprocal of x added to the reciprocal of 2 is $\dfrac{1}{x} + \dfrac{1}{2}$.

57.

$$\frac{20x}{3} + \frac{32x}{6} = 180$$

$$6\left(\frac{20x}{3} + \frac{32x}{6}\right) = 6(180)$$

$$6\left(\frac{20x}{3}\right) + 6\left(\frac{32x}{6}\right) = 6(180)$$

$$40x + 32x = 1080$$

$$72x = 1080$$

$$\frac{72x}{72} = \frac{1080}{72}$$

$$x = 15$$

$$\frac{20x}{3} = \frac{20(15)}{3} = 100$$

$$\frac{32x}{6} = \frac{32(15)}{6} = 80$$

The supplementary angles are 100° and 80°.

61.

$$\frac{4}{a^2 + 4a + 3} + \frac{2}{a^2 + a - 6} - \frac{3}{a^2 - a - 2} = 0$$

$$\frac{4}{(a + 3)(a + 1)} + \frac{2}{(a + 3)(a - 2)} - \frac{3}{(a - 2)(a + 1)} = 0$$

$$(a + 3)(a + 1)(a - 2)\left(\frac{4}{(a + 3)(a + 1)} + \frac{2}{(a + 3)(a - 2)} - \frac{3}{(a - 2)(a + 1)}\right) = (a + 3)(a + 1)(a - 2)(0)$$

$$(a + 3)(a + 1)(a - 2)\left(\frac{4}{(a + 3)(a + 1)}\right) + (a + 3)(a + 1)(a - 2)\left(\frac{2}{(a + 3)(a - 2)}\right)$$

$$- (a + 3)(a + 1)(a - 2)\left(\frac{3}{(a - 2)(a + 1)}\right) = 0$$

$$4(a - 2) + 2(a + 1) - 3(a + 3) = 0$$

$$4a - 8 + 2a + 2 - 3a - 9 = 0$$

$$3a - 15 = 0$$

$$3a = 15$$

$$a = 5$$

1.
$$3 \cdot \frac{1}{x} = 9 \cdot \frac{1}{6}$$
$$\frac{3}{x} = \frac{9}{6}$$
$$6x\left(\frac{3}{x}\right) = 6x\left(\frac{9}{6}\right)$$
$$18 = 9x$$
$$x = 2$$

The unknown number is 2.

5.
$$\frac{2}{x-3} - \frac{4}{x+3} = 8 \cdot \frac{1}{x^2-9}$$
$$(x-3)(x+3)\left(\frac{2}{x-3} - \frac{4}{x+3}\right) = (x-3)(x+3)\left(\frac{8}{x^2-9}\right)$$
$$(x-3)(x+3)\left(\frac{2}{x-3}\right) - (x-3)(x+3)\left(\frac{4}{x+3}\right) = 8$$
$$2(x+3) - 4(x-3) = 8$$
$$2x + 6 - 4x + 12 = 8$$
$$-2x + 18 = 8$$
$$-2x = -10$$
$$x = 5$$

The unknown number is 5.

9.

	Hours to Complete Total Job	Part of Job Completed in 1 Hour
Experienced	4	$\frac{1}{4}$
Apprentice	5	$\frac{1}{5}$
Together	x	$\frac{1}{x}$

$$\frac{1}{4} + \frac{1}{5} = \frac{1}{x}$$
$$20x\left(\frac{1}{4}\right) + 20x\left(\frac{1}{5}\right) = 20x\left(\frac{1}{x}\right)$$
$$5x + 4x = 20$$
$$9x = 20$$
$$x = \frac{20}{9} \qquad \text{or } 2\frac{2}{9}$$

The experienced surveyor and apprentice surveyor, working together, can survey the road bed in $2\frac{2}{9}$ hours.

13.

	Hours to Complete Total Job	Part of Job Completed in 1 Hour
Marcus	6	$\frac{1}{6}$
Tony	4	$\frac{1}{4}$
Together	x	$\frac{1}{x}$

$$\frac{1}{6} + \frac{1}{4} = \frac{1}{x}$$
$$12x\left(\frac{1}{6}\right) + 12x\left(\frac{1}{4}\right) = 12x\left(\frac{1}{x}\right)$$
$$2x + 3x = 12$$
$$5x = 12$$
$$x = \frac{12}{5} = 2\frac{2}{5}$$
$$45\left(\frac{12}{5}\right) = 108$$

Together, Marcus and Tony work for $2\frac{2}{5}$ hours at \$45 per hour. The labor estimate should be \$108.

17.

	Hours to Complete Total Job	Part of Job Completed in 1 Hour
First Pipe	20	$\frac{1}{20}$
Second Pipe	15	$\frac{1}{15}$
Third Pipe	x	$\frac{1}{x}$
3 Pipes Together	6	$\frac{1}{6}$

$$\frac{1}{20} + \frac{1}{15} + \frac{1}{x} = \frac{1}{6}$$

$$60x\left(\frac{1}{20}\right) + 60x\left(\frac{1}{15}\right) + 60x\left(\frac{1}{x}\right) = 60x\left(\frac{1}{6}\right)$$

$$3x + 4x + 60 = 10x$$
$$7x + 60 = 10x$$
$$60 = 3x$$
$$20 = x$$

It takes the third pipe 20 hours to fill the pond.

21.

distance	=	rate · time

First Portion	20	r	$\frac{20}{r}$
Cooldown Portion	16	$r - 2$	$\frac{16}{r-2}$

$$\frac{20}{r} = \frac{16}{r-2}$$
$$20(r - 2) = 16r$$
$$20r - 40 = 16r$$
$$-40 = -4r$$
$$r = 10 \text{ and } 4 - 2 = 10 - 2 = 8$$

His speed was 10 miles per hour during the first portion and 8 miles per hour during the cooldown portion.

25. Let w = the rate of the wind.

distance	=	rate · time

With the wind	48	$16 + w$	$\frac{48}{16+w}$
Into the wind	16	$16 - w$	$\frac{16}{16-w}$

$$\frac{48}{16 + w} = \frac{16}{16 - w}$$
$$48(16 - w) = 16(16 + w)$$
$$768 - 48w = 256 + 16w$$
$$512 = 64w$$
$$w = 8$$

The rate of the wind is 8 miles per hour.

29. $\dfrac{12}{4} = \dfrac{18}{x}$

$$12x = 72$$
$$x = 6$$

33. $\dfrac{16}{10} = \dfrac{34}{y}$

$$16y = 340$$
$$y = 21.25$$

37. $\dfrac{\frac{3}{4} + \frac{1}{4}}{\frac{3}{8} + \frac{13}{8}} = \dfrac{\frac{3+1}{4}}{\frac{3+13}{8}} = \dfrac{\frac{4}{4}}{\frac{16}{8}} = \dfrac{1}{2}$

41. Let r = Zanardi's speed.

distance	=	rate · time

Zanardi	1.952	r	$\frac{1.952}{r}$
Gugelmin	2	$r + 5.6$	$\frac{2}{r+5.6}$

$$\frac{1.952}{r} = \frac{2}{r + 5.6}$$
$$1.952(r + 5.6) = 2r$$
$$1.952r + 10.9312 = 2r$$
$$10.9312 = 0.048r$$
$$r \approx 227.7$$
$$r + 5.6 \approx 233.3$$

Zanardi's fastest lap speed was approximately 227.7 miles per hour, and Gugelmin's fastest lap speed was approximately 233.3 miles per hour.

Section 5.7

1. $\dfrac{\frac{1}{2}}{\frac{3}{4}} = \dfrac{1}{2} \cdot \dfrac{4}{3} = \dfrac{1 \cdot 2 \cdot 2}{2 \cdot 3} = \dfrac{2}{3}$

5. $\dfrac{\frac{-5}{12x^2}}{\frac{25}{16x^3}} = -\dfrac{5}{12x^2} \cdot \dfrac{16x^3}{25} = -\dfrac{5 \cdot 4 \cdot 4 \cdot x^2 \cdot x}{4 \cdot 3 \cdot x^2 \cdot 5 \cdot 5} = -\dfrac{4x}{15}$

9. $\dfrac{2 + \frac{7}{10}}{1 + \frac{3}{5}} = \dfrac{10\left(2 + \frac{7}{10}\right)}{10\left(1 + \frac{3}{5}\right)}$

$$= \dfrac{10(2) + 10\left(\frac{7}{10}\right)}{10(1) + 10\left(\frac{3}{5}\right)}$$

$$= \dfrac{20 + 7}{10 + 6}$$

$$= \dfrac{27}{16}$$

13. $\dfrac{\frac{1}{5} - \frac{1}{x}}{\frac{7}{10} + \frac{1}{x^2}} = \dfrac{10x^2\left(\frac{1}{5} - \frac{1}{x}\right)}{10x^2\left(\frac{7}{10} + \frac{1}{x^2}\right)}$

$$= \dfrac{10x^2\left(\frac{1}{5}\right) - 10x^2\left(\frac{1}{x}\right)}{10x^2\left(\frac{7}{10}\right) + 10x^2\left(\frac{1}{x^2}\right)}$$

$$= \dfrac{2x^2 - 10x}{7x^2 + 10}$$

$$= \dfrac{2x(x - 5)}{7x^2 + 10}$$

17. $\dfrac{\frac{4y - 8}{16}}{\frac{6y - 12}{4}} = \dfrac{4y - 8}{16} \cdot \dfrac{4}{6y - 12} = \dfrac{4(y - 2) \cdot 4}{4 \cdot 4 \cdot 6(y - 2)} = \dfrac{1}{6}$

21. $\dfrac{1}{2 + \frac{1}{3}} = \dfrac{3(1)}{3\left(2 + \frac{1}{3}\right)} = \dfrac{3(1)}{3(2) + 3\left(\frac{1}{3}\right)} = \dfrac{3}{6 + 1} = \dfrac{3}{7}$

25.

$$\dfrac{\dfrac{8}{x+4}+2}{\dfrac{12}{x+4}-2}=\dfrac{(x+4)\left(\dfrac{8}{x+4}+2\right)}{(x+4)\left(\dfrac{12}{x+4}-2\right)}$$

$$=\dfrac{(x+4)\left(\dfrac{8}{x+4}\right)+(x+4)(2)}{(x+4)\left(\dfrac{12}{x+4}\right)-(x+4)(2)}$$

$$=\dfrac{8+2x+8}{12-2x-8}$$

$$=\dfrac{16+2x}{4-2x}$$

$$=\dfrac{2(8+x)}{2(2-x)}$$

$$=\dfrac{8+x}{2-x}$$

29. answers may vary

33. Tie between *Eagles Greatest Hits* and *Born in the U.S.A.* for the fewest copies sold.

37.

$$\dfrac{1}{\dfrac{1}{R_1}+\dfrac{1}{R_2}}=\dfrac{R_1R_2(1)}{R_1R_2\left(\dfrac{1}{R_1}+\dfrac{1}{R_2}\right)}$$

$$=\dfrac{R_1R_2}{R_1R_2\left(\dfrac{1}{R_1}\right)+R_1R_2\left(\dfrac{1}{R_2}\right)}$$

$$=\dfrac{R_1R_2}{R_2+R_1}$$

41.

$$\dfrac{y^{-2}}{1-y^{-2}}=\dfrac{\dfrac{1}{y^2}}{1-\dfrac{1}{y^2}}$$

$$=\dfrac{y^2\left(\dfrac{1}{y^2}\right)}{y^2\left(1-\dfrac{1}{y^2}\right)}$$

$$=\dfrac{y^2\left(\dfrac{1}{y^2}\right)}{y^2(1)-y^2\left(\dfrac{1}{y^2}\right)}$$

$$=\dfrac{1}{y^2-1}$$

Chapter 5 Test

1. The rational expression is undefined when

$$x^2+4x+3=0$$
$$(x+3)(x+1)=0$$
$$x+3=0\quad\text{or}\quad x+1=0$$
$$x=-3\quad\text{or}\qquad x=-1$$

5. $\dfrac{x+6}{x^2+12x+36}=\dfrac{x+6}{(x+6)^2}=\dfrac{1}{x+6}$

9. $\dfrac{x^2-13x+42}{x^2+10x+21}\div\dfrac{x^2-4}{x^2+x-6}$

$$=\dfrac{x^2-13x+42}{x^2+10x+21}\cdot\dfrac{x^2+x-6}{x^2-4}$$

$$=\dfrac{(x-6)(x-7)\cdot(x+3)(x-2)}{(x+7)(x+3)\cdot(x+2)(x-2)}$$

$$\dfrac{(x-6)(x-7)}{(x+7)(x+2)}$$

13. $\dfrac{5a}{a^2-a-6}-\dfrac{2}{a-3}$

$$=\dfrac{5a}{(a-3)(a+2)}-\dfrac{2(a+2)}{(a-3)(a+2)}$$

$$=\dfrac{5a-2(a+2)}{(a-3)(a+2)}$$

$$=\dfrac{5a-2a-4}{(a-3)(a+2)}$$

$$=\dfrac{3a-4}{(a-3)(a+2)}$$

17.
$$\frac{4y}{y^2 + 6y + 5} - \frac{3}{y^2 + 5y + 4} = \frac{4y}{(y+5)(y+1)} - \frac{3}{(y+4)(y+1)}$$
$$= \frac{4y(y+4)}{(y+5)(y+1)(y+4)} - \frac{3(y+5)}{(y+4)(y+1)(y+5)}$$
$$= \frac{4y(y+4) - 3(y+5)}{(y+5)(y+1)(y+4)}$$
$$= \frac{4y^2 + 16y - 3y - 15}{(y+5)(y+1)(y+4)}$$
$$= \frac{4y^2 + 13y - 15}{(y+5)(y+1)(y+4)}$$

21.
$$\frac{10}{x^2 - 25} = \frac{3}{x+5} + \frac{1}{x-5}$$
$$(x+5)(x-5)\left(\frac{10}{(x+5)(x-5)}\right) = (x+5)(x-5)\left(\frac{3}{x+5} + \frac{1}{x-5}\right)$$
$$10 = (x+5)(x-5)\left(\frac{3}{x+5}\right) + (x+5)(x-5)\left(\frac{1}{x-5}\right)$$
$$10 = 3(x-5) + x + 5$$
$$10 = 3x - 15 + x + 5$$
$$10 = 4x - 10$$
$$20 = 4x$$
$$x = 5$$

In the original equation, 5 makes the denominator 0. This equation has no solution.

25.
$$\frac{8}{x} = \frac{10}{15}$$
$$8(15) = 10x$$
$$120 = 10x$$
$$12 = x$$

Chapter 6

SECTION 6.1

1.

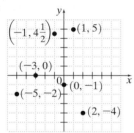

$(1, 5)$ is in quadrant I

$\left(-1, 4\frac{1}{2}\right)$ is in quadrant II

$(-5, -2)$ is in quadrant III

$(2, -4)$ is in quadrant IV

$(-3, 0)$ lies on the x-axis

$(0, -1)$ lies on the y-axis

5. $A: (0, 0)$ **9.** $E: (-2, -2)$

13. $B: (0, -3)$ **17.** $F: (-3, -1)$

21. a. $(1994, 578)$, $(1995, 613)$, $(1996, 654)$, $(1997, 675)$, $(1998, 717)$

b.

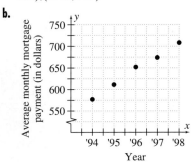

c. The scatter diagram shows a trend of increasing mortgage payments.

25. $x - 4y = 4$

Complete $(\ \ , -2)$:
$$y = -2$$
$$x - 4(-2) = 4$$
$$x + 8 = 4$$
$$x = -4$$
$(-4, -2)$

Complete $(4, \ \ )$:
$$x = 4$$
$$4 - 4y = 4$$
$$-4y = 0$$
$$y = 0$$
$(4, 0)$

29. $y = -7$

Complete $(11, \ \ )$:
$x = 11$, $y = -7$; $(11, -7)$
Complete $(\ \ , -7)$:
$y = -7$, $x = $ any x

33. $2x - y = 12$

Complete $(0, \ \ )$:
$$x = 0$$
$$2(0) - y = 12$$
$$-y = 12$$
$$y = -12$$
$(0, -12)$

Complete $(\ \ , -2)$:
$$y = -2$$
$$2x - (-2) = 12$$
$$2x + 2 = 12$$
$$2x = 10$$
$$x = 5$$
$(5, -2)$

Complete $(3, \ \ )$:
$$x = 3$$
$$2(3) - y = 12$$
$$6 - y = 12$$
$$-y = 6$$
$$y = -6$$
$(3, -6)$

x	y
0	-12
5	-2
3	-6

37. $x = 3$

All x table values are 3.

x	y
3	0
3	-0.5
3	$\frac{1}{4}$

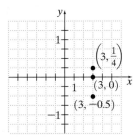

41. $y = 80x + 5000$

a. $x = 100$
$$y = 80(100) + 5000$$
$$= 8000 + 5000$$
$$= 13{,}000$$
$x = 200$
$$y = 80(200) + 5000$$
$$= 16{,}000 + 5000$$
$$= 21{,}000$$
$x = 300$
$$y = 80(300) + 5000$$
$$= 24{,}000 + 5000$$
$$= 29{,}000$$

x	100	200	300
y	13,000	21,000	29,000

b. $y = 8600$
$$8600 = 80x + 5000$$
$$3600 = 80x$$
$$x = 45 \text{ desks}$$

45. $2x + 4y = 5$
$$4y = 5 - 2x$$
$$y = \frac{5 - 2x}{4}$$

49. Plot the points:

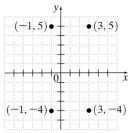

Rectangle is 9 units by 4 units, perimeter is $9 + 9 + 4 + 4 = 26$ units

53. answers may vary

SECTION 6.2

1. $x - y = 6$
$y = 0, x = 6; (6, 0)$
$x = 4$
$4 - y = 6$
$y = -2, (4, -2)$
$y = -1$
$x - (-1) = 6$
$x = 5; (5, -1)$

x	y
6	0
4	-2
5	-1

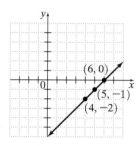

5. $y = \dfrac{1}{3}x$

$x = 0, y = 0; (0, 0)$
$x = 6, y = 2; (6, 2)$
$x = -3, y = -1; (-3, -1)$

x	y
0	0
6	2
-3	-1

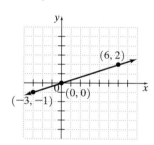

9. $x + y = 1$
Find 3 points:

x	y
0	1
1	0
-1	2

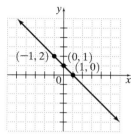

13. $x - 2y = 6$
Find 3 points:

x	y
0	-3
6	0
4	-1

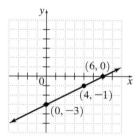

17. $x = -4$
$y = $ any value

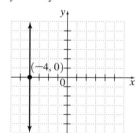

21. $y = x$
Find 3 points:

x	y
2	2
0	0
−2	−2

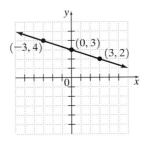

25. $x + 3y = 9$
Find 3 points:

x	y
0	3
3	2
−3	4

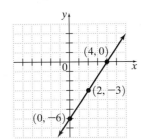

29. $3x - 2y = 12$
Find 3 points:

x	y
0	−6
4	0
2	−3

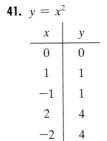

33. Find 3 points for each line:
$y = -2x$

x	y
2	−4
0	0
−2	4

$y = -2x - 3$

x	y
−2	1
−1	−1
0	−3

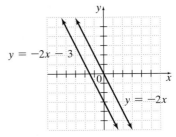

answers may vary

37. $x - y = -3$
$x = 0$
$0 - y = -3$
$\quad y = 3; (0, 3)$
$y = 0$
$x - 0 = -3$
$\quad x = -3; (-3, 0)$

x	y
0	3
−3	0

41. $y = x^2$

x	y
0	0
1	1
−1	1
2	4
−2	4

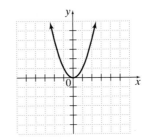

45. $y = 43x + 2035$

a. Find 3 points:

x	y
0	2035
4	2207
8	2379

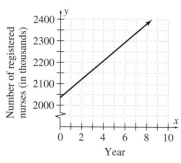

b. Yes; 8 years after 1997 (or in 2005) there will be 2379 thousand registered nurses.

SECTION 6.3

1. $x = -1$; $y = 1$;
$(-1, 0)$; $(0, 1)$

5. infinite

9. $x - y = 3$
If $x = 0$, then $y = -3$
If $y = 0$, then $x = 3$
Plot using $(0, -3)$ and $(3, 0)$:

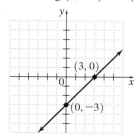

13. $-x + 2y = 6$
If $x = 0$, then $y = 3$
If $y = 0$, then $x = -6$
Plot using $(0, 3)$ and $(-6, 0)$:

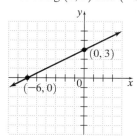

17. $x = 2y$
If $x = 0$, $y = 0$
Need another point:
If $y = 1$, $x = 2$
Plot using $(0, 0)$ and $(2, 1)$:

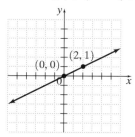

21. $x = y$
If $x = 0$, $y = 0$
Need another point:
If $x = 3$, $y = 3$
Plot using $(0, 0)$ and $(3, 3)$:

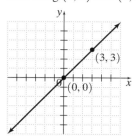

25. $5 = 6x - y$
If $x = 0$, $y = -5$
If $y = 0$, $x = \dfrac{5}{6}$

Plot using $(0, -5)$ and $\left(\dfrac{5}{6}, 0\right)$:

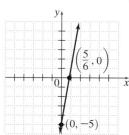

29. $x = -1$
For any y-value, x is -1.

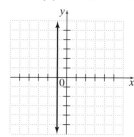

33. $y + 7 = 0$
$y = -7$
For any x-value, y is –7.

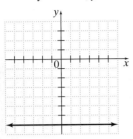

37. $\dfrac{-6 - 3}{2 - 8} = \dfrac{-9}{-6} = \dfrac{3}{2}$

41. $\dfrac{0 - 6}{5 - 0} = \dfrac{-6}{5} = -\dfrac{6}{5}$

45. $x = 3$
For any y-value, $x = 3$.
A

49. $3x + 6y = 1200$

　a. If $x = 0, 6y = 1200$
　　　　　　　$y = 200$
　(0, 200) corresponds to no chairs and 200 desks are
　manufactured.

　b. If $y = 0, 3x = 1200$
　　　　　　　$x = 400$
　(400, 0) corresponds to 400 chairs and no desks are
　manufactured.

　c. If $y = 50$,
　　$3x + 6(50) = 1200$
　　$3x + 300 = 1200$
　　　　　$3x = 900$
　　　　　$x = 300$
　300 chairs can be made

53. $y = 1.2x + 23.6$

　a. If $x = 0, y = 23.6$

　b. (0, 23.6)

　c. If 1995, U.S. farm expenses for livestock feed were
　$23.6 billion.

Section 6.4

1. $p_1 = (-1, 2)$; $p_2 = (2, -2)$
$$m = \frac{y_2 - y_1}{x_2 - x_1}$$
$$= \frac{-2 - 2}{2 - (-1)}$$
$$= -\frac{4}{3}$$

5. (0, 0) and (7, 8)
$$m = \frac{y_2 - y_1}{x_2 - x_1}$$
$$= \frac{8 - 0}{7 - 0}$$
$$= \frac{8}{7}$$

9. (1, 4) and (5, 3)
$$m = \frac{y_2 - y_1}{x_2 - x_1}$$
$$= \frac{3 - 4}{5 - 1}$$
$$= -\frac{1}{4}$$

13. (5, 1) and (−2, 1)
$$m = \frac{y_2 - y_1}{x_2 - x_1}$$
$$= \frac{1 - 1}{-2 - 5}$$
$$= \frac{0}{-7}$$
$$= 0$$

17. line 2 increases faster than line 1; line 2

21. $2x + y = 7$
$\quad\quad\ \ y = -2x + 7$
$\quad m = -2$

25. $x = 2y$
$\quad y = \dfrac{1}{2}x$
$\quad m = \dfrac{1}{2}$

29. $x = 1$
This is a vertical line, so it has an undefined slope.

33. $x - 3y = -6$　　　　$3x - y = 0$
$\quad\ 3y = x + 6$　　　　$\quad y = 3x$
$\quad\ y = \dfrac{1}{3}x + 2$　　　　$m = 3$
$\quad\ m = \dfrac{1}{3}$

$\dfrac{1}{3}(3) = 1 \neq -1$

neither

37. $6x = 5y + 1$　　　　$-12x + 10y = 1$
$\quad 5y = 6x - 1$　　　　$\quad\ 10y = 12x + 1$
$\quad\ y = \dfrac{6}{5}x - \dfrac{1}{5}$　　　　$\quad y = \dfrac{6}{5}x + \dfrac{1}{10}$
$\quad m = \dfrac{6}{5}$　　　　　　$m = \dfrac{6}{5}$

parallel

41. (−8, −4) and (3, 5)
$$m = \frac{y_2 - y_1}{x_2 - x_1}$$
$$= \frac{5 - (-4)}{3 - (-8)}$$
$$= \frac{9}{11}$$

　a. $\dfrac{9}{11}$　　　　　**b.** $-\dfrac{11}{9}$

45. $y - 1 = -6(x - (-2))$
$\quad y - 1 = -6(x + 2)$
$\quad y - 1 = -6x - 12$
$\quad\quad\ \ y = -6x - 11$

49. A vertical line has an undefined slope. B

53. $m = \dfrac{\Delta y}{\Delta x} = \dfrac{6}{10} = \dfrac{3}{5}$

57. 20 mpg

61. 86 to 87

65. $y = -1800x + 18{,}000$

   a. If $x = 0$,
$$y = -1800(0) + 18{,}000$$
$$y = 18{,}000$$
If $y = 0$,
$$0 = -1800x + 18{,}000$$
$$1800x = 18{,}000$$
$$x = 10$$
Intercepts: $(0, 18{,}000)$ and $(10, 0)$
In 1991, 18,000 Hepatitis B cases were reported in the United States. In $1991 + 10 = 2001$ the number of Hepatitis B cases reported in the United States will drop to zero.

   b. $m = -1800$

   c. The number of cases is decreasing by an average of 1800 per year.

SECTION 6.5

1. $m = 5, b = 3$
$$y = 5x + 3$$

5. $m = -\dfrac{1}{5}, b = \dfrac{1}{9}$
$$y = -\dfrac{1}{5}x + \dfrac{1}{9}$$

9. $y = \dfrac{2}{3}x + 5$

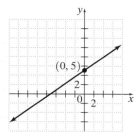

13. $4x + y = 6$
$$y = -4x + 6$$

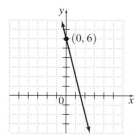

17. $m = 6; (2, 2)$
$$y - y_1 = m(x - x_1)$$
$$y - 2 = 6(x - 2)$$
$$y - 2 = 6x - 12$$
$$-6x + y = -10$$

21. $m = \dfrac{1}{2}; (5, -6)$
$$y - y_1 = m(x - x_1)$$
$$y - (-6) = \dfrac{1}{2}(x - 5)$$
$$y + 6 = \dfrac{1}{2}x - \dfrac{5}{2}$$
$$2y + 12 = x - 5$$
$$-x + 2y = -17$$
$$x - 2y = 17$$

25. $(3, 2)$ and $(5, 6)$
$$m = \dfrac{y_2 - y_1}{x_2 - x_1} = \dfrac{6 - 2}{5 - 3} = \dfrac{4}{2} = 2$$
$$y - y_1 = m(x - x_1)$$
$$y - 2 = 2(x - 3)$$
$$y - 2 = 2x - 6$$
$$2x - y = 4$$

29. $(2, 3)$ and $(-1, -1)$
$$m = \dfrac{y_2 - y_1}{x_2 - x_1} = \dfrac{-1 - 3}{-1 - 2} = \dfrac{-4}{-3} = \dfrac{4}{3}$$
$$y - y_1 = m(x - x_1)$$
$$y - (-1) = \dfrac{4}{3}(x - (-1))$$
$$y + 1 = \dfrac{4}{3}x + \dfrac{4}{3}$$
$$3y + 3 = 4x + 4$$
$$4x - 3y = -1$$

33. $(1, 32)$ and $(3, 96)$

   a. $m = \dfrac{s_2 - s_1}{t_2 - t_1} = \dfrac{96 - 32}{3 - 1} = \dfrac{64}{2} = 32$
$$s - s_1 = m(t - t_1)$$
$$s - 32 = 32(t - 1)$$
$$s - 32 = 32t - 32$$
$$s = 32t$$

   b. If $t = 4$ then $s = 32(4) = 128$ ft/sec

37. 2
$$(2)^2 - 3(2) + 1 = 4 - 6 + 1 = -1$$

41. No

45. $y = 2x + 1$
y-intercept $= (0, 1)$
slope $= 2$
B

49. $m = 3; (-1, 2)$
$$y - y_1 = m(x - x_1)$$
$$y - 2 = 3(x - (-1))$$
$$y - 2 = 3x + 3$$
$$3x - y = -5$$

SECTION 6.6

1. $\{(2, 4), (0, 0), (-7, 10), (10, -7)\}$
Domain: $\{-7, 0, 2, 10\}$
Range: $\{-7, 0, 4, 10\}$

5. Yes; each x-value is assigned to only one y-value.

9. No

13. Yes

17. 5:20 A.M.

21. 9:00 P.M.

25. Yes; it passes the vertical line test.

29.
$f(x) = x^2 + 2$
$f(-2) = (-2)^2 + 2 = 4 + 2 = 6$
$f(0) = 0^2 + 2 = 2$
$f(3) = 3^2 + 2 = 9 + 2 = 11$

33.
$f(x) = |x|$
$f(-2) = |-2| = 2$
$f(0) = |0| = 0$
$f(3) = |3| = 3$

37.
$h(x) = 2x^2 + 3$
$h(-1) = 2(-1)^2 + 3 = 2 + 3 = 5$
$h(0) = 2(0)^2 + 3 = 3$
$h(4) = 2(4)^2 + 3 = 32 + 3 = 35$

41.
$-x + 6 \leq 9$
$-x \leq 3$
$x \geq -3$

45. $f(x) = 2.59x + 47.24$

   a. $x = 46$
      $f(46) = 2.59(6) + 47.24 = 166.38$ cm

   b. $x = 39$
      $f(39) = 2.59(39) + 47.24 = 148.25$ cm

49.
$y = x + 7$
$f(x) = x + 7$

Section 6.7

1. $x - y > 3$
   $(0, 3): 0 - 3 > 3?$
           $-3 > 3?$
   No
   $(2, -1): -2 - (-1) > 3?$
              $-1 > 3?$
   No

5. $x < -y$
   $(0, 2): 0 < -2?$
   No
   $(-5, 1): -5 < -1?$
   Yes

9. $2x - y > -4$
   $2x - y = -4$
       $y = 2x + 4$

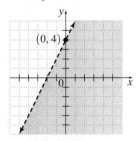

Test: $(0, 0$:
  $2(0) - 0 > -4?$
        $0 > -4?$
Yes; shade below the line.

13. $x \leq -3y$
   $x = -3y$
   $y = -\dfrac{1}{3}x$

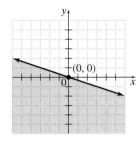

Test $(0, 1)$:
$0 \leq -3(1)?$
$0 \leq -3?$
No; shade below the line.

17. $y < 4$
   $y = 4$

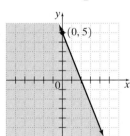

Test $(0, 0)$:
$0 < 4?$
Yes; shade below the line.

21. $5x + 2y \leq 10$
   $5x + 2y = 10$
       $2y = -5x + 10$
          $y = -\dfrac{5}{2}x + 5$

Test $(0, 0)$:
$5(0) + 2(0) \leq 10?$
        $0 \leq 10?$
Yes; shade below the line.

25. $x - y \le 6$
$x - y = 6$
$\quad\quad y = x - 6$

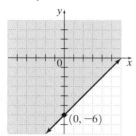

$(0, -6)$

Test $(0, 0)$:
$0 - 0 \le 6?$
$\quad\; 0 \le 6?$
Yes; shade above the line.

29. $2x + 7y > 5$
$2x + 7y = 5$
$\quad\quad 7y = -2x + 5$
$\quad\quad\; y = -\dfrac{2}{7}x + \dfrac{5}{7}$

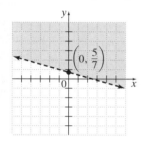

$\left(0, \dfrac{5}{7}\right)$

Test $(0, 0)$:
$2(0) + 7(0) > 5?$
$\quad\quad\quad\; 0 > 5?$
No; shade above the line.

33. $(-3, -1)$

CHAPTER 6 TEST

1. $12x - 7y = 5$
If $x = 1$, $12y - 7(1) = 5$
$\quad\quad\quad\quad\quad 12y = 12$
$\quad\quad\quad\quad\quad\; y = 1$
$(1, 1)$

5. $(6, -5)$ and $(-1, 2)$
$m = \dfrac{y_2 - y_1}{x_2 - x_1} = \dfrac{2 - (-5)}{-1 - 6} = \dfrac{7}{-7} = -1$

9. $2x + y = 8$
If $x = 0$, $y = 8$
If $y = 0$, $2x = 8$
$\quad\quad\quad\quad\; x = 4$
Plot using $(0, 8)$ and $(4, 0)$

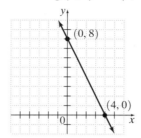

$(0, 8)$

$(4, 0)$

13. $5x - 7y = 10$
If $y = 0$, $5x = 10$
$\quad\quad\quad\quad x = 2$
If $y = -5$, $5x - 7(-5) = 10$
$\quad\quad\quad\quad\quad 5x + 35 = 10$
$\quad\quad\quad\quad\quad\quad\; 5x = -25$
$\quad\quad\quad\quad\quad\quad\;\; x = -5$
Plot using $(2, 0)$ and $(-5, -5)$

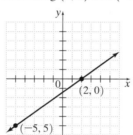

$(2, 0)$

$(-5, 5)$

17. $y = 2x - 6 \quad\quad\quad -4x = 2y$
$\quad m = 2 \quad\quad\quad\quad\quad\;\; y = -2x$
$\quad\quad\quad\quad\quad\quad\quad\quad\;\; m = -2$
neither

21. $m = \dfrac{1}{8}; b = 12$

$\quad\quad\quad y = \dfrac{1}{8}x + 12$
$\quad\quad\quad 8y = x + 96$
$\quad x - 8y = -96$

25. yes

Chapter 7

SECTION 7.1

1. a. $\quad x + y = 8 \quad\quad\quad\quad 3x + 2y = 21$
$\quad\quad\; 2 + 4 \overset{?}{=} 8 \quad\quad\quad 3(5) + 2(4) = 21$
$\quad\quad\quad\quad 6 = 8 \quad\quad\quad\quad\; 6 + 8 = 21$
$\quad\quad$ False $\quad\quad\quad\quad\quad\quad 14 = 21$
$\quad\quad\quad\quad\quad\quad\quad\quad\quad\quad\quad$ False
No, $(2, 4)$ is not a solution of the system.

b. $\quad x + y = 8 \quad\quad\quad\quad 3x + 2y = 21$
$\quad\quad\; 5 + 3 \overset{?}{=} 8 \quad\quad\quad 3(5) + 2(3) \overset{?}{=} 21$
$\quad\quad\quad\quad 8 = 8 \quad\quad\quad\quad 15 + 6 \overset{?}{=} 21$
$\quad\quad$ True $\quad\quad\quad\quad\quad\quad\quad 21 = 21$
Yes, $(5, 3)$ is a solution of the system.

5. a.

$$2y = 4x \qquad\qquad 2x - y = 0$$
$$2(-6) \stackrel{?}{=} 4(-3) \qquad 2(-3) - (-6) \stackrel{?}{=} 0$$
$$-12 = -12 \qquad\qquad -6 + 6 \stackrel{?}{=} 0$$
$$\text{True} \qquad\qquad\qquad 0 = 0$$
$$\text{True}$$

Yes, $(-3, -6)$ is a true solution of the system.

b.

$$2y = 4x \qquad\qquad 2x - y = 0$$
$$2(0) \stackrel{?}{=} 4(0) \qquad 2(0) - 0 \stackrel{?}{=} 0$$
$$0 = 0 \qquad\qquad 0 = 0$$
$$\text{True} \qquad\qquad \text{True}$$

Yes, $(0, 0)$ is a solution of the system.

9. $\begin{cases} x + y = 6 \\ -x + y = -6 \end{cases}$

Graph each linear equation on a single set of axes.

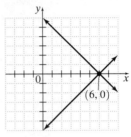

The solution is the intersection point of the two lines, $(6, 0)$.

13. $\begin{cases} y = x + 1 \\ y = 2x - 1 \end{cases}$

Graph each linear equation on a single set of axes.

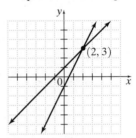

The solution is the intersection point of the two lines, $(2, 3)$.

17. $\begin{cases} y = -x - 1 \\ y = 2x + 5 \end{cases}$

Graph each linear equation on a single set of axes.

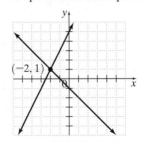

The solution is the intersection point of the two lines, $(-2, 1)$.

21. $\begin{cases} x + y = 5 \\ x + y = 6 \end{cases}$

Graph each linear equation on a single set of axes.

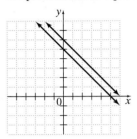

Since the lines are parallel, the system has no solution.

25. $\begin{cases} x - 2y = 2 \\ 3x + 2y = -2 \end{cases}$

Graph each linear equation on a single set of axes.

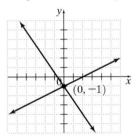

The solution is the intersection point of the two lines, $(0, -1)$.

29. $\begin{cases} x = 3 \\ y = -1 \end{cases}$

Graph each linear equation on a single set of axes.

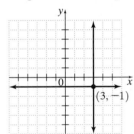

The solution is the intersection point of the two lines, $(3, -1)$.

33. $\begin{cases} 2x - 3y = -2 \\ -3x + 5y = 5 \end{cases}$

Graph each linear equation on a single set of axes.

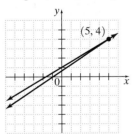

The solution is the intersection point of the two lines, $(5, 4)$.

37. answers may vary

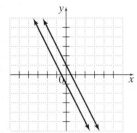

Any two parallel lines will meet the condition.

41. a. Each of the tables includes the point $(4, 9)$.
Therefore, $(4, 9)$ is a solution to the system.

b.

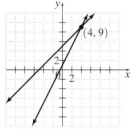

Yes, the lines intersect at the point $(4, 9)$.

45. $-y + 12\left(\dfrac{y-1}{4}\right) = 3$

$$-y + 3(y - 1) = 3$$
$$-y + 3y - 3 = 3$$
$$2y - 3 = 3$$
$$2y = 6$$
$$y = 3$$

The solution is $y = 3$.

49. answers may vary

SECTION 7.2

1. $\begin{cases} x + y = 3 \\ x = 2y \end{cases}$

Substitute $2y$ for x in the first equation. Then solve for y.

$$2y + y = 3$$
$$3y = 3$$
$$y = 1$$

Substitute 1 for y in the second equation.
Then solve for x.

$$x = 2(1)$$
$$x = 2$$

The solution is $(2, 1)$.

5. $\begin{cases} 3x + 2y = 16 \\ x = 3y - 2 \end{cases}$

Substitute $3y - 2$ for x in the first equation.
Then solve for y.

$$3(3y - 2) + 2y = 16$$
$$9y - 6 + 2y = 16$$
$$11y = 22$$
$$y = 2$$

Substitute 2 for y in the second equation.
Then solve for x.

$$x = 3(2) - 2$$
$$x = 6 - 2$$
$$x = 4$$

The solution is $(4, 2)$.

9. $\begin{cases} y = 3x + 1 \\ 4y - 8x = 12 \end{cases}$

Substitute $3x + 1$ for y in the second equation. Then
solve for x.

$$4(3x + 1) - 8x = 12$$
$$12x + 4 - 8x = 12$$
$$4x = 8$$
$$x = 2$$

Substitute 2 for x in the first equation. Then solve for
y.

$$y = 3(2) + 1$$
$$y = 6 + 1$$
$$y = 7$$

The solution is $(2, 7)$.

13. $\begin{cases} x + 2y = 6 \\ 2x + 3y = 8 \end{cases}$

Solve the first equation for x.

$$x = 6 - 2y$$

Substitute $6 - 2y$ for x in the second equation. Then
solve for y.

$$2(6 - 2y) + 3y = 8$$
$$12 - 4y + 3y = 8$$
$$-y = -4$$
$$y = 4$$

Substitute 4 for y in $x = 6 - 2y$. Then solve for x.

$$x = 6 - 2(4)$$
$$x = 6 - 8$$
$$x = -2$$

The solution is $(-2, 4)$.

17. $\begin{cases} 2y = x + 2 \\ 6x - 12y = 0 \end{cases}$

Solve the first equation for x.

$$x = 2y - 2$$

Substitute $2y - 2$ for x in the second equation.

$$6(2y - 2) - 12y = 0$$
$$12y - 12 - 12y = 0$$
$$-12 = 0$$

Since this is false, the system has no solution.

21. $\begin{cases} 2x - 3y = -9 \\ 3x = y + 4 \end{cases}$

Solve the second equation for y.

$$y = 3x - 4$$

Substitute $3x - 4$ for y in the first equation.
Then solve for x.

$$2x - 3(3x - 4) = -9$$
$$2x - 9x + 12 = -9$$
$$-7x = -21$$
$$x = 3$$

Substitute 3 for x in $y = 3x - 4$.
Then solve for y.

$$y = 3(3) - 4$$
$$y = 9 - 4$$
$$y = 5$$

The solution is $(3, 5)$.

25. $\begin{cases} 3x - y = 1 \\ 2x - 3y = 10 \end{cases}$

Solve the first equation for y.

$y = 3x - 1$

Substitute $3x - 1$ for y in the second equation. Then solve for x.

$2x - 3(3x - 1) = 10$

$2x - 9x + 3 = 10$

$-7x = 7$

$x = -1$

Substitute -1 for x in $y = 3x - 1$. Then solve for y.

$y = 3(-1) - 1$

$y = -3 - 1$

$y = -4$

The solution is $(-1, -4)$.

29. $\begin{cases} 5x + 10y = 20 \\ 2x + 6y = 10 \end{cases}$

Solve the first equation for x.

$x + 2y = 4$

$x = 4 - 2y$

Substitute $4 - 2y$ for x in the second equation. Then solve for y.

$2(4 - 2y) + 6y = 10$

$8 - 4y + 6y = 10$

$2y = 2$

$y = 1$

Substitute 1 for y in $x = 4 - 2y$. Then solve for x.

$x = 4 - 2(1)$

$x = 2$

The solution is $(2, 1)$.

33. $\begin{cases} \dfrac{1}{3}x - y = 2 \\ x - 3y = 6 \end{cases}$

Solve the second equation for x.

$x = 3y + 6$

Substitute $3y + 6$ for x in the first equation.

$\dfrac{1}{3}(3y + 6) - y = 2$

$y + 2 - y = 2$

$2 = 2$

Since this is always true, the system has an infinite number of solutions.

37. $3x + 2y = 6$

$-2(3x + 2y) = -2(6)$

$-6x - 4y = -12$

41. $3n + 6m$

$\underline{2n - 6m}$

$5n$

45. Simplify the first equation.

$-5y + 6y = 3x + 2(x - 5) - 3x + 5$

$y = 3x + 2x - 10 - 3x + 5$

$y = 2x - 5$

Simplify the second equation.

$4(x + y) - x + y = -12$

$4x + 4y - x + y = -12$

$3x + 5y = -12$

Solve the system

$\begin{cases} y = 2x - 5 \\ 3x + 5y = -12 \end{cases}$

Substitute $2x - 5$ for y in the second equation. Then solve for x.

$3x + 5(2x - 5) = -12$

$3x + 10x - 25 = -12$

$13x = 13$

$x = 1$

Substitute 1 for x in the first equation. Then solve for y.

$y = 2(1) - 5$

$y = 2 - 5$

$y = -3$

The solution is $(1, -3)$.

49. $\begin{cases} 3x + 2y = 14.05 \\ 5x + y = 18.5 \end{cases}$

Let $y_1 = \dfrac{14.05 - 3x}{2}$ and $y_2 = 18.5 - 5x$ and find the intersection.

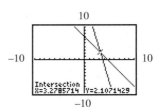

The approximate solution is $(3.28, 2.11)$.

Section 7.3

1. $\begin{cases} 3x + y = 5 \\ 6x - y = 4 \end{cases}$

$\overline{9x = 9}$

$x = 1$

Let $x = 1$ in the first equation.

$3(1) + y = 5$

$y = 2$

The solution is $(1, 2)$.

5. $\begin{cases} 3x + 2y = 11 \\ 5x - 2y = 29 \end{cases}$

$\overline{8x = 40}$

$x = 5$

Let $x = 5$ in the first equation.

$3(5) + 2y = 11$

$15 + 2y = 11$

$2y = -4$

$y = -2$

The solution is $(5, -2)$.

9. $\begin{cases} 3x + y = -11 \\ 6x - 2y = -2 \end{cases}$

$\begin{cases} 2(3x + y) = 2(-11) \\ 6x - 2y = -2 \end{cases}$

$\begin{cases} 6x + 2y = -22 \\ \underline{6x - 2y = -2} \end{cases}$

$\qquad 12x = -24$

$\qquad\quad x = -2$

Let $x = -2$ in the first equation.

$3(-2) + y = -11$

$\quad -6 + y = -11$

$\qquad\quad y = -5$

The solution is $(-2, -5)$.

13. $\begin{cases} 2x - 5y = 4 \\ 3x - 2y = 4 \end{cases}$

$\begin{cases} -3(2x - 5y) = -3(4) \\ 2(3x - 2y) = 2(4) \end{cases}$

$\begin{cases} -6x + 15y = -12 \\ \underline{6x - 4y = 8} \end{cases}$

$\qquad\quad 11y = -4$

$\qquad\qquad y = -\dfrac{4}{11}$

$\begin{cases} -2(2x - 5y) = -2(4) \\ 5(3x - 2y) = 5(4) \end{cases}$

$\begin{cases} -4x + 10y = -8 \\ \underline{15x - 10y = 20} \end{cases}$

$\qquad\quad 11x = 12$

$\qquad\qquad x = \dfrac{12}{11}$

The solution is $\left(\dfrac{12}{11}, -\dfrac{4}{11} \right)$.

17. $\begin{cases} 3x + y = 4 \\ 9x + 3y = 6 \end{cases}$

$\begin{cases} -3(3x + y) = -3(4) \\ 9x + 3y = 6 \end{cases}$

$\begin{cases} -9x - 3y = -12 \\ \underline{9x + 3y = 6} \end{cases}$

$\qquad\qquad 0 = -6$

Since this is false, the system has no solution.

21. $\begin{cases} \dfrac{2}{3}x + 4y = -4 \\ 5x + 6y = 18 \end{cases}$

$\begin{cases} 3\left(\dfrac{2}{3}x + 4y \right) = 3(-4) \\ -2(5x + 6y) = -2(18) \end{cases}$

$\begin{cases} 2x + 12y = -12 \\ \underline{-10x - 12y = -36} \end{cases}$

$\qquad\quad -8x = -48$

$\qquad\qquad x = 6$

Let $x = 6$ in the second equation.

$5(6) + 6y = 18$

$\quad 30 + 6y = 18$

$\qquad\quad 6y = -12$

$\qquad\quad y = -2$

The solution is $(6, -2)$.

25. $\begin{cases} 8x = -11y - 16 \\ 2x + 3y = -4 \end{cases}$

$\begin{cases} 8x + 11y = -16 \\ -4(2x + 3y) = -4(-4) \end{cases}$

$\begin{cases} 8x + 11y = -16 \\ \underline{-8x - 12y = 16} \end{cases}$

$\qquad\quad -y = 0$

$\qquad\qquad y = 0$

Let $y = 0$ in the first equation.

$8x = -11(0) - 16$

$8x = -16$

$\quad x = -2$

The solution is $(-2, 0)$.

29. $\begin{cases} \dfrac{x}{3} + \dfrac{y}{6} = 1 \\ \dfrac{x}{2} - \dfrac{y}{4} = 0 \end{cases}$

$\begin{cases} 6\left(\dfrac{x}{3} + \dfrac{y}{6} \right) = 6(1) \\ 4\left(\dfrac{x}{2} - \dfrac{y}{4} \right) = 4(0) \end{cases}$

$\begin{cases} 2x + y = 6 \\ \underline{2x - y = 0} \end{cases}$

$\qquad 4x = 6$

$\qquad\quad x = \dfrac{3}{2}$

To find y, we may multiply the second equation of the simplified system above by -1.

$\begin{cases} 2x + y = 6 \\ \underline{-2x + y = 0} \end{cases}$

$\qquad 2y = 6$

$\qquad\quad y = 3$

The solution is $\left(\dfrac{3}{2}, 3 \right)$.

33.
$$\begin{cases} \dfrac{x}{3} - y = 2 \\ -\dfrac{x}{2} + \dfrac{3y}{2} = -3 \end{cases}$$

$$\begin{cases} 3\left(\dfrac{x}{3} - y\right) = 3(2) \\ 2\left(-\dfrac{x}{2} + \dfrac{3y}{2}\right) = 2(-3) \end{cases}$$

$$\begin{cases} x - 3y = 6 \\ -x + 3y = -6 \end{cases}$$
$$0 = 0$$

Since this is always true, the system has an infinite number of solutions.

37. Let x = a number
$$20 - 3x = 2$$

41. a.
$$\begin{cases} x + y = 5 \\ 3x + 3y = b \end{cases}$$

The system will have an infinite number of solutions if the second equation is 3 times the first equation.
$$3(x + y) = 3(5)$$
$$3x + 3y = 15$$
Therefore, $b = 15$.

b. There are no solutions to the system if b is any real number except 15.

45. a.
$$\begin{cases} 21x - 7y = -3779 \\ -445x + 28y = 14{,}017 \end{cases}$$

$$\begin{cases} 4(21x - 7y) = 4(-3779) \\ -445x + 28y = 14{,}017 \end{cases}$$

$$\begin{cases} 84x - 28y = -15{,}116 \\ -445x + 28y = 14{,}017 \end{cases}$$
$$-361x = -1099$$
$$x \approx 3.04$$

Substitute 3.04 for x in the first equation.
$$21(3.04) - 7y = -3779$$
$$63.84 - 7y = -3779$$
$$7y = 3842.84$$
$$y \approx 548.98$$
The solution is approximately $(3, 549)$.

b. answers may vary

c. Since they were equal 3 years after 1988, then from $1988 + 4 = 1992$ to 1994 there were more UHF stations than VHF stations.

Section 7.4

1. Let x = one number
y = another number
$$\begin{cases} x + y = 15 \\ x - y = 7 \end{cases}$$

5. Let x = one number
y = another number
$$\begin{cases} x + y = 83 \\ x - y = 17 \end{cases}$$
$$2x = 100$$
$$x = 50$$
Let $x = 50$ in the first equation.
$$50 + y = 83$$
$$y = 33$$
The two numbers are 33 and 50.

9. Let x = number of points Cooper scored
y = number of points Bolton-Holifield scored
$$\begin{cases} x = 174 + y \\ x + y = 1068 \end{cases}$$
Substitute $174 + y$ for x in the second equation.
$$174 + y + y = 1068$$
$$2y = 894$$
$$y = 447$$
Let $y = 447$ in the first equation.
$$x = 174 + 447$$
$$x = 621$$
Cooper scored 621 points and Bolton-Holifield scored 447 points.

13. Let x = number of quarters
y = number of of nickels
$$\begin{cases} x + y = 80 \\ 0.25x + 0.05y = 14.6 \end{cases}$$
Substitute $80 - x$ for y in the second equation.
$$0.25x + 0.05(80 - x) = 14.6$$
$$0.25x + 4 - 0.05x = 14.6$$
$$0.2x = 10.6$$
$$x = 53$$
Let $x = 53$ in the first equation.
$$53 + y = 80$$
$$y = 27$$
There are 53 quarters and 27 nickels.

17. Let x = Pratap's rate in still water in miles per hour
y = rate of the current in miles per hour
$$\begin{cases} 2(x + y) = 18 \\ 4.5(x - y) = 18 \end{cases}$$
$$\begin{cases} 2x + 2y = 18 \\ 4.5x - 4.5y = 18 \end{cases}$$
$$\begin{cases} 2.25(2x + 2y) = 2.25(18) \\ 4.5x - 4.5y = 18 \end{cases}$$
$$\begin{cases} 4.5x + 4.5y = 40.5 \\ 4.5x - 4.5y = 18 \end{cases}$$
$$9x = 58.5$$
$$x = 6.5$$
Let $x = 6.5$ in the first equation.
$$2(6.5) + 2y = 18$$
$$2y = 5$$
$$y = 2.5$$
Pratap's rate in still water was 6.5 mph and the current's rate was 2.5 mph.

21. Let x = amount of 12% solution
 y = amount of 4% solution
$$\begin{cases} x + y = 12 \\ 0.12x + 0.04y = 0.09(12) \end{cases}$$
Substitute $12 - x$ for y in the second equation.
$$0.12x + 0.04(12 - x) = 0.09(12)$$
$$0.12x + 0.48 - 0.04x = 1.08$$
$$0.08x = 0.6$$
$$x = 7.5$$
Let $x = 7.5$ in the first equation.
$$7.5 + y = 12$$
$$y = 4.5$$
She needs $7\frac{1}{2}$ oz of 12% solution and $4\frac{1}{2}$ oz of

4% solution.

25. $4^2 = 16$

29. $(10y^3)^2 = (10)^2 \cdot y^{3 \cdot 2} = 100y^6$

33. Let x = first angle
 y = second angle
$$\begin{cases} x + y = 90 \\ y = 3x + 10 \end{cases}$$
Let $y = 3x + 10$ in the first equation.
$$x + 3x + 10 = 90$$
$$4x = 80$$
$$x = 20$$
Let $x = 20$ in the second equation.
$$y = 3(20) + 10$$
$$y = 70$$
The angles measure 20° and 70°.

37. Let x = width
 y = length
$$\begin{cases} 2x + y = 33 \\ y = 2x - 3 \end{cases}$$
Substitute $2x - 3$ for y in the first equation.
$$2x + 2x - 3 = 33$$
$$4x = 36$$
$$x = 9$$
Let $x = 9$ in the second equation.
$$y = 2(9) - 3$$
$$y = 15$$
The width is 9 feet and the length is 15 feet.

39. answers may vary

Chapter 7 Test

1. $\begin{cases} x - y = 2 \\ 3x - y = -2 \end{cases}$

Graph each linear equation on a single set of axes.

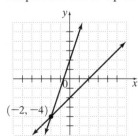

$(-2, -4)$

The solution is the intersection point of the two lines, $(-2, -4)$.

5. $\begin{cases} x - y = 4 \\ x - 2y = 11 \end{cases}$
Solve the first equation for x.
$$x = y + 4$$
Let $x = y + 4$ in the second equation
$$y + 4 - 2y = 11$$
$$-y = 7$$
$$y = -7$$
Let $y = -7$ in $x = y + 4$.
$$x = -7 + 4$$
$$x = -3$$
The solution is $(-3, -7)$.

9. $\begin{cases} 5x - 6y = 7 \\ 7x - 4y = 12 \end{cases}$
$$\begin{cases} -10x + 12y = -14 \\ 21x - 12y = 36 \end{cases}$$
$$\overline{11x = 22}$$
$$x = 2$$

Let $x = 2$ in the first equation.
$$5(2) - 6y = 7$$
$$-6y = -3$$
$$y = \frac{1}{2}$$

The solution is $\left(2, \frac{1}{2}\right)$.

13. Let x = amount of 30% solution
 y = amount of 70% solution
$$\begin{cases} x + y = 10 \\ 0.3x + 0.7y = 0.4(10) \end{cases}$$
Solve the first equation for x.
$$x = 10 - y$$
Let $x = 10 - y$ in the second equation.
$$0.3(10 - y) + 0.7y = 0.4(10)$$
$$3 - 0.3y + 0.7y = 4$$
$$0.4y = 1$$
$$y = 2.5$$
Let $y = 2.5$ in $x = 10 - y$.
$$x = 10 - 2.5$$
$$x = 7.5$$
There should be 7.5 liters of 30% solution and 2.5 liters of 70% solution.

Chapter 8

Section 8.1

1. $\sqrt{16} = 4$ because $4^2 = 16$ and 4 is positive.

5. $\sqrt{\dfrac{1}{25}} = \dfrac{1}{5}$ because $\left(\dfrac{1}{5}\right)^2 = \dfrac{1}{25}$ and $\dfrac{1}{5}$ is positive.

9. $\sqrt{-4}$ is not a real number because the index is even and the radicand is negative.

13. $\sqrt{\dfrac{9}{25}} = \dfrac{3}{5}$ because $\left(\dfrac{3}{5}\right)^2 = \dfrac{9}{25}$ and $\dfrac{3}{5}$ is positive.

17. $\sqrt[3]{64} = -4$ because $(-4)^3 = -64$

21. $\sqrt[3]{\dfrac{1}{8}} = \dfrac{1}{2}$ because $\left(\dfrac{1}{2}\right)^3 = \dfrac{1}{8}$.

25. answers may vary

29. $\sqrt[4]{-16}$ is not a real number.

33. $\sqrt[6]{1} = 1$ because $1^6 = 1$.

37. $\sqrt{136} \approx 11.662$

41. $\sqrt{z^2} = z$ because $(z)^2 = z^2$.

45. $\sqrt{9x^8} = 3x^4$ because $(3x^4)^2 = 9x^8$.

49. $50 = 25 \cdot 2$, 25 is a perfect square.

53. $28 = 4 \cdot 7$, 4 is a perfect square.

57. $\sqrt{\sqrt{81}} = \sqrt{9} = 3$

61. $y = \sqrt{x-2}$

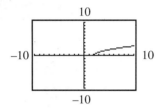

The graph starts at the point $(2, 0)$. $x - 2$ is greater than or equal to zero for $x \geq 2$.

65. $\sqrt[3]{195,112} = 58$ because $(58)^3 = 195,112$.
Each side would be 58 feet.

Section 8.2

1. $\sqrt{20} = \sqrt{4 \cdot 5} = \sqrt{4} \cdot \sqrt{5} = 2\sqrt{5}$

5. $\sqrt{50} = \sqrt{25 \cdot 2} = \sqrt{25} \cdot \sqrt{2} = 5\sqrt{2}$

9. $\sqrt{60} = \sqrt{4 \cdot 15} = \sqrt{4} \cdot \sqrt{15} = 2\sqrt{15}$

13. $\sqrt{52} = \sqrt{4 \cdot 13} = \sqrt{4} \cdot \sqrt{13} = 2\sqrt{13}$

17. $\sqrt{\dfrac{27}{121}} = \dfrac{\sqrt{27}}{\sqrt{121}} = \dfrac{\sqrt{9} \cdot \sqrt{3}}{11} = \dfrac{3\sqrt{3}}{11}$

21. $\sqrt{\dfrac{125}{9}} = \dfrac{\sqrt{125}}{\sqrt{9}} = \dfrac{\sqrt{25} \cdot \sqrt{5}}{3} = \dfrac{5\sqrt{5}}{3}$

25. $-\sqrt{\dfrac{27}{144}} = -\dfrac{\sqrt{27}}{\sqrt{144}} = -\dfrac{\sqrt{9} \cdot \sqrt{3}}{12} = -\dfrac{3\sqrt{3}}{12} = -\dfrac{\sqrt{3}}{4}$

29. $\sqrt{x^{13}} = \sqrt{x^{12} \cdot x} = \sqrt{x^{12}} \cdot \sqrt{x} = x^6\sqrt{x}$

33. $\sqrt{96x^4} = \sqrt{16x^4 \cdot 6} = \sqrt{16x^4} \cdot \sqrt{6} = 4x^2\sqrt{6}$

37. $\sqrt{\dfrac{9x}{y^2}} = \dfrac{\sqrt{9x}}{\sqrt{y^2}} = \dfrac{\sqrt{9} \cdot \sqrt{x}}{y} = \dfrac{3\sqrt{x}}{y}$

41. $\sqrt[3]{24} = \sqrt[3]{8 \cdot 3} = \sqrt[3]{8} \cdot \sqrt[3]{3} = 2\sqrt[3]{3}$

45. $\sqrt[3]{\dfrac{5}{64}} = \dfrac{\sqrt[3]{5}}{\sqrt[3]{64}} = \dfrac{\sqrt[3]{5}}{4}$

49. $\sqrt[3]{\dfrac{15}{64}} = \dfrac{\sqrt[3]{15}}{\sqrt[3]{64}} = \dfrac{\sqrt[3]{15}}{4}$

53. $6x + 8x = (6 + 8)x = 14x$

57. $9y^2 - 9y^2 = 0$

61. $\sqrt{x^2 + 4x + 4} = \sqrt{(x + 2)^2} = x + 2$

65. $C = 100\sqrt[3]{n} + 700$
$C = 100\sqrt[3]{1000} + 700$
$C = 100(10) + 700$
$C = 1000 + 700$
$C = 1700$
The cost is $1700.

Section 8.3

1. $4\sqrt{3} - 8\sqrt{3} = (4 - 8)\sqrt{3} = -4\sqrt{3}$

5. $6\sqrt{5} - 5\sqrt{5} + \sqrt{2} = (6 - 5)\sqrt{5} + \sqrt{2}$
$= \sqrt{5} + \sqrt{2}$

9. $2\sqrt{2} - 7\sqrt{2} - 6 = (2 - 7)\sqrt{2} - 6$
$= -5\sqrt{2} - 6$

13. $\sqrt{5} + \sqrt{5} = 1\sqrt{5} + 1\sqrt{5} = (1 + 1)\sqrt{5} = 2\sqrt{5}$

17. answers may vary

21. $\sqrt{45} + 3\sqrt{20} = \sqrt{9 \cdot 5} + 3\sqrt{4 \cdot 5}$
$= \sqrt{9} \cdot \sqrt{5} + 3\sqrt{4} \cdot \sqrt{5}$
$= 3\sqrt{5} + 3 \cdot 2\sqrt{5}$
$= 3\sqrt{5} + 6\sqrt{5}$
$= 9\sqrt{5}$

25. $4x - 3\sqrt{x^2} + \sqrt{x} = 4x - 3x + \sqrt{x}$
$= x + \sqrt{x}$

29. $3\sqrt{x^3} - x\sqrt{4x} = 3\sqrt{x^2 \cdot x} - x\sqrt{4 \cdot x}$
$= 3\sqrt{x^2} \cdot \sqrt{x} - x\sqrt{4} \cdot \sqrt{x}$
$= 3x\sqrt{x} - 2x\sqrt{x}$
$= x\sqrt{x}$

33. $\sqrt{8} + \sqrt{9} + \sqrt{18} + \sqrt{81}$
$= \sqrt{4} \cdot \sqrt{2} + 3 + \sqrt{9} \cdot \sqrt{2} + 9$
$= 2\sqrt{2} + 3 + 3\sqrt{2} + 9$
$= 5\sqrt{2} + 12$

37. $\sqrt{\dfrac{3}{4}} - \sqrt{\dfrac{3}{64}} = \dfrac{\sqrt{3}}{\sqrt{4}} - \dfrac{\sqrt{3}}{\sqrt{64}}$
$= \dfrac{\sqrt{3}}{2} - \dfrac{\sqrt{3}}{8}$
$= \dfrac{4\sqrt{3}}{8} - \dfrac{\sqrt{3}}{8}$
$= \dfrac{4\sqrt{3} - \sqrt{3}}{8}$
$= \dfrac{3\sqrt{3}}{8}$

41. $\sqrt{35} - \sqrt{140} = \sqrt{35} - \sqrt{4 \cdot 35}$
$= \sqrt{35} - \sqrt{4} \cdot \sqrt{35}$
$= \sqrt{35} - 2\sqrt{35}$
$= -\sqrt{35}$

45. $\sqrt{9x^2} + \sqrt{81x^2} - 11\sqrt{x} = 3x + 9x - 11\sqrt{x}$
$= 12x - 11\sqrt{x}$

49. $\sqrt{32x^2} + \sqrt{32x^2} + \sqrt{4x^2}$
$= \sqrt{16x^2} \cdot \sqrt{2} + \sqrt{16x^2} \cdot \sqrt{2} + 2x$
$= 4x\sqrt{2} + 4x\sqrt{2} + 2x$
$= 8x\sqrt{2} + 2x$

53. $(x + 6)^2 = x^2 + 2(x)(6) + 6^2$
$= x^2 + 12x + 36$

57. $\begin{cases} x = 2y \\ x + 5y = 14 \end{cases}$

Substitute $2y$ for x in the second equation.

$2y + 5y = 14$

$\quad\quad 7y = 14$

$\quad\quad\quad y = 2$

Let $y = 2$ in the first equation

$x = 2(2)$

$x = 4$

The solution is $(4, 2)$.

61. Area = area of two triangles + area of 2 rectangles

$= 2\left(\dfrac{3\sqrt{27}}{4}\right) + 2(8 \cdot 3)$

$= \dfrac{3\sqrt{9} \cdot \sqrt{3}}{2} + 48$

$= \dfrac{9\sqrt{3}}{2} + 48$

The total area is $\left(\dfrac{9\sqrt{3}}{2} + 48\right)$ sq. ft.

Section 8.4

1. $\sqrt{8} \cdot \sqrt{2} = \sqrt{8 \cdot 2} = \sqrt{16} = 4$

5. $\sqrt{6} \cdot \sqrt{6} = \sqrt{6 \cdot 6} = \sqrt{36} = 6$

9. $(2\sqrt{5})^2 = 2^2(\sqrt{5})^2 = 4 \cdot 5 = 20$

13. $\sqrt{3y} \cdot \sqrt{6x} = \sqrt{3y \cdot 6x}$

$\quad\quad\quad\quad = \sqrt{18xy}$

$\quad\quad\quad\quad = \sqrt{9} \cdot \sqrt{2xy}$

$\quad\quad\quad\quad = 3\sqrt{2xy}$

17. $\sqrt{2}(\sqrt{5} + 1) = \sqrt{2} \cdot \sqrt{5} + \sqrt{2} \cdot 1$

$\quad\quad\quad\quad = \sqrt{2 \cdot 5} + \sqrt{2}$

$\quad\quad\quad\quad = \sqrt{10} + \sqrt{2}$

21. $\sqrt{6}(\sqrt{5} + \sqrt{7}) = \sqrt{6} \cdot \sqrt{5} + \sqrt{6} \cdot \sqrt{7}$

$\quad\quad\quad\quad\quad = \sqrt{6 \cdot 5} + \sqrt{6 \cdot 7}$

$\quad\quad\quad\quad\quad = \sqrt{30} + \sqrt{42}$

25. $(\sqrt{3} + \sqrt{5})(\sqrt{2} - \sqrt{5})$

$= \sqrt{3} \cdot \sqrt{2} - \sqrt{3} \cdot \sqrt{5} + \sqrt{5} \cdot \sqrt{2} - \sqrt{5} \cdot \sqrt{5}$

$= \sqrt{3 \cdot 2} - \sqrt{3 \cdot 5} + \sqrt{5 \cdot 2} - \sqrt{5 \cdot 5}$

$= \sqrt{6} - \sqrt{15} + \sqrt{10} - \sqrt{25}$

$= \sqrt{6} - \sqrt{15} + \sqrt{10} - 5$

29. $(\sqrt{x} + 6)(\sqrt{x} - 6) = (\sqrt{x})^2 - 6^2 = x - 36$

33. $(\sqrt{6y} + 1)^2 = (\sqrt{6y})^2 + 2(\sqrt{6y})(1) + 1^2$

$\quad\quad\quad\quad = 6y + 2\sqrt{6y} + 1$

37. $\dfrac{\sqrt{21}}{\sqrt{3}} = \sqrt{\dfrac{21}{3}} = \sqrt{7}$

41. $\dfrac{\sqrt{75y^5}}{\sqrt{3y}} = \sqrt{\dfrac{75y^5}{3y}} = \sqrt{25y^4} = 5y^2$

45. $\dfrac{\sqrt{72y^5}}{\sqrt{3y^3}} = \sqrt{\dfrac{72y^5}{3y^3}} = \sqrt{24y^2} = \sqrt{4y^2} \cdot \sqrt{6} = 2y\sqrt{6}$

49. $\dfrac{\sqrt{3}}{\sqrt{5}} = \dfrac{\sqrt{3} \cdot \sqrt{5}}{\sqrt{5} \cdot \sqrt{5}} = \dfrac{\sqrt{15}}{5}$

53. $\dfrac{1}{\sqrt{6y}} = \dfrac{1 \cdot \sqrt{6y}}{\sqrt{6y} \cdot \sqrt{6y}} = \dfrac{\sqrt{6y}}{6y}$

57. $\sqrt{\dfrac{3}{x}} = \dfrac{\sqrt{3}}{\sqrt{x}} = \dfrac{\sqrt{3} \cdot \sqrt{x}}{\sqrt{x} \cdot \sqrt{x}} = \dfrac{\sqrt{3x}}{x}$

61. $\sqrt{\dfrac{2}{15}} = \dfrac{\sqrt{2}}{\sqrt{15}} = \dfrac{\sqrt{2} \cdot \sqrt{15}}{\sqrt{15} \cdot \sqrt{15}} = \dfrac{\sqrt{30}}{15}$

65. $\dfrac{3x}{\sqrt{2x}} = \dfrac{3x \cdot \sqrt{2x}}{\sqrt{2x} \cdot \sqrt{2x}} = \dfrac{3x\sqrt{2x}}{2x} = \dfrac{3\sqrt{2x}}{2}$

69. $\sqrt{\dfrac{y}{12x}} = \dfrac{\sqrt{y}}{\sqrt{12x}}$

$\quad\quad\quad = \dfrac{\sqrt{y}}{\sqrt{4} \cdot \sqrt{3x}}$

$\quad\quad\quad = \dfrac{\sqrt{y}}{2\sqrt{3x}}$

$\quad\quad\quad = \dfrac{\sqrt{y} \cdot \sqrt{3x}}{2\sqrt{3x} \cdot \sqrt{3x}}$

$\quad\quad\quad = \dfrac{\sqrt{3xy}}{2 \cdot 3x}$

$\quad\quad\quad = \dfrac{\sqrt{3xy}}{6x}$

73. $\dfrac{4}{2 - \sqrt{5}} = \dfrac{4(2 + \sqrt{5})}{(2 - \sqrt{5})(2 + \sqrt{5})}$

$\quad\quad\quad = \dfrac{4(2 + \sqrt{5})}{2^2 - (\sqrt{5})^2}$

$\quad\quad\quad = \dfrac{4(2 + \sqrt{5})}{4 - 5}$

$\quad\quad\quad = \dfrac{4(2 + \sqrt{5})}{-1}$

$\quad\quad\quad = -4(2 + \sqrt{5})$

$\quad\quad\quad = -8 - 4\sqrt{5}$

77. $\dfrac{\sqrt{3} + 1}{\sqrt{2} - 1} = \dfrac{(\sqrt{3} + 1)(\sqrt{2} + 1)}{(\sqrt{2} - 1)(\sqrt{2} + 1)}$

$\quad\quad\quad = \dfrac{\sqrt{6} + \sqrt{3} + \sqrt{2} + 1}{2 - 1}$

$\quad\quad\quad = \sqrt{6} + \sqrt{3} + \sqrt{2} + 1$

81. $\dfrac{3}{\sqrt{x} - 4} = \dfrac{3(\sqrt{x} + 4)}{(\sqrt{x} - 4)(\sqrt{x} + 4)} = \dfrac{3\sqrt{x} + 12}{x - 16}$

85. $\quad\quad 4z^2 + 6z - 12 = (2z)^2$

$\quad\quad\quad 4z^2 + 6z - 12 = 4z^2$

$4z^2 + 6z - 12 - 4z^2 = 4z^2 - 4z^2$

$\quad\quad\quad\quad\quad 6z - 12 = 0$

$\quad\quad 6z - 12 + 12 = 0 + 12$

$\quad\quad\quad\quad\quad\quad 6z = 12$

$\quad\quad\quad\quad\quad \dfrac{6z}{6} = \dfrac{12}{6}$

$\quad\quad\quad\quad\quad\quad z = 2$

89. Area = length $\cdot$ width

$= 13\sqrt{2} \cdot 5\sqrt{6}$

$= (13 \cdot 5)\sqrt{2 \cdot 6}$

$= 65\sqrt{12}$

$= 65\sqrt{4} \cdot \sqrt{3}$

$= 65 \cdot 2\sqrt{3}$

$= 130\sqrt{3}$

The area is $130\sqrt{3}$ sq. m.

93. $\dfrac{\sqrt{3}+1}{\sqrt{2}-1} = \dfrac{(\sqrt{3}+1)(\sqrt{3}-1)}{(\sqrt{2}-1)(\sqrt{3}-1)}$

$= \dfrac{3-1}{\sqrt{6}-\sqrt{2}-\sqrt{3}+1}$

$= \dfrac{2}{\sqrt{6}-\sqrt{2}-\sqrt{3}+1}$

Section 8.5

1. $\sqrt{x} = 9$

$(\sqrt{x})^2 = 9^2$

$x = 81$

5. $\sqrt{2x+6} = 4$

$(\sqrt{2x+6})^2 = 4^2$

$2x + 6 = 16$

$2x = 10$

$x = 5$

9. $3\sqrt{x} + 5 = 2$

$3\sqrt{x} = -3$

$\sqrt{x} = -1$

There is no solution since $\sqrt{x}$ cannot equal a negative number.

13. $\sqrt{2x+1} + 3 = 5$

$\sqrt{2x+1} = 2$

$(\sqrt{2x+1})^2 = 2^2$

$2x + 1 = 4$

$2x = 3$

$x = \dfrac{3}{2}$

17. $\sqrt{x+6} + 5 = 3$

$\sqrt{x+6} = -2$

There is no solution since the result of a square root cannot be negative.

21. $\sqrt{x} = \sqrt{3x-8}$

$(\sqrt{x})^2 = (\sqrt{3x-8})^2$

$x = 3x - 8$

$-2x = -8$

$x = 4$

25. $\sqrt{9x^2 + 2x - 4} = 3x$

$(\sqrt{9x^2 + 2x - 4})^2 = (3x)^2$

$9x^2 + 2x - 4 = 9x^2$

$2x = 4$

$x = 2$

29. $\sqrt{16x^2 + 2x + 2} = 4x$

$(\sqrt{16x^2 + 2x + 2})^2 = (4x)^2$

$16x^2 + 2x + 2 = 16x^2$

$2x = -2$

$x = -1$

A check shows that $x = -1$ is an extraneous solution. Therefore, there is no solution.

33. $\sqrt{x+7} = x + 5$

$(\sqrt{x+7})^2 = (x+5)^2$

$x + 7 = x^2 + 10x + 25$

$x^2 + 9x + 18 = 0$

$(x+6)(x+3) = 0$

$x + 6 = 0 \qquad \text{or} \qquad x + 3 = 0$

$x = -6 \text{ (extraneous)} \qquad\quad x = -3$

37. $\sqrt{2x+1} = x - 7$

$(\sqrt{2x+1})^2 = (x-7)^2$

$2x + 1 = x^2 - 14x + 49$

$0 = x^2 - 16x + 48$

$0 = (x-12)(x-4)$

$x - 12 = 0 \qquad \text{or} \qquad x - 4 = 0$

$x = 12 \qquad\qquad x = 4 \text{ (extraneous)}$

41. $\sqrt{1-8x} - x = 4$

$\sqrt{1-8x} = x + 4$

$(\sqrt{1-8x})^2 = (x+4)^2$

$1 - 8x = x^2 + 8x + 16$

$0 = x^2 + 16x + 15$

$0 = (x+15)(x+1)$

$x + 15 = 0 \qquad \text{or} \qquad x + 1 = 0$

$x = -15 \text{ (extraneous)} \qquad\quad x = -1$

45. $\sqrt{x-7} = \sqrt{x} - 1$

$(\sqrt{x-7})^2 = (\sqrt{x}-1)^2$

$x - 7 = x - 2\sqrt{x} + 1$

$2\sqrt{x} = 8$

$\sqrt{x} = 4$

$(\sqrt{x})^2 = 4^2$

$x = 16$

49. $\sqrt{x+8} = \sqrt{x} + 2$

$(\sqrt{x+8})^2 = (\sqrt{x}+2)^2$

$x + 8 = x + 4\sqrt{x} + 4$

$4 = 4\sqrt{x}$

$1 = \sqrt{x}$

$1^2 = (\sqrt{x})^2$

$1 = x$

$x = 1$

53. Let $x =$ width

$2x =$ length

$2(2x + x) = 24$

$4x + 2x = 24$

$6x = 24$

$\dfrac{6x}{6} = \dfrac{24}{6}$

$x = 4$

$2x = 8$

The length is 8 in.

57. answers may vary

A100

61. $-\sqrt{x+5} = -7x + 1$
$$y_1 = -\sqrt{x+5}$$
$$y_2 = -7x + 1$$

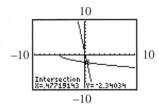

The solution is the x-value of the intersection, 0.48.

Section 8.6

1. $a^2 + b^2 = c^2$
$$2^2 + 3^2 = c^2$$
$$4 + 9 = c^2$$
$$13 = c^2$$
$$\sqrt{13} = c$$

The hypotenuse has a length of $\sqrt{13} \approx 3.61$.

5. $a^2 + b^2 = c^2$
$$7^2 + 24^2 = c^2$$
$$49 + 576 = c^2$$
$$625 = c^2$$
$$\sqrt{625} = c$$
$$25 = c$$

The hypotenuse has a length of 25.

9. $a^2 + b^2 = c^2$
$$4^2 + b^2 = 169$$
$$b^2 = 153$$
$$b = \sqrt{153}$$
$$b = 3\sqrt{17}$$

The unknown side has a length of $3\sqrt{17} \approx 12.37$.

13. $a^2 + b^2 = c^2$
$$a^2 + 2^2 = 6^2$$
$$a^2 + 4 = 36$$
$$a^2 = 32$$
$$a = \sqrt{32}$$
$$a = 4\sqrt{2} \approx 5.66$$

17. $a^2 + b^2 = c^2$
$$5^2 + 20^2 = c^2$$
$$25 + 400 = c^2$$
$$425 = c^2$$
$$\sqrt{425} = c$$
$$c \approx 20.6$$

The wire is approximately 20.6 feet long.

21. $b = \sqrt{\dfrac{3V}{h}}$
$$6 = \sqrt{\frac{3V}{2}}$$
$$6^2 = \frac{3V}{2}$$
$$2(36) = 3V$$
$$\frac{2(36)}{3} = V$$
$$V = 24$$

The volume is 24 cubic feet.

25. $r = \sqrt{2.5r}$
$$r = \sqrt{2.5(300)}$$
$$r = \sqrt{750}$$
$$r \approx 27$$

The car can travel at approximately 27 mph.

29. $9 = 3^2$
The number is 3.

33. $64 = 8^2$
The number is 8.

37.
$$a^2 + b^2 = c^2$$
$$[60(3)]^2 + [30(3)]^2 = c^2$$
$$(180)^2 + (90)^2 = c^2$$
$$32{,}400 + 8100 = c^2$$
$$40{,}500 = c^2$$
$$201 \approx c$$

They are approximately 201 miles apart.

41. answers may vary

Chapter 8 Test

1. $\sqrt{16} = 4$ because $4^2 = 16$ and 4 is positive.

5. $\sqrt[4]{-81}$ is not a real number because the index is even and the radicand is negative.

9. $\sqrt{y^7} = \sqrt{y^6 \cdot y} = \sqrt{y^6} \cdot \sqrt{y} = y^3\sqrt{y}$

13. $\sqrt{\dfrac{5}{16}} = \dfrac{\sqrt{5}}{\sqrt{16}} = \dfrac{\sqrt{5}}{4}$

17. $\sqrt{\dfrac{3}{4}} + \sqrt{\dfrac{3}{25}} = \dfrac{\sqrt{3}}{\sqrt{4}} + \dfrac{\sqrt{3}}{\sqrt{25}}$
$$= \frac{\sqrt{3}}{2} + \frac{\sqrt{3}}{5}$$
$$= \frac{5\sqrt{3}}{10} + \frac{2\sqrt{3}}{10}$$
$$= \frac{5\sqrt{3} + 2\sqrt{3}}{10}$$
$$= \frac{7\sqrt{3}}{10}$$

21. $\dfrac{\sqrt{50}}{\sqrt{10}} = \sqrt{\dfrac{50}{10}} = \sqrt{5}$

25. $\dfrac{8}{\sqrt{6} + 2} = \dfrac{8(\sqrt{6} - 2)}{(\sqrt{6} + 2)(\sqrt{6} - 2)}$
$$= \frac{8(\sqrt{6} - 2)}{6 - 4}$$
$$= \frac{8(\sqrt{6} - 2)}{2}$$
$$= 4(\sqrt{6} - 2)$$
$$= 4\sqrt{6} - 8$$

29. $\sqrt{2x - 2} = x - 5$
$$(\sqrt{2x - 2})^2 = (x - 5)^2$$
$$2x - 2 = x^2 - 10x + 25$$
$$0 = x^2 - 12x + 27$$
$$0 = (x - 9)(x - 3)$$
$$x - 9 = 0 \quad \text{or} \quad x - 3 = 0$$
$$x = 9 \qquad\qquad x = 3 \text{ (extraneous)}$$

Chapter 9

EXERCISE SET 9.1

1.
$$k^2 - 9 = 0$$
$$(k + 3)(k - 3) = 0$$
$$k + 3 = 0 \quad \text{or} \quad k - 3 = 0$$
$$k = -3 \qquad\qquad k = 3$$
The solutions are -3 and 3.

5.
$$2x^2 - 32 = 0$$
$$2(x^2 - 16) = 0$$
$$2(x + 4)(x - 4) = 0$$
$$x + 4 = 0 \quad \text{or} \quad x - 4 = 0$$
$$x = -4 \qquad\qquad x = 4$$
The solutions are -4 and 4.

9.
$$x^2 + 7x = -10$$
$$x^2 + 7x + 10 = 0$$
$$(x + 5)(x + 2) = 0$$
$$x + 5 = 0 \quad \text{or} \quad x + 2 = 0$$
$$x = -5 \qquad\qquad x = -2$$
The solutions are -5 and -2.

13. $x^2 = 21$
$$x = \pm\sqrt{21}$$
The solutions are $\pm\sqrt{21}$.

17. $x^2 = -4$
This equation has no real solution because the square root of -4 is not a real number.

21. $7x^2 = 4$
$$x^2 = \frac{4}{7}$$
$$x = \pm\sqrt{\frac{4}{7}}$$
$$x = \pm\frac{\sqrt{4}}{\sqrt{7}}$$
$$x = \pm\frac{2 \cdot \sqrt{7}}{\sqrt{7} \cdot \sqrt{7}}$$
$$x = \pm\frac{2\sqrt{7}}{7}$$
The solutions are $\pm\dfrac{2\sqrt{7}}{7}$.

25. answers may vary

29. $(x + 2)^2 = 7$
$$x + 2 = \pm\sqrt{7}$$
$$x = -2 \pm \sqrt{7}$$
The solutions are $-2 \pm \sqrt{7}$.

33. $(p + 2)^2 = 10$
$$p + 2 = \pm\sqrt{10}$$
$$p = -2 \pm \sqrt{10}$$
The solutions are $-2 \pm \sqrt{10}$

37. $(z - 4)^2 = -9$
This equation has no real solution because the square root of -9 is not a real number.

41.
$$(3x - 7)^2 = 32$$
$$3x - 7 = \pm\sqrt{32}$$
$$3x - 7 = \pm4\sqrt{2}$$
$$3x = 7 \pm 4\sqrt{2}$$
$$x = \frac{7 \pm 4\sqrt{2}}{3}$$
The solutions are $x = \dfrac{7 \pm 4\sqrt{2}}{3}$.

45. $x^2 - 4x + 4 = x^2 - 2(2)(x) + 2^2 = (x - 2)^2$

49. $A = \pi r^2$
$$36\pi = \pi r^2$$
$$36 = r^2$$
$$\sqrt{36} = r$$
$$6 = r$$
The radius is 6 in.

53. $(x - 1.37)^2 = 5.71$
$$x - 1.37 = \pm\sqrt{5.71}$$
$$x = 1.37 \pm \sqrt{5.71}$$
$$x = 1.37 - \sqrt{5.71} \qquad x = 1.37 + \sqrt{5.71}$$
$$x \approx -1.02 \qquad\qquad x \approx 3.76$$
The solutions are -1.02 and 3.76.

EXERCISE SET 9.2

1.
$$x^2 + 8x = -12$$
$$x^2 + 8x + 16 = -12 + 16$$
$$(x + 4)^2 = 4$$
$$x + 4 = \pm\sqrt{4}$$
$$x = -4 \pm 2$$
$$x = -6 \text{ or } x = -2$$
The solutions are -6 and -2.

5.
$$x^2 - 6x = 0$$
$$x^2 - 6x + 9 = 0 + 9$$
$$(x - 3)^2 = 9$$
$$x - 3 = \pm\sqrt{9}$$
$$x = 3 \pm 3$$
$$x = 0 \text{ or } x = 6$$
The solutions are 0 and 6.

9. $x^2 - 2x - 1 = 0$
$$x^2 - 2x = 1$$
$$x^2 - 2x + 1 = 1 + 1$$
$$(x - 1)^2 = 2$$
$$x - 1 = \pm\sqrt{2}$$
$$x = 1 \pm \sqrt{2}$$
The solutions are $1 \pm \sqrt{2}$.

13. $x^2 + 6x - 25 = 0$
$$x^2 + 6x = 25$$
$$x^2 + 6x + 9 = 25 + 9$$
$$(x + 3)^2 = 34$$
$$x + 3 = \pm\sqrt{34}$$
$$x = -3 \pm \sqrt{34}$$
The solutions are $-3 \pm \sqrt{34}$.

17.
$$4x^2 - 24x = 13$$
$$x^2 - 6x = \frac{13}{4}$$
$$x^2 - 6x + 9 = \frac{13}{4} + 9$$
$$(x-3)^2 = \frac{49}{4}$$
$$x - 3 = \pm\sqrt{\frac{49}{4}}$$
$$x = 3 \pm \frac{7}{2}$$

$$x = \frac{6-7}{2} \quad \text{or} \quad x = \frac{6+7}{2}$$
$$x = -\frac{1}{2} \qquad\qquad x = \frac{13}{2}$$

The solutions are $-\frac{1}{2}$ and $\frac{13}{2}$.

21.
$$2x^2 = 6x + 5$$
$$2x^2 - 6x = 5$$
$$x^2 - 3x = \frac{5}{2}$$
$$x^2 - 3x + \frac{9}{4} = \frac{5}{2} + \frac{9}{4}$$
$$\left(x - \frac{3}{2}\right)^2 = \frac{19}{4}$$
$$x - \frac{3}{2} = \pm\sqrt{\frac{19}{4}}$$
$$x = \frac{3}{2} \pm \frac{\sqrt{19}}{2}$$

The solutions are $\dfrac{3 \pm \sqrt{19}}{2}$.

25.
$$2y^2 + 8y + 5 = 0$$
$$2y^2 + 8y = -5$$
$$y^2 + 4y = -\frac{5}{2}$$
$$y^2 + 4y + 4 = -\frac{5}{2} + 4$$
$$(y+2)^2 = \frac{3}{2}$$
$$y + 2 = \pm\sqrt{\frac{3}{2}}$$
$$y = -2 \pm \frac{\sqrt{6}}{2}$$

The solutions are $-2 \pm \dfrac{\sqrt{6}}{2}$.

29.
$$\frac{3}{4} - \sqrt{\frac{25}{16}} = \frac{3}{4} - \frac{\sqrt{25}}{\sqrt{16}}$$
$$= \frac{3}{4} - \frac{5}{4}$$
$$= \frac{3-5}{4}$$
$$= -\frac{2}{4}$$
$$= -\frac{1}{2}$$

33. $\dfrac{6 + 4\sqrt{5}}{2} = \dfrac{6}{2} + \dfrac{4\sqrt{5}}{2} = 3 + 2\sqrt{5}$

37.
$$x^2 + kx + 16$$
$$\left(\frac{k}{2}\right)^2 = 16$$
$$\frac{k^2}{4} = 16$$
$$k^2 = 64$$
$$k = \pm\sqrt{64}$$
$$k = \pm 8$$

41.
$$x^2 + 8x = -12$$
$$y_1 = x^2 + 8x$$
$$y_2 = -12$$

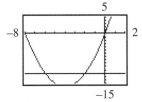

The solutions are the x-coordinates of the intersections, -6 and -2.

EXERCISE SET 9.3

1.
$$x^2 - 3x + 2 = 0$$
$$a = 1, b = -3, c = 2$$
$$x = \frac{-b \pm \sqrt{b^2 - 4ac}}{2a}$$
$$x = \frac{-(-3) \pm \sqrt{(-3)^2 - 4(1)(2)}}{2(1)}$$
$$x = \frac{3 \pm \sqrt{9 - 8}}{2}$$
$$x = \frac{3 \pm 1}{2}$$
$$x = 1 \text{ or } x = 2$$
The solutions are 1 and 2.

5.
$$49x^2 - 4 = 0$$
$$a = 49, b = 0, c = -4$$
$$x = \frac{-0 \pm \sqrt{0^2 - 4(49)(-4)}}{2(49)}$$
$$x = \frac{\pm\sqrt{784}}{98}$$
$$x = \pm\frac{28}{98}$$
$$x = \pm\frac{2}{7}$$

The solutions are $\pm\dfrac{2}{7}$.

9.
$$y^2 = 7y + 30$$
$$y^2 - 7y - 30 = 0$$
$$a = 1, b = -7, c = -30$$
$$y = \frac{-(-7) \pm \sqrt{(-7)^2 - 4(1)(-30)}}{2(1)}$$
$$y = \frac{7 \pm \sqrt{49 + 120}}{2}$$
$$y = \frac{7 \pm \sqrt{169}}{2}$$
$$y = \frac{7 \pm 13}{2}$$
$$y = -3 \text{ or } y = 10$$
The solutions are -3 and 10.

13.
$$m^2 - 12 = m$$
$$m^2 - m - 12 = 0$$
$$a = 1, b = -1, c = -12$$
$$m = \frac{-(-1) \pm \sqrt{(-1)^2 - 4(1)(-12)}}{2(1)}$$
$$m = \frac{1 \pm \sqrt{49}}{2}$$
$$m = \frac{1 \pm 7}{2}$$
$$m = -3 \text{ or } m = 4$$
The solutions are -3 and 4.

17.
$$6x^2 + 9x = 2$$
$$6x^2 + 9x - 2 = 0$$
$$a = 6, b = 9, c = -2$$
$$x = \frac{-9 \pm \sqrt{9^2 - 4(6)(-2)}}{2(6)}$$
$$x = \frac{-9 \pm \sqrt{129}}{12}$$
The solutions are $\dfrac{-9 \pm \sqrt{129}}{12}$.

21. $a^2 - 6a + 2 = 0$
$$a = 1, b = -6, c = 2$$
$$a = \frac{-(-6) \pm \sqrt{(-6)^2 - 4(1)(2)}}{2(1)}$$
$$a = \frac{6 \pm \sqrt{28}}{2}$$
$$a = \frac{6 \pm 2\sqrt{7}}{2}$$
$$a = 3 \pm \sqrt{7}$$
The solutions are $3 \pm \sqrt{7}$.

25.
$$3x^2 = 1 - 2x$$
$$3x^2 + 2x - 1 = 0$$
$$a = 3, b = 2, c = -1$$
$$x = \frac{-2 \pm \sqrt{2^2 - 4(3)(-1)}}{2(3)}$$
$$x = \frac{-2 \pm \sqrt{16}}{6}$$
$$x = \frac{-2 \pm 4}{6}$$
$$x = -1 \text{ or } x = \frac{1}{3}$$

The solutions are -1 and $\dfrac{1}{3}$.

29.
$$20y^2 = 3 - 11y$$
$$20y^2 + 11y - 3 = 0$$
$$a = 20, b = 11, c = -3$$
$$y = \frac{-11 \pm \sqrt{(11)^2 - 4(20)(-3)}}{2(20)}$$
$$y = \frac{-11 \pm \sqrt{361}}{40}$$
$$y = \frac{-11 \pm 19}{40}$$
$$y = -\frac{3}{4} \text{ or } y = \frac{1}{5}$$

The solutions are $-\dfrac{3}{4}$ and $\dfrac{1}{5}$.

33. $3p^2 - \dfrac{2}{3}p + 1 = 0$
$$9p^2 - 2p + 3 = 0$$
$$a = 9, b = -2, c = 3$$
$$p = \frac{-(-2) \pm \sqrt{(-2)^2 - 4(9)(3)}}{2(9)}$$
$$p = \frac{2 \pm \sqrt{-104}}{18}$$
There is no real solution.

37.
$$4p^2 + \frac{3}{2} = -5p$$
$$8p^2 + 10p + 3 = 0$$
$$a = 8, b = 10, c = 3$$
$$p = \frac{-10 \pm \sqrt{10^2 - 4(8)(3)}}{2(8)}$$
$$p = \frac{-10 \pm \sqrt{4}}{16}$$
$$p = \frac{-10 \pm 2}{16}$$
$$p = -\frac{3}{4} \text{ or } p = -\frac{1}{2}$$

The solutions are $-\dfrac{3}{4}$ and $-\dfrac{1}{2}$.

41. $28x^2 + 5x + \dfrac{11}{4} = 0$
$$112x^2 + 20x + 11 = 0$$
$$a = 112, b = 20, c = 11$$
$$x = \frac{-20 \pm \sqrt{20^2 - 4(112)(11)}}{2(112)}$$
$$x = \frac{-20 \pm \sqrt{-4528}}{224}$$
There is no real solution.

45. $y = -3$

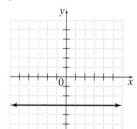

49. $a^2 + b^2 = c^2$
$$x^2 + 7^2 = 10^2$$
$$x^2 + 49 = 100$$
$$x^2 = 51$$
$$x = \sqrt{51}$$
The unknown side is $\sqrt{51}$ meters.

53. answers may vary

57. $7.3z^2 + 5.4z - 1.1 = 0$
$$a = 7.3, b = 5.4, c = -1.1$$
$$z = \frac{-5.4 \pm \sqrt{(5.4)^2 - 4(7.3)(-1.1)}}{2(7.3)}$$
$$z = \frac{-5.4 \pm \sqrt{61.28}}{14.6}$$
$$z \approx -0.9 \text{ or } z \approx 0.2$$
The approximate solutions are -0.9 and 0.2.

61.
$$y = -0.5x^2 + 12.5x + 82$$
$$150 = -0.5x^2 + 12.5x + 82$$
$$0 = -0.5x^2 + 12.5x - 68$$
$a = -0.5, b = 12.5, c = -68$
$$x = \frac{-12.5 \pm \sqrt{(12.5)^2 - 4(-0.5)(-68)}}{2(-0.5)}$$
$$x = \frac{-12.5 \pm \sqrt{20.25}}{-1}$$
$$x = 12.5 \pm 4.5$$
$$x = 8 \text{ or } x = 17$$
$$1994 + 8 = 2002$$
The first year that the sales will reach $150 billion is in 2002.

SECTION 9.4

1.

x	$y = 2x^2$
-2	8
-1	2
0	0
1	2
2	8

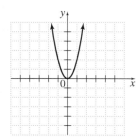

5.

x	$y = \frac{1}{3}x^2$
-5	$\frac{25}{3}$
-3	3
0	0
3	3
5	$\frac{25}{3}$

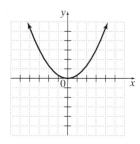

9. $y = x^2 + 4$
Find vertex.
$$x = \frac{-b}{2a} = -\frac{0}{2(1)} = 0$$
$$y = 0 + 4 = 4$$
vertex $= (0, 4)$
y-intercept $= (0, 4)$
Find x-intercepts.
$$0 = x^2 + 4$$
$$x^2 = -4$$
There are no x-intercepts because there is no solution to this equation.

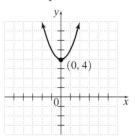

13. $y = x^2 + 2x - 8$
Find vertex.
$$x = -\frac{b}{2a} = -\frac{2}{2(1)} = -1$$
$$y = (-1)^2 + 2(-1) - 8 = -9$$
vertex $= (-1, -9)$
Find x-intercepts:
$$0 = x^2 + 2x - 8$$
$$0 = (x + 4)(x - 2)$$
$$x = -4 \text{ or } x = 2$$
x-intercepts: $(-4, 0), (2, 0)$
Find y-intercept.
$$y = 0^2 + 2(0) - 8 = -8$$
y-intercept $= (0, -8)$

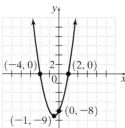

17. $y = x^2 + 5x + 4$
Find vertex.
$$x = \frac{-b}{2a} = \frac{-5}{2(1)} = -\frac{5}{2}$$
$$y = \left(-\frac{5}{2}\right)^2 + 5\left(-\frac{5}{2}\right) + 4 = -\frac{9}{4}$$
$$\text{vertex} = \left(-\frac{5}{2}, -\frac{9}{4}\right)$$
Find x-intercepts.
$0 = x^2 + 5x + 4$
$0 = (x + 4)(x + 1)$
$x = -4$ or $x = -1$
x-intercepts $= (-4, 0), (-1, 0)$
Find y-intercept.
$y = 0 + 5(0) + 4 = 4$
y-intercept $= (0, 4)$

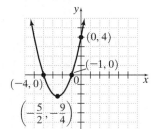

21. $\dfrac{\frac{1}{7}}{\frac{2}{5}} = \dfrac{1}{7} \cdot \dfrac{5}{2} = \dfrac{5}{14}$

25. $\dfrac{2x}{1 - \frac{1}{x}} = \dfrac{2x}{\frac{x-1}{x}} = \dfrac{2x}{1} \cdot \dfrac{x}{x-1} = \dfrac{2x^2}{x-1}$

29. $h = -16t^2 + 128t$

a. Find the value of h at vertex.
$$t = \frac{-b}{2a} = -\frac{128}{2(-16)} = 4$$
$$h = -16(4)^2 + 128(4) = 256$$
The fireball will reach a height of 256 feet.

b. Find the value of t at vertex.
$t = 4$
The fireball will reach a maximum height at 4 seconds.

c. Let $h = 0$. Solve for t.
$0 = -16t^2 + 128t$
$0 = -16t(t - 8)$
$t = 0$ or $t = 8$
The fireball will return to the ground after 8 seconds.

33. The graph opens up, $a > 0$, and it does not cross the x-axis. Graph C matches the description.

CHAPTER 9 TEST

1.
$$2x^2 - 11x = 21$$
$2x^2 - 11x - 21 = 0$
$(2x + 3)(x - 7) = 0$
$2x + 3 = 0$ or $x - 7 = 0$
$$x = -\frac{3}{2} \text{ or } \qquad x = 7$$

The solutions are $-\dfrac{3}{2}$ and 7.

5. $x^2 - 26x + 160 = 0$
$$x^2 - 26x = -160$$
$$x^2 - 26x + 169 = -160 + 169$$
$$(x - 13)^2 = 9$$
$$x - 13 = \pm\sqrt{9}$$
$$x = 13 \pm 3$$
$x = 10$ or $x = 16$.
The solutions are 10 and 16.

9.
$$(3x - 5)(x + 2) = -6$$
$$3x^2 + 6x - 5x - 10 = -6$$
$$3x^2 + x - 4 = 0$$
$$(3x + 4)(x - 1) = 0$$
$3x + 4 = 0$ or $x - 1 = 0$
$$x = -\frac{4}{3} \qquad\qquad x = 1$$

The solutions are $-\dfrac{4}{3}$ and 1.

13. $3x^2 - 7x + 2 = 0$

$a = 3, b = -7, c = 2$

$$x = \frac{-(-7) \pm \sqrt{(-7)^2 - 4(3)(2)}}{2(3)}$$

$$x = \frac{7 \pm \sqrt{25}}{6}$$

$$x = \frac{7 \pm 5}{6}$$

$$x = \frac{1}{3} \text{ or } x = 2$$

The solutions are $\dfrac{1}{3}$ and 2.

17. $y = x^2 - 4$
Find the vertex.
$$x = -\frac{b}{2a} = -\frac{0}{2(1)} = 0$$
$y = 0^2 - 4 = -4$
vertex $= (0, -4)$
y-intercept $= (0, -4)$
Find the x-intercepts.
$0 = x^2 - 4$
$x^2 = 4$
$x = \pm\sqrt{4}$
$x = \pm 2$
x-intercepts $= (-2, 0), (2, 0)$

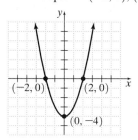

Index

Numerical coefficients, 215, 256, 287, 288
of term, 91, 176

Occupational Outlook Handbook Quarterly (1990–91), 580
OPEC (Organization of Petroleum Exporting Countries), 429
Operations, with real numbers on calculator, 50
Operation symbols, 13
Opposites (or additive inverses), 60, 78
finding, 25–26
Orbits, in shape of parabola, 664
Ordered pairs
plotting, 431–33, 518
and solutions, 535
Ordered pair solutions
completing, 434–36
found and plotted on graphic calculators, 450
Order of operations, 77
and exponents, 13–14
and grouping symbols, 14
Order property, for real numbers, 76
Origin, 431
Original inequality, 510

Paired data, 433, 497. *See also* Functions
Palatine Anthology, 404
Parabolas, 664, 683, 695
Parallel lines, A-2
cut by transversal, A-3
graphed on graphing calculator, 476
slopes of, 473–75, 487
Parallelogram, A-9
formula for area of, 145
Parentheses, 13, 14
distributive property for removal of, 93
and exponential expressions, 191
simplifying expressions containing, 93–94
Patterns, discovering on graphing calculator, 488
Percent, 178, R-1
written as decimal, R-32
Percent equations, solving, 151–52
Percents
and decimals, R-19–24, R-31–32
problem solving with, 178
written as decimals, R-23, R-32
Perfect squares, 597
Perfect square trinomials
defined, 337
factoring, 302–303, 337
recognizing, 301–302
Perpendicular lines, A-2
slope of, 473–75, 487

Pi (π), 6
Plane, A-1
Plane figures, A-2, A-9
Point-slope form, 520
of equation of line, 485
of linear equation, 487
to solve problems, 486–87
Polygons, A-9
defined, A-3
Polynomials, 189
adding and subtracting in one variable, 226
adding and subtracting in several variables, 226
addition of, 225, 257
defining types of, 216–17, 256
degree of, 216, 219, 257
dividing, 247–48, 258
dividing by polynomial other than monomial, 248–49, 258
evaluating, 217–18
factoring, 271–74, 295–96
modeling with, 255
multiplying, 231–32, 257
multiplying sum and difference of two terms, 239–40
multiplying vertically, 232
prime, 281
simplifying by combining like terms, 218–19
simplifying those with several variables, 219
special products, 237–38
steps in factoring, 311
subtraction of, 225–26, 257
terms and coefficients of, 215, 256
Positive integers, 5
Positive numbers, 25, 41, 48
Positive (or principal) square root, 595, 646
Positive slope, 470
Positive square root, 595
Power of a product rule, 256
for exponents, 194, 204
Power of a quotient rule, for exponents, 194, 204, 256
Power rule
for exponents, 193, 204, 256
for products and quotients, 194–95
Powers, 191
exponential expression raised to, 192, 193
real numbers raised to, 23
roots raised to, 593
Prices, problem solving about, 570–71
Prime factorization, R-30
writing, R-3–5
Prime numbers, 328, R-3
identifying, 318
Prime polynomials, 281
Problem solving, 1, 177

with addition, 25
formulas and, 141–44, 177
general strategy for, 129
modeled by proportions, 155–56
with percent, 152, 178
with proportions, 155–56
and quadratic equations, 323–27, 338
and radical equations, 635–37, 647, 648
and rational equations, 393–98, 417
with subtraction, 33–34
systems of linear equations and, 569–74, 584
translating and, 129–34
and translating words to symbols, 16–17
Product rule
for exponents, 192–93, 204, 256
for radicals, 603–604, 617, 646
for square roots, 603
Products, involving zero, 41, 78
Projectiles, in shape of parabola, 664
Properties
associative, 57
commutative, 57
distributive, 58–59
identity, 59
inverse, 60
Proportion, 178
defined, 153
solving, 153–54
solving problems modeled by, 155–56
Pyramid, volume formula for, 145
Pythagorean theorem, 325, A-6
for problem solving, 635–36

Quadrants, 431
Quadratic equations, 659
graphing in two variables, 683–87, 695
and problem solving, 323–27, 338
solved by quadratic formula, 673–76
solving by completing the square, 667–69, 694
solving by factoring, 313–16, 337
solving by quadratic formula, 694
solving by square root property, 661–63, 694
in standard form, 384
summary on, 681
Quadratic formula, 659, 681, 686
quadratic equations solved by, 673–76, 694
Quadrilateral, A-9
Quotient rule
for exponents, 195, 204, 256
for radicals, 604, 618, 646
for square roots, 604

Photo Credits